Methods in Molecular Biology

Series Editor
John M. Walker
School of Life and Medical Sciences
University of Hertfordshire
Hatfield, Hertfordshire, AL10 9AB, UK

For further volumes:
http://www.springer.com/series/7651

Systems Genetics

Methods and Protocols

Edited by

Klaus Schughart

Department of Infection Genetics, Helmholtz Centre for Infection Research & University of Veterinary Medicine Hannover, Braunschweig, Niedersachsen, Germany

Department of Microbiology, Immunology and Biochemistry, University of Tennessee Health Science Center, Memphis, TN, USA

Robert W. Williams

Department of Genetics, Genomics and Informatics, University of Tennessee Health Science Center, Memphis, TN, USA

Humana Press

Editors
Klaus Schughart
Department of Infection Genetics
Helmholtz Centre for Infection Research &
University of Veterinary Medicine Hannover
Braunschweig, Niedersachsen, Germany

Department of Microbiology, Immunology
and Biochemistry
University of Tennessee Health Science Center
Memphis, TN, USA

Robert W. Williams
Department of Genetics, Genomics
and Informatics
University of Tennessee Health Science Center
Memphis, TN, USA

ISSN 1064-3745 ISSN 1940-6029 (electronic)
Methods in Molecular Biology
ISBN 978-1-4939-6425-3 ISBN 978-1-4939-6427-7 (eBook)
DOI 10.1007/978-1-4939-6427-7

Library of Congress Control Number: 2016950574

Printed on acid-free paper

This Humana Press imprint is published by Springer Nature
The registered company is Springer Science+Business Media LLC New York
The registered company address is: 233 Spring Street, New York, NY 10013, U.S.A

Preface

Systems genetics is actually an old field with a new name. RA Fisher [1], S Wright [2-4], and JBS Haldane [5, 6]—the three leading figures of the *modern synthesis* who brought genetics into alignment with evolutionary biology—are the intellectual founders of what we would now call systems genetics. They used other terms—population genetics, statistical genetics, and quantitative genetics. We can add one more scientific progenitor, CH Waddington, a founder of what is now called systems biology and a key figure who helped align developmental biology with genetics [7].

The advantage of the term *systems genetics*, and the reason for its rapid rise in prominence, is that it emphasizes the concept "system" rather than the resource type (*population*), the measurement type (*quantitative*), or the method of analysis (*statistical*). Our colleague Grant Morahan coined the term in 2004 to refocus attention toward sets of related phenotypes, sets of gene variants, and sets of environmental factors and away from more restricted terms that were then in use—*genetical genomics*, *complex trait analysis*, and *QTL analysis* [8–10]. A short definition of systems genetics and its relations to other approaches may help.

Genetics can be divided roughly into three ways of looking at relations between genetic and phenotypic variation:

1. *One-to-one relations*—in other words, classical Mendelian genetics—the study of qualitative traits linked either to spontaneous mutations or to targeted modifications of genes.
2. *One-to-many relations* between single phenotypes and sets of loci or gene variants—in other words, QTL mapping, genome-wide association, and complex trait analysis.
3. *Many-to-many-to-many relations* among (a) sets of correlated and interacting phenotypes at different levels (metabolites, mRNAs, protein, organelles, cells, tissues, organ systems, and classic phenotypes and outcome measures), (b) sets of gene variants, and (c) sets of environmental factors and treatments.

The latter is the ultimate goal of systems genetics, but the reality is that we need to be working on problems at all three levels concurrently. No doubt about it: the amazing complexity and adaptability of biological systems needs to be dissected into manageable units for analytic and economic reasons. Results that make headlines and that are most highly rewarded tend to be the 1-to-1 simplifications—gene X causes aging, gene Y causes schizophrenia. But what is just as obvious now is that the yin of "dissection," "analysis," and "reduction" needs its complement—the yang of "assemble," "synthesis," and "integration."

The main motivation is not merely a scholastic intellectual balance—improved health care, agricultural productivity, and the design of robustly engineered biological systems absolutely require a deep understanding of the range of action of the whole.

The good news is that we finally have powerful tools both to dissect and to assemble biological systems with rapidly improving range, precision, and throughput. The duality of genetics can be balanced. Generating millions of precisely measured genotypes and molecular phenotypes—our biological parts list—is now practical for thousands of cases, in principle, under many conditions. Human cohorts of millions of subjects, all sequenced and accompanied with comprehensive health records, will soon be routine. For assembly and integration of these parts, we have the computer scientists, bioinformaticists, mathematicians, statisticians, and public funders to thank for every faster and more sophisticated ways to evaluate how best to put pieces together and how to predict outcomes with some level of precision. We now can even look forward with angst to *ab initio* creation—making new biological systems from scratch. We are on the cusp of amazing capabilities.

The chapters in this volume will give you a hands-on appreciation of the range of activity and methods in systems genetics. This volume does not cover the whole range of activity; our contributors are drawn from a small but vibrant community of rodent experimental geneticists. Most of us are focused on mouse models with the goal of translational impact to better understand and cure human diseases. Most of us grew up in this new genomics era of QTL mapping, and a dominant theme of many protocols is how best to track down genetic causes of heritable variation across a wide range of systems and traits. But if you stand back and envision the whole activity represented in this volume, you will see how protocols and results can be snapped together to build more holistic models in a true systems spirit. We are now well poised to implement ever more powerful methods and models.

We thank our many colleagues, collaborators, and the 100 contributors to this volume. Both of us were frankly surprised by the highly enthusiastic responses given to our requests for protocols in this new area—no thumbscrews required. That is an excellent sign. And in keeping with the theme of systems integration, we expect that there will be strength in numbers and complementarity—that readers will, we hope, find real synergy in using collections of these protocols.

Braunschweig, Germany ***Klaus Schughart***
Memphis, TN, USA ***Robert W. Williams***

References

1. Fisher R (1918) The correlation between relatives on the supposition of Mendelian inheritance. Phil Trans R Soc Edinburgh 52:399–433
2. Wright S (1921) Correlation and causation. J Agri Res 20:557–585
3. Wright S (1988) Surfaces of selective value revisited. Am Nat 131:115–123
4. Wright S (1990) The genetics of quantitative variability. Quantitative inheritance. Papers read at a colloquium held at the Institute of Animal Genetics Edinburgh University under the auspices of the Agricultural Research Council, 4–6 April 1950, pp 5–41
5. Haldane JB (1959) The theory of natural selection today. Nature 183(4663):710–713
6. Haldane JBS, Sprunt AD, Haldane NM (1915) Reduplication in mice (Preliminary Communication). Journal of Genetics 5(2):133–135. doi:110.1007/BF02985370
7. Waddington CH (1957) The strategy of the genes. A discussion of some aspects of theoretical biology. George Allen and Unwin Ltd, London
8. Morahan G, Williams RW (2007) Systems genetics: the next generation in genetics research? Novartis Found Symp 281:181–188, discussion 188–191, 208–189
9. Jansen RC, Nap JP (2001) Genetical genomics: the added value from segregation. Trends Genet 17(7):388–391, doi:S0168-9525(01)02310-1 [pii]
10. Threadgill DW (2006) Meeting report for the 4th annual Complex Trait Consortium meeting: from QTLs to systems genetics. Mamm Genome 17(1):2–4. doi:10.1007/s00335-005-0153-5

Contents

Contributors

ELEONORA ADAMI • *Max-Delbrück-Center for Molecular Medicine (MDC), Berlin, Germany*

MAHMOUD AGBARIA • *Department of Clinical Microbiology and Immunology, Faculty of Medicine, Tel-Aviv University, Tel-Aviv, Israel*

DANNY ARENDS • *Faculty of Life Sciences, Albrecht Daniel Thaer-Institut, Humboldt Universität zu Berlin, Berlin, Germany*

DAVID G. ASHBROOK • *Department of Computational and Evolutionary Biology, Faculty of Life Sciences, University of Manchester, Manchester, UK*

ERICH J. BAKER • *The Jackson Laboratory, Bar Harbor, ME, USA*

JACQUES BEHMOARAS • *Centre for Complement and Inflammation Research, Imperial College London, Hammersmith Hospital, London, UK*

CONNIE R. BEZZINA • *Department of Clinical and Experimental Cardiology, Academic Medical Centre (AMC), University of Amsterdam, Amsterdam, The Netherlands*

JUDITH A. BLAKE • *The Jackson Laboratory, Bar Harbor, ME, USA*

JAMES W. BOGENPOHL • *Department of Pharmacology and Toxicology, VCU Alcohol Research Center, Virginia Commonwealth University, Richmond, VA, USA*

LEONARDO BOTTOLO • *Department of Medical Genetics, University of Cambridge, Cambridge, UK; Department of Mathematics, Imperial College London, London, UK*

ALESSANDRA BRAGONZI • *Infections and Cystic Fibrosis Unit, Division of Immunology, Transplantation and Infectious Diseases, San Raffaele Scientific Institute, Milano, Italy*

GUDRUN A. BROCKMANN • *Faculty of Life Sciences, Albrecht Daniel Thaer-Institut, Humboldt Universität zu Berlin, Berlin, Germany*

JASON A. BUBIER • *The Jackson Laboratory, Bar Harbor, ME, USA*

CAROL J. BULT • *The Jackson Laboratory, Bar Harbor, ME, USA*

ELISSA J. CHESLER • *The Jackson Laboratory, Bar Harbor, ME, USA*

SUMANA R. CHINTALAPUDI • *Department of Anatomy and Neurobiology, Hamilton Eye Institute, University of Tennessee Health Science Center, Memphis, TN, USA*

METE CIVELEK • *Center for Public Health Genomics, Department of Biomedical Engineering, University of Virginia, Charlottesville, VA, USA*

WIM E. CRUSIO • *Institut de Neurosciences Cognitives et Intégratives d'Aquitaine, University of Bordeaux, Pessac, France; CNRS, Institut de Neurosciences Cognitives et Intégratives d'Aquitaine, Pessac, France*

YAN CUI • *Department of Microbiology, Immunology and Biochemistry, Center for Integrative and Translational Genomics, University of Tennessee Health Science Center, Memphis, TN, USA*

ANNA DELPRATO • *Institut de Neurosciences Cognitives et Intégratives d'Aquitaine, University of Bordeaux, Pessac, France; CNRS, Institut de Neurosciences Cognitives et Intégratives d'Aquitaine, Pessac, France*

AYCA DOGAN • *Faculty of Medicine, Department of Physiology, Istanbul Kemerburgaz, Istanbul, Turkey*

JANAN T. EPPIG • *The Jackson Laboratory, Bar Harbor, ME, USA*

ALLISON FABER • *Max-Delbrück-Center for Molecular Medicine (MDC), Berlin, Germany*
MARTIN T. FERRIS • *Department of Genetics, University of North Carolina, Chapel Hill, NC, USA*
JOHN E. FRENCH • *Center for Pharmacogenomics and Individualized Therapy, School of Pharmacy, University of North Carolina at Chapel Hill, Chapel Hill, NC, USA*
MICHAEL GALE JR. • *Department of Immunology, Center for Innate Immunity and Immune Diseases, University of Washington School of Medicine, Seattle, WA, USA*
DANIEL GATTI • *The Jackson Laboratory, Bar Harbor, ME, USA*
RICHARD GREEN • *Department of Immunology, Center for Innate Immunity and Immune Disease, University of Washington School of Medicine, Seattle, WA, USA*
REINMAR HAGER • *Department of Computational and Evolutionary Biology, Faculty of Life Sciences, University of Manchester, Manchester, UK*
RABEA A. HALL • *Department of Medicine II, Saarland University Medical Center, Homburg, Germany*
SEBASTIAAN VAN HEESCH • *Max-Delbrück-Center for Molecular Medicine (MDC), Berlin, Germany*
MATTHIAS HEINIG • *Helmholtz Zentrum München, Institute of Computational Biology (ICB), Neuherberg, Germany*
SEBASTIAN HEISE • *Faculty of Life Sciences, Albrecht Daniel Thaer-Institut, Humboldt Universität zu Berlin, Berlin, Germany*
PAULA HOFFMAN • *Department of Pharmacology, School of Medicine, University of Colorado Anschutz Medical Campus, Aurora, CO, USA*
YAEL HOURI-HADDAD • *Department of Prosthodontics, Dental School, The Hebrew University, Hadassah, Jerusalem, Israel*
NORBERT HUBNER • *Max-Delbrück-Center for Molecular Medicine (MDC), Berlin, Germany*
MARTHA IMPRIALOU • *Centre for Complement and Inflammation Research, Imperial College London, London, UK*
FUAD A. IRAQI • *Department of Clinical Microbiology and Immunology, Faculty of Medicine, Tel-Aviv University, Tel-Aviv, Israel*
MONICA M. JABLONSKI • *Department of Ophthalmology, Anatomy and Neurobiology, and Pharmaceutical Sciences, The University of Tennessee Health Science Center, Memphis, TN, USA*
LESLIE C. JELLEN • *Department of Genetics, Genomics, and Informatics, The University of Tennessee Health Science Center, Memphis, TN, USA*
BYRON C. JONES • *Department of Genetics, Genomics, and Informatics, The University of Tennessee Health Science Center, Memphis, TN, USA*
JAMES A. KADIN • *The Jackson Laboratory, Bar Harbor, ME, USA*
PATRICK W. KEELEY • *Department of Molecular, Cellular, and Developmental Biology, Neuroscience Research Institute, University of California, Santa Barbara, CA, USA*
FRANZISKA KREUCHWIG • *Max-Delbrück-Center for Molecular Medicine (MDC), Berlin, Germany*
FRANK LAMMERT • *Department of Medicine II, Saarland University Medical Center, Homburg, Germany*
MICHAEL A. LANGSTON • *The Jackson Laboratory, Bar Harbor, ME, USA*
ELISABETH M. LODDER • *Department of Clinical and Experimental Cardiology, Academic Medical Centre (AMC), University of Amsterdam, Amsterdam, The Netherlands*

MAARTEN LOOS • *Sylics (Synaptologics BV), Amsterdam, The Netherlands; Department of Molecular and Cellular Neurobiology, Center for Neurogenomics and Cognitive Research (CNCR), Neuroscience Campus Amsterdam, VU University Amsterdam, Amsterdam, The Netherlands*

NICOLA IVAN LORÈ • *Department of Infections and Cystic Fibrosis Unit, Division of Immunology, Transplantation and Infectious Diseases, San Raffaele Scientific Institute, Milano, Italy*

ALDONS J. LUSIS • *Department of Medicine, Division of Cardiology, University of California, Los Angeles, CA, USA*

HENRIKE MAATZ • *Max-Delbrück-Center for Molecular Medicine (MDC), Berlin, Germany*

PAUL L. MAURIZIO • *Department of Genetics, University of North Carolina, Chapel Hill, NC, USA*

KRISTIN M. MIGNOGNA • *Department of Psychiatry, VCU Alcohol Research Center, Virginia Commonwealth University, Richmond, VA, USA*

MICHAEL F. MILES • *Department of Pharmacology and Toxicology, VCU Alcohol Research Center, Virginia Commonwealth University, Richmond, VA, USA*

GRANT MORAHAN • *Centre for Diabetes Research, The Harry Perkins Institute of Medical Research, and Centre for Medical Research, The University of Western Australia, Nedlands, WA, Australia*

AIDA MORENA-MORAL • *Duke-NUS Medical School, Singapore, Singapore*

RICHARD MOTT • *Genetics Institute, University College London, London, UK*

KHYOBENI MOZHUI • *Preventive Medicine, University of Tennessee Health Science Center, Memphis, TN, USA*

MEGAN K. MULLIGAN • *Department of Genetics, Genomics, and Informatics, University of Tennessee Health Science Center, Memphis, TN, USA*

AYSAR NASHEF • *Department of Prosthodontics, Dental School, The Hebrew University, Hadassah, Jerusalem, Israel*

KAREN Y. ORÓSTICA • *Programa de Genética Humana ICBM, Facultad de Medicina, Universidad de Chile, Santiago, Chile*

RUPERT W. OVERALL • *CRTD-Center for Regenerative Therapies Dresden, Technische Universität Dresden, Dresden, Germany*

CALVIN PAN • *Department of Medicine, Division of Cardiology, University of California, Los Angeles, CA, USA*

FRANCESCO PESCE • *Imperial Centre for Translational and Experimental Medicine, National Heart and Lung Institute, Faculty of Medicine, Imperial College of London, London, UK*

ENRICO PETRETTO • *Duke-NUS Medical School, Singapore, Singapore*

SVITLANA PODLIESNA • *Department of Clinical and Experimental Cardiology, Academic Medical Centre (AMC), University of Amsterdam, Amsterdam, The Netherlands*

PJOTR PRINS • *University of Tennessee Health Science Center, Memphis, TN, USA*

RAMESH RAM • *Centre for Diabetes Research, The Harry Perkins Institute of Medical Research, and Centre for Medical Research, The University of Western Australia, Nedlands, WA, Australia*

CHRISTOPH D. RAU • *Department of Medicine, Division of Cardiology, University of California, Los Angeles, CA, USA*

BENJAMIN E. REESE • *Department of Psychological and Brain Sciences, Neuroscience Research Institute, University of California, Santa Barbara, CA, USA*

JOEL E. RICHARDSON • *The Jackson Laboratory, Bar Harbor, ME, USA*
MARTIN RINGWALD • *The Jackson Laboratory, Bar Harbor, ME, USA*
LAURA SABA • *Department of Pharmaceutical Sciences, Skaggs School of Pharmacy and Pharmaceutical Sciences, University of Colorado Anschutz Medical Campus, Aurora, CO, USA*
KLAUS SCHUGHART • *Department of Infection Genetics, Helmholtz Centre for Infection Research & University of Veterinary Medicine Hannover, Braunschweig, Niedersachsen, Germany; Department of Microbiology, Immunology and Biochemistry, University of Tennessee Health Science Center, Memphis, TN, USA*
ARIEL SHUSTERMAN • *Department of Prosthodontics, Dental School, The Hebrew University, Hadassah, Jerusalem, Israel*
AUGUST B. SMIT • *Department of Molecular and Cellular Neurobiology, Center for Neurogenomics and Cognitive Research (CNCR), VU University Amsterdam, Amsterdam, The Netherlands*
MAREN L. SMITH • *Department of Human and Molecular Genetics, VCU Alcohol Research Center, Virginia Commonwealth University, Richmond, VA, USA*
CYNTHIA L. SMITH • *The Jackson Laboratory, Bar Harbor, ME, USA*
SABINE SPIJKER • *Department of Molecular and Cellular Neurobiology, Center for Neurogenomics and Cognitive Research (CNCR), VU University Amsterdam, Amsterdam, The Netherlands*
BORIS TABAKOFF • *Department of Pharmaceutical Sciences, Skaggs School of Pharmacy and Pharmaceutical Sciences, University of Colorado Anschutz Medical Campus, Aurora, CO, USA*
RICARDO A. VERDUGO • *Programa de Genética Humana ICBM, Facultad de Medicina, Universidad de Chile, Santiago, Chile*
MATTHIJS VERHAGE • *Department of Functional Genomics, Center for Neurogenomics and Cognitive Research (CNCR), VU University Amsterdam, Amsterdam, The Netherlands; Department of Clinical Genetics, VU Medical Center, Amsterdam, The Netherlands*
IRENE E. WHITNEY • *Department of Molecular, Cellular and Developmental Biology, Neuroscience Research Institute, University of California, Santa Barbara, CA, USA*
ERVIN WIESS • *Department of Prosthodontics, Dental School, The Hebrew University, Hadassah Jerusalem, Israel*
COURTNEY WILKINS • *Department of Immunology, Center for Innate Immunity and Immune Disease, University of Washington School of Medicine, Seattle, WA, USA*
ROBERT W. WILLIAMS • *Department of Genetics, Genomics and Informatics, University of Tennessee Health Science Center, Memphis, TN, USA*
EVAN G. WILLIAMS • *Department of Biology, Institute of Molecular Systems Biology, Zürich, Switzerland*
LEAH C. SOLBERG WOODS • *Medical College of Wisconsin, Milwaukee, WI, USA*
JESSE D. ZIEBARTH • *Department of Microbiology, Immunology and Biochemistry, Center for Integrative and Translational genomics, University of Tennessee Health Science Center, Memphis, TN, USA*

Part I

Resources for Systems Genetics

Chapter 1

Resources for Systems Genetics

Robert W. Williams and Evan G. Williams

Abstract

A key characteristic of systems genetics is its reliance on populations that vary to a greater or lesser degree in genetic complexity—from highly admixed populations such as the Collaborative Cross and Diversity Outcross to relatively simple crosses such as sets of consomic strains and reduced complexity crosses. This protocol is intended to help investigators make more informed decisions about choices of resources given different types of questions. We consider factors such as costs, availability, and ease of breeding for common scenarios. In general, we recommend using complementary resources and minimizing depth of resampling of any given genome or strain.

Key words Genetic reference population (GRP), Recombinant inbred (RI), Collaborative Cross (CC), Congenic lines, Consomic and chromosome substitution lines, Recombinant congenic strains, RI intercross (RIX) and RI backcross (RIB) progeny, Heterogeneous stock (HS), Diversity outcross (DO), Hybrid diversity panel (HDP), Reduced complexity cross (RCC), Gene-by-environment interactions (G×E)

1 Introduction

A large number of innovative resources for systems genetics have been developed over the last 15 years [1]. There are at least two reasons for this burst of activity. The first catalyst was the introduction of far easier, cheaper, and more comprehensive methods of genotyping [2, 3] that we already take for granted. State-of-the-art genotyping for recombinant inbred (RI) strains consisted of ~1600 microsatellite markers (dinucleotide repeats) in 2001 [4]. Over the next 5 years this number increased to more than 10,000 SNPs [5], and we now rely on genotypes at more than 100,000 SNPs using Affymetrix or Illumina platforms [6, 7] at modest cost—well under $0.01 per marker. The second reason was rapid progress on ways to map quantitative traits with progressively higher precision and power [4, 8–16], culminating in the establishment of the Complex Trait Consortium in 2002 [17]. A good problem we now face is selecting wisely from the many options and resources that are now available. Any choice is a major commitment. This protocol highlights factors researchers should consider and balance.

Klaus Schughart and Robert W. Williams (eds.), *Systems Genetics: Methods and Protocols*, Methods in Molecular Biology, vol. 1488, DOI 10.1007/978-1-4939-6427-7_1, © Springer Science+Business Media New York 2017

2 Methods

2.1 Guidance on Using This Protocol

The goal of this protocol is to step through the decisions associated with selecting resources for both QTL mapping and systems genetics. The first issue is to define classes of questions. Different questions benefit from different types and mixtures of resources—the cliché "different horses for different courses" applies. In **Part 2.3** we review current murine resources used in QTL mapping and systems genetics. In **Part 2.4** we consider one multipurpose experimental design that will work reasonably well for a range of questions. Consider this design a starting point for your discussions and decisions. We provide some notes on the pros and cons of the resources, many in a simple question-and-answer format. Since everyone has their own biases, ask others for their opinions.

These are among the main considerations or themes that go into the choice of resources for systems genetics:

1. Cost and availability (strains, hybrids, cases).
2. Phenotype diversity, heritability, and genetic architecture.
3. Marker density, mapping precision, and power.
4. Sequence diversity and genetic blind spots.
5. Selective phenotyping or genotyping.
6. Complexity of QTL intervals.
7. Population structure, admixture, and analytic methods.
8. Depth of genetic, omics, and phenome data resources.
9. Robustness, replicability, extensibility, and translatability.

To foreshadow our conclusions: Most researchers currently rely on a single type of resource or cross, and while there are good historical reasons for this focus, this is no longer an optimal or advisable strategy. We now have such a range of powerful genetic resources optimized for different purposes that it makes sense to take advantage of combinations of complementary crosses and even multiple species [18–22]. Analytic methods do get more complex when using combinations of resources, but some of the same methods used to handle admixed human cohorts in genome-wide and phenome-wide association studies (GWAS and PheWAS) have now been adapted to handle combined experimental cohorts [23–25].

Our other conclusion is that a mapping resolution of about 1 Mb will usually be adequate to transition to validation, including translational analysis of human GWAS and PheWAS data sets [22, 26, 27], analysis of knockout (KO) and knockin (KI) phenotypes, bioinformatic and omics dissection, and pharmacological intervention. This is especially true in an era of super high precision but mechanistically unanchored GWAS. The need for high precision mapping in mouse has been supplanted by an acute need for powerful resources to understand and accurately predict genome-to-phenome (G2P) relations under a wide range of environments and treatments.

2.2 Types of Questions Guiding the Experimental Approach

We consider four main types of questions:

2.2.1 Type 1 Questions

The classic forward genetic question—what are the polymorphic genes and sequence variants that modulate a phenotype or disease risk? This is by far the most common question our research community has dealt with over the last two decades and will probably remain so for the next several decades. Almost all human GWASs have this same simple reductionist motivation—a generalization of the classic Mendelian approach but applied to messier and continuously variable quantitative traits.

The repeated mapping of large numbers of QTLs and their causal QT genes (QTGs) quickly leads to complex systems-level questions—a transition we now are beginning to see in human GWAS. This shift has happened gradually over the past decade. The pioneering work by Wakeland and colleagues on the family of gene variants that contribute to autoimmune disease is a fine example [28]. The work of Hunter and colleagues on metastasis networks [29, 30] and of Morahan and colleagues on type I diabetes [31] provide two other examples of this movement from QTL analysis to complex systems genetics. This shift is leading to the discovery of new biomarkers, diagnostics, mechanisms, and treatments.

Type 1 questions are usually approached in two steps: the first involves mapping QTLs to confidence intervals of 0.5–5 Mb, while the second and more problematic step involves proving to your own satisfaction (and that of reviewers and readers) that a polymorphic candidate gene has been validated as a source of trait variance [12, 15]. Almost all of the technical motivation and innovation in the late 1990s and early 2000s in the field of QTL mapping addressed mapping precision, with less explicit consideration given to statistical power. There was, and still is a good reason for this focus on precision: once the right gene has been identified, it becomes possible to switch from genetic causality defined by loci and LOD scores, to actionable molecular mechanisms modulated by differences of protein expression or sequence. Thanks to many human GWASs, we now understand much better how to control the risk of false discovery using populations that incorporate more and more recombinations and complex admixture. One goal of this protocol is to help you get to a sweet spot with a balance of power and precision. A second goal is to help ensure that the results are robust and translatable.

2.2.2 Type 2 Questions

Questions related to G×E and treatment effects on phenotypes. These types of questions will be crucial to those interested in systematic manipulations of diet, environmental stressors, age, pathogens, drug exposure, and differences in social interactions. Mice and other inbred and isogenic model organisms are extremely well suited to evaluate complex experimental effects in the context of QTL mapping. The ability to impose well-controlled perturbations across large cohorts is among the strongest motivations to use model organisms. This kind of design is already the most common and critical in agricultural genetics.

2.2.3 Type 3 Questions

Questions related to the global genetic modulation of single traits or of systems of correlated phenotypes. These types of large-scale questions often fall under the heading of "genetic architecture." This term encompasses the analysis of many components of heritable and nonheritable variation, particularly the number and effect sizes of loci, independence and interactions among loci, and the roles of the environment, epigenetics, parental effects, and developmental noise [32]. Oddly enough, before it became easy to map QTLs, these types of hard questions were at the heart of quantitative genetics. In fact, major branches of statistics had their birth in questions of genetic architecture, including ANOVA and path analysis [33, 34]. The diallel cross—the production of a matrix of F1 hybrids from inbred strains—is one of the mainstays of this type of quantitative genetics [35]. Recent examples include studies by Airey et al. [36], Crowley et al. [37], and Percival et al. [38] who have used diallel sets of RI strains and the founders of the Collaborative Cross (CC).

2.2.4 Type 4 Questions

Type 4 questions are related explicitly to predicting G2P relations. Given summed effects of gene variants (Type 1 questions), G×E interactions (Type 2), and the architecture of all sources of variance (Type 3), can we assemble predictive models of disease risk as a function of age, environment, diet, and drugs? This is the core question and quandary of precision health delivery. Precision medicine will have a short grace period, but if geneticists, molecular biologists, statisticians, and computational scientists have not delivered something impressive to match the hype, this term and the field risk being dismissed as a misnomer in the same way that *artificial intelligence* (AI) was dismissed and left unfunded for long periods. We need great experimental resources to generate and help validate predictions efficiently. The next section provides quick definitions and commentaries on the pros and cons of the important resources.

2.3 Pros and Cons of Resources and Crosses

We list of some of the major types of resources, from most simple to most complex in terms of level of genetic variation and complexity. The types of crosses and how they are generated are shown schematically in Fig. 1 with numbers that correspond to subsections.

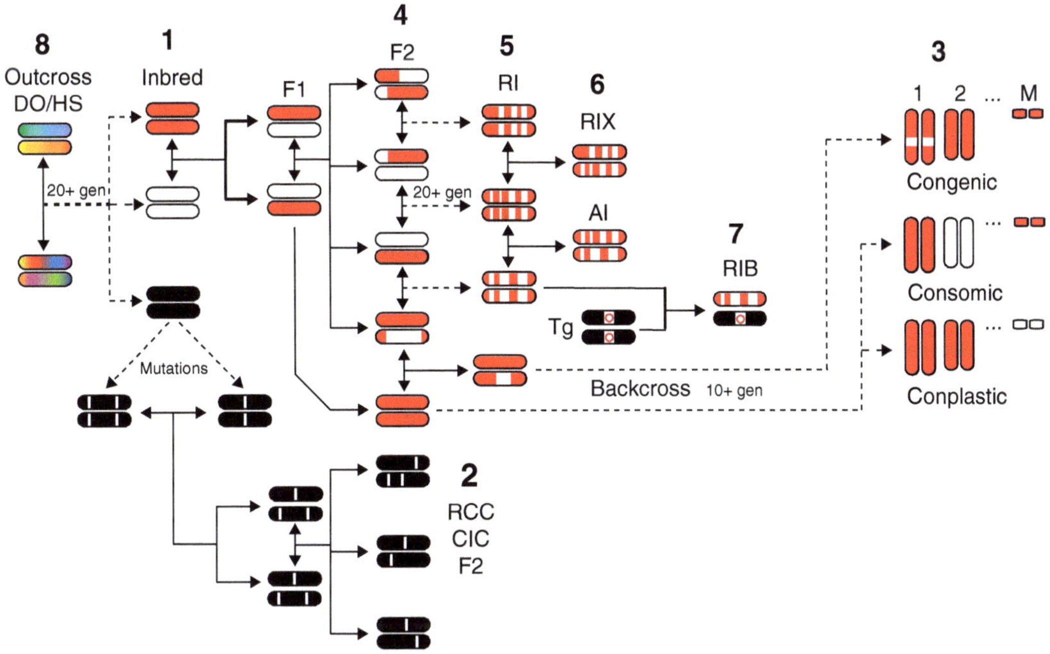

Fig. 1 Breeding schema for different genetic resources and populations. Breeding schemes used to generate the resources. Short bars symbolize pairs of chromosomes. The colors (here, arbitrarily chosen as *red*, *white*, or *black*) denote the haplotypes/genotypes of the chromosomes. The large numbers within the figure correspond to Section 2 subheadings. **Tg**—a transgenic line. Note that AI progeny are not generated from RI lines, but they have chromosomes with recombination patterns similar to those of RIX. For all other abbreviations see text of Section 23. Adapted from [1] with permission

2.3.1 Single Fully Inbred Strains (Fig. 1.1)

Single full inbred strains such as DBA/2 and C57BL/6, are often the starting point for in vivo studies. We usually do not think of inbred strains in isolation as a resource for systems genetics, but a family of knockouts can be bred into a single isogenic strain [39] or a single KO can be crossed into a hundred different inbred strains [40] to generate interesting cohorts.

Large sets of distinct inbred strains incorporate a great deal of genetic variation (three are shown in Fig. 1.1), and collectively may also be used as a core resource for systems genetics [41]. Genome sequence data are available for more than 36 inbred strains [42] (www.sanger.ac.uk/science/data/mouse-genomes-project) most of which are also part of the Mouse Phenome Project [43]. Such collections of inbred strains—often termed diversity panels—provide a quick and ready resource for profiling how traits vary across a wide range of genomes, but there are not enough easily available strains to map QTLs effectively. Power is low and false discovery rates (FDRs) are high. However, sets of common inbred strains combined with sets of RI strains are an excellent joint resource for systems genetics—a combination called a hybrid diversity panel to which we return below.

The most commonly used inbred strains have often been split into sets of substrains. These will carry different sets of a few spontaneous mutations that have been picked up over decades of maintenance in different colonies. In mice, C57BL/6J and C57BL/6N are the genetic backgrounds strains used for almost all KO, KI, and transgenic modifications (www.mousephenotype.org). Thanks to powerful sequencing technologies, sets of related substrains now provide an interesting new resource for G2P mapping. We describe this novel approach below.

2.3.2 Reduced Complexity Cross (RCC) or Coisogenic Cross (CIC) (Fig. 1.2)

Both RCCs and CICs are novel types of "postgenomic" intercrosses between very closely related substrains [44, 45] or even coisogenic pairs. For example, genomes of the C57BL/6J and C57BL/6N substrains differ at a total of about 36 known coding variants [42] but these substrains also differ for a surprisingly large numbers of phenotypes, including responses to several drugs and treatments [46–49]. BXD29/TyJ and BXD29-$\text{Tlr4}^{<\text{lps-2J}>}$/J constitute a coisogenic pair that differs at two or three loci [50]. How is it possible to map an F2 that has almost no sequence variants? Once two substrains have been sequenced deeply (>30-fold coverage), there will almost always be a large enough number of spontaneous noncoding mutations to assemble a sparse genome-wide panel of SNPs and indels for mapping sources of phenotypic differences.

While the mapping precision of an F2 RCC or CIC will be poor (intervals of 20 Mb or more), the small number of segregating variants within any interval means that it can be practical to identify candidate QTGs and even QT nucleotides (QTNs) efficiently [51]. Kumar used this approach to define a mutation in *Cyfip2* that controls response to cocaine and methamphetamine [44]. The utility of an RCC in mapping and even in systems genetics points out that the key variable in "cloning" QTLs is not mapping precision per se but the number of polymorphic genes and sequence variants within a QTL's confidence interval. A 5–10 Mb interval containing only a single sequence variant will be far more easily reduced to cause and mechanisms than a highly polymorphic 0.1 Mb QTL containing five genes and hundreds of sequence variants [44, 52].

2.3.3 Consomic and Congenic Whole Genome Panels (Fig. 1.3)

By backcrossing two inbred strains to each other while tracking genotypes of progeny over several generations, it is possible to effectively transplant whole chromosomes from donor strain *A* into recipient strain *B*. A full set of consomic strains will consist of 22 lines, each with one swapped chromosome plus the recipient control strains. There are now two sets of consomic strains—crosses of A/J or PWD/Ph into C57BL/6J [14, 53]. Buchner and Nadeau [54] have considered the pros and cons of consomic sets and their efficiency relative to other resources.

A whole genome congenic panel is basically a finer-grained version of a consomic set, but now each strain contains only a piece of a single donor chromosome [55]. The main utility of consomic and congenic sets is their high power to map phenotypes to single chromosomes. They have been used more recently to study epistasis and epigenetic effects [54]. Their main disadvantage is that mapping QTLs requires the production of a secondary F2 intercross or a set of interval-specific congenic strains. Whole chromosome effect sizes will almost inevitably decrease during this process [56].

2.3.4 Off-Target Mutations and Isogenic Strains

One important factor to consider before using congenic and consomic strain sets is their sensitivity to spontaneous mutations that will accumulate gradually and progressively on the recipient (non-transplanted) background chromosomes. Spontaneous mutations or allele conversion events that arise on these other 20 chromosomes can cause variant phenotypes, and these new phenotypes risk being misattributed to putative variants on the donor chromosome—essentially off-target effects [57]. It is therefore useful—sometimes even essential—to verify that traits map to the introgressed chromosome by making a small F2 from the consomic or congenic stock. Tracking down off-target variants is difficult because there are no known polymorphisms with which to map the other chromosomes. Sequencing consomic strains and using RCC methods is the obvious, but costly solution.

This raises a broad issue that applies to all crosses that are carried for many generations, including standard inbred strains, RI strains, AI progeny, and HS stock: what is the relative impact of inevitable de novo mutations on the measured phenotypes and results of different types? The good news is that for most of these resource types, new mutations will be unique to one strain or one case and will not segregate across the whole cross. Provided that the analysis and results are statistical collectives based on a large sample of strains or cases, then rare mutations, even those that are fixed in single strains, will simply be lumped as another source of error variance. In contrast, in situations in which mapping and other results depend on a single case and control—as when using congenic and consomic lines—there is a risk of misattribution of effects.

2.3.5 F2 Intercrosses and Backcrosses (Fig. 1.4)

The F2 intercross has been used widely in systems genetics, starting with the work of Damerval [58], Schadt, Lusis, and colleagues [59, 60]. Their main advantage is the ability to make large numbers of progeny quickly from almost any stock (usually inbred strains). F2 intercrosses and N2 backcrosses have a structure that makes mapping and the analysis of covariance among traits simple. There is no need to correct for population substructure (*see* **Note 1**)—a problem that arises in almost any multi-generation cross (e.g., heterogeneous stock (HS), AIs, and RI strains). It is practical to enhance the complexity and utility of an F2 intercross for systems genetics and for standard QTL mapping by making a four-way F2—for example by crossing A×B F1s to C×D F1s to produce AB×CD F2 progeny. This type of F2 is being used in an experimental study of life span in mice [61].

2.3.6 Advanced Intercrosses (Fig. 1.6)

AIs are simple extensions of F2s in which all subsequent generations are randomly bred, but with careful avoidance of sib matings [9, 10]. The number of recombination events per AI case climbs steadily as the depth of the pedigree increases. At the eighth generation (about 2 years of breeding), 100 AI progeny, if made correctly, will provide about the same mapping precision as 500 F2 for Mendelian traits [9]. The countervailing problems with AIs are (1) the more complex logistics of using more than 100 breeders for up to ten generations has a high cost, (2) the variable kinship among AI progeny needs to be factored into any kind of mapping or other statistical analysis, (3) the need for a significantly higher density of markers, and, perhaps most seriously (4) the loss of power associated with the increased number of recombinations per animal. A solution to some of these issues, first pointed out by Darvasi and Soller [9], is to generate RI strains from AI stock—the so-called Advanced RI (ARI) strains—and both the CC and many of the new BXD strains are actually ARIs.

Trade-Offs. There are important trade-offs between mapping precision and mapping power—the ability to detect QTLs with effects that account for a defined percent of the trait variance assuming a given sample size. As pointed out by Lander and Botstein [8], the longer the genetic map, the higher the thresholds for statistical significance. The relation is complex, but Table 1 provides a rough guide of tradeoffs. One column is marked **Recs/case** or recombinations per case, and a second column is marked **LOD Threshold**, or the linkage score that will often be needed to achieve genome-wide significance. **Recs/case** is an index of the potential precision of a resource, whereas the LOD score in this context is an inverse index of statistical power. High **Recs/case** are good for precision, but high LOD score requirements are bad for power.

The goal of course is precision with power. The simplest way to get both is to type larger and larger numbers of cases. A better solution is to combine complementary resources—one optimized for power such as a conventional F2 or conventional RI strains, and one optimized for precision—such as the Collaborative Cross (CC), a Hybrid Diversity Panel (HDP), AI, HS, or DO stock. The reason why joint resources are not used widely yet is because (1) many of the resources are new, and (2) the computational aspects of the analysis are more involved. But we now have powerful algorithms [23–25] that can handle dense genotypes and complex cohorts and covariates. Some of these are available online in the new version of GeneNetwork.

2.3.7 RI Strains (Fig. 1.5)

RI strains were originally made for mapping highly penetrant Mendelian traits [62, 63], but they were eventually adopted for the analysis of complex traits [64]. RIs are now a key resource in systems genetics. Their main advantage relative to F2s and HS is that

Table 1
Resources for systems genetics

Type of cross	Recs/case	LOD Threshold	$/Geno typing	$/Case[a]	Isogenic	Inbred	Phen-ome	GXE	Breeding	References
Consomic and congenic sets	1	1–2	0	140	Yes	Yes	Yes	Easy	Variable	[14, 55]
Reduced complexity cross	25	1–2	25	20	Almost	Almost	Hard	Hard	Easy	[44, 45]
F2 intercross, 2-way or 4-way	25	2.5–3	25	15	No	No	Hard	Hard	Easy	[8, 16]
Advanced intercross	100	4–5	100	100	No	No	Hard	Hard	Hard	[9, 10]
RI strains and advanced RI Strains	50 to 80	3–4	0	140	Yes	Yes	Yes	Easy	Variable	[4, 8, 22]
Advanced intercross RI strains	80	4–5	0	140	Yes	Yes	Yes	Easy	Variable	[4, 8]
RI Intercross F1s (RIX, RIB)	100 to 200	4–6	0	50	Yes	No	Hard	Easy	Easy	[36, 38, 40]
Hybrid diversity panel (HDP)	1000	6+	0	20–150	Yes	Yes	Yes	Yes	Easy	[18, 19]
Collaborative cross (8-way RI)	135	4–6	0	195	Yes	Yes	Yes	Easy	Variable	[13, 17]
Diversity outcross (DO HS)	400+	5–7	100	55	No	No	Hard	Hard	Easy	[84, 85]
Outbred stock (e.g., CD-1, CF-1)	1000	6+	100	7	No	No	Hard	Hard	Easy	[68, 79]

[a]Costs do not include shipping

each unique genometype (genetic individual) is represented by a stable inbred strain that can be replicated in large numbers—essentially a sexually reproducing clone. RIs are therefore an excellent resource for studies that benefit from replication across individuals (e.g., dosing and toxicity studies of genotypes) or across environments (i.e., studies on G × E), and for the gradual assembly of deep phenome data that can be used in G2P analysis. In mice, there are now sufficient numbers of RI strains to allow for comparatively precise and well-powered QTL mapping. There are currently two major types of RI strains in mice:

1. Classic two-parent RI strains. There are a total of about 340 of these types of mouse RI strains, including ~150 BXD available as live stock and many other small RI families: AXB/BXA (29 live), AKXD (20 cryopreserved), BXH (12 live), BRX58N (7 cryopreserved), CXB (12 cryopreserved), ILSXISS (60 cryopreserved), LGXSM (~18), NXSM (15 cryopreserved), SWXJ (13 cryopreserved).
2. The Collaborative Cross (CC). This is a unique eight-way RI set of about 100 strains that is now in widespread use for QTL analysis and systems genetics [13, 17]. These strains are available both from UNC Chapel Hill and the Jackson Laboratory.

Classic RI strains that are derived from standard F2 intercrosses harbor more recombinations per genome—about 40–50—than do backcrosses (10–15), or F2 intercrosses (20–30) and therefore deliver better QTL precision than one might expect even with modest samples size (Fig. 1, note the alternating red and white haplotype blocks that make up the chromosomes of the RI strains). The ability to resample individuals also reduces the impact of nongenetic trait variance—effectively boosting heritability [65]. Pandey and Williams [66] computed the empirical precision of *cis*-acting expression QTLs (*cis*-eQTLs) in the BXD family across the whole genome at different mean LOD scores and at different marker densities (their Fig. 8.6). With a cohort of 67 strains and using only two samples per strain, eQTLs with LOD scores of between 3 and 5 were located within ±2 Mb of the parent gene. Those with LOD scores above eight were typically within ±1 Mb. Corresponding empirical mapping precision based on *cis*-eQTLs can now be easily computed for many resource types across the whole genome using data sets and queries built into GeneNetwork ([67], this volume). Examples of doing this for a large AI ($n = 811$) and a well matched AI-derived RI set ($n = 40$) are given in **Note 2**.

The CC RI strains are capable of even better mapping precision than standard RIs for two reasons. First, the recombination load (the crossover probability) of CC strains is 1.75 times higher than that of typical two-parent RI strains due in part to the rounds of intercrossing required to merge all eight genomes

(Table 5 of ref. 16). Second, the inclusion of multiple parental genomes within the CC means that it is possible to carry out a fine-grained haplotype contrast analysis that can effectively reduce QTL intervals and numbers of QTG candidates [68]. Haplotype contrasts of the same general type can also be exploited using combinations of conventional RI families, inbred strains, and F2 crosses [18, 25, 63, 69].

The most important disadvantage of conventional RI strains and other standard two-parent crosses is that they segregate for only a fraction of all known polymorphisms. For example, the BXD family segregates for a total of ~5.2 million sequence variants—about 44% of common variants among standard inbred strains [70]. Some stretches of the genome will be almost completely identical by descent [6] and these regions will not normally contribute much to trait variance. This disadvantage however may also be viewed as an advantage when trying to dissect a QTL, since the load of polymorphisms within an interval will be about sixfold lower than that of the corresponding interval in the CC or DO stock, and thus the number of viable candidate genes may be much reduced. As shown by Li and colleagues, phenotypes that map into these genetic blindspots can be particularly easy to map to QTNs [52].

A practical disadvantage of RI strains is that they often have poor breeding performance compared to many F2s and outbred stock. While BXD strains average 4–5 pups per litter, some are hard to maintain and can be sensitive to housing conditions. Many CC lines have even lower fecundity. This is one reason why many inbred strains are so much more expensive than outcross or HS animals (Table 1) and why they are often cryopreserved rather than kept as live stock. This issue was also a factor motivating the creation of the DO: The DO provides a way to stabilize recombinations events that were at risk of extinction (Gary A. Churchill, personal communication). Speaking of the obvious, a final disadvantage of RI strains is that they are inbred—an anomalous genetic architecture that will not only decrease fitness but will often increase trait variance relative to isogenic F1 hybrids due to the loss of heterosis and allele buffering.

2.3.8 Advanced RI Lines

There are also several interesting variants of RI strains. The first of these are highly recombinant RI strains generated from AI progeny [9]. Many of the new BXD strains (BXD43 and higher) are AI-derived [4, 71], as are all of the LGXSM strains [72]. Instead of directly inbreeding siblings of an F2, progeny are crossed to avoid sib matings for as many as 30 generations, prior to the inbreeding phase (another 20 generations). The main benefit of using AI stock for making RI strains is a significant increase in potential QTL mapping precision (*see* **Note 2**), but as usual, with loss of power.

2.3.9 RI Backcrosses (Fig. 1.7)

The second variant involves making a set of F1 intercrosses between RI strains and a single inbred strain—usually one that carries interesting modifier alleles with a dominant or additive effect. For example Hunter and colleagues crossed 18 AKXD RIs to an FVB strain carrying a dominant cancer gene variant to map modifiers of metastasis [29]. They refer to this cross as an RI backcross (RIB) because the 18 sets of F1s are similar to a backcross—those chromosomes inherited from the RI parent are recombinant, whereas those inherited from the other strain are not. This idea can also be generalized across multiple RI sets and inbred strains. For example, Bennett and colleagues crossed an APOE transgenic strain to more than 31 common inbred strains and 66 BXD, AXB/BXA, BXH, CXB RI strains [40] to study the genetic architecture of atherosclerosis.

2.3.10 RIX Panels (Fig. 1.6)

RIX panels are a clever new extension of RI strains that have some interesting advantages over RI strains and HS. Given a set of 10 RI strains, it is simple to cross all of them to each other: 1 × 1, 2 × 1, 3 × 1, 3 × 2 and reciprocal crosses 1 × 2, 1 × 3, 2 × 3, and so on. From only 10 starting strains one can produce a full diallel set made up of 100 isogenic sets of F1. In a full diallel we do not gain much precision by resampling the same parental haploid genome in different combinations (1 × 2, 1 × 3, 1 × 4, etc.). While no new recombination event occur in making these F1s, one does expose an interesting range of phenotypes, such as those exploited by Rasmussen and colleagues [73] to develop mouse models of Ebola infection.

What makes RIX particularly attractive now for both mapping and systems genetics is that we have several large sets of RI strains—more than 100 BXDs and close to 100 CC lines. While it is not practical to generate or study a full 200 × 200 matrix of 40,000 RIX progeny and founders, it is practical to sample all 200 of these RI genomes by making 100 nonoverlapping sets of RIX litters: 1 × 2, 3 × 4, … 198 × 199, and 199 × 200. And two different RI sets can be crossed (e.g., BXD1 to CC001). A set of 100 disjoint (nonoverlapping) RIX progeny solves a number of problems—(1) efficient sampling of large RI families that exploits all recombination events in the parental RIs; (2) much lower inbreeding coefficients than inbred parents; (3) genetic complexity much more like that of human populations; (4) ability to study parent-of-origin and dominance effects; (5) fully defined genomes; (6) deep replication of any particular RIX to increase phenotype precision; (7) more direct analysis of G × E using precisely the same genometypes under two or more conditions; and as a (8) powerful resource to test predictive models of G2P relations.

Disadvantages of RIX sets include the following: (1) they can be costly to generate compared to HS or DO stock; (2) there will be a loss of genetic variance associated with the heterozygosity of RIX progeny compared to homozygous parents [74]; (3) breeding and cohort logistics are somewhat more complicated and expensive; and (4) it will be difficult for a community of researchers to define a single reference set of RIXs to use for collaborative phenotyping because there are such huge numbers of potential RIX that can be made.

2.3.11 Hybrid Diversity Panels (HDP)

An HDP is an aggregate of RI strains and common inbred strains that are usually phenotyped together and used as a single joint mapping resource [19, 69, 75, 76]. They are used for at least two reasons: (1) to achieve comparatively high mapping precision (intervals of 1–5 Mb) that can match those of HS and DO stock using as few as 100 inbred strains; (2) to make it possible to assemble large phenomes that can be used for G × E analysis. A HDP does not have a rigid definition, and a mouse HDP could and should include CCs, BXDs, and even RIX. Depending on its membership of isogenic genometypes, an HDP will share some of the same problems of any one RI family, but to a lesser degree. For example, the issue of genetic blind spots will be less serious except for a few regions of the genome that tend to be identical-by-descent even in the CC. The main problem of an HDP is the generally low to moderate fecundity of members and their high acquisition costs.

2.3.12 Outbred Stock (OS), Heterogeneous Stock (HS), and Diversity Outcross Stock (DO)

Outbred stock (OS)—often referred to as Swiss Webster stock [77, 78]—are the progeny of nine albinos (two males and seven females) imported from a colony in Lausanne to New York in 1926. They were subsequently distributed to researchers and commercial vendors worldwide as "standard laboratory" mice. As expected given this history, OS do not incorporate much genetic variation. Genomes of 66 OS colonies studied by Yalcin and colleagues [79] were heterozygous at no more than 34% of polymorphic loci, and a significant number of colonies were almost fully inbred. The theoretical attraction of some OS colonies is their potential high mapping precision with LD blocks that are only a few hundred kilobases.

HS and DO could be considered variants of OS, but here we use a modern definition of HS and DO as special stock generated from well-structured intercrosses and outcrosses among diverse sets of inbred progenitor strains. HS are almost always maintained using larger colonies—50 or more breeding cages—and breeding schemes that minimize mating of closely related individuals. One original motivation to make HS was to produce new models by intercrossing diverse strains, and then selectively breeding progeny for high and low phenotypes in responses to drugs, alcohol, and other treatments [80, 81]. The Northport HS (HS-Npt) made by intercrossing A/J, AKR/J, BALBc/J, CBA/J, C3H/HeJ, C57BL/6J, DBA/2J, and LP/J is a good example [82]. HS have also been used for high precision QTL mapping [83].

The DO is an example of a modern HS made by intercrossing early generations of the CC [84, 85]. DO mice are significantly more diverse even than HS-Npt or outbred stock for the simple reason that three of the progenitors of the DO and CC—PWK/Ph, CAST/Ei, and WSB/Ei—are inbred strains derived from highly diverse wild *Mus* species and subspecies. DO cohorts are now at the 22nd generation (G22) of outcrossing. The DO segregate for well over 40 million common sequence variants with minor allele frequencies above 10%. These animals breed well and incorporate four to sixfold more genetic variants than the number of common variants in human populations.

There are two key advantages of DO and HS: (1) they have a genetic complexity that equals or exceeds that of most human populations. They are excellent models for precision medicine; (2) like AI cohorts, they gradually accumulate large numbers of recombinations and therefore can resolve QTLs with high precision; (3) the high genetic diversity among parental strains ensures that phenotypes will be highly variable and that most regions of the genome will be polymorphic; and (4) they usually have excellent breeding performance, a feature that reduces costs.

The main disadvantages of HS and DO stock is the inevitable flip side: the high recombination load and map expansion will reduce statistical power per case and the high genetic complexity and numbers of haplotype can make it difficult to resolve single linked QTGs and QTNs. The last and most obvious experimental disadvantage is that HS and DO animals are genetically unique. This means that it is more difficult to acquire phenomes for these types of resources or to use them as effectively in G × E studies.

2.4 A Multipurpose Design for Systems Genetics

In this section we consider some of the designs that can now be used to address the four types of questions in Subheading 2.2. In the first section below (2.4.1) we consider Type 1 questions with a focus on mapping precision. In the second section (2.5), we start to wrap everything together by considering a single adaptable design for systems genetics that will be good for discussion purposes. We comment on ways to modify or extend this multipurpose design using a Question and Answer format. Much of the text is summarized in Tables 1 and 2.

2.4.1 Genotypes and Genetic Maps: What Mapping Resolution Is Needed?

The goal is usually to get down to about 1 Mb precision as efficiently as possible. Assume we are completely naive—we only know what traits interest us and that traits are somewhat variable among individual mice belonging to a few strains or stocks. We do not have estimates of heritability and we do not yet know what strains or crosses would be most useful.

One of the best resources in this situation is to study phenotypes in a small number of strains and F1 hybrids between these strains. This made sense several decades ago [75] and it makes even more sense today [86] because these initial "survey" data can eventually be wrapped into a mapping study with all other resources—whether

Table 2
A design for systems genetics using mouse resources

Group	Types	Type notes	N	Reps	M	F	Months
Group 1A	8	Fully inbred strains	48	6	3	3	2.4
Group 1B	4	4 F1s using 8 genometypes	24	6	3	3	1.2
Group 1C	8	Your choice	48	6	3	3	2.4
Group 2A	40	CC or BXD, AXB (exploratory)	160	4	2	2	8
Group 2B	40	CC or BXD, AXB (selective)	160	4	2	2	8
Group 3A	40	RIX (semi-exploratory)	160	4	2	2	8
Group 3B	40	RIX (selective)	160	4	2	2	8
Group 4A	100	DO or HS (predictive)	100	NA	50	50	5
Group 4B	100	DO or your choice	100	NA	50	50	5
Sums	380		960				48

HS, DO, CC, or RIX. For example, a study of 6 individuals each of 18 isogenic groups, such as sets of fully inbred strains some of their F1s, will answer questions related to trait heritability, trait dominance, and if you are lucky, even give you hints about genetic complexity and architecture. It may be possible to evaluate if the trait or disease phenotype is controlled by a small number of QTLs (the oligogenic model) or by hundreds of QTLs (the polygenic or "infinitesimal" model) [75]. This 120-case study will also enable you to perfect phenotyping and learn much more about sources of technical error, sex differences, and selecting better resources for the next stages.

The main risk in this type of pilot study is batch effect and phenotype drift. Systematically phenotyping strains *A* through *R* at a steady pace of one genotype per week over 4 months is a poor experimental design, since temporal variance and drift will masquerade as a heritable difference among lines. Interleave the phenotyping to study ten different genotypes with one or two individuals each for the first phase of the experiment and then repeat cycles as needed. An interleaved design may not be feasible in all situations, in which case consider re-phenotyping well-known strains throughout a study to check for drift.

2.4.2 Mapping Precision

While more mapping precision is always a good thing, there is not much justification to refine maps down to much less than confidence intervals of 1–2 Mb. Intervals of this size can now be efficiently dissected using an impressive and diverse array of data resources—including of course, full genome sequence for all genes in all strains. A small number of candidate genes and variants can now often be tested efficiently using genetically engineered mice, fish, flies, worms, or human GWAS data sets, in vitro analyses, or even phenome-wide association [22].

Another reason not to obsess about precision much below 1 Mb is the fuzzy functional definitions of genes. This is highlighted by a recent analysis of one of the strongest loci that modulates obesity in humans—SNPs within intron 1 of the *FTO* gene. While the position of linkage is not in question, these SNPs apparently tag variants in a long-range enhancer of *IRX3*—a small transcription factor 0.5 Mb distal [87]. This emphasizes that functional validation is critical, and that the law of diminishing return can kick in with some force under 1 Mb. We consider a 0.5–1 Mb as a reasonable goal that can usually be achieved efficiently using a combination of resources described below. This is not quite as precise as what can be achieved with large GWAS, but unlikely human studies we can efficiently transition to molecular mechanisms.

2.4.3 Assumptions

To develop this multipurpose design we assume almost nothing other than that the traits of interest are heritable and genetically complex, and that the initial focus is not on G×E or treatment effects, developmental stages and ages. We will come back to extensions that these types of questions toward the end of this section.

Sample size and costs of stock. As our starting parameter, we budget for 240 individuals per year over a 4-year period—960 cases total at a pace of 20 per month and 1 per day. This is a modest throughput that should be adaptable to almost any type of study, even electrophysiology, advanced imaging and behavioral methods. The cost of mice may range from as little as \$20 per case to as much as \$200. Standard inbred strains such as those used to generate the CC cost between \$20 (C57BL/6J) and \$200 per animal (WSB/EiJ) with an average of \$102. The average price for most of the resources discussed in this chapter is currently about \$150 per case. An experiment using 240 cases/year will typically require a budget of ~\$40,000/year. Housing costs are variable, but it is safe to assume 25–50 cages will incur a cost of \$10,000–\$20,000/year. If cases must be genotyped (e.g., F2, HS, and DO stock), then factor in a charge of as much as \$100 per case (Table 1).

Sex balance. Whenever possible males and females should be used in roughly equal numbers and concurrently. Not only is the use of both sexes becoming a mandate, but results will also be more interesting and robust in terms of their translational relevance. Finally, sex differences can provide mechanistic insight. The inclusion of both sexes in a design does not double the required sample size, even when using isogenic cohorts of RIs, RIXs, or HDPs. A balanced sample of just one or two males and females across multiple genometypes can be a powerful design to detect sex differences. Of course, sparse sampling does not address sex differences within any single strain, but this is a topic that may be worth revisiting in a second phase of work.

While it may look tidy in a Methods section, it is not necessary to get numbers of cases balanced precisely either by sex or genotype. Do not obsess about filling every cell in a design uniformly. If you must obsess about anything, make it (1) batch confounds, (2) drift in phenotyping standards, and (3) quality control for electronic records and case identifiers. When possible consider whether litter effects are a confounding factor in phenotype variation. This is a particular risk for RIX designs in which one single litter may be used for each genotype.

2.5 Experimental Design for Systems Genetics

2.5.1 Stage 1: Heritability, Technical Robustness of Assays, Effects of Sex, and Genetic Architecture (Table 2)

The main purpose of phase 1 is to make sure you understand more about the main sources of variance of phenotypes. It is well worth a 3–6 month pilot to make sure the phenotyping methods and assays work well. The data from this initial work will eventually be useful for mapping.

Group 1A: Six individuals each of eight inbred strains. It would make great sense to start with the parents of the CC. Depending on your field of study you could add or substitute AKR, BALB, DBA/2J, FVB, or other common strains.

$$n = 48$$

Error-checking: Since assignment errors can destroy your results, keep track of coat color, and even better, save tails of animals for *post hoc* genetic verification. This is important for all stages of the work.

Group 1B: Six individuals from each of four F1 hybrids made using strains A through H (AB, CD, EF, GH, or the reciprocals AB, BA, CD, DC). The parental strains for the F1s can be selected based either on greater genetic differences or on contrasting phenotypes.

$$n = 32$$

Group 1C: Six individuals from each of eight additional types based on the initial results above, or to encompass other interesting strains selected from the Mouse Phenome Project (phenome/jax.org) or based on any interest you have in RCC methods. You could also use this set of 48 cases to resolve problems or seize opportunities. This set could include F1 hybrids.

$$n = 48$$

Question 1: Is six samples per type really enough? ANSWER: If you are not examining different environmental factors, then yes. In fact, you probably should not do 6 per type at any one time or from only 1 or 2 litters, but break work into analysis of 2–4 cases for each of 12 types, and generate data over several batches. You may want to run pairs of males and females (littermates even) in single batches, since you are likely to be used paired t tests. If you find that the batch effects are large, then you have learned something important and may need to rethink the design of the larger study. If you find that there is variation as a function of age, you have also learned something important. Furthermore, after phenotyping six per type, you will have a good idea if any particular phenotype needs to be resampled to higher Ns. *See* **Note 3** that discusses some of the factors that should be considered when selecting number of biological replicates.

Question 2: Should I use wild strains such as PWD/PhJ, CAST/EiJ, or WSB/EiJ? ANSWER: Yes, unless there is some specific contraindication, such as cost, availability, or wildness. There is no reason to not expose yourself to the remarkably wide range of phenotypes at this stage. (Make sure you unbox wild strains carefully or you will have stories to tell.)

Question 3: Should I use HS or DO stock initially? ANSWER: No, not unless you have already used these types of resources or need them to address a specific hypothesis. You cannot estimate heritability from a single cohort of HS animals.

Question 4: Should I phenotype pairs of closely related substrains? ANSWER: Probably not at this stage unless you already know that there are significant differences in related phenotypes among substrains. If you are interested in exploiting RCC methods then include pairs or trios of substrains in Group 1C. Genetic variance will be lower in substrain contrasts, so you will need to increase sample size to 8–12 per type.

Question 5: Why are F1 hybrids useful? ANSWER: For at least these three reasons: (1) F1 hybrids are used to evaluate effects of gene variants on phenotypes in organisms with a more typical heterozygous genome. F1 hybrids are isogenic so they have many of the advantages of inbred strains. (2) F1 hybrids also enable us to evaluate whether phenotypes are dominant or recessive. (3) Reciprocal F1s can be used to study parent-of-origin effects on phenotypes. Note that some of these advantages do not apply to F1s between closely related substrains.

2.5.2 Stage 2: Low Resolution Mapping and Systems Genetics

The purpose is to understand the genetic complexity of phenotypes by low-resolution mapping but with good power. If there are a few QTLs with large effects then even a cross with 40 genometypes will highlight one or two loci. Since we rely on RI strains for this first analysis, it should be possible to compare all new data with all previously generated phenotypes and QTLs. We can be confident to find some interesting leads, generate new hypotheses, and perhaps even gain mechanistic insight.

Group 2A: Four each of 40 RI strains. Use four each if heritability is <0.4, otherwise consider using two each of 80 strains, particularly if you suspect that trait variance is controlled by a major effect locus. You can always return to the RI strains to boost your samples size.

$$n = 160$$

Group 2B: Same as above, but using a new set of 40 RI strains. You will now already know if you have detected suggestive or significant QTLs. If the answer is yes, then you can selectively phenotype those RI strains that have recombinations between the right haplotypes in the right regions. You might also want to replicate any outlier strains detected in Group 2A. If the results from Group 2A do not yet provide compelling candidates, then just forge ahead with more or different RI strains.

$$n = 160$$

Question 6: Could I not use RIX in Group 2B? ANSWER: Yes, since you will have RI strains available, this is an option. However, the RIX will not provide you much more genetic signal unless you use different RI parents to make the RIXs. RI and other fully homozygous strains have twice the genetic variance of F1 hybrids. This gives them a power advantage at early stages of mapping.

Question 7: Should I use BXDs, AXBs, or the CC strains? ANSWER: The CC will almost always be a good choice, as they are likely to exhibit the highest phenotypic variance in any target phenotype. BXDs and AXBs will provide better mapping power *per case* due to their lower genetic complexity, but this benefit can be neutralized by less phenotypic variance. If the parents of the RI panels differ markedly and your focus is more on systems genetics than mapping precision (e.g., C57BL/6J vs DBA/2J), then the BXD may be the best first choice for the simple reason that so much data has been accumulated for these strains. Availability of RI strains can sometimes be the main constraint.

Question 8: Can I mix CC stains with other RI panels? ANSWER: Yes, and this is precisely the motivation for resources such as the HDP. It is probably a good idea to sample at least 16 strains in any one RI set so that you can evaluate whether or not a locus is segregating and so that you can estimate trait covariance to some degree among phenotypes within single RI families.

Question 9: Should I use consomic or congenic panels for this work? ANSWER: No; not unless your screen in part 1 included PWD/Ph and A/J and suggested that these strains differed markedly from C57BL/6J. These are the strains that have been used to make consomic sets. Consomic strains can have good power if you sample each of 20 strains with 6 or more cases, but to achieve mapping precision (±5 Mb), you will have to generate your own derivative crosses, and effect sizes of loci can evaporate during the production of congenics [56].

Question 10: How do I handle outlier strains in the initial QTL analysis? ANSWER: Transform data so that outliers do not have an overwhelming effect on maps and other statistical results. You can winsorise high and low outliers or use a logarithm transform. Replicate outliers if you suspect technical error.

2.5.3 Stage 3: High-Resolution Mapping and More Systems Genetics

Group 3A: Four each of 40 sets of RIX progeny that are produced by crossing within or even across sets of RI strains. You will need 80 RI strains to make 40 nonoverlapping RIXs. Vendors may be willing to do this for you if the strains are not available to you. At this point you will almost surely have a small set of reasonably well mapped loci. You will also have enough data to decide if you want to reevaluate your questions. Are you really after QTGs, do you want to test a specific intervention, or do you want to try your luck at G2P prediction using a set of molecular and genetic biomarkers? This first set of 40 RIX progeny should enable you to do all three.

$$n = 160$$

Group 3B: Same as above but this set could be generated to test an intervention or age (using Group 3A as a control). Or this RIX group could be created selectively to test multilocus interactions or parent-of-origin effects.

$$n = 160$$

2.5.4 Stage 4: High-Resolution Mapping, Predictive Validation, and Systems Genetics

The combined results of the three stages should have left you with a set of loci mapped to less than 2 Mb. If that is not the case, then this final stage should help achieve that goal. Ideally, you might want to select DO stock on the basis of genotype, and that may be a service that will soon be available. This would be most useful if only one specific haplotype contrast is generating trait variance (e.g., a 1 vs 7 split of haplotype effects).

Group 4A: DO or HS. DO stock will probably be most accessible and also generally most suitable.

$$n = 100$$

Group 4B: Your wildcard. You could continue with a second set of 100 DO mice if the first results strengthened results. Or you could use the DO mice you still have to selectively cross animals with specific combinations of alleles. This would require selective genotyping of specific SNPs. DO mice are a wonderful source of genetic variance, but you may want to select or trim back some of those variants. This will position you well to predict phenotypes based on combinations of haplotypes at two or three loci.

Alternatively, use this group of cases for further studies on the effects of treatment, age or stage (see Group 3B).

$$n = 100$$

Question 11: How do I genotype DO or HS? **ANSWER**: Even in the most demanding situation of mapping DO, HS, and wild caught populations, markers need only be about 100 kb apart [79], and since the mouse genome is about 2.5 Gb, 100,000 well chosen markers will be more than adequate. Virtually any population, no matter how complex its genetic architecture can now be typed using the latest version of the mouse universal genotyping array (the GigaMUGA) or by sparse sequencing for about $100/case [88].

For selective genotyping of a handful of markers in DO or RCC F2 intercrosses you can use standard protocols that will probably require acquiring sets of PCR primers. Costs may be as high as $1/genotype/case. If you require a few hundred markers per case then a good ballpark cost for custom genotyping is under $0.10–0.20 per genotype per case—or $20–40 for 200–2000 markers for an F2 progeny. Finally almost all inbred, RI strains, an RIX progeny are already well typed and there is no cost at all.

Question 12: Is there a strong justification to use all of these types of resource—RIs, RIX and HS/DO? ANSWER: These resource types perform many of the same functions. However, G × E will be easier to study using RI and RIX. RIX progeny made using CC RIs are genetically similar to DO animals, but incorporate fewer recombinations per animal. Data from RIX cases can also be used to build up a phenome database and are potentially more useful for large collaborative teams, but this advantage may remain theoretical for the next several years. DO/HS animals are logistically far easier to obtain and provide you with access to the ultimate breadth of genetic and phenotypic diversity. They are the closest you can get to a wild-type mouse population short of capturing your own. If you results from Stages 1 to 3 are supported in DO populations, then you can be sure that results will have the maximum replicability and perhaps even translatability to human populations. You may also be able to computationally and genetically "extract" specific disease models from RI, CC, and DO stock.

3 Future Directions and Conclusions

Thanks to the massively reduced cost and increased scope of omics technologies, it is now feasible for small collaborative groups—and even single research groups—to execute large studies in systems genetics. We can anticipate that the use of this new systems paradigm will accelerate in the coming years with the advent of new and improved methods of quantifying an individual's proteomes, metabolomes, metagenomes, and epigenomes as a function of cell type, tissue, age, and state. It is great to have the core animal resources that are needed to take advantage of this rapidly expanding set of omics technologies.

What we have not considered in this chapter is the analytic and synthetic tools needed for high-content systems genetics. How do we actually map aggregated data from 1000 cases with complex substructure? How do we build predictive models and test their fit to empirical data? These questions are taken up in many of the chapters in this volume.

4 Notes

1. What is *population substructure* and how does it make statistical analysis and mapping trickier? We all have learned that observations used in many statistical tests should be independent. In genetic crosses all F2 progeny are usually treated as independent observations. But what if there are strong litter effects, or batch confounds due to technical errors. These effects can introduce variance into a cross that can obscure the detection of the genuine effects and produce spurious linkage. Similarly, in an AI cross, one mating pair may produce 50 siblings whereas another mating pair produces only 5. In this case we have known and unbalanced pedigree substructure that needs to be corrected even when doing something as simple as computing a correlation coefficient. Large GWASs sometimes combine data from different ethnicities and it is also essential to correct statistically for the kinship relations among members. In some cases we can use the genotypes of cases to compute a matrix of kinship similarity, and use this matrix to correct for the population substructure. If we know the litter and batch identifiers we can also adjust for these nuisance variables in a statistical model.

   In large RI sets such as the BXDs and CC, there is cryptic substructure that may not show up easily in genotypes but that that may still be important. The BXDs for example, were generated in multiple cohorts between 1970 and 2013 using the same parental strains—C57BL/6J and DBA/2J, but of course, 43 years of breeding history will add many new variants to both parents and some of these are already well known to have important effects [89].

2. To estimate empirical precision for QTLs across a population in GeneNetwork (www.genenetwork.org) you first need to select an expression data set from the pull-down menu. In this example, select **Species** = *Mouse*, **Group** = *B6D2 AI PSU*, **Type** = *Muscle mRNA*, and **Data Set** = *PSU B6D2 AI Muscle…*

   Enter this query into the **Get Any** box:

   cisLRS = (23 46 50)

   where cisLRS is the linkage statistic specifically for the *cis*-acting eQTLs. The first two values in parentheses are the minimum and maximum LRS values to return (LRS = LOD × 4.61), and the final number is the size in megabases of the acceptance window used to define how close a gene must be to the QTL peak to be considered *cis*-acting. In this case the acceptance window is very broad, and the peak LRS can be anywhere 50 Mb on either side of the gene.

   This search will generate 2086 hits. You can resort and download the results as an Excel table using the **Download Table** button. In this large F2 intercross with more than 800 cases generated by Ari Lionikas and colleagues between C57B/6J and DBA/2J, the mean offset between 2000 *cis*-eQTLs with LOD between 5 and 10 and their genes is 7.0 ± 0.21 Mb. If you try precisely the same set of operations with a matched BXD Advanced RI data set (*EPFL/LISP BXD CD+HFD... Exon Level*) you will find that the mean offset between 4400 *cis*-eQTL in this data set is 2.0 ± 0.06 Mb. The latter ARI data set is based on ~320 cases (1 array with 4–5 pooled samples for each of 40 strains under two conditions—high fat and standard chow diet [21].

3. Genetic studies usually benefit more by increasing the *n* of genometypes that are phenotyped than by increasing the *n* of replicates per type (e.g., Fig. 1b in ref. 90, and *see* ref. 65). All else being equal, a studying of 160 types without replication should be superior in terms of QTL results to one of 40 strains and 4 replicates of each. This is obvious for Mendelian traits such as coat color, but it also holds true for quantitative traits—even those with low heritability. However, at an early stage of a study it is vital to understand heritability and technical confounds and in some cases, replication is easy and cheap. For this reason, it is a good idea to begin work with six to eight replicates of a few "reference" genomes. When using isogenic cases we recommend two replicates minimum, one per sex. Bumping this up to two per sex per strain will improve the comfort level of many reviewers, although to keep them happy you will probably need 6–8 per group. There are also some good reasons to study six or more cases per genometype even after heritability is known: such as studies of genetics control of variation itself [91] or pharmacological effect thresholds.

   One way to think about the diminishing returns of high replication rates is to compare *t* scores and *z* scores required to achieve statistical significance for simple two-sample comparisons using different sample sizes. The *z* score assumes variance of the population is known and the critical value to reject the null at alpha 0.05 is $z = 1.96$. In contrast, the *t* score estimates variance from the sample itself, and the critical values start at a woefully high 12.71 for $n = 2$, but drops toward the asymptote of 1.96 very quickly: 3.182 for $n = 4$, 2.757 for $n = 6$, and 2.201 for $n = 12$.

Acknowledgments

We thank the support of the UT Center for Integrative and Translational Genomics, and funds from the UT-ORNL Governor's Chair.

References

1. Williams EG, Auwerx J (2015) The convergence of systems and reductionist approaches in complex trait analysis. Cell 162:23–32
2. Dietrich WF, Copeland NG, Gilbert DJ, Miller JC, Jenkins NA et al (1995) Mapping the mouse genome: current status and future prospects. Proc Natl Acad Sci U S A 92:10849–10853
3. Petkov PM, Cassell MA, Sargent EE, Donnelly CJ, Robinson P et al (2004) Development of a SNP genotyping panel for genetic monitoring of the laboratory mouse. Genomics 83:902–911
4. Williams RW, Gu J, Qi S, Lu L (2001) The genetic structure of recombinant inbred mice: high-resolution consensus maps for complex trait analysis. Genome Biol 2:RESEARCH0046
5. Shifman S, Bell JT, Copley RR, Taylor MS, Williams RW, Mott R, Flint J (2006) A high-resolution single nucleotide polymorphism genetic map of the mouse genome. PLoS Biol 4:e395
6. Yang H, Bell TA, Churchill GA, de Villena Pardo-Manuel F (2007) On the subspecific origin of the laboratory mouse. Nat Genet 39:1100–1107
7. Morgan AP, Fu CP, Kao CY, Welsh CE, Didion JP et al (2015) The mouse universal genotyping array: from substrains to subspecies. G3 (Bethesda) 6:263–279
8. Lander ES, Botstein D (1989) Mapping Mendelian factors underlying quantitative traits using RFLP linkage maps. Genetics 121:185–199
9. Darvasi A, Soller M (1995) Advanced intercross lines, an experimental population for fine genetic mapping. Genetics 141:1199–1207
10. Darvasi A (1998) Experimental strategies for the genetic dissection of complex traits in animal models. Nat Genet 18:19–24
11. Talbot CJ, Nicod A, Cherny SS, Fulker DW, Collins AC, Flint J (1999) High-resolution mapping of quantitative trait loci in outbred mice. Nat Genet 21:305–318
12. Complex Trait Consortium (2003) The nature and identification of quantitative trait loci: a community's view. Nat Rev Genet 4:911–916
13. Churchill GA, Airey DC, Allayee H, Angel JM, Attie AD et al (2004) The Collaborative Cross, a community resource for the genetic analysis of complex traits. Nat Genet 36:1133–1137
14. Singer JB, Hill AE, Burrage LC, Olszens KR, Song J et al (2004) Genetic dissection of complex traits with chromosome substitution strains of mice. Science 304:445–448
15. Flint J, Valdar W, Shifman S, Mott R (2005) Strategies for mapping and cloning quantitative trait genes in rodents. Nat Rev Genet 6:271–286
16. Broman KW (2005) The genomes of recombinant inbred lines. Genetics 169:1133–1146
17. Threadgill DW, Hunter KW, Williams RW (2002) Genetic dissection of complex and quantitative traits: from fantasy to reality via a community effort. Mamm Genome 13:175–178
18. Malmanger B, Lawler M, Coulombe S, Murray R, Cooper S, Polyakov Y, Belknap J, Hitzemann R (2006) Further studies on using multiple-cross mapping (MCM) to map quantitative trait loci. Mamm Genome 17:1193–1204
19. Ghazalpour A, Rau CD, Farber CR, Bennett BJ, Orozco LD et al (2012) Hybrid mouse diversity panel: a panel of inbred mouse strains suitable for analysis of complex genetic traits. Mamm Genome 23:680–692
20. Houtkooper RH, Mouchiroud L, Ryu D, Moullan N, Katsyuba E et al (2013) Mitonuclear protein imbalance as a conserved longevity mechanism. Nature 497:451–457
21. Williams EG, Mouchiroud L, Frochaux M, Pandey A, Andreux PA et al (2014) An evolutionarily conserved role for the aryl hydrocarbon receptor in the regulation of movement. PLoS Genet 10:e1004673
22. Wang X, Pandey AK, Mulligan MK, Williams EG, Mozhui K et al (2016) Joint mouse-human phenome-wide association to test gene function and disease risk. Nat Commun 7:10464
23. Lippert C, Listgarten J, Liu Y, Kadie CM, Davidson RI, Heckerman D (2011) FaST linear mixed models for genome-wide association studies. Nat Methods 8:833–835
24. Zhou X, Stephens M (2012) Genome-wide efficient mixed-model analysis for association studies. Nat Genet 44:821–824
25. Furlotte NA, Kang EY, Van Nas A, Farber CR, Lusis AJ et al (2012) Increasing association mapping power and resolution in mouse genetic studies through the use of meta-analysis for structured populations. Genetics 191:959–967
26. Koutnikova H, Markku L, Lu L, Combe R, Paananen J et al (2009) Identification of UBP1 as a critical blood pressure determinant. PLoS Genet 5:e1000591
27. Mozhui K, Wang X, Chen J, Mulligan MK, Li Z et al (2011) Genetic regulation of Nrxnl expression: an integrative cross-species analysis of schizophrenia candidate genes. Transl Psychiatry 1:e25

28. Subramanian S, Tus K, Li QZ, Wang A, Tian XH et al (2006) A Tlr7 translocation accelerates systemic autoimmunity in murine lupus. Proc Natl Acad Sci U S A 103:9970–9975
29. Hunter KW, Crawford NP (2008) The future of mouse QTL mapping to diagnose disease in mice in the age of whole-genome association studies. Annu Rev Genet 42:131–141
30. Hu Y, Wu G, Rusch M, Lukes L, Buetow KH et al (2012) Integrated cross-species transcriptional network analysis of metastatic susceptibility. Proc Natl Acad Sci U S A 109:3184–3189
31. Morahan G (2012) Insights into type 1 diabetes provided by genetic analyses. Curr Opin Endocrinol Diabetes Obes 19:263–270
32. Mackay TF (2001) The genetic architecture of quantitative traits. Annu Rev Genet 35:303–339
33. Fisher R (1918) The correlation between relatives on the supposition of Mendelian inheritance. Phil Trans R Soc Edin 52:399–433
34. Wright S (1921) Correlation and causation. J Agric Res 20:557–585
35. Lenarcic AB, Svenson KL, Churchill GA, Valdar W (2012) A general Bayesian approach to analyzing diallel crosses of inbred strains. Genetics 190:413–435
36. Airey DW, Lu L, Shou S, Williams RW (2002) Genetic sources of individual differences in cerebellum. Cerebellum 1:233–240
37. Crowley JJ, Kim Y, Lenarcic AB, Quackenbush CR, Barrick CJ et al (2014) Genetics of adverse reactions to haloperidol in a mouse diallel: a drug-placebo experiment and Bayesian causal analysis. Genetics 196:321–347
38. Percival CJ, Liberton DK, Pardo-Manuel de Villena F, Spritz R, Marcucio R et al (2016) Genetics of murine craniofacial morphology: diallel analysis of the eight founders of the Collaborative Cross. J Anat 228:96–112
39. Dowell RD, Ryan O, Jansen A, Cheung D, Agarwala S et al (2010) Genotype to phenotype: a complex problem. Science 328:469
40. Bennett BJ, Davis RC, Civelek M, Orozco L, Wu J et al (2015) Genetic architecture of atherosclerosis in mice: a systems genetics analysis of common inbred strains. PLoS Genet 11:e1005711
41. Bogue MA, Peters LL, Paigen B, Korstanje R, Yuan R et al (2014) Accessing data resources in the mouse phenome database for genetic analysis of murine life span and health span. J Gerontol A Biol Sci Med Sci 71:170–177
42. Keane TM, Goodstadt L, Danecek P, White MA, Wong K et al (2011) Mouse genomic variation and its effect on phenotypes and gene regulation. Nature 477:289–294
43. Bogue MA, Grubb SC (2004) The mouse phenome project. Genetica 122:71–74
44. Kumar V, Kim K, Joseph C, Kourrich S, Yoo SH et al (2013) C57BL/6N mutation in cytoplasmic FMRP interacting protein 2 (Cyfip2) regulates cocaine response. Science 342:1508–1512
45. Heiker JT, Kunath A, Kosacka J, Flehmig G, Knigge A et al (2014) Identification of genetic loci associated with different responses to high-fat diet-induced obesity in C57BL/6N and C57BL/6J substrains. Physiol Genomics 46:377–384
46. Khisti RT, Wolstenholme J, Shelton KL, Miles MF (2006) Characterization of the ethanol-deprivation effect in substrains of C57BL/6 mice. Alcohol 40:119–126
47. Mulligan MK, Ponomarev I, Boehm SL II, Owen JA, Levin PS et al (2008) Alcohol trait and transcriptional genomic analysis of C57BL/6 substrains. Genes Brain Behav 7:677–689
48. Simon MM, Greenaway S, White JK, Fuchs H, Gailus-Durner V et al (2013) A comparative phenotypic and genomic analysis of C57BL/6J and C57BL/6N mouse strains. Genome Biol 14:R82
49. Kirkpatrick SL, Bryant CD (2015) Behavioral architecture of opioid reward and aversion in C57BL/6 substrains. Front Behav Neurosci 8:450
50. Rosen GD, Azoulay NG, Griffin EG, Newbury A, Koganti L et al (2013) Bilateral subcortical heterotopia with partial callosal agenesis in a mouse mutant. Cereb Cortex 23:859–872
51. Cardin S, Scott-Boyer MP, Praktiknjo S, Jeidane S, Picard S et al (2014) Differences in cell-type-specific responses to angiotensin II explain cardiac remodeling differences in C57BL/6 mouse substrains. Hypertension 64:1040–1046
52. Li Z, Mulligan MK, Wang X, Miles MF, Lu L et al (2010) A transposon in Comt generates mRNA variants and causes widespread expression and behavioral differences among mice. PLoS One 5:e12181
53. Gregorova S, Divina P, Storchova R, Trachtulec Z, Fotopulosova V et al (2008) Mouse consomic strains: exploiting genetic divergence between Mus m. musculus and Mus m. domesticus subspecies. Genome Res 18:509–515
54. Buchner DA, Nadeau JH (2015) Contrasting genetic architectures in different mouse reference populations used for studying complex traits. Genome Res 25:775–791
55. Davis RC, Schadt EE, Smith DJ, Hsieh EW, Cervino AC et al (2005) A genome-wide set of congenic mouse strains derived from DBA/2J

on a C57BL/6J background. Genomics 86:259–270

56. Bryant CD, Kole LA, Guido MA, Sokoloff G, Palmer AA (2012) Congenic dissection of a major QTL for methamphetamine sensitivity implicates epistasis. Genes Brain Behav 11:623–632
57. Williams RW (1999) A targeted screen to detect recessive mutations that have quantitative effects. Mamm Genome 10:734–738
58. Damerval C, Maurice A, Josse JM, de Vienne D (1994) Quantitative trait loci underlying gene product variation: a novel perspective for analyzing regulation of genome expression. Genetics 137:289–301
59. Schadt EE, Monks SA, Drake TA, Lusis AJ, Che N et al (2003) Genetics of gene expression surveyed in maize, mouse and man. Nature 422:297–302
60. Schadt EE, Lamb J, Yang X, Zhu J, Edwards S et al (2005) An integrative genomics approach to infer causal associations between gene expression and disease. Nat Genet 37:710–717
61. Miller RA, Harrison DE, Astle CM, Floyd RA, Flurkey K et al (2007) An aging interventions testing program: study design and interim report. Aging Cell 6:565–575
62. Bailey DW (1971) Recombinant-inbred strains. An aid to finding identity, linkage, and function of histocompatibility and other genes. Transplantation 11:325–327
63. Taylor BA, Heiniger HJ, Meier H (1973) Genetic analysis of resistance to cadmium-induced testicular damage in mice. Proc Soc Exp Biol Med 143:629–633
64. Gora-Maslak G, McClearn GE, Crabbe JC, Phillips TJ, Belknp JK et al (1992) Use of recombinant inbred strains to identify quantitative trait loci in psychopharmacology. Psychopharmacology 104:413–424
65. Belknap JK (1998) Effect of within-strain sample size on QTL detection and mapping using recombinant inbred mouse strains. Behav Genet 28:29–38
66. Pandey A, Williams RW (2014) Genetics of gene expression in the CNS. Int Rev Neurobiol 116:195–231
67. Mulligan MK, Mozhui K, Prins P, Williams RW (2016) GeneNetwork – A toolbox for systems genetics. Methods Mol Biol (this volume)
68. Yalcin B, Flint J, Mott R (2005) Using progenitor strain information to identify quantitative trait nucleotides in outbred mice. Genetics 171:673–681
69. Williams RW, Strom RC, Goldowitz D (1998) Natural variation in neuron number in mice is linked to a major quantitative trait locus on Chr 11. J Neurosci 18:138–146
70. Roberts A, Pardo-Manuel de Villena F, Wang W, McMillan L, Threadgill DW (2007) The polymorphism architecture of mouse genetic resources elucidated using genome-wide resequencing data: implications for QTL discovery and systems genetics. Mamm Genome 18:473–481
71. Peirce JL, Lu L, Gu J, Silver LM, Williams RW (2004) A new set of BXD recombinant inbred lines from advanced intercross populations in mice. BMC Genet 5:7
72. Hrbek T, de Brito RA, Wang B, Pletscher LS, Cheverud JM (2006) Genetic characterization of a new set of recombinant inbred lines (LGXSM) formed from the inter-cross of SM/J and LG/J inbred mouse strains. Mamm Genome 17:417–429
73. Rasmussen AL, Okumura A, Ferris MT, Green R, Feldmann F et al (2014) Host genetic diversity enables Ebola hemorrhagic fever pathogenesis and resistance. Science 346:987–991
74. Hegmann JP, Possidente B (1981) Estimating genetic correlations from inbred strains. Behav Genet 11:103–114
75. Williams RW, Strom RC, Rice DS, Goldowitz D (1996) Genetic and environmental control of variation in retinal ganglion cell number in mice. J Neurosci 16:7193–7205
76. Overall RW, Kempermann G, Peirce J, Lu L, Goldowitz D et al (2009) Genetics of the hippocampal transcriptome in mouse: a systematic survey and online neurogenomics resource. Front Neurosci 3:55
77. Lynch CJ (1969) The so-called Swiss mouse. Lab Anim Care 19:214–220
78. Chia R, Achilli F, Festing MF, Fisher EM (2005) The origins and uses of mouse outbred stocks. Nat Genet 37:1181–1186
79. Yalcin B, Nicod J, Bhomra A, Davidson S, Cleak J et al (2010) Commercially available outbred mice for genome-wide association studies. PLoS Genet 6:e1001085
80. Kakihana R, Brown DR, McClearn GE, Tabershaw IR (1966) Brain sensitivity to alcohol in inbred mouse strains. Science 154:1574–1575
81. Holmes RS, Petersen DR, Deitrich RA (1986) Biochemical genetic variants in mice selectively bred for sensitivity or resistance to ethanol-induced sedation. Anim Genet 17:235–244
82. Hitzemann B, Dains K, Kanes S, Hitzemann R (1994) Further studies on the relationship between dopamine cell density and haloperidol-induced catalepsy. J Pharmacol Exp Ther 271:969–976

83. Valdar W, Soberg LC, Gaugiuier D, Burnett S, Klenerman P et al (2006) Genome-wide genetic association of complex traits in heterogeneous stock mice. Nat Genet 38:879–887
84. Churchill GA, Gatti DM, Munger SC, Svenson KL (2012) The diversity outbred mouse population. Mamm Genome 23:713–718
85. Svenson KL, Gatti DM, Valdar W, Welsh CE, Cheng R et al (2012) High-resolution genetic mapping using the mouse diversity outbred population. Genetics 190:437–447
86. Graham JB, Thomas S, Swarts J, McMillan AA, Ferris MT et al (2015) Genetic diversity in the collaborative cross model recapitulates human West Nile virus disease outcomes. MBio 6:e00493–15
87. Smemo S, Tena JJ, Kim KH, Gamazon ER, Sakabe NJ et al (2014) Obesity-associated variants within FTO form long-range functional connections with IRX3. Nature 507:371–375
88. Rat Genome Sequencing and Mapping Consortium, Baud A, Hermsen R, Guryev V, Stridh P et al (2013) Combined sequence-based and genetic mapping analysis of complex traits in outbred rats. Nat Genet 45:767–775
89. Anderson MG, Smith RS, Hawes NL, Zabaleta A, Chang B, Wiggs JL, John SW (2002) Mutations in genes encoding melanosomal proteins cause pigmentary glaucoma in DBA/2J mice. Nat Genet 30:81–85
90. Andreux PA, Williams EG, Koutnikova H, Houtkooper RH, Champy MF et al (2012) Systems genetics of metabolism: the use of the BXD murine reference panel for multiscalar integration of traits. Cell 150:1287–1299
91. Rönnegård L, Valdar W (2011) Detecting major genetic loci controlling phenotypic variability in experimental crosses. Genetics 188:435–447

Chapter 2

Heterogeneous Stock Populations for Analysis of Complex Traits

Leah C. Solberg Woods and Richard Mott

Abstract

Heterogeneous Stock (HS) populations allow for fine-resolution genetic mapping of a variety of complex traits. HS mice and rats were created from breeding together eight inbred strains, followed by maintaining the colony in a manner that minimizes inbreeding. After 50 or more generations of breeding, the resulting animals' chromosomes represent a genetic mosaic of the founders' haplotypes, with the average distance between recombination events in the centiMorgan range. This allows for genetic mapping to only a few Mb, a much smaller region than what can be identified using traditional F2 intercross or backcross mapping strategies. HS animals have been used to fine-map a variety of complex traits including anxiety and fear behaviors, diabetes, asthma, and heart disease, among others. Once a quantitative trait locus (QTL) has been identified, founder sequence and expression analysis can be used to identify underlying causal genes. In the following review, we provide an overview of how HS rats and mice have been used to identify genetic loci, and in some cases the causal genes, underlying complex traits. We discuss the creation and breeding strategies for both HS rats and mice. We then discuss the statistical analyses used to identify genetic loci, as well as strategies to identify causal genes underlying these loci. We end the chapter by discussing limitations faced when using HS populations, including several statistical challenges that have not been fully resolved.

Key words Resources for systems genetics, Genetic mapping, Outbred mice and rats, Expression analysis

1 Introduction

HS populations of mice and rats are created from breeding together eight inbred strains, followed by maintaining the colony in a manner that minimizes inbreeding (Fig. 1). After 50 or more generations of breeding, the resulting animals' chromosomes are a genetic mosaic of the founders' haplotypes, with the average distance between recombination events in the centiMorgan range allowing genetic mapping to only a few Mb [1, 2]. HS animals exhibit a high degree of both genetic and phenotypic diversity, allowing high-resolution genetic mapping for a wide variety of traits. While both HS rats and mice were originally created to be a resource population for experimental and selection studies [3, 4], Flint and

Klaus Schughart and Robert W. Williams (eds.), *Systems Genetics: Methods and Protocols*, Methods in Molecular Biology, vol. 1488, DOI 10.1007/978-1-4939-6427-7_2, © Springer Science+Business Media New York 2017

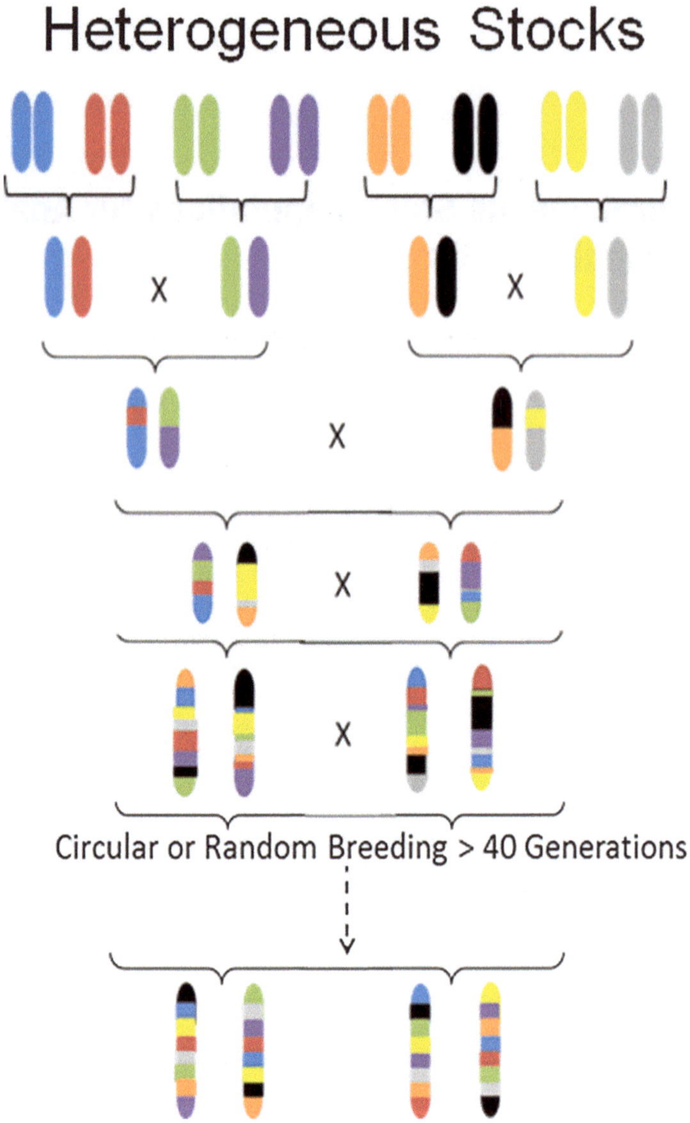

Fig. 1 Breeding scheme for heterogeneous stock (HS) populations. HS are created by breeding together eight inbred strains. Once all eight genomes are combined, the animals are bred using either a circular strategy or random breeding. Existing HS mouse and rat colonies have been bred for over 50 generations. Figure adapted from Solberg Woods, 2014

colleagues demonstrated in 1999 that these populations can be used to narrow a previously identified quantitative trait locus (QTL) for anxiety [5] to only 0.8 cM [6], a large improvement over mapping studies using traditional F2 intercross or backcross approaches which generally map to 30 cM or more. Since that time, multiple studies have used HS rats and mice for genetic mapping of complex traits. Whilst this chapter focuses on mouse and rat HS, similar types of populations—in which each individual's genome is a mosaic of the founders—have been made in these and other species [7–10]. Other outbred populations, such as advanced intercross lines [11, 12], the mouse Diversity Outcross (also created from eight inbred founder strains; [13]), and commercially available outbreds [14, 15], are also available and have been reviewed previously [16].

Upon demonstrating success of the HS strategy for fine-mapping a single locus for a behavioral trait, the Flint lab went on to use HS mice to conduct a large multi-phenotype study, including traits involved in fear and anxiety behaviors, diabetes and asthma, and others. The study included the largest cohort of mice at that time (1904 mice) and resulted in the identification of 843 QTL at 25 % false discovery rate (FDR) with an average confidence interval of only 2.8 Mb [17]. In separate studies, HS mice have also been used to fine-map traits such as fear [18, 19], ethanol-induced locomotor activity [20], and arthritis [21]. Two HS mouse colonies have been created: the Boulder HS [4] and the Northport HS [20]. Six of the inbred founder strains are shared between these stocks, namely A/J, AKR/J, BALB/cJ, C3H/HeJ, C57BL/6J, DBA/2J. The additional founders of the Boulder HS are the strains Is/Bi and RI11, while those of the Northport HS are CBA/2J and LP/J. Colonies were created and maintained with 24–40 breeder pairs using either a circular or random mating scheme.

With the success of HS mice, investigators began to also use HS rats for genetic fine-mapping. The HS rat colony (N:NIH-HS) was first established by the NIH in 1984 using the following inbred strains: ACI/N,BN/SsN,BUF/N, F344/N, M520/N, MR/N, WKY/N, WN/N [3]. The colony was maintained using 60 breeder pairs using a random mating strategy. Upon the retirement of the colony's originator, Dr. Carl Hansen, in 2006 the colony was transferred to the laboratory of Dr. Eva Redei at Northwestern University where it was maintained for 2 years. In 2006, Dr. Redei transferred breeder pairs to the Medical College of Wisconsin in the United States and the Autonomous University of Barcelona in Spain. The Medical College of Wisconsin currently maintains 64 breeder pairs and is using a random breeding strategy, using kinship coefficent to ensure closely related pairs are not bred together. A smaller colony is also currently being maintained at Barcelona (Fernandez Teruel, personal communication). The first genetic mapping study in HS rats fine-mapped a single locus for glucose tolerance from 60 Mb to only 2.4 Mb [22]. Using expression QTL analysis and founder sequence data, *Tpcn2* was identified as the likely causal gene within this region within only a few years [23]. Since that time, Flint and colleagues again conducted a large multi-phenotype study, including traits involved in anxiety, heart disease, and multiple sclerosis among others [24]. They were able to

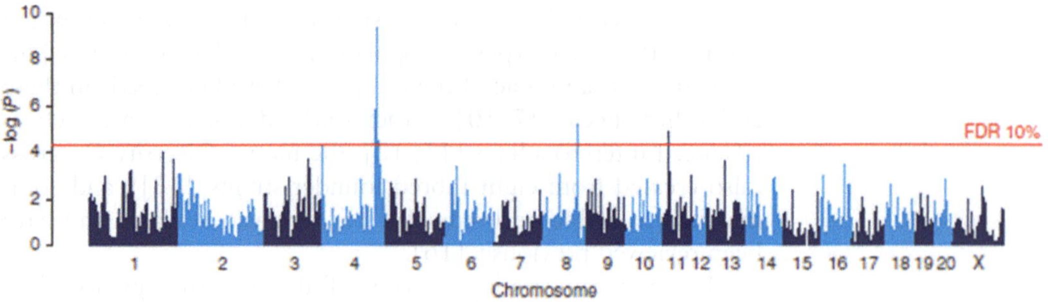

Fig. 2 Genome scan for platelet aggregation as shown in Baud et al., 2013. The scan shows the results of a haplotype-based mixed model. The *y*-axis shows the negative log *P* values for association with variation in platelet aggregation. The association peak on chromosome 4 harbors the von Willebrand factor gene that was identified through sequence analysis as the causative gene

fine-map 355 QTL for 122 traits at 10% FDR. An example of a scan for platelet aggregation is shown in Fig. 2. Using a merge analysis (described below) and protein modeling, they were able to identify 35 probable causative genes within these loci. These data are described in detail and publicly available at [25]. We note that different analysis methodologies were used in the mouse and rat HS experiments, reflecting methodological improvements (principally the development of mixed models, described below) in the interim. A recent study has also fine-mapped bone structure and strength in HS rats [26] and additional studies have demonstrated phenotypic variability, and thus suitability for future genetic studies, for additional phenotypic traits including kidney traits [27], drug abuse behaviors [28, 29], and behavioral and physiological responses to stress [30–32] and to ethanol [33–35]. Because of the rich history of the rat in behavioral studies, the HS rat will likely prove a useful model for genetic dissection of behaviors that are not easily modeled in the mouse (*see* ref. 36). The utility of the HS rat will also be enhanced by the recent availability of gene knockouts and other genetic manipulations now available in the rat [37, 38].

2 Methods: Statistical Analysis and Systems Genetics in HS Populations

In order to perform genetic mapping in HS populations, the underlying ancestral haplotype mosaics must first be determined. In this way, one can compute the probability that a particular locus in a given individual descended from which of the eight founder strains, thus providing increased information over simply analyzing genotype data (often based on biallelic single nucleotide polymorphisms or SNPs). HAPPY, a program developed by Mott and colleagues [2], uses a hidden Markov Model to determine the ancestral probabilities and has been shown to significantly improve genetic mapping results. Other programs, such as DOQTL which was more recently developed for use in the Diversity Outbred (DO) mouse population, a population of mice created from eight founder strains, very similar to the HS [39], can also be used to determine ancestral probabilities in the HS. Once probabilities are determined, regression modeling is conducted on the underlying mosaic structure to identify QTL.

2.1 Programs for QTL Analysis

Several programs are available for identifying QTL including Bagpipe [40], QTLRel [41], and DOQTL [39]. As with other highly recombinant populations, it is important to account for the complex family relationships within the HS population when conducting the analysis [24, 40, 42]. Most colonies are maintained using either a random or circular breeding strategy with anywhere from 40 to 60 breeding pairs, generally sufficient to minimize inbreeding and control genetic drift. Random mating strategies have the advantage of also avoiding reduced map expansion and therefore may be preferable to circular mating strategies [43]. As a result of the closed nature of the breeding strategy, animals within a colony are all related to each other but to differing degrees. If not accounted for, false positive QTL will be identified simply on the basis of relatedness, as opposed to actually pertaining to the phenotype. Moreover, in HS studies, in order to generate a large sample of animals for phenotyping, it is necessary to expand the colony size resulting in a large number of families. Thus the analysis of the phenotyped individuals acquires a mixture of linkage (due to family structure) and association (because all individuals are ultimately descended from eight founders). If the parents of the phenotyped generation are genotyped along with the phenotyped mice, then it is also possible to infer the maternal and paternal origin of the alleles, allowing the study of parent-of-origin effects [44].

There are several ways to account for unequal degrees of relatedness in outbred populations. These include mixed modeling approaches such as EMMA [45] or by resample model averaging [40]. When genome-wide genotype information is unavailable, and/or when the full pedigree is unknown, family can also be accounted for by including this as a random term in the model, as previously demonstrated for a single locus on rat chromosome 1 [22, 46]. When full pedigree information is known, QTLRel can be used to account for family relationships in highly recombinant animal populations such as the HS [41]. Resample model averaging approaches make use of genome-wide genotypes to determine genetic relatedness directly, and may prove advantageous under certain circumstances, particularly when pedigree information is unknown [40]. For mixed models, a kinship matrix is used to determine the genetic relatedness of each pair of individuals. This can be simply computed from SNP data (in the same way that it is computed in human studies) or from the ancestral haplotype mosaics [44]. Baud et al. [24] used a mixed model to control for differences in relatedness, whilst Valdar et al. 2006 [17] used resample model averaging (developed further in [40]). Each method has advantages and disadvantages. The mixed-model methodology is essentially equivalent to transforming both phenotypes and genotypes by multiplication by the square root of the inverse of the variance-covariance matrix, to create an uncorrelated dataset that can be analyzed by ordinary least squares. The method is well established and works well on phenotypes that are approximately normally distributed. Resampling methods, on the other hand, work well on phenotypes that are strongly skewed in distribution, but are slower than mixed models.

2.2 Identification of Causal Genes and Variants

Once QTLs are identified, there are several methods that can be used to identify the underlying causal gene(s) within the locus. These include a statistical merge analysis [47] (a form of genotype imputation), expression QTL mapping, and protein modeling. To date, complete genomes have been sequenced in the founder strains of the HS mice [48] and rats [24]. Relative to the respective reference genomes, more than four million SNPs per strain have been identified in the mouse [48] and more than two million SNPs per strain have been identified in the rat [24], in addition to structural variants and insertion/deletions. Because a repetitive portion of the genome (~15% in mouse and ~12% in rat) could not be reliably mapped to the reference strains, the number of variants identified is likely much larger than reported [24, 48]. The available sequence information can be used in several ways to identify candidate genes and/or variants within a fine-mapped QTL. By coupling founder sequences with relatively dense genotyping of the outbred population, it is possible to impute HS genotypes at all possible SNPs within a QTL. This can be followed by a merge analysis to narrow the potential causative variants within the QTL [47]. Briefly, a merge analysis uses probabilistically inferred descent to impute genotype dosages at unobserved loci, and then surveys those multiple imputed SNPs for their association with the phenotype. Using this method, two statistical models are compared: the haplotype model and the allelic model. In the haplotype model, the underlying ancestral probabilities at each SNP are used to model the QTL, with each founder haplotype permitted to carry a different phenotypic effect. This haplotype model is compared with one in which only the alleles for that SNP are used (allelic model). In the allelic model, the founder strain alleles are "merged" into two groups for each diallelic SNP: those containing allele "a" at a locus of interest and those containing allele "b" at this locus [47], with each group having a single phenotypic effect. Potentially causative variants are those in which the allelic model provides a better fit, that is, explaining the same amount of variation but with far fewer parameters, than the haplotype model (*see* ref. 21, 47).

This method has proven useful in narrowing the number of causative variants within QTL mapped in HS mice [48] and rats [24]. A single causal variant, however, is rarely identified and follow-up studies are often needed to identify the causal variant. In addition, the method works less well when multiple causal variants underlie a single QTL [24], but can be used to show that a QTL cannot be explained by a single biallelic variant (as was the case for about half the QTLs detected by Baud et al.). Merge analysis is most useful for excluding genetic variants that cannot be causal.

Transcript abundance levels, based on RNAseq or microarray data, can also be used to identify causal genes underlying a QTL, as well as identification of gene networks that play a role in a given trait. Expression QTL (eQTL) analysis in HS populations allows investigators to map both cis and trans-eQTL to within only a few Mb of the transcript itself [49]. Overlap of cis-eQTL with physiological QTL can then be used to identify candidate genes within an interval, as demonstrated in HS rats by Tsaih et al. [23]. Our group identified *Tpcn2* as a cis-eQTL within a physiological QTL for fasting glucose and insulin levels and demonstrated that glucose levels strongly correlated with *Tpcn2* expression levels in the HS rats. We then demonstrated that fasting glucose and insulin in response to a glucose challenge were altered in *Tpcn2* knock-out mice, and *Tpcn2* was nominally correlated with fasting insulin levels in humans, providing evidence that *Tpcn2* is the likely causal gene within this region. Transcript abundance levels can also be used to identify gene networks (groups of correlated transcripts) that may play a role in disease [50–53]. Although this strategy has not been applied specifically to HS populations to date, it offers a promising avenue of research for the future. Similarly, regions of the genome associated with open chromatin, such as DNAse-1 hypersensitive sites, identified in the mouse reference genome by the mouse ENCODE project [54] and across the HS founder strains [55] may be used to help identify causal variants and genes.

3 Further Considerations and Limitations

3.1 General Considerations

Using outbred HS populations offers several advantages over traditional F2 intercross or backcross strategies. The first is the ability to fine-map to only a few Mb, greatly decreasing the number of potential candidate genes within a given locus. Once loci are identified, full genome sequence is available for founder strains of the HS mouse [48] and HS rat [24] and this information has proven to be invaluable for identifying causative genes and variants within fine-mapped QTL. Despite these advantages, there are also several disadvantages that should be considered prior to embarking on a study with HS populations. Each animal is genetically and phenotypically distinct, so once a QTL, or even a gene, has been identified, there is no inbred model to go back to for functional testing, although the inbred founders might be used for this purpose. In addition, large numbers of animals are needed to have sufficient power for each mapping study and high density genotyping platforms (generally 10 thousand or greater) are required. Because each animal is unique, all animals need to be genotyped and phenotyped with each new study, as opposed to recombinant inbred lines where genotypes need be collected only once. As a result, it is beneficial to gather as much phenotype information as possible from the same group of animals, so that genotyping only needs to be done once and this information can be used to map multiple traits (e.g., [17, 24]). A further disadvantage is that these populations have been created through a single funnel (i.e., combining founder genomes only once), leading to loss of certain alleles and potentially unbalanced representation of the founder genomes. There are further considerations regarding confirmation of a potentially causal gene, and several statistical challenges that have not been resolved when using highly recombinant populations such as the HS.

3.2 Considerations in Determining Causality

Once a candidate gene is identified, follow-up studies are needed to confirm or disprove the role of that candidate in the trait. In addition to replication in a separate cohort, conducting a cross-species comparison can help provide support for the gene of interest. Of particular interest is human genome-wide association data which is often publicly available and can be mined to determine if a gene of interest falls just below the genome-wide significance threshold in humans. Once there is sufficient evidence to suspect a causal role for a specific gene, one of the most popular methods used to verify this gene is to study it in a knock-out model. Such methods have been available in the mouse since 1990 [56] and have recently become available in the rat [57]. Although popular because of their relative ease of constructing a knock-out, it is important to recognize that showing a change in phenotype in a knock-out model neither proves nor disproves a causative role of this gene at the QTL [58], particularly because there is no way to create a knock-out on the same background in which the QTL was identified. Methods such as quantitative complementation offer an alternative approach to test a causal role of the gene or variant [59, 60]. More importantly, new gene targeting approaches which allow for changes in single base pairs are now being used and offer a more realistic approach than a full gene knock-out (*see* ref. 61). The revolution in gene editing due to CRISPR/Cas9 technology (recently applied to rats in [38]) suggests that gene confirmation will become more straightforward in the future, at least for isolated coding variants. However, cases where multiple causal variants, carried on a single haplotype, are implicated will likely remain a challenge to prove causality, particularly if their effect is regulatory. It is therefore important for investigators to assess all available information, including expression, sequence, cross-species comparisons, results from a knock-out, allelic changes using CRISPR/Cas9 modifications, as well as possible in vitro studies to assess the potential causative role of a particular gene and/or variant.

3.3 Statistical Limitations

In addition to the above considerations, there are several statistical concerns that need to be taken into consideration when analyzing outbred HS populations. One of these is how best to determine significance thresholds. Cheng and Palmer [62] recently compared four different methods used in an advanced intercross population. They found that as long as an appropriate statistical model (i.e., one that takes into account the complex family relationships) is used, all methods worked relatively well, with gene-dropping (a simulation technique used in pedigree analysis) decreasing false QTL even when family is not taken into account. The best way to determine QTL confidence intervals is another challenge. Many studies use the 1.5 LOD drop method, (e.g., [46, 63, 64]), which was initially developed for use in F2 intercross populations. An alternative approach is to use nonparametric bootstrapping [65], in which the QTL is re-estimated under alternative datasets based on the original, with each alternative dataset created by resampling the individuals with replacement [66]. Although this method has been shown to be overly conservative [67], it does provide a complementary estimate of how sensitive the localization of the top QTL peak is to resampling, thus providing insight into whether more than one locus may underlie the QTL (*see* ref. 46).

Accurate determination of the joint effects of diplotypes (i.e., combinations of founder haplotypes—effectively any departure from the assumption of additivity in the haplotype effects) is also an on-going statistical challenge in outbred populations. Using the DO mouse population, investigators have looked at the effects of just the founder allele effects within the QTL [13, 63]. This has been useful in conducting haplotype analysis and narrowing the region of the QTL. However, within the HS or DO populations, there are in effect 36 possible diplotype combinations, and founder effects account for only eight of these. We [46] have recently published methods that accounts for all 36 possible diplotype effects and work in this area is on-going (*see* ref. 68).

A final statistical concern is that of statistical power. Although previous power calculations in multi-founder populations suggest that 1000–1500 animals provide sufficient power for mapping QTL explaining 5% of the variance [2, 69, 70], these simulations do not account for the confounding effects of relatedness (e.g., [40, 42]), or marker ascertainment (e.g., [71]), and are therefore likely overstated. A previous study in the diversity outbred mouse population used as few as 150 mice; however this study provided sufficient power to map only 11 of 113 traits that were measured [13]. Studies in HS populations have used over 1000 animals, successfully mapping most traits analyzed [17, 24], demonstrating the increased power of these studies. In order to have more accurate power estimates for future studies in these populations, power calculations will need to account for both family structure and polygenes. That said, increasing sample size has many benefits: there is greater power to detect QTLs of small effect, QTL effect sizes are less likely to be overestimated due to Winner's Curse, and confidence intervals are likely to be smaller and more accurate.

4 Conclusion

HS populations have proven useful for fine-mapping complex traits to only a few Mb, rapidly narrowing the number of potential candidate genes within the locus. Several strategies, including use of full founder sequence and expression QTL mapping, have been used to identify the underlying causal genes within these loci. It is important, however, to consider the cost and labor intensive challenges of working with HS populations, as well as the many unanswered statistical challenges that still remain. Despite these challenges, use of outbred models such as the HS has and will continue to enhance the knowledge of the genetic architecture of complex traits.

References

1. Mott R, Flint J (2013) Dissecting quantitative traits in mice. Annu Rev Genomics Hum Genet 14:421–439. doi:10.1146/annurev-genom-091212-153419
2. Mott R, Talbot CJ, Turri MG, Collins AC, Flint J (2000) A method for fine mapping quantitative trait loci in outbred animal stocks. Proc Natl Acad Sci U S A 97(23):12649–12654. doi:10.1073/pnas.230304397
3. Hansen C, Spuhler K (1984) Development of the National Institutes of Health genetically heterogeneous rat stock. Alcohol Clin Exp Res 8(5):477–479
4. McClearn GE, Wilson JR, Meredith W (1970) The use of isogenic and heterogenic mouse stocks in behavioral research. In: Lindzey G, Thiessen D (eds) Contributions to behavior-genetic analysis: the mouse as a prototype. Appleton Century Crofts, New York, pp 3–22
5. Flint J, Corley R, DeFries JC, Fulker DW, Gray JA, Miller S, Collins AC (1995) A simple genetic basis for a complex psychological trait in laboratory mice. Science 269(5229):1432–1435
6. Talbot CJ, Nicod A, Cherny SS, Fulker DW, Collins AC, Flint J (1999) High-resolution mapping of quantitative trait loci in outbred mice. Nat Genet 21(3):305–308. doi:10.1038/6825
7. Aylor DL, Valdar W, Foulds-Mathes W, Buus RJ, Verdugo RA, Baric RS, Ferris MT, Frelinger JA, Heise M, Frieman MB, Gralinski LE, Bell TA, Didion JD, Hua K, Nehrenberg DL, Powell CL, Steigerwalt J, Xie Y, Kelada SN, Collins FS, Yang IV, Schwartz DA, Branstetter LA, Chesler EJ, Miller DR, Spence J, Liu EY, McMillan L, Sarkar A, Wang J, Wang W, Zhang Q, Broman KW, Korstanje R, Durrant C, Mott R, Iraqi FA, Pomp D, Threadgill D, de Villena FP, Churchill GA (2011) Genetic analysis of complex traits in the emerging Collaborative Cross. Genome Res 21(8):1213–1222. doi:10.1101/gr.111310.110
8. Durrant C, Tayem H, Yalcin B, Cleak J, Goodstadt L, de Villena FP, Mott R, Iraqi FA (2011) Collaborative Cross mice and their power to map host susceptibility to Aspergillus fumigatus infection. Genome Res 21(8):1239–1248. doi:10.1101/gr.118786.110
9. Kover PX, Valdar W, Trakalo J, Scarcelli N, Ehrenreich IM, Purugganan MD, Durrant C, Mott R (2009) A multiparent advanced generation inter-cross to fine-map quantitative traits in Arabidopsis thaliana. PLoS Genet 5(7):e1000551. doi:10.1371/journal.pgen.1000551

10. Long AD, Macdonald SJ, King EG (2014) Dissecting complex traits using the Drosophila Synthetic Population Resource. Trends Genet 30(11):488–495. doi:10.1016/j.tig.2014.07.009
11. Darvasi A, Soller M (1995) Advanced intercross lines, an experimental population for fine genetic mapping. Genetics 141(3):1199–1207
12. Parker CC, Cheng R, Sokoloff G, Palmer AA (2012) Genome-wide association for methamphetamine sensitivity in an advanced intercross mouse line. Genes Brain Behav 11(1):52–61. doi:10.1111/j.1601-183X.2011.00747.x
13. Svenson KL, Gatti DM, Valdar W, Welsh CE, Cheng R, Chesler EJ, Palmer AA, McMillan L, Churchill GA (2012) High-resolution genetic mapping using the mouse diversity outbred population. Genetics 190(2):437–447. doi:10.1534/genetics.111.132597
14. Yalcin B, Flint J (2012) Association studies in outbred mice in a new era of full-genome sequencing. Mamm Genome 23(9–10):719–726. doi:10.1007/s00335-012-9409-z
15. Zhang W, Korstanje R, Thaisz J, Staedtler F, Harttman N, Xu L, Feng M, Yanas L, Yang H, Valdar W, Churchill GA, Dipetrillo K (2012) Genome-wide association mapping of quantitative traits in outbred mice. G3 (Bethesda) 2(2):167–174. doi:10.1534/g3.111.001792
16. Solberg Woods LC (2014) QTL mapping in outbred populations: successes and challenges. Physiol Genomics 46(3):81–90. doi:10.1152/physiolgenomics.00127.2013
17. Valdar W, Solberg LC, Gauguier D, Burnett S, Klenerman P, Cookson WO, Taylor MS, Rawlins JN, Mott R, Flint J (2006) Genome-wide genetic association of complex traits in heterogeneous stock mice. Nat Genet 38(8):879–887
18. Johannesson M, Lopez-Aumatell R, Stridh P, Diez M, Tuncel J, Blazquez G, Martinez-Membrives E, Canete T, Vicens-Costa E, Graham D, Copley RR, Hernandez-Pliego P, Beyeen AD, Ockinger J, Fernandez-Santamaria C, Gulko PS, Brenner M, Tobena A, Guitart-Masip M, Gimenez-Llort L, Dominiczak A, Holmdahl R, Gauguier D, Olsson T, Mott R, Valdar W, Redei EE, Fernandez-Teruel A, Flint J (2009) A resource for the simultaneous high-resolution mapping of multiple quantitative trait loci in rats: the NIH heterogeneous stock. Genome Res 19(1):150–158. doi:10.1101/gr.081497.108
19. Talbot CJ, Radcliffe RA, Fullerton J, Hitzemann R, Wehner JM, Flint J (2003) Fine scale mapping of a genetic locus for conditioned fear. Mamm Genome 14(4):223–230
20. Demarest K, Koyner J, McCaughran J Jr, Cipp L, Hitzemann R (2001) Further characterization and high-resolution mapping of quantitative trait loci for ethanol-induced locomotor activity. Behav Genet 31(1):79–91
21. Johnsen AK, Valdar W, Golden L, Ortiz-Lopez A, Hitzemann R, Flint J, Mathis D, Benoist C (2011) Genome-wide and species-wide dissection of the genetics of arthritis severity in heterogeneous stock mice. Arthritis Rheum 63(9):2630–2640. doi:10.1002/art.30425
22. Solberg Woods LC, Holl K, Tschannen M, Valdar W (2010) Fine-mapping a locus for glucose tolerance using heterogeneous stock rats. Physiol Genomics 41(1):102–108. doi:10.1152/physiolgenomics.00178.2009, 00178.2009 [pii]
23. Tsaih SW, Holl K, Jia S, Kaldunski M, Tschannen M, He H, Andrae JW, Li SH, Stoddard A, Wiederhold A, Parrington J, Ruas da Silva M, Galione A, Meigs J, Hoffmann RG, Simpson P, Jacob H, Hessner M, Solberg Woods LC (2014) Identification of a novel gene for diabetic traits in rats, mice, and humans. Genetics 198(1):17–29. doi:10.1534/genetics.114.162982
24. Baud A, Hermsen R, Guryev V, Stridh P, Graham D, McBride MW, Foroud T, Calderari S, Diez M, Ockinger J, Beyeen AD, Gillett A, Abdelmagid N, Guerreiro-Cacais AO, Jagodic M, Tuncel J, Norin U, Beattie E, Huynh N, Miller WH, Koller DL, Alam I, Falak S, Osborne-Pellegrin M, Martinez-Membrives E, Canete T, Blazquez G, Vicens-Costa E, Mont-Cardona C, Diaz-Moran S, Tobena A, Hummel O, Zelenika D, Saar K, Patone G, Bauerfeind A, Bihoreau MT, Heinig M, Lee YA, Rintisch C, Schulz H, Wheeler DA, Worley KC, Muzny DM, Gibbs RA, Lathrop M, Lansu N, Toonen P, Ruzius FP, de Bruijn E, Hauser H, Adams DJ, Keane T, Atanur SS, Aitman TJ, Flicek P, Malinauskas T, Jones EY, Ekman D, Lopez-Aumatell R, Dominiczak AF, Johannesson M, Holmdahl R, Olsson T, Gauguier D, Hubner N, Fernandez-Teruel A, Cuppen E, Mott R, Flint J (2013) Combined sequence-based and genetic mapping analysis of complex traits in outbred rats. Nat Genet 45(7):767–775. doi:10.1038/ng.2644
25. Baud A, Guryev V, Hummel O, Johannesson M, Flint J (2014) Genomes and phenomes of a population of outbred rats and its progenitors. Sci Data 1:140011. doi:10.1038/sdata.2014.11
26. Alam I, Koller DL, Canete T, Blazquez G, Mont-Cardona C, Lopez-Aumatell R, Martinez-Membrives E, Diaz-Moran S, Tobena A, Fernandez-Teruel A, Stridh P, Diez M, Olsson T, Johannesson M, Baud A, Econs MJ, Foroud T (2015) Fine mapping of bone structure and strength QTLs in heterogeneous stock rat. Bone 81:417–426. doi:10.1016/j.bone.2015.08.013

27. Solberg Woods LC, Stelloh C, Regner KR, Schwabe T, Eisenhauer J, Garrett MR (2010) Heterogeneous stock rats: a new model to study the genetics of renal phenotypes. Am J Physiol Renal Physiol 298(6):F1484–F1491. doi:10.1152/ajprenal.00002.2010, 00002.2010 [pii]

28. Richards JB, Lloyd DR, Kuehlewind B, Militello L, Paredez M, Solberg Woods L, Palmer AA (2013) Strong genetic influences on measures of behavioral-regulation among inbred rat strains. Genes Brain Behav 12(5):490–502. doi:10.1111/gbb.12050

29. Wang T, Han W, Wang B, Jiang Q, Solberg-Woods LC, Palmer AA, Chen H (2014) Propensity for social interaction predicts nicotine-reinforced behaviors in outbred rats. Genes Brain Behav 13(2):202–212. doi:10.1111/gbb.12112

30. Diaz-Moran S, Palencia M, Mont-Cardona C, Canete T, Blazquez G, Martinez-Membrives E, Lopez-Aumatell R, Tobena A, Fernandez-Teruel A (2012) Coping style and stress hormone responses in genetically heterogeneous rats: comparison with the Roman rat strains. Behav Brain Res 228(1):203–210. doi:10.1016/j.bbr.2011.12.002, S0166-4328(11)00844-8 [pii]

31. Lopez-Aumatell R, Guitart-Masip M, Vicens-Costa E, Gimenez-Llort L, Valdar W, Johannesson M, Flint J, Tobena A, Fernandez-Teruel A (2008) Fearfulness in a large N/Nih genetically heterogeneous rat stock: differential profiles of timidity and defensive flight in males and females. Behav Brain Res 188(1):41–55. doi:10.1016/j.bbr.2007.10.015, S0166-4328(07)00559-1 [pii]

32. Lopez-Aumatell R, Vicens-Costa E, Guitart-Masip M, Martinez-Membrives E, Valdar W, Johannesson M, Canete T, Blazquez G, Driscoll P, Flint J, Tobena A, Fernandez-Teruel A (2009) Unlearned anxiety predicts learned fear: a comparison among heterogeneous rats and the Roman rat strains. Behav Brain Res 202(1):92–101. doi:10.1016/j.bbr.2009.03.024, S0166-4328(09)00185-5 [pii]

33. Bice PJ, Liang T, Zhang L, Graves TJ, Carr LG, Lai D, Kimpel MW, Foroud T (2010) Fine mapping and expression of candidate genes within the chromosome 10 QTL region of the high and low alcohol-drinking rats. Alcohol 44(6):477–485. doi:10.1016/j.alcohol.2010.06.004, S0741-8329(10)00079-0 [pii]

34. Foroud T, Bice P, Castelluccio P, Bo R, Miller L, Ritchotte A, Lumeng L, Li TK, Carr LG (2000) Identification of quantitative trait loci influencing alcohol consumption in the high alcohol drinking and low alcohol drinking rat lines. Behav Genet 30(2):131–140

35. Spuhler K, Deitrich RA (1984) Correlative analysis of ethanol-related phenotypes in rat inbred strains. Alcohol Clin Exp Res 8(5):480–484

36. Parker CC, Chen H, Flagel SB, Geurts AM, Richards JB, Robinson TE, Solberg Woods LC, Palmer AA (2014) Rats are the smart choice: rationale for a renewed focus on rats in behavioral genetics. Neuropharmacology 76 Pt B:250–258. doi:10.1016/j.neuropharm.2013.05.047

37. Katter K, Geurts AM, Hoffmann O, Mates L, Landa V, Hiripi L, Moreno C, Lazar J, Bashir S, Zidek V, Popova E, Jerchow B, Becker K, Devaraj A, Walter I, Grzybowksi M, Corbett M, Filho AR, Hodges MR, Bader M, Ivics Z, Jacob HJ, Pravenec M, Bosze Z, Rulicke T, Izsvak Z (2013) Transposon-mediated transgenesis, transgenic rescue, and tissue-specific gene expression in rodents and rabbits. FASEB J 27(3):930–941. doi:10.1096/fj.12-205526

38. Zhao L, Oliver E, Maratou K, Atanur SS, Dubois OD, Cotroneo E, Chen CN, Wang L, Arce C, Chabosseau PL, Ponsa-Cobas J, Frid MG, Moyon B, Webster Z, Aldashev A, Ferrer J, Rutter GA, Stenmark KR, Aitman TJ, Wilkins MR (2015) The zinc transporter ZIP12 regulates the pulmonary vascular response to chronic hypoxia. Nature 524(7565):356–360. doi:10.1038/nature14620

39. Gatti DM, Svenson KL, Shabalin A, Wu LY, Valdar W, Simecek P, Goodwin N, Cheng R, Pomp D, Palmer A, Chesler EJ, Broman KW, Churchill GA (2014) Quantitative trait locus mapping methods for diversity outbred mice. G3 (Bethesda) 4(9):1623–1633. doi:10.1534/g3.114.013748

40. Valdar W, Holmes CC, Mott R, Flint J (2009) Mapping in structured populations by resample model averaging. Genetics 182(4):1263–1277. doi:10.1534/genetics.109.100727, genetics.109.100727 [pii]

41. Cheng R, Abney M, Palmer AA, Skol AD (2011) QTLRel: an R package for genome-wide association studies in which relatedness is a concern. BMC Genet 12:66. doi:10.1186/1471-2156-12-66, doi:1471-2156-12-66 [pii]

42. Cheng R, Lim JE, Samocha KE, Sokoloff G, Abney M, Skol AD, Palmer AA (2010) Genome-wide association studies and the problem of relatedness among advanced intercross lines and other highly recombinant populations. Genetics 185(3):1033–1044.

doi:10.1534/genetics.110.116863, genetics.110.116863 [pii]

43. Rockman MV, Kruglyak L (2008) Breeding designs for recombinant inbred advanced intercross lines. Genetics 179(2):1069–1078. doi:10.1534/genetics.107.083873
44. Mott R, Yuan W, Kaisaki P, Gan X, Cleak J, Edwards A, Baud A, Flint J (2014) The architecture of parent-of-origin effects in mice. Cell 156(1–2):332–342. doi:10.1016/j.cell.2013.11.043
45. Kang HM, Zaitlen NA, Wade CM, Kirby A, Heckerman D, Daly MJ, Eskin E (2008) Efficient control of population structure in model organism association mapping. Genetics 178(3):1709–1723. doi:10.1534/genetics.107.080101, doi:178/3/1709 [pii]
46. Solberg Woods LC, Holl KL, Oreper D, Xie Y, Tsaih SW, Valdar W (2012) Fine-mapping diabetes-related traits, including insulin resistance, in heterogeneous stock rats. Physiol Genomics 44(21):1013–1026. doi:10.1152/physiolgenomics.00040.2012
47. Yalcin B, Flint J, Mott R (2005) Using progenitor strain information to identify quantitative trait nucleotides in outbred mice. Genetics 171(2):673–681
48. Keane TM, Goodstadt L, Danecek P, White MA, Wong K, Yalcin B, Heger A, Agam A, Slater G, Goodson M, Furlotte NA, Eskin E, Nellaker C, Whitley H, Cleak J, Janowitz D, Hernandez-Pliego P, Edwards A, Belgard TG, Oliver PL, McIntyre RE, Bhomra A, Nicod J, Gan X, Yuan W, van der Weyden L, Steward CA, Bala S, Stalker J, Mott R, Durbin R, Jackson IJ, Czechanski A, Guerra-Assuncao JA, Donahue LR, Reinholdt LG, Payseur BA, Ponting CP, Birney E, Flint J, Adams DJ (2011) Mouse genomic variation and its effect on phenotypes and gene regulation. Nature 477(7364):289–294. doi:10.1038/nature10413
49. Huang GJ, Shifman S, Valdar W, Johannesson M, Yalcin B, Taylor MS, Taylor JM, Mott R, Flint J (2009) High resolution mapping of expression QTLs in heterogeneous stock mice in multiple tissues. Genome Res 19(6):1133–1140. doi:10.1101/gr.088120.108
50. Chen Y, Zhu J, Lum PY, Yang X, Pinto S, MacNeil DJ, Zhang C, Lamb J, Edwards S, Sieberts SK, Leonardson A, Castellini LW, Wang S, Champy MF, Zhang B, Emilsson V, Doss S, Ghazalpour A, Horvath S, Drake TA, Lusis AJ, Schadt EE (2008) Variations in DNA elucidate molecular networks that cause disease. Nature 452(7186):429–435. doi:10.1038/nature06757, nature06757 [pii]
51. Ghazalpour A, Doss S, Zhang B, Wang S, Plaisier C, Castellanos R, Brozell A, Schadt EE, Drake TA, Lusis AJ, Horvath S (2006) Integrating genetic and network analysis to characterize genes related to mouse weight. PLoS Genet 2(8):e130. doi:10.1371/journal.pgen.0020130, 06-PLGE-RA-0128R2 [pii]
52. Keller MP, Choi Y, Wang P, Belt Davis D, Rabaglia ME, Oler AT, Stapleton DS, Argmann C, Schueler KL, Edwards S, Steinberg HA, Chaibub Neto E, Kleinhanz R, Turner S, Hellerstein MK, Schadt EE, Yandell BS, Kendziorski C, Attie AD (2008) A gene expression network model of type 2 diabetes links cell cycle regulation in islets with diabetes susceptibility. Genome Res 18(5):706–716. doi:10.1101/gr.074914.107, gr.074914.107 [pii]
53. Schadt EE, Lamb J, Yang X, Zhu J, Edwards S, Guhathakurta D, Sieberts SK, Monks S, Reitman M, Zhang C, Lum PY, Leonardson A, Thieringer R, Metzger JM, Yang L, Castle J, Zhu H, Kash SF, Drake TA, Sachs A, Lusis AJ (2005) An integrative genomics approach to infer causal associations between gene expression and disease. Nat Genet 37(7):710–717. doi:10.1038/ng1589, ng1589 [pii]
54. Stamatoyannopoulos JA, Snyder M, Hardison R, Ren B, Gingeras T, Gilbert DM, Groudine M, Bender M, Kaul R, Canfield T, Giste E, Johnson A, Zhang M, Balasundaram G, Byron R, Roach V, Sabo PJ, Sandstrom R, Stehling AS, Thurman RE, Weissman SM, Cayting P, Hariharan M, Lian J, Cheng Y, Landt SG, Ma Z, Wold BJ, Dekker J, Crawford GE, Keller CA, Wu W, Morrissey C, Kumar SA, Mishra T, Jain D, Byrska-Bishop M, Blankenberg D, Lajoie BR, Jain G, Sanyal A, Chen KB, Denas O, Taylor J, Blobel GA, Weiss MJ, Pimkin M, Deng W, Marinov GK, Williams BA, Fisher-Aylor KI, Desalvo G, Kiralusha A, Trout D, Amrhein H, Mortazavi A, Edsall L, McCleary D, Kuan S, Shen Y, Yue F, Ye Z, Davis CA, Zaleski C, Jha S, Xue C, Dobin A, Lin W, Fastuca M, Wang H, Guigo R, Djebali S, Lagarde J, Ryba T, Sasaki T, Malladi VS, Cline MS, Kirkup VM, Learned K, Rosenbloom KR, Kent WJ, Feingold EA, Good PJ, Pazin M, Lowdon RF, Adams LB (2012) An encyclopedia of mouse DNA elements (Mouse ENCODE). Genome Biol 13(8):418. doi:10.1186/gb-2012-13-8-418
55. Hosseini M, Goodstadt L, Hughes JR, Kowalczyk MS, de Gobbi M, Otto GW, Copley RR, Mott R, Higgs DR, Flint J (2013) Causes and consequences of chromatin variation between inbred mice. PLoS Genet 9(6):e1003570. doi:10.1371/journal.pgen.1003570

56. Thomas KR, Capecchi MR (1990) Targeted disruption of the murine int-1 proto-oncogene resulting in severe abnormalities in midbrain and cerebellar development. Nature 346(6287):847–850. doi:10.1038/346847a0
57. Geurts AM, Cost GJ, Freyvert Y, Zeitler B, Miller JC, Choi VM, Jenkins SS, Wood A, Cui X, Meng X, Vincent A, Lam S, Michalkiewicz M, Schilling R, Foeckler J, Kalloway S, Weiler H, Menoret S, Anegon I, Davis GD, Zhang L, Rebar EJ, Gregory PD, Urnov FD, Jacob HJ, Buelow R (2009) Knockout rats via embryo microinjection of zinc-finger nucleases. Science 325(5939):433. doi:10.1126/science.1172447, doi:325/5939/433 [pii]
58. Flint J, Eskin E (2012) Genome-wide association studies in mice. Nat Rev Genet 13(11):807–817. doi:10.1038/nrg3335
59. Long AD, Mullaney SL, Mackay TF, Langley CH (1996) Genetic interactions between naturally occurring alleles at quantitative trait loci and mutant alleles at candidate loci affecting bristle number in Drosophila melanogaster. Genetics 144(4):1497–1510
60. Yalcin B, Willis-Owen SA, Fullerton J, Meesaq A, Deacon RM, Rawlins JN, Copley RR, Morris AP, Flint J, Mott R (2004) Genetic dissection of a behavioral quantitative trait locus shows that Rgs2 modulates anxiety in mice. Nat Genet 36(11):1197–1202
61. Adams DJ, van der Weyden L (2008) Contemporary approaches for modifying the mouse genome. Physiol Genomics 34(3):225–238. doi:10.1152/physiolgenomics.90242.2008
62. Cheng R, Palmer AA (2013) A simulation study of permutation, bootstrap, and gene dropping for assessing statistical significance in the case of unequal relatedness. Genetics 193(3):1015–1018. doi:10.1534/genetics.112.146332
63. Logan RW, Robledo RF, Recla JM, Philip VM, Bubier JA, Jay JJ, Harwood C, Wilcox T, Gatti DM, Bult CJ, Churchill GA, Chesler EJ (2013) High-precision genetic mapping of behavioral traits in the diversity outbred mouse population. Genes Brain Behav 12(4):424–437. doi:10.1111/gbb.12029
64. Samocha KE, Lim JE, Cheng R, Sokoloff G, Palmer AA (2010) Fine mapping of QTL for prepulse inhibition in LG/J and SM/J mice using F(2) and advanced intercross lines. Genes Brain Behav 9(7):759–767. doi:10.1111/j.1601-183X.2010.00613.x, GBB613 [pii]
65. Hastie T, Tibshirani R, Friedman J (2009) The elements of statistical learning: data mining, inference, and prediction, 2nd edn. Springer, New York
66. Visscher PM, Thompson R, Haley CS (1996) Confidence intervals in QTL mapping by bootstrapping. Genetics 143(2):1013–1020
67. Manichaikul A, Dupuis J, Sen S, Broman KW (2006) Poor performance of bootstrap confidence intervals for the location of a quantitative trait locus. Genetics 174(1):481–489. doi:10.1534/genetics.106.061549, genetics.106.061549 [pii]
68. Durrant C, Mott R (2010) Bayesian quantitative trait locus mapping using inferred haplotypes. Genetics 184(3):839–852. doi:10.1534/genetics.109.113183, genetics.109.113183 [pii]
69. Mott R, Flint J (2002) Simultaneous detection and fine mapping of quantitative trait loci in mice using heterogeneous stocks. Genetics 160(4):1609–1618
70. Valdar WS, Flint J, Mott R (2003) QTL finemapping with recombinant-inbred heterogeneous stocks and in vitro heterogeneous stocks. Mamm Genome 14(12):830–838. doi:10.1007/s00335-003-3021-1
71. Valdar W, Flint J, Mott R (2006) Simulating the collaborative cross: power of quantitative trait loci detection and mapping resolution in large sets of recombinant inbred strains of mice. Genetics 172(3):1783–1797. doi:10.1534/genetics.104.039313, doi:genetics.104.039313 [pii]

Part II

Tools for Analysis and Integration in Systems Genetics

Chapter 3

Mouse Genome Informatics (MGI): Resources for Mining Mouse Genetic, Genomic, and Biological Data in Support of Primary and Translational Research

Janan T. Eppig, Cynthia L. Smith, Judith A. Blake, Martin Ringwald, James A. Kadin, Joel E. Richardson, and Carol J. Bult

Abstract

The Mouse Genome Informatics (MGI), resource (www.informatics.jax.org) has existed for over 25 years, and over this time its data content, informatics infrastructure, and user interfaces and tools have undergone dramatic changes (Eppig et al., Mamm Genome 26:272–284, 2015). Change has been driven by scientific methodological advances, rapid improvements in computational software, growth in computer hardware capacity, and the ongoing collaborative nature of the mouse genomics community in building resources and sharing data. Here we present an overview of the current data content of MGI, describe its general organization, and provide examples using simple and complex searches, and tools for mining and retrieving sets of data.

Key words Resources for systems genetics, Database, Genome informatics

1 Introduction and Background (Box 1)

The Mouse Genome Informatics (MGI) resource (www.informatics.jax.org) [1] represents a collection of highly interactive integrated database projects with the shared mission of facilitating the use of mouse as a model for studies of human biology and disease. The core component of MGI is the Mouse Genome Database (MGD) [2], which provides the gold-standard, authoritative source, for many mouse genome data types, including the unified genome features catalog and genome locations, functional annotations for mouse protein-coding genes, the comprehensive compendium of mouse mutants and their phenotypes, strains, and variants, and experimentally defined models of human genetic diseases. The Gene Expression Database (GXD, www.informatics.jax.org/expression.shtml) [3] component of MGI provides extensive data on expression assays, results, and images, with an emphasis on expression during mouse embryonic development for both wild-type and

Klaus Schughart and Robert W. Williams (eds.), *Systems Genetics: Methods and Protocols*, Methods in Molecular Biology, vol. 1488, DOI 10.1007/978-1-4939-6427-7_3,

mutant genotypes. Researchers planning conditional mutagenesis experiments will find the Cre Portal (www.creportal.org) [4] web site essential for examining Cre line specificity data that enables optimal selection of Cre lines for conditional mutagenesis experiments. The Mouse Tumor Biology Database (MTB, www.tumor.informatics.jax.org) [5] component provides mouse cancer model data to aid in selection of appropriate strains for experimentation, and provides the ability to review and compare tumor incidence and patterns of mutations in specific cancer models. Recent work incorporates PDX (Patient-Derived Xenograft) data derived from engrafting human tumors into NSG™ immune-deficient mice [6]. The International Mouse Strain Resource (IMSR, www.findmice.org) [7] facilitates users' ability to find and order mouse strains and mutants (live, cryopreserved embryos or sperm, or in ES cell lines) from repositories worldwide. The newest addition to MGI resources is the Human–Mouse: Disease Connection portal (HMDC, www.diseasemodel.org) [8] a tool allowing translational and clinical researchers to explore established and candidate relationships between mouse phenotypes and human diseases, and providing links for obtaining mouse models available from repositories.

Box 1. Acronyms, Used

Acronym	Acronym meaning	URL for data source or definition
APF	Australian Phenomics Facility	http://www.apf.edu.au
API	Application Programming Interface	https://en.wikipedia.org/wiki/Application_programming_interface
BioGrid	Biological General Repository for Interaction Datasets	http://thebiogrid.org
CNV	Copy number variation	https://en.wikipedia.org/wiki/Copy-number_variation
CrePortal	Recombinase (cre) Activity Portal	http://www.informatics.jax.org/recombinase.shtml
CvDC	Cardiovascular Development Consortium	http://www.benchtobassinet.com/About/AboutCvDC.aspx
dbSNP	Single-Nucleotide Polymorphism database	http://www.ncbi.nlm.nih.gov/SNP/index.html
EMAP	Edinburgh Mouse Atlas , Project, Mouse Developmental Anatomy	http://www.emouseatlas.org/emap/ema/DAOAnatomyJSP/anatomy.html
EMAPA	Edinburgh Mouse Atlas Project, Abstract Mouse, with developmental stages grouped	http://www.emouseatlas.org/emap/ema/theiler_stages/StageDefinition/AnatomyOntology/AbstractMouseontology.html
FAQ	Frequently Asked Questions	

Acronym	Acronym meaning	URL for data source or definition
FTP	File Transfer Protocol	https://en.wikipedia.org/wiki/File_Transfer_Protocol
GAF	GO Annotation Format	http://geneontology.org/page/go-annotation-file-formats
GO	Gene Ontology	http://geneontology.org
GOA	Gene , Ontology Annotation Project (EBI)	http://www.ebi.ac.uk/GOA
GXD	Gene Expression Database	http://www.informatics.jax.org/expression.shtml
Havana	Human and Vertebrate Analysis and Annotation	http://www.sanger.ac.uk/science/groups/vertebrate-annotation
HCOP	HGNC Comparison of Orthology Predictions	http://www.genenames.org/cgi-bin/hcop
HGNC	HUGO , Gene Nomenclature Committee	http://www.genenames.org
HMDC	Human–Mouse Disease Connection	http://www.diseasemodel.org
Homologene	NCBI's automated system for sequence-based homology grouping in eukaryotes	http://www.ncbi.nlm.nih.gov/homologene
HPO	Human Phenotype Ontology	http://www.human-phenotype-ontology.org
IKMC	International , Knockout Mouse Consortium	http://www.mousephenotype.org/about-ikmc
IMPC	International Mouse Phenotyping Consortium	http://www.mousephenotype.org
IntAct	Interaction Database	http://www.ebi.ac.uk/intact/
InterPro	Protein Sequence Analysis & Classification	http://www.ebi.ac.uk/interpro/
MGD	Mouse Genome Database	http://www.informatics.jax.org
MGI	Mouse , Genome Informatics	http://www.informatics.jax.org
miRBase	microRNA Database	http://www.mirbase.org
MP	Mammalian Phenotype Ontology	http://www.informatics.jax.org/searches/MP_form.shtml
MPD	Mouse Phenome Database	http://phenome.jax.org
MTB	Mouse Tumor Biology Database	http://tumor.informatics.jax.org/mtbwi/index.do

Acronym	Acronym meaning	URL for data source or definition
Mutagenetix	ENU mutagenesis program at University of Texas Southwestern	https://mutagenetix.utsouthwestern.edu
NCBI	National Center for , Biotechnology Information	http://www.ncbi.nlm.nih.gov
OMIM	Online Mendelian Inheritance in Man	http://www.omim.org
Panther	Protein Analysis Through Evolutionary Relationships	http://www.pantherdb.org
PDX	Patient-Derived Xenograft	http://tumor.informatics.jax.org/mtbwi/pdxSearch.do;jsessionid=A95DE8D26ACA65271B5A8C9C232A936A
QTL	Quantitative Trait Loci	https://en.wikipedia.org/wiki/Quantitative_trait_locus
RefSeq	NCBI's Reference , Sequence Database	http://www.ncbi.nlm.nih.gov/refseq/
SNPs	Single-Nucleotide Polymorphisms	http://ghr.nlm.nih.gov/handbook/genomicresearch/snp
SO	Sequence Ontology	http://www.sequenceontology.org
UniGene	Unified view of the transcriptome	http://www.ncbi.nlm.nih.gov/unigene
UniProt KB	Universal Protein Resource Knowledgebase	http://www.uniprot.org
URL	Uniform Resource Identifier	https://en.wikipedia.org/wiki/Uniform_Resource_Locator
VCF	, Variant Call Files	http://samtools.github.io/hts-specs/VCFv4.2.pdf
VEGA	Vertebrate Genome Annotation Database	http://vega.sanger.ac.uk/index.html
VLAD	Visual Annotation Display	http://proto.informatics.jax.org/prototypes/vlad/

2 Methods

2.1 Types of Data and Principle Analysis Tools

2.1.1 Data Types

MGI includes a significant breadth and depth of genetic, genomic, and biological data for the laboratory mouse. As the central mission of MGI is to support the use of mouse as a model for human biology and disease, significant effort goes to standardizing and curating, data so that they can be used as an integrated whole. Table 1 provides an overview of major data

Table 1
Major data areas of MGI and MGI's role in providing these data to the greater scientific community

Data class	Data description	Sources/methods
Genomic	Unified nonredundant catalog of genome features and locations, with links to providers and sequences	With each new genome build, gene predictions from NCBI, Ensembl, and Havana/VEGA are downloaded and equivalent gene models among their assemblies are algorithmically determined using MGI's fjoin algorithm [13]. These are compared and integrated with existing MGI genome features and the MGI unified catalog updated as required [14]. Ongoing updates to the mouse reference sequence occur as NCBI, Ensembl and Havana/VEGA release annotation updates and as curators from these groups and from MGI continue to analyze complex and/or conflicting gene prediction overlap regions. MGI's unified catalog is the authoritative listing of mouse genome features used by NCBI Gene, IMPC, MGI, and other resource providers
	Strain specific genome features	MGI integrates genome features and genes not present in the C57BL/6J reference genome, but found in other mouse strains
	Conflict in genome feature types	Genome feature type call conflicts (mostly gene vs. pseudogene) occur among NCBI, Ensembl and Havana/VEGA sources and these conflicts are stored and displayed in MGI gene pages
	Genes to sequence associations	Mapping genes to their nucleotide sequences is a co-curation effort of the Mouse Genome Annotation groups, including MGI, NCBI, Ensembl, and VEGA; mappings of genes to their protein sequences is a co-curation effort of MGI, UniProt KB and Protein Ontology groups
	Nomenclature for genes and genome features	MGI is the web host for nomenclature rules set by the International Committee on Standardized Genetic Nomenclature for Mice. MGI implements the policies and serves as the authoritative source for mouse symbols, names, and synonyms for genes and genome features. Nomenclature also is coordinated with human and rat resources
	Sequence Ontology (SO) annotations	MGI provides primary curation for associations between mouse genome features and SO terms, working with SO on defining new terms as needed
Comparative	Mouse-human orthologs	MGI maintains an algorithmic comparison and rule-based representation unifying orthology data from NCBI's Homologene and the Human Gene Nomenclature Committee's HCOP (HGNC Comparison of Orthology Predictions) resource. These mouse-human orthology results are used throughout MGI
	Mouse-vertebrate homologs	MGI represents data incorporated from NCBI's Homologene for selected vertebrate genomes

(continued)

Table 1 (continued)

Data class	Data description	Sources/methods
Functional	Gene Ontology (GO) annotations to mouse genes and gene products	MGI develops the definitive nonredundant mouse GO annotation set that is available from the MGI website or downloadable; and provides these data to the Gene Ontology Consortium website. MGI provides primary curation of GO terms to mouse genes and gene products and integrates efforts from other mouse annotation groups such as UniProtKB/GOA and via orthology-based inferences
Expression	Gene expression data, emphasizing endogenous expression during mouse development	Detailed expression data are standardized and integrated from large-scale projects and curated from scientific literature. Assay types currently supported include RNA in situ hybridization, immunohistochemistry, in situ reporter (knock-in), RT-PCR, Northern and Western blots, and RNase and Nuclease S1 protection assays. Records include digitized images, as available. Inclusion of high-throughput sequencing expression data is under development
	Mouse Anatomy Ontologies	Developed in collaboration with EMAP, the Mouse Developmental Anatomy Ontology provides a standardized nomenclature for anatomical terms and the developmental stages when the anatomical structures are present. A developmental-stage independent version, EMAPA, is concurrently maintained
	Gene expression literature index for mouse development	The literature index for gene expression in mouse development contains, for each article, what genes and ages were analyzed and what assay types were used. The user presentation provides a high-level overview and rapid access to detailed expression data
Recombinase/cre	Recombinase containing knock-in alleles and transgenes	MGI develops the complete list of recombinase containing knock-in alleles and transgenes used for conditional mutagenesis experiments. Data include nomenclature, identifiers, descriptions of the recombinase-containing construct, cre driver(s) and inducers (if applicable)
	Recombinase/cre specificity	Knowing where and when recombinase activity occurs is key to selecting the right recombinase-bearing line for experimentation. Reports of recombinase/cre characterization for on- and off-target sites is curated and integrated by MGI from scientific literature and downloaded from projects doing large-scale generation of new recombinase/cre containing mice

Variant/mutant	SNPs	MGI provides comprehensive information on reference SNPs including the reference flanking sequence, assays that define the SNP, and gene/marker associations. SNP web pages link to popular genome browsers including MGI's Mouse JBrowse
	Mutant alleles and genome rearrangements	MGI develops a complete catalog of mutations, with unique identifiers, description of mutant construction and inheritance, coordinating with major mutagenesis projects (IKMC, IMPC, Mutagenetix, APF, CvDC, etc.)
	QTL	MGI assigns nomenclature for QTL, assigns maximum location range in the genome for QTL with defined flanking markers, and annotates QTL trait data
Phenotypes	Phenotype annotations for mouse genotypes using Mammalian Phenotype Ontology (MP)	Phenotype data are integrated from large-scale phenotyping projects including Europhenome, Sanger Institute Mouse Genetics Program, the International Mouse Phenotyping Consortium, the Cardiovascular Disease Consortium and others, as well as data incorporated by MGI curators from the scientific literature [15]
	Mammalian Phenotype Ontology (MP)	MGI is the primary developer and central site for the Mammalian Phenotype Ontology. New terms are added as required by researcher's phenotyping projects and through ongoing phenotype curation [16, 17]. The organizational structure of the MP branches are reviewed by community experts
Mouse strains	Catalog of strains	MGI provides IDs and standard nomenclature for strain designations, implementing the rules of the International Committee on Standardized Genetic Nomenclature for Mice.
	International Mouse Strain Resource (IMSR, www.findmice.org)	MGI produces a website consolidating holdings of mouse resources (live, cryopreserved embryos and sperm, ES cells) from repositories worldwide. The site is updated weekly and links are provided to repository sites for strain data and ordering and to MGI's phenotype pages
Tumors	Data on endogenous spontaneous and induced mouse models for human cancer	A spectrum of data including tumor frequency and latency data, genetic strain definition (inbred, hybrid, mutant, genetically engineered strains), pathology images and diagnostic reports, are integrated using standardized tissue and tumor type vocabularies [5]
	Patient-Derived Xenograft (PDX) data	Current PDX data are from the Jackson Laboratory, but will expand to integrate PDX data from other sources. Data include de-identified clinical information, annotated histopathology images and diagnostic markers, copy number variation (CNVs), transcription profiling and targeted exome sequencing data [5]

(continued)

Table 1
(continued)

Data class	Data description	Sources/methods
Disease models	Associations between mouse genotypes and human diseases	MGI staff curate mouse model data from the scientific literature and integrate research submissions. Mouse model data are annotated to mouse genotypes. Users view these data at the genotype level or at the gene level (when they can be algorithmically aggregated to the gene level), depending on the query method applied by the user [8, 18]
	Associations between human genes and human diseases	MGI requires these associations to relate human data to mouse disease models. Association of human diseases with human genes is provided by OMIM, with additional data from NCBI's Gene Review and Gene Test data collections

areas covered by MGI and the role of MGI in analyzing and assembling these data for the biomedical community. For a fuller enumeration of the contents of MGI, consult the MGI statistics page at http://www.informatics.jax.org/mgihome/homepages/stats/all_stats.shtml.

Data (whether from published scientific articles, electronic submissions from researchers, or large-scale resource data downloads) are processed and/or curated for nomenclature, anatomy, phenotype, disease, function, and genome feature terminology using authoritative standards and ontologies. Ultimately this enables the many facets of robust searching, retrieving, displaying, and analyzing data in MGI.

2.1.2 Web Access and Search Results

Table 2 describes some of the ways in which users can access MGI data. Of particular note is the continual development of the MGI website to increasingly accommodate customized searches and the downloading or forwarding of these results to users desktops or to other analysis tools (note that the MouseMine [9] tool, described below also enables users to save data sets to the MGI servers for future use).

2.1.3 Analysis Tools and Bulk Data Extraction from MGI

MGI continues to add to the analysis tools available to users. Table 3 briefly describes some of these tools. In addition to maintaining MGI's gold-standard data sets and providing easy web access, we are committed to providing ways for users to capture data sets of interest to use in their own analysis. It should be noted that many MGI results pages now include the ability to forward the data retrieved to MouseMine or MGI's Batch Query (see below) or to download the data retrieved in text or Excel formats.

2.2 Examples (Step-by-Step)

In this section, we describe general principles and a few common examples for searching, viewing, and mining data from MGI. A comprehensive delineation of all searching and analysis options is beyond the scope of a single chapter. Detailed user help documentation is available at MGI online, along with direct contacts to our User Support staff, (email mgi-help@jax.org) to assist users with web navigation, data search and retrieval, and analysis tools.

In addition, data submission systems are in place for laboratories wishing to contribute their data (published or unpublished), as well as to confirm nomenclature for alleles, genes, and mouse strains, and obtain MGI IDs that can be included in publications. Data may be submitted prior to publication and held confidentially until the publication is public. See http://www.informatics.jax.org/submit.shtml for a direct link to submission forms or follow the "Submit Data" link in the navy blue navigation bar present at the top of most MGI web pages.

Table 2
Common methods of accessing MGI data using search forms[a]

Start from...	Use to search...	Results include...	Access from...
Quick search	When the desired search is broad and inclusive, for nomenclature, vocabulary/ontology terms, and annotations	Genes and genome features and vocabulary terms for phenotype, disease, function, anatomy, and protein-domains, with links to data details (*see* Fig. 2 for a Quick Search results example)	• The top of the topics list on the MGI homepage (www.informatics.jax.org); • Or in the upper right corner of other MGI web pages
Genes and markers query	Search for genes and genome features using various parameters: nomenclature, feature type (e.g., protein-coding gene), genome location, and vocabulary annotations	Genes and genome features addressing your query, and including genome location and links to each gene's detail page (*see* Fig. 3 for a Genes and Markers Query and results example) Gene detail pages (Fig. 4) include graphical maps, human and vertebrate homologs, annotations to human disease, mutants and alleles, phenotypes, GO annotations, expression profiles, gene interactions, sequence and protein links, reagents, and references	• Use the Search pull-down on the navigation bar strip at the top of any MGI page (select All Search Tools or Genes); • Or select Genes from the topics list on the MGI homepage (www.informatics.jax.org); • Or use the Genes tab above the navigation bar on most MGI web pages
Phenotypes, alleles and diseases query	Search for mutant or genetically engineered alleles, transgenes, QTL, etc. using search parameters: phenotype or disease annotations, gene or mutant allele nomenclature, genome location, allele generation method and/or allele attributes, or allele project collections	Summary of alleles for specified parameters and display of allele attributes, high level systems phenotypes, and human diseases modeled. Links to detail page for the mutant allele, including mutation's molecular detail, phenotype data and disease annotations, and links to obtaining mice, and allele-specific references	• Use the Search pull-down on the navigation bar strip at the top of any MGI page (select All Search Tools or Phenotypes); • Or select Phenotypes and Mutant Alleles from the topics list on the MGI homepage (www.informatics.jax.org); • Or use the Phenotypes tab above the navigation bar on most MGI web pages

Human–mouse disease connection (HMDC)	Search for mouse models of human disease and potential candidate genes using search parameters: mouse or human genes, mouse or human genome locations, phenotypes and disease terms. Searches also accept mouse or human VCF (variant call format) files and text files of genes or gene IDs as input	Grid and table views of the results can be toggled using the tabs. Data include mouse/human orthologs, phenotype classes, and human diseases/disease models fitting the parameters entered. Grid cells are active links to data details. Other links go to MGI homology pages, gene detail pages, references, and IMSR (International Mouse Strain Resource) (*see* Fig. 8 for a HMDC search result example)	• Use the Search pull-down on the navigation bar strip at the top of any MGI page (select Human Disease); • Or select Human–Mouse: Disease Connection from the topics list on the MGI homepage (www.informatics.jax.org); • Or use the Human Disease tab above the navigation bar on most MGI pages NOTE: this portal also has its own URL: www.diseasemodel.org
Gene expression data query	Search for detailed expression results using various parameters: nomenclature, gene annotations, genome location, anatomy, developmental stage, assay type	Gene, assay type, age, anatomical structure, images, tissue x stage matrix, tissue x gene matrix, links to additional experimental detail, and references	• Use the Search pull-down on the navigation bar strip at the top of any MGI page (select Expression); • Or select Gene Expression Database from the topics list on the MGI homepage (www.informatics.jax.org); • Or use the Expression tab above the navigation bar on most MGI pages
Recombinase (cre)	Search for transgenes and knock-ins carrying a recombinase by specifying where (an anatomical structure) the recombinase is active or what the cre driver is	A summary table of drivers, transgene and knock-in alleles, tissue systems in which the transgene or knock-in is active or is not active, common nomenclature synonyms, inducing agent (if applicable), and links to IMSR and references. Links to detail pages for the transgenes and knock-ins provide a tissue x age matrix displaying cre activity and phenotype information. Links from the cre activity matrix go to details of cre activity for specific tissues and subtissues	• Use the Search pull-down on the navigation bar strip at the top of any MGI page (select Recombinase (cre)); • Or select Recombinase (cre) from the topics list on the MGI homepage (www.informatics.jax.org); • Or use the Recombinases tab above the navigation bar on most MGI pages • NOTE: this portal also has its own URL: www.creportal.org

(continued)

Table 2
(continued)

Start from…	Use to search…	Results include…	Access from…
Function (GO)	Search for or select GO terms using MGI's GO Browser	A hierarchical display of GO terms showing the term of choice relative to its parent terms, synonyms, term definition, and links to all mouse GO annotations to the term and children of the term and showing type of supporting evidence, annotation context, and references	• Use the Search pull-down on the navigation bar strip at the top of any MGI page (select Function); • Or select Function from the topics list on the MGI homepage (www.informatics.jax.org); • Or use the Function tab above the navigation bar on most MGI pages
SNP query	Search for SNPs by gene or among strains or by genome location	A summary table of SNP IDs, genome location, functional class, variation type, and strains with SNP calls. Links are provided to MGI's SNP detail pages, dbSNP, and MPD as well as to MGI's gene detail pages (as appropriate)	• Use the Search pull-down on the navigation bar strip (select Strains/SNPs, then SNP Query); • Or select Strains, SNPs and Polymoprhisms from the topics list on the MGI homepage (www.informatics.jax.org) and follow the SNP Query link; • Or use the Strains/SNPs tab above the navigation bar on most MGI pages
Mouse tumor biology	Search by organ of tumor origin, tumor classification or name, strain, gene, somatic genetic mutation type, pathology images, references	Summary of tumor instance and frequency records filling the query parameters, including tumor name, organ, treatment type, strains, frequency, metastases, and with links to strain tumor summaries and images	• Use the Search pull-down on the navigation bar strip (select Tumors); • Or select Mouse Models of Human Cancer from the topics list on the MGI homepage (www.informatics.jax.org); • Or use the Tumors tab above the navigation bar on most MGI pages
PDX (patient-derived xenografts)	Search by cancer site, diagnosis, gene variants, gene expression or by gene amplification/ deletion	Summary tables of models satisfying the query with tumor site and type, patient sex and age, and additional data available for each model. Links go to specific model's data with engraftment data and model characterizations	• Go to the Mouse Tumor Biology page (see row above) and select PDX Model Search listed in the Additional Resources in the left toolbar

Mouse genome browser	Search by chromosome and genome coordinates	Optionally turn on tracks for mutant alleles, SNPs, QTL, phenotypes, or switch to viewing human GRCh38 build or pseudo genomes for strains other than the C57BL/6J reference genome	• Use the Search pull-down on the navigation bar strip (select Mouse Genome Browser); • Or select Genes from the topics list on the MGI from the homepage (www.informatics.jax.org) and then select the Mouse Genome Browser option
Vocabulary browsers	Search or browse vocabulary terms and select a term of interest. Mouse Developmental Anatomy and Adult Anatomy, GO, and Phenotype Ontology terms are displayed hierarchically; OMIM terms are displayed alphabetically	Developmental Anatomy terms provide Theiler stage and parent term links. GO and MP terms provide definitions, synonyms. Each vocabulary term links to all annotations for the selected term. OMIM terms link to MGI Disease Model pages and to OMIM entries	• Use the Search pull-down on the navigation bar strip and, hovering over the Vocabularies section, select GO, Human Disease (OMIM), Mammalian Phenotype (MP), Mouse Developmental Anatomy, or Adult Mouse Anatomy browser; • Or, select a topic area among Phenotypes and Mutant Alleles, Gene Expression, or Function from the topics list on the MGI homepage (www.informatics.jax.org) and follow links to the relevant vocabulary

[a]This is not an exhaustive list of search methods or data that can be retrieved from MGI. Users are encouraged to explore MGI, visit the "Getting Started" section on the homepage (www.informatics.jax.org), the Help, and FAQ sections in the upper left of each web page, or use the "Contact Us" link in the navy blue navigation bar at the top of MGI web pages or send email to mgi-help@jax.org for questions and assistance

Table 3
MGI tools for bulk data access and analysis[a]

Tool	What users can do...	Unique features...
MGI batch query	Enter or load a file of gene symbols, or IDs from a variety of resources and request additional data and annotations for these genes (e.g., genome locations, mutant alleles, IDs from other resources and annotations for these genes for GO, phenotype, expression, human diseases, SNPs, nucleotide and protein sequences). Results can be downloaded or forwarded to MouseMine	• Provides an easy "ID translation" among major data resource providers (e.g., MGI ID -to- NCBI Gene ID); • Quickly returns annotation sets for all genes queried; • Interface is intuitive and easy to understand
MouseMine	Data warehouse product for MGI data that allows users to build queries that are difficult to answer via MGI web pages [9]. There is great flexibility in the reports generated and data can be filtered, sorted, and saved for future use in analyses and comparisons	• Allows users to combine data available on standard MGI web searches with MGI data that are not readily searchable/viewable via the web; • MouseMine also includes some data sets external to MGI (currently genome interactions from BioGrid and IntAct and homology data from Panther); • MouseMine supports intersection/union of different data sets and performs enrichment analyses
Visual annotation display (VLAD)	Provides enrichment analysis of data using GO or Mammalian Phenotype Ontology or Anatomy ontologies for mouse data [11]. Other ontology sources may be used if supplied by the user. Evaluates data for positive or negative directions or can be used to compare data set 1 with data set 2 for relative enrichment	• Provides graphical results, that when a comparison is involved also displays in a relative bar diagram how the data sets compare; • Provides a table with statistical results and relative numbers (in the case of comparison data), and links to gene pages (as appropriate)
Human–mouse disease connection (HMDC)	Specific tool for exploration of human disease and mouse model relationships, including experimentally determined models of human disease, mouse phenotype profiling, and suggestions for candidate disease relationships	• Displays explicit experimental mouse models for human disease; • Exposes cases where a mouse model exists, but the human gene is not associated with the disease (potential candidate gene for a patient with a disease, where known gene associations are negative); • Conversely, exposes cases where the human gene and human disease are associated, but no mouse model exists (potential model development opportunity for genetically engineering a new mouse model)

FTP site	MGI maintains a public FTP site with more than 60 weekly updated data reports, each containing different data sets and combinations of data elements from MGI. These reports are free for downloading, mostly as text files, for use by other resource consolidators and for individual data mining and analysis projects	• These reports are used extensively in the community and are regularly reviewed for relevance to add or remove reports based on community use and need • New FTP reports can be requested as a one-time report or as an ongoing report by contacting User Support at mgi-help@jax.org
API	MGI provides an API (application program interface) through MouseMine that automates data retrieval, list creation, user profiles, and data model introspection	• MGI takes advantage of this feature of InterMine software [10] to provide an API for data without having to develop our own de-novo API

[a]Increasingly MGI is adding options to download data sets from web summary and results pages generated by user searches or to forward these data sets to other analysis tools

2.2.1 Beginning with the Homepage (www.informatics.jax.org)

The MGI Homepage (Fig. 1) provides the initial entrée into the database and its tools. The major parts of this web page include: (1) in the left column, the Quick Search box and listing of major topic areas of MGI that are active links to relevant specific search and analysis tools, (2) in the right column a mission statement, an "About Us", "MGI Publications", Facebook, and Twitter links; a carousel of news items (new MGI data, new/updated displays, search options and tools, upcoming conferences with MGI talks/posters, etc.) and other MGI news items. Links to MGI statistics and software releases, as well as to MGI tutorials, are provided.

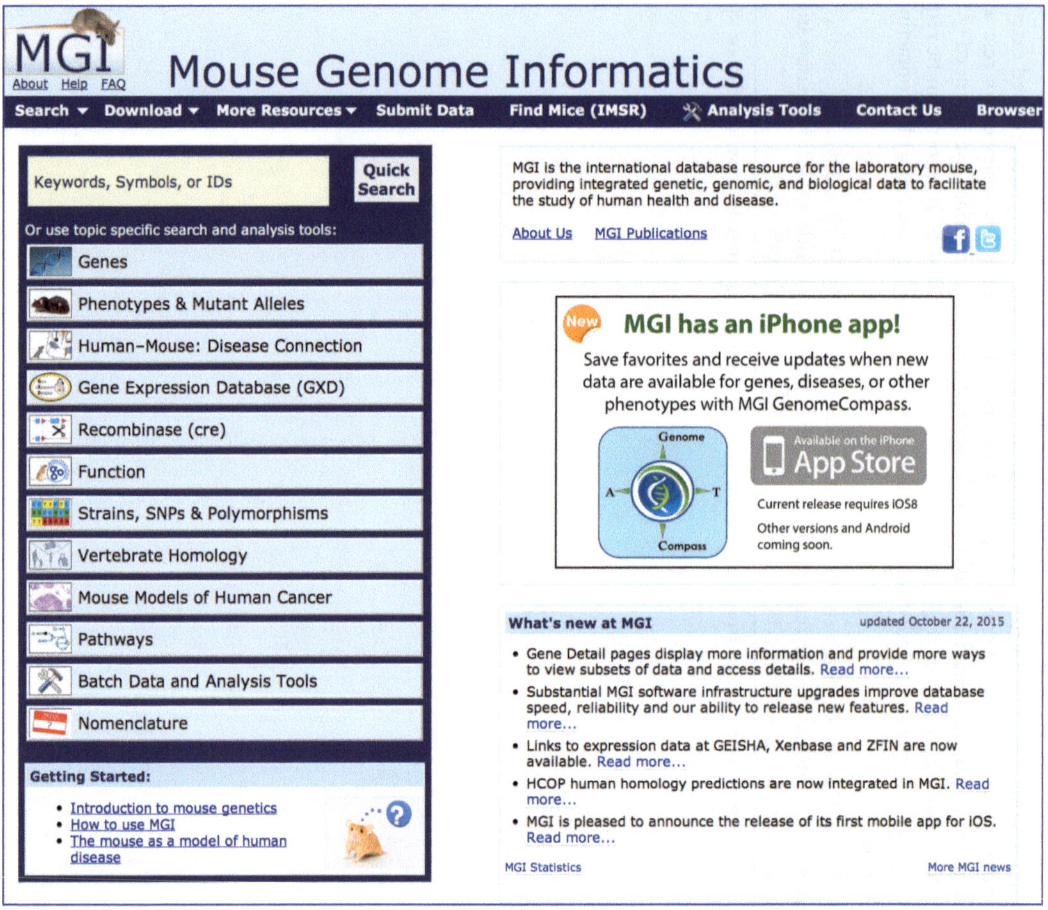

Fig. 1 Mouse Genome Informatics (MGI) Homepage (www.informatics.jax.org). The Homepage provides an entrée into the many data areas and tools provided at the MGI site. The main topics listing in the left column provides immediate access to the Quick Search function (*top*), as well as to pages designed for access to search forms, and topic-specific FAQs, statistics, and help. The *right portion* of the Homepage includes a mission statement, social media links, a rotating carosel of news, information, and recent MGI publication links, and a "What's new in MGI" section delineating recently added MGI features and data, as well as links to MGI Statistics. Below (not shown) are items of Community Interest, including MGI workshop information and recent research highlights

2.2.2 Using the MGI Quick Search Function

The MGI Quick Search is best used for two purposes: (1) by a new user or a user wishing to get a broad view on what MGI has on subject "X", where "X" may be either a nomenclature term (e.g., gene symbol, gene name, gene family or synonym, or human or vertebrate homolog of a mouse gene) or any annotation term (currently these include phenotype, disease, anatomy, function (GO, Ontology), or protein domain terms). Such searches may return thousands of items (e.g., a search for ATP in the Quick Search retrieves 4755 Genome Feature results and 1347 Vocabulary Term results (search done on 10/30/2015)); and (2) by a user who is well-acquainted with MGI and wants to jump quickly to the gene, alleles, or disease of interest (e.g., what are the alleles known for the *Pax3* gene? (search for *Pax3*); what genes/alleles in MGI are associated with Crouzon Syndrome, ? (search for Crouzon; or search for the full term in quotes "Crouzon Syndrome"). An example search result is shown in Fig. 2. Currently, Genome Feature results can be forwarded to the MGI Batch Query using the "Get more data" button (see Using the MGI Batch Query below).

2.2.3 Using Topic-Area Search Forms

The topic-specific search forms are designed to provide access to data in particular areas, and using parameter searching that may be topic-specific. For example, Fig. 3a shows the current search form for Genes and Markers. Here users can more precisely define the data that they seek as contrasted with the Quick Search. For the Genes and Markers search, users can define the type of genome feature (e.g., protein-coding gene), genome location by chromosome or by genome coordinates, and specify parameters from the GO vocabulary, InterPro domains, or mutant phenotype(s), or disease (OMIM) terms. Figure 3b shows the results of a Genes and Markers search. Note that export tools are available for these search results, either for download in text or Excel formats or results can be forwarded to the MGI Batch Query or MouseMine for adding additional annotation data to the results or for enrichment analysis. Similar search paradigms exist for other topic-area search forms.

2.2.4 Gene Detail Pages

The Gene Detail Pages are a favorite landing site for many MGI queries. Here the multitude of information about what a gene does is synthesized for the user, with many links leading deeper into the database for additional detail on particular aspects. Figure 4, shows a portion of the gene detail page for mouse gene *Fgfr2* (fibroblast growth factor receptor 2) as of Oct. 28, 2015. The new format of this page includes the ability to customize one's view to see more or less information within each categorical stripe, grid displays summarizing annotations for phenotypes, function (GO), and expression for the gene, and references grouped by topical area.

Fig. 2 Quick Search Results page example. The Quick Search tool is found at the *top left* of the MGI Homepage or on other MGI pages in the *upper right* corner. In this example, cleft palate was entered as the search term. 2929 results appear in the Genome Features section and 251 results appear in the Vocabulary Terms section. The results are returned in best-match order, indicated by four stars, with the one star matches at the end. The best matches will be those data that have "cleft palate" (both words occurring together) as a primary match in the symbol name or term annotation or its children term annotations. At the one star end of the matches, each word will be searched for independently, so matches will include terms matching only "cleft", for example. For multiword terms such as cleft palate using quotes restricts the search to only return items matching "cleft palate" as a phrase. The search done in this manner returns 554 results in the Genome Features section and 38 results in the Vocabulary Terms section (search done 10/30/2015). From the Quick Search Results page, the symbols in the Genome Features section link to the relevant MGI Gene Detail page or Mutant Allele Detail page. In the Vocabulary Terms section, the term links to the relevant term page: for phenotype terms, to the Mammalian Phenotype Ontology Browser; for disease terms, to the MGI Human Disease and Mouse Model Detail page; for an anatomical terms to the Mouse Developmental Anatomy Browser; for functional terms to the Gene Ontology Browser; for a protein family term to the MGI Protein Superfamily Detail page. The Associated Data column links to the underlying annotations and brings the user to the relevant annotation detail page

2.2.5 Using the MGI Batch Query

The Batch Query performs as its name suggests. The Batch Query takes input of gene symbols (either current symbols or including synonyms and homologs) or various ID types (e.g., MGI gene IDs, NCBI's Gene IDs, Ensembl IDs, VEGA IDs, UniGene IDs, miRBase IDs, GenBank/RefSeq IDs, UniProt IDs, GO IDs, RefSNP

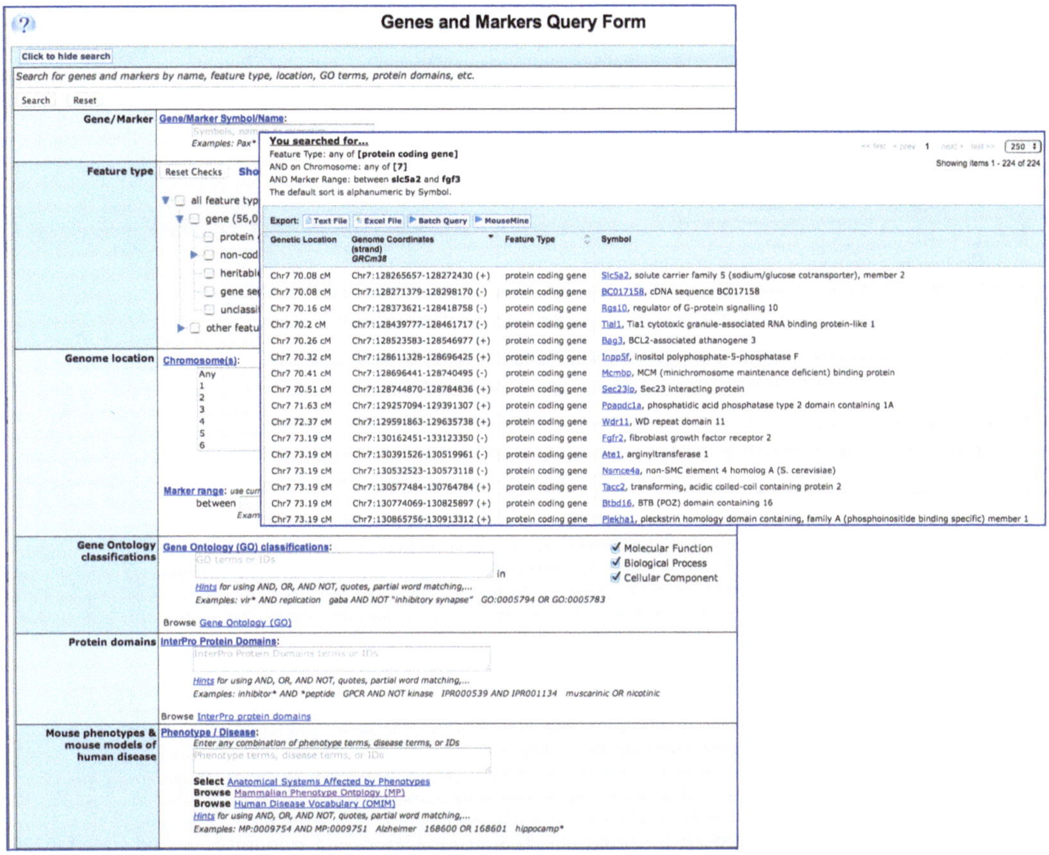

Fig. 3 The Genes and Markers Query Form and Results Page example. *Panel A* shows the Genes and Markers Query Form where users may specify as many search parameter as desired. A very straightforward search might be to simply enter a gene symbol, such as *Pax6*, with no additional parameters specified. *Panel B* shows the results of a search where the Feature type "protein coding gene" was selected and the Genome location of Chromosome "7" and Marker range between "*Slc5a2*" and "*Fgf3*" was specified. This search returned 224 genes. The genes were then sorted by Genome Coordinates using the arrowhead toggle in that column. Note the "You searched for…" feature at the *top*, which reminds the user what the search parameters were and the "Export" utilities for downloading results as text or Excel files or forwarding the results to the Batch Query tool or to MouseMine

IDs), which can be entered into the Query Form, or uploaded as a file. Users can then select attributes and annotation types as desired for their submitted set of genes. Selectable gene attributes include nomenclature (MGI ID, current gene symbol and name, and genome feature type), genome location, Ensembl ID, NCBI Gene ID, and VEGA ID. In addition, multiple-value annotation data can be selected including Gene Ontology, Mammalian Phenotype Ontology, Human Disease, Alleles, Gene Expression, RefSNP IDs, GenBank/RefSeq IDs, and UniProt IDs. A table is returned with all data requested and links to the relevant MGI gene detail pages. Note that this tool is particularly useful as a ID conversion tool, so that, for example, entering a set of Ensembl gene IDs as input, one could request corresponding MGI IDs and VEGA IDs in the table returned. The Batch Query results also can be downloaded as a text

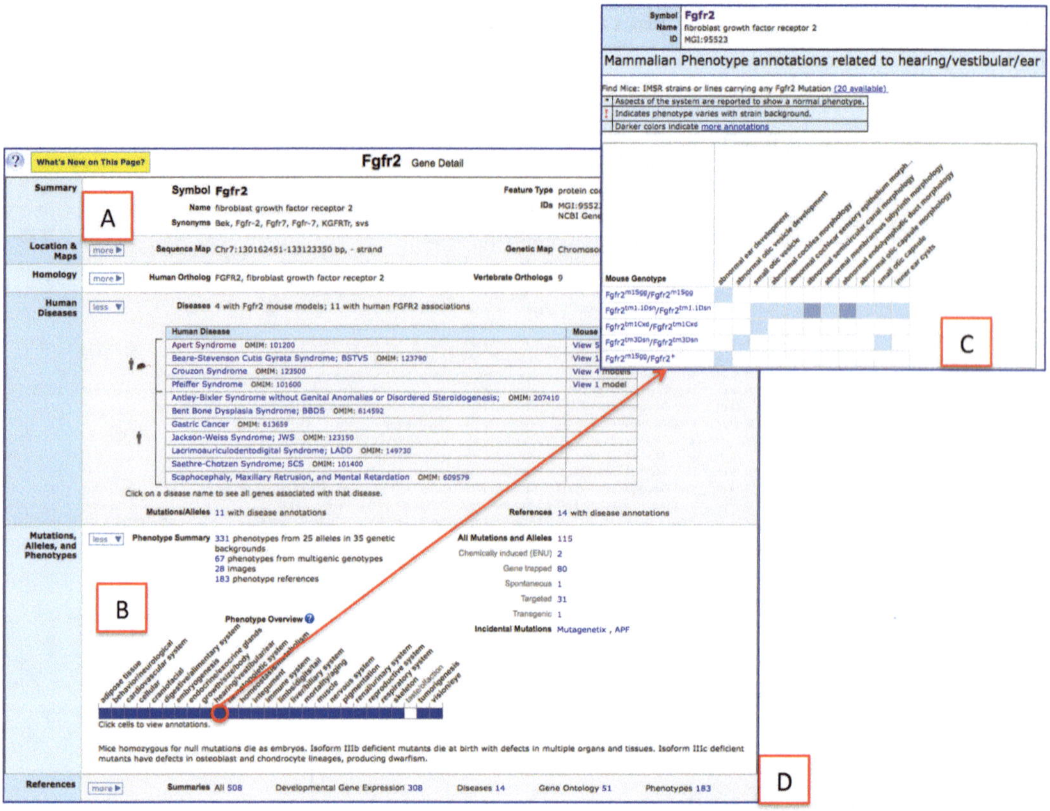

Fig. 4 Portion of the Gene Detail Page for *Fgfr2*. The Gene Detail Pages are frequently visited sites for MGI users, as they summarize the salient features of what a gene does. Recent revisions to this page have pulled forward additional data, making the composite information available richer in data content. *Panel A* shows the *top portion* of the Gene Detail page for the *Fgfr2* gene. Important new features include the ability to customize by toggling "more" or "less" data to be viewed in each data stripe. Here the Location and Maps data stripe and the Homology data stripe are toggled closed; while the Human Diseases and the Mutations, Alleles, and Phenotypes data stripes are toggled open. Additional stripes on this Gene Detail page (not shown) include Gene Ontology (GO) Classifications, Expression, Interactions, Sequences and Gene Models, Polymorphisms, Protein Information, Molecular Reagents, Other Accession IDs. The References data stripe appearing at the bottom of the page is shown. *Panel B* highlights the Mutations, Alleles, and Phenotypes data stripe. Here a number of links to summary data sets are enumerated. A significant addition to data availability is the Phenotype Overview graphic. Here 27 high-level system terms are presented, with filled cells indicating aberrant phenotypes are reported for homozygous or heterozygous mutations in one or more mutant alleles of *Fgfr2*. Each filled cell is an active link to underlying data. *Panel C* shows the detail available by clicking on the filled cell for the "hearing/vestibular/ear" phenotype. Here users can see that five distinct genotypes with *Fgfr2* mutations have hearing/vestibular/ear phenotypes and also can note the more specific annotation terms represented. Within this more detailed view, the filled cells again are active links to more detailed data (not shown) and the darker shading of the colored cells indicates more annotations. *Panel D* shows the revised Reference data stripe, where references are not only available as "all" references for *Fgfr2*, but also are displayed categorically based on the types of data provided by subsets of the references (e.g., Expression, Diseases, GO, Phenotypes)

or Excel file, or data can be forwarded to MouseMine for further manipulation or analysis. Figure 5 shows a Batch Summary where a set of 402 MGI gene IDs were forwarded from a search done using the Genes and Markers Query form for protein-coding genes AND the Gene Ontology term "DNA repair" (data retrieved 10/30/2015). The request from the Batch Query was to return, for these 402 IDs, nomenclature, Ensembl IDs, Entrez Gene IDs, and Human Disease annotations.

2.2.6 Using the MouseMine Warehouse Tool

MouseMine [9] is accessible directly at www.mousemine.org or via the MGI web page navigation bar (use the Search pull-down menu) or by following the Analysis Tools link. Based on Intermine [10], MouseMine is a new data warehouse for storing and accessing MGI data. MouseMine allows users to construct powerful queries of

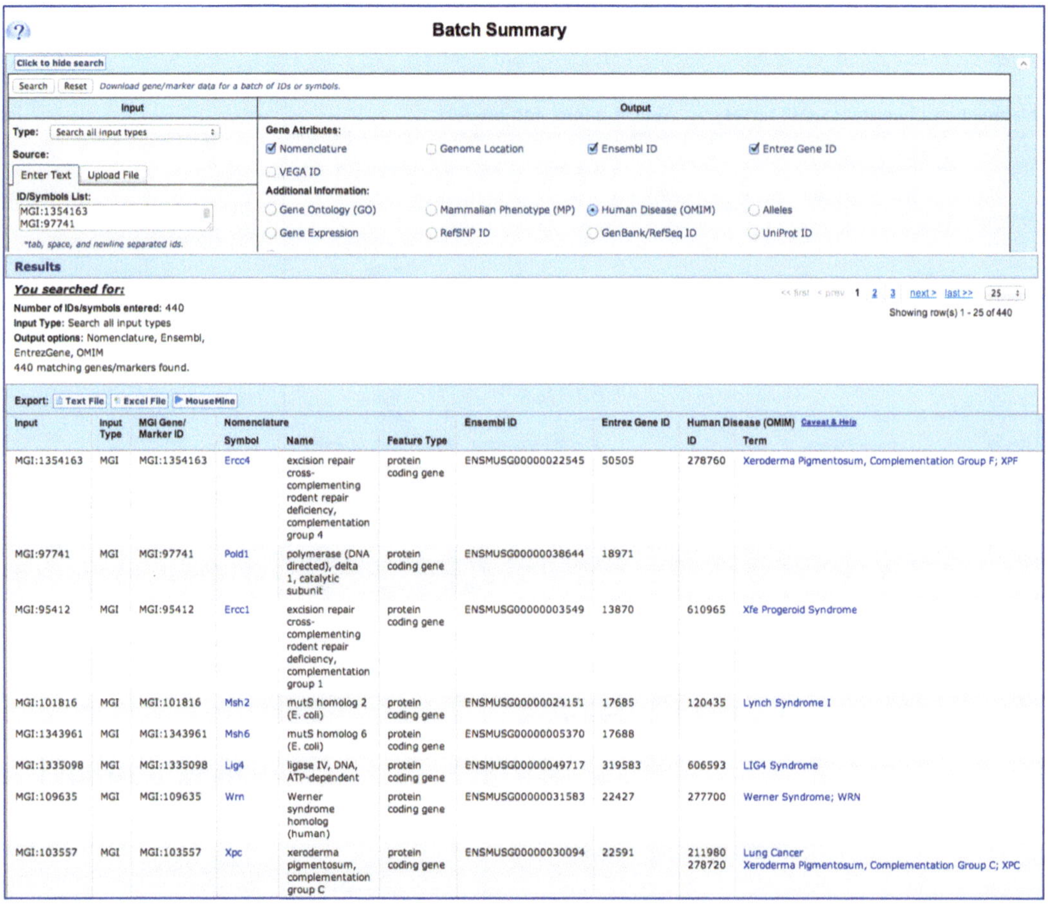

Fig. 5 Batch Query and Results. Here a file of 402 MGI Gene IDs was uploaded into the *Input Source* (*upper left*) and the *Output requested* (*upper right*) included Nomenclature, Ensembl ID, Entrez (NCBI) Gene ID and Human Disease (OMIM) annotations. The results returned are a tabular display of all data matching the user request and can be downloaded as text or Excel files or forwarded to MouseMine for further analysis. Links in the Nomenclature Symbol column take the user to the Gene Detail page for that gene; and links in the Disease (OMIM) Term column take the user to the MGI Human Disease and Mouse Model Detail page

MGI content, develop customized data sets, store and combine results of different searches, save or download results, and use MouseMine's built in enrichment tools. In particular, users can query MGI in ways not possible through the web interface (e.g., query fields that are displayed through MGI, but are not searchable such as retrieving the set of heterozygous genotypes for gene X). In addition MouseMine incorporates and integrates some data external to MGI, such as interaction data from BioGrid and IntAct and homology data from Panther. MouseMine, also takes advantage of ready-made API (Applications Programming Interface) utilities provided as part of the InterMine software to provide an API to MGI data. While the MouseMine interface is not as intuitive as the MGI web interface, it is a powerful tool for informaticians and those willing to spend a little learning time in exchange for being able to extensively explore and manipulate MGI data. Figure 6 shows a MouseMine search and example comparison result.

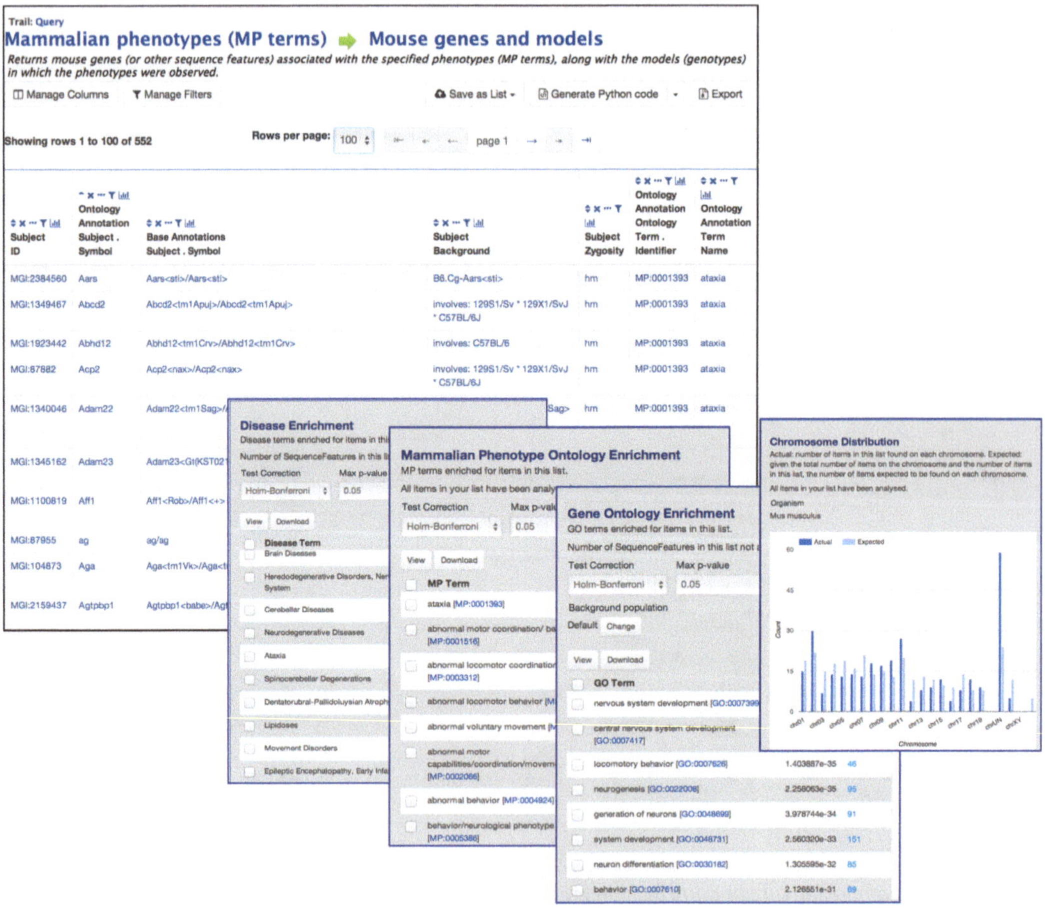

Fig. 6 MouseMine Results. A simple search for the Mammalian Phenotype term "ataxia" returns genotypes where this phenotype is observed. 552 results were returned by this search on 10/30/2015 (large table). Overlaying this table (*lower right*) are shown three enrichment analyses (for disease, phenotype, and GO) using these results and a chromosome distribution analysis of the actual vs. expected genomic location distribution of these genes

2.2.7 Visual Annotation Display (VLAD) Tool

VLAD, developed by MGI's Joel Richardson [11], is an exploratory tool for term enrichment for mouse data. One of VLAD's key qualities is its flexibility in being able to use any ontology and any annotation set. Although VLAD provides the choice of using the Gene Ontology or Mammalian Phenotype Ontology available at MGI, it is able to use any ontology on the Open Biomedical Ontologies site (http//www.obofoundry.org). Annotation sets associating gene-to-annotated term are supported in GAF format (GO Annotation Format, http://geneontology.org/page/go-annotation-file-gaf-format-10). VLAD can also analyze more than one gene set at a time and the user can control the appearance of the graphical display. Figure 7 shows an example output from VLAD.

2.2.8 The Human–Mouse Disease Connection (HMDC) Tool

The Human–Mouse Disease Connection [8], available at www.diseasemodel.org or from the MGI homepage, is a tool for exploring the phenotype-disease relationships for mouse and

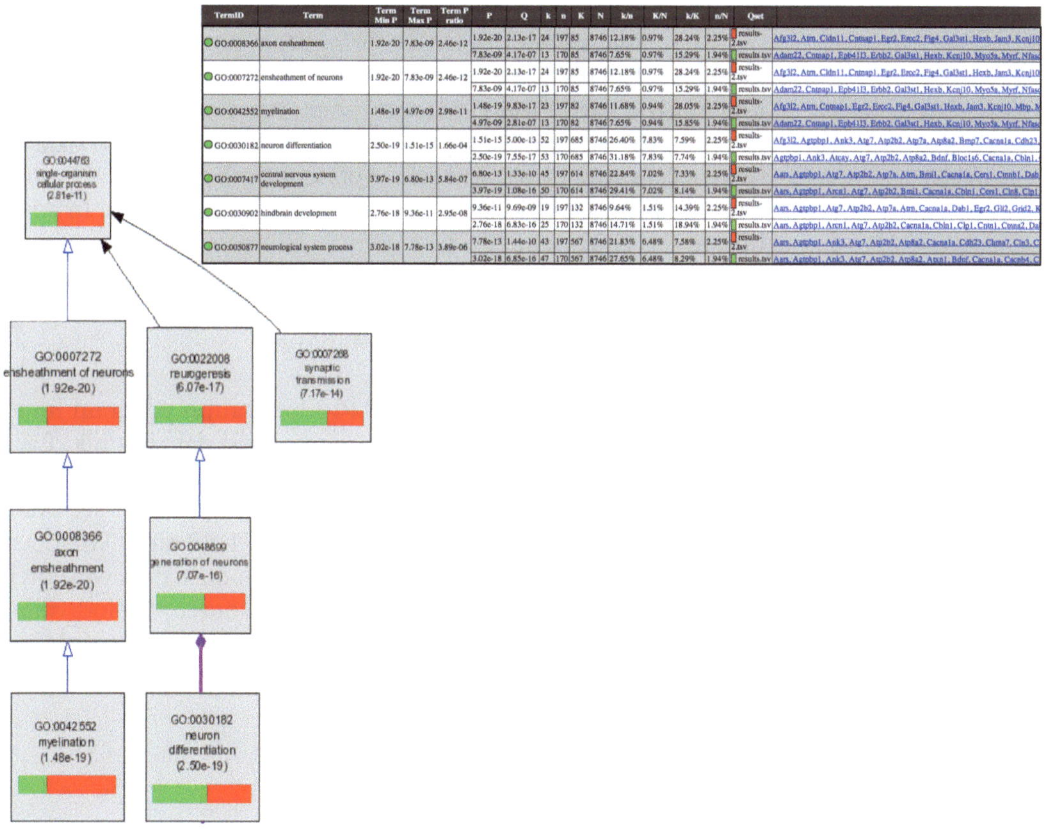

TermID	Term	Term Min P	Term Max P	Term P ratio	P	Q	k	n	K	N	k/n	K/N	k/K	n/N	Qset	
GO:0008366	axon ensheathment	1.92e-20	7.83e-09	2.46e-12	1.92e-20	2.13e-17	24	197	85	8746	12.18%	0.97%	28.24%	2.25%	results-2.tsv	Afg3l2, Atm, Cldn11, Cntnap1, Egr2, Ercc2, Fig4, Gal3st1, Hexb, Jam3, Kcnj10
					7.83e-09	4.17e-07	13	170	85	8746	7.65%	0.97%	15.29%	1.94%	results.tsv	Adam22, Cntnap1, Epb41l3, Erbb2, Gal3st1, Hexb, Kcnj10, Myo5a, Myrf, Nfasc
GO:0007272	ensheathment of neurons	1.92e-20	7.83e-09	2.46e-12	1.92e-20	2.13e-17	24	197	85	8746	12.18%	0.97%	28.24%	2.25%	results-2.tsv	Afg3l2, Atm, Cldn11, Cntnap1, Egr2, Ercc2, Fig4, Gal3st1, Hexb, Jam3, Kcnj10
					7.83e-09	4.17e-07	13	170	85	8746	7.65%	0.97%	15.29%	1.94%	results.tsv	Adam22, Cntnap1, Epb41l3, Erbb2, Gal3st1, Hexb, Kcnj10, Myo5a, Myrf, Nfasc
GO:0042552	myelination	1.48e-19	4.97e-09	2.98e-11	1.48e-19	9.83e-17	23	197	82	8746	11.68%	0.94%	28.05%	2.25%	results-2.tsv	Afg3l2, Atm, Cntnap1, Egr2, Ercc2, Fig4, Gal3st1, Hexb, Jam3, Kcnj10, Mbp, M
					4.97e-09	2.81e-07	13	170	82	8746	7.65%	0.94%	15.85%	1.94%	results.tsv	Adam22, Cntnap1, Epb41l3, Erbb2, Gal3st1, Hexb, Kcnj10, Myo5a, Myrf, Nfasc
GO:0030182	neuron differentiation	2.50e-19	1.51e-15	1.66e-04	1.51e-15	5.00e-13	52	197	685	8746	26.40%	7.83%	7.59%	2.25%	results-2.tsv	Afg3l2, Agtpbp1, Ank3, Atg7, Atp2b2, Atp7a, Atp8a2, Bmp7, Cacna1a, Cdh23,
					2.50e-19	7.55e-17	53	170	685	8746	31.18%	7.83%	7.74%	1.94%	results.tsv	Agtpbp1, Ank3, Atcay, Atg7, Atp2b2, Atp8a2, Bdnf, Bloc1s6, Cacna1a, Cbln1,
GO:0007417	central nervous system development	3.97e-19	6.80e-13	5.84e-07	6.80e-13	1.33e-10	45	197	614	8746	22.84%	7.02%	7.33%	2.25%	results-2.tsv	Aars, Agtpbp1, Atg7, Atp2b2, Atp7a, Atm, Bmi1, Cacna1a, Cers1, Ctnnb1, Dab
					3.97e-19	1.08e-16	50	170	614	8746	29.41%	7.02%	8.14%	1.94%	results.tsv	Aars, Agtpbp1, Arcn1, Atg7, Atp2b2, Bmi1, Cacna1a, Cbln1, Cers1, Cln8, Clp1
GO:0030902	hindbrain development	2.76e-18	9.36e-11	2.95e-08	9.36e-11	9.69e-09	19	197	132	8746	9.64%	1.51%	14.39%	2.25%	results-2.tsv	Aars, Agtpbp1, Atg7, Atp2b2, Atp7a, Atm, Cacna1a, Dab1, Egr2, Gli2, Grid2, K
					2.76e-18	6.83e-16	25	170	132	8746	14.71%	1.51%	18.94%	1.94%	results.tsv	Aars, Agtpbp1, Arcn1, Atg7, Atp2b2, Cacna1a, Cbln1, Clp1, Cntn1, Ctnna2, Da
GO:0050877	neurological system process	3.02e-18	7.78e-13	3.89e-06	7.78e-13	1.44e-10	43	197	567	8746	21.83%	6.48%	7.58%	2.25%	results-2.tsv	Aars, Agtpbp1, Ank3, Atg7, Atp2b2, Atp8a2, Cacna1a, Cdh23, Chrna7, Cln3, C
					3.02e-18	6.85e-16	47	170	567	8746	27.65%	6.48%	8.29%	1.94%	results.tsv	Aars, Agtpbp1, Ank3, Atg7, Atp2b2, Atp8a2, Atxn1, Bdnf, Cacna1a, Cacnb4, C

Fig. 7 Output from VLAD. *Panel A* is a small portion of the graphical view produced by a VLAD analysis. Here the distribution of positive and negative expression of genes is indicated by the proportion of *green* and *red* in the bars. In a larger graphic it is easy to see what branches of the ontologies (whether representing function, phenotype, etc.) are overrepresented in one's experiment. *Panel B* shows a small portion of the data analysis output table, with various calculated ratios, and colored indicator of data directionality, and the list of genes to which each row belongs

human. In particular, it is designed to highlight both the known experimental mouse disease models, and to graphically show (1) where a mouse model of a disease has been studied, but the human homolog is not identified (yet) as disease causing in human; thus suggesting where one might look for a potential causative mutation with a patient not fitting currently known human disease etiology; and (2) where a mouse model has not been identified for a particular disease-gene connection that is known in humans, suggesting that genetically engineering a mouse mutant in that gene may create a model system for studying this human disease. Searching HMDC can be done from either the mouse or human perspective. Specifically, searchable parameters include: mouse or human gene(s), mouse or human genome, location(s), and phenotypes and disease terms. Phenotypes at present are limited to mouse, but the human phenotype data, via the Human Phenotype Ontology (HPO) [12] annotations will be added soon. Searches also can be initiated using mouse or human VCF (variant call format) files or text files of gene symbols or gene IDs.

The initial search of HMDC brings the user to a Grid of data satisfying the search parameters. Human and mouse orthologs are listed at the left of the Grid, with phenotypes and diseases across the top. Mouse data are represented by blue cells; human data by orange cells; and bi-colored cells represent both mouse and human data. Cells in the Grid are active links to additional data. In addition to the Grid view, tabs take users to alternate displays based on genes or diseases.

Figure 8 illustrates the Grid results from a search of the Human–Mouse Disease Connection portal.

2.2.9 FTP and API Access

FTP (file transfer protocol) and API (application program interface) provide an additional way to access MGI data in bulk. Each week, MGI generates more than 60 specific data reports combining various data elements of MGI and creates files that are available for downloading via the MGI public FTP site. These are used largely by other resources to integrate mouse data with their applications, but can also serve as a data source for individual analysis projects. API access is provided through MouseMine. MGI takes advantage of the supplied API that comes with the InterMine software [10].

Fig. 8 (continued) to human disease association, but with no reported mouse model, thus suggesting genes that could be engineered to potentially model the human mutation/human disease; and (3) *bi-colored cells* represent cases where both the human gene mutation/human disease association is known and where mouse mutations in the orthologous gene(s) phenotypically model the human disease characteristics

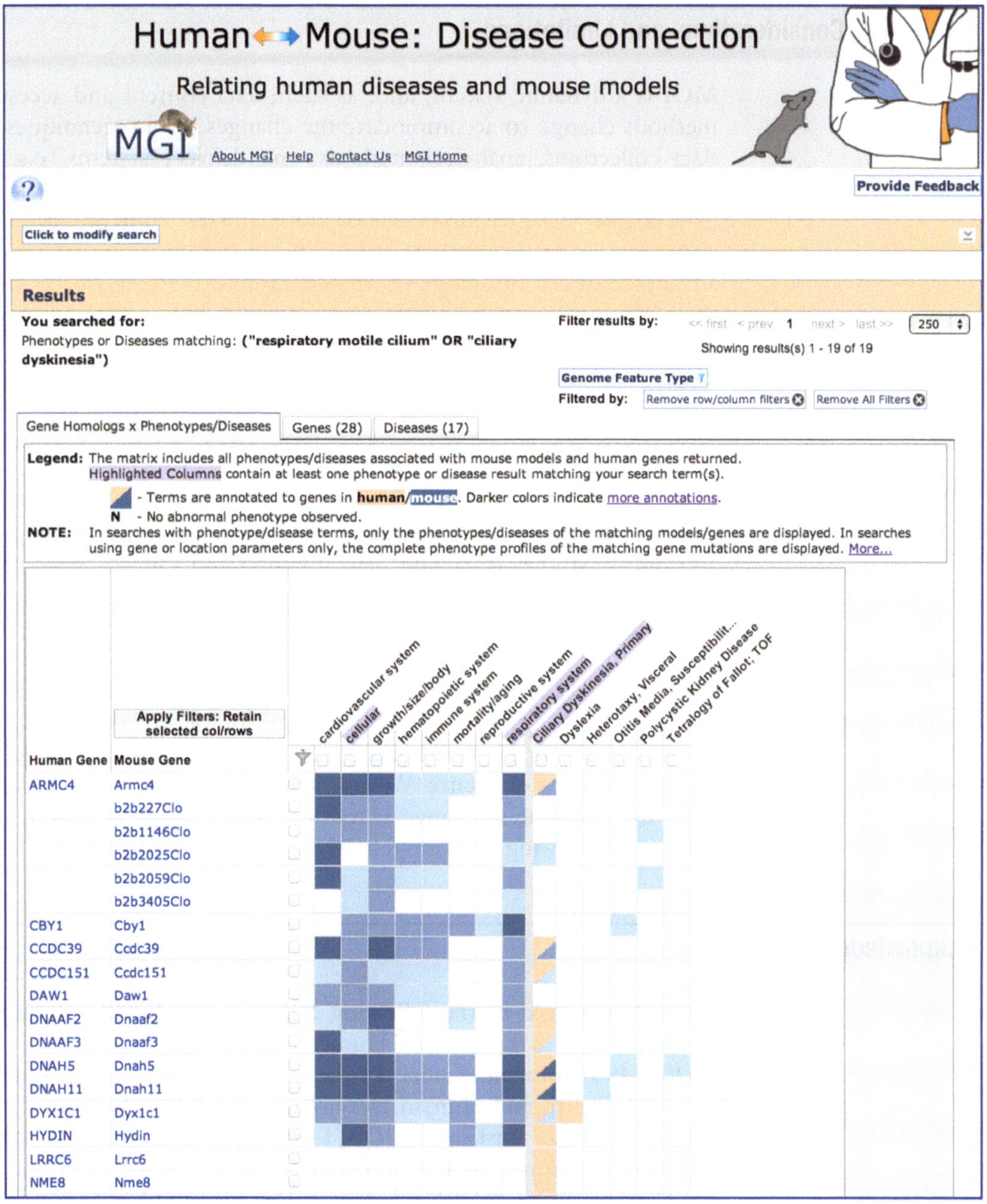

Fig. 8 Example output from a Human–Mouse: Disease Connection search. A search for phenotypes "respiratory motile cilium" OR "ciliary dyskinesia" in the HMDC Search produced this resulting Grid. Most human and mouse genes satisfying this search have clear orthologs (*left rows*), with the mouse genes/markers in rows 2–6 representing ENU mutations that thus far not associated with known genes. The columns represent phenotype annotations (*left portion*) and disease associations (*right portion*). In the phenotype portion of this Grid (*left portion*), mouse phenotypes are indicated by *blue-shaded cells*, with more annotations attributed to darker cells. In the disease portion of the Grid (*right portion*): (1) *blue cells* show diseases where the mutant mouse gene is associated with disease models, but no human gene is thus far associated with the disease (thus a candidate for patient mutations where existing etiologies do not fit); (2) *orange-shaded cells* show human gene

3 Further Considerations and Limitations

MGI is a dynamic system, and, as such, data content and access methods change to accommodate the changes in bio-techniques, data collections, analytical methods, and delivery systems (e.g., web interfaces, APIs, batch, queries, and bulk data downloads). This chapter has provided a cursory look at MGI content, access, and tools as of November 2015. We encourage users to take advantage of the full range of MGI data, to provide feedback on their experiences, and to help us keep the "community" in this community resource through data submission (see www.informatics.jax.org/submit.shtml).

4 Outlook

The mouse as a model organism and surrogate for human biology and disease studies is gaining new attention and value as genome sequencing costs decrease and as our ability to do comparative genomic and phenotypic analyses increases. In particular, mouse is proving to be a critical model for drug studies, cancer biology, rare disease analysis, and the understanding of complex and chronic diseases. MGI will continue to be a major contributor , to our collective knowledge about the mouse and continue to evolve with the ongoing advancement of science. We anticipate the shape and content of MGI will continue to change as research needs change and as new analysis and viewing tools are required in the era of "Big Data".

Acknowledgements

MGI is supported by the following NIH grants: HG000330 and HG002273 from the National Human Genome Research Institute; HD062499 from the Eunice Kennedy Shriver National Institute of Child Health and Human Development; CA089713 from the National Cancer Institute; OD011190 from the Office of the Director, Division of Comparative Medicine; and NS082666 from the National Institute of Neurological Disorder and Stroke.

References

1. Eppig JT, Richardson JE, Kadin JA, Ringwald M, Blake JA, Bult CJ (2015) Mouse Genome Informatics (MGI): reflecting on 25 years. Mamm Genome 26:272–284
2. Eppig JT, Richardson JE, Kadin JA, Smith CL, Blake JA, Bult CJ, MGD Team (2015) Mouse Genome Database: from sequence to phenotypes and disease models. Genesis 53:458–473
3. Finger JH, Smith CM, Hayamizu TF, McCright IJ, Xu J, Eppig JT, Kadin JA, Richardson JE, Ringwald M (2015) The mouse

Gene Expression Database (GXD): new features and how to get the most out of them. Genesis 53:510–522

4. Murray SA, Eppig JT, Smedley D, Simpson EM, Rosenthal N (2012) Beyond knockouts: cre resources for conditional mutagenesis. Mamm Genome 23:587–599
5. Bult CJ, Krupke DM, Begley DA, Richardson JE, Neuhauser SB, Sundberg JP, Eppig JT (2015) Mouse Tumor Biology (MTB): a database of mouse models for human cancer. Nucleic Acids Res 43:D818–D824
6. Shultz LD, Lyons BL, Burzenski LM, Gott B, Chen X, Chaleff S, Kotb M, Gillies SD, King M, Mangada J, Greiner DL, Handgretinger R (2005) Human lymphoid and myeloid cell development in NOD/LtSz-scid IL2R gamma null mice engrafted with mobilized human hemopoietic stem cells. J Immunol 174:6477–6489
7. Eppig JT, Motenko H, Richardson JE, Richards-Smith B, Smith CL (2015) The International Mouse Strain Resource (IMSR): cataloging worldwide mouse and ES cell line resources. Mamm Genome 26:448–455
8. Eppig JT, Blake JA, Bult CJ, Kadin JA, Richardson JE, Mouse Genome Database Group (2015) The Mouse Genome Database (MGD): facilitating mouse as a model for human biology and disease. Nucleic Acids Res 43:D726–D736
9. Motenko H, Neuhauser SB, O'Keefe M, Richardson JE (2015) MouseMine: a new data warehouse for MGI. Mamm Genome 26:325–330
10. Smith RN, Aleksic J, Butano D, Carr A, Contrino S, Hu F, Lyne M, Lyne R, Kalderimis A, Rutherford K, Stepan R, Sullivan J, Wakeling M, Watkins X, Micklem G (2012) InterMine: a flexible data warehouse system for the integration and analysis of heterogeneous biological data. Bioinformatics 28:3163–3165
11. Richardson JE, Bult CJ (2015) Visual annotation display (VLAD): a tool for finding functional themes in lists of genes. Mamm Genome 26:567–573
12. Köhler S, Doelken SC, Mungall CJ, Bauer S, Firth HV, Bailleul-Forestier I, Black GC, Brown DL, Brudno M, Campbell J, FitzPatrick DR, Eppig JT, Jackson AP, Freson K, Girdea M, Helbig I, Hurst JA, Jähn J, Jackson LG, Kelly AM, Ledbetter DH, Mansour S, Martin CL, Moss C, Mumford A, Ouwehand WH, Park SM, Riggs ER, Scott RH, Sisodiya S, Van Vooren S, Wapner RJ, Wilkie AO, Wright CF, Vulto-van Silfhout AT, de Leeuw N, de Vries BB, Washingthon NL, Smith CL, Westerfield M, Schofield P, Ruef BJ, Gkoutos GV, Haendel M, Smedley D, Lewis SE, Robinson PN (2014) The Human Phenotype Ontology project: linking molecular biology and disease through phenotype data. Nucleic Acids Res 42:D966–D974
13. Richardson JE (2006) fjoin: simple and efficient computation of feature overlaps. J Comput Biol 13:1457–1464
14. Zhu Y, Richardson JE, Hale P, Baldarelli RM, Reed DJ, Recla JM, Sinclair R, Reddy TBK, Bult CJ (2015) A unified gene catalog for the laboratory mouse reference genome. Mamm Genome 26:295–304
15. Blake JA, Bult CJ, Eppig JT, Kadin JA, Richardson JE, Mouse Genome Database Group (2014) The Mouse Genome Database: integration of and access to knowledge about the laboratory mouse. Nucleic Acids Res 42:D810–D817
16. Smith CL, Eppig JT (2012) The Mammalian Phenotype Ontology as a unifying standard for experimental and high-throughput phenotyping data. Mamm Genome 23:653–668
17. Smith CL, Eppig JT (2015) Expanding the mammalian phenotype ontology to support automated exchange of high throughput mouse phenotyping data generated by large-scale mouse knockout screens. J Biomed Semantics 6:11
18. Bello SM, Eppig JT, MGI Software Group (2016) Inferring gene-to-phenotype and gene-to-disease relationships at mouse genome informatics: challenges and solutions. J Biomed Semantics 7:14

Chapter 4

GeneNetwork: A Toolbox for Systems Genetics

Megan K. Mulligan, Khyobeni Mozhui, Pjotr Prins, and Robert W. Williams

Abstract

The goal of systems genetics is to understand the impact of genetic variation across all levels of biological organization, from mRNAs, proteins, and metabolites, to higher-order physiological and behavioral traits. This approach requires the accumulation and integration of many types of data, and also requires the use of many types of statistical tools to extract relevant patterns of covariation and causal relations as a function of genetics, environment, stage, and treatment. In this protocol we explain how to use the GeneNetwork web service, a powerful and free online resource for systems genetics. We provide workflows and methods to navigate massive multiscalar data sets and we explain how to use an extensive systems genetics toolkit for analysis and synthesis. Finally, we provide two detailed case studies that take advantage of human and mouse cohorts to evaluate linkage between gene variants, addiction, and aging.

Key words WebQTL, Interval mapping, Pair scan, Principal component analysis, Test cross, Recombinant inbred strain, Reverse genetics, dbSNP, GeneWeaver, BioGPS, NCBI, GeneRIF, UCSC Genome Browser, Gemma, GEO, Allen Brain Atlas, GWAS Catalog, GTEx, WebGestalt, PLINK, Manhattan plot, eQTL analysis, R/qtl, WGCNA, Proteomics, Metabolomics, Metagenomics

1 Introduction

GeneNetwork (www.genenetwork.org, GN) is a web service for systems genetics. It started in 2001 as *WebQTL*—an online version of Ken Manly's *Map Manager QT* program [1] combined with data sets in the *Portable Dictionary of the Mouse Genome* [2]. GN is a data repository and analytic platform for systems genetics that integrates large and diverse molecular and phenotype data sets. Just over 1400 papers listed in Google Scholar have used GN in many different ways.

GN was initially used as a traditional forward genetics tool to map quantitative trait loci (QTLs) and expression QTLs (eQTLs) in sets of recombinant inbred (RI) strains and standard genetic test crosses, including F2 intercrosses and backcrosses [3]. As the number and variety of data types grew it became practical to implement multivariate analysis in GN to study genetic covariation among large numbers of phenotypes [4–6]. This kind of assembly,

Klaus Schughart and Robert W. Williams (eds.), *Systems Genetics: Methods and Protocols*, Methods in Molecular Biology, vol. 1488, DOI 10.1007/978-1-4939-6427-7_4, © Springer Science+Business Media New York 2017

analysis, and integration of sets of phenotypes and even entire phenomes is a hallmark of systems genetics and is the forerunner and experimental companion of personalized health genomics and precision medicine. Thanks to recent breakthroughs in sequencing technology, GN can now also be used for novel reverse genetics approaches such as phenome-wide association studies (PheWAS). In a typical reverse genetics approach, gene function is determined through manipulation, either by gene deletion (knockout), addition of altered sequence (knock-in), silencing (RNA interference or RNAi), or gene editing (e.g. clustered regularly-interspaced short palindromic repeats or CRISPRs). Similar to these more traditional approaches, a PheWAS begins with known genes and sequence variants and then tracks down sets of linked biomarkers and phenotypic consequences [7–9].

At its most basic level, GN is a tool for studying covariation and causal connections among traits and DNA variants. This sounds simple enough, but it can be challenging to know how to get started and how to navigate and use the many program modules and options. Here we provide detailed instructions for using GN along with "worked" examples and some test questions (and answers) that should ease entry into this resource. All examples and figures were taken from the production version 1 of GN (late 2015). While the interface may change in the next few years (GN version 2, GN2), all of the logic, data types, and procedures described here will still be applicable.

The potential scope of GN analysis tools is broad—well-organized collections of genetic, genomic, and trait data from different species can be integrated easily—either as private or open data. At this point, GN includes curated data sets for a variety of model organisms and plant species, including humans, monkeys, rodents, *Drosophila*, and *Arabidopsis*, soy, and barley. Data are usually open and exportable, and data typically include information for hundreds to thousands of individuals with matched genotypes for thousands to millions of markers (usually SNPs), array or RNA-sequencing (RNA-seq) data for tens of thousands of transcripts, and in a growing number of cases, proteomic, metabolomic, metagenomic, behavioral, and morphological data.

Massive omics data sets are unwieldy to access, normalize, and analyze. Even those skilled in bioinformatics spend more than half of their time simply wrangling, reformatting, and error checking data sets to match the requirements of different workflows. GN spares the user most of these problem. Data are formatted and normalized, and usually come with good metadata (often in the form of links to more information). This greatly simplifies QTL and eQTL analysis, candidate gene discovery, coexpression analysis, and hypothesis testing [3, 10]. The GN toolkit includes many search functions, tools to study correlation and partial correlation, multiple QTL mapping methods (including R/qtl, PLINK, and

GEMMA, and FaST-LMM in GN2), and powerful dimension-reduction techniques (principle component analysis and weighted gene coexpression analysis), network construction, enrichment analysis, variant analysis, and links to key informatics resources such as NCBI (www.ncbi.nlm.nih.gov), the UCSC Genome Browser (genome.ucsc.edu), BioGPS (biogps.org), the GWAS Catalog (www.ebi.ac.uk), Gemma (www.chibi.ubc.ca), the Allen Brain Atlas (www.brain-map.org), and GeneWeaver (GeneWeaver.org).

In this chapter we introduce the basic architecture of GN (Subheading 2) and work through two detailed cases studies (Subheadings 4.1 and 4.2) that analyze both mouse and human data sets. We also explain how GN links to other web sites that provide complementary resources and analysis tools (Subheading 3). Throughout the chapter we provide a series of questions that can be used to test your proficiency. Answers are provided at the end of the protocol in the **Notes** section. Both **Case Studies** in Subheadings 4.1 and 4.2 provide detailed protocols needed to exploit GN data resources and to test specific hypotheses. Work through both of these examples and use the notes to gain an excellent understanding of the range of applications and types of questions that can be addressed and often answered using a systems genetics approach.

2 Organization of GeneNetwork

The first challenge in using GN is to locate cohorts (groups of subjects or samples) and associated data sets. The hierarchical organization of GN's main **Select and Search** menu is simple and makes it relatively easy to find relevant data sets (Fig. 1). To get data, after opening the browser, select the most appropriate **Species** from the dropdown menu. For an open-ended search of phenotypes you can also select **All Species** at the bottom of the menu. The next steps are to select the **Group**, **Type**, and **Data Set** from the drop-down menus. For many groups, a combination of phenotypes, genotypes, and molecular data are available. This makes it possible to perform QTL mapping and the analysis of trait and gene covariation. Table 1 provides a sample of human and rodent data sets that are amenable to these types of analyses.

As a navigation aid in this protocol, all active links in GN (buttons and linked text) and all data that you type into search fields such as **Get Any** are displayed using bold italic font. In contrast, page names, titles, column headers, and static menu items are displayed using **bold** font.

2.1 Types of Data

Almost all human data sets in GN include gene expression measurements (Table 1). In addition, several data sets also include genotypes and can therefore also be used for eQTL analyses. Examples include all of the human GTEx data sets and several

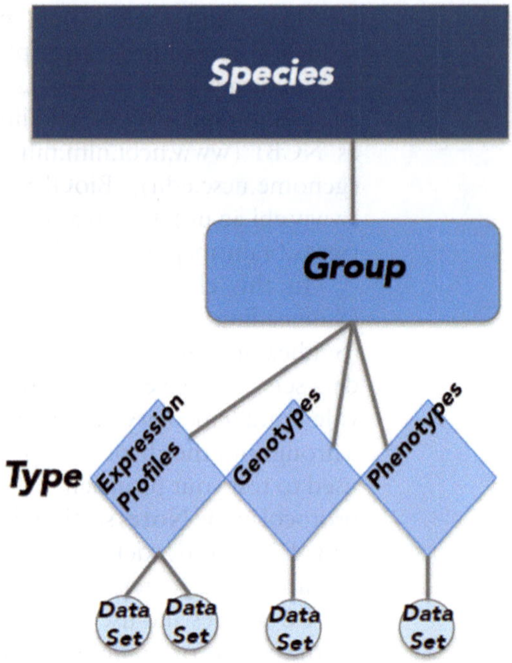

Fig. 1 Organization of data sets in GeneNetwork

human brain and liver expression data sets: *Brain, Aging: AD, Normal Gene Expression with Genotypes (Myers)*, and *Liver: Normal Gene Expression with Genotypes (Merck)* (Table 1). Concerns about subject confidentiality sometimes limit the amount of data available for human cohorts. Nonhuman cohorts, such as rodent populations, do not suffer from these restrictions and often contain more levels of data (Table 1). The rodent cohort with the most extensive data collection is currently the BXD family of strains derived from a cross between C57BL/6J (B) and DBA/2J (D) [11]. Inbred panels and RI strains represent stable populations that allow for deep resampling of individual genotypes and the accumulation of many different levels of data over time, and across laboratories and research communities enabling replication of research and the study of the pleiotropic actions of variants. The BXD set, for example, includes a wide variety of trait measurements collected over the last four decades [12]. Other populations commonly used for quantitative genetics and systems genetics include F2 intercrosses and outbred populations such as heterogeneous stock (HS) mice. For most F2, outbred, and human populations, each individual is truly unique and collecting multiple levels of data and studying gene-by-environmental (GXE) interactions and lab-to-lab replication is usually not practical.

Table 1
A sample of well-characterized human and mouse data sets

	Group	Genotypes	Molecular traits	Higher-order phenotype traits	Description and usage	References
Human	All Tissue, RNA-Seq GTEx v5	Yes	Expression profiles (RNA-seq) from 30 peripheral tissues and 11 brain subregions	No	Massive collaborative effort to explore associations between genotype and gene expression across tissues collected from up to 1,000 individuals. Suitable for quantitative genetics (QTL mapping) and systems genetics, including correlation and network analysis to compare associations between tissues and between other human or rodent data sets	[25]
	Brain, Aging: AD, Normal Gene Expression with genotypes (Myers)	Yes	Expression profiles (Agilent microarray) from cerebellum, prefrontal cortex, and primary visual cortex	No	A study of cortical gene expression for normal aged and Alzheimer's disease cases with ~176 cases and 187 controls per tissue. Suitable for quantitative genetics (QTL mapping) and systems genetics, including correlational and network analysis to compare associations between tissues and disease state or between other human or rodent data sets	[26]

(continued)

Table 1
(continued)

Group	Genotypes	Molecular traits	Higher-order phenotype traits	Description and usage	References
Liver: Normal Gene Expression with Genotypes (Merck)	Yes	Expression profiles (Rosetta/Merck Human 44K 1.1 microarray) from liver	Metabolic traits	Gene expression profiles from 427 human liver samples that includes measurements of activity for nine enzymes Suitable for quantitative genetics (QTL mapping) and systems genetics, including correlational and network analysis to compare associations between tissues and disease state or between other human or rodent data sets	[20]
Brain: Normal Gene Expression (NIH/Gibbs)	No	Expression profiles (Illumina humanRef-8 v2.0 expression beadchip microarray) from cerebellum, caudal pons, frontal cortex, and temporal cortex	No	Study that includes brain expression profiles from 147 individuals. Suitable for correlational and network analysis to compare associations between tissues and between other human or rodent data sets	[27]
Brain, Aging: AD, HD, Normal Gene Expression (Harvarid/Merk)	No	Expression profiles (Agilent microarray) from cerebellum, prefrontal cortex, and primary visual cortex	No	A study that includes brain expression profiles from 307 Alzheimer's disease cases, 152 Huntington's disease cases, and 132 controls. Suitable for correlational and network analysis to compare associations between tissues and disease state and between other human or rodent data sets. Tissues provided by the Harvard Brain Tissue Resource Center (www.brainbank.mclean.org)	[28]

Mouse	Brain, Aging: AD, Normal Gene Expression (Liang)	No	Expression profiles (Affymetrix Human genome U133 Plus 2.0 microarray) from entorhinal cortex, hippocampus, medial temporal gyrus, posterior cingulate cortex, superior frontal gyrus, primary visual cortex. Expression profiles for peripheral tissue (adipose, adrenal gland, bone, cartilage, eye, and retina, gastrointestinal tract, kidney, liver, lung, muscle, and spleen), brain tissue (whole brain, amygdala, cerebellum, hippocampus, hypothalamus, midbrain, neocortex, nucleus accumbens, pituitary, prefrontal cortex, and striatum), and cell type (hematopoietic cells, hepatocytes, hippocampal precursor cells, and T-cells) measured on multiple microarray platforms and using RNA-sequencing. Some proteome data from liver is also available.	No	A survey of gene expression across six brain regions for normal aged and Alzheimer's disease cases with ~ 14 biological replicates per tissue and condition. Suitable for correlational and network analysis to compare associations between tissues and disease state or between other human or rodent data sets	[29]
	BXD	Yes		Behavioral, Metabolic, Morphological, Pharmacological, Toxicology	Recombinant inbred genetic reference population (GRP) derived by crossing a C57BL/6J (B) female with a DBA/2J (D) male. The BXD set was derived from three separate crosses of B and D parental strains in early 1970's, late 1990's, and early 2000's Data collection is part of a massive collaborative effort from multiple investigators. Suitable for quantitative genetics (QTL mapping) and systems genetics, including correlation and network analysis to compare associations between tissues and between other rodent or human data sets	[11]

(continued)

Table 1
(continued)

Group	Genotypes	Molecular traits	Higher-order phenotype traits	Description and usage	References
Mouse Diversity Panel	Yes	Expression profiles for bone, dorsal root ganglion, hippocampus, and liver measured using microarray platforms.	Behavioral, Metabolic, Morphological, Pharmacological, and Toxicology	The Mouse Diversity Panel (MDP) is represented by multiple and genetically divergent inbred strains. This panel has a higher recombination rate, level of genetic variation, and phenotypic diversity than crosses derived from two parental inbred strains but demonstrates significant population structure. Suitable for quantitative genetics (QTL mapping) and systems genetics, including correlation and network analysis to compare associations between tissues and between other rodent or human data sets	[30]
BHF2 (Apoe Null) UCLA	Yes	Expression profiles for adipose, brain, liver, and muscle measured using the agilent microarray platform.	Metabolic	This data set features a large F2 cross derived from C57BL/6J and C3H/HeJ (BHF2) of 334 individuals. Both inbred progenitors were null for ApoE resulting in a population of genetically diverse F2 individuals that lack ApoE. Loss of this gene recapitulates some of the phenotypes associated with metabolic syndrome. The F2 population was fed a high-fat diet from 8 to 24 weeks of age. Suitable for quantitative genetics (QTL mapping) and systems genetics of metabolism	[31]

Heterogeneous Stock	Yes	Expression profiles for hippocampus, liver, and lung using the Illumina Mouse WG-6 v1, v1.1 microarray platform.	Morphological	Heterogeneous Stock (HS) mice are derived from eight different inbred strains (A/J, AKR/J, BALBc/J, CBA/J, C3H/HeJ, C57BL/6J, DBA/2J, and LP/J). This panel has a higher recombination rate, level of genetic variation, and phenotypic diversity than crosses derived from two parental inbred strains but can demonstrate significant population structure. Suitable for high-resolution quantitative genetics (QTL mapping) and systems genetics, including correlation and network analysis to compare associations between tissues and between other rodent or human data sets	[32]

Many of the **Data Sets** are amenable to systems genetics mapping and other methods and are accessible at GeneNetwork. The **Description and Usage** column provides details about the data set and potential usage. Note that only the first three human data sets have both genotype and gene expression data and only the third data set features genotypes, gene expression, and higher-order trait data in the form of metabolic phenotypes

2.2 Starting an Analysis

The main GN search page and an overview of a typical workflow are shown in Fig. 2. Data sets are selected based on **Species**, **Group**, and **Type** (Fig. 1). Detailed information and metadata can often be reached by clicking the Info buttons to the left (Fig. 2). **Data Sets** are queried using either **Get Any** or the **Combined** options. Searching with **Get Any** performs matches to entered text using the logical OR operator. For example, if the term "*alcohol ethanol*" is entered into **Species = Mouse, Group = BXD, Type = Phenotypes**, and **Data Set = BXD Published Phenotypes**, then the search will return all matches for "*alcohol*" or for "*ethanol*" (>500 results). In contrast a search for "*alcohol consumption*" in the **Combined** search option (Fig. 2b) uses the logical AND operation and generates far fewer results. Very long lists of gene symbols or probe set IDs—a thousand or more—will fit into these search boxes.

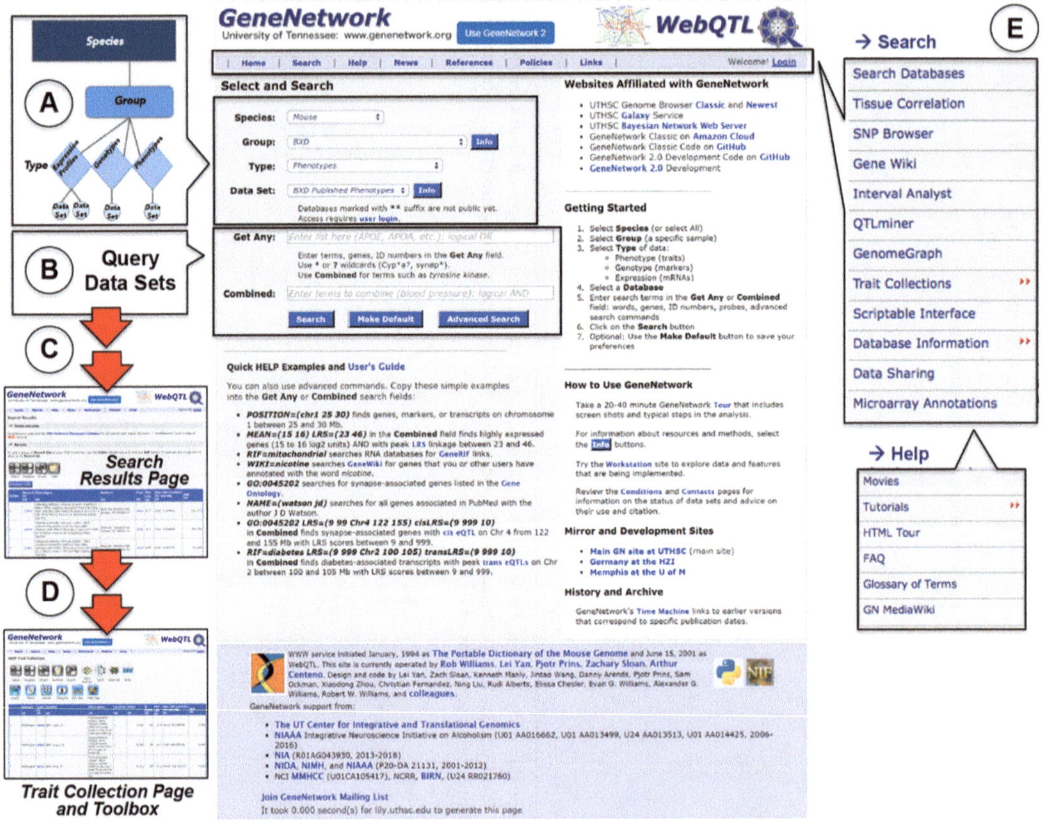

Fig. 2 GeneNetwork main search page and organization. Most analyses in GeneNetwork will follow the steps shown in *panels A* through *D*. In this workflow, a data set is selected (*A*) and mined for traits of interest based on user search queries (*B*). Traits are then selected from the search (*C*) and placed in a collection for further inspection and quantitative analysis (*D*). The banner menu contains additional search options and helpful resources under the Search and Help tab, respectively (*E*)

To get started, experiment with the **Quick HELP Examples** located just below the **Combined** search option (Fig. 2, center). Test whether you can find all genes on human chromosome (Chr) 21 that have high expression (>4.0 $\log_2$ RPKM) in the frontal cortex (*see* **Note 1**). Search queries are dependent on the **Data Set** type. For example, genotype data sets can be searched by marker name or marker position; phenotype data sets can be searched by phenotype description or authors' names; gene and protein expression data sets can be searched based on expression level, gene location, gene symbol, a Gene Ontology category (GO), or even by NCBI *Gene Reference into Function* (GeneRIF) text string.

To compare and improve compatibility across data sets, most array data have been $\log_2$ transformed and rescaled to an average of 8 and a standard deviation of ±2 units. This is true of Affymetrix and Illumina array data. However, Agilent data report gene expression as the $\log_{10}$ of the ratio between a specific tissue compared against a reference pool of multiple tissues (mlratio). RNA-seq data is usually normalized to $\log_2$ (RPKM + 1).

Many of GN data sets can be searched for traits or transcripts based on QTL position and significance levels (LRS or LOD score). Transcripts or proteins that are controlled by variants in or near their parent gene produce so-called cis-acting expression QTLs (cis-eQTLs) whereas those that are controlled by a more distant locus, usually on a different chromosome, produce trans-acting QTLs (trans-eQTLs) (Fig. 3). Test whether you can find a set of proteins in the mouse liver that are strongly controlled by trans-eQTLs (*see* **Note 2**).

2.3 Create a Trait Collection

Once you have selected a **Data Set** and submitted a search, results will appear in a **Search Results** page (Figs. 2c and 4). From this page, select the individual traits, transcripts, or gene markers for additional analysis by adding them to a **Trait Collection** (Fig. 2d). Do this from the **Search Results** page either by selecting all rows with the Select icon (Fig. 4a) or by selecting a subset of rows with the **Add** icon (Fig. 4b). Trait collections are usually restricted to a single species and group. Comparisons across groups and species are possible, but in most cases this involves assembling several **Trait Collections**—one for each group.

From either **Search Results** or from a **Trait Collection** (Fig. 5) you can inspect traits in greater detail by clicking on their ***Record ID*** or ***Trait ID***. This will direct you to the **Trait Data and Analysis** page (Fig. 6) that contains links to other web resources and GN tools.

The GN banner Search pull-down lists additional options, each of which is reviewed briefly below (Fig. 2e). Search Databases and Trait Collections are simply navigations aids to quickly get back to these two pages.

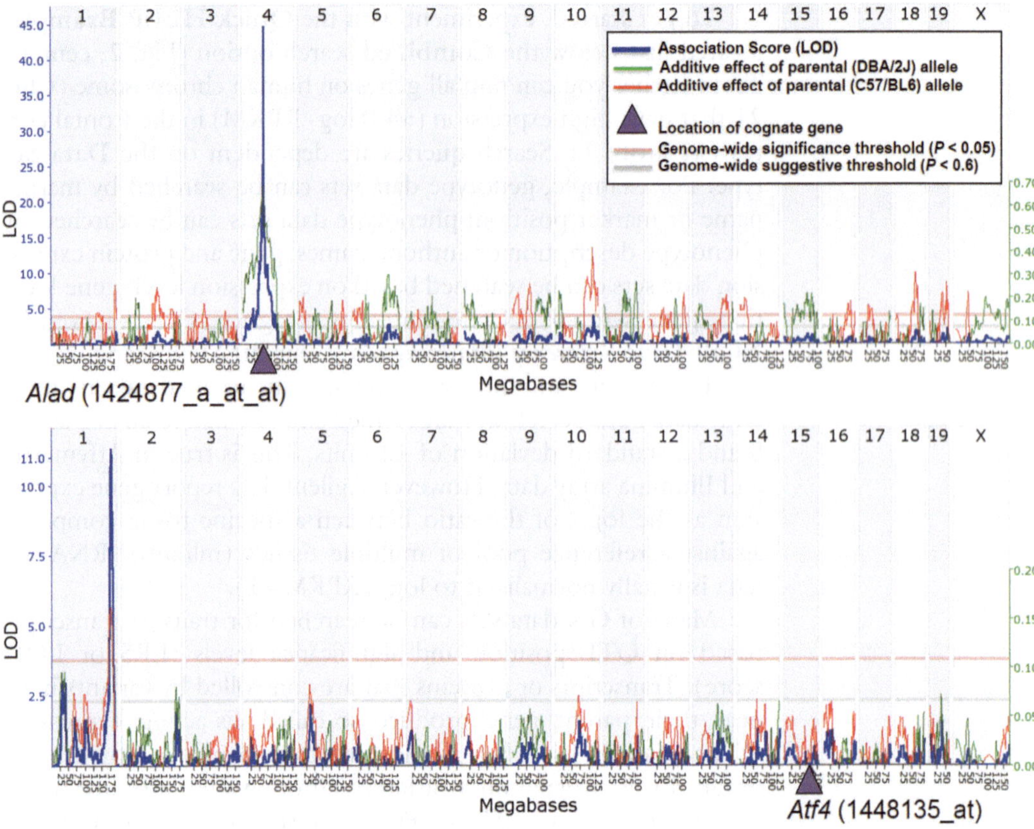

Fig. 3 Local or distant modulation of gene expression in the hippocampus of BXD strains. QTL maps are shown for *Alad* and *Atf4* in the *top* and *bottom panels* with the association score (LOD) plotted on the *Y* axis across the genome (*X*-axis). Chromosomes and megabase position are shown at the *top* and *bottom of the graph*, respectively. Expression of *Alad* is modulated by a local cis-eQTL whereas expression of *Atf4* is modulated by a distant trans-eQTL. The sequence variant underlying expression of *Alad* is actually a copy number variant such that the parental DBA/2J strain and BXD strains that have inherited the *D allele* at this locus have additional copies of the gene and higher expression (indicated by the *green line* associated with the QTL peak in *blue*). The expression of Atf4 is modulated from a distal region on Chr 1. BXD strains that have inherited the *B* allele from the C57BL/6J parent at the Chr 1 locus have higher expression of *Atf4*. This distal region on Chr 1 (often referred to as QTL rich region 1 or *QRR1*) is a major regulatory locus of many expression and behavioral traits. The additive effect is shown in *green to the right*. The expression data can be accessed using **Mouse** *Species*: **Mouse**, *Group*: **BXD Phenotypes**, *Type*: **BXD** Data Set: **Hippocampus Consortium M430v2 (Jun06) RMA** and entering the probe set IDs in the **Get Any** search option

Tissue Correlation computes correlations of gene expression level across sets of 26 different tissues or 32 different brain regions from inbred (isogenic) strains of mice. Variation in expression is purely due to differences among cell and organ systems rather than being due to genetic or environmental factors. The output tables and graphs are particularly useful when studying genes with minimal annotation or when testing the hypothesis that expression of two or more genes is jointly regulated across tissues.

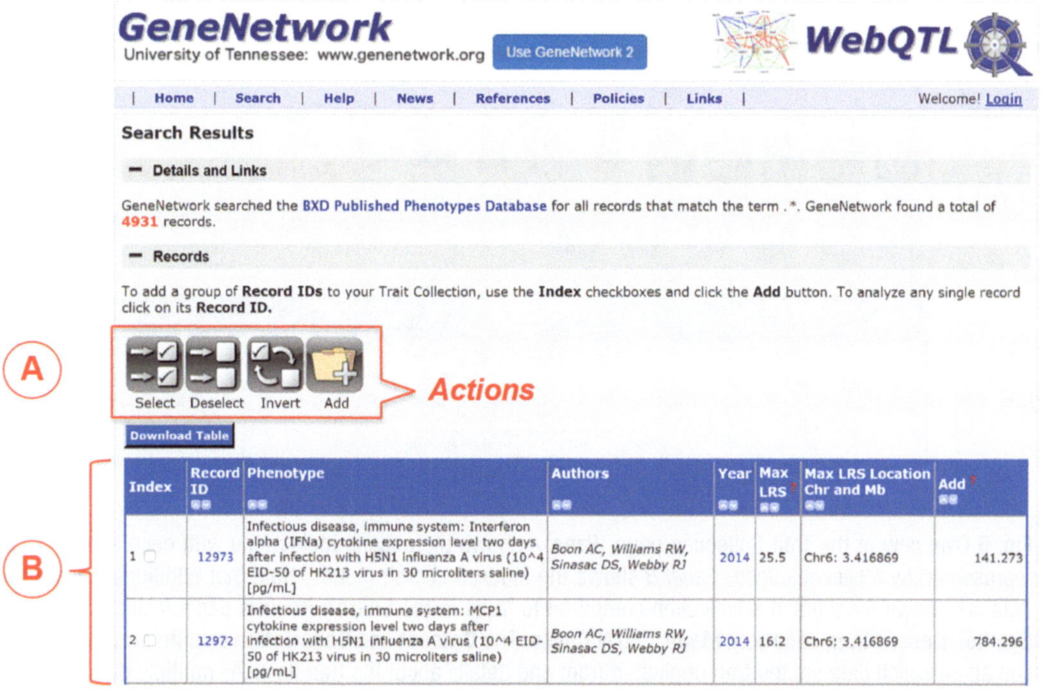

Fig. 4 Overview of Search Results page. *Panel A* indicates actions and *panel B* shows indexed search results. Number of records that match search term are shown in the **Details and Links** section at the top of the page. Note that this page was generated using the **Mouse** (*Species*), **BXD** (*Group*) **Phenotypes** (*Type*) **BXD Published Phenotypes** Data **Set** and entering the wild card character (*asterisk*) using the **Get Any** option. Summarized information for each trait varies based on data set type but, in general, **Record ID** gives a unique identifier for each data set, (e.g. a number for phenotype data sets and a probe set identifier for expression data sets), **Max LRS** and **MAX LRS Location Chr and Mb** give the maximum association score for each trait, and associated peak chromosome and megabase position, respectively. **Add** gives the additive allele effect, which is the estimated effect on trait expression associated with inheritance of the maternal or paternal allele. Positive or negative values indicate higher or lower expression associated with inheritance of the paternal or maternal allele, respectively. From the **Search Results** page additional information about individual traits can be accessed by clicking the **Record ID**. Multiple traits can be selected (or deselected) using the actions options Select, Deselect, and Invert. Selected traits can be added to a **Trait Collection** for further analysis using the Add option. The *red question marks* are links to additional information about column headings

SNP Browser, ***Interval Analyst***, and ***QTLminer*** all provide three different ways to screen for genes and gene variants within defined genomic regions—but currently only for the mouse genome. QTLminer is the most comprehensive of the three tools, and takes advantage of the many levels of data available in GN. This tool can provide output tables that includes many types of QTL information, data on gene expression, and genetic variation across multiple mouse groups [13].

Gene Wiki allows anyone to add notes on genes, transcripts, or proteins to GN. It is essentially an open public notepad with good search functions. GeneWiki incorporates current NCBI GeneRIF annotations.

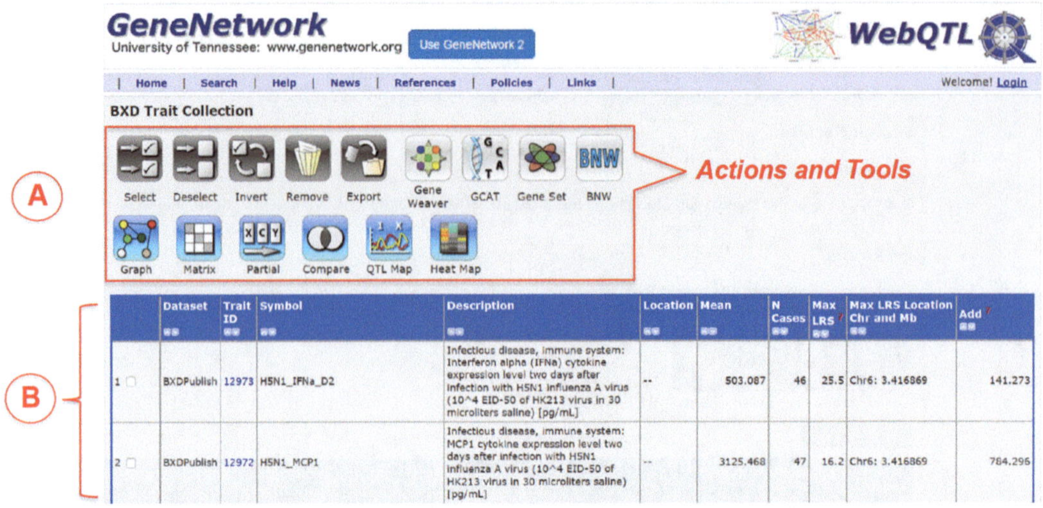

Fig. 5 Overview of the Trait Collection page. *Panel A* shows the actions tools menu with each action or tool represented by a clickable icon. *Panel B* shows the indexed search results. Note that additional columns of data are shown for traits in a collection compared to traits in the **Search Results** page, including **Dataset**, **Symbol**, **Description**, **Location**, **Mean**, and **N Cases**. The **Dataset** and **Description** column provide information about which data set the trait originated from and details about the trait itself. As multiple different types of data can be added to the same **Group** collection it is useful to keep track of which data set the trait originated from, especially if exploring the expression of the same gene across tissue types. For phenotype data sets, detailed descriptions are provided about trait measurement and for gene expression data sets, the full gene name is given along with information about the probe set used to measure the expression of that gene. The **Symbol** column gives the gene symbol for expression data sets and an abbreviated name for phenotypes. **Location** and **Mean** give the location of the gene for expression data sets and average trait expression, respectively. **N Cases** shows the number of individuals that were included in the trait measurement. The *red question marks* are links to additional information about column headings

GenomeGraph provides a way to review global genetic modulation for many gene expression data sets. This tool plots the physical position of each gene against the position of the highest linkage score for the corresponding transcript or probe (this function is not yet available for human data sets). *GenomeGraph* provides two complementary overviews (see the tabs) of the distribution of cis- and trans-eQTLs. One of these is suitable for figures, while the other is interactive and enables zooming and clicking on individual transcript/marker coordinates. The *GenomeGraph* is used to detect both the cis-acting eQTLs and prominent trans-eQTL bands—loci that modulate the expression of large numbers of transcripts or proteins [14]. Can you use this tool to check for trans-eQTL bands in mouse liver (*see* **Note 3**)?

Scriptable Interface is a more complex option that enables direct queries of GN databases using a set of keywords and commands—an application programming interface (API) that can be used to link one web resource with another. It is possible to access

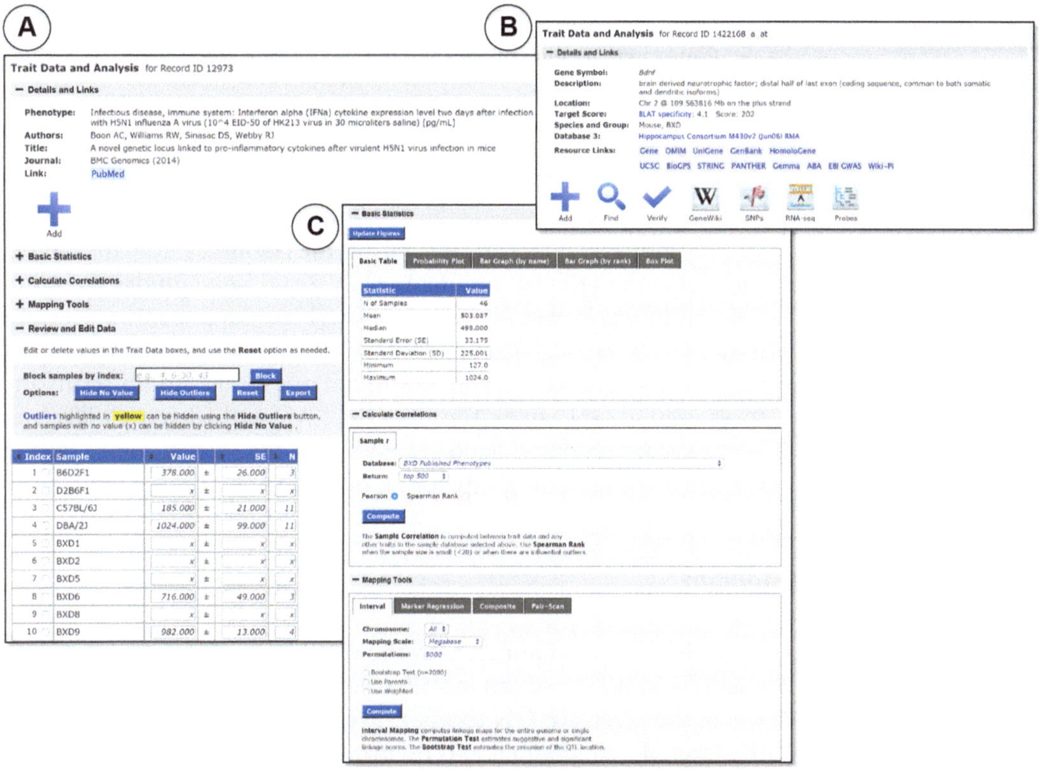

Fig. 6 Layout of Trait Data and Analysis page. Users can explore individual traits in detail in the **Trait Data and Analysis** page. In the **Details and Links** track, a full description of the trait and associated actions and tools are shown. Actions and tools vary slightly depending on whether the trait is from a phenotype (*A*) or gene expression (*B*) Data Set. The results in *B* can be generated by selecting **Mouse** (*Species*), **BXD** (*Group*), **Hippocampus mRNA** (*Type*), **Hippocampus Consortium M430v2 (Jun06) RMA** (Data Set) and entering the gene symbol "*Bdnf*" using the **Get Any** option. Multiple links to outside resources (shown as **Resource Links**) are provided for gene expression data in addition to the GeneNetwork actions and tools Add, Find, Verify, GeneWiki, SNPs, RNA-seq, and Probes. Both traits have a common set of tools shown in *Panel C* as the **Basic Statistics**, **Calculate**, **Correlations**, and **Mapping Tools** tracks. Each track gives the user options to graph the trait distribution, correlate expression of the trait with all other traits in a Data Set from the same Group, or perform QTL mapping for the trait, respectively. Actual trait values are shown in the **Review and Edit Data** track

or download data and tools using R, Python or other code and scripts. The API consists of a query that returns results in a JSON format that is easily loaded locally. The R/qtl package, for example, can read GN REST API data by default. Examples of such functionality are:

1. Fetch all genotype data belonging to a cross or sample.
2. Fetch all phenotype data belonging to an experiment or population.
3. Get the genome scan results for a particular phenotype.

4. Get a list of phenotype correlates and their correlations.
5. Get a list of phenotypes with a QTL in a given interval.
6. Get a list of genes matching a QTL in a given interval.

The final three pull-down items—Database Information, Data Sharing, and Annotations—provide documentation and download tools.

In addition there are several useful resources available under the Help tab in the banner menu (Fig. 2e). Useful guides and tutorials outlining how to use the GN web resource can be accessed under the Movies, Tutorials, and HTML Tour options. Extremely useful explanations to frequently asked questions and for terms and tools used in GN can be found in the FAQ and Glossary of Terms. The glossary has been hand curated since the inception of GN and is a great companion guide for all new users.

3 The GeneNetwork Toolbox

Now that you are familiar with the organization of data and typical search workflows, we can introduce resources available for trait analysis in the extensive GN toolbox. We will explore these tools first at the level of a single trait, and then at the level of multiple traits.

3.1 Tools for Single Trait Analysis

The **Trait Data and Analysis** page is key to using GN and includes many useful tools for studying single traits (Fig. 6). Options differ by data type and species. A trait such as body weight has very different **Resource Links** than mRNA, protein, metabolite, and genotype data. Most data sets that include transcript or protein assay measurements include links to resources that provide information about function, homology, expression across tissues, and genomic location. These include Gene pages at NCBI, ***OMIM***, ***HomoloGene***, ***UCSC*** Genome Browser, and BioGPS. Other links are focused on protein structure and function, including STRING, PANTHER, and Wiki-PI. Gemma and ABA provide access and analysis of thousands of transcriptome and in situ expression data sets, respectively. EBI GWAS searches human genome-wide association studies for matches to selected transcripts or proteins.

The row of icons labeled Add, Find, Verify, GeneWiki etc. link to large GN database resources. The Add icon is used to build up collections of traits for network analysis in a **Trait Collection**. Find locates similar expression traits in other data sets and other species. GeneWiki provides a summary of gene and protein function based on notes made by GN users and published data. It is simple to add your own notes to GN by selecting GeneWiki and then New GeneWiki Entry. SNPs links to a **Variant Browser** that is identical to the SNP Browser accessed from the GN banner under the

Search tab. Verify, RNA-seq, and Probes provide quality control information about transcripts and peptides. Both Verify and RNA-seq link to GN mirrors of the Genome Browser.

The Verify and RNA-seq tools use the transcript, peptide, or probe sequence to align against the reference genome. The BLAT reanalysis results and annotations at the top of the Trait Data and Analysis page should match, but mismatches are frequent and arise from poor annotation, poor sequence selection, or ambiguous alignment. The RNA-seq tool performs the same type of BLAT alignment but includes tracks with data on all genomic variants segregating between the parents of the BXD mouse cohort [15], and expression profiles from whole brain [7] and striatum [16] generated by RNA-seq. Sequence variants are displayed in the **DBA/2J Sequence and Structural Variation** track and RNA-seq data from brain (B, D, and BXD strains) and striatum (B and D strains) are displayed in the **RNA-seq: Brain (BR) ABI, N tags/nt, adjusted** track and the **RNA-seq: Striatum (STR) ILM, N tags/nt, adjusted** track, respectively. These data are useful for visualizing variants within genes that may affect expression, and can also be used to determine whether variants overlap probe sequences. Array platforms have all been designed based on the genome of a single reference genome (C57BL/6J in the case of mice, Brown-Norway in the case of rats). The use of a single genome for design purposes can result in biased hybridization in array studies and biased alignment in RNA-seq studies [17]. The RNA-seq data is also useful for validating expression differences detected using array platforms. The related Probes tool is useful only for Affymetrix data sets and is used to evaluate the performance of individual array probes.

3.2 Analysis and Mapping Methods for Single Trait Analysis

The lower set of four panels (Fig. 6c) on the **Trait Data and Analysis** page include the core computational functions of GN—**Basic Statistics**, **Calculate Correlations**, **Mapping Tools**, and **Review and Edit Data**.

Basic Statistics is used to summarize statistical properties of single (univariate) traits. Open this section (click on the bar) and select the **Basic Table** tab or **Probability Plot** or **Bar Graph** tabs. These options are reviewed below in detail in **Case Studies** in Subheadings 4.1 and 4.2.

Calculate Correlations is used to compute the bivariate correlations between the reference trait and any other set of traits that has been measured in the same **Group**. Open this section and select a target **Database**, the number of correlations to **Return** (default is **top 500**, but the range is between **100** and **20,000**), and the method of correlation—**Pearson** or **Spearman Rank**. Note the tabs: GN can compute three types of correlation—**Sample r**, **Literature r**, and **Tissue r. Sample r** does what you expect. It computes correlations using values listed at the bottom of the page.

Literature r computes correlations between genes based on their shared vocabularies in PubMed. The same method is applied when using the GCAT tool (http://binfl.memphis.edu/gcat/help.html, [18]). Finally, **Tissue r** computes correlations based on variation in expression of genes across about 30 tissues and organs in mouse (identical to the Tissue Correlation tool). All correlation output results are displayed in a **Correlation Table**. Any of the rows in these tables can be evaluated in their own **Trait Data and Analysis** page by selecting the Record ID, or large sets of rows and covariates can be analyzed as a group using tools at the top of **Correlation Table** page. Use either the Index check boxes or the Select, Deselect, Invert, and Add icons to move traits into a collection.

Mapping Tools includes a number of on-line "live" QTL mapping methods. The association function in PLINK is currently the default for human GWAS. **Interval** mapping is the default for almost all plant and nonhuman cohorts. Interval mapping exploits Haley-Knott regression equations to evaluate the linkage across all autosomes and chromosome (Chr) X. Linkage is displayed either as a likelihood ratio statistic (LRS) or the log of the odds ratio (LOD). Both scores provide an estimate of the statistical strength of linkage and the LRS is derived from the LOD score by multiplying by 4.61. A linkage probability of 0.001 is roughly equivalent to a LOD of 3 and an LRS of 13.8. Genome-wide association studies (GWAS) in humans often use a $-\log_{10}(P)$ value where P is the probability of linkage between differences in genotype and differences in trait or disease severity.

Mapping Tools also include **Marker Regression**, a very simple method that computes statistics only for individual marker genotypes. Composite interval mapping (Composite) is a variant of simple interval mapping that enables control for one or more other markers. It is equivalent to mapping the results of a partial correlation. Pair-Scan is an experimental mapping option implemented for larger RI sets (samples of 50 or more strains) that searches for epistatic interactions among loci.

Review and Edit Data contains a working copy of the trait values for each case. Outliers, if any, are highlighted in yellow. Users can manually change trait values, select subsets of individuals for further analysis, exclude outlier values, export values for analysis offline, or reset to the original values.

3.3 Tools for Multiple Trait Analysis

A key feature of GN is access to several different levels of data that all originate from well-defined groups of subjects or cases. The levels can range from genotypes to behavior, but can also include different treatments, developmental stages, or laboratory settings. Users can assemble computationally coherent collections of traits to explore joint gene control, gene-by-treatment, gene-by-lab, and gene-by-environmental interactions. Users may want to examine expression for a single gene, gene families, or members of a biological pathway across multiple tissues. To accomplish these tasks it is necessary to find the data types and then assemble them into a single collection. This is done using the **Search Results** page, the **Trait Data and Analysis** page, and several other tables generated by tools in GN, particularly **Correlation Tables**. Once these multiscalar data sets have been assembled, a number of new tools are available for joint analysis from the **Trait Collection** (Fig. 4). Basic actions are similar to those found in the **Search Results** page, including Select, Deselect, and Invert. Other actions include Remove and Export.

Analysis tools that are optimized for large collections of genes and proteins include Gene Weaver (discussed in detail in the chapter 6), GCAT, Gene Set analysis (WebGestalt), and BNW (Bayesian Network Webserver). GCAT uses text mining to determine if a list is functionally coherent and related based on the literature [18]. Gene Set searches for significant enrichment based on GO categories (functional annotations describing gene function or location) and Graph, Matrix, Partial, and Compare are tools that leverage correlations to identify patterns and relations among traits. The Graph tool is used to construct and visualize correlation networks from selected traits. The lines or edges connecting trait nodes can be filtered and exported to the open source Cytoscape software platform or graph images can be reconfigured and saved as a PDF. Matrix generates correlation matrices from any number of traits using both Pearson and Spearman coefficients. Scatter plots can be generated for each pairwise comparison. Principal component analysis (PCA), a data reduction and pattern detection technique, is also performed and eigenvectors are generated for the principal components that capture the majority of the variation in expression of selected traits. Eigenvector values can be added to the **Trait Collection** and are handled by GN in the same way as other traits. The pattern of expression captured across cases by each eigenvector trait can be used for mapping, to find additional correlates, or to check for technical artifacts.

The Partial correlation tool computes correlation between traits after controlling for other traits, markers, or cofactors such as age or sex. Partial correlations can be calculated for a subset of traits in a **Trait Collection** or against an entire data set. Select at least one **Primary** trait (X), one or more **Target** traits (Y), and a set of **Control** traits (Z). Again you have the option of computing either Pearson's r or Spearman's rho partial correlations.

The final correlation tool is Compare. This tool is used to identify intersecting sets of traits across data sets from the same **Group** that are correlated with selected traits in the **Trait Collection** based on a user defined threshold. It will essentially compute the intersecting values of a Venn diagram using 2–20 or more variables in the collection.

Tools for exploring the genetic control and mapping of multiple traits from the same collection include QTL Map and Heat Map. The QTL Map tool allows users to compare QTLs for up to ten traits globally or by single chromosome. This tool is useful to visually explore traits that may be modulated by the same chromosomal position. The Heat Map tool is used to compare global patterns of genetic modulation for up to 500 traits at a time. Individual traits are represented by columns with genomic position shown by row. Significant QTLs are indicated for each trait as intense blue or red bands depending on whether expression is increased by the maternal or paternal allele (blue and red respectively for the BXD RI set).

The tools available for individual or multiple trait analysis in GN are designed for users to explore data sets and detect relations among traits that are driven by genetic and nongenetic factors. The underlying genetic variants responsible for some of these associations and their potential impact on higher-order phenotypic variation can then be evaluated. We provide two case studies below that put these tools and data sets into context, and that illustrate how they can be used in a systems genetics approach.

4 Case Studies and Workflows

In this section we have provided case studies for both mouse and human data sets that illustrate the utility of GN. Other case and use studies can be found in this book and other publications [19].

4.1 Mouse Case Study

The BXD family of strains and their parents—C57BL/6J (B) and DBA/2J (D)—differ greatly in their preference and sensitivity to alcohol and many other drugs. As a result, the BXDs have been used as a genetic model system to map loci and define gene variants that may be involved in addiction. Using data and tools in GN we can ask whether there are any gene variants associated with addiction and whether gene expression varies as a function of strain and genotype. We can also test the possible causes and consequences of variation in gene sequence and gene expression. This case study takes you through the main steps in this process.

1. Navigate to the **Select and Search** page at www.genenetwork.org.
2. Choose an expression database by picking the following options. **Species** = Mouse, **Group** = BXD, **Type** = Hippocampus mRNA, **Data Set** = Hippocampus Consortium M430v2 (Jun06) RMA (the third data set in this menu). For this example we will use an Affymetrix hippocampus expression data set that uses the RMA normalization method. This is the most commonly used normalization method for Affymetrix arrays and is therefore the best choice for comparing across tissue and even species data

sets. The hippocampus is one of many brain regions important for episodic memory formation and spatial navigation. It is also particularly sensitive to many types of environmental and pharmacological perturbations. For more information (metadata) about how this and other data sets were generated, click the Info button to the right of the data set name.

3. Search for genes. Enter the following search string in the **Combined** option: "*Mean*=*(8 16) cisLRS*=*(10 99 10) RIF*=*addiction*" (remove the double quotes). This search will return all transcripts (in this case also called probe sets) that have a mean $\log_2$ expression between 8 and 16 units and whose expression is modulated by a cis-acting eQTL with an LRS between 10 and 99 that have also been linked to addiction. By using the **Combined** search field, all three components of the query have been combined automatically using a Boolean AND operator. The first component—*Mean*=*(8 16)*—limits the search to transcripts that have moderate to very high expression level. Eight is the average $\log_2$ expression level for most array expression data sets in GN while 16 is very high. Typically, a trait with an average $\log_2$ expression value less than 6 is not considered expressed.
The second component of the query—*cisLRS*=*(10 999 10)*—limits the search to those transcripts associated with a cis-eQTL LRS value between 10 and 99. An LRS score of 10 corresponds to a LOD of 2.2 and is roughly associated with a nominal (point-wise) p value of 0.01. Similarly, an LRS of 99 is equivalent to a LOD of 21.5. The third parameter (also 10) included in the query limits how far the eQTL location can be from the corresponding gene associated with the mRNA. In this case we set a 10 Mb exclusion limit. Finally, the third query term—*RIF*=*addiction*—limits the search to genes that have been annotated with the term "*addiction*" in NCBI GeneRif collection.

4. Click on the Search button to explore the results of this query. The search returns 31 records (November 2015). The **Symbol** and **Description** columns provide the gene symbol and full name. The **Record ID** column gives the probe, exon, or transcript ID that has been used to measure expression. The particular part of the mRNA that is the target of the assay is often listed in the **Description** column after the gene name (e.g., "distal 3′ UTR)". Gene location is given in the **Location Chr and Mb** column, whereas the location of highest LRS associated with the trait is given in the **Max LRS Location Chr and Mb** column. The last **Add** column lists the additive effect of alleles at the **Max LRS Location**. In this case, the positive and negative values of **Add** indicate that expression is increased by the paternal (*D*) or maternal (*B*) allele, respectively. All of these **Search Result** columns can be sorted. Initially the list is sorted

alphabetically by **Symbol** but can also be sorted by probe set genomic location (**Location Chr and Mb**) or by eQTL strength (**Max LRS**). The top ten unique genes sorted by **Max LRS** include *Rb1*, *Csnk1e*, *Cntnap2*, *Cdkn1b*, *Mpdz*, *Gria1*, *Comt*, *Gabra2*, *Kcnj3*, and *Slc1a2*. Select all and then click Add to move all of the search results into a **BXD Trait Collection** for further analysis.

5. To study the expression of the *Rb1* transcript in greater detail, select its **Record ID** or **Trait ID** (1417850_at) to navigate to the **Trait Data and Analysis** page (Fig. 7). Each trait can be examined in more detail in this manner, whether it is a transcript, peptide, metabolite, genotype, or behavioral trait. There are a number of tools for single trait analysis on the **Trait Data and Analysis** page. We now will take you through many of these in the next few steps.

6. Examine the expression of *Rb1* across all of the BXD family members included in the data set using the **Basic Statistics** track. Expand the track by clicking the "+" symbol or in the gray bar. Under the **Include** drop-down menu select "BXD Only". The **Basic Table** provides simple univariate statistics such as **N**

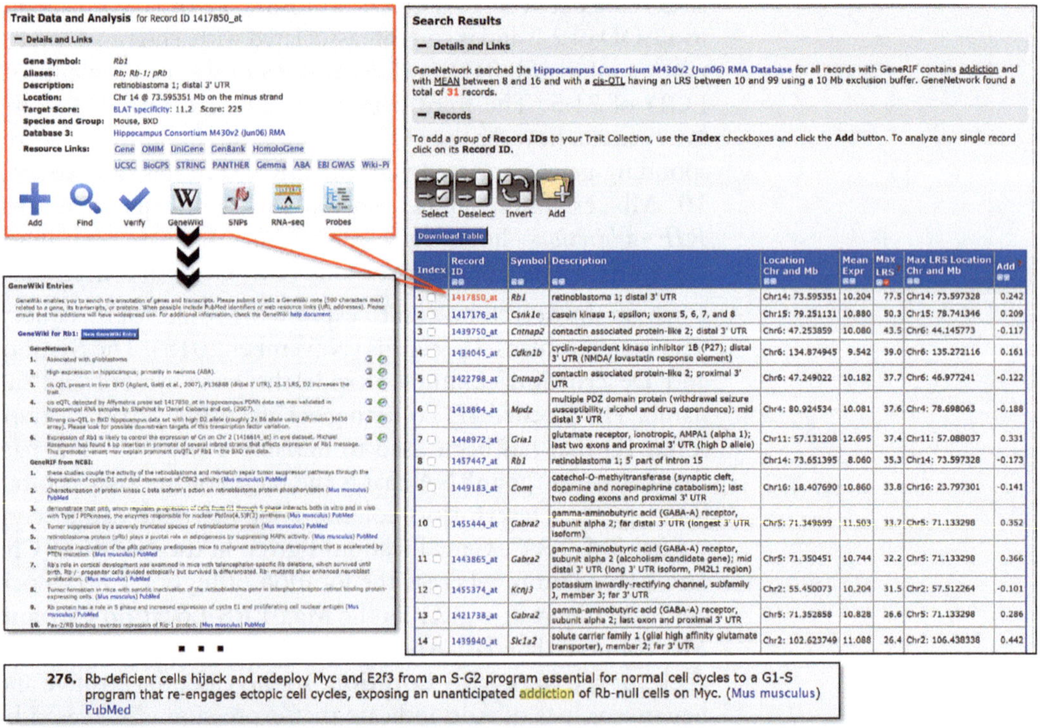

Fig. 7 Exploring the function of *Rb1*. An unusual use of the term addiction in NCBI GeneRIF lead to the inclusion of *Rb1* in our search for addiction-related genes whose expression is modulated by a strong cis-eQTL

of Samples, **Mean**, and **Range**. This particular data set includes 71 samples with a **Range (fold)** of 2.34 fold on this $\log_2$ scale.

The **Probability Plot** tab is a critical tool for detecting outliers and for reviewing the distribution of trait values. If the distribution is close to normal then the observed **Trait values** on the *Y*-axis will line up well with the **Expected Z scores** on the *X*-axis. Deviations from the expected straight line of normality—an S-shape, a set of abrupt breaks (as here), or a set of ripples—indicate that one or more large effects may be influencing the distribution. A strong QTL or a sex difference can produce such effects. For an example of a sex effect (and potential confounder), review the expression of *Xist:* probe set 1436936_s_at.

Another means to visualize data distributions are with **Bar Graph (by rank)** and **Bar Graph (by name)**. By selecting **Bar Graph (by rank)** you can see that expression of *Rb1* is reasonably close to expectation (a normal distribution), although there are two or three small breaks. This could indicate the presence of one or more loci that have a modest impact on expression and that are segregating among the BXD family members. In this case there are no outliers.

Had outliers been detected it would have been necessary to handle them in the **Review and Edit Data** section toward the bottom of the page. This part of the **Trait Data** page contains a working copy of the data values. Values can be deleted or blocked with an *X*. Data can be modified, winsorized, or truncated to make them less extreme. Even a single outlier can have a very adverse impact on genetic mapping—often increasing the risk of false-positive QTLs and producing Pearson correlations that are inflated. The original values can be Reset or downloaded using the Export function.

7. Perform QTL mapping using the **Mapping Tools** track, below the basic statistics and calculate correlations tracks. Very fast interval mapping is a powerful feature of GN that makes it possible to carry out complex trait analysis of most cohorts in real time. Click on the Compute button under the **Interval** tab using the default options. We already noted that the distribution of *Rb1* expression had some breaks. We can now explore possible causes of these disruptions to the expected normal distribution by mapping trait variance.

   The results of whole genome interval mapping are displayed as a graphical map with chromosome number and megabase position displayed at the top and bottom of the map, respectively. You can change to a genetic map measured in centimorgans (cM), but this is rarely useful when a physical map is available. The LRS linkage score is displayed on the left *Y*-axis. **Blue**, **red,** and **green** lines plot the **LRS**, the additive coefficient for the *B* allele (inherited by roughly half of the strains from C57BL/6J) and *D* allele across the genome, respectively.

The horizontal red and grey lines show the threshold for significant and suggestive linkage scores based on mapping 5000 permutations (see the **Histogram of Permutation Test**). A permutation is simply the random rearrangement of elements in an ordered list (in this case a list of genotypes and associated trait values). A permutation test is a method for evaluating statistical significance by randomly reshuffling and recomputing scores for list elements. To achieve a significance of $p=0.05$, the original association score between genotype and trait expression must be greater than at least 95% of all permuted associations. All of these calculations, including the default 5001 genome scans, and the display, usually take less than a minute to generate.

The visual display of the graph can be altered by changing the attributes in the box above the graph. Note the purple arrowhead at the bottom of the *X*-axis that indicates the position of the cognate gene. Here we see strong and highly significant linkage between expression of *Rb1* and a locus on Chr 14 that overlaps the physical location of the *Rb1* gene, a cis-eQTL. Change the units to **LOD** in the attribute box above the map and click on the Chr 14 icon to zoom in and replot the map using a LOD score scale.

To look at the relationship between gene expression, genotype, and the segregation pattern of parental alleles in greater detail, check the **Haplotype Analyst** box and change the **View** to 70–80 Mb in the attributes box and then select Remap. This will zoom in and show the pattern of inheritance for each BXD strain with the location of gene models shown at the top of the plot followed by a map of the chromosome for each strain (strain name to the right) and the corresponding trait value sorted from highest to lowest (value to the right of the strain name). The vertical black lines represent the location of genotyped markers that reveal whether that position in the genome was inherited from the maternal or paternal strain (the corresponding marker names are shown at the bottom of the chromosome map). Similar genotypes across a set of adjacent markers define a haplotype and are represented here as large blocks of green (inherited from the paternal strain) and red (inherited from the maternal strain) with intervening undefined grey regions. Somewhere within the grey interval a recombination event occurred and more markers will be needed to resolve the haplotype blocks more completely. Blue areas are or were heterozygous when the strains were genotyped last. You may have already noticed the striking segregation of green haplotype blocks at the top and red haplotype blocks to the bottom of the chromosome map. Parental alleles at this locus are strongly associated with expression variation and this can be seen here as BXD strains that have inherited the paternal *D allele* (in green) have high expression of *Rb1* and those strains that have inherited the maternal *B allele* (in red) have lower expression (expression values shown for each strain at the far right).

It is often useful to define a confidence interval in which the candidate variant or gene driving trait variation is likely to be located based on the mapping results. One rough estimate of the confidence interval is the 1.5 LOD drop-off which is defined as the interval bordered to the left and right of the peak QTL in which the LOD score (represented by the blue line) drops by 1.5 LOD units. In this example, that would be the point on the blue line to the left and right of the peak that represents a value of 15.5 LOD. This can be roughly approximated visually from the graph such that the 1.5 LOD confidence interval defining the cis-eQTL is roughly between 73 and 75 Mb on Chr 14.

To view the precise association score for any single marker and the corresponding chromosomal position, click the 'Download result in a tab-delimited text format' link toward the top left side of the **Map Viewer** page. Note that the peak marker is rs3701623 located on Chr 14 at 73.597 Mb. To estimate the amount of trait variance that is genetic and captured by this single QTL, navigate back to the main GN **Select and Search** page (use the Search Databases option under the **Search** dropdown in the banner or click on GeneNetwork in the top left corner of the browser window). Enter the marker 'rs3701623' using the **Get Any** query under **Group** = BXD, **Type** = Genotypes, **Data Set** = BXD Genotypes and select Search. This query will return information about genotypes at this marker. Select the marker and Add it to the **Trait Collection**. The collection should now contain all 31 genes from the previous search results and the marker rs3701623. Select the marker and the *Rb1* probe set, and then choose the Matrix tool. We will learn more about the matrix tool later, but for now we have just generated the Pearson (left value) and Spearman Rank (right value) correlation coefficient for our expression trait and marker. The Pearson r is 0.83 and the corresponding r^2 is ~0.7. In other words, about 70% of the variation in hippocampal *Rb1* expression among BXD strains is explained by a cis-eQTL.

8. Verify that *Rb1* is linked in to addiction or substance abuse in GeneWiki. *Rb1* is a tumor suppressor with high expression in hippocampus. But is there a link to addiction of the type we expect? From the Wiki pages perform a search for the work "addiction". This will highlight entry 276. However, *Rb1* is linked to addiction in a different context: the acute need of cells for *Myc* expression to survive. Try this using another gene from the original list—*Cdkn1b* (*see* **Note 4**).

9. As shown above, quality control is critical. Both the Verify and the RNA-seq tools on the **Trait Data and Analysis** page are used to confirm the correct identity of probe sequences and detect possible problems associated with local sequence variants. Probe set 1450486_a_at (*Oprl1*) is a good example of how sequence variants can interfere with expression measurements. Select *Oprl1* probe set 1450486_a_at from the **Trait Collection** and link to the corresponding **Trait Data and Analysis** page.

   Confirm involvement of this gene in addiction by clicking the GeneWiki link and performing the same analysis as in **step 8**. Note that the term "addiction" appears in three separate GeneRIF entries. From the **Trait Data and Analysis** page perform quality control by selecting the RNA-seq tool. This tool is similar to Verify in that it uses UCSC BLAT to align the probe set to the reference genome. The **BLAT Search Results** page (Fig. 8) summarizes alignment scores. Click on the far left browser link of the top row.

   The RNA-seq browser page displays many tracks (Fig. 8 bottom). These include the alignment of the 11 probes (black rectangles), the region of the gene targeted by the probes (the 3′ UTR, exons, or in rare cases, the introns), DBA/2J sequence variants, and RNA-seq expression measurements. Confirm that the probes target the right gene (*Oprl1*) and determine if any variants overlap probes and might interfere with expression measurements (Fig. 8).

   Note that the probe set targets *Oprl1* correctly. However, several probes overlap SNPs (probes 299709 452573; Fig. 8). These SNPs could impact measurements of expression in strains that inherit the *D* allele. To check whether or not expression differs between probes that overlap SNPs, use the Probes tool in the **Trait Data and Analysis** page for *Oprl1* (probe set 1450486_a_at). Affymetrix microarrays feature multiple probes whose expression is then summarized to get a measure of cognate gene expression. The Probes tool allows you to explore individual probe expression, genetic mapping, and covariation. In the case of the M430 array used here, expression is based on hybridization of 11 perfect match (PM) and 11 mismatch (MM) probes (Fig. 9). Use the Select PM button to select the perfect match probes and then select the Heat Map icon to look at the eQTL profile for all 11 probes (Fig. 9). The heat map shows the location and strength of eQTLs for each probe. A strong cis-eQTL indicating higher expression in BXD strains that have inherited the *B* allele of *Oprl1* (blue, Fig. 9) is only associated with probes overlapping SNPs (299709 and 452573). The strong cis-eQTL detected for *Oprl1* is actually a technical artifact caused by sequence variants that disrupt the hybridization of probes to their target RNA sequence in strains

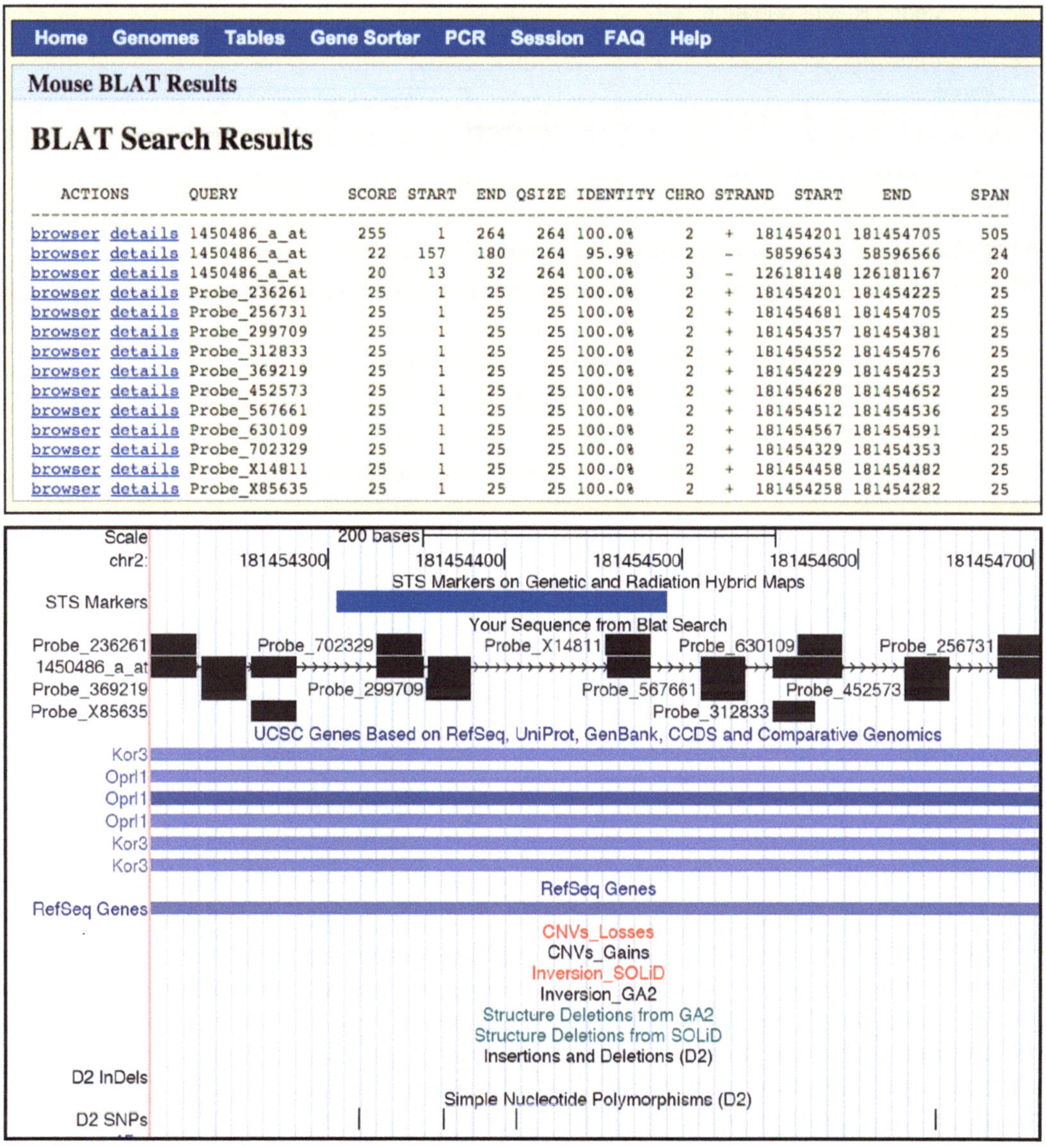

Home Genomes Tables Gene Sorter PCR Session FAQ Help

Mouse BLAT Results

BLAT Search Results

ACTIONS	QUERY	SCORE	START	END	QSIZE	IDENTITY	CHRO	STRAND	START	END	SPAN
browser details	1450486_a_at	255	1	264	264	100.0%	2	+	181454201	181454705	505
browser details	1450486_a_at	22	157	180	264	95.9%	2	-	58596543	58596566	24
browser details	1450486_a_at	20	13	32	264	100.0%	3	-	126181148	126181167	20
browser details	Probe_236261	25	1	25	25	100.0%	2	+	181454201	181454225	25
browser details	Probe_256731	25	1	25	25	100.0%	2	+	181454681	181454705	25
browser details	Probe_299709	25	1	25	25	100.0%	2	+	181454357	181454381	25
browser details	Probe_312833	25	1	25	25	100.0%	2	+	181454552	181454576	25
browser details	Probe_369219	25	1	25	25	100.0%	2	+	181454229	181454253	25
browser details	Probe_452573	25	1	25	25	100.0%	2	+	181454628	181454652	25
browser details	Probe_567661	25	1	25	25	100.0%	2	+	181454512	181454536	25
browser details	Probe_630109	25	1	25	25	100.0%	2	+	181454567	181454591	25
browser details	Probe_702329	25	1	25	25	100.0%	2	+	181454329	181454353	25
browser details	Probe_X14811	25	1	25	25	100.0%	2	+	181454458	181454482	25
browser details	Probe_X85635	25	1	25	25	100.0%	2	+	181454258	181454282	25

Fig. 8 Probe set quality control. The RNA-seq button performs alignment of a probe set sequence against the appropriate reference genome using UCSC Genome Browser's BLAST-like alignment tool (BLAT). The results are shown for probe set 1450486_a_at in the *top panel*. The SCORE is a function of the size and match. For large sequences a perfect score is 255. START, END, and QSIZE provide information about the size in base pairs of the query sequence. IDENTITY provides information about the match with 100 % indicating a perfect match to the reference C57BL/6J genome. The location and span of the match are given by CHRO (chromosome) STRAND, START, END, and SPAN. Note that both the probe set and the 11 perfect match probes that comprise the probe set are shown and that the best match for the individual probes and entire probe set is on the positive strand on Chr 2 around 181.45 Mb. Clicking the browser link for the best match directs to a graphical display of the probe set alignment, shown in the *bottom panel*. The genome browser display can be cluttered for the uninitiated. The basic layout is a display of several different *Tracks* of information. These tracks can be modified by scrolling down to the track tables at the *bottom of the page*. The display in the above panel was generated by selecting the hide option for all tracks EXCEPT the **Mapping and Sequencing**, **Genes and Gene Prediction**, and the **DBA/2J Sequence and Structural Variation** tracks. The position of all 11 probes and the composite probe set are shown in the *bottom panel in black* with the corresponding IDs shown to the *left*. The *arrowheads* designate the alignment of the probe set on the positive (or sense) strand. The targeted gene (*Oprl1*) is shown below and indicates that the probe set is designed to target the 3′ UTR according to the UCSC gene model. The location of sequence variants in the DBA/2J strain relative to the C57BL/6J reference genome are shown in the last two tracks (D2 InDels and D2 SNPs). Note probes 299709 and 452573 overlap a DBA/2J SNP

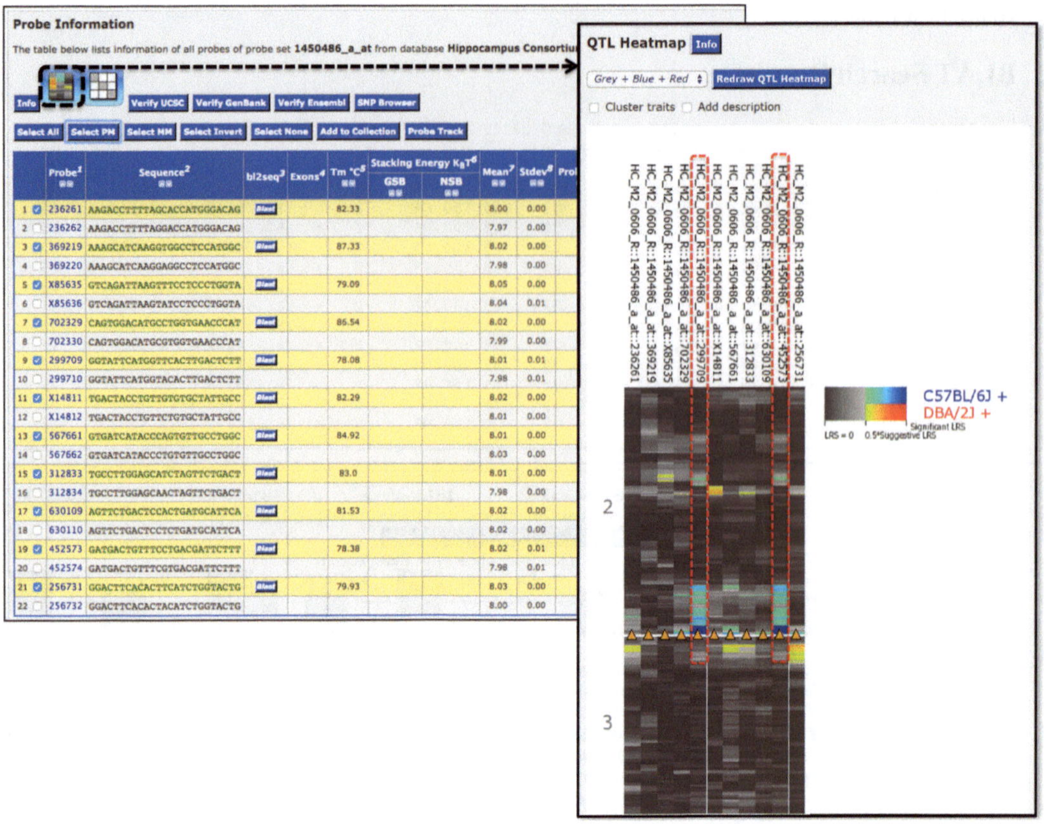

Fig. 9 Impact of variants overlapping probe sets in microarray data sets. SNPs overlapping *Oprl1* probe set 1450486_a_at (perfect match or PM probes 299709 and 452573) lead to expression measurements that are higher in BXD strains that have inherited the *B allele* and lower in strains that have inherited the *D allele*. The **QTL Heatmap** reveals a strong eQTL with higher expression associated with inheritance of the B allele at the *Oprl1* locus (*blue*) only for the probes that overlap SNPs. The *arrowhead* indicates the genomic position of the probes. No other probes demonstrate a strong association between inheritance of alleles at this locus and gene expression. This analysis reveals that the strong cis-eQTL detected for *Oprl1* is actually the result of a technical artifact resulting from sequence variants that disrupt the hybridization of probes to their target RNA sequence in strains other than the reference B6 strain (in this case the D2 strain)

other than those with the reference *B* haplotype. When exploring eQTLs it is good practice to determine: (1) That the assay targets the right genes, and (2) Whether or not measurements might be impacted by sequence variants. Try this analysis on *Kcnj3*, probe set 1455374_at (*see* **Note 5**).

Thus far we have searched and returned a list of genes whose expression is likely modulated by local sequence variants segregating in the BXD cohort that may play a role in addiction. We identified two genes (*Rb1* and *Oprl1*) whose presence on the list is due to different types of technical errors. What about the remaining genes? Are these genes connected in any other way?

10. Select the top nine genes from our **Search Results** page (1417176_at, 1434045_at, 1422798_at, 1418664_at, 1448972_at, 1449183_at, 1421738_at, 1439940_at, 1437920_at, 1421202_at) and Add them to the **Trait Collection** (Fig. 10).

    We can now explore whether these traits are connected at the level of genetic regulation or gene expression. Select all traits and then select the Matrix tool. The output is a correlation matrix comprised of pair-wise correlations for each selected probe set (Fig. 11) and the results of a PCA that will be described below (Fig. 12). From the correlation matrix at the top of the page, we can explore whether the expression of these traits are correlated in the hippocampus of 71 BXD strains. With this number of individuals, a correlation of ~|0.3| will be significant at a *p*-value less than 0.01, however, only correlation coefficients greater than |0.5| are highlighted in the matrix. For each pair-wise correlation, it is possible to generate a scatterplot that also displays the associated *p*-value by clicking on each correlation (Fig. 11). Note that nine pairwise correlations are significant ($p < 0.01$) within this gene set.

    Embedded in the Matrix tool is a module to compute principal components (PCs) and eigenvector scores. PCA is used to extract shared patterns of variation from larger numbers of traits that covary for different reasons. For example, the first PC

BXD Trait Collection

	Dataset	Trait ID	Symbol	Description	Location	Mean	N Cases	Max LRS	Max LRS Location Chr and Mb	Add
1	HC_M2_0606_R	1421202_at	*Chrna4*	cholinergic receptor, nicotinic, alpha 4 (high affinity nicotine receptor with Chrnb2); last two exons and proximal 3' UTR	Chr2: 180.759407	8.310	99	23.4	Chr2: 179.257463	0.096
2	HC_M2_0606_R	1417176_at	*Csnk1e*	casein kinase 1, epsilon; exons 5, 6, 7, and 8	Chr15: 79.251131	10.880	99	50.3	Chr15: 78.741346	0.209
3	HC_M2_0606_R	1422798_at	*Cntnap2*	contactin associated protein-like 2; proximal 3' UTR	Chr6: 47.249022	10.182	99	37.7	Chr6: 46.977241	-0.122
4	HC_M2_0606_R	1418664_at	*Mpdz*	multiple PDZ domain protein (withdrawal seizure susceptibility, alcohol and drug dependence); mid distal 3' UTR	Chr4: 80.924534	10.081	99	37.6	Chr4: 78.698063	-0.188
5	HC_M2_0606_R	1448972_at	*Gria1*	glutamate receptor, ionotropic, AMPA1 (alpha 1); last two exons and proximal 3' UTR (high D allele)	Chr11: 57.131208	12.695	99	37.4	Chr11: 57.088037	0.331
6	HC_M2_0606_R	1449183_at	*Comt*	catechol-O-methyltransferase (synaptic cleft, dopamine and norepinephrine catabolism); last two coding exons and proximal 3' UTR	Chr16: 18.407690	10.860	99	33.8	Chr16: 23.797301	-0.141
7	HC_M2_0606_R	1421738_at	*Gabra2*	gamma-aminobutyric acid (GABA-A) receptor, subunit alpha 2; last exon and proximal 3' UTR	Chr5: 71.352858	10.828	99	26.6	Chr5: 71.133298	0.286
8	HC_M2_0606_R	1439940_at	*Slc1a2*	solute carrier family 1 (glial high affinity glutamate transporter), member 2; far 3' UTR	Chr2: 102.623749	11.088	99	26.4	Chr2: 106.438338	0.442
9	HC_M2_0606_R	1437920_at	*Epha5*	Eph receptor A5; proximal 3' UTR of a short message	Chr5: 84.496134	9.874	99	25.8	Chr5: 78.347362	0.132

Choose File No file chosen append Load Collection Save Collection

Fig. 10 Top cis-modulated genes associated with addiction

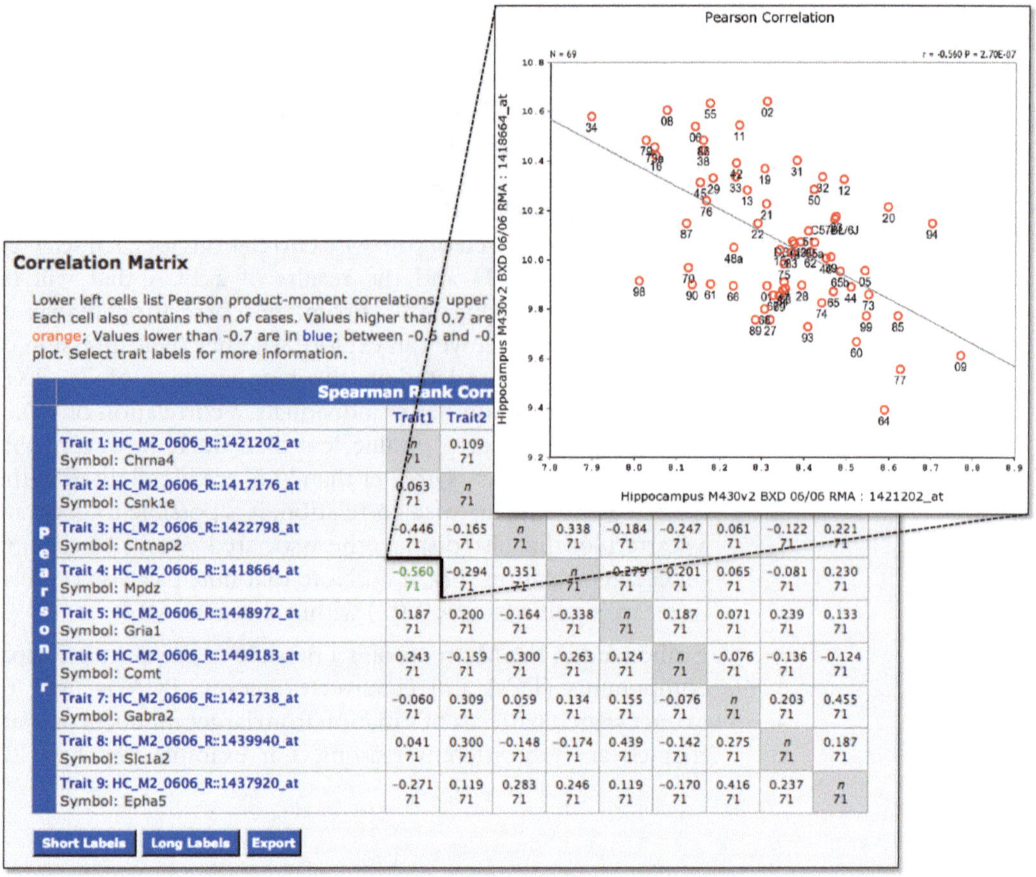

Fig. 11 Exploring covariation. The matrix function allows users to investigate covariation between genes (or probe sets) in the **Trait Collection**. To display the gene symbols along with the probe set IDs, use the Short Labels button to redraw the correlation matrix. The matrix displays the correlation for each pair of genes (or probe sets) with the spearman correlation coefficient shown to the *right* of the diagonal and the Pearson Correlation Coefficient shown to the *left* (the diagonal is indicated by *grey shading* and would normally be represented as a 1, or the correlation of each probe set with itself). Scatterplots can be generated by clicking the correlation in the matrix. The scatterplot can be customized by selecting the Show Options icon, adjusting the settings, and replotting

could represent a technical error or batch effect, a second PC could correspond to sex differences, and a third PC could correspond to variation produced by a gene variant. In many cases, PCs will not correspond to any obvious single source of variance. Scores can be assigned to each subject in the analysis for each of the PCs. These PC scores (also known as eigenvector scores or even "eigengene" score in transcriptome studies) are similar to residuals and have a mean of 0. The **Scree Plot** describes the fraction of variance that is explainable by each of the PCs in descending order. For a set of randomly selected transcripts as much as 25 % of the variance may be described by the first PC—often an indicator of an uncorrected batch effect.

The **Factor Loadings Plot** describes how each trait loads onto, or is correlated with the first and second PCs (Fig. 12). In this example the first factor, or PC1, explains ~28 % of the variance in expression of the nine top transcripts from our search. The PC scores can be used as composite traits and entered into GN collections and workflows just like any other trait. To perform mapping and analysis of the PC scores, select the PCA Traits link under **PCA Traits** (e.g., **PC01**) then review the scores in the corresponding **Trait Data and Analysis** page (Fig. 12). In this example two PCs capture most of the variation in expression. Use the **Interval** tab in the **Mapping Traits** track to perform standard QTL interval mapping. This common source of variation is not derived from a single genetic locus as there are no strong QTLs modulating either PC.

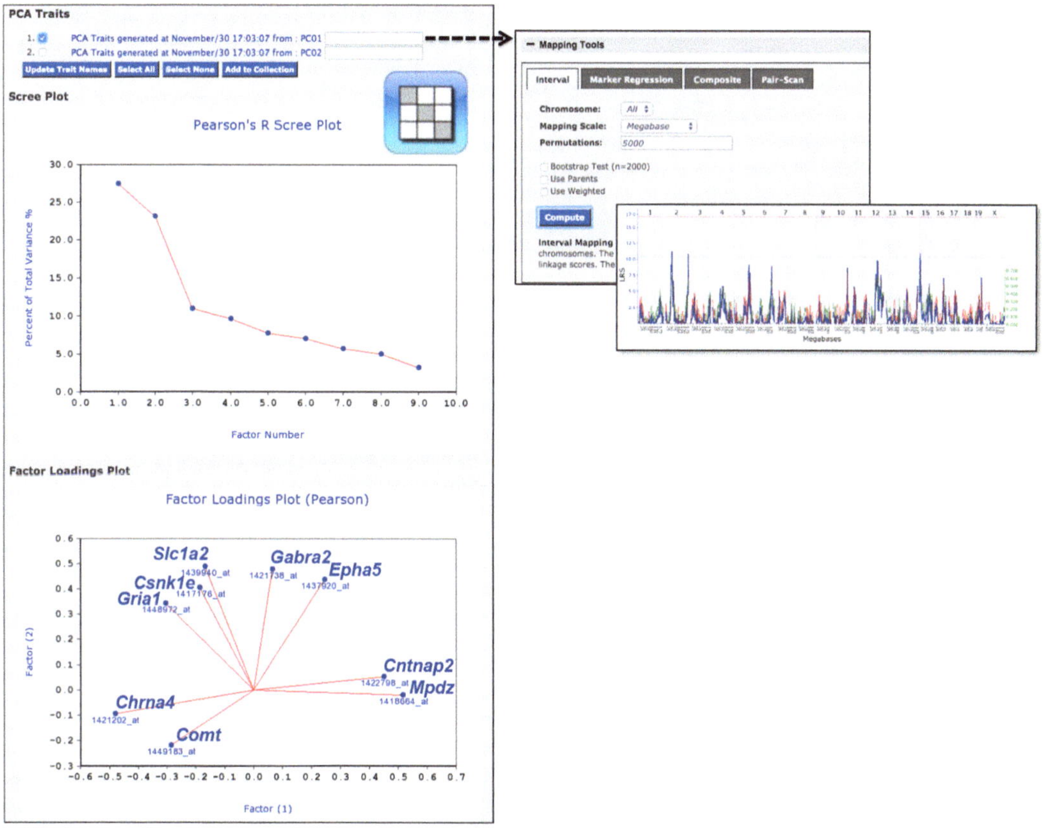

Fig. 12 Principal component analysis (PCA). As part of the matrix tool, a PCA is performed on the selected traits. The **Scree Plot** (*left panel*) plots each principal component (PC) based on the amount of variance each PC or factor explains. The **Factor Loadings Plot** displays the loading (the correlation) between each treat (the measured variable) and the factor or PC (latent variable). Each PC can be treated as a trait. If selected the same basic functions and tools for individual trait analysis can be used for the PC. QTL mapping is shown for PC1 in the *top right panel*. Interval mapping does not suggest strong genetic control originating from a single locus for PC1

11. Construct a network graph from the **Trait Collection** using the Graph tool.

    Additional tools are available in the **Trait Collection** to analyze relations among the top genes (probe sets) in our list. Select all nine traits and the Graph tool. This tool constructs a network graph that shows all possible correlations among selected traits at a given threshold (Fig. 13a). Users can control the way the graph is displayed using the options provided. The type of network can be changed using the **Select Graph Method** dropdown menu. In addition, line color and style, correlation type and threshold, and node label, font, and shape

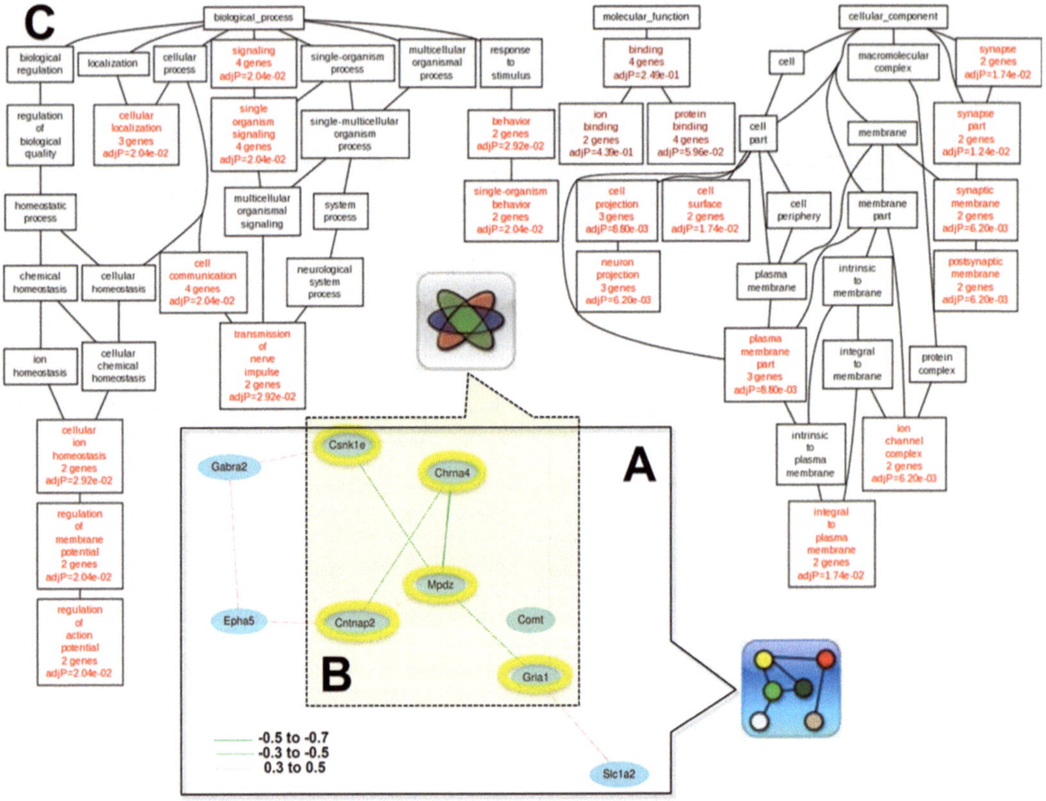

Fig. 13 Creating networks and analysis of biological enrichment. From the **Trait Collection** a network graph depicting relations between gene set members can be constructed using the Graph tool. Display and correlation threshold can be adjusted using the Network Graph interface. Each node represents a gene (probe set) and the edge indicates the correlation (*green* for negative correlations and *red* for positive correlations). In this case the network shown in *A* was given a threshold of $r = |0.3|$ as this represents a significant correlation ($p < -0.01$) in this data set. Based on the network, a subset of genes (shown in the *yellow panel in B*) can be selected for enrichment analysis. Select the subset in the **Trait Collection** and select the Gene Set tool. Enrichment analysis is shown in the background (*C*), with significant (adjusted *p*-value or AdjP < 0.05) enrichment of biological function (based on GO annotations) shown in *red*

are all customizable. High-quality PDF or GIF files can also be generated. In our example, *Mpdz* is the highest connected gene in the network and has four connections at a correlation of |0.3| or better (Fig. 13a), in contrast, *Comt* is not connected at all. Highly connected genes, sometimes called network hubs or hub genes, are thought to have important biological roles, although this is a topic of much debate in systems biology. In less complex systems (flies, worms, and yeast), such hub genes are often essential genes required for survival. However, in higher organisms the role of such hub genes is less clear. Note, that our network of nine genes (or nodes) is much too small to make grand biological conclusions, but is sufficient for an exploratory analysis and tutorial.

12. Test whether a subset of selected expression traits is enriched for biological function using the Gene Set tool. Variation or covariation, such as that observed using the Matrix (pair-wise correlations) and PCA (data reduction and pattern analysis) or the Graph tool (covariation) can indicate underlying genetic control or shared biological function. The Gene Set tool in the **Trait Collection** page can be used to investigate whether selected sets of genes share common biological functions. Select *Mpdz* and its correlates (*Chrna4*, *Gria1*, *Csnk1e*, and *Cntnap2*) and the Gene Set tool (Fig. 13b). This tool uses WebGestalt to compare functional GO anotations within the selected genes compared to a background gene list that includes all of the genes (probe sets) included on the M430 microarray used to generate this data set. Select View results to display a directed acyclic graph of significantly enriched functional categories (Fig. 13c). Even though the gene list submitted is quite small (only five genes), several categories are enriched at an adjusted *p*-value less than 0.05. These categories include signaling (*Chrna4*, *Cntnap2*, *Mpdz*, and *Csnk1e*), part of neuron projection (*Chrna4*, *Cntnap2*, and *Mpdz*), and regulation of action potential (*Chrna4* and *Mpdz*). Click on the Trait ID of each gene in the **Trait Collection** and use the GeneWiki tool to explore their function in more detail. These genes function in overlapping biological pathways, play a critical role in synaptic and intracellular signaling, and have been linked to addiction. In addition, expression of all genes is correlated and the expression of each is variable in BXD hippocampus—likely due to the presence of local sequence variants that modulate expression.
13. Perform a reverse systems genetics analysis to dissect the consequences of genomic variation on higher-order traits by selecting the link for Trait ID 1449183_at (*Comt*) to navigate to the **Trait Data and Analysis** page.

Now that we have initiated a functional search and explored variation and covariation among sets of genes, let us use the vast data resources available in GN to perform a reverse systems genetics analysis to dissect the consequences of genomic variation on higher-order traits. From the **Trait Data and Analysis** page for *Comt*, navigate to the GeneWiki entry. This gene has been extensively studied in human populations and in the BXD cohort. A common polymorphism in humans results in the substitution of the amino acid valine (*Val*) to methionine (*Met*), and a decrease in activity. *COMT* is involved in the degradation of catecholamines, including the neurotransmitters adrenaline, noradrenaline, and dopamine. *COMT* alleles have been associated with subtle differences in risk of psychiatric disease and difference in cognition and attention. A *Comt* polymorphism also segregates among the BXD population such that the maternal strain and those BXD progeny that have inherited the *B* allele have a ~200 bp insertion (a type of mutational event in which additional DNA is added to the genomic sequence) in the 3′ UTR that leads to truncation when compared to the paternal haplotype (*D* allele) [7]. Interestingly, for some *Comt* probe sets (1449183_at) this mutation leads to higher expression in those strains that have inherited the *B* allele, unless the probe sets target the most distal part of the 3′ UTR (1418701_at) that is not expressed in those cases. In the latter case, higher expression is observed in those strains that have inherited the *D* allele. To look at this interesting discordance between probe sets, use the Find tool to identify probe sets targeting *Comt* in multiple expression data sets from BXD. Using the tools introduced to you earlier in this case study, compare where each *Comt* probe set (1418701_at and 1449183_at) aligns to the reference genome, the strain distribution of expression for each probe set, and the difference in cis-eQTL mapping (*see* **Note 6**). Note the different Record IDs for *Comt* that correspond to different probes or probe sets across different microarray platforms. Different regions of the *Comt* gene are being targeted by each probe or probe set, and this is generally true for most genes and microarray platforms. The Find tool can also be used to find corresponding probe sets for the same gene in human and rat data sets.

We know that the expression of *Comt* varies across the BXD set and we now know from GeneWiki that the causal mutation underlying this variation is an insertion. We can use GN data sets to determine the functional consequences of this variation. In other words, we can ask what phenotypes are controlled by the genetic variation at the *Comt* locus. To do this we can navigate back to the **Select and Search** page and identify phenotypes from the BXD Phenotypes BXD Published Phenotypes data set that map back to the *Comt* locus. In the

Combined search option enter "*LRS=(9 99 chr16 16 22)*" to identify all phenotypes that have a peak QTL located within 2 Mb of the *Comt* locus on Chr 16 at 18.4 Mb. This should return at least 12 traits that we can add to our collection. Do the traits returned make sense given the role of *Comt* in the regulation of catecholamine (epinephrine, norepinephrine, and dopamine) levels? The expression of these phenotypes is controlled by a QTL that precisely overlaps the location of *Comt*. To compare the overlap in QTL mapping among these phenotypes and with the *Comt* probe set, select all phenotype traits and the expression trait in the **Trait Collection** and select the Heat Map tool. For finer mapping resolution up to ten traits can be mapped together using the QTL Map tool.

In many cases this type of a reverse genetic analysis is complicated by the linkage disequilibrium inherent in the BXD population, which has an average haplotype block of about 50 Mb and an eQTL mapping resolution of around 1 Mb. This often results in the presence of several genes and variants within a QTL confidence interval that could control trait expression. In our case, *Comt* is the only gene within a 4 Mb interval that contains a variant. Thus, traits that map back to this locus are controlled by the variation in *Comt*. You can also use this same search query in different BXD expression data sets to find downstream expression traits (probe sets that map back to the *Comt* locus or are controlled by a trans-eQTL that originates from the *Comt* locus) or to find phenotypes or expression traits that correlate with *Comt* expression.

In the preceding series of examples we have illustrated how to query the GN database and use some of the many tools available to perform systems-level analyses, including genetic mapping, exploring patterns of covariation and performing a reverse genetics systems analysis to uncover the functional impact of sequence variation. All examples rely on a large and well-characterized genetic reference population, the BXD cohort. In the next example we will explore some of the ways to search human data sets available in GN.

4.2 Human Case Study

In this example we will make use of a publicly available multilevel data set collected from a human cohort. As in the mouse case study, navigate to the **Select and Search** page and this time select **Species**=Human, and **Group**=Liver: Normal Gene Expression with Genotype (Merck). Clicking on the Info button will show that this data set was originally published in 2008 [20] and then in 2010 [21] and was specifically used to examine gene expression and cytochrome P450 activity in human liver. Click on the **Type** dropdown menu to see the types of data that are available for this group. You will see that there are two data types available for this group. The Phenotypes data set (named as HLC Published

Phenotypes) consists of phenotypes collected from this population that can be used for genetic mapping. Additionally, for some of the human cohorts including this particular group, the Phenotypes category can also include some individual-level demographic data such as age, race, socioeconomic status, etc. The other data type for this group is microarray gene expression data for the liver (Liver mRNA). Additionally, there is genotype data available for this cohort and users can perform basic genetic association analysis within GN using PLINK.

Using a simple workflow, we will demonstrate how functions in GN enable secondary analysis of published human data. We start out with basic demographic data—the age of subjects—and examine what we can learn about age-related gene expression changes in the liver.

1. Select **Type** = Phenotype and enter the wildcard symbols * or ? in the Get Any search box. These wildcards will retrieve all records available for this cohort in the database. As of November 2015, there are 17 records in the Phenotype category for this group and include three demographic variables, 12 metabolic and physiologic traits, and two morphometric traits. Can you now use the Matrix and Graph tools that were described in the above mouse case study to inspect the correlation structure among these demographic variables and the different phenotypes (*see* **Note** 7)?
2. Click on the Record ID 10001 (Demographics, age: Age [year]) to open the Trait **Data and Analysis** page for the age data. Notice that the layout of the page is similar to that of the expression traits described in the mouse case study, but without the **Resource Links** and probe tools that are relevant to gene expression traits. Examine the descriptive statistics and distribution profiles for this data using the **Basic Statistics** track. You will see that the mean age is about 50 years (±17 SD) and ranges from 1 to 94 years.
3. Given this wide range in sample age, we can now query if age is associated with differences in gene expression in the liver. Open the **Calculate Correlations track and** Select **Database** = GSE9588 Human Liver Normal (Mar11) Both Sexes. It is also possible to stratify the analysis by sex by choosing either the male or female expression data. For this example, we will retrieve the top 500 transcripts that have the highest correlation with age in both sexes. Select **Pearson** and click Compute. The result of this analysis will be displayed in the **Correlation Table** page. The top of this page will display actions and tools as in the **Trait Collection** page (Fig. 5). The main correlation results are in the **Sample r** and **Sample p(r)** columns (Pearson correlations and *p*-values, respectively) (Fig. 14a). To access individual correlation plots, click on an *r* value and this will display a **Sample Correlation Scatterplot**

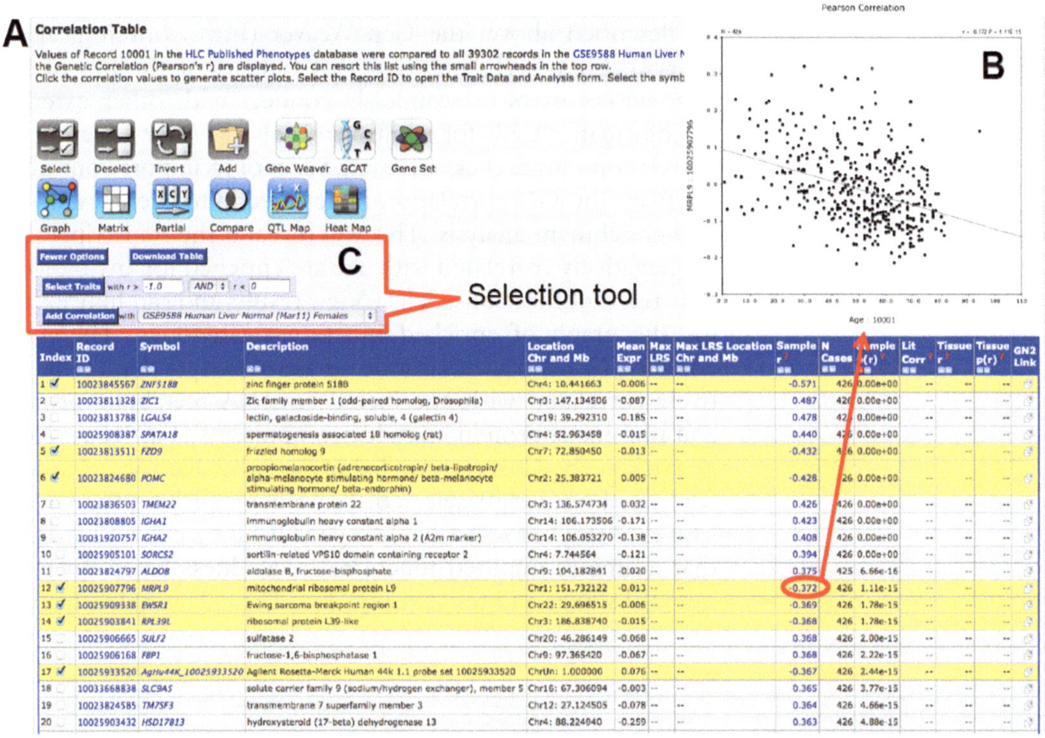

Fig. 14 Correlation table and correlation scatter plot. (*A*) The Correlation Table displays the results of a correlation analysis between a trait or data of interest and other traits collected from the same cohort. In this case, the correlation analysis is between the demographic age data and gene expression in the liver. (*B*) Individual scatter plots can be displayed by clicking on correlation values found in the **Sample r** column. This example shows a significant negative correlation between the expression of a mitochondrial ribosomal protein gene, *MRPL9*, and age. (*C*) Users can select transcripts in the table by setting the correlation criteria using AND/OR operators

with the trait on the *X*-axis (in this case, age) and the mRNA expression on the *Y*-axis (Fig. 14b). For this example, click on the correlation (*r* value) for the 12th transcript in the list (mitochondrial ribosomal protein L9, *MRPL9*) and we see that the expression of this mitochondrial ribosomal protein (MRP) gene is negatively correlated with age. You can customize the scatterplot by selecting Show Options in the **Sample Correlation Scatterplot** and setting your own preferences. For instance, in this example (Fig. 14b), the axes have been renamed from the default and the sample ID tag hidden.

4. The entire correlation results table can be exported by clicking on the Download Table button. Additionally, you can also select a set of records based on correlation values using AND/OR operators by clicking the More Options button and setting the selection criteria. In the example in Fig. 14, all transcripts that are negatively correlated with age are selected by setting the Pearson correlations to range between $r > -1.0$ AND $r < 0$ (Fig. 14c).

5. As described above, the GeneWeaver (http://ontologicaldiscovery.org), GCAT, and Gene Set buttons at the top of the page allows users to seamlessly connect with other external bioinformatics tools for additional analyses. After selecting by correlation range, click the Gene Set tool to import your gene list from the GN correlation table directly to WebGestalt for GO enrichment analysis. This will reveal if the transcripts that are negatively correlated with age are enriched for any biologically relevant functions. Select View results and carefully examine the graph of enriched functional categories. The most enriched GO categories in this list of transcripts that are negatively correlated with age include mRNA metabolic process and ribonucleoprotein complex components (Fig. 15a). Now go back to the **Correlation Table** page that has the negatively correlated transcripts selected. From here, clicking the GCAT icon exports your selections as a gene list for a network analysis that examines imputed functional relatedness based on published abstracts and text mining (Fig. 15b). This quick analysis

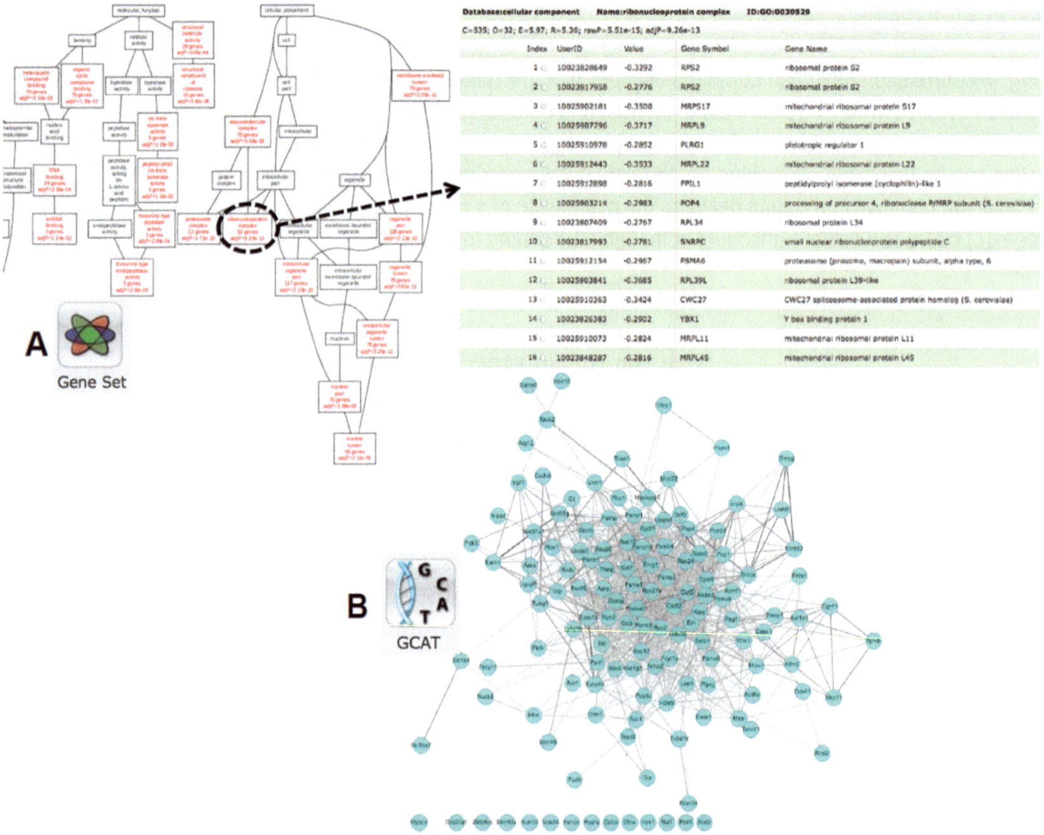

Fig. 15 Biological enrichment and network analysis. Gene lists can be sent directly from gene network to other external websites for (*A*) Gene Ontology, and (*B*) functional network analysis

indicates that ribosomal genes are downregulated in expression during aging. The negative correlation between *MRP* genes and age is striking, and members of this family of genes modulate aging and lifespan in mice and *C. elegans* [22].

Now that we have performed a GO analysis of the transcripts that are negatively correlated with age, repeat the analysis above with transcripts that are positively correlated and demonstrate increased expression with age (*see* **Note 8**).

While mapping functions in GN are better optimized for model organisms and standard test crosses, GN also provides an interface to PLINK for performing simple GWAS in humans. Below we conclude this case study with a demonstration of this mapping tool.

6. So far, we have used a wildcard search key to retrieve all the trait data available for the Merck liver cohort and examined gene expression changes associated with age. Now to perform a genetic association analysis using the phenotype data, open the **Trait Data and Analysis** page for Record ID 10015. This is CYP2C8 enzymatic activity measured in 362 cases.
7. Using the **Basic Statistics** track, note that unlike the age data, which had a normal distribution, this phenotype has a highly skewed distribution. This phenotype provides an example in which the choice between Pearson and Spearman Rank in the **Calculate Correlation** section has a significant impact on the resulting list of correlated genes. First, perform a Pearson correlation and retrieve just the top 100 correlates from the GSE9588 Human Liver Normal (Mar11) Both Sexes data. Perform the same analysis but this time select the Spearman Rank option. Compare the two correlation tables. Note that while the top gene for the Pearson correlation is *TOMM40L* (*ID 10023831160*), the top transcript computed using Spearman rho is *CYP2C8* itself (*10033668843*). The scatter plots for the Pearson *r* and Spearman rho reveals why the Spearman rank correlation is better suited for this CYP2C8 enzymatic activity data and, from the Spearman correlation table, we find that CYP2C8 enzymatic activity is correlated with the expression of a number of other cytochrome P450 genes.
8. Now we test whether variation in CYP2C8 enzyme activity and *CYP2C8* expression share common genetic causes. From the **Trait Data and Analysis** page for record *ID 10015*, navigate to the **Mapping Tools** section. This tool provides a quick but basic interface to PLINK [23]. Note that you can set the thresholds for the minor allele frequency and as well as the *p* value. The current version of this function in GN allows only the basic genetic association tests and users cannot set the threshold for Hardy-Weinberg equilibrium or include other covariates for population structure or demographic covariates

in the association model. So use this tool with these caveats in mind (and compare with GN2 which does include some of these important functions). To initiate the genetic association test, click the Compute Using PLINK button and **Keep** outliers for this preliminary test. Perform the same analysis for *CYP2C8* expression (10033668843) to identify eQTLs.

9. The mapping result will be displayed as a Manhattan plot with chromosomal location on the *X*-axis and the $-\log_{10}(p)$ values on the *Y*-axis (Fig. 16). For the enzyme activity phenotype, the

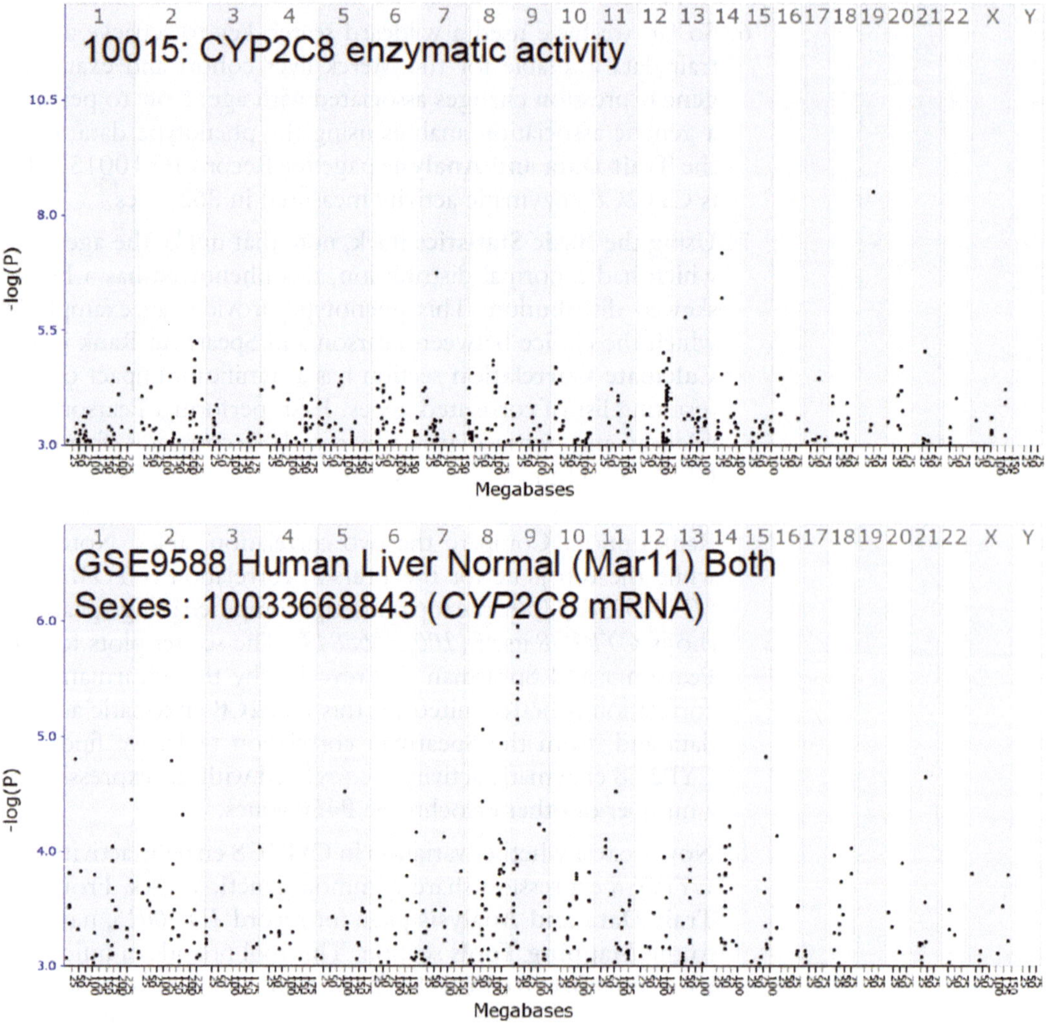

Fig. 16 Manhattan Plots. Basic genetic association test is performed within GeneNetwork using PLINK and result is displayed as a standard Manhattan plot. Comparing between the GWAS results for the (*top*) CYP2C8 enzyme activity (Record ID *10015*), and (*bottom*) expression of *CYP2C8* gene in liver (*GSE9588 Human Liver Normal (Mar11) Both Sexes: 10033668843*), we find no common genetic modulator of the two related traits

top significant association ($p < 0.0000001$) is with SNP rs6508937 chromosome 19. Clicking on the **SNP Name (rsID)** will take you to NCBI's dbSNP page for that particular SNP which will contain additional information on the type of variation, the ancestral allele, minor allele frequency, etc. For the expression trait, the most significant association is with SNP rs10964657 on Chr 9 ($p < 0.0001$). Surprisingly, in this case, the comparison of the two Manhattan plots does not flag any common SNPs and therefore does not provide support for the hypothesis that covariation in expression of transcripts and enzymes are due to shared genetic causes.

5 Future Directions and Conclusions

One of the main values of GN is its vast resource of data that enables both exploratory data-mining as well as specific hypothesis testing and cross-correlations between phenotypes at many scales. At the end of 2015, GN contained 578 systems genetics data sets for eight species and well over 70 different cell, tissue, and organ types making it a 160 GB database of genotypes and well-structured phenotypes. The amount of data in GN is growing rapidly: 255 datasets have been added in the past 2 years, compared to ~100 in the preceding decade. With this volume of data, search is a key feature for analysis and exploration. GN allows searching through genomic, genetic, and phenotype data contained in the database. Users can then select multiple datasets and perform analysis on selected genes, traits, and collections. The web-browser interface allows for interactive exploration of GN resources and the use of built-in analysis tools. This allows biomedical researchers to explore the data without training in more advanced bioinformatics programming languages, such as R and Python.

GN started out as a simple database and web site that was used primarily for analysis of mouse, rat, and human genes, chromosomes, and linked phenotypes. GN has now transformed into a service for on-line QTL mapping, eQTL analysis, and systems genetics. GN allows researchers to upload and store their own research data, run analyses—including QTL mapping, GWAS, and network analysis, generate publishable figures, compare results with those of other datasets, and explore relations between QTLs, genes, and phenotypes.

In this chapter we have highlighted the potential of GN by discussing built-in functionality and providing a few use cases. GN is an evolving service. The goals and challenges are to integrate new and sophisticated mapping and analysis features while maintaining an easy user interface in a structured environment and providing a powerful REST programming interface for power users. The new version of GN (GN2) will provide greater flexibility and additional features such as the use of generalized linear mixed models (LMMs), QTL mapping with covariates, and Weighted Gene Coexpression

Analysis (WGCNA) [24]. These tools are already available in the beta-release of the next generation of GN2 (Fig. 2). There are many packages and web services available that can do individual components of a quantitative genetics or systems genetics analysis well, such as QTL mapping or data reduction and organization. However, there are no other resources that provide both a data repository and an integrated set of tools and services for systems genetics. Because GN and its environment consist of free and open source software, the whole system is easily installed and deployed locally allowing for coexistence of both a public data resource (the heart of GN) and local (private) data. It is even possible to rebrand the webserver and make it outward facing for new projects or institute.

6 Notes

1. Select **Species = Human, Group = All Tissues... Type = Frontal Cortex mRNA**. Click on the Default button to lock-in these settings. Now review the **Quick HELP Examples and User's Guide**. The final query string should be entered into the **Combined** search. It should look like this: POSITION = (chr21 0 1000) MEAN = (4 1000) and should generate 28 hits. To focus on genes involved in Down syndrome, also known as Trisomy 21, add RIF = trisomy. This will trim the set down to four hits.
2. Select **Species = Mouse, Group = BXD, Type = Liver Proteome, Data Set = EPFL/ETHZ BXD Liver, Chow Diet....** Click on the Default button to lock-in these settings. Review the **Quick HELP Examples and User's Guide**. The query string should be entered into the **Get Any** search. It should look like this: transLRS = (20 999 10) and should generate ~136 hits. This search will return all trans-QTLs with an LRS between 20 and 999 using a 10 Mb window. Sort the results by the **Max LRS Location** column and look for patterns in the types of proteins that map to the same eQTL location; e.g. Chr 5 at about 127–128 Mb and Chr 10 at 107 Mb. These are potential trans-regulatory regions.
3. Yet another way to visualize whole data sets and search for regulatory regions would be to select GenomeGraph from the Search tab in the banner menu. Select the **"EPFL..."** data set described above in *see* **Note 2** and choose the Mapping option. This should generate a graph that shows genome location on the *X*-axis (each block is a chromosome) and position of the gene on the *Y*-axis. Each red cross represents a significant association at a false discovery rate (FDR) less than 0.2 (default is set to 0.2 or a FDR of 20%). Note the vertical bands (or trans-bands) that indicate a number of significant associations on several chromosomes, including Chr 5. In contrast to trans-eQTLs,

cis-eQTLs are indicated as a red cross on the diagonal (a significant association that corresponds to the location of the gene).

4. Check the function of *Cdkn1b* by selecting Gene Wiki from the dropdown menu under the Search tab in the banner menu and entering the gene name in the box and selecting submit. Inspect the entries and then perform a search for the term "*addiction*". Again, the term addiction (entry 799) is used in an interesting way, "*Data indicate that the addiction of MYCC-amplified ovarian cancer cells to MYCC differs*...".
5. The probe set for *Kcnj3* appears to align far beyond (distal to) the known limits of the gene. To verify this, perform the RNA-seq BLAT alignment, click on the browser link (far left), and then click on the zoom out 10× button twice. Note that the RNA-seq tracks (blue and red) show intense expression in the region well beyond the standard model 3′ UTR. This is not unusual; 3′ UTRs are often not well annotated. The probe set actually does target the gene, and does so at the distal part of the 3′ UTR. Two of the probes overlap SNPs (736871 and 725381), and both are associated with strong cis-eQTL artifacts (see Heat Map).
6. To get to the Find tool you must navigate to the **Trait Data and Analysis** page for the gene (or probe set) of interest. Use this page or an existing **BXD Trait Collection** if active from the mouse case study. Alternatively, start over from the main search page by searching for *Comt* in most open BXD or even human lymphoblastoid and some aging brain expression data sets (**Groups** from Meyers and Liang). For *Comt*, the Find tool will return a number of results from four human data sets, four rat data sets and over 20 mouse data sets. For many of these data sets the expression of *Comt* is measured from multiple probes. For mouse and human data sets, the expression of each probe set appears to vary despite targeting the same gene (see **Mean Expr** or Mean Expression column). Note the large number of probes for data sets annotated with the term exon; exon-level microarrays have probes designed to target each feature of a transcript (UTRs, introns, and exons). Explore each probe set for *Comt* by clicking on the Record ID for 1418701_at and 1449183_at using **Tissue = Hippocampus** and **Dataset = Hippocampus Consortium M430 (Jun06) RMA**. You will be redirected to the **Trait Data and Analysis** page for each record where you can compare the *Comt* transcript feature targeted by each probe set using the Verify or RNA-seq tool, explore the distribution of expression across BXD strains using the **Basic Statistics** track and **Probability Plot** and **Bar Graph (by rank)** options, and compare allelic effects and cis-eQTL mapping using the **Mapping Tools** track and the **Interval** mapping option. You should see that probe sets targeting the distal end of the *Comt* transcript (the distal 3′ UTR,

probe set 1418701_at) have a very different pattern of expression across inbred strains of mice and the BXD panel when compared to probe sets that target coding exons or more proximal regions of the 3′ UTR (probe set 1449183_at).

7. Start by selecting all 17 records from the **Search Results** page and Add to **Trait Collection**. From the **Trait Collection** page, select all 17 records and then click the Matrix tool. For a graphical visualization of the correlation among the different variables select the Graph tool. Try any of the network methods from the Select Graph Method and set the correlation to |0.25| with Pearson as **Correlation Type**. By examining the correlation matrix and the network graphs, you will learn that the various enzyme activity traits form a correlated network. While ethnicity and sex show no correlation with any of the traits, age is positively correlated with weight, which, as might be expected, has a strong positive correlation with BMI.

8. Transcripts from the **Correlation Table** that are positively correlated with age can be selected by setting the More Options track to $r>0$ AND $r<1.0$. Alternatively, a quicker way is to simply click the Invert Select option. Send this list of genes to WebGestalt by clicking the Gene Set tool. Note that the enrichment p values for this set of genes positively correlated with age are not as significant as for those that are negatively correlated with age.

Acknowledgment

We thank Lei Yan, Arthur Centeno, and Zachary Sloan, for their many contributions to building and maintaining GN over the past decade. GN code has benefited greatly from contributions by Jintao Wang, Sam Ockman, Xiaodong Zhou, Ning Liu, and Alex G. Williams, and Drs. Rudi Alberts, Arends, Elissa J. Chesler, Kenneth Manly, Danny and Evan G. Williams. We also thank M. Trevor Houseal and Austin Kimes for their help in editing this chapter. Support for GeneNetwork has been provided by NIH grants U01AA013499, U01AA16662, U01AA014425, P20DA21131, U01CA105417, and U24 RR021760. GN is also generously supported by the UT Center for Integrative and Translational Genomics, and funds from the UT-ORNL Governor's Chair.

References

1. Manly KF, Olson JM (1999) Overview of QTL mapping software and introduction to map manager QT. Mamm Genome 10(4):327–334
2. Williams RW (1994) The portable dictionary of the mouse genome: a personal database for gene mapping and molecular biology. Mamm Genome 5(6):372–375
3. Chesler EJ, Lu L, Shou S, Qu Y, Gu J, Wang J, Hsu HC, Mountz JD, Baldwin NE, Langston MA, Threadgill DW, Manly KF, Williams RW (2005) Complex trait analysis of gene expression uncovers polygenic and pleiotropic networks that modulate nervous system function. Nat Genet 37(3):233–242. doi:10.1038/ng1518

4. Andreux PA, Williams EG, Koutnikova H, Houtkooper RH, Champy MF, Henry H, Schoonjans K, Williams RW, Auwerx J (2012) Systems genetics of metabolism: the use of the BXD murine reference panel for multiscalar integration of traits. Cell 150(6):1287–1299. doi:10.1016/j.cell.2012.08.012
5. Chesler EJ, Wang J, Lu L, Qu Y, Manly KF, Williams RW (2003) Genetic correlates of gene expression in recombinant inbred strains: a relational model system to explore neurobehavioral phenotypes. Neuroinformatics 1(4):343–357. doi:10.1385/NI:1:4:343
6. Wang J, Williams RW, Manly KF (2003) WebQTL: web-based complex trait analysis. Neuroinformatics 1(4):299–308. doi:10.1385/NI:1:4:299
7. Li Z, Mulligan MK, Wang X, Miles MF, Lu L, Williams RW (2010) A transposon in comt generates mRNA variants and causes widespread expression and behavioral differences among mice. PLoS One 5(8):e12181. doi:10.1371/journal.pone.0012181
8. Williams EG, Mouchiroud L, Frochaux M, Pandey A, Andreux PA, Deplancke B, Auwerx J (2014) An evolutionarily conserved role for the aryl hydrocarbon receptor in the regulation of movement. PLoS Genet 10(9):e1004673. doi:10.1371/journal.pgen.1004673
9. Wang X, MK M, Pandey A, Williams EG, Mozhui K et al (2016) Joint mouse-human phenome-wide association to test gene function and disease risk. Nat Commun 7:10464
10. Chesler EJ, Lu L, Wang J, Williams RW, Manly KF (2004) WebQTL: rapid exploratory analysis of gene expression and genetic networks for brain and behavior. Nat Neurosci 7(5):485–486. doi:10.1038/nn0504-485
11. Peirce JL, Lu L, Gu J, Silver LM, Williams RW (2004) A new set of BXD recombinant inbred lines from advanced intercross populations in mice. BMC Genet 5:7. doi:10.1186/1471-2156-5-7
12. Taylor BA, Heiniger HJ, Meier H (1973) Genetic analysis of resistance to cadmium-induced testicular damage in mice. Proc Soc Exp Biol Med 143(3):629–633
13. Alberts R, Schughart K (2010) QTLminer: identifying genes regulating quantitative traits. BMC Bioinformatics 11:516. doi:10.1186/1471-2105-11-5161471-2105-11-516
14. Overall RW, Kempermann G, Peirce J, Lu L, Goldowitz D, Gage FH, Goodwin S, Smit AB, Airey DC, Rosen GD, Schalkwyk LC, Sutter TR, Nowakowski RS, Whatley S, Williams RW (2009) Genetics of the hippocampal transcriptome in mouse: a systematic survey and online neurogenomics resource. Front Neurosci 3:55. doi:10.3389/neuro.15.003.2009
15. Wang XS, Agarwala R, Capra JA, Chen ZG, Church DM, Ciobanu DC, Li ZS, Lu L, Mozhui K, Mulligan MK, Nelson SF, Pollard KS, Taylor WL, Thomason DB, Williams RW (2010) High-throughput sequencing of the DBA/2J mouse genome. BMC Bioinformatics 11:O7. doi:10.1186/1471-2105-11-S4-O7
16. Bottomly D, Walter NA, Hunter JE, Darakjian P, Kawane S, Buck KJ, Searles RP, Mooney M, McWeeney SK, Hitzemann R (2011) Evaluating gene expression in C57BL/6J and DBA/2J mouse striatum using RNA-Seq and microarrays. PLoS One 6(3):e17820. doi:10.1371/journal.pone.0017820
17. Ciobanu DC, Lu L, Mozhui K, Wang X, Jagalur M, Morris JA, Taylor WL, Dietz K, Simon P, Williams RW (2010) Detection, validation, and downstream analysis of allelic variation in gene expression. Genetics 184(1):119–128. doi:10.1534/genetics.109.107474, doi:genetics.109.107474 [pii]
18. Homayouni R, Heinrich K, Wei L, Berry MW (2005) Gene clustering by latent semantic indexing of MEDLINE abstracts. Bioinformatics 21(1):104–115. doi:10.1093/bioinformatics/bth464
19. Williams RW, Mulligan MK (2012) Genetic and molecular network analysis of behavior. Int Rev Neurobiol 104:135–157. doi:10.1016/B978-0-12-398323-7.00006-9
20. Schadt EE, Molony C, Chudin E, Hao K, Yang X, Lum PY, Kasarskis A, Zhang B, Wang S, Suver C, Zhu J, Millstein J, Sieberts S, Lamb J, GuhaThakurta D, Derry J, Storey JD, Avila-Campillo I, Kruger MJ, Johnson JM, Rohl CA, van Nas A, Mehrabian M, Drake TA, Lusis AJ, Smith RC, Guengerich FP, Strom SC, Schuetz E, Rushmore TH, Ulrich R (2008) Mapping the genetic architecture of gene expression in human liver. PLoS Biol 6(5):e107. doi:10.1371/journal.pbio.0060107
21. Yang X, Zhang B, Molony C, Chudin E, Hao K, Zhu J, Gaedigk A, Suver C, Zhong H, Leeder JS, Guengerich FP, Strom SC, Schuetz E, Rushmore TH, Ulrich RG, Slatter JG, Schadt EE, Kasarskis A, Lum PY (2010) Systematic genetic and genomic analysis of cytochrome P450 enzyme activities in human liver. Genome Res 20(8):1020–1036. doi:10.1101/gr.103341.109
22. Houtkooper RH, Mouchiroud L, Ryu D, Moullan N, Katsyuba E, Knott G, Williams

RW, Auwerx J (2013) Mitonuclear protein imbalance as a conserved longevity mechanism. Nature 497(7450):451–457. doi:10.1038/nature12188

23. Purcell S, Neale B, Todd-Brown K, Thomas L, Ferreira MA, Bender D, Maller J, Sklar P, de Bakker PI, Daly MJ, Sham PC (2007) PLINK: a tool set for whole-genome association and population-based linkage analyses. Am J Hum Genet 81(3):559–575. doi:10.1086/519795
24. Langfelder P, Horvath S (2008) WGCNA: an R package for weighted correlation network analysis. BMC Bioinformatics 9:559. doi:10.1186/1471-2105-9-559
25. GTEx Consortium (2015) Human genomics. The Genotype-Tissue Expression (GTEx) pilot analysis: multitissue gene regulation in humans. Science 348(6235):648–660. doi:10.1126/science.1262110
26. Webster JA, Gibbs JR, Clarke J, Ray M, Zhang W, Holmans P, Rohrer K, Zhao A, Marlowe L, Kaleem M, McCorquodale DS 3rd, Cuello C, Leung D, Bryden L, Nath P, Zismann VL, Joshipura K, Huentelman MJ, Hu-Lince D, Coon KD, Craig DW, Pearson JV, NACC-Neuropathology Group, Heward CB, Reiman EM, Stephan D, Hardy J, Myers AJ (2009) Genetic control of human brain transcript expression in Alzheimer disease. American Journal of Human Genetics 84(4):445–458
27. Gibbs JR, van der Brug MP, Hernandez DG, Traynor BJ, Nalls MA, Lai SL, Arepalli S, Dillman A, Rafferty IP, Troncoso J, Johnson R, Zielke HR, Ferrucci L, Longo DL, Cookson MR, Singleton AB (2010) Abundant quantitative trait loci exist for DNA methylation and gene expression in human brain. PLoS Genetics 6(5), e1000952
28. Liang WS, Reiman EM, Valla J, Dunckley T, Beach TG, Grover A, Niedzielko TL, Schneider LE, Mastroeni D, Caselli R, Kukull W, Morris JC, Hulette CM, Schmechel D, Rogers J, Stephan DA (2008) Alzheimer's disease is associated with reduced expression of energy metabolism genes in posterior cingulate neurons. Proc Natl Acad Sci U S A 105(11):4441–4446. doi:10.1073/pnas.0709259105
29. Zhang B, Gaiteri C, Bodea LG, Wang Z, McElwee J, Podtelezhnikov AA, Zhang C, Xie T, Tran L, Dobrin R, Fluder E, Clurman B, Melquist S, Narayanan M, Suver C, Shah H, Mahajan M, Gillis T, Mysore J, MacDonald ME, Lamb JR, Bennett DA, Molony C, Stone DJ, Gudnason V, Myers AJ, Schadt EE, Neumann H, Zhu J, Emilsson V (2013) Integrated systems approach identifies genetic nodes and networks in late-onset Alzheimer's disease. Cell 153(3):707–720. doi:10.1016/j.cell.2013.03.030
30. McClurg P, Janes J, Wu C, Delano DL, Walker JR, Batalov S, Takahashi JS, Shimomura K, Kohsaka A, Bass J, Wiltshire T, Su AI (2007) Genomewide association analysis in diverse inbred mice: power and population structure. Genetics 176(1):675–683
31. Wang S, Yehya N, Schadt EE, Wang H, Drake TA, Lusis AJ (2006) Genetic and genomic analysis of a fat mass trait with complex inheritance reveals marked sex specificity. PLoS Genet 2(2), e15
32. Huang GJ, Shifman S, Valdar W, Johannesson M, Yalcin B, Taylor MS, Taylor JM, Mott R, Flint J (2009) High resolution mapping of expression QTLs in heterogeneous stock mice in multiple tissues. Genome Res 19(6):1133–1140. doi:10.1101/gr.088120.108

Chapter 5

Complex Trait Analyses of the Collaborative Cross: Tools and Databases

Ramesh Ram and Grant Morahan

Abstract

The Collaborative cross (CC) is a powerful mouse resource for investigating complex genetic traits. Here we discuss various tools and techniques for gene mapping and identification using the CC. The data analyses procedures are illustrated with examples.

Key words Analysis tools in systems genetics, Collaborative Cross, Quantitative trait loci (QTL) mapping methods, Haplotype analysis, Imputation, Mixed model, Whole-genome association

1 Introduction

Complex trait analysis using the Collaborative Cross (CC) [1] has gained popularity among researchers worldwide due to the power of this resource, including the genetic and phenotypic variability exhibited by the CC strains. This diversity was achieved by selecting eight diverse founder strains [2] that harbored over 90% of the common genetic variants of the mouse species. The CC resource was produced at three sites [3–5]; the Australia CC strains are also termed the Gene Mine [3] due to its ability to support identification of genes mediating complex traits. Recombination events that occurred during the breeding program are "archived" such that each strain has a mosaic genome of chromosome segments derived from the founders. The genome architecture of the CC [2] enables definition of narrower phenotypic association regions than conventional mapping [6]. The genome intervals of the CC strains can be readily assigned to each of the eight founders [7]. This enables the application of classical linkage mapping approaches for unbiased discovery of genetic polymorphisms that determine various complex traits.

A gene mapping analysis in the CC can be achieved in just two steps: (a) phenotyping (*see* ref. 8, 9 for examples) and (b)

Klaus Schughart and Robert W. Williams (eds.), *Systems Genetics: Methods and Protocols*, Methods in Molecular Biology, vol. 1488, DOI 10.1007/978-1-4939-6427-7_5, © Springer Science+Business Media New York 2017

genetic analyses using tools and software for gene identification [7]. In this chapter, we describe how to use a set of bioinformatic tools that are collectively known as The GeneMiner [7]. These tools are accessible online via the following URL: www.sysgen.org/GeneMiner.

Genotyping is an essential step in conventional genetic analyses, but is not required to be performed by users because the CC strains have already been genotyped. This was accomplished using a microarray technology marketed as the Mouse Universal Genotyping Array (MUGA) [10] and its subsequent derivatives, MegaMuga and GigaMuga. The MegaMuga genotyping platform can type 77,808 SNPs located on chromosomes 1 through X. Of these SNPs, 68,903 are reliably homozygous among the eight founders, and with at least one founder having the nonreference allele. In addition there are 26 and 23 SNPs in the Y and mitochondria chromosomes respectively. The GigaMuga array can genotype 143,259 SNPs per strain and includes 66,992 of the MegaMuga SNPs; however, we found the call rates of the newly introduced SNPs to be lower than expected (data not shown). Alternatively, the Mouse Diversity Genotyping Arrays have been used to genotype at 623,000 SNPs, but only 170,935 of these were usable for downstream analysis [11].

Effective phenotypic screening of the CC strains can be achieved by prioritizing lines that are available from the breeding program at any given time. The CC strains are inbred, so the same genetically defined individuals can be phenotyped. This ensures more confidence in assignment of phenotypic scores than conventional analyses that rely on a single individual for each genome tested.

We have implemented convenient web interfaces for various tools to perform step-by-step statistical data analytic tasks linking genotypes and phenotypes facilitating gene mapping and identification. The tools implemented online allow various analyses including simple genome-wide association studies (GWAS) of genotyped SNPs with results displayed as Manhattan plots; linkage-style analysis using inferred founder haplotypes with results displayed as interactive logarithm of odds (LOD) score plots; and association-style analysis of imputed SNPs and indels that are predicted to impact protein sequence or gene expression [12]. Accessory tools have also been implemented such as a regional founder haplotype visualizer; a founder effect analyzer, and LOD-drop region calculators. With these tools, researchers are able to perform systems genetics analysis of CC lines.

2 Materials

Usage of web-tools for gene identification requires access to a computer with the following simple requirements.

2.1 Equipment

1. Fast Internet access.
2. Google Chrome Internet browser (google.com/chrome).
3. Spreadsheet software.
4. Basic R packages installed (cran.r-project.org).

2.2 Genotype and Phenotype Data

1. All the latest CC strain genotyping data are already preloaded and made available on the server side.
2. Prepare the CC strains phenotypic data in one spreadsheet and calculate mean and standard errors for each strain. Strains with large standard errors are not reliable and should be used for gene mapping with caution.
3. Mapping may be performed on normalized data or using the data "as is". In order to decide whether to use the normalize option, data can be imported into R and a Shapiro-Wilk test of normality can be used. Additionally, a QQ-plot of the data can be visualized using R functions *qqnorm* and *qqline*. Delete outliers if appropriate (*see* **Note 1**).

3 Methods

3.1 GeneMiner Phenotypic Input

All methods are performed online through this URL: www.sysgen.org/GeneMiner.

1. Prepare the phenotype data in a spreadsheet with two columns: "strain name" and "mean phenotype value" of the strain replicates and ensure there are no duplicate data present. Covariates, if any, should be regressed out prior to the analysis phase (*see* **Note 2**).
2. There are options available to normalize a nonnormally distributed quantitative phenotype. When the methods with "normalize" options are selected (as opposed to "as is"), the phenotype values are rank-averaged and then replaced by normal quantiles. The quantiles will then be used in all subsequent analyses.
3. When the phenotype data is nominal/qualitative (refer to *see* **Note 3**), the data needs to be specified so that strains are assigned to one of up to six groups. The groups can be labeled with text. Methods for discretization of data values are discussed in *see* **Note 4**.
4. The server performs quality control on the user-provided input, checking that: it does not contain missing or nonnumeric values; duplicate strains are not present; there are less than seven categories in a nominal trait; the strain name entered is correct and does exist in the database. Any errors will be reported and appropriate measures such a deleting or correcting those lines at the input should enable resubmission of the task.

5. There are 14 mapping options available. These are discussed below sequentially as per their appearance in the online interface. Depending on the nature of the phenotype being quantitative or qualitative, an appropriate option is chosen.

3.2 Analysis Method Options Using Genotype SNP Data

1. Option 1 and 2 are GWAS style analysis options making use of the raw genotyping data obtained using the MUGA platform. The method implements a simple fast association analysis and returns the result in the form of a GWAS style Manhattan plot (*see* **Note 5**) and SNPs with $P<1\times10^{-5}$ are output. The threshold of $P<5\times10^{-8}$ can be considered as significant (refer to *see* **Note 6**).
2. Option 3 is a fast pair-wise epistatic analysis for quantitative traits, again making use of the raw genotyping data (refer to *see* **Note 7**). A Bonferroni correction threshold is applied to determine significant interactions at >95 % confidence. The result is shown in a circular network plot and SNP pairs with adjusted $P<0.05$ are output in table format.
3. Options 4 and 5 provide efficient mixed model association (EMMA) [13] based GWAS analyses. A genotypic kinship matrix and first two principal components are used in this method (refer to *see* **Note 8**). The result is presented in the form of a GWAS-style Manhattan plot and SNPs that are significant are output along with their *P*-values (*see* **Note 6**).
4. Option 6 is for performing GWAS style analysis when the phenotype data is either binary (1/0) or categorized into as many as six groups. The results are presented similar to that explained in **step 3**.

3.3 Analysis Method Options Using Inferred Haplotype Data

A step-by-step approach to haplotype inference was presented previously ([4], *see Supplementary Methods*). In these steps, HAPPY [14] was used to infer haplotype probabilities. Other methods such as RABBIT [15] can be used for further refinement. The results of the haplotype analysis of all the GeneMine strains are already available online at GeneMiner to test against phenotypes. Options 7 and higher make use of such haplotype data. For faster results, haplotypes tested for associations at 7802 chromosomal positions are called "MugaQTL" and the full (and longer) scan is called "MMugaQTL." The results obtained are not likely to vary, as the haplotypes in both cases are predetermined using the full set of genotypes. Analysis steps are illustrated in Fig. 1.

1. Option 7 performs mapping using inferred founder haplotypes for a binary or multinominal trait (*see* **Note 3**). The result is displayed as a LOD-score plot (*see* **Note 9**). The method employed is a multinominal logistic regression (*see* **Note 10**). The plot is displayed in an interactive fashion so that the user may click on the chromosome of interest, to load up an interactive LOD-score

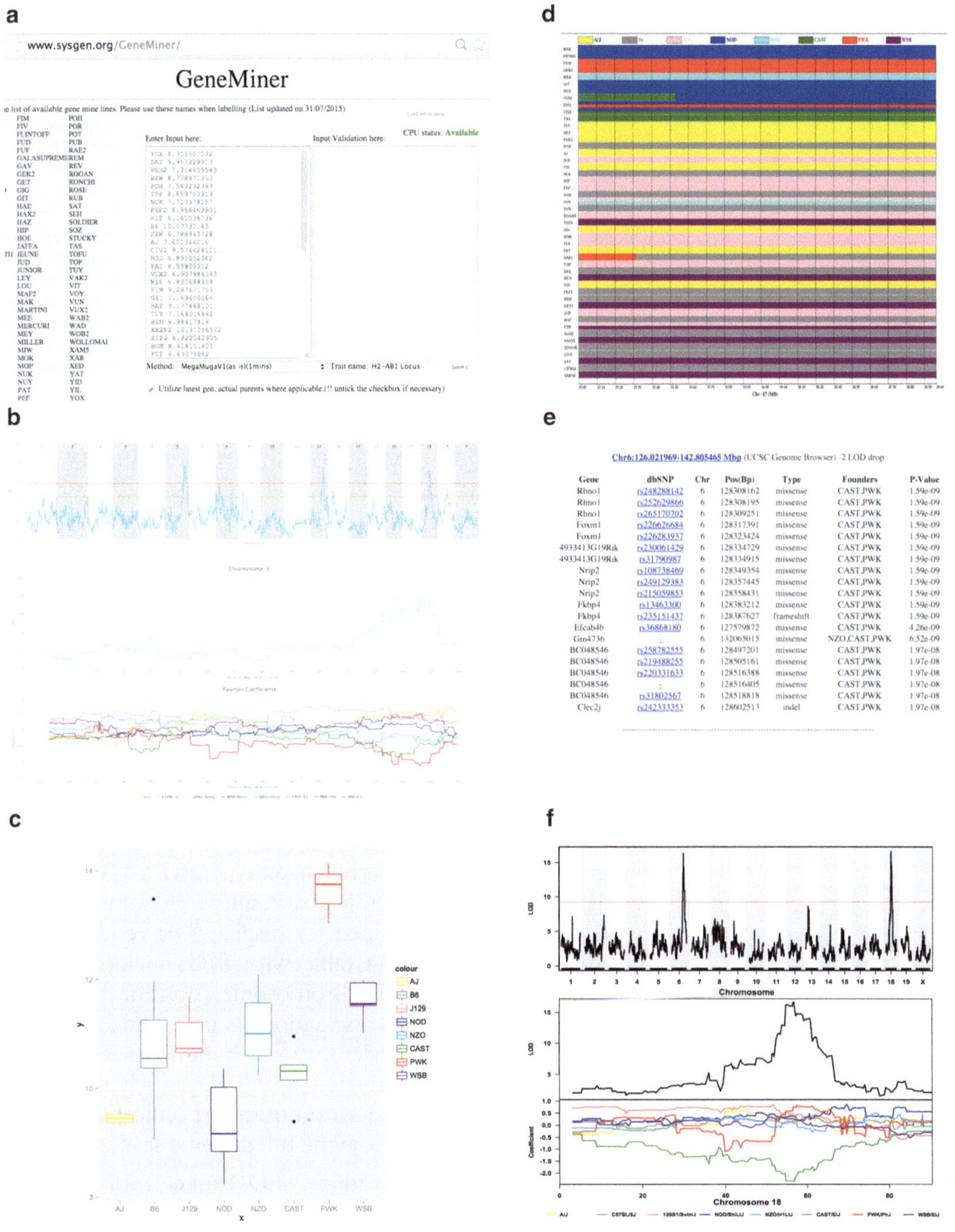

Fig. 1 Step-by-step systems genetics analyses using GeneMiner. (**a**) Step 1. The input phenotype data is entered in the box provided. Step 2. From a selection of 14 different methods, an appropriate method for performing analysis is selected, depending on considerations such as whether the phenotype is quantitative or binary. (**b**) Step 3. The analysis is performed by programs on the server and results are displayed as interactive plots where the mouse genome is on the *x*-axis and significance (LOD) scores on the *y*-axis. The areas of high significance are of importance. Step 4. The chromosomes with the most interesting results can be clicked and viewed in more detail. (**c**) Step 5. The basis for high significance at a chromosome position is explained using a box plot comparing the 8 Gene Mine founders. Extreme differences in the box plots can be seen and noted at this step. (**d**) Step 6. The founder haplotype segregation is visualized at a chromosome position where high significance is conferred. (**e**) Step 7. A list of candidate genes at a region of high significance is displayed. (**f**) Step 8. Figures and tables suitable for publications can be obtained for download

plot for the chromosome as well as a plot showing the eight founders' effects as a coefficient plot (*see* **Note 11**).

2. Option 8 and 9 perform mapping using inferred founder haplotypes for a quantitative trait. The method employed is a R/QTL regression [16] (*see* **Note 12**). The result displayed is in the same format as that of option 7. In these scans, the heterozygous region optimization is performed where required (*see* **Note 13**).
3. Option 10 and 11 perform mapping of quantitative traits using inferred founder haplotype probabilities instead of fixed best haplotypes. Hence, this option cannot perform optimization in any remnant heterozygous regions.
4. Options 12–14 perform the same functions as options 8–11 but employ a full MegaMuga haplotype scan at 68,903 chromosomal positions where SNPs were genotyped.
5. Option 7–14 all perform 1000 permutations to determine three thresholds: 95% (full), 90% (approaching) and 63% (suggestive) significance. These thresholds are indicated by lines drawn on the LOD-score QTL plots. Regions identified by these peaks can be considered to be associated with the phenotype with the indicated degree of confidence.
6. Association of imputed SNPs and indels is performed when selecting options 7–14. Some 101,232 functional variants (missense, indel, frameshift, etc where at least one of the eight founders had the variant) were downloaded from the Sanger Centre [12] and preloaded into GeneMiner. Variants are imputed based on the inferred haplotype and a GWAS-style association is performed. Individual results with nominal unadjusted $P<0.0001$ are retained for display. The results of the analysis are presented in a table with fields including gene name, imputed SNP rsid, chromosome, position, founders that have minor allele, and association P value. The SNP rsid is linked to NCBI dbSNP (www.ncbi.nlm.nih.gov/SNP). Appropriate results are extracted and displayed when the user clicks on a coordinate of a chromosome specific LOD-score plot, which is displayed below the full genome scan.
7. A region may be defined within 2 LOD units from any user-selected chromosomal peak position ("a LOD 2 drop interval"). The region's coordinates are displayed in UCSC genome browser (genome.ucsc.edu) format with linked to that website. This allows for genes in the LOD 2-drop interval to be viewed.
8. Founder coefficient plots are used to visualize founder effects at the loci of interest. The coefficients are shown as negative, zero, or positive values. Founders with a zero coefficient have phenotypic values close to the mean phenotypic value, while negative coefficient founders have high phenotypic values and positive coefficient founders have low phenotypic values. In addition, a

box plot of the founder effects at a selected position can also be visualized where the actual phenotype values instead of coefficients are displayed.

9. All resulting webpages have a top panel that contains options for a haplotype viewer and a result downloader. The haplotype viewer will create an image of the underlying haplotype of strains tested (sorted in ascending order of the phenotype). This allows users to visualize each strain's genome at the linkage region. The result downloader allows downloading all results in text format. Publication quality figures may also be requested.

4 Notes

1. The Shapiro-Wilk test is a simple test for normality of phenotype data. If $P < 0.001$ then the phenotype is not normal and will require normalization. The *qqnorm* and *qqline* functions in R allow plotting the phenotype data against theoretical quantiles along with a regression line between the two. Outlier data points are those that are further away from this regression line and can be excluded from downstream analysis. Alternatively, the winsorize method [17] of outlier removal can be employed.
2. It is important that covariate adjustments are performed for covariates (if any) prior to testing traits using GeneMiner. Covariate adjustment can be achieved by using the linear regression function *lm* in R. First, the phenotype can be regressed on the covariate variable/s and then the residuals can be obtained and used as the adjusted phenotype. Additionally it is good practice also to regress out covariate interaction terms.
3. In some situations the phenotype data can be binary, e.g. affected or unaffected. In other situations the phenotype can be multinominal, e.g. high, medium-high, medium-low, and low. In these situations the phenotype data can be coded as numeric groups such as group 1, 2, 3, and so on. Multinomial phenotype mapping can sometimes reveal genes that are otherwise confounded by quantitative variations.
4. Discretization of a quantitative trait can be achieved by the use of the minimum description length (MDL) method implemented in the R package *discritization*.
5. A Manhattan plot is a plot of $-\log_{10}(P)$ against genomic position, where P is the association p-value.
6. The commonly accepted thresholds of GWAS significances are $P < 1 \times 10^{-5}$ (suggestive) and $P < 5 \times 10^{-8}$ (fully significant). However a false discovery rate (FDR) correction of the P-values can be also performed. For this, download the results and import the p-values in R and use the function *p.adjust* with

methods such as "fdr." This will yield FDR adjusted *P*-values wherein a threshold of 0.05 can be applied.

7. Prior to GWAS-style epistatic analysis, SNPs with missing genotypes >10% and minor allele frequency (MAF) less than 0.05 are removed. Then, after performing epistasis tests, a Bonferroni correction is applied by dividing the *P*-value with the total number of tests performed.
8. The kinship is calculated using *emma.kinship* function in EMMA R package. The PCs are calculated using PLINK "—pca" option.
9. A LOD-score plot is one where the *x*-axis corresponds to the genome chromosomes 1–*X* and the *y*-axis correspond to the Logarithm of odds ratio (LOD). LOD score is calculated as $\chi^2/(2\log_e 10)$ where χ is either the model chi-square value or the Likelihood Ratio Statistic (LRS) [16].
10. Multinomial logistic regression is performed using the function multinom inside R package *nnet.* An Anova chi-square test is performed to obtain the LOD score.
11. Founder coefficients are the beta-coefficients of the fitted regression model. In the case of the multinomial regression, the coefficients are also log odds of the founders. Otherwise, coefficients are simple effect estimates. Where significant peaks occur, one or more founder coefficients deviate away from zero. Such founders can be referred to as causal founders.
12. The method implemented for testing haplotypes is that of the function *scanone* in the R/QTL R package that performs a genome scan of single QTL models.
13. Heterozygous region optimization is a process to improve statistical power in small sample phenotype data. At any given chromosomal position, heterozygous strains are isolated from the strains that are homozygous. The homozygous strains are first tested against the phenotype, then each individual heterozygous strain is added to the model as homozygous of one of the two founders that provide the heterozygosity, such that the chosen founder maximizes association strength and aligns well with the purely homozygous data. This optimization is performed only when less than 10% of the strains are heterozygous and the strains that are homozygous cover all eight-founder haplotypes.

Acknowledgments

This work was supported by grants from the Diabetes Research Foundation of Western Australia and the National Health and Medical Research Council (1069173), Australia.

References

1. Threadgill DW, Miller DR, Churchill GA, de Villena FP (2011) The collaborative cross: a recombinant inbred mouse population for the systems genetic era. ILAR J 52:24–31
2. Collaborative Cross Consortium (2012) The genome architecture of the Collaborative Cross mouse genetic reference population. Genetics 190:389–401
3. Morahan G, Balmer L, Monley D (2008) Establishment of "The Gene Mine": a resource for rapid identification of complex trait genes. Mamm Genome 19:390–393
4. Chesler EJ, Miller DR, Branstetter LR, Galloway LD, Jackson BL, Philip VM, Voy BH, Culiat CT, Threadgill DW, Williams RW, Churchill GA, Johnson DK, Manly KF (2008) The Collaborative Cross at Oak Ridge National Laboratory: developing a powerful resource for systems genetics. Mamm Genome 19:382–389
5. Iraqi FA, Churchill G, Mott R (2008) The Collaborative Cross, developing a resource for mammalian systems genetics: a status report of the Wellcome Trust cohort. Mamm Genome 19:379–381
6. Threadgill DW, Churchill GA (2012) Ten years of the Collaborative Cross. Genetics 190:291–294
7. Ram R, Mehta M, Balmer L, Gatti DM, Morahan G (2014) Rapid identification of major-effect genes using the collaborative cross. Genetics 198:75–86
8. Ferguson B, Ram R, Handoko HY, Mukhopadhyay P, Muller HK, Soyer HP, Morahan G, Walker GJ (2015) Melanoma susceptibility as a complex trait: genetic variation controls all stages of tumor progression. Oncogene 34:2879–2886
9. Weerasekera LY, Balmer L, Ram R, Morahan G (2015) Characterization of retinal vascular and neural damage in a novel model of diabetic retinopathy. Invest Ophthalmol Vis Sci 56:3721–3730
10. Didion JP, Buus RJ, Naghashfar Z, Threadgill DW, Morse HC, de Villena FP (2014) SNP array profiling of mouse cell lines identifies their strains of origin and reveals cross-contamination and widespread aneuploidy. BMC Genomics 15:847
11. Durrant C, Tayem H, Yalcin B, Cleak J, Goodstadt L, de Villena FP, Mott R, Iraqi FA (2011) Collaborative Cross mice and their power to map host susceptibility to Aspergillus fumigatus infection. Genome Res 21:1239–1248
12. Yalcin B, Wong K, Agam A, Goodson M, Keane TM, Gan X, Nellåker C, Goodstadt L, Nicod J, Bhomra A, Hernandez-Pliego P, Whitley H, Cleak J, Dutton R, Janowitz D, Mott R, Adams DJ, Flint J (2011) Sequence-based characterization of structural variation in the mouse genome. Nature 477:326–329
13. Kang HM, Zaitlen NA, Wade CM, Kirby A, Heckerman D, Daly MJ, Eskin E (2008) Efficient control of population structure in model organism association mapping. Genetics 178:1709–1723
14. Mott R, Talbot CJ, Turri MG, Collins AC, Flint J (2000) A method for fine mapping quantitative trait loci in outbred animal stocks. Proc Natl Acad Sci 97:12649–12654
15. Zheng C, Boer MP, Eeuwijk FA (2015) Reconstruction of genome ancestry blocks in multiparental populations. Genetics 200:1073–1087
16. Broman KW, Sen S (2009) A guide to Qtl mapping with R/Qtl. Statistics for biology and health. Springer, Dordrecht
17. Fernández JR, Etzel C, Beasley TM, Shete S, Amos CI, Allison DB (2002) Improving the power of sib pair quantitative trait loci detection by phenotype winsorization. Hum Hered 53:59–67

Chapter 6

Integrative Functional Genomics for Systems Genetics in GeneWeaver.org

Jason A. Bubier, Michael A. Langston, Erich J. Baker, and Elissa J. Chesler

Abstract

The abundance of existing functional genomics studies permits an integrative approach to interpreting and resolving the results of diverse systems genetics studies. However, a major challenge lies in assembling and harmonizing heterogeneous data sets across species for facile comparison to the positional candidate genes and coexpression networks that come from systems genetic studies. GeneWeaver is an online database and suite of tools at www.geneweaver.org that allows for fast aggregation and analysis of gene set-centric data. GeneWeaver contains curated experimental data together with resource-level data such as GO annotations, MP annotations, and KEGG pathways, along with persistent stores of user entered data sets. These can be entered directly into GeneWeaver or transferred from widely used resources such as GeneNetwork.org. Data are analyzed using statistical tools and advanced graph algorithms to discover new relations, prioritize candidate genes, and generate function hypotheses. Here we use GeneWeaver to find genes common to multiple gene sets, prioritize candidate genes from a quantitative trait locus, and characterize a set of differentially expressed genes. Coupling a large multispecies repository curated and empirical functional genomics data to fast computational tools allows for the rapid integrative analysis of heterogeneous data for interpreting and extrapolating systems genetics results.

Key words IT-tools for systems genetics, GeneWeaver data base, Data mining, QTL candidate gene

1 Introduction

Systems genetics studies generate large volumes of gene expression networks, and positional candidate genes. Resolving and prioritizing these results requires refinement of the causal variants, functional role of genes and gene products and relationships of gene coexpression networks to mechanistic biology. A wealth of data exists to perform this refinement, and importantly, this data can be drawn from other populations and species, thereby providing

Electronic supplementary material: The online version of this chapter (doi:10.1007/978-1-4939-6427-7_6) contains supplementary material, which is available to authorized users.

Klaus Schughart and Robert W. Williams (eds.), *Systems Genetics: Methods and Protocols*, Methods in Molecular Biology, vol. 1488, DOI 10.1007/978-1-4939-6427-7_6, © Springer Science+Business Media New York 2017

orthogonal information to break apart systems genetic network structure driven by population structure. With the growth of functional genomics and systems genetics, research groups have rapidly amassed large collections of genome-wide functional genomics datasets. The primary analysis of these datasets generally reports the strongest and most novel features of the data, but the remainder of the data is often deposited in a database, and many early studies have been simply reported through manuscript supplements, rarely to be used again. Analyzing your own data in light of the vast amount of recent and historical data can enable the discovery of highly supported connections among genetic variants, gene products, biological molecules, and complex phenotypes. There are many resources for annotating the results of a genome-wide study, all located in diverse databases and other web content. Having the ability to harmonize and analyze historic data, together with highly curated public resource data such as that found in model organism databases, adds tremendous depth and orthogonal information sources to prioritize and refine the results of genetic analysis. In developing GeneWeaver, we have brought together primary data with curated ontology annotations including Mammalian Phenotype Ontology [1], Human Phenotype Ontology [2], Gene Ontology [3] and pathway resources such as Kyoto Encyclopedia of Genes and Genomes [4] and Pathway Commons [5]. These datasets are harmonized for analysis using set intersection-based analysis tools implemented with fast graph algorithms, for the rapid integration of large collections of gene sets in real time.

GeneWeaver is an online database of gene sets coupled to a suite of analysis tools that allow for hypothesis generation through the integration of heterogeneous functional genomics data [6, 7]. GeneWeaver is unique in that it supports dynamic analysis of user provided and stored gene sets from multiple species (currently supporting ten species of eukaryotes) using cross-platform gene identifier mappings. This identifier mapping and homology determination is done automatically, mapping each uploaded gene-identifier to a unique ID cluster. Users have the option of including homology or limiting their analysis to a species of interest. GeneWeaver also stores the results of all analysis so you can return and retrieve them at a later date. Users may also share their data with other users selectively, make it public, or keep it restricted to a private account. Data can be imported by users, uploading their gene set data directly or exporting to GeneWeaver from within another online resource such as Neuro Informatics Framework (NIF) [8], Grappa [9], Mouse Phenome Database (MPD) [10] or GeneNetwork [11]. These datasets can then be added to your collection to be analyzed together with other gene sets retrieved from the GeneWeaver database.

To begin a GeneWeaver analysis a user must collect "GeneSets" together in a "Project". From the user's project page, the sets of interest are selected and the tools are executed. Each tool can be used to answer different questions emerging from systems genetics as will be described below. Additional use cases can be found in Bubier et al. [7]. Once complete, results can be exported as images, genes or gene sets can be exported as text files, or results can be used interactively through link-out to resources with more information about a gene of interest at locations such as Entrez [12], Ensembl [13], Allen Brain Atlas [14], MGI, comparative toxicogenomics database (CTD) [15] or String [16].

2 Materials

1. GeneWeaver works with any modern browser provided javascript and session cookies are enabled.
2. Though not required, it is recommended that all users create an account to store and share gene sets and results. Having an account will also allow users to create groups or communities with access to shared datasets.

3 Methods

3.1 How to Navigate GeneWeaver and Find Genes Common to Multiple Gene Sets

When performing systems genetics analysis one often wants to determine what genes are common to a group of gene sets of interest. Here we are asking the question "What genes are found to be commonly regulated by the four types of chemicals associated with water wells contaminated by hydraulic fracturing compounds BTEX (benzene, toluene, ethylbenzene and xylene)."[17] This query makes use of data within GeneWeaver from the Comparative Toxicogenomics Database.

1. Go to www.geneweaver.org, and create a user account, or log on if you already have an account, using the links in the upper right corner.
2. Locate the Search Box.
3. Search for just GeneSets (uncheck the other boxes) and type the phrase "benzene OR toluene OR ethylbenzene OR xylene".
4. Click on the *Search* Icon.
5. A list of results will be displayed (Fig. 1). On the general tab on the left side, the results are broken down by tiers (*see* **Note 1**), species and attribution.
6. To restrict the analysis to data curated by the CTD select CTD under Attribution (on the General Tab) and unselect the other boxes (e.g. GO Annotations or No attribution).
7. Explore one of the gene sets in detail by clicking on the GeneSet ID or name. This will take you to a description of the set and display the Gene List (Fig. 2). In the Gene List is the gene sym-

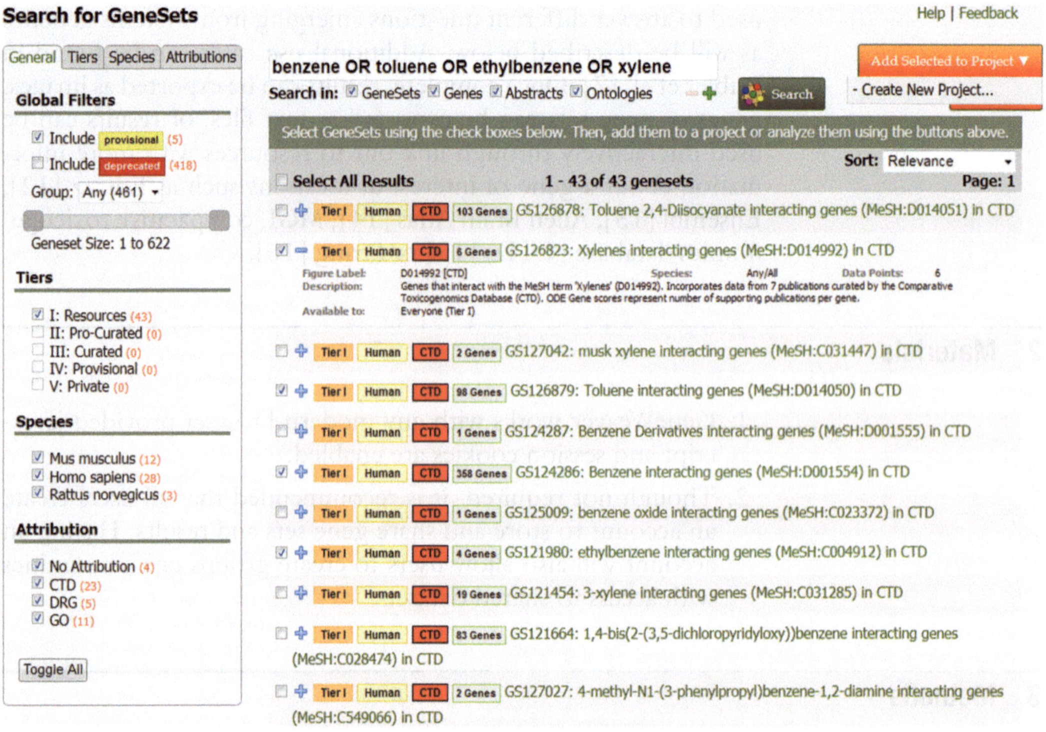

Fig. 1 Search for GeneSets Results. This page displays the tier, species, and attribution for each GeneSet, the number of genes in that set, the GeneSet ID and the name. Faceted search enables filtering of query results. GeneSets can be selected by clicking in the box to the left of each set for addition into user specified "Projects" for analysis. Additional information about the set is revealed by expanding the label using the "+" icon

bol for the associated species displayed, for these CTD sets the species is *Homo sapiens*. Beside the gene symbols is a column of links to additional bioinformatics resources associated with each gene, and a continuous or binary score for the association of each gene to the gene list. The data values for the CTD reflect the number of publications that each gene association is drawn from. In other gene sets in GeneWeaver the data value will represent a *p*-value, *q*-value, correlation, or effect size. The score information and inclusion criteria are provided in the GeneSet description. If you are interested in this set of genes, there is also an *Export Data* link above the Gene List which will export a plain text file of the GeneSet contents.

8. From this page you can also identify other gene sets in the database that are similar in composition to this gene set by clicking on *View Similar GeneSets*. What results from this is a list of gene sets and their Jaccard similarity score (1 identical, 0 no overlap).
9. Clicking the browser back button twice will return you to your search results.

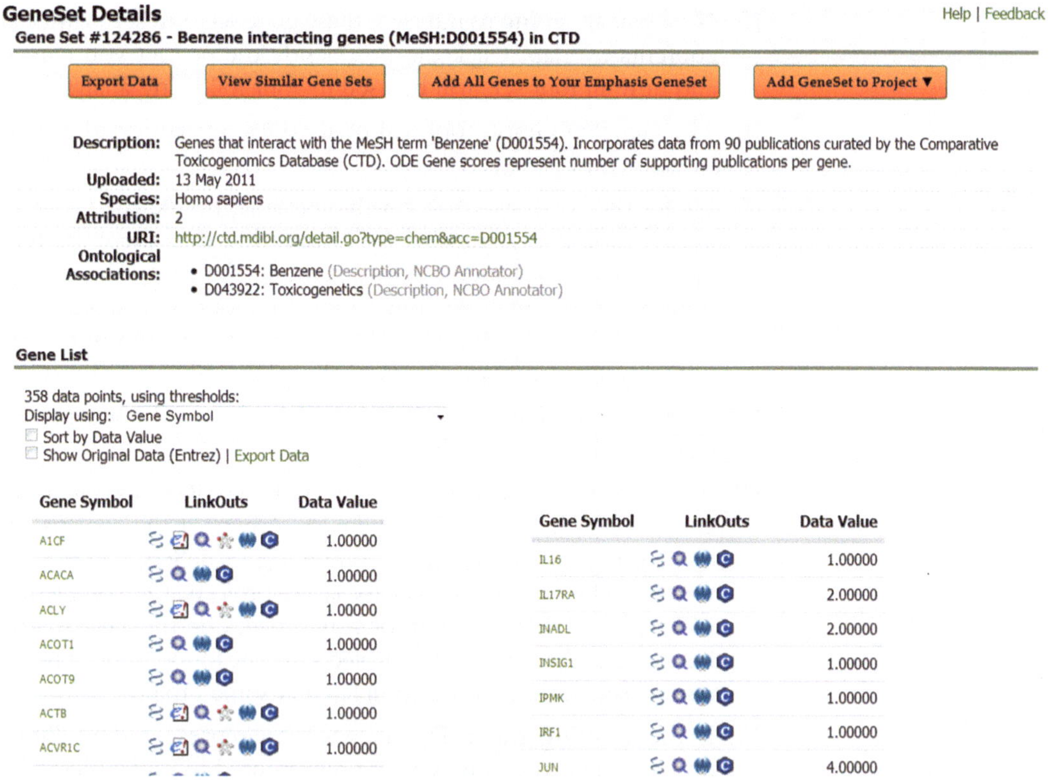

Fig. 2 The GeneSet detail page displays a description of the set, figure label, date added, species, and publication metadata. Below the header is the GeneList, containing the gene symbol, data value, and link outs to various external resources

10. Click on the "+" next to the GeneSet ID of one of the sets to reveal more information about the set (Fig. 1). This metadata includes the description, the figure label (what the set will be labeled in the graphics produced in GeneWeaver), as well as additional information about the gene sets such as the number of genes, the species, and the tier (items also summarized by the tags next to the GeneSet ID on the search results page).
11. Look over the list and select the check box if it says "Xylenes interacting genes", "Ethylbenzene interacting genes" "Toluene interacting genes" or "Ethylbenzene interacting genes". The four sets you select will be GS126823, GS121980, GS126879, GS124286 (Fig. 1).
12. Next add these gene sets to a project by clicking on the *Add Selected to Project* drop-down arrow and selecting and –*Create New Project*—.
13. Now enter a name for your project (e.g. BTEX) and click ok.
14. This will now direct you to the My Projects Page where your gene sets will be expanded. You can click on the "-" to collapse the file folder contents or access additional sets at this time.

15. Now you are going to intersect these gene sets and see how their contents overlap. Click the box beside the project you named and now all four sets will be used in your following analysis.
16. The GeneSet Graph tool will be used for a partitioned display of genes and gene sets.
17. Under the Analysis Tools on the left side, beside the GeneSet Graph tool icon is another "+" indicating that there are tool options you can modify before running the tool. The default options are to include homology, to suppress disconnected gene sets from the display and to determine the minimum size of the graph automatically. We will use the default settings.
18. Click on the GeneSet graph icon (Fig. 3), on the left underneath Analysis tools to initiate the tool.
19. As the tool runs you will be directed to a Running GeneSet Graph Tool Status Screen, where the events of the analysis are logged as they occur, e.g. "Combining GeneSets", "Drawing Graph". This will take longer for larger projects. On completion you are taken to your result, or you can click on the *View Results* button. Results of very large projects may be found under a user's results page once they are completed at some time the future.
20. In the resulting graph (Fig. 4), genes are represented by elliptical nodes and the gene sets are represented in rectangles. The least-connected genes are displayed on the left, followed by the

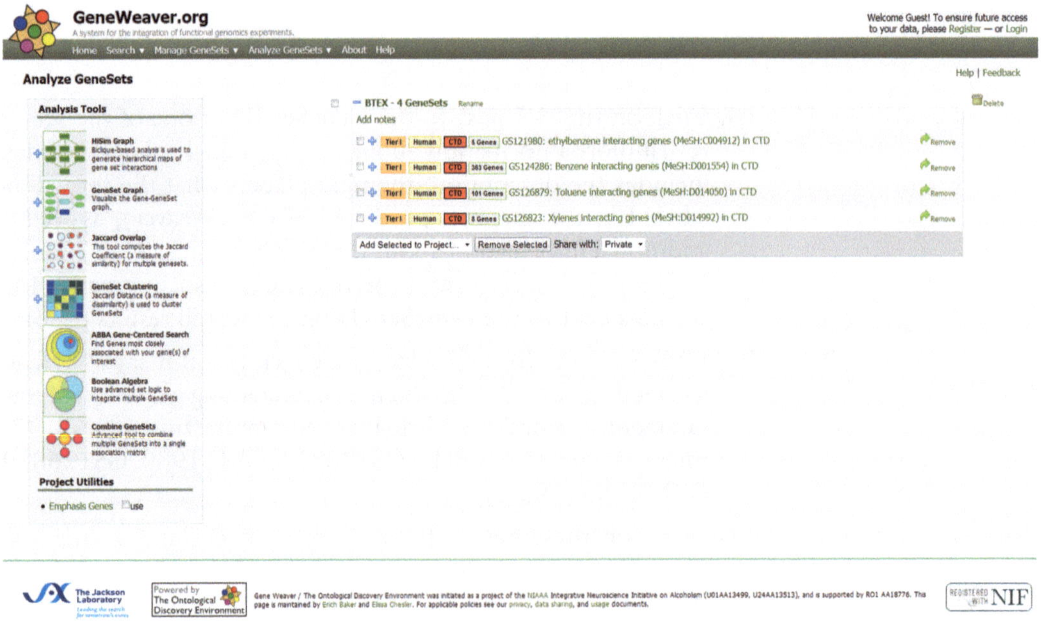

Fig. 3 The analysis tools are located on the left side of the analyze GeneSets page. To execute an an analysis, click on the icon for that tool. To modify the tool settings, click the "+" to view all tool options and set additional parameters before clicking the icon to execute

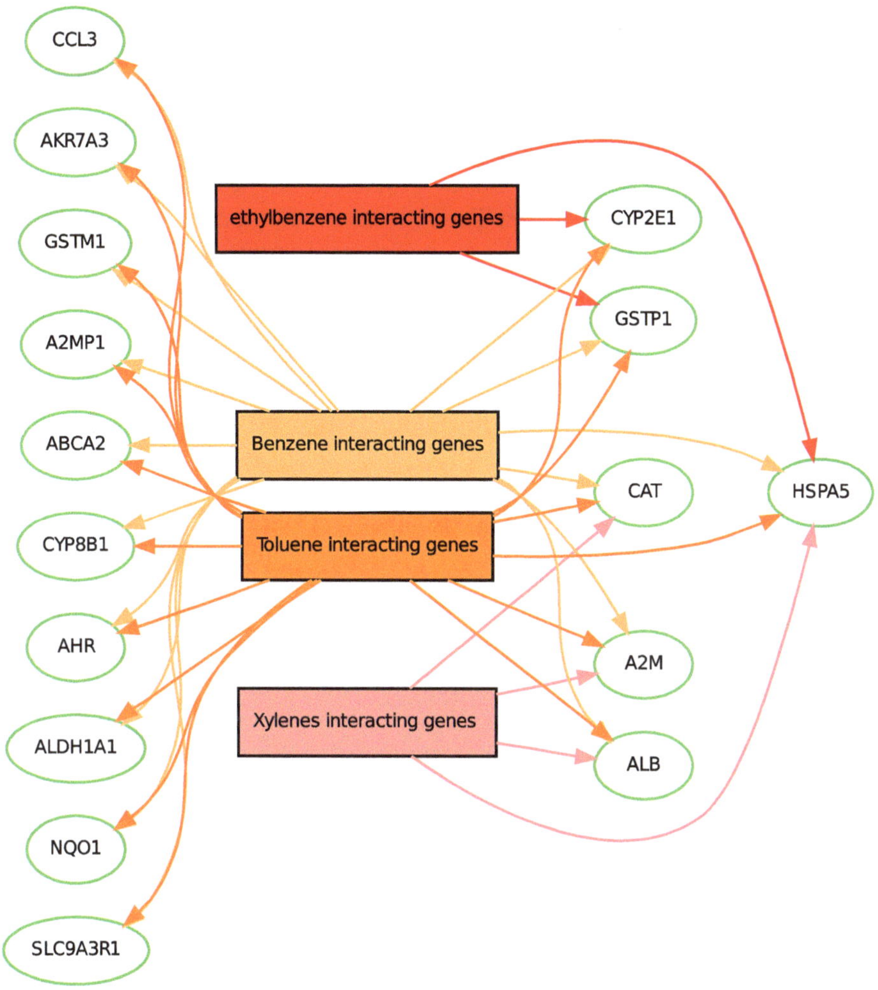

Fig. 4 The GeneSet graph constructed from the genes interacting with each of the four BTEX compounds as identified by the CTD. The most highly connected genes are to the far right and each column moving leftward is connected to a lesser degree. HSPA5 is the most highly connected and common to all four sets of genes

gene sets, then the more-connected genes in increasing order to the right. Genes and gene sets are connected by colored lines to show what genes are in which gene sets. In this way, the GeneSet Graph displays the bipartite graph of the genes and gene sets, but modifies the display of the gene partition to make it easier to visually interpret.

21. You can interact with the graph by clicking on different elements. If you click on the gene sets you will be taken to the GeneSet detail page. If you click on a gene, GeneWeaver will launch a search for all the gene sets that contain that gene, and display that result. The graph can be exported as a pdf by clicking on the *Export link* beneath the graph. Click on the gene to the far right, common to all gene sets, HSPA5 (Fig. 4). A search for all records within the database containing HSPA5 is launched.

22. HSPA5 is found in over 900 GeneWeaver GeneSets at the time of this writing. Central nervous systems and respiratory effects are common results of consuming the water contaminated by these four compounds [18]. To limit our search so that we can specifically explore the basis of these effects we can type in the complete phrase "HSPA5 AND nervous OR respiration", and click *Search*. This reduces the number of gene sets returned. On the general tab, under attributions select just the mammalian phenotype (MP) or the GO (Gene Ontology). You will see that mice with mutations in this gene have defects in respiratory function (GS164339) and the gene is associated with the central nervous system (in both mouse GS193432 and human GS210376). This result suggests that if the BTEX compounds are working through a common mechanism to affect human health, a common pathway would be through interaction with HSPA5.

3.2 Prioritizing Candidate Genes from Quantitative Trait Loci

Current and historic genetic mapping studies aim to identify quantitative trait loci. These loci are often large intervals containing numerous candidate genes. GeneWeaver is a powerful tool for helping to prioritize the candidate genes within the interval by integrating convergent findings, from heterogeneous data types. The tools in GeneWeaver allow for easy visualization of the genes with the most supporting evidence.

1. To begin an analysis, a QTL of interest must be identified. This can be done by uploading your own list of genes, or searching for a QTL that has already been uploaded into GeneWeaver. Alternatively you can export a QTL region from GeneNetwork.org, which will be illustrated here or skip to **step 6** below, to use the gene set already in GeneWeaver.org.
2. Search GeneNetwork.org for "nociception", in the Get Any box (Fig. 5). Click *Search*. Select and click on record ID 11821, Hargraves test males and females (Fig. 6) of the 33 records returned.
3. Use the mapping tools to map the QTL (*see* **Note 2**). The genome scan for 5000 permutations is shown in (Fig. 7).
4. Zoom in on the Chromosome 4 QTL, by clicking on the Chromosome 4 region of the genome scan (Fig. 8). Accessing the mapping data by *Downloading* the tab-delineated file (Fig. 8, arrow), calculate the 1.5 LOD drop, in this case 46.553–65.048 MBP. Use View box (Fig. 8, arrow) to set the interval to "46.553"–"65.048" Mb and click on *remap*.
5. Below the graph, select all the genes in the interval by clicking on the select icon (Fig. 9) To bring your data to GeneWeaver, click on the GeneWeaver icon, making sure to be previously login to your GeneWeaver account. You will be brought to the GeneSet upload page with the Genes Uploaded and the

Fig. 5 Default settings at GeneNetwork.org are set to search "Mouse", "Phenotypes", from among the "BXD Published Phenotypes" data set. Here the term nociception was searched for

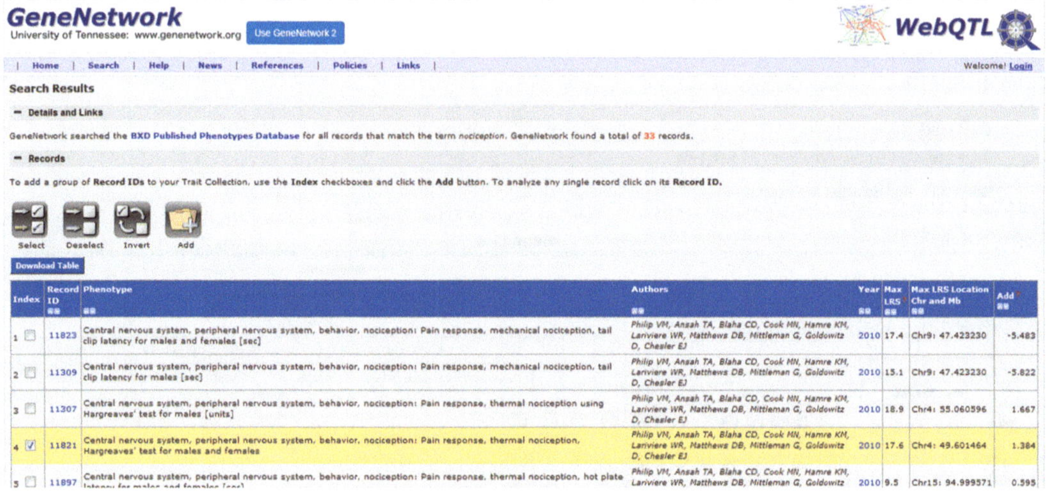

Fig. 6 The search results page in GeneNetwork showing the 33 records retrieved from the phenotype search for nociception. Select the Pain QTL on Chr 4, from Record ID 11821

metadata filled out. You should add additional details explaining the dataset accordingly. Click on *Upload GeneSet*, then select "Add GeneSet to Project" and create and name a new project e.g. "Pain QTL Candidate Genes." Skip to **step 7**.

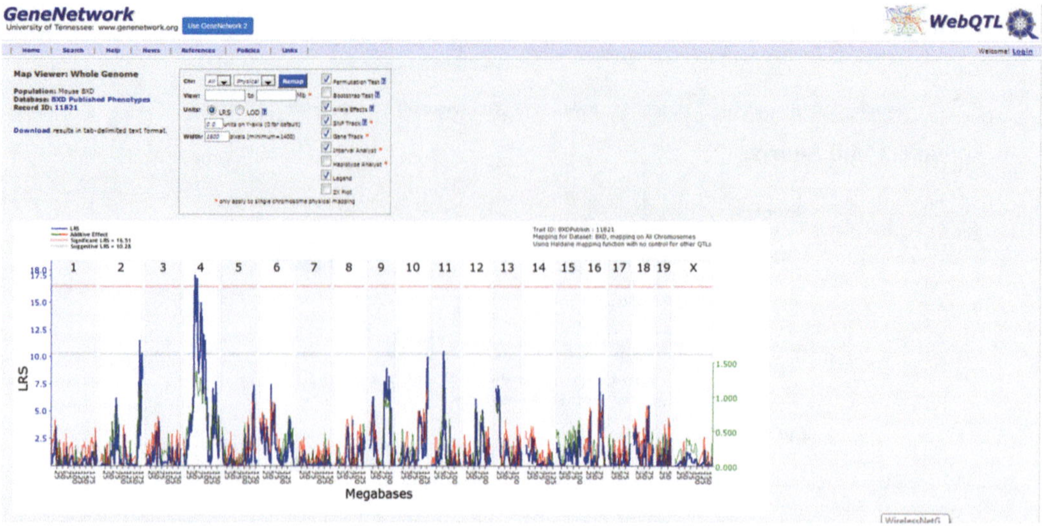

Fig. 7 Genome Scan for the Hargreaves nociception assay in BXD mice. A significant QTL $p<0.05$ is seen on chromosome 4

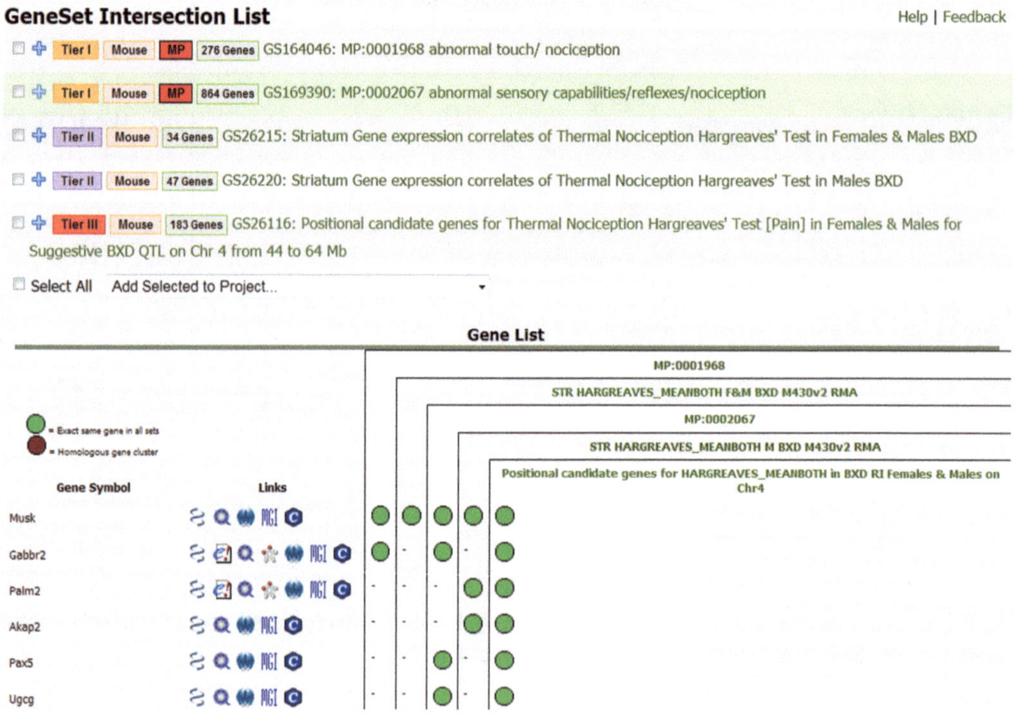

Fig. 8 The map viewer detailed Chr 4 plot, available by clicking on the chromosome 4 region in the whole genome scan is shown

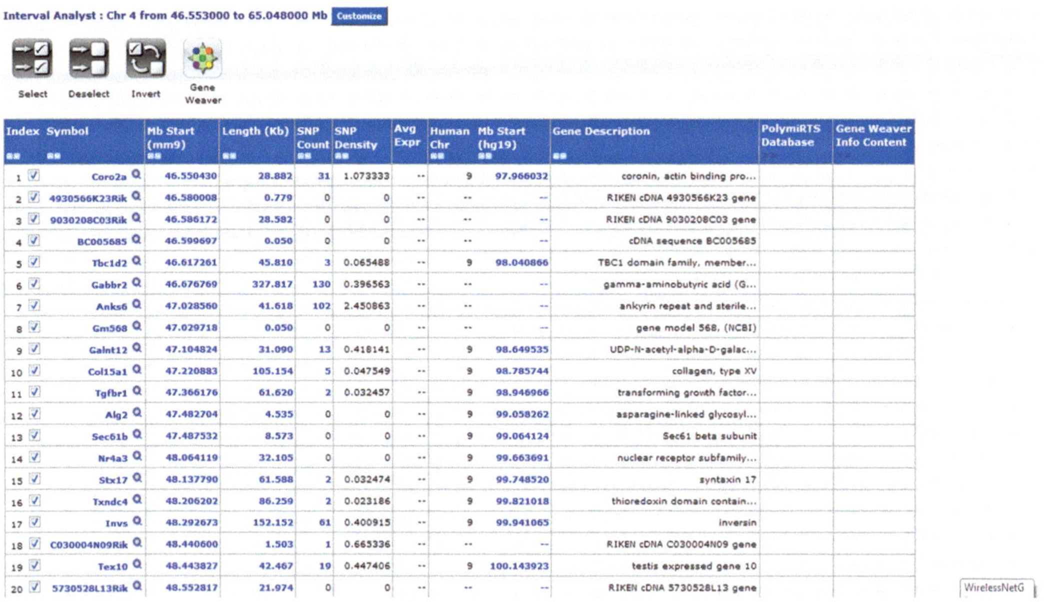

Index		Symbol	Mb Start (mm9)	Length (Kb)	SNP Count	SNP Density	Avg Expr	Human Chr	Mb Start (hg19)	Gene Description	PolymiRTS Database	Gene Weaver Info Content
1	☑	Coro2a	46.550430	28.882	31	1.073333	--	9	97.966032	coronin, actin binding pro...		
2	☑	4930566K23Rik	46.580008	0.779	0	0	--	--	--	RIKEN cDNA 4930566K23 gene		
3	☑	9030208C03Rik	46.586172	28.582	0	0	--	--	--	RIKEN cDNA 9030208C03 gene		
4	☑	BC005685	46.599697	0.050	0	0	--	--	--	cDNA sequence BC005685		
5	☑	Tbc1d2	46.617261	45.810	3	0.065488	--	9	98.040866	TBC1 domain family, member...		
6	☑	Gabbr2	46.676769	327.817	130	0.396563	--	--	--	gamma-aminobutyric acid (G...		
7	☑	Anks6	47.028560	41.618	102	2.450863	--	--	--	ankyrin repeat and sterile...		
8	☑	Gm568	47.029718	0.050	0	0	--	--	--	gene model 568, (NCBI)		
9	☑	Galnt12	47.104824	31.090	13	0.418141	--	9	98.649535	UDP-N-acetyl-alpha-D-galac...		
10	☑	Col15a1	47.220883	105.154	5	0.047549	--	9	98.785744	collagen, type XV		
11	☑	Tgfbr1	47.366176	61.620	2	0.032457	--	9	98.946966	transforming growth factor...		
12	☑	Alg2	47.482704	4.535	0	0	--	9	99.058262	asparagine-linked glycosyl...		
13	☑	Sec61b	47.487532	8.573	0	0	--	9	99.064124	Sec61 beta subunit		
14	☑	Nr4a3	48.064119	32.105	0	0	--	9	99.663691	nuclear receptor subfamily...		
15	☑	Stx17	48.137790	61.588	2	0.032474	--	9	99.748520	syntaxin 17		
16	☑	Txndc4	48.206202	86.259	2	0.023186	--	9	99.821018	thioredoxin domain contain...		
17	☑	Invs	48.292673	152.152	61	0.400915	--	9	99.941065	inversin		
18	☑	C030004N09Rik	48.440600	1.503	1	0.665336	--	--	--	RIKEN cDNA C030004N09 gene		
19	☑	Tex10	48.443827	42.467	19	0.447406	--	9	100.143923	testis expressed gene 10		
20	☑	5730528L13Rik	48.552817	21.974	0	0	--	--	--	RIKEN cDNA 5730528L13 gene		

Fig. 9 Chrosmomse 4 nociception QTL positional candidates with 1.5 LOD confidence interval (46.553–65.048 MBP). These are selected and exported to GeneWeaver as a new GeneSet using the icon at the *top*

6. If not exporting from GeneNetwork, this pain QTL candidate gene set can be retrieved by searching for "Pain QTL Chr4" to retrieve, among other sets, GS26116 [19], which corresponds to what was just exported from GeneNetwork. Select this GeneSet and create a new project, such as "Pain QTL Candidate Genes" to add it to.
7. Since we are narrowing an interval we want to highlight only gene set intersections that contain genes within the interval. We will use the "Emphasis GeneSet" feature within GeneWeaver. Click on the Gene Set, GS26166 (or your exported set) and select "Add all Genes to Your Emphasis GeneSet" (Fig. 10, *see* **Note 3**).
8. You are brought to an Emphasis Genes page displaying all the genes within the QTL GeneSet. Return to the Search page by clicking on Search. We want to add additional studies to this project. This particular QTL was mapped in the BXD Recombinant Inbred mouse population and the data is available within GeneNetwork.org. Gene expression correlates of this trait have already been exported from GeneNetwork. Search for the measured trait "HARGREAVES_MEANBOTH", to retrieve the gene expression correlates from multiple tissues on multiple sequencing platforms. Select all these records and add them to your project.
9. We want to determine if there is a likely candidate gene, already known in this interval. If there is, then there should be a tier 1

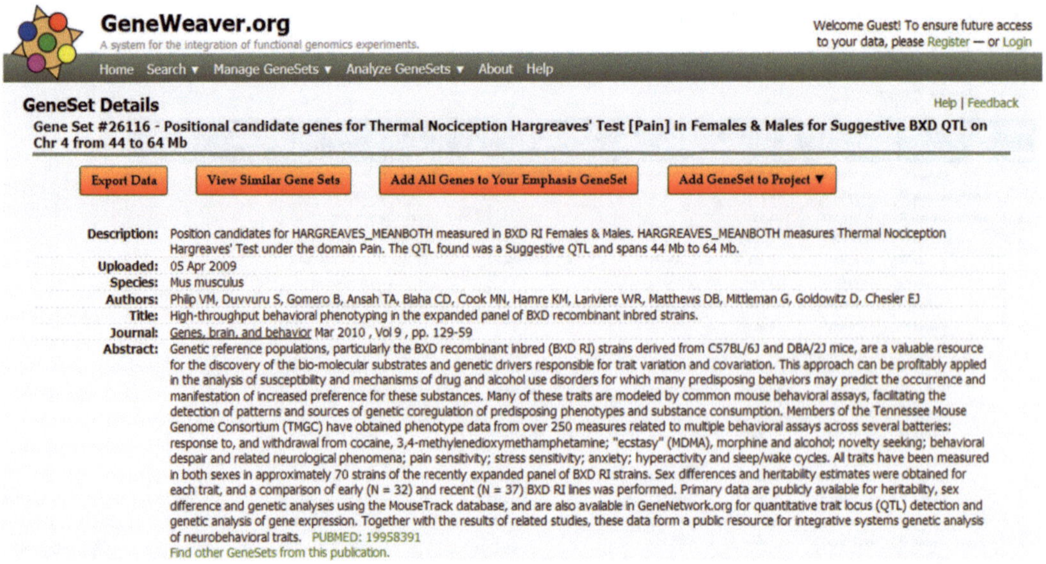

Fig. 10 The GeneSet detail page shows metadata for the GeneSet including a description, the species, the data uploaded, and any associated publication. In addition, you can "Export Data", "View Similar GeneSets", "Add the GeneSet to a Project", or as in this example, "Add All Genes to Your Emphasis Gene Set"

resource data containing the gene which could be retrieved by searching for the term nociception. Click on Search and search for "nociception." After the results are returned, select from the tab on the left "Tier 1:Resources." This brings the results down to a smaller size. Within the remaining 18 gene sets, select GeneSet GS164046 "MP:0001968 abnormal touch/nociception" and another broad term GeneSet GS169390 "MP:0002067 abnormal sensory capabilities/reflexes/nociception". Add these selected sets to your project.

10. Select all the gene sets in your project "Pain QTL Candidate Genes" by clicking the box next to the GeneSet name. To highlight positional candidate genes, use the "Emphasis genes" feature by checking the "use" box beside "Emphasis Genes" in the Projects Utilities side bar, beneath the tool options. Execute the tool HiSim graph using default settings by clicking on HiSim Graph Icon (Fig. 3). The HiSim Graph is a tool for grouping functional genomic datasets based on the genes they contain. HiSim will display a graph of gene set intersections of very high order. In algorithmic terms, these intersections are created from the overlap of maximal bicliques in discrete bipartite structures.

11. You will be brought to the Running HiSim Graph page where the progress of the analysis is displayed. The steps as they occur will be shown e.g. "Combining GeneSets", "Integrating Homologous Genes", "Running Biclique Algorithm", "Determining Subset Relationships", "Prepressing Nodes",

"Building Trees", "Outputting Nodes", "Outputting Edges", "Graph Output Finished" and finally "Drawing trees". You will automatically be taken to the resulting graph or you can click on the *View Graph* link.

12. The HiSim Graph opens in a new tab (Figs. 11 and 12), and can be interactively panned and zoomed using the mouse. More details of each intersection can be viewed by clicking on the individual nodes in the tree. A link at the bottom of the frame allows download of the pdf. In terms of gene sets, the smallest intersections (fewest gene sets, most genes) are at the lower levels, and the largest intersections (most gene sets, fewest genes) are at the top. When thinking about the genes in all the gene sets, the roles are reversed (smallest number of genes at the top, largest number of genes at the bottom).
13. The HiSim Graph can be modified with tools available beneath the graph under the "Show Display Options" link. The number of gene sets displayed on the graph, the GeneSet ID labels, and publication and gene counts can be toggled off and on. The graph is interactive and the nodes can be dragged around and clicked on. Click on the highest blue node at the top of the graph. This node, being blue, will contain genes that are QTL positional candidates (emphasis genes) as well as any gene set containing genes that overlap.
14. This brings up the GeneSet Intersection list (Fig. 13) displaying the genes in the six most highly connected gene sets and the genes that connect them. The most highly connected gene within the interval and this project, is also associated with nociception, *Musk*. Clicking out to Entrez reveals several PubMed associations to "neuromuscular junction". *Musk* could be a potential candidate gene from this QTL worth of further genetic investigation. At this point we have demonstrated functional but not genetic sufficiency of the candidate. e.g. are there causative B6 vs D2 SNPs in *Musk*, or a *cis*-eQTL that could support its role in genetic variation in nociception. Additional

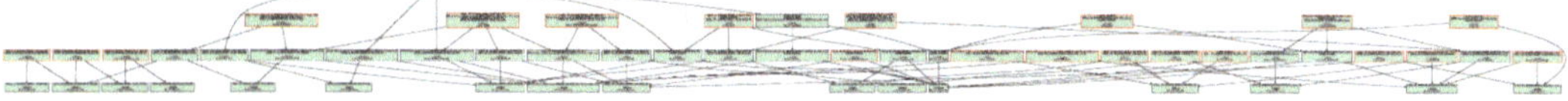

Fig. 11 Output of the hierarchical similarity graph tool. Individual gene sets are displayed at the *bottom* of the graph, intersections are represented as parents of individual gene sets

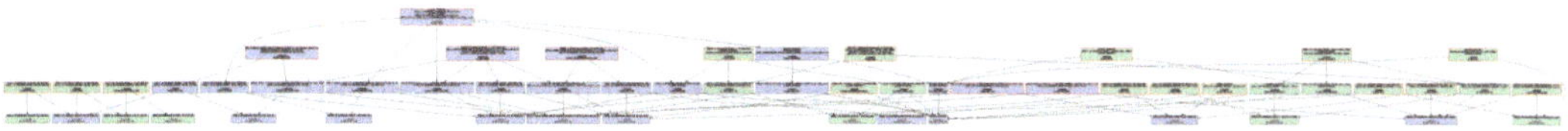

Fig. 12 A zoomed in portion of a HiSim graph with the emphasis genes indicated in the *purple nodes*

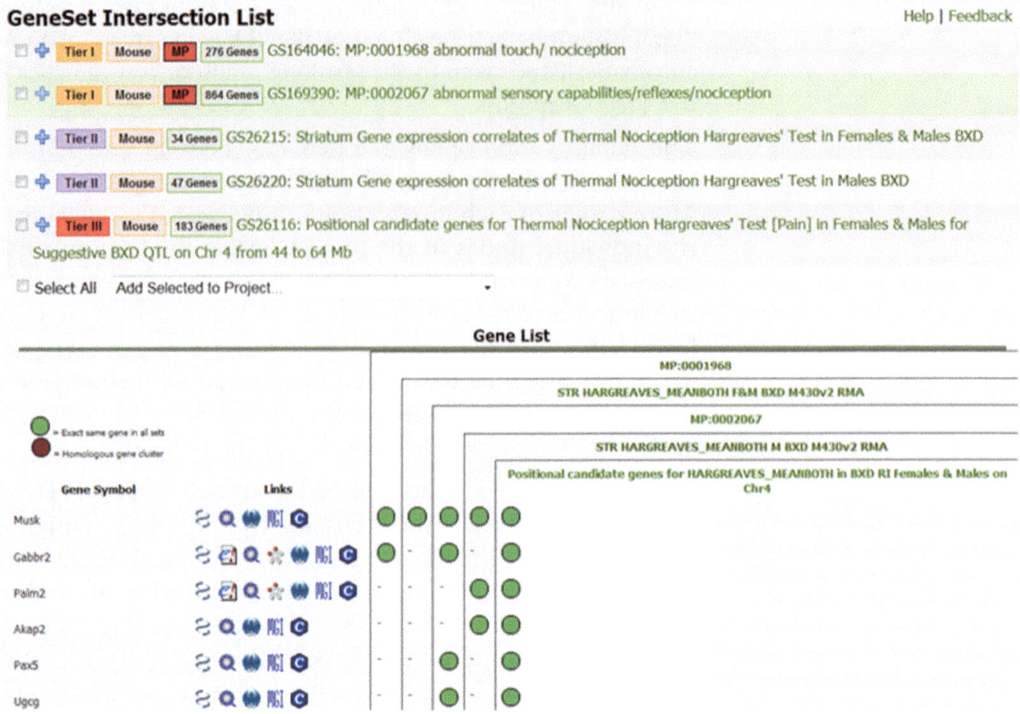

Fig. 13 The GeneSet intersection page is displayed when you click on any of the nodes in a HiSim graph. The genes found to be at the intersection of these genesets are shown at the *top* of the list, and genes with fewer intersections found below

genes such as *Rab23b*, *Eif3a*, and *Gabbr2* are also identified in some of the three and four way intersections with our list of candidate genes and may be more strongly supported by the genetic evidence in this and other mapping populations.

3.3 Understanding Differentially Expressed Genes and Coexpression Networks

Studies of differentially expressed genes previously measured using microarrays and now commonly assayed by RNA-Seq technologies result in large datasets that need interpretation. There are many approaches that can be taken to explore these datasets. Using GeneWeaver, we will integrate data from multiple heterogeneous studies to explore a set of chronic cocaine regulated genes from Feng et al. [20].

1. Locate the supplemental material for the publication by Feng et al. [20]. The set of genes we are interested in can be found in the hyperlinked Table S1 of Feng et al. [20] (http://www.genomebiology.com/content/supplementary/gb-2014-15-4-r65-s3.xlsx). From this file we will select tab = "chronic.gene.diff". We want to extract just the Gene Symbols and *q*-values from this file using copy and paste to create a new file with these two columns (http://www.geneweaver.org/docs/DiffChronic.xlsx).

2. You will now begin uploading your gene set by clicking on upload GeneSet, under Manage GeneSets on the main menu (To add multiple gene sets at once *see* **Note 4**).
3. On this page (Fig. 14) fill in the Metadata for this gene set. We have a sample of how to format your gene set metadata according to the standards documentation (http://geneweaver.org/

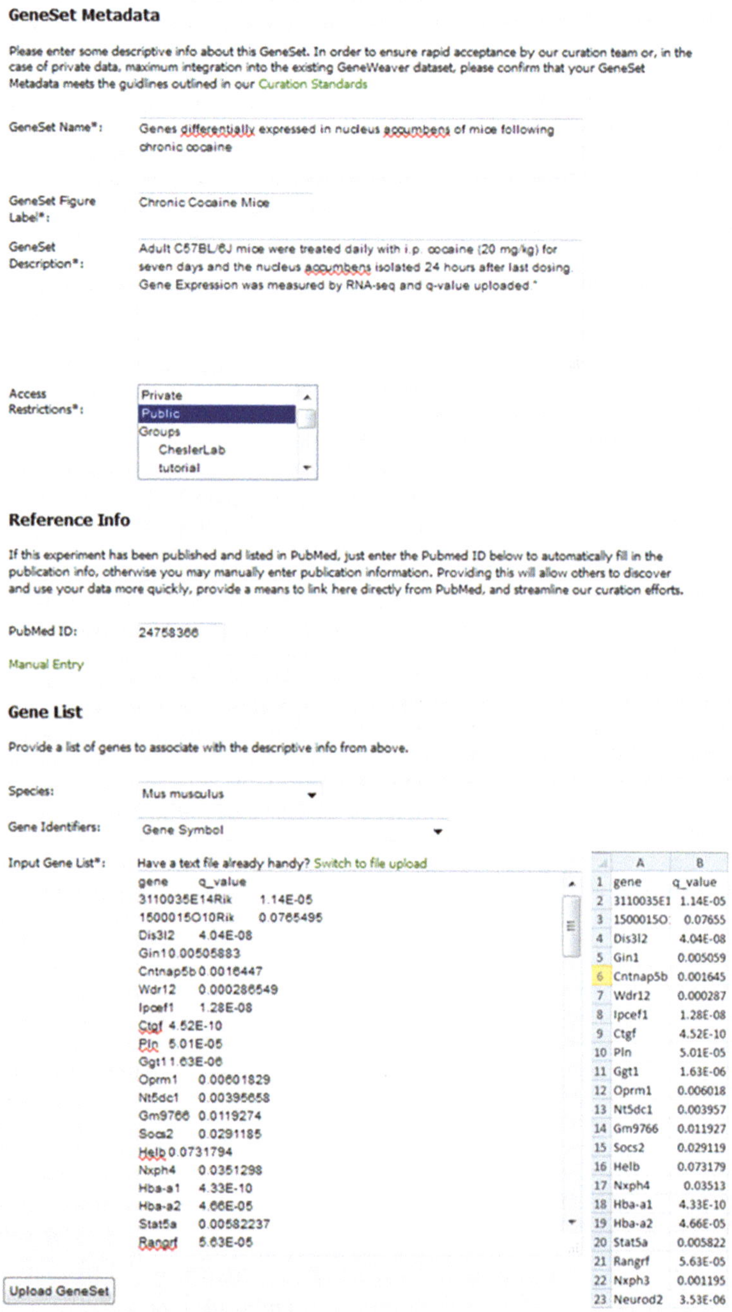

Fig. 14 GeneSet upload page

wiki/index.php?title=Curation_Standards). For this gene set a good name may be "Genes differentially expressed in nucleus accumbens of mice following chronic cocaine." For the Gene Set Figure label use something concise and clear: "Chronic Cocaine Mice" For the gene set description include relevant experimental details, such as treatment paradigm and mouse strain as well as the value type uploaded in the second column: fold change, *p*-value, *q*-value, etc. For this gene set we will use the following "Adult C57BL/6J mice were treated daily with i.p. cocaine (20 mg/kg) for 7 days and the nucleus accumbens isolated 24 h after last dosing. Gene Expression was measured by RNA-seq and *q*-value uploaded."

4. After filling out the description you can decide if you want the file to be public, a tier IV record that will be reviewed by a curator before being upgraded to tier III, or to be a Private Record (tier V), visible only to you and groups you share it with.
5. To populate the record with the abstract, title, journal, and authors from the publication, enter the PMID 24758366 in the box within the reference information section.
6. Entering the Gene List section, select the species (*Mus musculus*) and what gene identifier is used. Gene Identifiers can be tricky, for a more detailed discussion on identifier mapping *see* **Note 5** and Jay, 2012 [21]. For this gene set we have no choice but to use the provided gene symbols. Copy the two columns from the spreadsheet (Fig. 14 inset) and paste them into the sheet. Alternatively the spreadsheet can be saved as a .txt file and uploaded by clicking on "Switch to file upload." Once complete click on upload GeneSet.
7. Once completed you are taken to the GeneSet detail page. If there are errors in your uploaded data you can correct them by clicking on "Edit".
8. Use the *Add Selected to Project*, and create a new project, e.g. "Chronic Cocaine".
9. Now using the Search function populate this project with additional gene sets related to this study trying Queries such as "Cocaine Addiction", "Chronic Cocaine". The sets we have chosen are in Table 1.
10. To look for pairwise overlap between sets we are going to run the Jaccard Overlap. Click on your project containing the newly uploaded gene set and the sets described in Table 1. Now select Jaccard similarity tool and you will be taken to the "Running Jaccard Similarity" screen, where the progress will be displayed e.g. "Combining GeneSets", "Integrating Homologous Genes", "Performing Pairwise Counting". When completed you will automatically be taken to the results or you can click on *View result*.

Table 1
Selected gene sets retrieved from the GeneWeaver database related to chronic cocaine addiction in mouse, rat, and human

GeneSet ID	Description	Number of genes in set	Species	Reference
GS170435	Enhanced behavioral response to cocaine	25	Mouse	MP:0009754
GS87427	Genes significantly changed after cocaine CPP treatment in the hippocampus	43	Rat	17640290
GS178912	Response to cocaine	23	Mouse	GO:0042220
GS136985	Genes with significant differential expression (FDR < 0.2) observed in chronic cocaine-addicted individuals	22	Human	21464311
GS86746	List of cocaine-treated WT vs. Saline-treated WT significantly regulated genes	571	Mouse	17988634
GS1256	Gene expression in human hippocampus from cocaine addicts	48	Human	18000554
GS87302	Transcripts differentially expressed in the majority of cocaine users	50	Human	15009677

11. The results are presented as a pairwise matrix (Fig. 15). Each Venn diagram represents the pairwise gene overlap between the two gene sets depicted for each row and column. Text overlays show the exact gene counts, Jaccard similarity coefficient, and *p*-value for every pair. The Jaccard similarity matrix can be scrolled and browsed with the mouse, and more details of each intersection can be viewed by clicking on the Venn diagrams in the display. This tool can be very useful in identifying the presence of duplicate datasets within a project, or to remove datasets that do not overlap any of the other datasets.
12. Now close out that window and return to the browser open to the *Analyze GeneSets* page. We can look for what genes underlie the set overlap observed in the Jaccard summary. Click on the GeneSet Graph Icon again using the default settings.
13. The resulting graph (Fig. 16) shows a number of two way intersections involving our gene set (on the left), and one three way intersect for the gene *Drd2* which was not found in our set (the right side). Three genes from this study, *Scube1*, *Clql3*, and *Hs3st2* overlap with another mouse study of differential expression following chronic cocaine exposure [22]. Two mouse gene homologs in human, KL and SPTBN1, were differentially expressed in a comparison of ten cocaine users compared to drug free controls [23]. One mouse gene homolog, human CTGF was found differentially expressed in a different

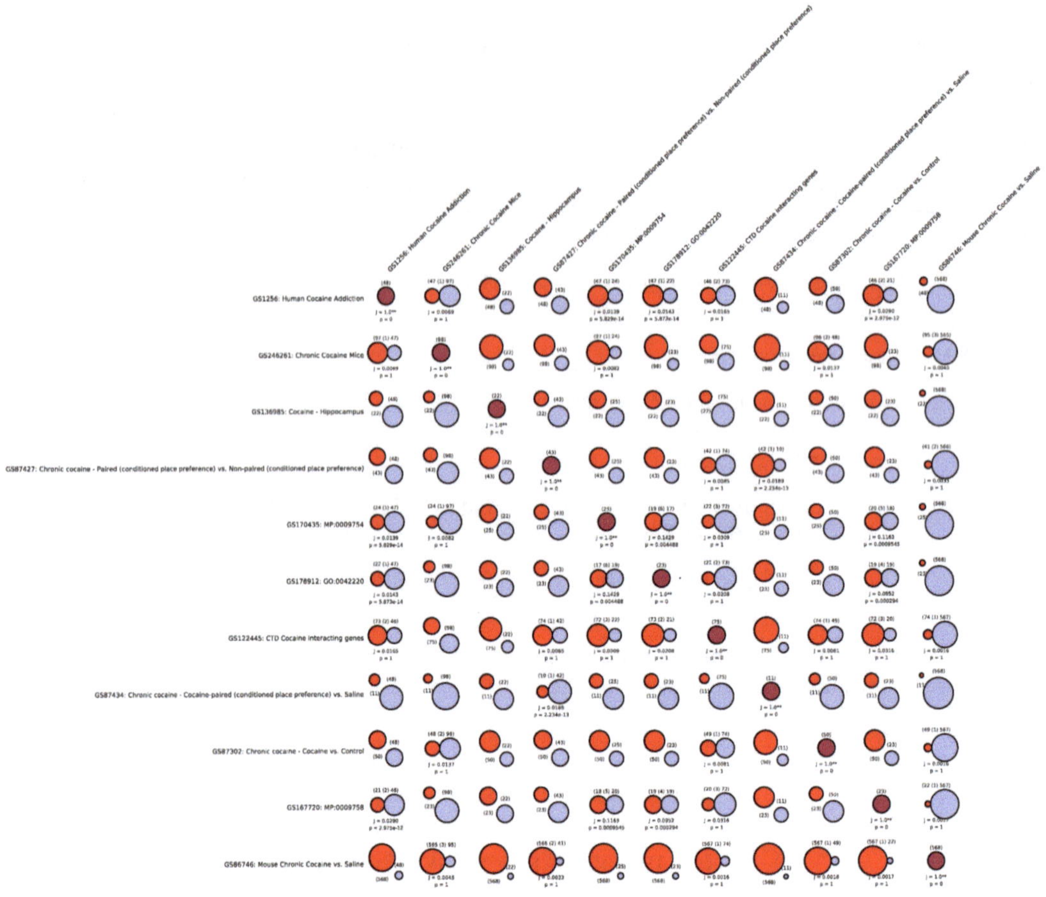

Fig. 15 Venn diagrams of GeneSet overlap using jaccard simlarity tool

set of 10 cocaine users compared to 11 controls [24]. Finally knock-out studies in mouse have shown that one of the overlapping genes, *Oprm1*, has an enhanced behavioral response to cocaine and is annotated to MP:0009754 from [25].

4 Notes

1. The tiered structure in GeneWeaver allows the organization of the data types. Each gene set is assigned a tier, I-V. Tiers I, II, and III represent public resources, machine-generated resources, and human-curated data sets respectively. Tiers IV and V represent data submissions from users that are either pending curatorial review or stored for private use.
2. For detailed instruction on using GeneNetwork and all options and datasets available *see* Williams et al. [11].
3. If you have previously used the emphasis GeneSet feature, you will need to clear your emphasis genes before adding new empha-

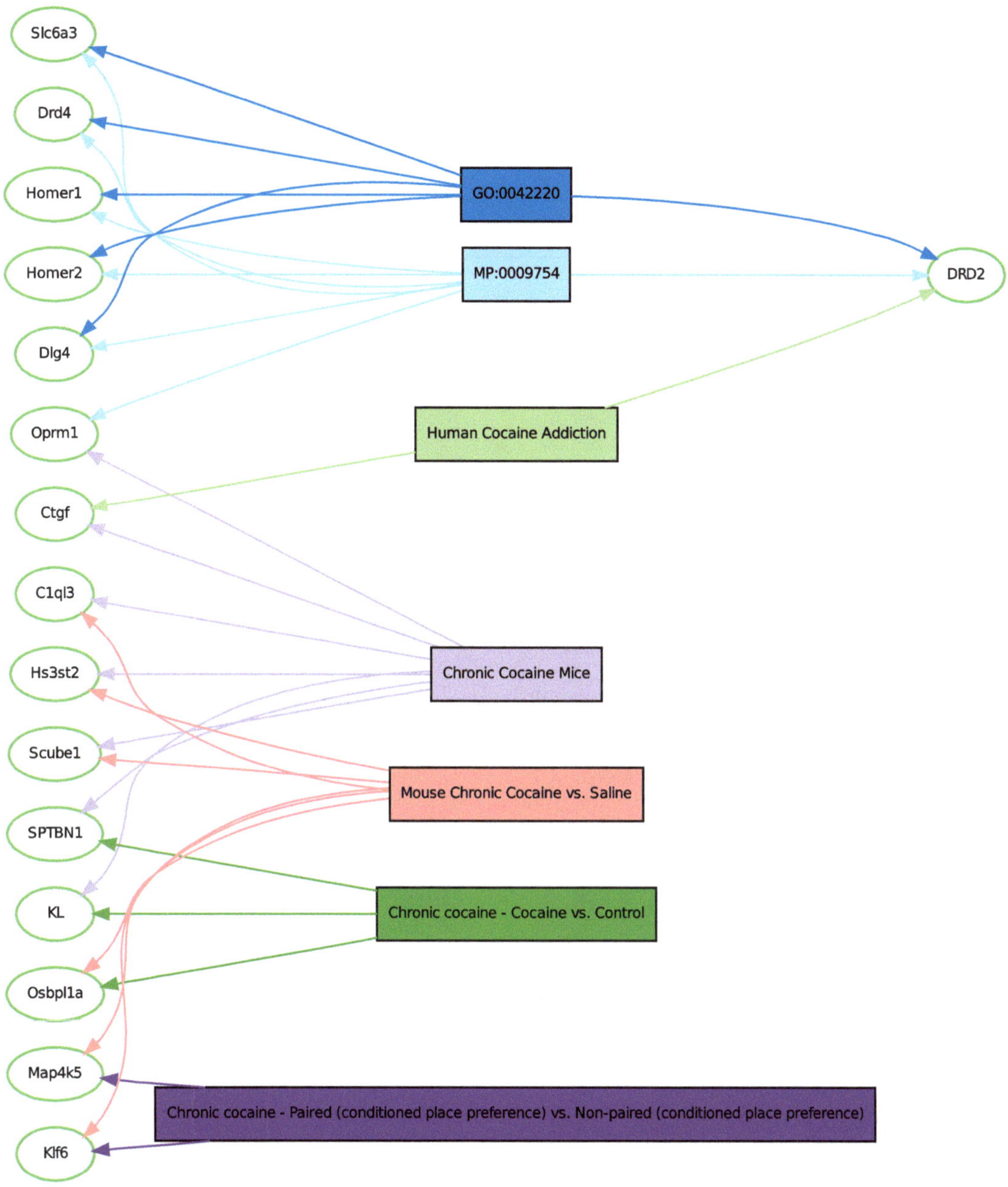

Fig. 16 GeneSet graph of the differentially expressed genes from chronic cocaine exposure in mice

sis genes. This is done on the Analyze GeneSets, MyProjects Page, clicking on Emphasis genes, and scrolling to the bottom to click on "clear" to empty the Emphasis Genes clipboard.

4. The "Batch GeneSet Upload" allows the user to upload multiple gene sets at one time. It uses a tab delimited format, a sample of which is displayed when you click Batch GeneSet Upload under Manage GeneSets.

5. GeneWeaver currently supports 63 different popular gene identifiers from microarray probes, genome browser names as well as the official model organisms symbols. If you cannot find the correct identifier or your identifier is not supported try converting at a website such as NIAID's DAVID website (https://david.ncifcrf.gov/) which has a nice ID conversion tool [26].

Acknowledgements

GeneWeaver is currently supported by NIH AA18776 jointly funded by NIAAA/NIDA.

References

1. Smith CL, Eppig JT (2012) The Mammalian Phenotype Ontology as a unifying standard for experimental and high-throughput phenotyping data. Mamm Genome 23(9–10):653–668. doi:10.1007/s00335-012-9421-3
2. Kohler S, Doelken SC, Mungall CJ, Bauer S, Firth HV, Bailleul-Forestier I, Black GC, Brown DL, Brudno M, Campbell J, FitzPatrick DR, Eppig JT, Jackson AP, Freson K, Girdea M, Helbig I, Hurst JA, Jahn J, Jackson LG, Kelly AM, Ledbetter DH, Mansour S, Martin CL, Moss C, Mumford A, Ouwehand WH, Park SM, Riggs ER, Scott RH, Sisodiya S, Van Vooren S, Wapner RJ, Wilkie AO, Wright CF, Vulto-van Silfhout AT, de Leeuw N, de Vries BB, Washingthon NL, Smith CL, Westerfield M, Schofield P, Ruef BJ, Gkoutos GV, Haendel M, Smedley D, Lewis SE, Robinson PN (2014) The Human Phenotype Ontology project: linking molecular biology and disease through phenotype data. Nucleic Acids Res 42(Database issue):D966–D974. doi:10.1093/nar/gkt1026
3. Blake JA, Dolan M, Drabkin H, Hill DP, Li N, Sitnikov D, Bridges S, Burgess S, Buza T, McCarthy F, Peddinti D, Pillai L, Carbon S, Dietze H, Ireland A, Lewis SE, Mungall CJ, Gaudet P, Chrisholm RL, Fey P, Kibbe WA, Basu S, Siegele DA, McIntosh BK, Renfro DP, Zweifel AE, Hu JC, Brown NH, Tweedie S, Alam-Faruque Y, Apweiler R, Auchinchloss A, Axelsen K, Bely B, Blatter M, Bonilla C, Bouguerleret L, Boutet E, Breuza L, Bridge A, Chan WM, Chavali G, Coudert E, Dimmer E, Estreicher A, Famiglietti L, Feuermann M, Gos A, Gruaz-Gumowski N, Hieta R, Hinz C, Hulo C, Huntley R, James J, Jungo F, Keller G, Laiho K, Legge D, Lemercier P, Lieberherr D, Magrane M, Martin MJ, Masson P, Mutowo-Muellenet P, O'Donovan C, Pedruzzi I, Pichler K, Poggioli D, Porras Millan P, Poux S, Rivoire C, Roechert B, Sawford T, Schneider M, Stutz A, Sundaram S, Tognolli M, Xenarios I, Foulgar R, Lomax J, Roncaglia P, Khodiyar VK, Lovering RC, Talmud PJ, Chibucos M, Giglio MG, Chang H, Hunter S, McAnulla C, Mitchell A, Sangrador A, Stephan R, Harris MA, Oliver SG, Rutherford K, Wood V, Bahler J, Lock A, Kersey PJ, McDowall DM, Staines DM, Dwinell M, Shimoyama M, Laulederkind S, Hayman T, Wang S, Petri V, Lowry T, D'Eustachio P, Matthews L, Balakrishnan R, Binkley G, Cherry JM, Costanzo MC, Dwight SS, Engel SR, Fisk DG, Hitz BC, Hong EL, Karra K, Miyasato SR, Nash RS, Park J, Skrzypek MS, Weng S, Wong ED, Berardini TZ, Huala E, Mi H, Thomas PD, Chan J, Kishore R, Sternberg P, Van Auken K, Howe D, Westerfield M (2013) Gene Ontology annotations and resources. Nucleic Acids Res 41(Database issue):D530–D535. doi:10.1093/nar/gks1050
4. Kanehisa M, Goto S, Sato Y, Kawashima M, Furumichi M, Tanabe M (2014) Data, information, knowledge and principle: back to metabolism in KEGG. Nucleic Acids Res 42(Database issue):D199–D205. doi:10.1093/nar/gkt1076
5. Cerami EG, Gross BE, Demir E, Rodchenkov I, Babur O, Anwar N, Schultz N, Bader GD, Sander C (2011) Pathway Commons, a web resource for biological pathway data. Nucleic Acids Res 39(Database issue):D685–D690. doi:10.1093/nar/gkq1039
6. Baker EJ, Jay JJ, Bubier JA, Langston MA, Chesler EJ (2012) GeneWeaver: a web-based system for integrative functional genomics. Nucleic Acids Res 40(Database issue):D1067–D1076. doi:10.1093/nar/gkr968
7. Bubier JA, Phillips CA, Langston MA, Baker EJ, Chesler EJ (2015) GeneWeaver: find-

ing consilience in heterogeneous cross-species functional genomics data. Mamm Genome 26(9–10):556–566. doi:10.1007/s00335-015-9575-x

8. Gardner D, Akil H, Ascoli GA, Bowden DM, Bug W, Donohue DE, Goldberg DH, Grafstein B, Grethe JS, Gupta A, Halavi M, Kennedy DN, Marenco L, Martone ME, Miller PL, Muller HM, Robert A, Shepherd GM, Sternberg PW, Van Essen DC, Williams RW (2008) The neuroscience information framework: a data and knowledge environment for neuroscience. Neuroinformatics 6(3):149–160. doi:10.1007/s12021-008-9024-z
9. Pesch R, Lysenko A, Hindle M, Hassani-Pak K, Thiele R, Rawlings C, Kohler J, Taubert J (2008) Graph-based sequence annotation using a data integration approach. J Integr Bioinform 5(2). doi:10.2390/biecoll-jib-2008-94
10. Grubb SC, Bult CJ, Bogue MA (2014) Mouse phenome database. Nucleic Acids Res 42(Database issue):D825–D834. doi:10.1093/nar/gkt1159
11. Williams RW, Mulligan MK (2012) Genetic and molecular network analysis of behavior. Int Rev Neurobiol 104:135–157. doi:10.1016/B978-0-12-398323-7.00006-9
12. Geer LY, Marchler-Bauer A, Geer RC, Han L, He J, He S, Liu C, Shi W, Bryant SH (2010) The NCBI BioSystems database. Nucleic Acids Res 38(Database issue):D492–D496. doi:10.1093/nar/gkp858
13. Cunningham F, Amode MR, Barrell D, Beal K, Billis K, Brent S, Carvalho-Silva D, Clapham P, Coates G, Fitzgerald S, Gil L, Giron CG, Gordon L, Hourlier T, Hunt SE, Janacek SH, Johnson N, Juettemann T, Kahari AK, Keenan S, Martin FJ, Maurel T, McLaren W, Murphy DN, Nag R, Overduin B, Parker A, Patricio M, Perry E, Pignatelli M, Riat HS, Sheppard D, Taylor K, Thormann A, Vullo A, Wilder SP, Zadissa A, Aken BL, Birney E, Harrow J, Kinsella R, Muffato M, Ruffier M, Searle SM, Spudich G, Trevanion SJ, Yates A, Zerbino DR, Flicek P (2015) Ensembl 2015. Nucleic Acids Res 43(Database issue):D662–D669. doi:10.1093/nar/gku1010
14. Lein ES, Hawrylycz MJ, Ao N, Ayres M, Bensinger A, Bernard A, Boe AF, Boguski MS, Brockway KS, Byrnes EJ, Chen L, Chen TM, Chin MC, Chong J, Crook BE, Czaplinska A, Dang CN, Datta S, Dee NR, Desaki AL, Desta T, Diep E, Dolbeare TA, Donelan MJ, Dong HW, Dougherty JG, Duncan BJ, Ebbert AJ, Eichele G, Estin LK, Faber C, Facer BA, Fields R, Fischer SR, Fliss TP, Frensley C, Gates SN, Glattfelder KJ, Halverson KR, Hart MR, Hohmann JG, Howell MP, Jeung DP, Johnson RA, Karr PT, Kawal R, Kidney JM, Knapik RH, Kuan CL, Lake JH, Laramee AR, Larsen KD, Lau C, Lemon TA, Liang AJ, Liu Y, Luong LT, Michaels J, Morgan JJ, Morgan RJ, Mortrud MT, Mosqueda NF, Ng LL, Ng R, Orta GJ, Overly CC, Pak TH, Parry SE, Pathak SD, Pearson OC, Puchalski RB, Riley ZL, Rockett HR, Rowland SA, Royall JJ, Ruiz MJ, Sarno NR, Schaffnit K, Shapovalova NV, Sivisay T, Slaughterbeck CR, Smith SC, Smith KA, Smith BI, Sodt AJ, Stewart NN, Stumpf KR, Sunkin SM, Sutram M, Tam A, Teemer CD, Thaller C, Thompson CL, Varnam LR, Visel A, Whitlock RM, Wohnoutka PE, Wolkey CK, Wong VY, Wood M, Yaylaoglu MB, Young RC, Youngstrom BL, Yuan XF, Zhang B, Zwingman TA, Jones AR (2007) Genome-wide atlas of gene expression in the adult mouse brain. Nature 445(7124):168–176. doi:10.1038/nature05453
15. Davis AP, Grondin CJ, Lennon-Hopkins K, Saraceni-Richards C, Sciaky D, King BL, Wiegers TC, Mattingly CJ (2015) The comparative toxicogenomics database's 10th year anniversary: update 2015. Nucleic Acids Res 43(Database issue):D914–D920. doi:10.1093/nar/gku935
16. Szklarczyk D, Franceschini A, Kuhn M, Simonovic M, Roth A, Minguez P, Doerks T, Stark M, Muller J, Bork P, Jensen LJ, von Mering C (2011) The STRING database in 2011: functional interaction networks of proteins, globally integrated and scored. Nucleic Acids Res 39(Database issue):D561–D568. doi:10.1093/nar/gkq973
17. Hildenbrand ZL, Carlton DD, Fontenot BE, Meik JM, Walton JL, Taylor JT, Thacker JB, Korlie S, Shelor CP, Henderson D, Kadjo AF, Roelke CE, Hudak PF, Burton T, Rifai HS, Schug KA (2015) A comprehensive analysis of groundwater quality in the Barnett Shale region. Environ Sci Technol 49(13):8254–8262. doi:10.1021/acs.est.5b01526
18. U.S. Department of Health and Human Services (2004) Interaction profile for: benzene, toluene, ethylbenzene, and xylenes (BTEX). Agency for Toxic Substances and Disease Registry, Atlanta, GA
19. Philip VM, Duvvuru S, Gomero B, Ansah TA, Blaha CD, Cook MN, Hamre KM, Lariviere WR, Matthews DB, Mittleman G, Goldowitz D, Chesler EJ (2010) High-throughput behavioral phenotyping in the expanded panel of BXD recombinant inbred strains. Genes Brain Behav 9(2):129–159. doi:10.1111/j.1601-183X.2009.00540.x
20. Feng J, Wilkinson M, Liu X, Purushothaman I, Ferguson D, Vialou V, Maze I, Shao N, Kennedy P, Koo J, Dias C, Laitman B,

Stockman V, LaPlant Q, Cahill ME, Nestler EJ, Shen L (2014) Chronic cocaine-regulated epigenomic changes in mouse nucleus accumbens. Genome Biol 15(4):R65. doi:10.1186/gb-2014-15-4-r65

21. Jay JJ (2012) Cross species integration of functional genomics experiments. Int Rev Neurobiol 104:1–24. doi:10.1016/B978-0-12-398323-7.00001-X

22. Renthal W, Maze I, Krishnan V, Covington HE III, Xiao G, Kumar A, Russo SJ, Graham A, Tsankova N, Kippin TE, Kerstetter KA, Neve RL, Haggarty SJ, McKinsey TA, Bassel-Duby R, Olson EN, Nestler EJ (2007) Histone deacetylase 5 epigenetically controls behavioral adaptations to chronic emotional stimuli. Neuron 56(3):517–529. doi:10.1016/j.neuron.2007.09.032

23. Albertson DN, Pruetz B, Schmidt CJ, Kuhn DM, Kapatos G, Bannon MJ (2004) Gene expression profile of the nucleus accumbens of human cocaine abusers: evidence for dysregulation of myelin. J Neurochem 88(5):1211–1219

24. Mash DC, ffrench-Mullen J, Adi N, Qin Y, Buck A, Pablo J (2007) Gene expression in human hippocampus from cocaine abusers identifies genes which regulate extracellular matrix remodeling. PLoS One 2(11):e1187. doi:10.1371/journal.pone.0001187

25. Yoo JH, Yang EM, Lee SY, Loh HH, Ho IK, Jang CG (2003) Differential effects of morphine and cocaine on locomotor activity and sensitization in mu-opioid receptor knockout mice. Neurosci Lett 344(1):37–40

26. da Huang W, Sherman BT, Lempicki RA (2009) Systematic and integrative analysis of large gene lists using DAVID bioinformatics resources. Nat Protoc 4(1):44–57. doi:10.1038/nprot.2008.211

Chapter 7

A Suite of Tools for Biologists That Improve Accessibility and Visualization of Large Systems Genetics Datasets: Applications to the Hybrid Mouse Diversity Panel

Christoph D. Rau, Mete Civelek, Calvin Pan, and Aldons J. Lusis

Abstract

In this chapter we address the recent explosion in large multilevel population studies such as the METSIM study in humans as well as large panels of animal models such as the Hybrid Mouse Diversity Panel or the BXD set of recombinant inbred strains. These studies have harnessed the increasing affordability of large-scale high-throughput profiling to gather massive quantities of data. These datasets, spread across different -omics levels (genome, transcriptome, etc.), different tissues (e.g. heart, plasma, bone) and different environmental factors (e.g. diet, drugs) each individually have led to a number of novel findings relevant to a variety of complex diseases and other phenotypes. The analysis of these results, however, is often limited to individuals with a comprehensive understanding of database languages such as SQL. In this chapter, we describe the development of a GUI-based database analysis suite, using the Hybrid Mouse Diversity Panel as an example to lay out a series of methods for visualization and integration of large systems genetics datasets. The database is based on the Shiny suite of tools in R, and is transferrable to other SQL-based datasets.

Key words Analysis tools in systems genetics, GUI-based database analysis suite, Multilevel population studies, Hybrid mouse diversity panel, BxD recombinant inbred strains, METSIM in humans

1 Introduction and Background

Systems genetics deals with the analysis of massive datasets typically gathered by large, multicenter studies. In order to overcome hurdles ranging from the need to amass large sample sizes to performing data analysis to dealing with administrative issues such as informed consent or obtaining funding, these studies often require teams of individuals who, working together, are able to generate data which must subsequently be sifted through to find notable results. As the field has matured, increasingly ambitious studies have

Electronic supplementary material The online version of this chapter (doi:10.1007/978-1-4939-6427-7_7) contains supplementary material, which is available to authorized users.

Klaus Schughart and Robert W. Williams (eds.), *Systems Genetics: Methods and Protocols*, Methods in Molecular Biology, vol. 1488, DOI 10.1007/978-1-4939-6427-7_7, © Springer Science+Business Media New York 2017

been designed and implemented by building off of the framework of other studies. For instance, a recent study, the METabolic Syndrome in Men (METSIM) [1], is collecting multiple -omics levels of data from over 10,000 men and intends to do follow-up studies with these same -omes over the next several decades. Similarly, large panels of inbred animal models, such as the Hybrid Mouse Diversity Panel (HMDP) [2], the BxD set of recombinant inbred strains, or the Collaborative Cross [3] have been constructed which allow for a continually expanding set of diseases and phenotypes to be explored on genetically identical individuals. These animal models are able to avoid the issues of environmental confounders and informed consent, while still providing powerful insights into the underlying mechanisms of complex phenotypes and diseases.

These large studies have had considerable successes in discovering genes and pathways which are implicated in the regulation of phenotypes and disease progression. In the process, they have generated massive datasets which span different tissue types, environmental conditions and -omics levels. The analysis of these large datasets, however, can prove challenging. On the one hand, individuals without a sufficient background in programming may have difficulty navigating the (often SQL) databases in which these data are typically deposited and/or be unable to visualize and interpret these results even if they were able to access them. On the other hand, the scale of the data generated means that a comprehensive analysis is often beyond the capabilities of a few members of a team who do possess the ability to access, visualize and understand these data after they are generated. This chapter lays out the implementation of a suite of point-and-click tools for the visualization and interpretation of large systems genetics datasets, designed to allow researchers who do not deal with computational techniques on a regular basis to access and interpret these data in an intuitive way. The database accessibility suite has been coded using the Shiny R package [4], therefore each tool can be modified to interface with any SQL-based database by a programmer knowledgable about the struture of the database.

2 Methods

2.1 Type of Data and Principle Analysis Tools

2.1.1 The Hybrid Mouse Diversity Panel

The dataset we will be using throughout this chapter is the Hybrid Mouse Diversity Panel, a set of over 150 unique inbred mouse lines. These lines have been extensively studied, and at present the database consists of mice studied under five different environmental conditions or genetic stressors (Table 1) and ten different tissues from which transcriptomes and other -omics studies have been performed. In total, over 300 clinical traits, 40,000 transcripts, 350 metabolites and 3500 protein fragments have been queried in one or more HMDP study (Figs. 1 and S1). Through a variety of systems biology techniques, many candidate genes and

Table 1
Experimental models in the HMDP database

Environmental condition/stressor	Primary traits [references]
1. Low-fat chow diet	Plasma lipids, adiposity [5] Bone density [6, 7] Behavior [8] Blood cell levels [9] Proteomics [10] Macrophage inflammation [11] Metabolomics (hepatic) [12] DNA methylation [13]
2. High-fat, high sucrose diet	Dietary responsiveness [14] Diabetes/insulin resistance [15] Gut microbiota [14, 16] Bone marrow stem cells (ALLAYEE, LUSIS) Fatty liver [17]
3. Isoproterenol treatment	Heart failure [18]
4. High-fat, high-cholesterol diet and ApoE-Leiden, CETP transgenes	Atherosclerosis [19]
5. Low-fat chow diet, auditory stressors	Hearing [20, 21]

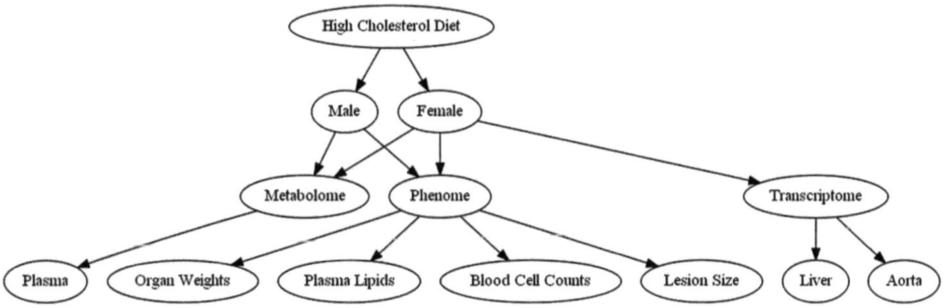

Fig. 1 A subset of the data available in the HMDP Database. Depicted here is a subset of the total HMDP database, organized with study name (in this case, a high-cholesterol diet) at the *top level*, followed by gender, then -omics level, and finally by individual tissues or phenotypes. A full depiction of the contents of the HMDP Database can be found in Fig. S1

pathways [5, 9, 14, 18] have been identified using this panel. While past studies have generally focused on data from an individual experimental model/condition, integration across models can also be fruitful. Because the mice studied under various conditions are identical in terms of their genetic backgrounds, a variety of combined analyses across multiple tissues or studies are possible; for example, the data can be used for the discovery of novel cross-tissue or cross-conditional relationships.

Below, we describe the suite of accessibility tools we have developed to assist in these sorts of HMDP analyses. Broadly speaking, our database implements two categories of analysis: The Visualization of previously generated data (e.g., the creation of a Manhattan plot) and the Discovery of new relationships between these data (e.g., identifying correlations of genes and phenotypes across multiple tissues/studies).

2.1.2 Overall Design

It was important that the tools we created for accessing the database operated similarly to one another. Each tool (Figs. S2, S3, S4, S5, S6, S7, and S8) asks users for inputs on the left side of the screen using drop-down menus, checkboxes and places for users to input text. As many of our studies examine similar measurements, each drop-down menu is dependent on the selection of the menus above it. For instance, in Fig. S2, the plotted Manhattan plot was generated from clinical trait data from the atherosclerosis study using female mice and the adiposity phenotype. Similarly, in Fig. S3, the displayed beeswarm plot was generated from clinical data from one of the Chow studies, using males from the first chow study and visualizing the effects on HDL of the SNP rs31423553. On the right hand side of each tool is the output, with the figure or table at the top followed by a button to download the results from the database to one's own computer.

2.2 Visualization Tools

2.2.1 Generating a Manhattan Plot

A classic tool for the analysis of GWAS results is the Manhattan plot, which visualizes genome-wide association. The locations of single nucleotide polymorphisms (SNPs) are plotted on the *X*-axis and the strength of their association with the trait of interest as $-\log_{10}$ (*p* value) on the *Y*-axis. The HMDP contains over 400,000 individual quantitative traits. For each trait, the database allows for a Manhattan plot (using the qqman package [22]) to be generated either over the entire genome (Fig. 2a S2a) or in greater detail at an individual chromosome (Fig. S2b) or at an even narrower scope to look at a more localized region. At distances of less than 10 Mb, the UCSC genome browser is linked to the output (Fig. S2c), allowing for direct identification of possible candidate genes. Additionally, at any point the data displayed may be downloaded for later analysis.

2.2.2 Generating a Beeswarm Plot

Whereas a Manhattan plot displays the results of an association study across a range of SNPs, it is often desirable to examine the distribution of individual samples across a single polymorphism. In this case, a "beeswarm" plot is often used. Our database is capable of quickly generating such a plot for any phenotype/SNP pairing (Figs. 2b and S3) through the use of the Beeswarm package [23].

2.2.3 Visualizing Values Across Strains and Tissues

A frequently asked question in animal research is whether a given animal model is the best model to explore a particular phenotype of interest. Since the strains which comprise the HMDP are

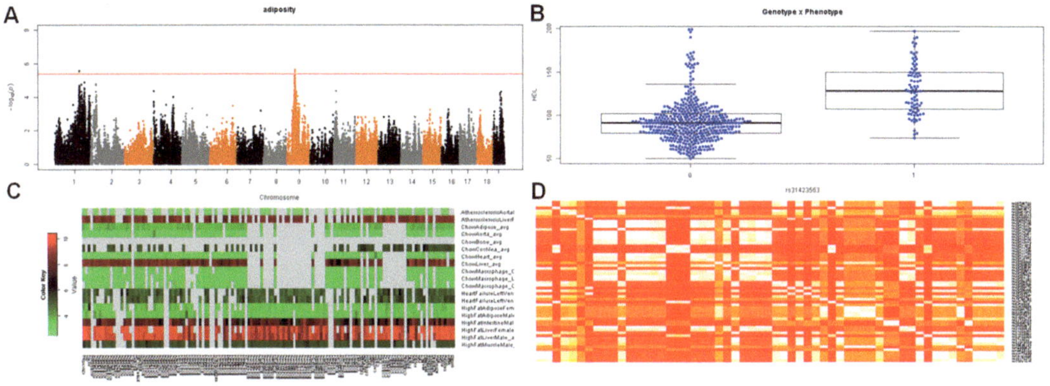

Fig. 2 Visualization tool outputs in the HMDP Database GUI. (**a**) A Manhattan plot for adiposity reveals a significant association on chromosome 9. (**b**) A beeswarm plot demonstrates the effect of SNP rs31423553 on HDL levels in plasma. (**c**) The expression of the gene *Abcc6* is plotted across the different studies of the HMDP, revealing high expression in liver compared to other tissues. (**d**) A LD plot of chromosome 3 between 10 and 11 Mb shows evidence of a small LD block from 10.1 to 10.5 Mb

publically available, one of the benefits of the HMDP as a model is that it can answer these questions and provide researchers with the ideal mouse strain for further research. By visualizing the phenotype or gene expression value across all or a subset of strains and all or a subset of studies/tissues (Figs. 2c and S4), our database allows for quick analysis and subsequent download of any gene expression or phenotypic value it contains. For example, we can see in Fig. 2c that the gene *Abcc6* is highly expressed in the liver and moderately expressed in the intestine, but weakly expressed in other tissues.

2.2.4 Linkage Disequilibrium

While linkage disequilibrium (LD) in humans is typically quite small, the LD structure in mouse panels is broader, ranging from 1 Mb to up to 10 Mb in the HMDP. Consequently, identifying candidate genes near significantly associated SNPs involves examining all genes which lie within the LD of the peak SNP, rather than simply examining the one or two nearest genes to the peak SNP as is often done in human studies. Our database allows researchers to either specify an individual SNP, in which case a proposed LD block around the SNP will be provided, or provide a particular window to examine, in which case the LD structure between each pair of SNPs in the window will be displayed (Figs. 2d and S5).

2.3 Discovery Tools

2.3.1 Identifying Nonsynonymous SNPs Within a Gene

Prioritizing candidate genes at a locus can be difficult, as some genes in a locus can be poorly annotated or described. One common technique used to identify genes with a greater likelihood to be implicated in the phenotype of interest is to examine the gene for nonsynonymous and splice mutations which may act to disrupt the structure and function of a gene at the locus without affecting its expression. Our database makes use of the Wellcome Trust

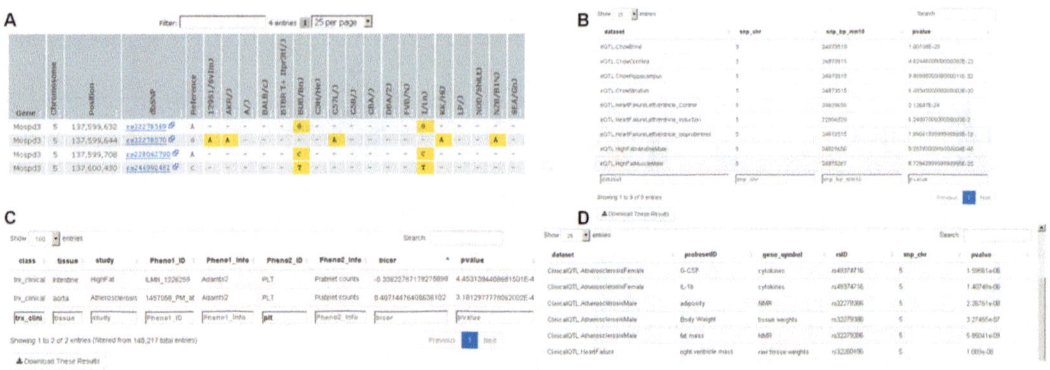

Fig. 3 Discovery Outputs of the HMDP Database GUI. (**a**) Querying the Wellcome Trust Mouse Genomes resource reveals a number of nonsynonymous mutations in the gene *Mospd3*. (**b**) *cis*-eQTLs located near *Prkag2* show strong local regulation in multiple tissues (**c**) Correlation of *Adamts2* with other phenotypes (*see* Fig. S8 for additional correlations) reveals a previously unappreciated correlation with platelet counts in multiple studies. (**d**) Examination of the *Mospd3* locus for RV weight reveals additional significant loci across the HMDP studies

Mouse Genomes Project [24] which contains the full sequences of 18 of the inbred lines of the HMDP and, therefore, the vast majority of all sequence variations within the panel. The output allows researchers to identify which strains have mutations within them, what sort of mutations they are and where in the gene they are located (Figs. 3a and S6).

2.3.2 Identifying Local-eQTLs Within Tissues

If a candidate gene's physical amino acid sequence is not altered, another means by which a SNP may affect a gene's function is by altering its expression. SNPs residing near a gene whose expression is associated with that SNP are commonly termed local or *cis*-eQTLs. Our database allows users to search across all HDMP studies and tissues to identify *cis*-eQTLs by either probe ID or gene symbol, returning the best *p*-value and location for a SNP within a user-defined window (default 2 Mb) (Figs. 3b and S7). The table can be searched quickly within the GUI or downloaded for more detailed analysis.

2.3.3 Find Correlations Within and Across Studies, Tissues and Conditions

The HMDP is an expanding resource, where each subsequent study builds on prior research. Finding relationships between genes of interest in one dataset and genes or phenotypes of interest in another, however, can be challenging. We have implemented a searchable unified correlation table, allowing users to examine how genes and phenotypes might correlate to one another across tissues and experimental conditions. For example, we can observe that *Adamts2* expression in the heart is linked to changes in heart weight after isoproterenol stimulation (Figs. 3c and S8). However, we can also see that *Adamts2* expression in the aorta and intestine are linked to plasma platelet counts, suggesting a role for the gene in clotting or wound repair. It is notable that while both aortic and intestinal expression of *Adamts2* is linked to plasma platelets, they are linked in opposite directions: Higher expression in the intestines in mice with a high-fat diet is linked to lower platelets, while higher expression in the aorta in mice induced to develop atherosclerosis is linked to increased platelet counts. Such inter-study observations can provide additional clues to the roles genes play in physiologic/disease traits.

2.3.4 Identify Overlapping Loci

In addition to exploring how genes and phenotypes are correlated to one another across different tissues and conditions in the HMDP, we can use the large number of associations available in the panel to look for loci which are shared between multiple phenotypes. This could have a number of applications, from the validation of *cis* or *trans*-eQTLs in multiple tissues to exploring how a complex phenotype or set of phenotypes, identified across multiple tissues, may be regulated by a single locus or set of loci. Our database allows researchers to look at and download either all phenotypes which are associated with a SNP at a specific *p*-value, or to examine all phenotypes which have a *p*-value of a given significance within a user-defined window. For example, a locus on chromosome 5 near the gene *Mospd3* has been linked to increased right ventricular weight after isoproterenol stimulation [18]. By examining this same locus across all of our tissues, we can see that we also have links to food intake on a high-fat diet, levels of the IL-1b cytokine, levels of the metabolite hexanoyl-carnitine, and the abundance of several proteins in the liver (Figs. 3d and S9).

2.3.5 Availability

The HMDP database may be accessed at systems.genetics.ucla.edu. The most up-to-date version of the code for the implementation of the database can be found at https://github.com/ChristophRau/HMDPDatabase.

3 Further Considerations and Limitations

Other databases do exist for the analysis of mouse data, the most notable of which is the Mouse Genome Informatics (MGI) database provided by Jackson Labs (informatics.jax.org). There it is possible to download phenotypes and genotypes of a number of strains from a variety of different studies as well as visualize the results of each individual study in terms of the phenotypes observed as well as information on a gene of interest. One strength of the HMDP over the studies typically found at this database is the scale of the study and the comparability across studies. While most HMDP studies involve over 100 strains of mice, studies in the MGI repository are typically much smaller. At the same time, few studies involve the same strain of mice as others, making it difficult for researchers to compare results between individual studies. Additionally, few studies involve the study of multiple -omics layers, and differences in housing, diet and other environmental factors make it difficult to directly compare studies to one another. Moreover, the tools made available in the MGI database to query these studies are generally designed for the smaller, less systems-wide studies that make up the bulk of the repository.

It is always possible to perform more nuanced and complicated analyses of the HDMP database using direct SQL queries. While results gathered by these sorts of analyses may be stronger and more meaningful than the results obtainable by our analysis suite, it requires the ability to navigate the SQL databases directly. As mentioned in the introduction, we believe that an ultimately more fruitful approach is to open up access to such databases to anyone with a grasp of the concepts involved, but perhaps not the technical skill to access the data directly.

4 Outlook

The HMDP is a constantly evolving tool with ongoing studies in several laboratories and an expanding database. Our graphical interface is designed to be able to access any data which is added to the database without the need to manually update a number of tables. We plan to continue to add modules to the database, for instance allowing users to select a series of phenotypes, genes, metabolites and proteins and receive in return graphical and tabular outputs of SNPs on the genome which are associated with one or more of these inputs, a tool which will complement our currently implemented ability to study a given genomic location for association overlaps. Additionally, we plan to improve the relationships of the modules to one another, allowing a user to start in one module, get a result, click on that result and then be taken to another module for additional analyses. Finally, we plan to expand our results to interface with some of the gene-centric databases which currently exist (e.g. NCBI Gene), to allow researchers to seamlessly travel from phenotype to locus to gene. The code for each of these updates will be made available and like the rest of the interface, will be designed to be easily modifiable to interface with other researchers' SQL databases.

Acknowledgments

The HMDP database was developed and is currently maintained through support from NIH grants HL30568, HL28481.

5 Appendix: Code for Database Algorithm v0.7

The Shiny package, developed by Rstudio allows users to implement a graphical user interface with an R-based backend. A GUI created with Shiny requires two R scripts to properly function. The first, Server.R, is the part of the code which actually performs the various analyses, interfaces with the SQL database and does all the things that a standard R script would do. The second, ui.R, controls the layout and structure of the GUI itself and is responsible for sending inputs to and receiving outputs from Server.R. Updates to this code may be found on GitHub at: https://github.com/ChristophRau/HMDPDatabase.

6 Server.R

```
# A Graphical User Interface for querying a genetics SQL Database using Shiny in R
# Version: 0.7
# Last Modified: 12/9/15
#
# The following is an implementation of a GUI using the Shiny package in Rstudio.  Shiny
programs have two scripts associated with them.  This one, server.R acts as the "brains"
# of the code and contains all of the functions which actually compute results.  The other
script, ui.R, controls the appearance of the GUI and provides inputs and displays outputs from
# Server.R

#####Startup Stuff#####
options(shiny.maxRequestSize = 50*1024^2)
options(stringsAsFactors=FALSE)
#Scripts and packages required for operation
source("manhattan.R")
if (!require("beeswarm")) install.packages("beeswarm")
if (!require("gplots")) install.packages("gplots")
library(beeswarm)
library("gplots")

#Initialize the database reader and get a list of all relevant tables
print("Initializing Database")
library(RODBC)
dbhandle <- odbcDriverConnect('driver={SQL Server};server=JLUSISDB;database=HM
DP;trusted_connection=true')
FullTables=sqlTables(dbhandle)

#A section to define limited Tables if you want to password protect the full data.
limitedTables=FullTables  #no password protection

limitedTables=FullTables[grep("Chow",FullTables[,3]),] #limit to a subset of data

allTables = limitedTables


#The following section goes through all of the tables in the database and creates a
master list which maps gene symbols to probe IDs.  This takes some time and it is often
#    easier to generate this file separately and read it in with read.csv
#####Create the Master table of Gene names and IDs#####
print("Creating Genes Table… this could take up to 5 minutes…")
# AFFY="Affymetrix_HT_MG-430A_v33"
# ILMN="Illumina_MouseRef8_v2_R3"
#
```

```
# #Load Affy Data
#
# query <- paste("Select gene_symbol, probesetID from [TranscriptAnnotation].[",AFFY,"]",sep="")
# temp=sqlQuery(dbhandle, query)
#
# #Load Illumina Data
#
# query <- paste("Select Symbol, probesetID from [TranscriptAnnotation].[",ILMN,"]",sep="")
# temp2=sqlQuery(dbhandle, query)
# colnames(temp)=c("Symbol", "probesetID")
# all_genes=rbind(temp,temp2)
# all_genes[,1]=toupper(all_genes[,1])
#
#
# #Creating the Search_List.  This can take a moment…
# geneSearchList<-do.call(rbind,
#              by(all_genes,all_genes$Symbol, function(x)
#               with (x,
#                  data.frame(
#                   Symbol=unique(Symbol),
#                   probeIDs=paste(probesetID,collapse=",")
#                  )
#               )
#              )
# )
# geneSearchList=as.matrix(geneSearchList)
# geneSearchList=paste(geneSearchList[,1],geneSearchList[,2],sep=",")
#or just read it in.
geneSearchList=as.matrix(read.delim(file="GeneSearchList.csv"))
allStrains=as.matrix(read.csv(file="Strains.csv")) #reads in all the strains (individuals) used in the study
#####At this point the server is initialized and ready to launch#####
print("Launching!")
shinyServer(function(input, output,session) { #basic implementation of a
shinyServer


#####Suggestions#####
 #This first part of the code implements a simple suggestion .csv for recording bugs and/or suggestions
  Suggest <- reactive({
   outfile=file("Suggestions.csv","a")
   name=input$Suggestion_Name
   value=input$Suggestion_Report
   outrow=paste(name,value,sep=",")
   cat(outrow,file=outfile,sep="\n")
   close(outfile)
  })
  output$Suggestion_Text <-  renderText({
   if(input$Suggestion_Button==0){
     return (NULL)} else{
       isolate(Suggest())
       val="Thanks!"
       return(val)
     }

  })

  #####Login#####
```

```
  #A way to implement some degree of password protection on your data, say if
you have some public and some private information.
  Password="Password"

  Login <- reactive({
   if(input$Password_Go==0)
   {return("")} else {
     isolate({
       if(input$Password == Password){
         allTables <<- FullTables
         return("Login Successful")
       } else {
         return("Login Failed")
       }
     })
 }
   })

  output$PassOK <- renderText({
   LogVal=Login()
   print(LogVal)})



  #####ProbeID_Lookup#####
  #This section allows for rapid conversion of Probe_IDs to Gene Symbols and vice versa.
Works in batch mode.

  Lookup <- reactive({
   inFile <- input$Lookup_Batch
   if(is.null(inFile)){  #If no batch file uploaded
     if(is.null(input$Lookup_One)){ return(NULL)} else{  #if nothing entered,
return nothing
       val=input$Lookup_One
       print(val)
       entry=geneSearchList[grep(paste0(val,"(,|$)"),geneSearchList, ignore.
case= TRUE)]  #find gene in master table
       print(entry)
       entry=strsplit(entry,",")[[1]]
       print(entry)
       out=c()
       for(i in 2:length(entry)){
         out=rbind(out,c(entry[1],entry[i])) #generate output for gene
       }
       colnames(out)=c("Gene Name","Probe ID")
       print(out)
       return(out)
     }

    } else { #If batch file uploaded
     vals=read.csv(inFile$datapath,header=F) #read the file
     out=c()
     for(i in 1:nrow(vals)){ #for each gene
       val=vals[i,1]
       entry=geneSearchList[grep(paste0(val,"(,|$)"),geneSearchList, ignore.
case= TRUE)] #find in master table
       entry=strsplit(entry,",")[[1]]
```

```
        for(j in 2:length(entry)){  #append results to output
          out=rbind(out,c(entry[1],entry[j]))
        }
      }
      colnames(out)=c("Gene Name","Probe ID")
      return(out)
      }
  })

  output$Lookup_Table <-  renderDataTable({  #This tells the ui how to output the data.
    if(input$Lookup_Button==0){
      return (NULL)} else{
        isolate(Lookup())
      }

  })

  #####Data Vizualization Section####
  #This section creates Manhattan Plots for any study in the database

  #This function populates the possible studies to be drawn from based on the type of data being mapped
  output$DataViz_StudyUI<- renderUI({
    if(input$DataViz_DataType=="Clinical"){
      DV_StudyChoices=allTables[allTables[,2]=="ClinicalTraitAnnotation",3]
    }
    if(input$DataViz_DataType=="Expression"){
      DV_temp=allTables[allTables[,4]=="TABLE",]
      DV_StudyChoices=DV_temp[DV_temp[,2]=="TranscriptAbundance",3]
    }
    if(input$DataViz_DataType=="Metabolite"){
      DV_StudyChoices=allTables[allTables[,2]=="MetaboliteAnnotation",3]
    }
    if(input$DataViz_DataType=="Protein"){
      DV_StudyChoices=allTables[allTables[,2]=="ProteinAnnotation",3]
    }
    selectInput("DataViz_Study", "Select Study", DV_StudyChoices ) #the select
input to be placed into the UI
  })

  #After a study has been selected, this function populates the possible phenotypes to select from
  output$DataViz_PhenotypeUI <- renderUI({
    cur_table=input$DataViz_Study   #Get which study is being examined
    if(input$DataViz_DataType=="Clinical"){
      query=paste("SELECT distinct trait_name  FROM HMDP.ClinicalTraitAnnotati
on.",input$DataViz_Study,"",paste="")
      DV_PhenoChoices=as.vector(sqlQuery(dbhandle, query)[[1]])
    }
    if(input$DataViz_DataType=="Expression"){ #My original idea was to do a
drop-down menu for all phenotypes, including genes/probes.  This proved too
taxing and instead I've implemented a simple text entry box.
      # The code for the drop-down menu is below.
      #  query=paste("Select Top 1 HMDP.TranscriptAbundance.",input$DataViz_
Study,".* FROM HMDP.TranscriptAbundance.",input$DataViz_Study,sep="")
      # DV_TempExpressionQuery=sqlQuery(dbhandle,query)
      #  query=paste("SELECT distinct probesetID ",colnames(DV_
TempExpressionQuery)[2]," FROM HMDP.TranscriptAbundance.",input$DataViz_Study,
```

```
      #            " WHERE ",colnames(DV_TempExpressionQuery)[2],"='",DV_
TempExpressionQuery[2][[1]],"'",sep="")
      #  DV_PhenoChoices=as.vector(sqlQuery(dbhandle, query)[[1]])
    }
    if(input$DataViz_DataType=="Metabolite"){
      query=paste("SELECT distinct metabolite_name  FROM HMDP.MetaboliteAnnota
tion.",input$DataViz_Study,"",paste="")
      DV_PhenoChoices=as.vector(sqlQuery(dbhandle, query)[[1]])
    }
    if(input$DataViz_DataType=="Protein"){
      query=paste("SELECT distinct gene_symbol  FROM HMDP.ProteinAnnotation.",
input$DataViz_Study,"",paste="")
      DV_PhenoChoices=as.vector(sqlQuery(dbhandle, query)[[1]])
    }
    if(input$DataViz_DataType=="Expression"){  #Text entry for expression
      textInput("DataViz_Pheno", "Please Enter your probesetID")
    } else{ # Select Input for phenotype
      selectInput("DataViz_Pheno", "Select Phenotype", DV_PhenoChoices )}
  })

  #This function differentiates between sub-studies (for instance, male vs female mice)
  output$DataViz_FinalTableSelectUI <-renderUI({
    if(input$DataViz_DataType=="Clinical"){
      DV_temp=allTables[allTables[,4]=="VIEW",]
      DV_FinalTableChoices=DV_temp[DV_temp[,2]=="ClinicalQTL",3]
    }
    if(input$DataViz_DataType=="Expression"){
      DV_temp=allTables[allTables[,4]=="VIEW",]
      DV_FinalTableChoices=DV_temp[DV_temp[,2]=="expressionQTL",3]
    }
    if(input$DataViz_DataType=="Metabolite"){
      DV_temp=allTables[allTables[,4]=="VIEW",]
      DV_FinalTableChoices=DV_temp[DV_temp[,2]=="MetaboliteQTL",3]
    }
    if(input$DataViz_DataType=="Protein"){
      DV_temp=allTables[allTables[,4]=="VIEW",]
      DV_FinalTableChoices=DV_temp[DV_temp[,2]=="ProteinQTL",3]
    }
    DV_FinalTableContenders=DV_FinalTableChoices[grep(input$DataViz_Study,DV_
FinalTableChoices)]

    selectInput("DataViz_ExactView", "Select Table", DV_FinalTableContenders )
  })

  #Finally, now that we have pinpointed the exact phenotype/study combination
desired, we get the data from the server
  DV_GetData <- reactive({  #A reactive function only triggers if a variable
within it changes (in this case, if the button 'DataViz_Calulate' is pressed)
    if(input$DataViz_Calculate==0)
    {return("NULL")} else {
      isolate({  #Nothing within the isolate function "counts" for the reac-
tive function above.  This allows the user to modify what they are looking for
without constantly telling the program to start interacting with the database
        if(input$DataViz_DataType=="Clinical"){
          Group="ClinicalQTL"
        }
```

```
        if(input$DataViz_DataType=="Expression"){
          Group="expressionQTL"
        }
        if(input$DataViz_DataType=="Metabolite"){
          Group="MetaboliteQTL"
        }
        if(input$DataViz_DataType=="Protein"){
          Group="ProteinQTL"
        }
        #We are now going to construct the query to the SQL server.
        query <- paste("SELECT trait_name,rsID,snp_chr,snp_bp_mm10,pvalue  FROM
HMDP.",Group,".",input$DataViz_ExactView," WHERE trait_name='",input$DataViz_
Pheno,"'",sep="")
        print("Query Constructed")
        #and here we actually run the query
        DV_Data=sqlQuery(dbhandle, query)
      })
    }
  })
  #This function makes the Manhattan Plot.  It is separate from the above
function to allow users to modify their Manhattan plot parameters (eg look at
specific chromosomes)
  #without having to re-download the data
  DV_MakeManhattan <- reactive({

    withProgress(message="Constructing Query…",value=0,{ #withProgress allows
for the creation of a progress bar in the GUI.  In this case, its reporting
that the query is being constructed
      DV_Data=DV_GetData()}) #and then getting the data
    print("Data Aquired")
    #print(dim(DV_Data))
    if(DV_Data[1]!="NULL"){ #If there is data…
      print("Running")
      withProgress(value=.5, message="Processing Results…",{ #another update
on the current progress (50% complete)
        subset=DV_Data[,c(3,4,5)] #get relevant values from the output (chro-
mosome, position, p-value)
        if(input$DataViz_Chromosome!="All"){ #if we are NOT looking at all
chromosomes, we need to filter our data
          subset=subset[subset[,1]==input$DataViz_Chromosome,] #Limit to just
the chromosome of interest
          positions=as.numeric(subset[,2])
          #And Limit to the region on the chromosome of interest
          subset=subset[positions>as.numeric(input$DataViz_Lower_
Bound)*1000000,]
          positions=as.numeric(subset[,2])
          subset=subset[positions<as.numeric(input$DataViz_Upper_Bound)*1000000,]
        }
        #Tweak the X and Y chromosome to Chromosomes '20' and '21' respective-
ly
        levels(subset[,1])[levels(subset[,1])=="X"]="20"
        levels(subset[,1])[levels(subset[,1])=="Y"]="21"
        subset=apply(subset,2,as.numeric)
        colnames(subset)=c("CHR","BP","P")
        subset=as.data.frame(subset)
        #At this point we have our final data table.
      })
```

```
      withProgress(value=.9,message="Constructing Plot…",{ #Finally, construct the plot
        print("Constructing Plot")
        DV_Button_Value=DV_Button_Value+1
        #This function is provided in Manhattan.R from http://
GettingGeneticsDone.blogspot.com.  In this case, we are alternating color on
each chromosome and have set the genome-wide
        #singificance like to 10^-5.387, which is the accepted significance line of the HMDP.
        manhattan(subset, colors=c("black","#666666","#CC6600"), main=DV_
Data[1,1], pch=20, genomewideline=5.387, suggestiveline=F)
      })
    }
  })

  #This function is a simple wrapper which takes the output of the above func-
tions (which are reactive and cannot be directly interfaced with the UI) and
makes a continually updating plot for the UI
  DV_Button_Value=0
  output$DataViz_Manhattan <-renderPlot({

    DV_MakeManhattan()
  })

  #This function takes the GWAS results for the phenotype of interest and
packages it for download for later off-line analysis.
  output$DataViz_Download<-downloadHandler(
    filename = "results.gwas", #the default filename
    content = function(file) {
      write.table(DV_GetData(), file,row.names=F,sep="\t")
    })

  #Finally, this function displays the UCSC genome browser in a frame for regions
of less than 10MB in length
  output$DV_GenomeBrowser <-renderUI({
    if(input$DataViz_Chromosome=="All" || input$DataViz_Upper_Bound-
input$DataViz_Lower_Bound>10 )  #If we are looking at more than 10 MB of ge-
nome
    {return(tags$iframe(src="",seamless=T,height=800,width="100%"))} else {
#Return an empty frame.  Otherwise…
      options(scipen=999)    #turn off scientific notation
      #This creates a link to the Mouse UCSC Genome broser for the chromosome and
region of interest.  It will need to be changed for other species, of course.
      temp=paste("http://genome.ucsc.edu/cgi-bin/hgTracks?hgHubConnect.destUrl",
                 "=..%2Fcgi-bin%2FhgTracks&clade=mammal&org=Mouse&db=mm10&position=chr",
                  input$DataViz_Chromosome,"%3A",input$DataViz_Lower_Bound*1000000,
                  "-",input$DataViz_Upper_Bound*1000000,
                  "&hgt.positionInput=enter+position%2C+gene+symbol+or+search+t
erms&knownGene=pack",
                  "&ensGene=hide&xenoRefGene=hide&refGene=hide&ucscRetroAli6=hi
de&mrna=hide",
                 "&intronEst=hide&snp138Common=dense&rmsk=dense&Submit=submit",sep="")
      options(scipen=0) #Turn back on Scientific Notation
      tags$iframe(src=temp,seamless=T,height=800,width="100%") #create a frame
which displays the link above
    }
  })
```

```
  #The following section allows users to create a 'beeswarm' plot to examine
the effect of a single SNP on a phenotype
  #####Beeswarm Section#####
  #these first sections are identical to those for the Manhattan Plots above,
but due to the weirdness of Shiny, are simply repeated here.
  output$Beeswarm_StudyUI<- renderUI({
    if(input$Beeswarm_DataType=="Clinical"){
      BS_StudyChoices=allTables[allTables[,2]=="ClinicalTraitAnnotation",3]
    }
    if(input$Beeswarm_DataType=="Expression"){
      BS_temp=allTables[allTables[,4]=="TABLE",]
      BS_StudyChoices=BS_temp[BS_temp[,2]=="TranscriptAbundance",3]
    }
    if(input$Beeswarm_DataType=="Metabolite"){
      BS_StudyChoices=allTables[allTables[,2]=="MetaboliteAnnotation",3]
    }
    if(input$Beeswarm_DataType=="Protein"){
      BS_StudyChoices=allTables[allTables[,2]=="ProteinAnnotation",3]
    }
    selectInput("Beeswarm_Study", "Select Study", BS_StudyChoices )
  })
  output$Beeswarm_FinalTableSelectUI <-renderUI({
    if(input$Beeswarm_DataType=="Clinical"){
      BS_temp=allTables[allTables[,4]=="VIEW",]
      BS_FinalTableChoices=BS_temp[BS_temp[,2]=="ClinicalTraits",3]
    }
    if(input$Beeswarm_DataType=="Expression"){
      BS_temp=allTables[allTables[,4]=="VIEW",]
      BS_FinalTableChoices=BS_temp[BS_temp[,2]=="TranscriptAbundance",3]
    }
    if(input$Beeswarm_DataType=="Metabolite"){
      BS_temp=allTables[allTables[,4]=="VIEW",]
      BS_FinalTableChoices=BS_temp[BS_temp[,2]=="MetaboliteAbundance",3]
    }
    if(input$Beeswarm_DataType=="Protein"){
      BS_temp=allTables[allTables[,4]=="VIEW",]
      BS_FinalTableChoices=BS_temp[BS_temp[,2]=="ProteinQTL",3]
    }
    # print(input$Beeswarm_Study)
    BS_FinalTableContenders=BS_FinalTableChoices[grep(input$Beeswarm_Study,BS_
FinalTableChoices)]
    # print("Done")
    selectInput("Beeswarm_ExactView", "Select Table", BS_FinalTableContenders )
  })
  output$Beeswarm_PhenotypeUI <- renderUI({
    cur_table=input$Beeswarm_Study
    if(input$Beeswarm_DataType=="Clinical"){
      query=paste("SELECT distinct trait_name  FROM HMDP.ClinicalTraitAnnotati
on.",input$Beeswarm_Study,"",paste="")
      DV_PhenoChoices=as.vector(sqlQuery(dbhandle, query)[[1]])
    }

    if(input$Beeswarm_DataType=="Expression"){ #once again, a drop-down menu
isn't practical here.
      #query=paste("Select probesetID FROM HMDP.TranscriptAbundance.",input$Be
eswarm_ExactView,sep="")
      #DV_TempExpressionQuery=sqlQuery(dbhandle,query)
      #DV_PhenoChoices=as.vector(sqlQuery(dbhandle, query)[[1]])
    }
```

```
    if(input$Beeswarm_DataType=="Metabolite"){
      query=paste("SELECT distinct metabolite_name  FROM HMDP.MetaboliteAnnota
tion.",input$Beeswarm_Study,"",paste="")
      DV_PhenoChoices=as.vector(sqlQuery(dbhandle, query)[[1]])
    }
    if(input$Beeswarm_DataType=="Protein"){
      query=paste("SELECT distinct gene_symbol  FROM HMDP.ProteinAnnotation.",
input$Beeswarm_Study,"",paste="")
      DV_PhenoChoices=as.vector(sqlQuery(dbhandle, query)[[1]])
    }
    if(input$Beeswarm_DataType=="Expression"){
      textInput("Beeswarm_Pheno", "Please Enter your probesetID")
    } else{
      selectInput("Beeswarm_Pheno", "Select Phenotype", DV_PhenoChoices )}

  })


  #With all the inputs set, this actually creates the output once the button
'Beeswarm_Calculate' is pressed.
  BS_GetData <- reactive({
    if(input$Beeswarm_Calculate==0)
    {return("NULL")} else {
      isolate({
        if(input$Beeswarm_DataType=="Clinical"){
          Group="ClinicalTraits"
        }
        if(input$Beeswarm_DataType=="Expression"){
          Group="TranscriptAbundance"
        }
        if(input$Beeswarm_DataType=="Metabolite"){
          Group="MetaboliteAbundance"
        }
        if(input$Beeswarm_DataType=="Protein"){
          Group="ProteinAbundance"
        }
        if(input$Beeswarm_DataType=="Expression"){
          #Expression Data is a little bit different from the other datatypes
in terms of how it is stored, so it is collected in its own conditonal phrase.
          query<- paste("SELECT * FROM HMDP.",Group,".",input$Beeswarm_
ExactView," WHERE probesetID='",input$Beeswarm_Pheno,"'",sep="")
          print("Query Constructed")
          print(query)
          #Get Expression Data
          Pheno_Data=sqlQuery(dbhandle, query)
          query <- paste("SELECT * FROM HMDP.genotypes.MouseDivArray_genotype_
calls_emma_format WHERE rsID='",input$Beeswarm_rsID,"'", sep="")
          print("Query Constructed")
          #Get SNP Info
          Geno_Data=sqlQuery(dbhandle,query)
          Mapping=match(colnames(Pheno_Data),colnames(Geno_Data))
          temp=is.na(Mapping)
          Mapping[temp]=1
          BS_Data=cbind(t(Geno_Data[,Mapping]),t(as.matrix(Pheno_Data)))
        }
        else {
          #The Althersclerosis data is slightly different from the other data,
so this corrects for a difference in column names. (Strain vs Maternal_strain)
```

```
          if(length(grep("Atherosclerosis",input$Beeswarm_ExactView))>0)
{query <- paste("SELECT Maternal_strain, \"",input$Beeswarm_Pheno,"\" FROM
HMDP.",Group,".",input$Beeswarm_ExactView,sep="")}
          else{query <- paste("SELECT Strain, \"",input$Beeswarm_Pheno,"\"
FROM HMDP.",Group,".",input$Beeswarm_ExactView,sep="")}
          print("Query Constructed")
          #Run the query and get the Phenotype
          Pheno_Data=sqlQuery(dbhandle, query)
          #get the SNP data at that particular RSID
          query <- paste("SELECT * FROM HMDP.genotypes.MouseDivArray_genotype_
calls_emma_format WHERE rsID='",input$Beeswarm_rsID,"'", sep="")
          print("Query Constructed")
          Geno_Data=sqlQuery(dbhandle,query)
          Mapping=match(Pheno_Data[,1],colnames(Geno_Data))
          temp=is.na(Mapping)
          Mapping[temp]=1
          BS_Data=cbind(t(Geno_Data[,Mapping]),as.numeric(as.matrix(Pheno_
Data)[,2]))
        }
      })
    }
  })

  #This section actually creates the Beeswarm Plot from the data gathered
above
  output$BS_Plot <- renderPlot({
    if(input$Beeswarm_Calculate==0){
      return ("NULL")} else{
        isolate({ #Nothing after this matters for updating, so it will only
update if the button is pressed.
          BS_Data=BS_GetData()
          print("Data Aquired")
          if(BS_Data[1]!="NULL"){
            print("Running")
            #Remove missing values…
            temp0=!is.na(BS_Data[,1])
            BS_Data=BS_Data[temp0,]
            #Identify which strains are WT or SNP for that rsID
            temp1=BS_Data[,1]==0
            temp2=BS_Data[,1]==1
            BS_Data=rbind(BS_Data[temp1,],BS_Data[temp2,])
            BS_Data=apply(BS_Data,2,as.numeric)
            #calculate a p-value for that rsID and phenotype
            p_val=t.test(as.numeric(BS_Data[BS_Data[,1]==0,2]),as.numeric(BS_
Data[BS_Data[,1]==1,2]))$p.value
            #Create a simple boxplot of data
            boxplot(BS_Data[,2]~BS_Data[,1],xlab=input$Beeswarm_
rsID,ylab=input$Beeswarm_Pheno,main="Genotype x Phenotype",outline=FALSE)
            #overlay it with the beeswarm
            beeswarm(BS_Data[,2]~BS_Data[,1],col = 4, pch = 16,add=TRUE)

          }
        })
      }
  })
```

```
  #The following section identifies the most significant eQTL within a specific
user-defined window of the gene in question.  Mostly useful for answering 'does
this gene have a cis-eQTL?'
  #####ciseQTL section#####
  #While this function isn't specifically needed (the outputed UI could have just
been included in the main UI script, it is here so it can be expanded if needed)
  output$ciseQTL_PhenotypeUI <- renderUI({
    # Needs to be probesetID
    textInput("ciseQTL_Pheno", "Please Enter your probesetID")
  })

  #Once again, a workhorse function to actually get the data only when the
button is pushed.
  CE_GetData <- reactive({
    if(input$ciseQTL_Calculate==0)
    {return("NULL")} else {
      isolate({

        Group="expressionQTL"
        Gene=input$ciseQTL_Pheno
        #form the SQL Query.  This first one simply gets the information about
the gene, namely its location on the genome
        query <-paste("SELECT TOP 1 [dataset],[probesetID],[gene_chr],[gene_
start_bp],[gene_end_bp],[snp_chr],[snp_bp_mm10],[pvalue] from HMDP.Unified.QTL_
AllInfo WHERE probesetID='",Gene,"'",sep="")
        Pheno_Data=sqlQuery(dbhandle,query)
        Gene_chr=Pheno_Data[,3]
        SNP_Lower=Pheno_Data[,4]-(input$ciseQTL_Window*1000000)
        SNP_Upper=Pheno_Data[,5]+(input$ciseQTL_Window*1000000)
        #This next query grabs all SNPs for that particular gene within that
window from all studies
        query <-paste("SELECT [dataset],[probesetID],[snp_chr],[snp_bp_
mm10],[pvalue] from HMDP.Unified.QTL_AllInfo WHERE probesetID='",Gene,"' and
snp_chr='",Gene_chr,
                      "' and snp_bp_mm10>",SNP_Lower," and snp_bp_mm10<",SNP_Upper,sep="")
        #A progress bar to get the data
        withProgress(value=0,message="Getting Data",{
          Pheno_Data=sqlQuery(dbhandle,query)
        })
        #Now that we have the data, we need to find the best eQTL for each
study
        withProgress(value=.2,message="Generating Table",{
          table_names=names(table(Pheno_Data[,1]))  #get study names
          output=c()
          temp_count=0
          for(i in table_names){ #for each study
            temp_count=temp_count+1
            incProgress(amount=temp_count/length(table_names)*.8) #slowly
increment the % complete from 0% to 80%
            temp=grep(i,Pheno_Data[,1])
            temp_data=Pheno_Data[temp,c(1,3,4,5)] #extract study from master
table
            temp=which.min(as.numeric(temp_data[,4])) #find minimum value
            output=rbind(output,temp_data[temp,]) #add minimum row to total
output
          }
          rownames(output)=NULL
        })
```

```
        output #return output.
      })
    }
  })
  #Creates the actual output table for the cis_eQTL data
  output$ciseQTL_Table <-  renderDataTable({
    if(input$ciseQTL_Calculate==0){
      return (NULL)} else{
        isolate(CE_GetData())
      }
  })
  #Creates a downloadable table with the cis_eQTL data
  output$ciseQTL_Download<-downloadHandler(
    filename = "results.txt",
    content = function(file) {
      write.table(isolate(CE_GetData()), file,row.names=F,sep="\t")
    })

  #This section looks for all loci (for different phenotypes/tissues/studies)
which overlap a particular region.  Useful for identifying relationships be-
tween different phenotypes
  #####Overlapping Loci section#####

  #There are two means of looking for overlap which are supported.  The first is
to provide a genomic interval of interest, the second to provide a specific rsID.
  #The following two lines are only rendered in the UI if it is in "window"
format.
  output$Overlap_chr <-renderUI({if(input$Overlap_window_or_rsID=="window")
{selectInput("Overlap_Chr","Select Chromosome",c(1:19,"X"))}})
  output$Overlap_LB  <-renderUI({if(input$Overlap_window_or_rsID=="window")
{numericInput("Overlap_LB","Lower Bound (in MB)",value=10,min=0)}})
  #This last one is *either* the upper bound OR the rsID.
  output$Overlap_additional <-renderUI({
    if(input$Overlap_window_or_rsID=="window"){
      numericInput("Overlap_UB","Upper Bound (in MB)",value=15,min=0)
    } else {
      textInput("Overlap_rsID","Please enter a SNP rsID")
    }
  })

  #The Workhorse function for this section
  OL_GetData <- reactive({
    if(input$Overlap_Calculate==0)
    {return("NULL")} else { #if the button is pushed…
      isolate({
        print("Generating Overlap Table, Please Wait…")
        if(input$Overlap_window_or_rsID=="window"){ #If we are in Window format
          withProgress(value=0,message="Obtaining Results…",{
            #There is an option to either include eQTLs or exclude them.  It
is significantly faster to NOT include eQTLs.
            if(input$Overlap_includeGenes){  #if we are including eQTLs, we
draw from the entire unified QTL table.
              query <- paste("Select [dataset],[probesetID],[gene_
symbol],[rsID],[snp_chr],[snp_bp_mm10],[LD_block_start_mm10],",
                             "[LD_block_end_mm10],[pvalue] from Unified.QTL_
AllInfo WHERE snp_chr='",input$Overlap_Chr,"' and snp_bp_mm10>",
                             input$Overlap_LB*1000000," and snp_bp_
mm10<",input$Overlap_UB*1000000," and pvalue<",input$Overlap_threshold,sep="")
```

```
              Pheno_Data=sqlQuery(dbhandle, query)
            } else { #if we are excluding eQTLs, we filter out the eQTLs as
part of the query
              query <- paste("Select [dataset],[category],[probesetID],[gene_
symbol],[rsID],[snp_chr],[snp_bp_mm10],[LD_block_start_mm10],",
                              "[LD_block_end_mm10],[pvalue] from Unified.QTL_
AllInfo WHERE snp_chr='",input$Overlap_Chr,"' and snp_bp_mm10>",
                              input$Overlap_LB*1000000," and snp_bp_
mm10<",input$Overlap_UB*1000000," and pvalue<",input$Overlap_threshold,
                              " and category!='expression QTL'",sep="")
              Pheno_Data=sqlQuery(dbhandle, query)
              Pheno_Data=Pheno_Data[,-2] #To make the rest of the code work,
we remove the "category" column from the data.
            }


            print("Data Aquired, Analyzing.  Please Wait…")
            all_names=apply(Pheno_Data[,c(1:2)],1,paste,collapse="") #We wish
to find all of the unique peaks in the returned data
            unique_names=names(table(all_names))
          })
          print(paste("There are ",length(unique_names)," peaks!",sep=""))

          #We now process the results to find the minimum p-value for each phe-
notype peak.
          output=c()
          withProgress(value=0,message="Processing Peaks…",{
            for(i in 1:length(unique_names)){
              # print(i)
              if(i%%50==0){incProgress(50/length(unique_names),detail=paste(i,"
Peaks Processed",sep=""))}
              cur_name=unique_names[i]  #for the ith unique name
              temp=grep(cur_name,all_names) #find those names in the main out-
put (non-unique)
              temp_array=Pheno_Data[temp,] #form a subset from just those rows
              temp=which.min(as.numeric(temp_array[,9])) #find the minimum
pvalue
              winner=temp_array[temp,] #declare THAT particular SNP the 'win-
ner'
              winner=winner[,c(1:4,9)] #extract the important information
(study, probeID/identifier, gene symbol, rsID, p value)
              output=rbind(output,winner) #add the winner to the final output

            }
          })

        } else { #if we are looking at a specific rsID
          withProgress(value=0,message="Obtaining Results…",{
            #We first find the precomputed linkage disequilibrium block for that
SNP
            query <- paste("Select top 1 [dataset],[probesetID],[gene_
symbol],[rsID],[snp_chr],[snp_bp_mm10],[LD_block_start_mm10],",
                            "[LD_block_end_mm10],[pvalue] from Unified.QTL_
AllInfo WHERE rsID='",input$Overlap_rsID,"'",sep="")
            #print(query)
            Pheno_Data=sqlQuery(dbhandle, query)
            print(Pheno_Data)
            incProgress(.5,detail="part 2")
```

```
          snp_chr=Pheno_Data[5] #chr of LD block
          lower=Pheno_Data[7] #lower bound of LD block
          upper=Pheno_Data[8] #upper bound of LD block
         #And now we basically do the same thing as when we were doing the window.
          query <- paste("Select [dataset],[category],[probesetID],[gene_
symbol],[rsID],[snp_chr],[snp_bp_mm10],[LD_block_start_mm10],",
                         "[LD_block_end_mm10],[pvalue] from Unified.QTL_
AllInfo WHERE snp_chr='",snp_chr,"' and snp_bp_mm10>",
                         lower," and snp_bp_mm10<",upper," and
pvalue<",input$Overlap_threshold,sep="")
          #print(query)
          Pheno_Data=sqlQuery(dbhandle, query)
        })
        if(input$Overlap_includeGenes){
          Pheno_Data=Pheno_Data[,-2]
        } else {
          tokeep=Pheno_Data[,2]!="expression QTL"
          Pheno_Data=Pheno_Data[tokeep,]
          Pheno_Data=Pheno_Data[,-2]
        }
        print("Data Aquired, Analyzing.  Please Wait…")
        withProgress(value=0,message="Processing Peaks…",{
          all_names=apply(Pheno_Data[,c(1:2)],1,paste,collapse="")
          unique_names=names(table(all_names))
          print(paste("There are ",length(unique_names)," peaks!",sep=""))
          output=c()
          for(i in 1:length(unique_names)){
            # print(i)
            if(i%%50==0){incProgress(50/length(unique_names),detail=paste(i,"
Peaks Processed",sep=""))}
            cur_name=unique_names[i]
            temp=grep(cur_name,all_names)
            temp_array=Pheno_Data[temp,]
            temp=which.min(as.numeric(temp_array[,9]))
            winner=temp_array[temp,]
            winner=winner[,c(1:5,9)]
            output=rbind(output,winner)

          }
        })
      }
      #Finally, we prepare the data for output by tweaking the significant
digits and scientific notation of the p values.
      withProgress(value=.9,message="Outputing…",{
        vals=as.matrix(output[,5])
        vals=as.numeric(vals)
        format(vals,digits=3,scientific=T)
        vals=as.character(vals)
        output[,5]=vals
      })
      #and we return that final output
      output


    })}
  })
```

```
  #This function simply takes the output of the workhorse function above and
repackages it in a form acessible to the GUI.
  output$Overlap_Table <-  renderDataTable({

    if(input$Overlap_Calculate==0){
      return (NULL)} else{

        isolate(OL_GetData())
      }

  })

  #This function creates a downloadable file of the table generated above
  output$Overlap_Download<-downloadHandler(
    filename = "results.txt",
    content = function(file) {
      write.table(isolate(OL_GetData()), file,row.names=F,sep="\t")
    })


  #This section creates a visual depiction of the Linkage Disequilibrium of
a specific region of the genome or, alternately, provides the precomputed LD
block around a provided rsID.
  #####LD Block Section#####
  #Like the section above, this is designed to take either a genomic interval
of interest OR a specific rsID.
  output$LD_chr <-renderUI({if(input$LD_window_or_rsID=="window")
{selectInput("LD_Chr","Select Chromosome",c(1:19,"X"))}})
  output$LD_LB  <-renderUI({if(input$LD_window_or_rsID=="window")
{numericInput("LD_LB","Lower Bound (in MB)",value=10,min=0)}})
  output$LD_additional <-renderUI({
    if(input$LD_window_or_rsID=="window"){
      numericInput("LD_UB","Upper Bound (in MB)",value=15,min=0)
    } else {
      textInput("LD_rsID","Please enter a SNP rsID")
    }
  })
  output$LD_MAFCutoff <-renderUI({if(input$LD_window_or_rsID=="window")
{numericInput("LD_MAF","Minor Allele Frequency Cutoff",value=.05,min=0)}})

  #First function for the 'simple' case where we are dealing with a rsID.
  LD_GetData_rsID <- reactive({
    if(input$LD_Calculate==0 || input$LD_window_or_rsID=="window")  #if we are
looking for the LD within an interval, return nothing.
    {return ("NULL")} else { #otherwise… all we need to do is grab that specific
SNP from our database…
      query <- paste("Select top 1 [dataset],[rsID],[snp_chr],[snp_bp_mm10],[LD_
block_start_mm10],",
                     "[LD_block_end_mm10] from Unified.QTL_AllInfo WHERE
rsID='",input$LD_rsID,"'",sep="")
      #print(query)
      withProgress(value=0,message="Getting Data",{
        Pheno_Data=sqlQuery(dbhandle, query)
        val=Pheno_Data[2]
        print(val)
        val=val[[1]]
      })
```

```
      #and print out specific values found within its entry
      outtext=paste("The SNP ",val," Located at Chr",Pheno_Data[3],":",Pheno_
Data[4]," Has a proposed LD window of ",
                    (as.numeric(Pheno_Data[6])-as.numeric(Pheno_Data[5]))," bp
spanning from ",Pheno_Data[5]," to ",Pheno_Data[6],sep="")
      outtext
    }
  })

  #The more complicated situation is where instead of looking at a particular
rsID we are interested in visualizing the LD structure within a particular re-
gion of the genome.
  LD_GetData_Window <- reactive({
    if(input$LD_Calculate==0 || input$LD_window_or_rsID=="rsID") #if we are
looking at just one rsID… output nothing.
    {return ("NULL")} else {
      print("Beginning")
      options(scipen=999) #turn off scientific notation
      #construct our query to extract out all the SNPs (but not the geno-
types!) within the region
      query<- paste("Select [snp_chr],[rsID],[snp_bp_mm10] from genotypes.
MouseDivArray_genotype_calls_plink_format where snp_chr='",
                    input$LD_Chr,"' and snp_bp_mm10>'",input$LD_LB*1000000,"'
and snp_bp_mm10<'",input$LD_UB*1000000,"'",sep="")
      Pheno_Data=sqlQuery(dbhandle,query)
      options(scipen=0) #return scientific notation to normal

      SNPs=Pheno_Data[,2] #get SNP names
      SNPs=paste(SNPs,collapse="' OR rsID='") #create a master search entry
which looks like 'rsID="SNPA" OR rsID="SNPB" OR…'

      print("Getting SNPs")
      withProgress(value=0,message="Getting SNPS",{
        #Here we actually get out the genotypes for the SNPs identified above
(this is done to save a significant amount of time)
       query<- paste("Select genotypes.MouseDivArray_genotype_calls_emma_format.*,
rsID AS Expr1 from genotypes.MouseDivArray_genotype_calls_emma_format WHERE ",
                      "rsID='",SNPs,"'",sep="")
        #massage the data into the right format.
        Pheno_Data=sqlQuery(dbhandle,query)
        Pheno_Data=Pheno_Data[,-1]
        Pheno_Data=Pheno_Data[,-ncol(Pheno_Data)]
        positions=Pheno_Data[,1]
        Pheno_Data=Pheno_Data[,-1]
        #calculate the minor allele frequency of each SNP and remove those who
whose MAFs are less than the predefinited cutoff
        sums=apply(Pheno_Data,1,sum,na.rm=TRUE)
        ngoodcol=table(is.na(Pheno_Data[1,]))[1]
        MAFS=sums/ngoodcol
        tokeep=MAFS>input$LD_MAF
        Pheno_Data=Pheno_Data[tokeep,]
        positions=positions[tokeep]
      })
      #Now that we have the exact phenotypes we care about, we can calulate
the relationship of each of these SNPs to one another
      withProgress(value=.5,message="Calculating Correlations",{
        print("Calculating Correlations")
```

```
        cortable=corFast(t(Pheno_Data),use="pairwise.complete.obs")
        cortable[which(is.na(cortable))]=0
        cortable2=cortable^2
        rownames(cortable2)=positions
        colnames(cortable2)=positions
        #cortable2=1-cortable2
      })
      #and finally create the output PDF, which is simply a heatmap of the cor-
relations of each SNP to each other SNP within the window.
      print("Creating Plot")
      title=paste("LD Block Structure: Chr ",input$LD_Chr," ",input$LD_LB, "
to ",input$LD_UB," Mb",sep="")
      heatmap.2(cortable2, Rowv=FALSE,Colv=FALSE, dendrogram="none", col=heat.
colors(75), scale="none",
                key=FALSE, symkey=FALSE, density.info="none", trace="none", ce
xRow=0.5,cexCol=.15,main=title)

    }
  })

  #Two output functions, one for the rsID version and one for the window version.
  output$LD_rsIDOut <- renderText({
    if(input$LD_Calculate==0 || input$LD_window_or_rsID=="window"){return("")}
else{isolate(LD_GetData_rsID())}
  })

  output$LD_windowOut <- renderPlot({
    if(input$LD_Calculate==0 || input$LD_window_or_rsID=="rsID"){return(NULL)}
else {isolate(LD_GetData_Window())}
  })

  #A very simple section which takes a gene name as input and opens up a
browser window to the Wellcome Trust Mouse Genomes SNP Query site for that
gene and the strains in the HMDP
  #Originally, this actually opened in a frame, but recent changes to the
Wellcome Trust site mean that a new tab/window is now necessary.
  #####NONSYNNONYMOUS SNP SECTION#####
  output$NonSynnon_Result <-renderUI({
    if(input$NonSynnon_Calculate==0){return(NULL)} else {
      isolate({
        temp=paste("http://www.sanger.ac.uk/sanger/Mouse_SnpViewer/rel-
1505?gene=",
                   input$NonSynnon_Gene,"&context=0&loc=&release=rel-
1505&sn=frameshift_variant&sn=missense_variant&",
                   "sn=splice_region_variant&sn=stop_gained&sn=stop_
lost&sv=complex_events&sv=copy_number_gain&sv=",
                   "deletion&sv=insertion&sv=inversion&st=129s1_
svimj&st=a_j&st=akr_j&st=balb_cj&st=btbr_t__itpr3tf_j",
                   "&st=bub_bnj&st=c3h_hej&st=c57l_j&st=c58_j&st=cba_j&st=dba_
2j&st=fvb_nj&st=i_lnj&st=kk_hij&st=lp_j",
                   "&st=nod_shiltj&st=nzb_b1nj&st=sea_gnj",
                   sep="")
        browseURL(temp)
        #tags$iframe(src=temp,seamless=F,height=1024,width=1600)
      })
    }
  })
```

```
  #This section creates a heatmap of all the values for a particular gene
across all the studies/strains of the HMDP
  #####Vizualize Values Across Strains Section#####
  output$VVAS_StudyUI<- renderUI({
    if(input$VVAS_DataType=="Phenotype"){
      VVAS_StudyChoices=c("Not", "Implemented","Yet")  #There are some significiant
challenges here.  See systems.genetics.ucla.edu for an eventual update…
      selectInput("VVAS_Pheno", "Select Study", VVAS_StudyChoices )
    }
    if(input$VVAS_DataType=="Gene"){
      textInput("VVAS_Gene", "Please Enter your probesetID OR Gene Name")
    }
  })

  #This section of the UI lets users select a subset of all of the studies to examine
  output$VVAS_SelectExperimentsUI <-renderUI({

    selectors=allTables[allTables[,2]=="TranscriptAbundance",]
    selectors=selectors[selectors[,4]=="VIEW",]
    selectors=selectors[,3]


    checkboxGroupInput("VVAS_Experiments", "Select Experiments to Include",
selectors, selected=selectors)
  })
  #Workhorse function for this section
  VVAS_Output <-reactive({
    if(input$VVAS_Calculate==0 || input$VVAS_DataType=="Phenotype")
    {return ("NULL")} else {
      withProgress(value=0,message="Setting up…",{
        #prepare the specific studies we are interested in…
        Table_subset=input$VVAS_Experiments
        Table_subset=paste0("HMDP.TranscriptAbundance.",Table_subset)
        Table_subset=as.matrix(Table_subset)
        })
      withProgress(value=0,message="Generating Table…",{
       #here we are creating the first query of our output, to which the rest will be added.
        gene_query=paste("(ProbesetID='",input$VVAS_Gene,"' OR gene_
symbol='",input$VVAS_Gene,"')",sep="")

       query <- paste("SELECT * FROM ",Table_subset[1]," WHERE ", gene_query,paste="" )
      })
      print(paste0("Fetching Data From ",Table_subset[1]))
      withProgress(value=1/length(Table_subset),message=paste0("Fetching Data
From ",Table_subset[1]),{
        #Actually get the data for the first row and process
        #The eventual result should have the gene name as a row and all
strains of interest as columns.  If more than one probeset is returned (if us-
ing gene name instead of probesetID)
        #then the most highly expressed value will be added.
        Pheno_Data=as.matrix(sqlQuery(dbhandle, query))
        Pheno_Data=Pheno_Data[,-c(1:2)]
        Pheno_Data=as.matrix(Pheno_Data)
        if(ncol(Pheno_Data)==1){
          Pheno_Data=t(Pheno_Data)
        }
```

```
        if(nrow(Pheno_Data)>1){
          Pheno_Data=apply(Pheno_Data,2,as.numeric)
          averages=apply(Pheno_Data[,-c(1:2)],1,mean,na.rm=TRUE)
          Pheno_Data=Pheno_Data[which.max(averages),]
          Pheno_Data=t(as.matrix(Pheno_Data))
        }
        rownames(Pheno_Data)=strsplit(strsplit(Table_subset[1],".",fixed=T)
[[1]][3],"_",fixed=T)[[1]][1]
        gene_data=Pheno_Data
      })
      #now we do the same thing for every other study of interest, merging the
results with the growing master output table
      for(i in 2:length(Table_subset)){
       query <- paste("SELECT * FROM ",Table_subset[i]," WHERE ", gene_query,paste="" )
        print(paste0("Fetching Data From ",Table_subset[i]))
        withProgress(value=i/length(Table_subset),message=paste0("Fetching
Data From ",Table_subset[i]),{
          Pheno_Data=as.matrix(sqlQuery(dbhandle, query))
          Pheno_Data=Pheno_Data[,-c(1:2)]
          Pheno_Data=as.matrix(Pheno_Data)
          if(ncol(Pheno_Data)==1){Pheno_Data=t(Pheno_Data)}

          if(nrow(Pheno_Data)==0){ #we need a special case if the gene/probe
isn't found in the study's array.  In this case, we just add a row of NAs.
            new_val= strsplit(Table_subset[i],".",fixed=T)[[1]][3]
            temp=c(rownames(gene_data),new_val)
            new_row=rep(NA,ncol(gene_data))
            gene_data=rbind(gene_data,new_row)
            rownames(gene_data)=temp
          } else {
            if(nrow(Pheno_Data)>1){
              Pheno_Data=apply(Pheno_Data,2,as.numeric)
              averages=apply(Pheno_Data[,-c(1:2)],1,mean,na.rm=TRUE)
              Pheno_Data=Pheno_Data[which.max(averages),]
              Pheno_Data=t(as.matrix(Pheno_Data))
            }
            rownames(Pheno_Data)=strsplit(Table_subset[i],".",fixed=T)[[1]][3]
            temp=c(rownames(gene_data),rownames(Pheno_Data))
            gene_data=merge(gene_data,Pheno_Data,all=TRUE,sort=FALSE)
            rownames(gene_data)=temp
          }
        })
      }
      #and finally, we return the combined data
      gene_data


    }
  })

  #Unused in the final code, this allows for testing of which samples will be
included for the strain select portion of the UI.
  output$TEST_Checkbox <-renderText({
    res=input$VVAS_SelectStrains
    strain_classes=input$VVAS_SelectStrains
    strains=c()
    for(q in 1:length(strain_classes)){
```

```
      temp=allStrains[allStrains[,2]==strain_classes[q],1]
      strains=c(strains,temp)
    }

    print(strains)
    })


  #This section actually outputs the heatmap of all the expression values
  output$VVAS_Plot <- renderPlot({
    if(input$VVAS_Calculate==0){
      return ("NULL")} else{

        isolate({
          VVAS_Data=VVAS_Output() #get the data…
          print("Data Aquired")
          if(VVAS_Data[1]!="NULL"){ #if there is data…
            print("Running")

            strain_classes=input$VVAS_SelectStrains #get which strain classes
(mouse panels) we are interested in from the GUI.
            strains=c()
          for(q in 1:length(strain_classes)){ #create a master list of strains of interest.
              temp=allStrains[allStrains[,2]==strain_classes[q],1]
              strains=c(strains,temp)
            }
            #Filter output for only the strains of interest
            temp=match(strains,colnames(VVAS_Data))
            temp=temp[!is.na(temp)]
            print(temp)
            VVAS_Data=VVAS_Data[,temp]

            #and now actually generate the heatmap.
            withProgress(value=0,message="Generating Figure…",{

              gene_data=VVAS_Data
              pheno_names=rownames(gene_data)
              strain_names=colnames(gene_data)
              gene_data=gene_data[,order(strain_names)]
              strain_names=strain_names[order(strain_names)]

              gene_data=apply(gene_data,2,as.numeric)
              rownames(gene_data)=pheno_names
            heatmap.2(gene_data,Rowv=FALSE,Colv=FALSE,dendrogram="none",trace="none",
                      col=greenred(100),na.color="grey",keysize=1.2,density.
info="none",margins=c(5,9))
            })
          }
        })
      }
  })

  #This function allows users to download the values plotted in the heatmap.
  output$VVAS_Download<-downloadHandler(
    filename = "results.txt",
```

```
    content = function(file) {
      write.table(isolate(VVAS_Output()), file,row.names=T,sep="\t")
    })


  #This section examines the entirety (or a subset) of the data currently
available to find significant/suggestive correlations between a phenotype of in-
tetrest and other phenotypes, studies, tissues, etc.
  #####Find Correlations#####
  #This first function allows the user to select a subset of the entire data
to look for correlations in.  Obviously, the fewer experiments, the faster it
goes.
  output$FC_SelectExperimentsUI <-renderUI({

    selectors=allTables[allTables[,2]=="Correlations",]  #find all correlations
in allTables
    selectors=selectors[selectors[,4]=="VIEW",] #find all of those correlations
which are views (to avoid duplicates)
    selectors=selectors[,3] #get names

    #We have to clean up the names a little bit (namely remove anything after
the first "_"), so here is a quick function to do so.
    retElement <- function(x,num){
      temp=strsplit(x,"_")
      temp=temp[[1]][num]
      return(temp)

    }
    selectors=sapply(selectors,retElement,1)
    selectors=names(table(selectors))

    checkboxGroupInput("FC_Experiments", "Select Experiments to Include", se-
lectors, selected=selectors)
  })

  #The workhorse function which actually finds the correlations
  FC_GetResults<- reactive({
    experiments=allTables[allTables[,2]=="Correlations",] #get all correlations
    experiments=experiments[experiments[,4]=="VIEW",]
    experiments=experiments[,3]
    if(!input$FC_Include_Probes){ #we have the option to keep or remove all the eQTLs.
     temp=sapply(experiments,retElement,3)!="trx" #if we don't want the eQTLs, we filter them out.

      countElement <-function(x){
        temp=strsplit(x,"_")
        temp=length(temp[[1]])
        return(temp)
      }
      t2=sapply(experiments,countElement)!=4

      temp= temp | t2
      experiments=experiments[temp]
    }
    all_experiments=experiments
```

```
    #We now filter ALL experiments by the ones we selected above that we wish to keep
    experiments=c()
    for(i in 1:length(input$FC_Experiments)){
      temp=input$FC_Experiments[i]
      temp=paste(temp,"_",sep="")
      temp=grep(temp,all_experiments,fixed=T)
      temp=all_experiments[temp]
      experiments=c(experiments,temp)
    }

    #Trim off the 'AllInfo' part.

    for(i in 1:length(experiments)){
      temp=strsplit(experiments[i],"_")[[1]]
      temp=temp[-length(temp)]
      temp=paste(temp,collapse="_")
      experiments[i]=temp
    }

    #Make the MASSIVE experiment filter for the eventual query

    exp_filter="(dataset='"
    for(i in 1:(length(experiments)-1)){
      temp=experiments[i]
      exp_filter=paste0(exp_filter,temp,"' OR dataset='")
    }
    exp_filter=paste0(exp_filter,experiments[length(experiments)],"')")

   # print(exp_filter)
    #The much smaller phenotype_filter
    pheno_query=paste("(ProbesetID_1='",input$FC_Input,"' OR gene_
symbol='",input$FC_Input,"' OR clinical_trait_1='",input$FC_Input,"' OR
metabolite_1='",input$FC_Input,"' OR protein_1='",input$FC_Input,"')",sep="")

   # print(pheno_query)
    #The tiny pvalue filter
    pval_filter=paste0("pvalue<='",input$FC_threshold,"'")
   # print(pval_filter)
    #and finally we combine everything together to create our master SQL query.
    final_query=paste0("SELECT * FROM Unified.Correlations_AllInfo WHERE ",exp_
filter," AND ",pheno_query," AND ",pval_filter)

    print(final_query)
    withProgress(value=0,message="Generating Results… this may take some
time.",{ #It really might.  Working on a way to improve speed now.
      FC_Data=sqlQuery(dbhandle, final_query) #and here we actually are getting the results
    })


    outdata=c()
    withProgress(value=0,message="Processing Results…",{
      #our correlations can be with all sorts of different things… a gene, a
phenotype, a metabolite, a protein, etc, etc.  Our initial input can be any of
those things as wel
```

```
      #as a result, we have to figure out which entries in our unified correla-
tion database is actually filled
      for(i in 1:nrow(FC_Data)){ #for each row of the correlations we've downloaded
        incProgress(1/nrow(FC_Data))
        cur_row=FC_Data[i,] #extract that row
       tokeep=c(2:4) #keep a few columns that are always needed (type of correlation,
tissue of interest, study) and then look to see which other columns are filled
        if(!is.na(cur_row[5])){ tokeep=c(tokeep,5,6)} #gene 1
        if(!is.na(cur_row[11])){ tokeep=c(tokeep,11,12)} #gene 2
        if(!is.na(cur_row[17])){ tokeep=c(tokeep,17,18)} #phenotype 1
        if(!is.na(cur_row[20])){ tokeep=c(tokeep,20,21)} #phenotype 2
        if(!is.na(cur_row[23])){ tokeep=c(tokeep,23,24)} #metabolite 1
        if(!is.na(cur_row[25])){ tokeep=c(tokeep,25,26)} #metabolite 2
        if(!is.na(cur_row[27])){ tokeep=c(tokeep,27,27)} #protein 1
        if(!is.na(cur_row[28])){ tokeep=c(tokeep,28,28)} #protein 2
        tokeep=c(tokeep,29,30) #and we want to keep the last two values as
well (correlation score and pvalue)
        cur_row=cur_row[tokeep] #and now we actually filter the row to the val-
ues we care about
        names(cur_row)=c("class","tissue","study","Pheno1_ID","Pheno1_
Info","Pheno2_ID","Pheno2_Info","bicor","pvalue") #add names to those values
        outdata=rbind(outdata,cur_row) #and add it to our master output.
      }
    })
    #this output is a condensed form of the SQL query which removes empty
spaces and is better for visualization
    outdata

  })

  #the output function for the data above.
  output$FC_Output <-renderDataTable({
    if(input$FC_Calculate==0){
      return (NULL)} else{

        isolate(FC_GetResults())
      }

  })
  #and a downloader to allow downloading of all correlations.
  output$FC_Download<-downloadHandler(
    filename = "results.txt",
    content = function(file) {
      write.table(isolate(FC_GetResults()), file,row.names=F,sep="\t")
    })
})
#####EXTRAS#####
#Takes a string x, splits it and reurns  the num-th element.
retElement <- function(x,num){
  temp=strsplit(x,"_")
  temp=temp[[1]][num]
  return(temp)
}
```

7 ui.R

```
# A Graphical User Interface for querying a genetics SQL Database using Shiny in R
# Version: 0.7
# Last Modified: 12/9/15
#
# The following is an implementation of a GUI using the Shiny package in
Rstudio.  Shiny programs have two scripts associated with them.  This script,
ui.R, controls the appearance of
# The GUI and provides inputs to and displays outputs from Server.R which con-
tains the actual functions.
#for details on how this page's layout works, please see http://shiny.rstudio.
com/tutorial/ and http://shiny.rstudio.com/reference/shiny/latest/
shinyUI(fluidPage(
  titlePanel("Welcome to the HMDP Database Shiny Server v0.7"),
  tabsetPanel(
    tabPanel("Start Here/Login",
             h6("Welcome to the first iteration of the searchable HMDP Database.
Please click a relevant tab to begin."),
             textInput("Password","Enter password for full access"),
             actionButton("Password_Go","Login"),
             textOutput("PassOK")),
    tabPanel("Visualize GWAS Result",
             sidebarLayout(
               sidebarPanel(
      selectInput("DataViz_DataType",label=h3("Select a type of data"),choices = c
("Clinical","Expression","Metabolite","Protein")), #a selectInput is a dropdown menu
            htmlOutput("DataViz_StudyUI"), #an 'htmlOutput' is actually a way
to create dynamic inputs.  In this case, Server.R is taking the selection from
above and creating a new
                                              #selectInput populated with all the
studies which have that type of data
      htmlOutput("DataViz_FinalTableSelectUI"), #then this one is allowing for
fine tuning of the selection (typically selecting which gender of mice to examine)
      htmlOutput("DataViz_PhenotypeUI"), #and finally this one gives you a list
of all possible phenotypes that can be used.
            actionButton("DataViz_Calculate","Create Manhattan Plot"),  #This
is a button which, when clicked, tells Server.R to start calculating.

      selectInput("DataViz_Chromosome",label="Which Chromosome?",choices=c("Al
l",c(1:19),"X"),selected="All"), #another select input
            numericInput("DataViz_Lower_Bound","Lower Bound (In MB)",1,min=0),
#a numeric input which will take any number
            numericInput("DataViz_Upper_Bound","Upper Bound (In MB)",999,min=0)
               ),
              mainPanel(downloadButton('DataViz_Download', 'Download These
Results') , #creates a download button which takes a created file from Server.R
                          plotOutput('DataViz_Manhattan'), #creates a plot
                          h5("At distances of less than 10Mb, the UCSC Genome
Browser will Appear Below."),
                          htmlOutput("DV_GenomeBrowser"))) #once again an ht-
mlOutput, but in this case it really is an output, namely a visualization of
the UCSC genome browser
    ),
    tabPanel("Create Beeswarm Plot",
             sidebarLayout(
               sidebarPanel(
```

```
                selectInput("Beeswarm_DataType",label=h3("Select a type of
data"),choices = c("Clinical","Expression","Metabolite","Protein")),
                htmlOutput("Beeswarm_StudyUI"),
                htmlOutput("Beeswarm_FinalTableSelectUI"),
                htmlOutput("Beeswarm_PhenotypeUI"),
                textInput("Beeswarm_rsID","Enter your SNP of
choice",value=""), #will take any string as an input
                actionButton("Beeswarm_Calculate","Create Plot") ),
              mainPanel(plotOutput('BS_Plot'))
              )
            ),
    tabPanel("Visualize Values Across Strains and Tissues",
            sidebarLayout(
             sidebarPanel(
               selectInput("VVAS_DataType",label=h3("Select a type of
data"),choices = c("Phenotype","Gene")),
               htmlOutput("VVAS_StudyUI"),
               htmlOutput("VVAS_SelectExperimentsUI"),
             checkboxGroupInput("VVAS_SelectStrains",label="Strain Groups",choices=c("In
bred","AxB","BxA","BxD","BxH","CxB"),selected=c("Inbred","AxB","BxA","BxD","BxH","CxB")),
               actionButton("VVAS_Calculate","Create Plot") ),
             mainPanel(downloadButton('VVAS_Download', 'Download These
Results'),
                       plotOutput('VVAS_Plot')
                       #,textOutput("TEST_Checkbox")
                       )
            ) ),
    tabPanel("Nonsynnonymous SNPs",
            sidebarLayout(
              sidebarPanel(
                textInput("NonSynnon_Gene","Enter Gene (SYMBOL FOR NOW)"),
                actionButton("NonSynnon_Calculate","Run")
                ),
              mainPanel(htmlOutput("NonSynnon_Result")))),
    tabPanel("cis-eQTLs",
            sidebarLayout(
              sidebarPanel(
                htmlOutput("ciseQTL_PhenotypeUI"),
                numericInput("ciseQTL_Window","Size of cis-eQTL window in
MB",min=0,value=2),
                actionButton("ciseQTL_Calculate","Create Table")
                ),
              mainPanel(dataTableOutput('ciseQTL_Table'),
                    downloadButton('ciseQTL_Download', 'Download These Results')))
              ),
    tabPanel("Gene/Phenotype Correlations",sidebarLayout(
      sidebarPanel(
        textInput("FC_Input","Enter your gene or phenotype name"),
        htmlOutput("FC_SelectExperimentsUI"),
        numericInput("FC_threshold","P-value
threshold",min=0,value=.0000042,max=1),
        checkboxInput("FC_Include_Probes","Include Genes?",value=TRUE), #a simple
checkbox for TRUE/FALSE statments.  In this case, should genes be included when
calculating correlations?
        actionButton("FC_Calculate","Create Table")
      ),
      mainPanel(dataTableOutput("FC_Output"),
               downloadButton('FC_Download', 'Download These Results'))
    )),
```

```
    tabPanel("Overlapping Loci",
             sidebarLayout(
               sidebarPanel(
                 selectInput("Overlap_window_or_rsID","Please select to
begin",c("window","rsID"))
                ,htmlOutput("Overlap_chr"),
                htmlOutput("Overlap_LB"),
                htmlOutput("Overlap_additional"),

                numericInput("Overlap_threshold","P-value threshold",min=0,val
ue=.0000042,max=1),
               checkboxInput("Overlap_includeGenes","Include eQTLs?",value=FALSE),
                actionButton("Overlap_Calculate","Create Table")
               ),
               mainPanel(dataTableOutput('Overlap_Table'),
                       downloadButton('Overlap_Download', 'Download These Results'))
               )),
    tabPanel("Generate LD Plot",
             sidebarLayout(
               sidebarPanel(
              selectInput("LD_window_or_rsID","Please select to begin",c("window","rsID"))
                 ,htmlOutput("LD_chr"),
                 htmlOutput("LD_LB"),
                 htmlOutput("LD_additional"),
                 htmlOutput("LD_MAFCutoff"),
                 actionButton("LD_Calculate","Calculate!")
               ),
               mainPanel(textOutput("LD_rsIDOut"),plotOutput("LD_windowOut"))
             )),
    tabPanel("Gene Name Conversions",
             sidebarLayout(
               sidebarPanel(
                textInput("Lookup_One","Please enter a gene name or probesetID"),
                fileInput("Lookup_Batch","Or upload a file for batch conversion"),
                actionButton("Lookup_Button", "Convert!")
                 ),
               mainPanel(
                 dataTableOutput("Lookup_Table")
                 )
               )),
    tabPanel("More Tools To Come!",h3("Soon…")),
    tabPanel("Bugs/Suggestions",
             textInput("Suggestion_Name","Name"),
             textInput("Suggestion_Report","Suggestion/Bug"),
             tags$style(type='text/css', "#Suggestion_Report { height: 300px;
width: 600px; }"),
             actionButton("Suggestion_Button","Suggest!"),
             textOutput("Suggestion_Text"),
             h3("Planned Changes:"),
             h4("Make it Faster (Especially correlations)"),
             h4("Eliminate Bugs"),
             h4("Make it Look Nice")
             )
    )))
```

References

1. Stancakova A, Javorsky M, Kuulasmaa T, Haffner SM, Kuusisto J, Laakso M (2009) Changes in insulin sensitivity and insulin release in relation to glycemia and glucose tolerance in 6,414 Finnish men. Diabetes 58(5):1212–1221. doi:10.2337/db08-1607
2. Ghazalpour A, Rau CD, Farber CR, Bennett BJ, Orozco LD, van Nas A, Pan C, Allayee H, Beaven SW, Civelek M, Davis RC, Drake TA, Friedman RA, Furlotte N, Hui ST, Jentsch JD, Kostem E, Kang HM, Kang EY, Joo JW, Korshunov VA, Laughlin RE, Martin LJ, Ohmen JD, Parks BW, Pellegrini M, Reue K, Smith DJ, Tetradis S, Wang J, Wang Y, Weiss JN, Kirchgessner T, Gargalovic PS, Eskin E, Lusis AJ, LeBoeuf RC (2012) Hybrid mouse diversity panel: a panel of inbred mouse strains suitable for analysis of complex genetic traits. Mamm Genome 23(9–10):680–692. doi:10.1007/s00335-012-9411-5
3. Threadgill DW, Miller DR, Churchill GA, de Villena FP (2011) The collaborative cross: a recombinant inbred mouse population for the systems genetic era. ILAR J 52(1):24–31
4. Chang W, Cheng J, Allaire J, Xie Y, McPherson J (2015) Shiny: web application framework for R. http://cran.r-project.org/package=shiny
5. Bennett BJ, Farber CR, Orozco L, Kang HM, Ghazalpour A, Siemers N, Neubauer M, Neuhaus I, Yordanova R, Guan B, Truong A, Yang WP, He A, Kayne P, Gargalovic P, Kirchgessner T, Pan C, Castellani LW, Kostem E, Furlotte N, Drake TA, Eskin E, Lusis AJ (2010) A high-resolution association mapping panel for the dissection of complex traits in mice. Genome Res 20(2):281–290. doi:10.1101/gr.099234.109
6. Calabrese G, Bennett BJ, Orozco L, Kang HM, Eskin E, Dombret C, De Backer O, Lusis AJ, Farber CR (2012) Systems genetic analysis of osteoblast-lineage cells. PLoS Genet 8(12): e1003150. doi:10.1371/journal.pgen.1003150
7. Farber CR, Bennett BJ, Orozco L, Zou W, Lira A, Kostem E, Kang HM, Furlotte N, Berberyan A, Ghazalpour A, Suwanwela J, Drake TA, Eskin E, Wang QT, Teitelbaum SL, Lusis AJ (2011) Mouse genome-wide association and systems genetics identify Asxl2 as a regulator of bone mineral density and osteoclastogenesis. PLoS Genet 7(4):e1002038. doi:10.1371/journal.pgen.1002038
8. Park CC, Gale GD, de Jong S, Ghazalpour A, Bennett BJ, Farber CR, Langfelder P, Lin A, Khan AH, Eskin E, Horvath S, Lusis AJ, Ophoff RA, Smith DJ (2011) Gene networks associated with conditional fear in mice identified using a systems genetics approach. BMC Syst Biol 5:43. doi:10.1186/1752-0509-5-43
9. Davis RC, van Nas A, Bennett B, Orozco L, Pan C, Rau CD, Eskin E, Lusis AJ (2013) Genome-wide association mapping of blood cell traits in mice. Mamm Genome 24(3–4):105–118. doi:10.1007/s00335-013-9448-0
10. Ghazalpour A, Bennett B, Petyuk VA, Orozco L, Hagopian R, Mungrue IN, Farber CR, Sinsheimer J, Kang HM, Furlotte N, Park CC, Wen PZ, Brewer H, Weitz K, Camp DG II, Pan C, Yordanova R, Neuhaus I, Tilford C, Siemers N, Gargalovic P, Eskin E, Kirchgessner T, Smith DJ, Smith RD, Lusis AJ (2011) Comparative analysis of proteome and transcriptome variation in mouse. PLoS Genet 7(6):e1001393. doi:10.1371/journal.pgen.1001393
11. Orozco LD, Bennett BJ, Farber CR, Ghazalpour A, Pan C, Che N, Wen P, Qi HX, Mutukulu A, Siemers N, Neuhaus I, Yordanova R, Gargalovic P, Pellegrini M, Kirchgessner T, Lusis AJ (2012) Unraveling inflammatory responses using systems genetics and gene-environment interactions in macrophages. Cell 151(3):658–670. doi:10.1016/j.cell.2012.08.043
12. Ghazalpour A, Bennett BJ, Shih D, Che N, Orozco L, Pan C, Hagopian R, He A, Kayne P, Yang WP, Kirchgessner T, Lusis AJ (2014) Genetic regulation of mouse liver metabolite levels. Mol Syst Biol 10:730. doi:10.15252/msb.20135004
13. Orozco LD, Morselli M, Rubbi L, Guo W, Go J, Shi H, Lopez D, Furlotte NA, Bennett BJ, Farber CR, Ghazalpour A, Zhang MQ, Bahous R, Rozen R, Lusis AJ, Pellegrini M (2015) Epigenome-wide association of liver methylation patterns and complex metabolic traits in mice. Cell Metab 21(6):905–917. doi:10.1016/j.cmet.2015.04.025
14. Parks BW, Nam E, Org E, Kostem E, Norheim F, Hui ST, Pan C, Civelek M, Rau CD, Bennett BJ, Mehrabian M, Ursell LK, He A, Castellani LW, Zinker B, Kirby M, Drake TA, Drevon CA, Knight R, Gargalovic P, Kirchgessner T, Eskin E, Lusis AJ (2013) Genetic control of obesity and gut microbiota composition in response to high-fat, high-sucrose diet in mice. Cell Metab 17(1):141–152. doi:10.1016/j.cmet.2012.12.007
15. Parks BW, Sallam T, Mehrabian M, Psychogios N, Hui ST, Norheim F, Castellani LW, Rau CD, Pan C, Phun J, Zhou Z, Yang WP, Neuhaus I, Gargalovic PS, Kirchgessner TG, Graham M, Lee R, Tontonoz P, Gerszten RE, Hevener AL, Lusis AJ (2015) Genetic architecture of insulin resistance in the mouse. Cell Metab 21(2):334–346. doi:10.1016/j.cmet.2015.01.002

16. Org E, Parks BW, Joo JW, Emert B, Schwartzman W, Kang EY, Mehrabian M, Pan C, Knight R, Gunsalus R, Drake TA, Eskin E, Lusis AJ (2015) Genetic and environmental control of host-gut microbiota interactions. Genome Res 25(10):1558–1569. doi:10.1101/gr.194118.115
17. Hui ST, Parks BW, Org E, Norheim F, Che N, Pan C, Castellani LW, Charugundla S, Dirks DL, Psychogios N, Neuhaus I, Gerszten RE, Kirchgessner T, Gargalovic PS, Lusis AJ (2015) The genetic architecture of NAFLD among inbred strains of mice. Elife 4:e05607. doi:10.7554/eLife.05607
18. Rau CD, Wang J, Avetisyan R, Romay MC, Martin L, Ren S, Wang Y, Lusis AJ (2015) Mapping genetic contributions to cardiac pathology induced by Beta-adrenergic stimulation in mice. Circ Cardiovasc Genet 8(1):40–49. doi:10.1161/CIRCGENETICS.113.000732
19. Bennett BJ, Davis RC, Civelek M, Orozco L, Wu J, Qi HX, Pan C, Packard RR, Eskin E, Yan M, Kirchgessner T, Wang Z, Li X, Gregory JC, Hazen SL, Gargalovic P, Lusis AJ (2015) Genetic architecture of atherosclerosis in mice: a systems genetics analysis of common inbred strains. PLoS Genet 11:e1005711
20. Crow AL, Ohmen J, Wang J, Lavinsky J, Hartiala J, Li Q, Li X, Salehide P, Eskin E, Pan C, Lusis AJ, Allayee H, Friedman RA (2015) The genetic architecture of hearing impairment in mice: evidence for frequency specific genetic determinants. G3 (Bethesda) 5:2329–2339. doi:10.1534/g3.115.021592
21. Ohmen J, Kang EY, Li X, Joo JW, Hormozdiari F, Zheng QY, Davis RC, Lusis AJ, Eskin E, Friedman RA (2014) Genome-wide association study for age-related hearing loss (AHL) in the mouse: a meta-analysis. J Assoc Res Otolaryngol 15(3):335–352. doi:10.1007/s10162-014-0443-2
22. Turner S (2014) qqman: Q-Q and Manhattan plots for GWAS data. http://cran.r-project.org/package=qqman
23. Eklund A (2015) Beeswarm: the Bee Swarm plot, an alternative to Stripchart. http://cran.r-project.org/package=beeswarm
24. Yang H, Wang JR, Didion JP, Buus RJ, Bell TA, Welsh CE, Bonhomme F, Yu AH, Nachman MW, Pialek J, Tucker P, Boursot P, McMillan L, Churchill GA, de Villena FP (2011) Subspecific origin and haplotype diversity in the laboratory mouse. Nat Genet 43(7):648–655. doi:10.1038/ng.847

Chapter 8

Expression QTLs Mapping and Analysis: A Bayesian Perspective

Martha Imprialou, Enrico Petretto, and Leonardo Bottolo

Abstract

The aim of expression Quantitative Trait Locus (eQTL) mapping is the identification of DNA sequence variants that explain variation in gene expression. Given the recent yield of trait-associated genetic variants identified by large-scale genome-wide association analyses (GWAS), eQTL mapping has become a useful tool to understand the functional context where these variants operate and eventually narrow down functional gene targets for disease. Despite its extensive application to complex (polygenic) traits and disease, the majority of eQTL studies still rely on univariate data modeling strategies, i.e., testing for association of all transcript-marker pairs. However these "one at-a-time" strategies are (1) unable to control the number of false-positives when an intricate Linkage Disequilibrium structure is present and (2) are often underpowered to detect the full spectrum of *trans*-acting regulatory effects. Here we present our viewpoint on the most recent advances on eQTL mapping approaches, with a focus on Bayesian methodology. We review the advantages of the Bayesian approach over frequentist methods and provide an empirical example of polygenic eQTL mapping to illustrate the different properties of frequentist and Bayesian methods. Finally, we discuss how multivariate eQTL mapping approaches have distinctive features with respect to detection of polygenic effects, accuracy, and interpretability of the results.

Key words IT-tools for systems genetics, Polygenic eQTL, *trans*-eQTLs, LASSO, Penalized-regression, Bayesian variable selection

1 Introduction

Genetics shape the landscape of phenotypic variation between humans through changes in the mechanisms regulating gene transcription and, consequently, gene expression. Detecting genetic drivers of gene expression can help understand the functional effects of DNA sequence variations at the cellular level. In particular, with the growing number of genetic variants associated with complex traits and diseases by genome-wide association studies (GWAS), understanding how these variants act though changes to the transcriptome might help elucidating their cellular context and prioritize functional gene targets [1].

Klaus Schughart and Robert W. Williams (eds.), *Systems Genetics: Methods and Protocols*, Methods in Molecular Biology, vol. 1488, DOI 10.1007/978-1-4939-6427-7_8,

Expression Quantitative Trait Loci (eQTLs) are genetic loci that control variation in the expression level of a gene (or transcript) in a given tissue or cell-type. In the literature eQTLs are distinguished by their relative position to the gene they regulate, as *cis*- (or proximal) and *trans*-acting. This distinction is important, as it can be informative on the mechanisms underlying variation in gene expression. For example, *cis*-eQTLs can be located within the promoter or enhancer region of the gene, and thus indicate interactions with the gene's own regulatory elements. Typically, *cis*-eQTLs are more easily detected and are of large genetic effect, whereas *trans*-eQTLs have relatively smaller effects and can reveal secondary regulatory mechanisms of gene expression. While it has been reported that a substantial fraction of observed *trans*-eQTL associations can be explained by *cis*-mediation [2], the identification of large clusters of *trans*-eQTLs can be informative of coordinated genetic regulation of gene expression and regulatory networks underlying complex traits [3–6].

The classical set-up of an eQTL mapping study involves quantifying the expression levels of selected genes or of the whole transcriptome using microarrays or RNA-sequencing analysis, and then treating each expression level as a quantitative trait to be mapped against a set of genetic markers. The goal is to estimate the number, effect size, and kind (i.e., *cis*- or *trans*-acting) of eQTLs in a given tissue or cell-type. eQTLs can be detected using linkage or association mapping, much the same as in GWAS for quantitative traits. Linkage mapping is typically used to detect genetic linkages in pedigrees of related individuals for highly penetrant phenotypes with a few major effect genes (or under monogenic control), while association is more powerful when working with traits determined by many small-effect variants (i.e., polygenic) and in populations of unrelated individuals. There is vast literature on linkage-based eQTL mapping in inbred populations, families as well as in experimental model systems; however, in this review we restrict our attention to association mapping, as it is more relevant to the interpretation of GWAS signals in common disease.

Due to the complex genetic architecture of expression traits, statistical power is key when choosing an eQTL-mapping strategy. Contemporary eQTL studies are characterized by the "large p, small n paradigm", as the number of predictors (genetic markers) is orders of magnitude larger than the number of genotyped samples. Typically, the contribution of most predictors to the expression trait is negligible, so most experiments aim to discover the few SNPs with substantial effects and use separate analyses to detect *cis*- and *trans*-effects. In this, *cis*-eQTLs are usually investigated by analyzing only the SNPs located nearby the gene, therefore reducing the need for multiple testing adjustments. However, "one at-a-time" models, which estimate individual SNP's contribution to the gene expression, are less capable of identifying the full spectrum of (*cis*- and *trans*-acting) eQTLs in the genome, giving way to multivariate selection approaches.

A wide range of genetic mapping programs, tailored for eQTL analysis, is currently available, using either frequentist or Bayesian inference [7–20]. These methods vary greatly in terms of statistical power to detect associations, interpretability of results and computational efficiency and the choice between different approaches is usually influenced by the trade-off between these three factors. Frequentist univariate models, for example, are fast and usually come with straightforward conversions to false-positive rates and false discovery rates (FDR), but have limited ability to detect small-effect *trans*-eQTLs and polygenic contributions to gene expression. Multivariate selection models (using penalization on the regression coefficients or sparsity prior on their number) are substantially more powerful than univariate approaches since they are able to decrease the uncertainty of the results by selecting (noncollinear) independent predictor variables avoiding at the same time over-fitting. However these advantages do come at a price: these methods are computationally more demanding and less efficient to deal with genome-wide eQTL-mapping experiments.

Another problem of frequentist univariate models relates to their ability to distinguish a tissue-specific eQTL (i.e., a genetic marker linked to gene expression in a specific tissue or cell-type) from an eQTL that is conserved across tissues. In contrast, the simultaneous and multivariate eQTL mapping of expression levels across tissues has been shown to increase power to detect common *trans*-eQTLs [20–22] in comparison with a naïve intersection of eQTLs mapped separately within individual tissues.

Here we review statistical methodologies that are most commonly used for the discovery of eQTLs. In this, after introducing eQTL mapping that use the frequentist approach, we focus on Bayesian approaches and appraise their advantages and distinctive features. For illustrative purposes, we report an example of eQTL mapping of simultaneous *cis*- and *trans*-effects (i.e., polygenic control of gene expression) as well as the extension to multiple tissues, to illustrate features specific to each eQTL mapping method.

2 Frequentist eQTL Mapping

In classical statistics, the observed data are considered an instance of infinitely many possible independent samples, while the tested hypothesis h, and any model parameters, are fixed and unknown. Hypothesis testing aims at deciding to accept or reject the null hypothesis with a high probability, which amounts to estimating the likelihood of observing the current instance of the data (or any function of it) under the null hypothesis. The p-value, a measure of the probability of observing under the same experimental conditions future samples equal or more extreme than the observed data, is used to decide on a hypothesis, based on whether it is smaller than an arbitrary significance level (typically $<5\%$ when a single hypothesis is tested).

2.1 Simple Parametric Models

Early attempts of eQTL mapping were predominantly frequentist, and utilized mapping strategies that were used in ordinary linkage of GWAS analysis settings. Most of these methods test the association of the expression level of each transcript to each marker independently, partitioning the samples in groups based on their genotype—e.g., in isogenic populations this is essentially differential expression analysis using the allele as grouping variable [23, 24] while in multi-allelic data an ANOVA test is performed with genotypes as grouping variable. Since both *t*-test and ANOVA can be seen as special cases of the linear regression model, several software packages implementing simple linear regression for eQTL mapping are available [7, 10, 14, 15].

Here we introduce the basic principles of the linear regression approach in eQTL mapping. Let's assume an expression profiling experiment with n samples that are genotyped at p markers which are the predictor variables. Without loss of generality, here we also assume that $n > p$. The expression of one transcript can be described as:

$$y = \alpha + x_1\beta_1 + \ldots + x_p\beta_p + \varepsilon, \varepsilon \sim \mathrm{N}\left(0,\sigma^2\right),$$

where y is the $n \times 1$ vector of expression levels, α is a constant, $x_j = \left(x_{1,j},\ldots,x_{n,j}\right)$, $j = 1,\ldots,p$, is the $n \times 1$ predictor vector which corresponds to the sample genotypes at the jth marker and ε is the normally distributed error term, centered in zero with residual variance $\sigma^2 1_n$. The regression coefficients $\beta = \left(\beta_1,\ldots,\beta_p\right)^T$, which encode the contribution of each marker to the gene expression y, can be estimated by minimizing the sum of squared residuals using Ordinary Least Squares (OLS), i.e., by solving:

$$\beta^{\wedge} = \arg\min_{\beta}\left\{\sum_{i=1}^{n}\left(y_i - \alpha - \sum_{j=1}^{p} x_{ij}\beta_j\right)^2\right\}.$$

A hypothesis test can be set-up testing whether all β regression coefficients are zero (null hypothesis) or at least one is not zero, in which case an eQTL association is detected.

$$H_0 : \beta = 0$$

$$H_1 : \text{at least } \beta_j \neq 0,\ j = 1,\ldots,p.$$

Different statistics can be used to test this hypothesis, each employed by different methods in the frequentist eQTL literature: the t-statistic [24] if each βj, $j = 1,\ldots,p$, is tested independently, the F-statistic [7, 15, 25] if all β are tested simultaneously, the Pearson's r [7] or the, closely related, Likelihood Ratio test [10,

12]. These linear regression models are quite flexible and can be extended in several ways, for example by including sex, age, batch effects, population structure, etc., or considering confounders as fixed effects (covariates), by combining additive, recessive, and dominant effects of the genotypes or by adding a random effect that, for instance, can be used to account for family/pedigree structure [26]. Since in typical eQTL mapping experiments the number of markers is much larger than the number of observations, $p \gg n$, (also known as the "large *p*, small *n*" paradigm), linear regression models cannot be used straightforwardly with the whole set of markers. To overcome this problem, simple univariate strategies have been proposed where all possible transcript-markers pairs are tested for association. However, these procedures are suboptimal since they are not able to control the number of false-positive associations when an intricate Linkage Disequilibrium (LD) structure with correlated markers is present.

2.2 Nonparametric Models

When the assumptions of normality and/or linearity are not guaranteed, nonparametric models, based on the Wilcoxon rank-sum test [23, 27, 28], a nonparametric version of the *t*-test, or Spearman's rank correlation [9] have been proposed and employed to map eQTLs, in particular in simple model organisms [23]. Sometimes nonparametric models are used in conjunction with linear models to help establish a significance threshold, especially in the presence of outliers.

Both parametric and nonparametric frequentist approaches based on the "one at-a-time" strategy are widely adopted because of their computational performance—many employ efficient memory allocation techniques [11] or minimize the number of required operations [7, 29]. The appealing "simplicity" and widespread use of the *p*-value is another attractive feature of these approaches, as it allows for straightforward control of family-wise error rate (FWER) and FDR (e.g., using for instance the Benjamini-Hochberg method [30]) although both procedures assume the independence of the statistical tests that are rarely met in practice due to LD structure in the genetic markers.

Despite its extensive use, the *p*-value as a measure of association is based only on the null distribution and it cannot control the power, which depends on the alternative hypothesis. The lack of power control provided by *p*-values is particularly undesirable in typical eQTL studies based on linear regression models since it is hard to detect associations with small effect sizes, such as those observed for *trans*-eQTLs. For instance, it has been shown that with 5 M SNPs a sample size of at least 200 is required to detect common (i.e., minor allele frequency, MAF > 20%) *trans*-eQTLs and over 500 is required to detect rare (MAF < 5%) variants [31]. Reaching this sample size requirement can be difficult in many eQTL-mapping experiments since relevant tissue for expression profiling is difficult to obtain, in particular in human eQTL analyses.

2.3 Penalized-Regression Models

Penalized-regression methods such as ridge regression [14, 32], the LASSO [33–36], Elastic Net [37] and Group Lasso [38] have been proposed to address the limitations of classical regression-based eQTL mapping methods. This class of approaches tries to account for a sparse representation of the genetic markers that contribute to the expression of the gene when $p \gg n$ and for the presence of blocks of LD between genetic markers. In penalized-regression approaches the output consists of a sparse set of predictors (genetic markers) that are obtained by shrinking the majority of regression coefficients towards zero. Here, we focus on the LASSO [33–36], as it is one of the most widely used method in eQTL mapping, and it is a key component of a larger class of penalized-based approaches [39–46]. In LASSO, shrinkage is achieved by restricting the OLS solution such that the absolute sum of the regression coefficients (L^1-norm) does not exceed a threshold *t*:

$$\sum_{j=1}^{p} \left|\beta_j\right| \le t$$

which is equivalent to solve

$$\beta^{\wedge} = \arg\min_{\beta} \left\{ \sum_{i=1}^{n} \left(y_i - \alpha - \sum_{j=1}^{p} x_{ij}\beta_j \right)^2 + \lambda \sum_{j=1}^{p} \left|\beta_j\right| \right\}.$$

The parameter λ is called penalty, which is typically selected by cross-validation, such that it minimizes the off-sample prediction mean square error. However imposing a L^1-norm restriction on the effects, the nonzero regression coefficient estimates become biased.

The eQTL mapping by penalized regression-based approaches typically leads to the identification of a few genetic markers as eQTLs, implicitly assuming that the majority of markers in the genome have negligible effects of gene expression. While this hypothesis is plausible from a biological viewpoint, the interpretation of the results can be sometimes difficult, as the nonzero regression coefficients are not informative about the genome-wide significance of the eQTL results, and their estimate cannot be used straightforwardly to control the FWER or FDR.

To overcome this limitation, additional resampling-based approaches such as stability selection [47, 48] (which accounts for the number of times a genetic marker is selected by a LASSO-type algorithm during the resampling procedure) provides a selection frequency (posterior probability) for each predictor, that can be used to control the FWER, but not the FDR. Another limitation of this approach is that current strategies to calibrate the penalty parameter λ are not robust: in general there is no optimal strategy for the tuning of the parameter λ, while standard calibration strategies may lead to inconsistent prediction with either too many false-positives or false-negatives [49]. This is particularly important in the presence of moderately correlated predictors, which is usually the case in eQTL mapping studies due to the underlying LD structure in the genome [44].

In the presence of a group of highly correlated variables, the LASSO tends to select one variable from a group and ignore the others. To overcome this limitation, Elastic Net [37] has been proposed. This method adds an extra penalty (L^2-norm) which, when used alone, corresponds to the ridge regression:

$$\breve{\beta} = \arg\min_{\beta}\left\{\sum_{i=1}^{n}\left(y_i - \alpha - \sum_{j=1}^{p} x_{ij}\beta_j\right)^2 + \lambda_1\sum_{j=1}^{p}\left|\beta_j\right| + \lambda_2\sum_{j=1}^{p}\left(\left|\beta_j\right|^2\right)^{\frac{1}{2}}\right\}.$$

It is well known that including groups of correlated predictors in the sparse solution (i.e., the set of eQTLs) can produce large variance in the final parameter estimates since the determinant operator required in the OLS solution is close to zero, which makes the linear algebra operator "ill-conditioned", and therefore the matrix inversion cannot be performed with as much precision (i.e., large variances). However, adding an extra penalty regularizes the matrix inversion, reducing the variance of the nonzero effects. Although the resulting nonzero regression coefficient estimates are biased, the expected mean squared error is lower than OLS since the bias is largely compensated by a smaller variance. Despite the theoretical and intuitive arguments in favor of the Elastic Net, the choice of the penalty parameters λ_1 and λ_2 by cross-validation is computationally time consuming since the optimization should be done in a two-dimensional grid. Moreover the optimal solution for λ_1 and λ_2 can lie in a very small interval that is not covered by the user-defined grid of penalty parameters, with the risk of producing a sub-optimal solution.

A concise list of the most commonly used frequentist eQTL mapping methods and their software implementation is reported in Table 1.

3 Bayesian eQTL Mapping

3.1 Concepts of Bayesian Modeling

Bayesian methods are becoming increasing popular in modern genetics [51], possibly as a consequence of recent more efficient algorithmic/computational implementations, cheaper high-performance computing solutions, and in general less computational constraints in their genome-wide applications. Unlike frequentist approaches, which try to infer the value of fixed models parameters from random data, in Bayesian inference the data are

Table 1
Frequentist eQTL mapping approaches

Strategy	Method, [ref] and availability	Statistic	Additional features	Multi-tissue	Genetic models
One at-a-time test	ANOVA	F-statistic	–	Yes	Additive
One at-a-time test	Krux [13] https://github.com/tmichoel/krux	Kruskal-Wallis	–	No	Additive
One at-a-time test	Matrix-eQTL [7] http://www.bios.unc.edu/research/genomic_software/Matrix_eQTL	Pearson's r	• Heteroskedastic error term (for correlated transcripts) • Slice computations in small matrices for computational efficiency	No	• Additive • Dominant
One at-a-time test	Genevar [9] https://www.sanger.ac.uk/resources/software/genevar/	Spearman's ρ	–	No	Additive
One at-a-time regression	R/QTL [10] http://www.rqtl.org	*t*-test	–	No	Additive
Multiple regression	snpMatrix [11] http://www.bioconductor.org/packages/2.3/bioc/html/snpMatrix.html	Chi-squared	• Generalized linear models • SNP conditioning search	No	Additive
Multiple regression	eMap [12] http://www.mybiosoftware.com/emap-1-2-eqtl-analysis.html	Likelihood Ratio	Inclusion of covariates via backward selection	No	• Additive • Dominant
Multiple regression	HEFT [14] http://mezeylab.cb.bscb.cornell.edu/Software.aspx	*t*-test	• Ridge regression • Detection of hidden covariates by factor analysis	No	Additive
One at-a-time regression	SNPTEST [15] https://mathgen.stats.ox.ac.uk/genetics_software/snptest/snptest.html	F-statistic	Frequentist and Bayesian analysis (*see* Table 2)	Yes	• Additive • Dominant • Recessive • Heterozygote
Multiple regression	Glmnet [50] https://cran.r-project.org/web/packages/glmnet/index.html	–	• LASSO • Elastic Net	Yes	Additive

treated as a fixed quantity (since there is no randomness after observing the data) while the parameters are treated as random variables. This allows researchers to assign to parameters (and models) probabilities, making the inferential framework more intuitive and straightforward. Here we introduce a few general concepts that are at the core of the Bayesian paradigm. Denoting the parameters by θ and the observed data by D, the Bayes theorem allows to write:

$$\pi(\theta \mid D) = \frac{\ell(D \mid \theta)\pi(\theta)}{\ell(D)} = \frac{\ell(D \mid \theta)\pi(\theta)}{\int \ell(D \mid \theta)\pi(\theta)\mathrm{d}\pi},$$

where $\pi(\theta \mid D)$ is the posterior distribution, $\ell(D \mid \theta)$ is the likelihood (conditionally on some parameters' value), $\pi(\theta)$ is the prior distribution on the parameters and $l(D)$ the marginal likelihood. In a nutshell, the equation above states that the Bayesian paradigm provides a distribution regarding what it has been learned about the parameter from the data. In contrast to the frequentist approach, where only a point estimate (MLE) and a standard error (SE) are obtained from the inferential process, in the Bayesian paradigm the whole distribution of the parameters is available.

Similarly, the Bayesian model selection is obtained by assigning a distribution of probability over alternative competing models and, after observing the data, selecting the most promising model as the one with the largest posterior probability. The assignment of probabilities to model parameters is made using both the information captured by the data D and prior knowledge (or beliefs) about the structure of the model, which is encoded by the prior probability $\pi(M)$ of a model M. Then, a typical Bayesian experiment updates the prior distribution $\pi(M)$ to the posterior $\pi(M \mid D)$ by multiplying the likelihood $\ell(D \mid M)$ with the prior probability of the model $\pi(M)$, using the Bayes theorem:

$$\pi(M \mid D) = \frac{\ell(D \mid M)\pi(M)}{\pi(D)} = \frac{\ell(D \mid M)\pi(M)}{\sum_i \ell(D \mid M_i)\pi(M_i)},$$

where $\ell(D \mid M)$ is the conditional probability of observing the data under the model and $\pi(D)$ is the probability of the data, which can be computed by summing over the conditionals of all possible models.

An alternative way to evaluate which model is most supported by the data D, between two alternative models M_1 and M_2, is to calculate the so-called Bayes Factor (BF) [52]:

$$BF(M_1, M_2) = \frac{\ell(D \mid M_1)}{\ell(D \mid M_2)} = \frac{\dfrac{\pi(M_1 \mid D)}{\pi(M_1)}}{\dfrac{\pi(M_2 \mid D)}{\pi(M_2)}} = \frac{\dfrac{\pi(M_1 \mid D)}{\pi(M_2 \mid D)}}{\dfrac{\pi(M_1)}{\pi(M_2)}}$$

which is the ratio between posterior odds $\pi(M_1 \mid D)/\pi(M_2 \mid D)$ and prior odds $\pi(M_1)/\pi(M_2)$. The BF can also be interpreted as a Likelihood Ratio test between two competing models M_1 and M_2 when all the uncertainty about nuisance parameters η (i.e., parameters that are of no direct interest but are specified in the model) has been marginalized (integrated) out

$$\frac{\ell(D \mid M_1)}{\ell(D \mid M_2)} = \frac{\int \ell(D \mid M_1,\eta)\pi(\eta)\mathrm{d}\eta}{\int \ell(D \mid M_2,\eta)\pi(\eta)\mathrm{d}\eta}$$

without conditioning as in frequentist approaches

In Bayesian eQTL mapping the observed data typically include a $n \times 1$ vector of outcomes y (i.e., gene expression levels) and a $n \times p$ matrix of predictor variables X (i.e., genetic markers). The set of model parameters, their prior distribution, and hence the joint posterior distribution may vary between approaches [53]. The Bayesian models presented here attempt to infer the posterior distribution of the vector of regression coefficients $\beta = (\beta_1,\ldots,\beta_p)^{\mathrm{T}}$, which encodes the effect of markers to the gene expression level, i.e., the eQTLs.

3.2 Univariate Regression Models

One class of Bayesian eQTL approaches associates the outcome with one marker "at-a-time", by computing the BF for each SNP (instead of the frequentist p-value) [15, 19, 20, 54]. This approach is computationally efficient since only two alternative models M_1 and M_2 are compared each time, i.e., M_1 and M_2, encoding for the inclusion/exclusion of the marker, respectively. In this framework the BF is further simplified

$$BF(M_1,M_2) = \frac{\pi(M_1 \mid D)}{1-\pi(M_1 \mid D)}\frac{1-\pi(M_1)}{\pi(M_1)},$$

where $\pi(M_1)$ and $\pi(M_1 \mid D)$ are the prior and posterior probability, respectively, that the marker is an eQTL. Markers whose BF exceeds a certain threshold are therefore defined as eQTL (the general criteria for setting the optimal BF threshold based on number of predictors can be found in [15, 55]). Beyond setting the BF threshold, it has been shown that using the BF is superior to conventional p-value since the Bayesian-inferred associations can benefit from the elicitation of "biologically primed" informative priors [56, 57], which in some cases can improve power [58].

SNPTEST [15] performs a single-marker eQTL association analysis, i.e., implementing a "one at-a-time" strategy, which incorporates both frequentist and Bayesian association tests. In its Bayesian form, SNPTEST fits a linear regression model that computes the posterior odds of including marker j in the linear regression model $y = \alpha + \beta x_j + \varepsilon$. The error term is normally distributed

$\varepsilon \sim \mathrm{N}\left(0,\sigma^2\right)$, while the model parameters β, σ^2 are given a conjugate Normal-Inverse-Gamma prior set-up. This regression approach can be extended to map eQTL under dominant or recessive inheritance models: the prior distributions remain the same, but the genotype vector is modified and recoded to reflect dominant or recessive inheritance model. A normal prior is used on β with larger variance, reflecting the assumption that dominant or recessive alleles contribute differently to the phenotypic (i.e., gene expression) variance.

3.3 Bayesian Variable Selection Methods

Initially, most "one at-a-time" Bayesian strategies were applied to GWAS of clinical traits or disease, in which the phenotype (disease trait) was analyzed against a genome-wide panel of genetic predictors (SNPs). When applied to these data and especially to data problem that is typical of eQTL mapping (i.e., large number of *both* expression phenotypes *and* genetic markers), these methods display similar problems as the simple frequentist models described in Section 2; namely, an inflated number of false-positive associations and loss of power due to setting arbitrary study-wide significance thresholds [59].

Similarly to penalized-regression models, Bayesian Variable Selection (BVS) methods have been developed for eQTL mapping to analyze jointly the whole set of markers. However, differently from penalized-regression, BVS is able to perform model choice (select the markers that are likely to influence the expression of the gene) and provide parameters estimate (the regression coefficients of the active markers) at the same time [16–18, 60–64]. With both these quantities available, genome-wide significance can be obtained by controlling the FDR level [65]. Here, we describe the major components of BVS and introduce a few computational implementations of this class of approaches.

BVS—Prior set-up: similar to penalized-regression methods, BVS models try to choose few important markers with large effects. Unlike LASSO-type regressions, in BVS sparsity is not only controlled by the prior distribution on the regression coefficients (i.e., the L^p-norm penalty in the frequentist approach), but by specifying an *a priori* number of eQTLs encoded in a latent binary vector $\gamma = \left(\gamma_1, \ldots, \gamma_p\right)^{\mathrm{T}}$, where $\gamma_j = 1$ if $\beta_j \neq 0$ and $\gamma_j = 0$ if $\beta_j = 0$. In a nutshell, BVS can control *both* the level of shrinkage and *the* number of nonzero eQTL effects that can be detected. These two tasks are tightly connected in BVS given the prior specification of the regression coefficients [66]:

$$\beta \mid \gamma, \sigma^2 \sim N\left(0, g\sigma^2 \left(X_\gamma^{\mathrm{T}} X_\gamma\right)^{\lambda}\right).$$

The equation above states that for the selected markers, i.e., for markers with $\gamma_j = 1$ since $\beta_j \neq 0$, the prior distribution on the vector of regression coefficients is normal distributed and centered in zero. The covariance matrix can be the unit diagonal matrix, giving rise to the so-called independent prior, if $\lambda = 0$ or the inverse of the covariance matrix which characterize the so-called *g*-prior if $\lambda = -1$, multiplied by a constant *g* and the residual variance σ^2. Under this specification the linear regression model becomes $y = \alpha + X_\gamma \beta_\gamma + \varepsilon$, $\varepsilon \sim N(0, \sigma^2)$, where γ, the vector of binary values indicating which markers are selected and therefore their number, receives a Binomial prior distribution

$$\pi(\gamma) = \text{Bin}(p, \theta)$$

where θ can be a fixed parameter or a further level of hierarchy can be specified [1]. The prior distribution for the model parameters β, σ^2 usually follows a Normal-Inverse-Gamma set-up [16, 17, 21] with a different specification for the power-prior λ. The choice of $\lambda = -1$ is particularly appealing in eQTL studies since *g*-prior "discourages" highly collinear predictors to enter the models simultaneously by inducing a negative correlation between the coefficients, therefore controlling LD structure automatically. On the opposite side with $\lambda = 0$, the regression coefficients are *a priori* mutually independent [15] although, given the influence of the likelihood, this does not hold *a posteriori*. It turns out that *a priori* independent prior is less capable in handling intricate correlations between markers, but its use is encouraged because it induces, like the ridge estimator, an absolute shrinkage to the regression coefficients, i.e., it shrinks greatly in directions of small eigenvalues, whereas the *g*-prior proportional shrinkage retains much more of the OLS estimator in ill-conditioned directions [67].

The specification of the prior distribution for the model parameters β, σ^2 is an active area of research in Bayesian statistics, since different prior set-ups imply different levels of shrinkage. Sparsity-inducing prior set-ups include the Laplace prior [68], the spike-and-slab priors [61, 62], the horseshoe shrinkage prior [69] and local adaptation priors [17, 63]—analyzing the different features of these set-ups in detail goes beyond the scope of this review. However, here we mention that in the piMASS eQTL mapping method [17] a novel prior set-up links the expected genetic effect sizes with the model size. In particular, the effect size prior demonstrates the biologically primed idea that if the model size is small, then the few associated markers will have large effect sizes—the opposite is expected when then model size is large. This exemplifies how the Bayesian setting can effectively leverage "biologically informed" priors to improve and refine eQTL detection.

BVS—Model selection and posterior computation: in typical genomics and eQTL mapping experiments the number of predictor variables is too large to enumerate all possible combinations of latent binary vector γ. Therefore search algorithms are used to explore the model space. Most methods use Markov Chain Monte Carlo (MCMC) [17, 18, 61–63], a sampling technique in which the posterior distribution $\pi(\gamma \mid D)$ is simulated using Markov Chain algorithms. The idea behind is that not all the $2p$ possible models (i.e., combination of markers) need to be simulated, since the majority of them are unable to explain the data with $\pi(\gamma \mid D) \approx 0$. On the contrary, it is more efficient to concentrate the markers' space exploration on important models with large $\pi(\gamma \mid D)$. In a post-processing analysis, $\pi(\gamma \mid D)$ can be used to rank the visited models and decide which one to report. The output provided by MCMC algorithm is very rich since the sampled distribution of $\pi(\gamma \mid D)$ is available (apart from noninteresting models with $\pi(\gamma \mid D) \approx 0$). However, this comes at a price since MCMC algorithms are computational intensive and, as for frequentist penalized-regression methods, rather time consuming. If the goal of the analysis is to report only the top visited model, alternative faster sampling algorithms based on the expectation maximization (EM) algorithm [70] have been recently proposed [71].

There is a very large literature on the MCMC sampling schemes that can be used to sample realizations of $\pi(\gamma \mid D)$. The simplest MCMC algorithm that can be implemented is the Gibbs sampling [72], which is particular suitable when spike-and-slab priors are specified for the regression coefficients [18, 61, 62]. Since spike-and-slab priors can be seen as two-point mixture distribution, once conditioning on value of *j*th binary latent variable *γj*, the posterior distribution of the *j*th "spike" or the "slab" is ready available and it is relatively simple to simulate from. The drawback of this approach is that it tends to mix slowly when there are correlated predictors (e.g., in the presence of LD between the markers), since the posterior distribution of regression coefficient for the *j*th predictor depends on the neighbor predictors and if a marker has been selected, $\gamma_j = 1$, markers in strong LD with it will be selected as well. It turns out that the algorithm may be stuck in a particular configuration of γ for many iterations of the MCMC algorithm (slow mixing) without being able to detect the optimal combination of predictive markers. To overcome this problem, MCMC algorithms that explore more efficiently the model space have been proposed. For instance, using a "shotgun" stochastic search [73] one can explore the entire neighbor of the current model and randomly pick up with a non-uniform probability a model from that list. For instance in piMASS [17], once a model has been selected, the next active marker that can be included in the model is the one that shows the (residual) highest absolute correlation with the phenotype so that correlated predictors are less likely to be included in the model. The Evolutionary Stochastic Search method [16, 60, 74, 75], which we will discuss in detail below, has been designed for a more efficient and far-reaching exploration of the model space. It runs several parallel MCMC samplers that swap information about the different configurations of markers selected in each chain and therefore avoiding the slow mixing phenomenon described above.

BVS—Posterior summary and interpretation: a large number of MCMC iterations are generally required in order to match the frequency a particular model has been sampled, $\breve{\pi}(\gamma \mid D)$, with the theoretical posterior probability of that model, $\pi(\gamma \mid D)$. In that case the algorithm is said to have reached convergence. From a practical point of view, assessing convergence of MCMC is not easy and many diagnostic measures can be applied to detect any anomalous behavior of the algorithm. Moreover the initial draws of the algorithm (burn-in phase) are usually discarded because it may be possible that the models are sampled with the wrong frequency compared with the correct theoretical probability with some models over-represented or *vice versa* during the initial phase. All the models visited by the search algorithm after the burn-in are kept and summarized into a marginal posterior probability of inclusion (MPPI) $\breve{\pi}\left(\gamma_j = 1 \middle| D, \gamma|_j\right)$, which indicates the frequency the *j*th marker has been selected in all models visited by the search algorithm. Despite its straightforward interpretation (MPPI = the probability that the marker *j* explains the variation of the gene expression given all other markers), the use of MPPI alone in variable selection by setting a threshold is not recommended apart for prediction [76], as there is no direct interpretation of it with respect to effect size. However the classification of the MPPI into two groups will allow the assessment of their genome-wide significance. Specifically, one can employ the EM algorithm to fit a mixture of two beta distributions and then use the classification probabilities to derive the FDR, as described in [77]. Alternatively, versatile R packages that can estimate local (tail-area) FDR from the posterior distribution [78], are also available [79].

piMASS [17] is a BVS algorithm for eQTL mapping with a new regression coefficients' prior variance that allows either models with a large number of predictors with a small proportion of variance explained (PVE) or a small number of predictors with a large PVE. This prior set-up is in tune with what is expected in typical eQTL mapping experiments, where few *cis*-eQTLs are present with large effects and large PVE, whereas many *trans*-eQTLs have relatively smaller effects and smaller PVE. Its implementation is based on a single MCMC chain with a sampling strategy that explores models made by faraway and/or uncorrelated genetic markers.

Another BVS algorithm is Evolutionary Stochastic Search (ESS) [16, 21, 60, 75] in which the level of sparsity can be controlled directly by the user specifying the *a priori* expected number of predictors to be included in the model and its variance. Moreover given the prior structure on the regression coefficients that can be thought as a mixture of *g*-priors and an Inverse-Gamma prior [80], the level of proportional shrinkage automatically adapts to different real data scenarios. ESS uses an advanced stochastic search algorithm in which multiple models are explored by parallel MCMC samplers. Specifically, at each iteration, each chain locally selects a different model using local moves based on the Gibbs sampler [72] or a fast version of the Metropolis-Hasting algorithm [81]. Global moves, which allow the exchange of information between parallel chains about the models selected, are also implemented, using a MCMC version of genetic algorithms [82]. The combination of local and global moves allows the efficient exploration of the model space and prevents the algorithm from getting stuck to a sub-optimal model made by highly correlated predictors (i.e., genetic markers in high LD).

A concise list of the most commonly used Bayesian eQTL mapping methods and their software implementation is reported in Table 2.

4 Multi-tissue Extensions

Transcriptomic studies can assess gene expression levels in multiple tissues or cell-types in order to understand the mechanism of gene regulation at the systems-level, including mapping of eQTLs in multiple systems [84]. While expression of certain genes and pathways can be conserved across different tissues, intersecting results from several single-tissue eQTL analyses (for instance by imposing the same FDR threshold in each eQTL study) may be too conservative and can lead to inflated false-negative rate [83]. In contrast, utilizing a cross-tissue analysis of eQTLs by jointly mapping gene expression profiles from multiple tissues, has been shown to increase power to detect small effect eQTLs (specifically, *trans*-eQTLs) [20–22].

Several eQTL mapping approaches, including some discussed above, have been extended to allow eQTL mapping of tissue-consistent QTLs (i.e., eQTLs that are detected across multiple tissues), by allowing BVS models to analyze multivariate outcomes. Thus, assuming an experiment with n samples, p predictors and q outcomes (tissues or cell-types), the multiple outcome linear regression model can be written as:

$$Y = A + x_1 B_1 + \ldots + x_p B_p + E, E \sim MN(0,,I_n,,\Sigma),$$

where Y is a $n \times q$ matrix of outcomes, A is a $n \times q$ matrix of intercepts, xj is the jth predictor encoded in a $n \times 1$ vector and $B_j = (\beta_{j,1},,\ldots,,\beta_{j,q})$ is the vector of regression coefficients that links the jth predictor with the multiple outcomes Y. Finally, E is the $n \times q$ matrix of errors that is distributed as a matrix-variate normal distribution centered in zeros, with the matrix Σ that controls the residual correlation between the q outcomes.

Table 2
Bayesian eQTL mapping approaches

Strategy	Method [ref] and availability	Statistic	Additional features	Multi-tissue	Genetic models
Univariate	SNPTEST [15] https://mathgen.stats.ox.ac.uk/	BF	• Bayesian and frequentist and analysis (*see* Table 1) • Covariates can be included in the model • Imputation of missing genotypes	Yes	• Additive • Dominant • Recessive • Heterozygote • General
	Sherlock [19] http://sherlock.ucsf.edu/	BF	• Integration of known GWAS hits	No	Additive
	eQTLBMA [20] https://github.com/timflutre/eqtlbma	BF	Multiple tissues, while allowing different eQTLs per tissue	Yes	Additive
BVS	ESS [16, 21, 60, 75] www.bgx.org.uk/software/guess.html (command-line implementation) https://cran.r-project.org/package=R2GUESS (R implementation)	• MPPI • Best models visited	• Covariates can be included in the model • FDR control • Extension for eQTLs hotspots [74] • Extension for eQTLs hotspots in multiple tissues [83]	Yes	Additive
	piMASS [17] http://www.haplotype.org/pimass.html	MPPI	• Linear and logistic regression	No	Additive
	iBMQ [18] https://www.bioconductor.org/packages/release/bioc/html/iBMQ.html	MPPI	• FDR control • Extension for eQTLs hotspots	No	Additive

The above equation can be seen as the multiple-outcome extension of the linear model and both SNPTEST [15] and ESS [21, 75] come with this multivariate outcome extensions. Both algorithms use a similar prior set-up, modeling the matrix of regression coefficients $B = \left(B_1,,\ldots,,B_p\right)^{\mathrm{T}}$ by a matrix-variate normal prior $|\Sigma \sim \mathrm{MN}\left(g\left(X_\gamma{}^{\mathrm{T}}X_\gamma\right)^{\lambda},\Sigma\right)$, where $(X\gamma^{\mathrm{T}}X\gamma)\lambda$ is the correlation matrix between the selected markers with $\lambda = 0$ in SNPTEST and $\lambda = -1$ in ESS and Σ is the $q \times q$ matrix modeling the correlation between outcomes (i.e., gene expression levels in different tissues). The model is further specified by placing an Inverse-Wishart prior on Σ, $\Sigma \sim IW(c,Q)$, where c indicates the degrees of freedom and Q is proportional to the expected residual variance. eQTLBMA [20] is another eQTL mapping method that is designed to handle multi-tissue eQTLs, again using a matrix-variate normal prior set-up—but it also uses a hierarchical model which permits heterogeneity between tissues, to allow the estimate of genetic effects both between- and within-tissues.

Frequentist approaches for multiple-tissue eQTL analysis have also been implemented: for example the multivariate version of the ANOVA model (MANOVA) or the Wilks' test statistic [85], a generalization of the F-statistic for multivariate random variables [86]. Multiple-outcome penalized-regression approaches have also been proposed [43, 87], while the R package glmnet [50] includes options that fit multiple-outcome Gaussian models. However, controlling for FWER and FDR is more challenging than in the case of univariate penalized linear regression, and extensions of stability selection [47] for the multivariate problem are still in the stage of development. As a result, interpretation of the multi-tissue eQTL results from multivariate penalized-regression has to be based on the value of regression coefficients, and so thresholding can be a challenge.

5 Empirical Comparison of Frequentist and Bayesian eQTL Mapping

In this section we present an illustrative example of previously reported eQTL mapping for the *Hopx* gene, which in the rat has been shown to be under control by two loci on chromosome 14 (*cis*-eQTL) and chromosome 2 (*trans*-eQTL), respectively; where both *cis*- and *trans*-eQTLs have been experimentally validated [21]. Rather than providing a comprehensive comparison of eQTL mapping methods (systematic simulation studies that compare methods in a variety of scenarios can be found in [75]), our purpose here is to use this empirical eQTL mapping example to facilitate discussion on the comparison between frequentist and Bayesian eQTL mapping approaches. In this eQTL mapping exercise, we have used microarray gene expression data for the *Hopx* gene in two tissues (heart and fat) from 29 recombinant inbred (RI) rat strains (generated by sibling-mating the offspring of a genetic cross until the progenies are inbred), genotyped at 1307 SNPs. Since rats within an RI strain have complete homozygosity at each locus in the genome, each genetic marker allows splitting the rat population in two groups. We considered (1) a single-tissue example using gene expression data from the heart only and (2) a multi-tissue example using gene expression data from both tissues.

5.1 Single-Tissue Example

We mapped genome-wide eQTLs for the heart gene expression data using three frequentist (Matrix-eQTL [7], Kruskal-Wallis test [8] and LASSO from the R package glmnet [50]) and three Bayesian methods (SNPTEST [15], ESS [16, 60], piMASS [17])—*see* Tables 1 and 2 for reference. The parameters and eQTL analysis details for all methods are reported in Table 3.

All six approaches detected a clear *cis*-QTL signal on rat chromosome 14 (close to the location of *Hopx* gene), although for the

Table 3
Details genome-wide eQTL analysis—part 1

Method	Genome-wide eQTL analysis details
Matrix-eQTL	We used the linear additive model as the genotypes are binary. *p*-values were adjusted using the Benjamini-Yekutieli FDR method [88]. We selected eQTL associations at 1 % FDR
Kruskal-Wallis test	The test is the nonparametric equivalent of a one-way ANOVA. We used the kruskal.test function in R to extract *p*-values, and selected eQTLs at 1 % FDR employing Benjamini-Yekutieli method
Glmnet-LASSO	We performed nine-fold cross-validation using the function cv.glmnet, setting alpha = 1 and family = "gaussian". After obtaining estimates on the regression coefficients, these were transformed in posterior probabilities by using stability selection method, implemented in the R package stabs [89]. We declared significance with a threshold of 0.2 on the posterior probabilities
SNPTEST	We ran the Bayesian version of SNPTEST-v.2.5.2 with $\beta \sim N(0, 0.02\sigma^2), \sigma^2 \sim IG(3,2)$ as priors (-prior_qt_mean_b 0 -prior_qt_V_b 0.02 -prior_qt_a 3 -prior_qt_b 2). We called eQTLs at $\log_{10} BF \geq 0.25$
piMASS	We ran piMASS-v.0.90 setting the -prior probability that a SNP is truly associated with the phenotype to range between 1 and 56 (-pmin 1 -pmax 56) and the model size to range from 1 to 100 (-smin 1 –smax 100). We did not impose constraints on the hyperparameter *h* and to the minor allele frequency (-exclude_maf 0). The burn-in phase was set to 10^6 iterations, followed by 10^7 sampling iterations, while only one every ten models considered by the sampling steps was recorded (-w 1,000,000 –s 10,000,000 –num 10). We computed the FDR on the marginal posterior probabilities of inclusion (MPPI) by fitting a mixture of beta distributions, as described in [90]
ESS	We ran GUESS-v.1.1, setting the *a priori* expected model size to $E = 5, S = 3$ (-Egam 5 -Sgam 3) and ran 25,000 steps of which the first 5000 as burn-in (-nsweep 25,000 -burn_in 5000). We computed FDR on the MPPI provided by ESS in the same way as described above for piMASS algorithm

ANOVA and SNPTEST the level of significance reached at the *cis*-eQTL is only a little higher than the rest of the genome. In this example, Glmnet-LASSO and ESS are the only methods that unambiguously detect a *trans*-eQTL signal on chromosome 2. However, Glmnet-LASSO is also picking an additional eQTL signal on chromosome 3. Therefore, in this example, the classic method that implements penalization (Glmnet-LASSO) and one of the Bayesian approaches that uses sparsity (ESS), show good performance in detecting both the *cis*- and *trans*-eQTL signals (however, Glmnet-LASSO is also picking an comparable eQTL signal on chromosome 3). The most striking observation that we can derive from this empirical analysis is that widely used frequentist methods (e.g., Matrix-eQTL) which employ a "one at-a-time" strategy were not able to detect the *trans*-eQTL signal on rat chromosome 2 (with both the *cis*- and *trans*-eQTL signals experimentally validated, as previously reported in [21]), therefore highlighting an important limitation of this approach.

5.2 Multiple-Tissue Example

For the second illustrative example, we ran multivariate ANOVA, Glmnet-LASSO and ESS to jointly map eQTLs for *Hopx* gene expression levels across heart and fat tissues from the 29 rat RI strains used for single-tissue eQTL analysis. The parameters and eQTL analysis details for all methods are reported in Table 4.

Similarly to the results of eQTL analysis in the single tissue, all three methods unambiguously identified a strong *cis*-effect on rat chromosome 14, which therefore suggests the presence of a common *cis*-eQTL in heart and fat tissues. However, only the Glmnet-LASSO and ESS methods were able to identify an additional *trans*-eQTL on rat chromosome 2, suggestive of common *trans*-regulation between the two tissues (as previously shown for other *trans*-eQTL signals conserved across multiple tissues in this genetic system [90]). Glmnet-LASSO identifies the two eQTLs without identifying false-positives, although the signal from the *trans*-effect is much weaker than that of the *cis-effect*. One important issue with Glmnet-LASSO is in the output provided by the algorithm:

Table 4
Details genome-wide eQTL analysis—part 2

Method	Genome-wide eQTL analysis details
MANOVA	We ran a Wilks' test using the R function wilks.test setting method = "rank", and selected associations at 1 % FDR employing Benjamini-Yekutieli method
Glmnet-LASSO	We set parameters to cv.glmnet in the same way as in the single-tissue analysis, but specified family = "mgaussian" to perform multivariate analysis
ESS	The prior set-up was the same as in single-tissue analysis described in the table above, but we instead ran 110,000 sampling steps, of which 10,000 were burn-in. No further specification for multi-outcome analysis is required by ESS that automatically recognizes the multivariate nature of the matrix Υ

although the regression coefficients of the selected markers for the two tissues are clearly reported there is not a simple way to combine them and to transform the tissue-specific effects into a posterior probability, for instance, by the stability selection procedure. In the same eQTL example ESS picks up with a very low MPPI an additional signal from chromosome 10, which is likely to be a false-positive. A noteworthy observation that can be derived from the results of the ESS analysis is that the MPPI of the *trans*-effect is almost doubled in multiple-tissues compared to the single tissue analysis, highlighting the advantage of combining information from multiple sources (in this case tissues). From a biological viewpoint, when compared the MPP of the same *trans*-eQTL detected in the single-tissue analysis (Fig. 1), the signal in the multi-tissue maybe reflects a potential pleiotropic nature of this eQTL.

6 Discussion and Outlook

We discussed the challenges in eQTL mapping and reviewed several commonly used approaches, including their advantages and advantages. In particular, we emphasized the useful features provided by the Bayesian methods. Using a simple yet informative example of polygenic regulation of gene expression in the rat, we illustrated the major differences between frequentist and Bayesian eQTL mapping approaches. In this, we first focused on single-tissue eQTL mapping (Fig. 1), where both *cis*- and *trans*-signals have been previously experimentally validated [21]. We used this demonstrative example to show that frequentist approaches based on a computational efficient strategy that tests for association all transcript-marker pairs ("one at-a-time") were not suitable to detected polygenic control of gene expression. In contrast, methods based on multivariate models, either frequentist (LASSO) or Bayesian (ESS), were able to detect both eQTLs, although ESS performed marginally better as it eliminated possible false-positive associations identified by the LASSO-based approach. We then extended this example to include gene expression data from two tissues for the same gene: the eQTL results were very similar to what observed in the two single tissue cases, with the Bayesian variable selection method detecting unambiguously both *cis*- and *trans*-eQTLs (Fig. 2). This example also highlighted the benefits of using multiple tissues for simultaneous eQTL mapping since, by joint modeling the dependence between tissues, it further increased the power to detect (small-effect) *trans*-eQTLs compared to the single-tissue experiment [20–22].

For Bayesian approaches, the ability to handle the whole set of predictors (genome-wide genetic markers) and model their correlation (i.e., accounting for LD structure) as well as providing the whole posterior distribution of the parameters come at a price.

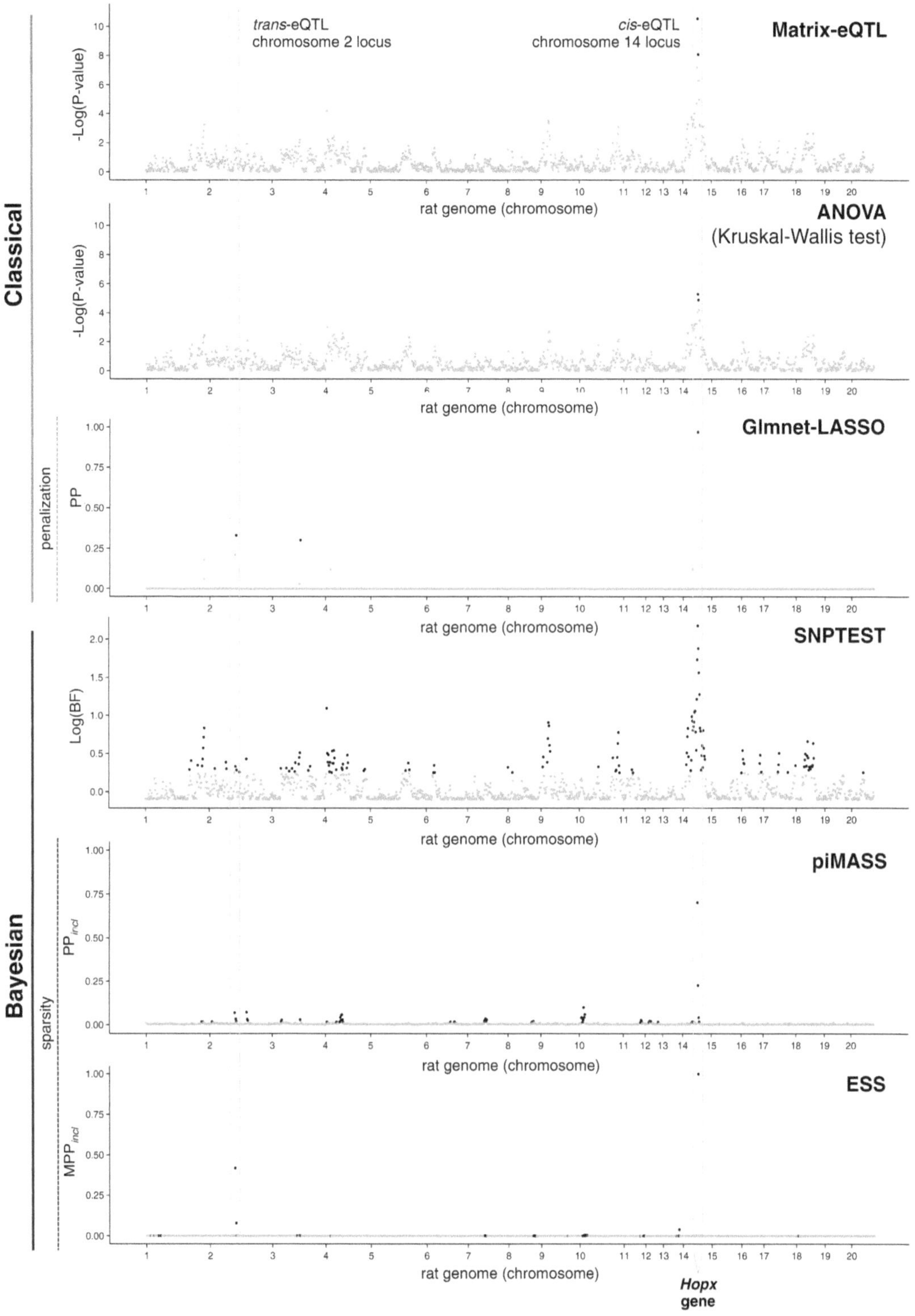

Fig. 1 For each SNP genotyped in the rat genome (*x*-axis) and for each method we report the evidence in support of genetic regulation of *Hopx* gene expression in the heart tissue (*y*-axis). The input consisted of $n \times 1$ expression values and a $n \times p$ matrix of predictors (genome-wide SNPs), where $n = 29$ and $p = 1307$. *Black dots*, associations called at 1 % FDR. *Boxes* highlight the chromosomal locations where the *cis-* and *trans-*eQTLs are located, respectively

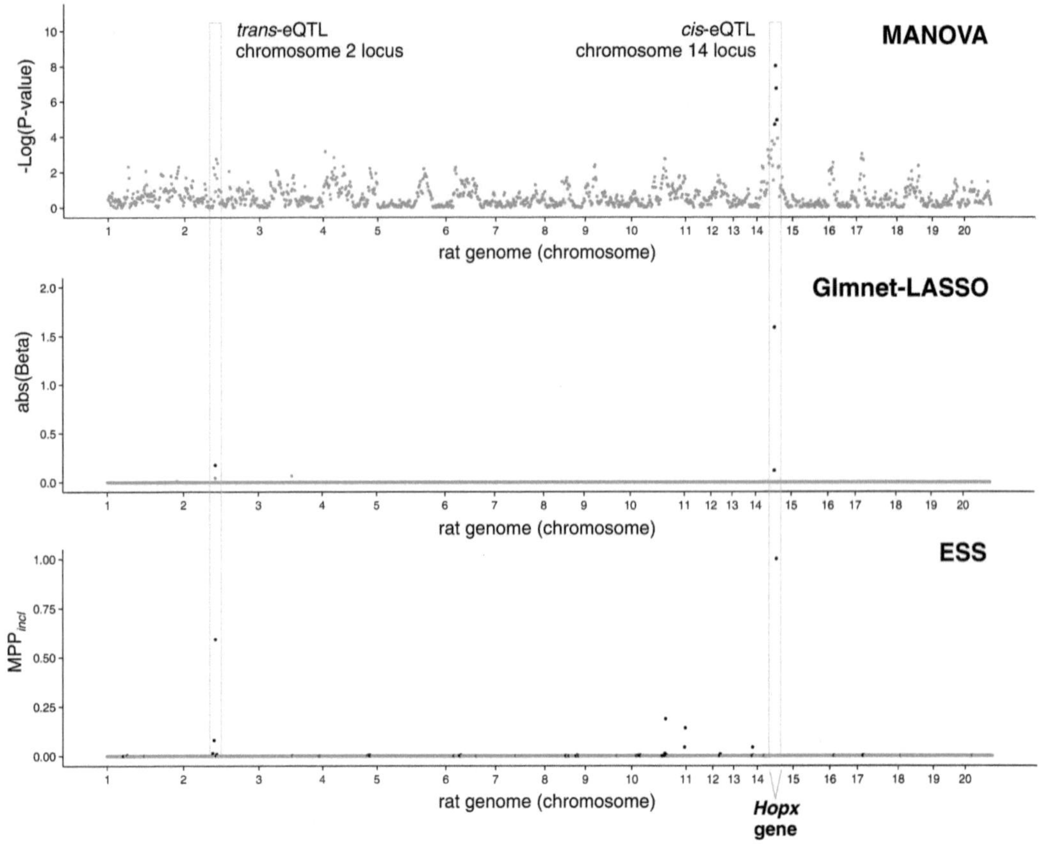

Fig. 2 For each SNP genotyped in the rat genome (*x*-axis) and for each method we report the evidence in support of genetic regulation of *Hopx* gene expression simultaneously in the heart and fat tissues (*y*-axis). The input consisted of $n \times 2$ expression values (fat and heart, respectively) and a $n \times p$ matrix of predictor variables (genome-wide SNPs), where $n = 29$ and $p = 1307$. *Black dots*, associations called at 1 % FDR. For the Glmnet-LASSO, the *blue* and *black dots* indicate the absolute values of β-coefficients estimated in fat and heart tissues, respectively. *Boxes* highlight the chromosomal locations where the *cis*- and *trans*-eQTLs are located, respectively

The more traditional (frequentist) eQTL approaches (such as Matrix-eQTL [7]) have the attractive feature of computational efficiency compared to the more demanding BVS methods. This might account for the common application of frequentist eQTL mapping methods in biomedical research. However, as highlighted in our illustrative examples, the high computational efficiency of frequentist approaches might miss the polygenic control of gene expression. This can have important implications when both *cis*- and *trans*-eQTLs are investigated at the genome-wide level, usually resulting in a smaller fraction of "replicable" *trans*-eQTLs as compared with *cis*-eQTLs, and advocating the use of larger populations to boost detection of small *trans*-effects [91].

However, recent advancements in high-performance computing have rendered the application of MCMC methods feasible even for hundreds of thousands of predictors in hundreds (if not thousands) of individuals [75]—which now justifies the increasing popularity of the Bayesian eQTL mapping methods. In contrast, although recent advances in the computational aspects of the LASSO solution [92], frequentist penalized-regression methods still need time-consuming cross-validation procedure to estimate the penalty parameter λ. In the case of Elastic Net a two-dimensional grid is required in order to select the optimal λ_1, λ_2 penalties. Selecting the optimal parameters, however, necessitates a very fine-grained grid of penalties to be analyzed, which is even more computationally expensive.

Regarding interpretation of the eQTL results, in BVS approaches all the models visited by the search algorithm (after the burn-in) are kept and summarized into a marginal posterior probability of inclusion (MPPI). Penalized-regression models usually output the estimates of regression coefficient values, which can vary largely between experiments and therefore are less safe for declaring eQTL associations consistently across studies. Moreover, estimation of the FDR from the regression coefficients is not possible, so one is limited to controlling family-wise error rates, a more conservative approach that can lead to false-negatives. In contrast, several techniques that control the FDR from the MPPI are now available, making the genome-wide control of the significance level less of a problem for Bayesian eQTL methods. In addition, although not directly investigated in our illustrative examples, the Bayesian prior set-up offers more flexibility to consider (and explore) different eQTL models, for example by specifying the number of expected eQTLs and their effect size or by using genomic locations of the transcripts to improve the accuracy of the posterior distribution for the location of the eQTL [93].

In summary, we advocate that Bayesian approaches are in general more flexible to analyze complex genetic regulation of expression than frequentist methods. In particular, Bayesian eQTL mapping strategies can adapt naturally to a wider range of applications, such as (1) detection of polygenic effects on gene expression [21], (2) epistatic eQTL interactions [62], (3) eQTLs hotspots [74] and (4) eQTLs and eQTLs hotspots across multiple-tissues [21, 74, 83]. We also argue that using Bayes Factors might provide a more objective way to call statistically significant eQTLs [55, 94] and compare them across studies. Conversely, using computationally inexpensive *p*-values generated by frequentist approaches to call significant eQTLs requires a threshold for genome-wide significance that can varies largely with sample size as well as with other study-specific factors. While this issue is well known and yet often ignored, it is likely to be highly relevant to the development of reference eQTL databases and resources. Since eQTL analyses have been proved useful in the identification of molecular pathways affecting disease susceptibility, e.g., [6, 90, 95, 96], it is generally advisable to use truly multivariate eQTL mapping strategies that can provide more flexibility in modeling complex data structures and can have enhanced interpretability of the results. In this respect, Bayesian mapping approaches now provide a valid alternative to traditional "one at-a-time" frequentist methods and a richer and easy to interpret output than penalized-regression methods.

Acknowledgments

We acknowledge funding from Medical Research Council Grant G 1002319 (L.B.), MR/M013138/1 (L.B.), MR/M004716/1 (M.I. and E.P.) and Duke-NUS Graduate Medical School Singapore (E.P.).

References

1. Guo H, Fortune MD, Burren OS et al (2015) Integration of disease association and eQTL data using a Bayesian colocalisation approach highlights six candidate causal genes in immune-mediated diseases. Hum Mol Genet 24:3305–3313. doi:10.1093/hmg/ddv077
2. Pierce BL, Tong L, Chen LS et al (2014) Mediation analysis demonstrates that trans-eQTLs are often explained by cis-mediation: a genome-wide analysis among 1,800 South Asians. PLoS Genet 10:e1004818. doi:10.1371/journal.pgen.1004818
3. Kang H, Kerloc'h A, Rotival M et al (2014) Kcnn4 is a regulator of macrophage multinucleation in bone homeostasis and inflammatory disease. Cell Rep 8:1210–1224. doi:10.1016/j.celrep.2014.07.032
4. Rotival M, Zeller T, Wild PS et al (2011) Integrating genome-wide genetic variations and monocyte expression data reveals trans-regulated gene modules in humans. PLoS Genet 7:e1002367. doi:10.1371/journal.pgen.1002367
5. Fehrmann RSN, Jansen RC, Veldink JH et al (2011) Trans-eQTLs reveal that independent genetic variants associated with a complex phenotype converge on intermediate genes, with a major role for the HLA. PLoS Genet 7:e1002197. doi:10.1371/journal.pgen.1002197
6. Small KS, Hedman AK, Grundberg E et al (2011) Identification of an imprinted master trans regulator at the KLF14 locus related to multiple metabolic phenotypes. Nat Genet 43:561–564. doi:10.1038/ng.833
7. Shabalin AA (2012) Matrix eQTL: ultra fast eQTL analysis via large matrix operations. Bioinformatics 28:1353–1358. doi:10.1093/bioinformatics/bts163
8. MacDonald JH (2009) Kruskal-Wallis test. Biol Handb Stat 165–172. doi:10.1002/9780470479216.corpsy0491
9. Yang T-P, Beazley C, Montgomery SB et al (2010) Genevar: a database and Java application for the analysis and visualization of SNP-gene associations in eQTL studies. Bioinformatics 26:2474–2476. doi:10.1093/bioinformatics/btq452
10. Broman KW, Wu H, Sen S, Churchill GA (2003) R/qtl: QTL mapping in experimental crosses. Bioinformatics 19:889–890
11. Clayton D, Leung H-T (2007) An R package for analysis of whole-genome association studies. Hum Hered 64:45–51. doi:10.1159/000101422
12. Sun W (2009) eQTL analysis by Linear Model. In: http://www.bios.unc.edu/~weisun/software/eMap.pdf. Accessed 20 Oct 2015
13. Qi J, Asl HF, Björkegren J, Michoel T (2014) kruX: matrix-based non-parametric eQTL discovery. BMC Bioinformatics 15:11. doi:10.1186/1471-2105-15-11
14. Gao C, Tignor NL, Salit J et al (2014) HEFT: eQTL analysis of many thousands of expressed genes while simultaneously controlling for hidden factors. Bioinformatics 30:369–376. doi:10.1093/bioinformatics/btt690
15. Marchini J, Howie B, Myers S et al (2007) A new multipoint method for genome-wide association studies by imputation of genotypes. Nat Genet 39:906–913. doi:10.1038/ng2088
16. Bottolo L, Chadeau-hyam M, Hastie DI et al (2011) ESS++: a C++ objected-oriented algorithm for Bayesian stochastic search model exploration. Bioinformatics 27:587–588. doi:10.1093/bioinformatics/btq684
17. Guan Y, Stephens M (2011) Bayesian variable selection regression for genome-wide association

studies and other large-scale problems. Ann Appl Stat 5:1780–1815
18. Scott-Boyer MP, Imholte GC, Tayeb A et al (2012) An integrated hierarchical Bayesian model for multivariate eQTL mapping. Stat Appl Genet Mol Biol. doi: 10.1515/1544-6115.1760
19. He X, Fuller CK, Song Y et al (2013) Sherlock: detecting gene-disease associations by matching patterns of expression QTL and GWAS. Am J Hum Genet 92:667–680. doi:10.1016/j.ajhg.2013.03.022
20. Flutre T, Wen X, Pritchard J, Stephens M (2013) A statistical framework for joint eQTL analysis in multiple tissues. PLoS Genet 9:e1003486. doi:10.1371/journal.pgen.1003486
21. Petretto E, Bottolo L, Langley SR et al (2010) New insights into the genetic control of gene expression using a Bayesian multi-tissue approach. PLoS Comput Biol 6:e1000737. doi:10.1371/journal.pcbi.1000737
22. Sul JH, Han B, Ye C et al (2013) Effectively identifying eQTLs from multiple tissues by combining mixed model and meta-analytic approaches. PLoS Genet 9:e1003491. doi:10.1371/journal.pgen.1003491
23. Brem RB, Yvert G, Clinton R, Kruglyak L (2002) Genetic dissection of transcriptional regulation in budding yeast. Science 296:752–755. doi:10.1126/science.1069516
24. Dudoit S, Yang YH, Callow MJ, Speed TP (2002) Statistical methods for identifying differentially expressed genes in replicated cDNA microarray experiments. Stat Sin 12:111–139. doi:10.1146/annurev.psych.53.100901.135153
25. Gerrits A, Li Y, Tesson BM et al (2009) Expression quantitative trait loci are highly sensitive to cellular differentiation state. PLoS Genet 5:e1000692. doi:10.1371/journal.pgen.1000692
26. Zhou X, Stephens M (2014) Efficient multivariate linear mixed model algorithms for genome-wide association studies. Nat Methods 11:407–409. doi:10.1038/nmeth.2848
27. Narahara M, Higasa K, Nakamura S et al (2014) Large-scale East-Asian eQTL mapping reveals novel candidate genes for LD mapping and the genomic landscape of transcriptional effects of sequence variants. PLoS One 9:e100924. doi:10.1371/journal.pone.0100924
28. Duggal G, Wang H, Kingsford C (2014) Higher-order chromatin domains link eQTLs with the expression of far-away genes. Nucleic Acids Res 42:87–96. doi:10.1093/nar/gkt857
29. Gatti DM, Shabalin AA, Lam T-C et al (2009) FastMap: fast eQTL mapping in homozygous populations. Bioinformatics 25:482–489. doi:10.1093/bioinformatics/btn648
30. Benjamini Y, Hochberg Y (1995) Controlling the false discovery rate: a practical and powerful approach to multiple testing. J R Stat Soc Ser B Methodol 57:289–300. doi:10.2307/2346101
31. GTEx Consortium (2013) The genotype-tissue expression (GTEx) project. Nat Genet 45:580–585. doi:10.1038/ng.2653
32. Hoerl AE, Kennard RW (1970) Ridge regression: biased estimation for nonorthogonal problems. Technometrics 12:55–67. doi:10.1080/00401706.1970.10488634
33. Tibshirani R (2011) Regression shrinkage and selection via the lasso: a retrospective. J R Stat Soc Ser B Stat Methodol 73:273–282. doi:10.1111/j.1467-9868.2011.00771.x
34. Wu TT, Chen YF, Hastie T et al (2009) Genome-wide association analysis by lasso penalized logistic regression. Bioinformatics 25:714–721. doi:10.1093/bioinformatics/btp041
35. Zou H (2006) The adaptive lasso and its oracle properties. J Am Stat Assoc 101:1418–1429. doi:10.1198/016214506000000735
36. Tibshirani R, Saunders M, Rosset S et al (2005) Sparsity and smoothness via the fused lasso. J R Stat Soc Ser B Stat Methodol 67:91–108. doi:10.1111/j.1467-9868.2005.00490.x
37. Zou H, Hastie T (2005) Regularization and variable selection via the elastic net. J R Stat Soc Ser B Stat Methodol 67:301–320. doi:10.1111/j.1467-9868.2005.00503.x
38. Yuan M, Lin Y (2006) Model selection and estimation in regression with grouped variables. J R Stat Soc Ser B 68:49–67. doi:10.1111/j.1467-9868.2005.00532.x
39. Kim S, Xing EP (2009) Statistical estimation of correlated genome associations to a quantitative trait network. PLoS Genet 5:e1000587. doi:10.1371/journal.pgen.1000587
40. Wang W, Zhang X (2011) Network-based group variable selection for detecting expression quantitative trait loci (eQTL). BMC Bioinformatics 12:269
41. Lee S, Xing EP (2012) Structured input-output Lasso, with application to eQTL mapping, and a thresholding algorithm for fast estimation. Available at: https://arxiv.org/abs/1205.1989
42. Cheng W, Zhang X, Guo Z et al (2014) Graph-regularized dual Lasso for robust eQTL mapping. Bioinformatics 30:139–148. doi:10.1093/bioinformatics/btu293
43. Kim S, Xing EP (2012) Tree-guided group lasso for multi-response regression with structured sparsity, with an application to eQTL mapping. Ann Appl Stat 6:1095–1117. doi:10.1214/12-AOAS549

44. Leng C, Lin Y, Wahba G (2006) A note on the lasso and related procedures in model selection. Stat Sin 16:1273–1284
45. Rakitsch B, Lippert C, Stegle O, Borgwardt K (2013) A Lasso multi-marker mixed model for association mapping with population structure correction. Bioinformatics 29:206–214. doi:10.1093/bioinformatics/bts669
46. Brown AA, Richardson S, Whittaker J (2011) Application of the Lasso to expression quantitative trait loci mapping. Stat Appl Genet Mol Biol 10:1–35. doi:10.2202/1544-6115
47. Meinshausen N, Bühlmann P (2010) Stability selection. J R Stat Soc Ser B 72:417–473. doi:10.1111/j.1467-9868.2010.00740.x
48. Shah RD, Samworth RJ (2013) Variable selection with error control: another look at stability selection. J R Stat Soc Ser B 75:55–80. doi:10.1111/j.1467-9868.2011.01034.x
49. Waldmann P, Mészáros G, Gredler B et al (2013) Evaluation of the lasso and the elastic net in genome-wide association studies. Front Genet 4:270. doi:10.3389/fgene.2013.00270
50. Friedman J, Hastie T, Tibshirani R (2010) Regularization paths for generalized linear models via coordinate descent. J Stat Softw 33:1–22. doi:10.18637/jss.v033.i01
51. Beaumont MA, Rannala B (2004) The Bayesian revolution in genetics. Nat Rev Genet 5:251–261. doi:10.1038/nrg1318
52. Kass RE, Raftery AE (1995) Bayes factors. J Am Stat Assoc 90:773–795. doi:10.2307/2291091
53. O'Hara RB, Sillanpää MJ (2009) A review of Bayesian variable selection methods: what, how and which. Bayesian Anal 4:85–117
54. Servin B, Stephens M (2007) Imputation-based analysis of association studies: candidate regions and quantitative traits. PLoS Genet 3:1296–1308. doi:10.1371/journal.pgen.0030114
55. Stephens M, Balding DJ (2009) Bayesian statistical methods for genetic association studies. Nat Rev Genet 10:681–690. doi:10.1038/nrg2615
56. Lee S-I, Dudley AM, Drubin D et al (2009) Learning a prior on regulatory potential from eQTL data. PLoS Genet 5:e1000358. doi:10.1371/journal.pgen.1000358
57. Das A, Morley M, Moravec CS et al (2015) Bayesian integration of genetics and epigenetics detects causal regulatory SNPs underlying expression variability. Nat Commun 6:8555. doi:10.1038/ncomms9555
58. Balding DJ (2006) A tutorial on statistical methods for population association studies. Nat Rev Genet 7:781–791. doi:10.1038/nrg1916
59. Kendziorski CM, Chen M, Yuan M et al (2006) Statistical methods for expression quantitative trait loci (eQTL) mapping. Biometrics 62:19–27. doi:10.1111/j.1541-0420.2005.00437.x
60. Bottolo L, Richardsony S (2010) Evolutionary stochastic search for bayesian model exploration. Bayesian Anal 5:583–618. doi:10.1214/10-BA523
61. Zhang M, Montooth KL, Wells MT et al (2005) Mapping multiple quantitative trait loci by Bayesian classification. Genetics 169:2305–2318. doi:10.1534/genetics.104.034181
62. Zhang M, Zhang D, Wells MT (2008) Variable selection for large p small n regression models with incomplete data: mapping QTL with epistases. BMC Bioinformatics 9:251. doi:10.1186/1471-2105-9-251
63. Liu J, Liu Y, Liu X, Deng H-W (2007) Bayesian mapping of quantitative trait loci for multiple complex traits with the use of variance components. Am J Hum Genet 81:304–320. doi:10.1086/519495
64. Chun H (2009) Expression quantitative trait loci mapping with multivariate sparse partial least squares regression. Genetics 182:79–90. doi:10.1534/genetics.109.100362
65. Chen W, Ghosh D, Raghunathan TE, Sargent DJ (2009) Bayesian variable selection with joint modeling of categorical and survival outcomes: an application to individualizing chemotherapy treatment in advanced colorectal cancer. Biometrics 65:1030–1040. doi:10.1111/j.1541-0420.2008.01181.x
66. Chipman H, George EI, McCulloch RE (2001) The practical implementation of Bayesian model selection. Institute of Mathematical Statistics, Beachwood, OH, pp 65–116
67. Brown PJ, Vannucci M, Fearn T (2002) Bayes model averaging with selection of regressors. J R Stat Soc Ser B 64:519–536. doi:10.1111/1467-9868.00348
68. Park T, Casella G (2008) The Bayesian Lasso. J Am Stat Assoc 103:681–686. doi:10.1198/016214508000000337
69. Carvalho CM, Polson NG, Scott JG (2010) The horseshoe estimator for sparse signals. Biometrika 97:465–480. doi:10.1093/biomet/asq017
70. Dempster AP, Laird NM, Rubin DB (1977) Maximum likelihood from incomplete data via the EM algorithm. J R Stat Soc Ser B 39:1–38, 10.1.1.133.4884
71. Ročková V, George EI (2014) EMVS: the EM approach to Bayesian variable selection. J Am Stat Assoc 109:828–846. doi:10.1080/01621459.2013.869223
72. Gelfand AE, Smith AFM (2012) Sampling-based approaches to calculating marginal densities. J Am Stat Assoc 85:398–409. doi:10.1080/01621459.1990.10476213

73. Hans C, Dobra A, West M (2007) Shotgun stochastic search for “Large p ” regression. J Am Stat Assoc 102:507–516. doi:10.1198/016214507000000121
74. Bottolo L, Petretto E, Blankenberg S et al (2011) Bayesian detection of expression quantitative trait loci hot spots. Genetics 189:1449–1459. doi:10.1534/genetics.111.131425
75. Bottolo L, Chadeau-Hyam M, Hastie DI et al (2013) GUESS-ing polygenic associations with multiple phenotypes using a GPU-based evolutionary stochastic search algorithm. PLoS Genet 9:e1003657. doi:10.1371/journal.pgen.1003657
76. Barbieri MM, Berger JO (2015) Optimal predictive model selection. Ann Stat 32:870–897
77. Broët P, Lewin A, Richardson S et al (2004) A mixture model-based strategy for selecting sets of genes in multiclass response microarray experiments. Bioinformatics 20:2562–2571. doi:10.1093/bioinformatics/bth285
78. Efron B (2008) Microarrays, empirical bayes and the two-groups model. Stat Sci 23:1–22. doi:10.1214/08-STS236REJ
79. Strimmer K (2008) fdrtool: a versatile R package for estimating local and tail area-based false discovery rates. Bioinformatics 24:1461–1462. doi:10.1093/bioinformatics/btn209
80. Zellner A, Siow A (1980) Posterior odds ratios for selected regression hypotheses. Trab Estad Y Investig Oper 31:585–603. doi:10.1007/BF02888369
81. Hastings WK (1970) Monte Carlo sampling methods using Markov chains and their applications. Biometrika 57:97–109. doi:10.2307/2334940
82. Eiben AE, Raué PE, Ruttkay Z (1994) Genetic algorithms with multi-parent recombination. In: Parallel problem solving from nature — PPSN III. Springer, Heidelberg, pp 78–87
83. Lewin A, Saadi H, Peters JE et al (2016) MT-HESS: an efficient Bayesian approach for simultaneous association detection in OMICS datasets, with application to eQTL mapping in multiple tissues. Bioinformatics 32:523–32. doi:10.1093/bioinformatics/btv568
84. Ardlie KG, Deluca DS, Segre AV et al (2015) The genotype-tissue expression (GTEx) pilot analysis: multitissue gene regulation in humans. Science 348:648–660. doi:10.1126/science.1262110
85. Todorov V, Filzmoser P (2010) Robust statistic for the one-way MANOVA. Comput Stat Data Anal 54:37–48. doi:10.1016/j.csda.2009.08.015
86. Kim S, Becker J, Bechheim M et al (2014) Characterizing the genetic basis of innate immune response in TLR4-activated human monocytes. Nat Commun 5:5236. doi:10.1038/ncomms6236
87. Chen X, Shi X, Xu X et al (2012) A two-graph guided multi-task Lasso approach for eQTL mapping. ece.ubc.ca XX:208–217
88. Benjamini Y, Yekutieli D (2001) The control of the false discovery rate in multiple testing under dependency. Ann Stat 29:1165–1188. doi:10.1214/aos/1013699998
89. Hofner B, Boccuto L, Göker M (2015) Controlling false discoveries in high-dimensional situations: boosting with stability selection. BMC Bioinformatics 16:144
90. Heinig M, Petretto E, Wallace C et al (2010) A trans-acting locus regulates an anti-viral expression network and type 1 diabetes risk. Nature 467:460–464. doi:10.1038/nature09386
91. Grundberg E, Small KS, Hedman ÅK et al (2012) Mapping cis- and trans-regulatory effects across multiple tissues in twins. Nat Genet 44:1084–1089. doi:10.1038/ng.2394
92. Wu TT, Lange K (2008) Coordinate descent algorithms for lasso penalized regression. Ann Appl Stat 2:224–244
93. Gelfond JAL, Ibrahim JG, Zou F (2007) Proximity model for expression quantitative trait loci (eQTL) detection. Biometrics 63:1108–1116. doi:10.1111/j.1541-0420.2007.00778.x
94. Wakefield J (2009) Bayes factors for genome-wide association studies: comparison with P-values. Genet Epidemiol 33:79–86. doi:10.1002/gepi.20359
95. Emilsson V, Thorleifsson G, Zhang B et al (2008) Genetics of gene expression and its effect on disease. Nature 452:423–428. doi:10.1038/nature06758
96. Westra H-J, Peters MJ, Esko T et al (2013) Systematic identification of trans eQTLs as putative drivers of known disease associations. Nat Genet 45:1238–1243. doi:10.1038/ng.2756

Chapter 9

Epigenetics and Control of RNAs

Henrike Maatz*, Sebastiaan van Heesch*, Franziska Kreuchwig, Allison Faber, Eleonora Adami, Norbert Hubner, and Matthias Heinig

Abstract

Histone modifications are epigenetic marks that fundamentally impact the regulation of gene expression. Integrating histone modification information in the analysis of gene expression traits (eQTL mapping) has been shown to significantly enhance the prediction of eQTLs. In this chapter, we describe (1) how to perform quantitative trait locus (QTL) analysis using histone modification levels as traits and (2) how to integrate these data with information on RNA expression for the elucidation of the epigenetic control of transcript levels. We will provide a comprehensive introduction into the topic, describe in detail how ChIP-seq data are analyzed and elaborate on how to integrate ChIP-seq and RNA-seq data from a segregating disease animal model for the identification of the epigenetic control of RNA expression.

Key words Histone modifications, RNA expression, ChIP-seq, Integrative analysis, Recombinant inbred panel, QTL mapping

1 Introduction and Background

Histones, the DNA-packaging proteins, are the main component of chromatin within the eukaryotic nucleus. Two of each H2A, H2B, H3, and H4 histones form compact protein cores around which the DNA is wrapped resulting in nucleosomes, the basic repeating unit of DNA packaging. The histone tails, in particular the N-terminal tails that reach out of the histone multimer, are frequently posttranslationally modified by addition or removal of acetyl, phosphoryl and methyl groups. With the advent of high-throughput sequencing technologies histone modification patterns have been extensively characterized at high resolution on a genome-wide scale [1–3]. This has become feasible by combining chromatin immunoprecipitation using histone modification-specific antibodies with high-throughput sequencing (ChIP-seq) of the co-immunoprecipitated

*These two authors have contributed equally for this chapter.

Klaus Schughart and Robert W. Williams (eds.), *Systems Genetics: Methods and Protocols*, Methods in Molecular Biology, vol. 1488, DOI 10.1007/978-1-4939-6427-7_9,

histone-bound DNA. These studies show that recurrent combinations of histone modifications are associated with defined chromatin states corresponding to repressed, poised, and active promoters, strong and weak enhancers, as well as transcribed and repressed regions of the genome. Histone modifications thus constitute a regulatory code that plays a critical role in genome organization and in turn the epigenetic regulation of gene expression.

Given the strong correlation of histone modifications with gene expression, histone modification information can be used to enhance the discovery of gene expression traits whose variation is attributed to genetic factors. Segregating populations such as the rat HXB/BXH recombinant inbred (RI) panel are especially suited for this integrative analysis. The HXB/BXH RI panel is a well-characterized model system for the metabolic syndrome [4–6]. The panel was generated by breeding spontaneously hypertensive rats (SHR/Ola, referred to as SHR) and normotensive Brown-Norway rats (BN-*Lx*/Cub, referred to as BN) to produce an F2 population. This F2 population was subsequently brother–sister mated for currently beyond 80 generations [7]. A total of 30 RI strains were derived from crosses of female SHR and male BN rats (HXB strains, $n=20$) or female BN and male SHR rats (BXH strains, $n=10$). This RI panel has been extensively used to study the inheritance of gene expression and other (molecular) phenotypic traits. Genome-wide gene expression profiles have been collected and combined with available genotype data for the identification of underlying expression quantitative trait loci (eQTLs) [5].

Recently it was shown that genetic variation also affects histone modifications and thus shapes the landscape of transcriptionally competent chromatin [8–12]. Most genetic variants linked to disease by large-scale genome-wide association studies map to the noncoding regions of the genome and may exert their effect by influencing chromatin conformation and subsequently gene expression. In the HXB/BXH RI panel, histone modifications vary extensively in methylation levels among the strains. Therefore they can be considered a quantitative trait [10]. HistoneQTL mapping in the RI strains can provide information on the extent to which *cis*- and *trans*-acting factors affect chromatin status. In addition, histoneQTLs have been shown to be important predictors of gene expression changes.

Here we will describe how histoneQTL mapping and subsequent integration with expression data can increase the detection of eQTLs in RI panels using ChIP-seq, RNA-seq and genotype data. Recently, quality standards for assaying peak-like features in ChIP-seq experiments have been established and will be discussed as they are key steps to define high-quality intermediate phenotypes. Calling of broad histone modification marks such as H3K27me3 and H4K20me1 is not trivial and requires specific considerations, because most software for the identification of modified regions is designed for narrow peak-like features [13].

2 Materials

Our analysis employs high-resolution ChIP-seq data for histone marks including H3K4me3 and H3K27me3 with genome–wide RNA-seq and genotype data. ChIP-seq data sets can be generated from mono-nucleosome preparations as described [1, 10] or by any other protocol that is suitable to produce high-resolution ChIP-seq data. For many RI panels such as the HXB/BXH panel, high-throughput expression profiles and genotype data exist and may not need to be generated anew. Instead, these data can be downloaded from sources such as EBI's ArrayExpress (http://www.ebi.ac.uk/arrayexpress) [14] or the European Nucleotide Archive (ENA; http://www.ebi.ac.uk/ena) [15]. For HXB/BXH transcriptome and genotype data, the respective accession numbers are E-MTAB-1102 (Array express) and ERP001430 (ENA). Where no such data exist, the Illumina TruSeq kit is a good option to produce rRNA-depleted poly(A)$^+$ mRNA as described in Rintisch et al. [10]. Genotype data can be generated using genotyping arrays (if variant positions are known) or by applying whole genome resequencing using for example Illumina's Nextera DNA kit or Illumina's TruSeq DNA PCR-Free kit, followed by sequence read alignment and SNP calling (not further discussed in this chapter).

3 Methods

We will start our ChIP-seq data analysis with the raw sequence data. We assume that you have obtained one file in fastq format for each of your samples. **Steps 1–3** must be performed on each of the files. **Steps 4–6** identify and quantify epigenetic traits that will be used as intermediate phenotypes in the QTL analysis (**steps 7–8**). Finally, results from multiple phenotypes such as levels of histone modifications and gene expression will be integrated in **step 9**. Especially for the data integration of multiple traits it is advisable to stick to a uniform file-naming convention; a distinct prefix that denotes the name of the trait and common suffixes that denote the data type. For many of the steps, a wide variety of tools exist allowing for variations in the protocol.

1. Sequence level quality control

   The first step is to assess the base calling quality and base pair composition biases in the raw sequencing reads. FastQC [16] is one of the many tools for this purpose. It generates an HTML page summarizing basic statistics with diagnostic plots.

   ```
   fastqc sample1.fastq
   ```

2. Alignment

Align the reads to the genome using bowtie [17] using only reads that are uniquely mapping to the genome, then save alignments in sam format [18].

```
bowtie -m 1 --best --strata -S path/to/genome_index \
sample1.fastq sample1.sam
```

Convert the alignments to binary sam format (bam) and then sort and index for faster file access.

```
samtools view -b -o sample1.bam -S sample1.sam
samtools sort sample1.bam sample1_sorted
mv sample1_sorted.bam sample1.bam
samtools index sample1.bam
```

Remove duplicated reads, which are likely PCR artifacts, and index the cleaned file. This file will be the basis for the downstream analysis.

```
samtools rmdup sample1.bam sample1_rmdup.bam
samtools index sample1_rmdup.bam
```

3. Alignment level QC

First we check basic statistics of the alignment such as the total number of reads sequenced, the number of uniquely mapped reads and the number of uniquely mapped reads after duplicate removal.

```
samtools flagstat sample1.bam > sample1.stat
samtools flagstat sample1-rmdup.bam > sample-rmdup.stat
```

Next, check the quality of the alignments based on the cross-correlation of reads mapping to the forward and the reverse strand using spp_nodups from the *phantompeakqualtools* package [19].

```
run_spp_nodups.R -c=sample1_rmdup.bam -savp \
-out=sample1_qc.txt
```

After performing this step for each of the samples, the individual results are summarized in a table and barplots to ensure that all samples have been sequenced with comparable depth and quality. Here we will use functions from the R package seqQTL [20].

```
qc = spp.qc.summary("./bam/", pattern="_qc.txt$")
plot.spp.qc(qc)
```

4. Identification/definition of modified regions

De novo identification of modified regions is necessary when you are dealing with peak-like features without any prior knowledge about the location of these features, as is the case with enhancer marks. The identification of modified regions is performed using peak calling algorithms [21].

```
macs callpeak -t sample1_rmdup.bam -c input_rmdup.bam \
--name=macs-sample1 --format=BAM
```

De novo identification of modified regions is especially challenging for histone marks with broad genomic footprints, such as the polycomb associated H3K27me3 and heterochromatin associated H3K9me3. Specialized peak callers such as histone-HMM can be used to call these large modified domains [13]. As additional input the tool requires a file with chromosome length information as provided by the UCSC data base. The "-t" option selects just one chromosome for parameter estimation to speed up the analysis.

```
histoneHMM_call_regions.R -c chromInfo.txt -t chr20 \
-o histoneHMM-H3K27me3-sample1 H3K27me3-sample1_rmdup.bam
```

Modified regions identified in individual strains are then summarized. To assure identification of high-quality regions, we consider only regions that have been found in a minimum number of strains or (biological) replicates.

```
peaks = merge.peaks(dir="./bam/", FDR = 5, pattern="macs-.*")
save(peaks, file="peaks.RData")
GR2gff(peaks, file="peaks.gff")
```

Other histone marks have well known genomic positions because they are associated with specific gene features. For example, H3K4me3 marks active promoter regions and H3K36me3 and H3K79me2 mark transcribed regions. Because the positions of these marks can be obtained from gene structure annotation databases, we can create cumulative coverage plots that visualize their patterns over the specified gene features.

```
genes = gtf2GR("annotation.gtf", "gene_id")
genes = genes[values(genes)[, "type"] == "gene"]
TSS = promoters(genes, 2000, 2000)
cvg = coverageBamInGRanges("sample1-rmdup.bam", TSS)
plot(-2000:2000, colMeans(cvg))
```

From the coverage plots it becomes clear which regions to use for the quantification of histone modifications. Also, modified regions can be identified as annotated features that satisfy a certain minimal read coverage (**steps 5–6**).

5. Quantification of regions

   Next we count the reads within the regions defined in **step 4**. Here we will use functions from the R package seqQTL [20]. The result will be a feature count matrix, which is the basis for defining our intermediate quantitative phenotypes.

```
counts = get.count.matrix(TSS, dir="bam/", \
pattern="H3K4me3.*-rmdup.bam$")
colnames(counts) = gsub("-rmdup.bam$", "",
colnames(counts))
write.table(counts, file="H3K4me3-counts.txt",
sep="\t", \
quote=F)
```

6. Filtering and normalization

   Filtering and normalization are important steps to avoid artifacts from very low and thus unreliable read counts. Systematic read count differences are caused by differences in sequencing depth in the individual samples and outliers that might be present in the data. The procedure described here is based on the recommendations for RNA-seq-based eQTL mapping developed in the GTEx consortium [22].

First we ensure that the sequencing depth, especially the coverage in the regions of interest, is comparable between samples by creating boxplots of the raw, log-transformed count matrix. Note that we also add a pseudo count of "one" to retain the zero counts in the plot after the log transformation.

```
boxplot.matrix(log10(counts + 1))
```

Depending on whether we use regions of fixed width (such as 2 kb around the transcriptional start site (TSS)) or variable width (such as the results from peak calling or coverage in gene bodies), the read counts need to be width-normalized to obtain reads per kilobase. To account for different total sequencing depth, counts are further normalized to reads per kilobase per million (RPKM) sequenced.

```
annotation = gtf2GR("gene_annotation.gtf", c("gene_id"))
annotation = annotation[values(annotation)[, "type"] == "exon"]
rpkm = get.rpkm(annotation, counts)
```

Inspect the distribution of RPKM values for a bimodal pattern. This can be used to derive thresholds for discriminating modified from unmodified regions.

```
hist(log10(rpkm + 1), breaks=50)
```

In the GTEx protocol, all regions that have greater than zero counts in at least $p = 50\%$ of the samples are used.

```
p = 0.5
```

In a recombinant inbred scenario with 100 % penetrance we expect a segregation pattern of around 50 % of strains showing the modification and the other 50 % showing no modification (low counts). A suitable percentage threshold (p) corresponding to a significance level (alpha) can be computed from the binomial distribution. Alpha is the probability of observing no counts in at least the fraction p of the strains given that the true segregation pattern is 50:50.

```
N = ncol(rpkm)
alpha = 0.01
p = qbinom(alpha, prob=0.5, size=N) / N
```

In both cases the actual filtering process is performed like this.

```
expressed = apply(rpkm > 0, 1, function(x) sum(x) /
ncol(rpkm) > p)
filtered = rpkm[expressed,]
```

Filtered genes are then quantile normalized.

```
qnormalized = normalize.quantile(filtered)
```

The counts per trait are then further normalized towards a standard normal distribution to remove the effect of outliers, which can be extreme in sequencing data. The resulting matrix ("pheno") contains our intermediate phenotypes for the QTL analysis.

```
pheno = t(apply(qnormalized, 1, function(x)
return(qnorm((rank(x, ties="random")) / (length(x) + 1)))))
rownames(pheno) = rownames(filtered)
colnames(pheno) = colnames(filtered)
phenotype.file = "phenotypes.txt"
write.table(pheno, phenotype.file, sep="\t", quote=F)
```

7. QTL analysis

There are many approaches to perform the actual QTL analysis. The classical approach is to use linear trait—marker regression models in combination with a permutation procedure. Recent advances in vectorized computation have produced considerable gains in speed [23]. This allows for the computation of 10,000 permutations for millions of trait—marker pairs on a high-performance computing cluster (HPC) in a matter of a few days. The R package *eQTLpipeline* [24] implements this permutation strategy for a wide range of popular HPC systems supported by the *BatchJobs* R package [25]. Since the size of most data tables are in the range of Gigabytes, we will use the *data.table* R package [26] allowing for efficient file access by memory mapping.

First, we load the required R packages.

```
library(data.table)
library(eQTLpipeline)
```

Then, we define the location of the files containing the genomic coordinates of markers and genes. The file formats are described in the package documentation.

```
snps_location_file_name = "snp_positions.txt"
gene_location_file_name = "gene_positions.txt"
snp.pos = as.data.frame(fread(snps_location_file_name))
positions = read.csv(gene_location_file_name, sep="\t",
stringsAsFactors=F)
```

Load available covariates, including technical covariates from the sequencing or biological covariates such as sex. The covariates should be in a tab-separated file, with each sample in a column with covariates as row names.

```
covariates.file = "covariates.txt"
covar = read.csv(file=covariates.file, sep="\t")
```

Define the location of the genotype file. Genotypes should be encoded as allele dosage. The file is tab-separated with sample IDs as column names and genetic markers as rows.

```
genotypes.file = "imputed_dosage.txt"
```

Now we start the actual QTL analysis. Note that the *compute.all* flag causes all results to be returned. The default behavior is to return only results with nominal *P*-values below a certain threshold. Because we determine the thresholds by permutation, all results are required. Also note that this step can be omitted, as it will automatically be performed within the permutation analysis. We give it here for completeness or in case a permutation analysis is not desired.

```
actual.eqtls = eqtl(phenotype.file, genotypes.file,
covariates.file, positions, snp.pos, prefix="phenotype_
name", compute.all=T)
```

Start the permutation analysis. Note that if we do not specify the *actual.eqtls* argument here, the initial eQTL analysis will be performed automatically.

```
esnps = find.eSNPs(pheno, covar, positions, snp.pos, genotypes.
file, dir="eqtls", min.perm=1000, max.perm=10000, exit.
criterion=15, actual.eqtls=actual.eqtls)
write.table(esnps, "esnps.txt", sep="\t", quote=F,
row.names=F)
```

Finally, obtain the trait-level summary of the eQTL results, which is called *eGenes* in the literature.

```
egenes = get.egenes.from.esnps (esnps, fdr=0.05)
write.table(egenes, "egenes.txt", sep="\t", quote=F)
```

8. QC of the QTL analysis

   To control the quality of the QTL analysis we will use a quantile–quantile plot (QQ plot). It visualizes the relation between the observed and the theoretical distribution of our test statistics, the –log10 (*P*-value). Under the null hypothesis of no association, *P*-values follow a uniform distribution on the interval from zero to one.

```
qquniform(egenes[,"p.value"])
```

9. Integration of genetic, epigenetic and gene expression data

   Integration of epigenetic and gene expression data is performed on the gene level requiring an assignment of histone modification traits to genes. Such an assignment is easily obtained when histone modification traits are based on gene annotations. The use of de novo epigenetic features requires more complicated strategies [27], which will not be covered here.

   In the following steps, we assume that we have data for two histone modifications, H3K4me3 and H3K27me3, as well as gene expression and genotype data. The first step of gene-level data integration is to combine all QTL results, genotypes and quantitative traits into a single data frame using functions from the GraphicalModels R package [10]. The prefixes refer to the phenotype prefixes used in the individual QTL analyses.

```
H3K4me3.prefix = "H3K4me3"
H3K27me3.prefix = "H3K27me3"
expr.prefix = "RNA"
strainlist = scan("strain_names.txt",
what=character())
data = integrate(traits=list(H3K4me3=H3K4me3.prefix,
H3K27me3=H3K27me3.prefix, expr=expr.prefix), strain-
list, cutoff=0.05)
```

This data frame contains data for all gene–marker pairs, where at least one of the molecular traits has a significant QTL. For each gene we can now determine the relation between the genotype and the traits using likelihood-based model selection [28]. This procedure is implemented in the *GraphicalModels* R package [10]. These models consist of a graph (V, E) and an associated set of conditional distributions (P). V is the set of random variables that corresponds to the genotypes and quantitative traits and E is a set of directed edges that encodes the dependency structure between variables. Variables can be either discrete or continuous. All parents of discrete variables in G must also be discrete. Discrete variables are distributed according to a multinomial distribution whereas continuous variables are modeled by a normal distribution. The mean is linearly dependent on the values of the parent nodes: $f(x_i|pa(i)) = N(\beta_0 + \Sigma_{j \text{ in } pa(i)} \beta_j x_j, \sigma)$. Here $pa(i)$ is the set of parent nodes of node i and $N(\mu, \sigma)$ is the density function of the Gaussian distribution with mean μ and standard deviation σ. Whether a variable is discrete or continuous is determined automatically from the data frame, where factor columns are treated as discrete and numeric columns as continuous. The graphs of all models that are to be compared have to be specified by edgelists.

```
nodes = c("snp", "h3k4", "h3k27", "expr")
models = list(snp.h3k4.expr=matrix(
                c("snp", "h3k4",
                  "h3k4", "expr"), byrow=T),
              snp.h3k27.expr=matrix(
                c("snp", "h3k27",
                  "h3k27", "expr"), byrow=T))
```

Once all model structures have been defined, the most likely dependency structure can be identified by comparing model fits according to the Akaike information criterion (AIC).

```
model.selection = single.locus.bootstrap
(genes, models, data, q.cols, fname, n.bootstrap=100)
```

The resulting models can then be visualized.

```
instance = 100
selected.model = models[[model.selection[instance,
"model"]]]
selected.data = data[data[,"trait"] ==
model.selection[instance, "gene"] &
data[,"marker"] == model.selection[instance, "mark-
er"], ]
plot.model(selected.model, selected.data)
```

4 Expected Results

The first result we need to judge is the quality of the sequencing. We expect a high sequencing quality throughout the length of the sequencing read (Fig. 1).

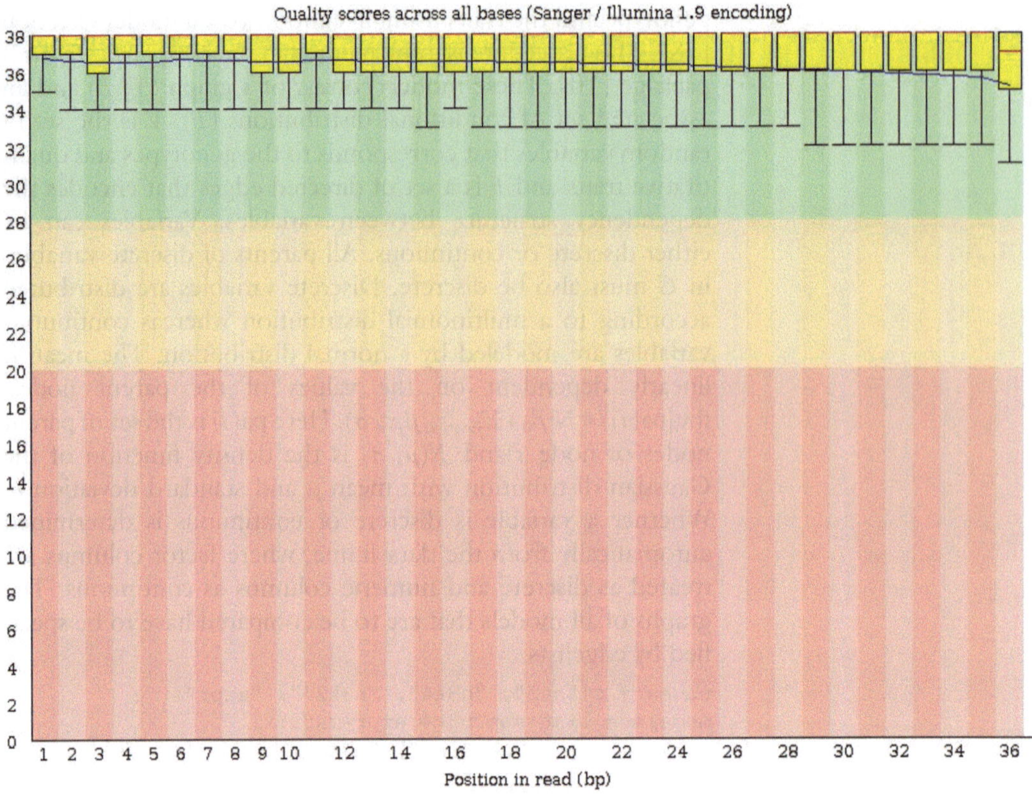

Fig. 1 Sequence level QC plot. The plot shows the sequence quality score distribution (*y*-axis) as boxplots by position in the read (*x*-axis). In this example there is a consistently high sequencing quality with a slight decrease towards the end of the sequenced read

A slight decrease in sequencing quality toward the end of the sequences is expected as a result of the sequencing process. Positions with severe deterioration of sequencing quality should be trimmed off. Base pair composition should be close to the genome-wide GC content and uniform along the whole sequencing read. If mono-nucleosomal chromatin was obtained by enzyme digestion there might be a bias in the base pair composition at the start of the reads that reflect the enzyme recognition sequence.

After alignment, we expect at least 10 million uniquely mapping reads for peak-like features and at least 40 million uniquely mapping reads for broader histone marks, because these domains span a large proportion of the genome. In the strand cross-correlation plot we expect a strong correlation at the expected fragment size. If experimentally determined fragment sizes are available, they should be matching with the estimated fragment sizes. In the case of histone ChIP-seq data, the sequencing libraries should be prepared from mono-nucleosomal fragments (*see* Subheadings 2 and 5). Therefore the fragment size is expected to be roughly

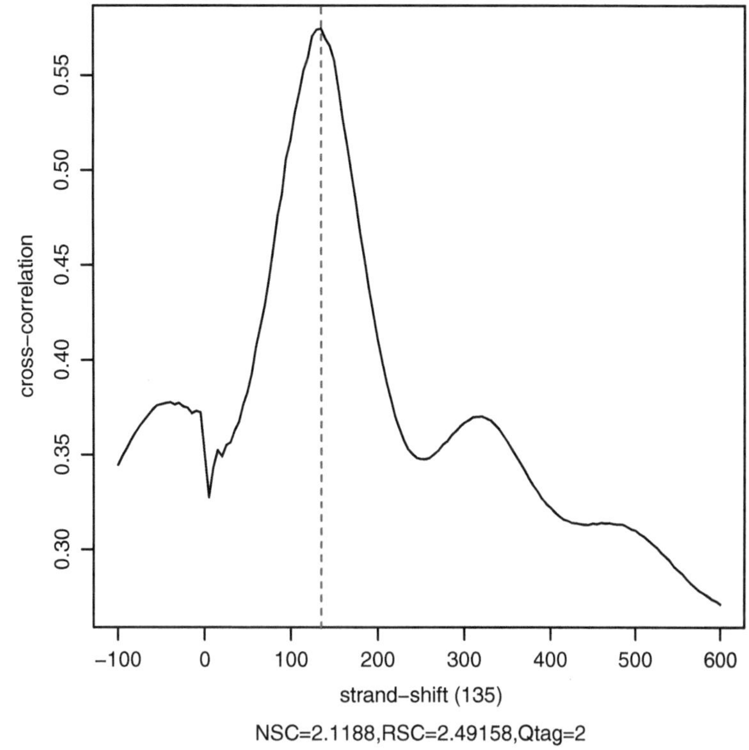

Fig. 2 Alignment level QC plot. The plot shows the correlation between the number of reads aligned at position *x* on the forward strand and the number of reads that aligned at position '*x*+ s' on the reverse strand on the *y*-axis. Here "s" denotes the strand shift that is shown on the *x*-axis. NSC denotes the normalized strand cross-correlation, RSC denotes the relative strand cross-correlation [29] and Qtag is a discretized quality tag based on RSC where the value of 2 is best. The maximum cross-correlation marked by the *dashed line* is attained at a fragment length of 135 which is close to the 146–147 bp of DNA that is wrapped around a single nucleosome

around 146–147 bp as this is the number of base pairs that are wrapped around a single nucleosome. Figure 2 shows an estimated fragment size of 135 and a periodic pattern that represents fragments with two and three nucleosomes.

For the promoter-associated H3K4me3 marks we expect a bimodal distribution of read coverage around the TSS with peaks at the average positions of the nucleosomes upstream and downstream of the TSS, surrounding the nucleosome-free region (NFR). Figure 3 shows that most reads fall within 2 kb of the TSS, allowing for an annotation-based quantification of H3K4me3 traits. The relation between high levels of H3K4me3 modification around the TSS and transcriptional activity becomes clear when coverage plots are generated for expressed and nonexpressed genes separately (Fig. 3). Differing from H3K4me3 peaks, H3K27me3 marks span large genomic domains, which can contain multiple genes.

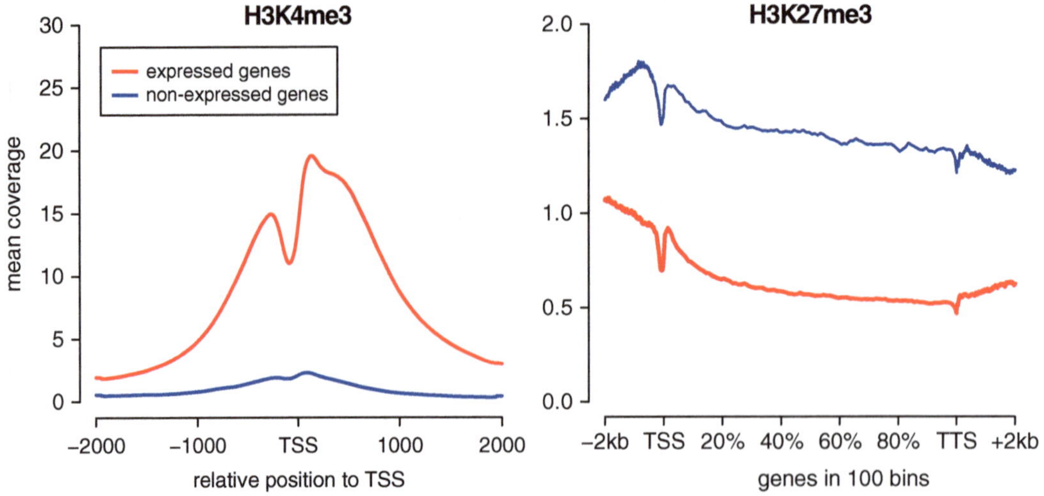

Fig. 3 Coverage plots. Coverage plots show the average coverage (*y*-axis) aggregating over genomic features defined based on gene annotation. Here genes are separated according to their expression levels into expressed genes (*red*) and nonexpressed genes (*blue*). The *left plot* shows occupancy patterns of H3K4me3 in 2 kb regions around the TSS. The *right plot* shows occupancy patterns of H3K27me3 across gene bodies, from the TSS to the transcription termination site (TTS). As genes vary in length, the *x*-axis shows relative positions by binning positions along each gene into 100 bins

Coverage is present across the gene body with a slight decay towards the end of the gene with a particular concentration around the TSS. Therefore gene bodies can be used as counting units for H3K27me3 traits. H3K27me3 is associated with polycomb-mediated silencing thus higher coverage is observed on non-expressed genes.

The histogram of RPKM values for features defined from gene annotations is expected to display two modes, one for unmodified regions and one for modified regions. Visual inspection or fitting of a mixture model can be used to determine an RPKM threshold for the filtering step. Figure 4 indicates that an RPKM threshold of 1 yields a good separation for H3K4me3 marks. If the input for this graph is based on the output of a peak calling algorithm, the distribution will not necessarily be bimodal because regions without peaks, such as inactive TSSs, are not considered.

Depending on the sample size and coverage depth of the histone modifications, many hundreds to thousands of histoneQTLs are expected. An example of a histoneQTL is shown in Fig. 5. To generate such figures we recommend using a genome browser such as IGV [30] or web based tools like JBrowse [31], both capable of exporting images in the modifiable vector-based SVG format.

Quantile–quantile plots (QQ plots) are the recommended quality control of the QTL analysis. The observed quantiles of the test statistic should be equal to the theoretical quantiles for low (nonsignificant) values of the test statistic; lying on the diagonal of the QQ plot (Fig. 6). For high (significant) values of the test

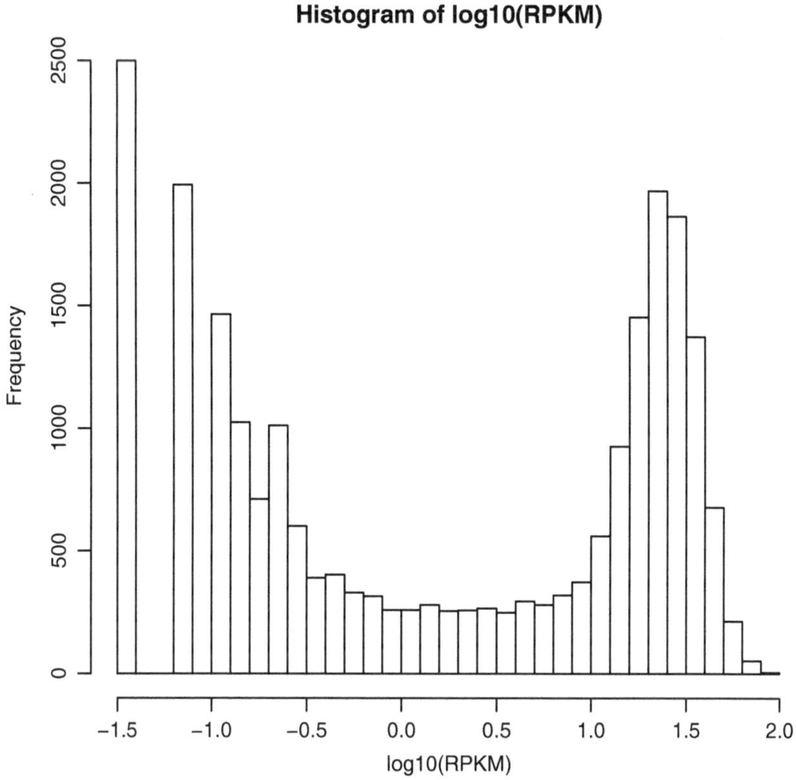

Fig. 4 Histogram of RPKM values. The histogram of RPKM values for H3K4me3 counted in TSS regions from −2 kb to +2 kb shows a clear bimodal distribution representing modified (RPKM > 1) and unmodified regions (RPKM < 1). Note that a pseudo count of one has been added to each counting bin so that values of zero are not lost in the log transformation

statistic we expect a deviation from the diagonal with the observed quantiles being larger than the theoretical quantiles.

The integrated analysis of histone modifications and gene expression data is expected to improve the detection rate of genes whose expression is determined by sequence variants. This is because the integrated analysis also allows for the identification of indirect connections, where histone modification levels are strongly associated to the sequence variants and correlated to gene expression levels. In our study we were able to increase the number of detected genes by about 15 %. Moreover graphical models allow for the characterization of the interplay between traits in particular the direction of the effects such as up- or downregulation, either by visual inspection (Fig. 7) or by systematic comparison of the estimated model parameters. Taking into account the strength of the association between the SNP and each of the quantitative traits helps to define via which regulatory mechanism gene expression or histone modification is affected. For example, a SNP can affect H3K4me3 levels, associated with transcriptional activation and

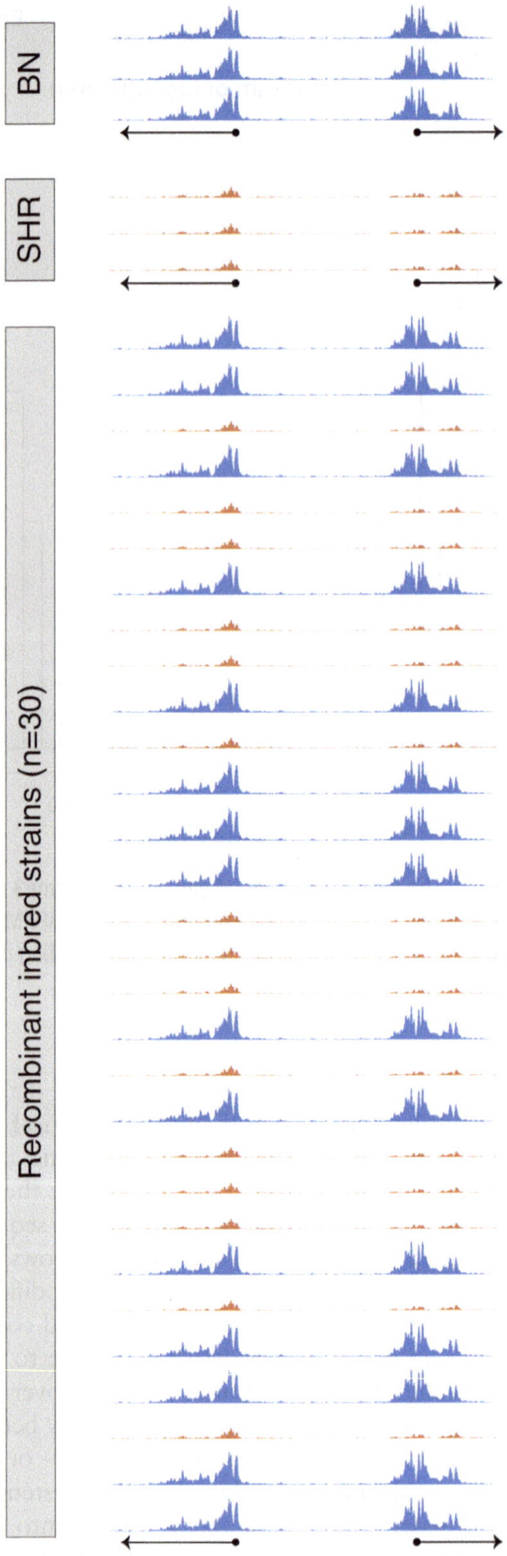

Fig. 5 Illustrative example of a histoneQTL. Three replicate samples for the parental strains (*top*) are displayed, alongside all 30 RI strains. For this exemplary genomic region, histone methylation levels are regulated in *cis*. The H3K4me3 enrichment is allele specific with those RI strains carrying a local BN genotype (*blue*) showing more enrichment for both peaks than the RI strains carrying the SHR allele (*orange*)

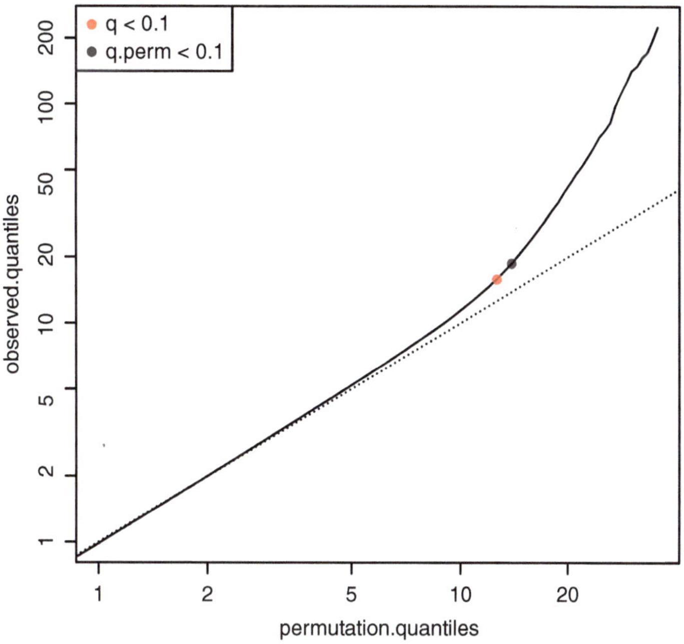

Fig. 6 QQ Plot of QTL $-\log_{10}$(*P*-values). The plot shows the quantiles of the theoretical distribution on the *x*-axis and the observed quantiles on the *y*-axis. The *dashed line* marks the diagonal where theoretical and observed quantiles are equal

change expression levels concordantly, without interfering with H3K27me3 levels. Second, the transcriptional activity of the gene can be increased when a SNP negatively influences the positioning of the repressive H3K27me3 mark. Alternatively, a SNP can be positively associated with changes in both histone marks (H3K4me3 and H3K27me3), without affecting nearby gene expression levels. Examples of various possible graphical models are shown in Fig. 7.

5 Further Considerations and Limitations

Below we discuss strengths and weaknesses of both the system and the procedures described in this chapter, outlining what is worth considering for a proper experimental design.

1. **Considerations on the resolution of QTL mapping in recombinant inbred systems**

   The HXB/BXH recombinant inbred panel consists of 30 lines, each carrying a unique mixture of the BN and SHR genotype. Because these animals are fully inbred, having 30 lines is sufficient for the reliable detection of both local (*cis*) and distant (*trans*) effects. However, the number of recombination events that occurred in the initial cross limits the resolution of QTL mapping. In the HXB/BXH panel the genome is

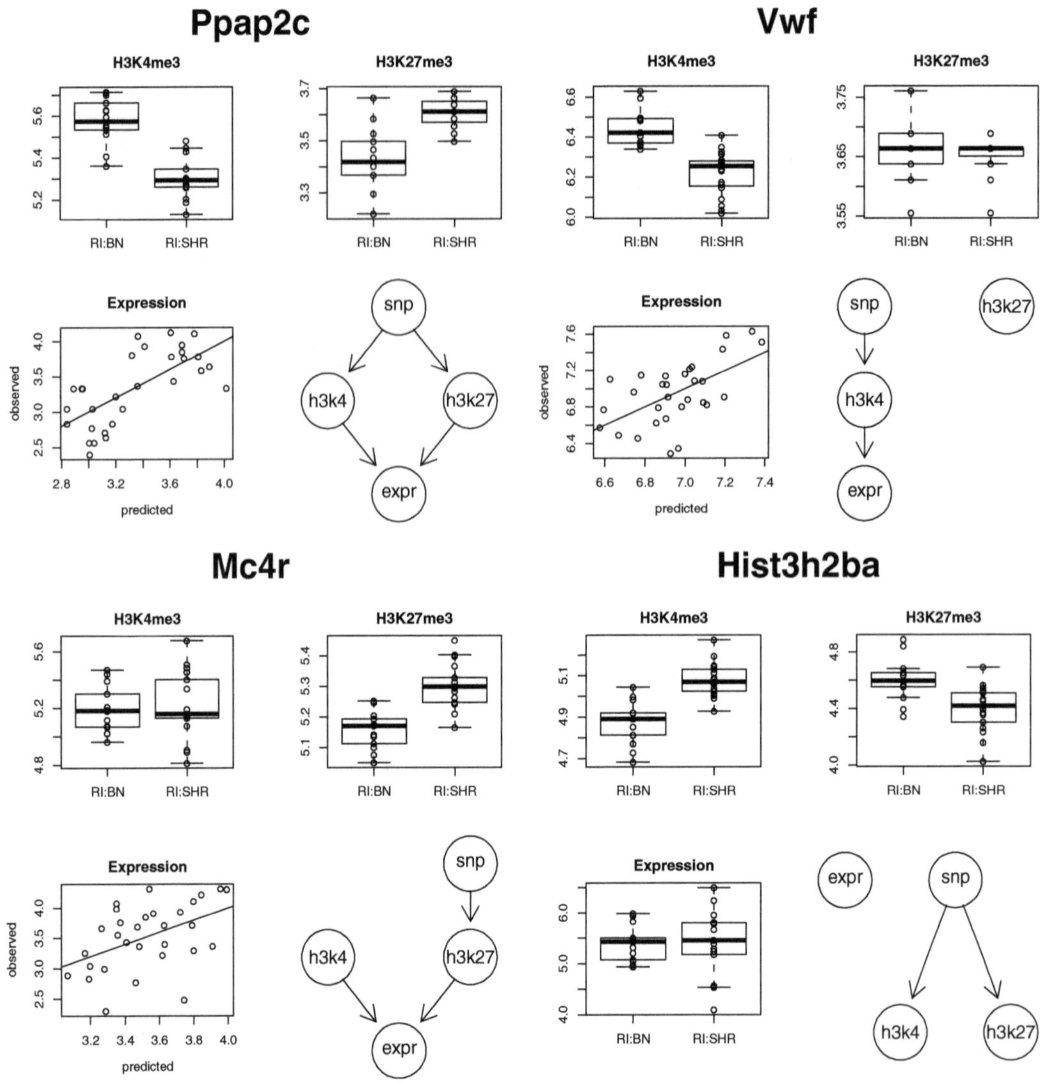

Fig. 7 Examples of graphical models. Each panel uses one gene to illustrate a selected graphical model using histone modification and gene expression level data

partitioned in 1348 recombination events between the BN and SHR genomes, as defined by variable strain distribution patterns (SDPs) of the markers used for genotyping [10]. This means that significant marker-trait associations can still result in megabase-sized regions, which challenges the identification of the causal genetic variant and the interpretation the mechanistic consequences of these variants.

The use of other (outbred) genetic model systems such as heterogeneous stocks (HS) [32–34], including the NIH HS rats (NIH-HS) [35], can improve the resolution of QTL mapping down to the single gene level, but comes with disadvantages as well. Heterogeneous stocks have been randomly crossed for approximately 50 generations from a larger number of progenitor strains that are models for multiple complex traits. The recombination pattern after that many generations reflects the genetic diversity of a human population well, making it a versatile tool for the fine-mapping of QTLs. Nevertheless, genetic mapping in such systems is increasingly difficult because of the presence of a mixture of heterozygous and homozygous alleles, requiring precise genotyping and separation of haplotypes. Also, variability in the rate of kinship (genetic similarity) between samples complicates QTL mapping, whereas kinship is near equal across the rat HXB/BXH animals where each animal shares approximately 50% of its genome. In HS systems, much larger numbers of animals are required to be phenotyped and sampled (1407 for the NIH-HS) and since each animal is unique it is impossible to renew source material.

2. **Considerations of cost and replicate number in large animal panels**

   A potential other limitation of performing histoneQTL mapping and integration with RNA-seq data in genetic systems is the high number of samples that need to be processed, which currently makes this a costly approach. Since we require high-quality RNA-seq and ChIP-seq data, deep sequencing is required (*see* Subheading 4). With respect to the use of replicates, using recombinant inbred panels for QTL mapping comes with an advantage here. Even though replicates for the RI animals are desirable they are not absolutely necessary for mapping QTLs since each local genotype (either from the BN or SHR background) can function as a replicate. For example, if at a given position in the genome 10 of the HXB/BXH animals carry the BN allele and 20 carry the SHR allele, these would function as replicates for each genotype. Additionally, permutation testing reduces the number of false-positive identifications that arise by chance. For the two parental strains, in our case BN and SHR, we do ideally require five biological replicates each.

3. **Influence of different ChIP-seq procedures on quantifying traits.**

   There are different ways to prepare mono-nucleosomal chromatin for ChIP-seq analysis. The two most often utilized methods either use micrococcal nuclease (MNase) digestion without cross-linking to fragment the chromatin (referred to as native ChIP) or chromatin fragmentation can be achieved by sonication, which requires prior cross-linking (referred to as X-ChIP). Both methods have different impact on the final quality and interpretation of the data. While cross-linking using formaldehyde may prevent changes in nucleosome positions and histone modifications during sample preparation, these may be biased by subsequent reverse cross-linking steps.

Also, since DNA is so tightly bound to the nucleosomes, DNA-protein cross-linking for histone modification IPs is not absolutely necessary—other than for transcription factor ChIP-seq. The native ChIP procedure on the other hand has been associated with a more pronounced sequence bias [36] and may lead to undesired post-sampling chromatin remodeling or modification. However, MNase digestion in native ChIP has the great advantage to remove linker-DNA in between nucleosomes more effectively than sonication and thus allows for a reliable and reproducible single-nucleosome resolution. This subsequently results in more precise mapping of immunoprecipitated reads and ultimately high-resolution ChIP-seq data.

The detection of differential histone marks not only relies on high resolution of the ChIP-seq data, but also on the type of histone modification that is being assessed. Whereas some activating marks like H3K4me3 have tight, near single-nucleosome distributions around the transcriptional start site of genes, other heterochromatin-associated marks like H3K27me3 and H3K9me3 have broad distributions covering entire gene bodies and beyond. The latter type of modification is extremely challenging to identify with traditional peak calling algorithms, as those are mostly optimized to call clearly distinguishable peak features. To still be able to reliably discriminate broader enriched regions and quantitative differences therein, the histoneHMM method, which uses a bivariate Hidden Markov Model, was developed [13]. HistoneHMM uses an unsupervised classification procedure of aggregated short reads across larger genomic regions, after which it compares samples and provides a classification of each region as either being not modified at all, modified in all samples or differentially modified within the sample group. For the use of broad histone marks for histoneQTL mapping, precise definition of the quantitative trait used in the segregation analysis is pivotal, such as those obtained from histoneHMM or gene annotations.

4. **Outlook**

As the cost of sequencing declines and technological developments increase the resolution of data tremendously, QTL mapping becomes an attractive method for many larger study designs, including patient cohorts. For example, large amounts of data can be generated from a human population, including whole genome sequencing to determine variant positions and RNA-seq or ChIP-seq to define expression levels or chromatin states in a given tissue or cell type [27]. Also, the recent development of single-cell technologies such as single-cell mRNA-seq [37] and single-cell ChIP-seq [38] make it likely that the near future will show follow-ups of such QTL studies in patient cohorts. Using single-cell techniques, the diseased tissue could be sampled directly to assess cellular heterogeneity and ultimately assign molecular phenotypes to specific subpopulations of cell types. That way, the mechanistic consequences of genetic variants can be studied directly in the cell type relevant to the disease and indirect (*trans*) effects between cell types responding to the diseased situation can be assessed as well. However, these methodologies are currently still challenged by technological difficulties that need to be overcome before high-resolution QTL analyses comparable to the approach described in this chapter can be performed.

Acknowledgments

This work was supported by funding from the European Union EURATRANS award (HEALTH-F4-2010-241504 to N.H.), the Helmholtz Alliance ICEMED, the Deutsche Forschungsgemeinschaft (Forschergruppe 1054, HU 1522/1-1) to N.H, and an EMBO Long-Term Fellowship (ALTF 186-2015) and Marie Curie Actions (LTFCOFUND2013, GA-2013-609409) to SvH.

References

1. Barski A, Cuddapah S, Cui K, Roh TY, Schones DE, Wang Z, Wei G, Chepelev I, Zhao K (2007) High-resolution profiling of histone methylations in the human genome. Cell 129(4):823–837. doi:10.1016/j.cell.2007.05.009
2. Mikkelsen TS, Ku M, Jaffe DB, Issac B, Lieberman E, Giannoukos G, Alvarez P, Brockman W, Kim TK, Koche RP, Lee W, Mendenhall E, O'Donovan A, Presser A, Russ C, Xie X, Meissner A, Wernig M, Jaenisch R, Nusbaum C, Lander ES, Bernstein BE (2007) Genome-wide maps of chromatin state in pluripotent and lineage-committed cells. Nature 448(7153):553–560. doi:10.1038/nature06008
3. Ernst J, Kheradpour P, Mikkelsen TS, Shoresh N, Ward LD, Epstein CB, Zhang X, Wang L, Issner R, Coyne M, Ku M, Durham T, Kellis M, Bernstein BE (2011) Mapping and analysis of chromatin state dynamics in nine human cell types. Nature 473(7345):43–49. doi:10.1038/nature09906
4. Pravenec M, Churchill PC, Churchill MC, Viklicky O, Kazdova L, Aitman TJ, Petretto E, Hubner N, Wallace CA, Zimdahl H, Zidek V, Landa V, Dunbar J, Bidani A, Griffin K, Qi N, Maxova M, Kren V, Mlejnek P, Wang J, Kurtz TW (2008) Identification of renal Cd36 as a determinant of blood pressure and risk for hypertension. Nat Genet 40(8):952–954. doi:10.1038/ng.164
5. Hubner N, Wallace CA, Zimdahl H, Petretto E, Schulz H, Maciver F, Mueller M, Hummel O, Monti J, Zidek V, Musilova A, Kren V, Causton H, Game L, Born G, Schmidt S, Muller A, Cook SA, Kurtz TW, Whittaker J, Pravenec M, Aitman TJ (2005) Integrated transcriptional profiling and linkage analysis for identification of genes underlying disease. Nat Genet 37(3):243–253. doi:10.1038/ng1522
6. Printz MP, Jirout M, Jaworski R, Alemayehu A, Kren V (2003) Genetic Models in Applied Physiology. HXB/BXH rat recombinant inbred strain platform: a newly enhanced tool for cardiovascular, behavioral, and developmental genetics and genomics. J Appl Physiol 94(6):2510–2522. doi:10.1152/japplphysiol.00064.2003
7. Simonis M, Atanur SS, Linsen S, Guryev V, Ruzius FP, Game L, Lansu N, de Bruijn E, van Heesch S, Jones SJ, Pravenec M, Aitman TJ, Cuppen E (2012) Genetic basis of transcriptome differences between the founder strains of the rat HXB/BXH recombinant inbred panel. Genome Biol 13(4):r31. doi:10.1186/gb-2012-13-4-r31
8. Kasowski M, Kyriazopoulou-Panagiotopoulou S, Grubert F, Zaugg JB, Kundaje A, Liu Y, Boyle AP, Zhang QC, Zakharia F, Spacek DV, Li J, Xie D, Olarerin-George A, Steinmetz LM, Hogenesch JB, Kellis M, Batzoglou S, Snyder M (2013) Extensive variation in chromatin

states across humans. Science 342(6159):750–752. doi:10.1126/science.1242510

9. McVicker G, van de Geijn B, Degner JF, Cain CE, Banovich NE, Raj A, Lewellen N, Myrthil M, Gilad Y, Pritchard JK (2013) Identification of genetic variants that affect histone modifications in human cells. Science 342(6159):747–749. doi:10.1126/science.1242429
10. Rintisch C, Heinig M, Bauerfeind A, Schafer S, Mieth C, Patone G, Hummel O, Chen W, Cook S, Cuppen E, Colome-Tatche M, Johannes F, Jansen RC, Neil H, Werner M, Pravenec M, Vingron M, Hubner N (2014) Natural variation of histone modification and its impact on gene expression in the rat genome. Genome Res 24(6):942–953. doi:10.1101/gr.169029.113
11. Heinz S, Romanoski CE, Benner C, Allison KA, Kaikkonen MU, Orozco LD, Glass CK (2013) Effect of natural genetic variation on enhancer selection and function. Nature 503(7477):487–492. doi:10.1038/nature12615
12. Kilpinen H, Waszak SM, Gschwind AR, Raghav SK, Witwicki RM, Orioli A, Migliavacca E, Wiederkehr M, Gutierrez-Arcelus M, Panousis NI, Yurovsky A, Lappalainen T, Romano-Palumbo L, Planchon A, Bielser D, Bryois J, Padioleau I, Udin G, Thurnheer S, Hacker D, Core LJ, Lis JT, Hernandez N, Reymond A, Deplancke B, Dermitzakis ET (2013) Coordinated effects of sequence variation on DNA binding, chromatin structure, and transcription. Science 342(6159):744–747. doi:10.1126/science.1242463
13. Heinig M, Colome-Tatche M, Taudt A, Rintisch C, Schafer S, Pravenec M, Hubner N, Vingron M, Johannes F (2015) histoneHMM: differential analysis of histone modifications with broad genomic footprints. BMC Bioinformatics 16:60. doi:10.1186/s12859-015-0491-6
14. Parkinson H, Kapushesky M, Shojatalab M, Abeygunawardena N, Coulson R, Farne A, Holloway E, Kolesnykov N, Lilja P, Lukk M, Mani R, Rayner T, Sharma A, William E, Sarkans U, Brazma A (2007) ArrayExpress--a public database of microarray experiments and gene expression profiles. Nucleic Acids Res 35(Database issue):D747–D750. doi:10.1093/nar/gkl995
15. Leinonen R, Akhtar R, Birney E, Bower L, Cerdeno-Tarraga A, Cheng Y, Cleland I, Faruque N, Goodgame N, Gibson R, Hoad G, Jang M, Pakseresht N, Plaister S, Radhakrishnan R, Reddy K, Sobhany S, Ten Hoopen P, Vaughan R, Zalunin V, Cochrane G (2011) The European Nucleotide Archive. Nucleic Acids Res 39(Database issue):D28–D31. doi:10.1093/nar/gkq967
16. Andrews S (2010) FastQC: a quality control tool for high throughput sequence data
17. Langmead B, Salzberg SL (2012) Fast gapped-read alignment with Bowtie 2. Nat Methods 9(4):357–359. doi:10.1038/nmeth.1923
18. Li H, Handsaker B, Wysoker A, Fennell T, Ruan J, Homer N, Marth G, Abecasis G, Durbin R, Genome Project Data Processing Subgroup (2009) The sequence alignment/map format and SAMtools. Bioinformatics 25(16):2078–2079. doi:10.1093/bioinformatics/btp352
19. Marinov GK, Kundaje A, Park PJ, Wold BJ (2014) Large-scale quality analysis of published ChIP-seq data. G3 (Bethesda) 4(2):209–223. doi:10.1534/g3.113.008680
20. Heinig M (2015) Toolset for preprocessing sequencing based trais for QTL analysis. https://github.com/matthiasheinig/seqQTL
21. Zhang Y, Liu T, Meyer CA, Eeckhoute J, Johnson DS, Bernstein BE, Nusbaum C, Myers RM, Brown M, Li W, Liu XS (2008) Model-based analysis of ChIP-Seq (MACS). Genome Biol 9(9):R137. doi:10.1186/gb-2008-9-9-r137
22. Consortium GT (2015) Human genomics. The Genotype-Tissue Expression (GTEx) pilot analysis: multitissue gene regulation in humans. Science 348(6235):648–660. doi:10.1126/science.1262110
23. Shabalin AA (2012) Matrix eQTL: ultra fast eQTL analysis via large matrix operations. Bioinformatics 28(10):1353–1358. doi:10.1093/bioinformatics/bts163
24. Heinig M (2015) Toolset to perform eQTL analysis on HPC systems. https://github.com/matthiasheinig/eQTLpipeline
25. Bischl B (2015) BatchJobs and BatchExperiments: abstraction mechanisms for using R in batch environments. http://www.jstatsoft.org/article/view/v064i11
26. Dowle M, Srinivasan A, Short T, Lianoglou S with contributions from Saporta R, Antonyan E (2015) Extension of Data.frame. https://github.com/Rdatatable/data.table/wiki
27. Waszak SM, Delaneau O, Gschwind AR, Kilpinen H, Raghav SK, Witwicki RM, Orioli A, Wiederkehr M, Panousis NI, Yurovsky A, Romano-Palumbo L, Planchon A, Bielser D, Padioleau I, Udin G, Thurnheer S, Hacker D, Hernandez N, Reymond A, Deplancke B, Dermitzakis ET (2015) Population variation and genetic control of modular chromatin architecture in humans. Cell 162(5):1039–1050. doi:10.1016/j.cell.2015.08.001
28. Schadt EE, Lamb J, Yang X, Zhu J, Edwards S, Guhathakurta D, Sieberts SK, Monks S,

Reitman M, Zhang C, Lum PY, Leonardson A, Thieringer R, Metzger JM, Yang L, Castle J, Zhu H, Kash SF, Drake TA, Sachs A, Lusis AJ (2005) An integrative genomics approach to infer causal associations between gene expression and disease. Nat Genet 37(7):710–717. doi:10.1038/ng1589

29. Encode Project Consortium (2012) An integrated encyclopedia of DNA elements in the human genome. Nature 489(7414):57–74. doi:10.1038/nature11247
30. Robinson JT, Thorvaldsdottir H, Winckler W, Guttman M, Lander ES, Getz G, Mesirov JP (2011) Integrative genomics viewer. Nat Biotechnol 29(1):24–26. doi:10.1038/nbt.1754
31. Skinner ME, Uzilov AV, Stein LD, Mungall CJ, Holmes IH (2009) JBrowse: a next-generation genome browser. Genome Res 19(9):1630–1638. doi:10.1101/gr.094607.109
32. Mott R, Talbot CJ, Turri MG, Collins AC, Flint J (2000) A method for fine mapping quantitative trait loci in outbred animal stocks. Proc Natl Acad Sci U S A 97(23):12649–12654. doi:10.1073/pnas.230304397
33. Valdar W, Solberg LC, Gauguier D, Burnett S, Klenerman P, Cookson WO, Taylor MS, Rawlins JN, Mott R, Flint J (2006) Genome-wide genetic association of complex traits in heterogeneous stock mice. Nat Genet 38(8):879–887. doi:10.1038/ng1840
34. Rat Genome Sequencing and Mapping Consortium, Baud A, Hermsen R, Guryev V, Stridh P, Graham D, McBride MW, Foroud T, Calderari S, Diez M, Ockinger J, Beyeen AD, Gillett A, Abdelmagid N, Guerreiro-Cacais AO, Jagodic M, Tuncel J, Norin U, Beattie E, Huynh N, Miller WH, Koller DL, Alam I, Falak S, Osborne-Pellegrin M, Martinez-Membrives E, Canete T, Blazquez G, Vicens-Costa E, Mont-Cardona C, Diaz-Moran S, Tobena A, Hummel O, Zelenika D, Saar K, Patone G, Bauerfeind A, Bihoreau MT, Heinig M, Lee YA, Rintisch C, Schulz H, Wheeler DA, Worley KC, Muzny DM, Gibbs RA, Lathrop M, Lansu N, Toonen P, Ruzius FP, de Bruijn E, Hauser H, Adams DJ, Keane T, Atanur SS, Aitman TJ, Flicek P, Malinauskas T, Jones EY, Ekman D, Lopez-Aumatell R, Dominiczak AF, Johannesson M, Holmdahl R, Olsson T, Gauguier D, Hubner N, Fernandez-Teruel A, Cuppen E, Mott R, Flint J (2013) Combined sequence-based and genetic mapping analysis of complex traits in outbred rats. Nat Genet 45(7):767–775. doi:10.1038/ng.2644
35. Johannesson M, Lopez-Aumatell R, Stridh P, Diez M, Tuncel J, Blazquez G, Martinez-Membrives E, Canete T, Vicens-Costa E, Graham D, Copley RR, Hernandez-Pliego P, Beyeen AD, Ockinger J, Fernandez-Santamaria C, Gulko PS, Brenner M, Tobena A, Guitart-Masip M, Gimenez-Llort L, Dominiczak A, Holmdahl R, Gauguier D, Olsson T, Mott R, Valdar W, Redei EE, Fernandez-Teruel A, Flint J (2009) A resource for the simultaneous high-resolution mapping of multiple quantitative trait loci in rats: the NIH heterogeneous stock. Genome Res 19(1):150–158. doi:10.1101/gr.081497.108
36. Tolstorukov MY, Kharchenko PV, Goldman JA, Kingston RE, Park PJ (2009) Comparative analysis of H2A.Z nucleosome organization in the human and yeast genomes. Genome Res 19(6):967–977. doi:10.1101/gr.084830.108
37. Tang F, Barbacioru C, Wang Y, Nordman E, Lee C, Xu N, Wang X, Bodeau J, Tuch BB, Siddiqui A, Lao K, Surani MA (2009) mRNA-Seq whole-transcriptome analysis of a single cell. Nat Methods 6(5):377–382. doi:10.1038/nmeth.1315
38. Rotem A, Ram O, Shoresh N, Sperling RA, Goren A, Weitz DA, Bernstein BE (2015) Single-cell ChIP-seq reveals cell subpopulations defined by chromatin state. Nat Biotechnol 33:1165–1172. doi:10.1038/nbt.3383

Chapter 10

Integrating Multidimensional Data Sources to Identify Genes Regulating Complex Phenotypes

Rupert W. Overall

Abstract

Phenotypes collected with a view to quantitative trait locus mapping can be augmented with compatible whole-transcriptome expression data and information from several other sources. These different data sources can be assembled into multidimensional network models which allow the identification of key genes potentially driving the phenotype of interest. The following chapter describes this approach using an example workflow. Several alternatives and potential limitations are discussed to aid the researcher when applying these techniques to their own work.

Key words Multilayer networks, Data integration, Complex traits, Network theory, Gene–gene interactions

1 Introduction

Systems genetics is concerned with understanding the genetic regulation of complex phenotypes. "Complex" in this sense indicating that the phenotype is under the control of many genes, rather than exhibiting monogenic Mendelian inheritance. The polygenic nature of complex phenotypes means that the effects of any single gene polymorphism will contribute to only a small part of the total phenotypic response. Attempting to directly associate a highly polygenic phenotype with segregating genomic markers, as in traditional QTL mapping, can often be unproductive due to the numerous molecular steps intervening between polymorphism and high-level phenotype. In such situations, one can take advantage of transcript expression data, as well as other forms of gene–gene interaction data, to infer possible pathways and help build a link from the phenotype back to a causal gene. In many cases, such interaction data are already available in public archives and need not be generated anew by the researcher [1]. These different sources of interaction data can be collated into network models (*see* **Note 1**) which allow analysis using techniques borrowed from graph theory.

Klaus Schughart and Robert W. Williams (eds.), *Systems Genetics: Methods and Protocols*, Methods in Molecular Biology, vol. 1488, DOI 10.1007/978-1-4939-6427-7_10,

An important advantage of a network representation over a simple listing of genes correlating to a phenotype is that the interactions between the genes are also taken into account. This allows closely related clusters of genes to be identified which may work together to modulate the phenotype of interest. In this way, some genes—for which sufficient data are not available—might still be assigned a function, or effect on the phenotype, through "guilt by association". A role for such genes being inferred from the fact that they are tightly linked to other, better characterized, genes.

The networks generated from each of the data sources can then be merged to create a multidimensional, or multilevel, network. In graph theory parlance, this is also termed a multigraph. Many of the tools used for network analysis can be directly applied to either the multigraph or to a collapsed "averaged" version of it. This allows genes and modules to be selected which are likely to influence regulation of the phenotype of interest. Multidimensional network models offer the additional advantages of reproduction—the support of individual links from different data sources; and completion—where interactions missing in one resource may be provided by another (Fig. 1a). Reliance on a single data source presents two key limitations which can be ameliorated by the use of multidimensional networks. Firstly, technical variation can influence the correlations between phenotypes, and thus the weights of

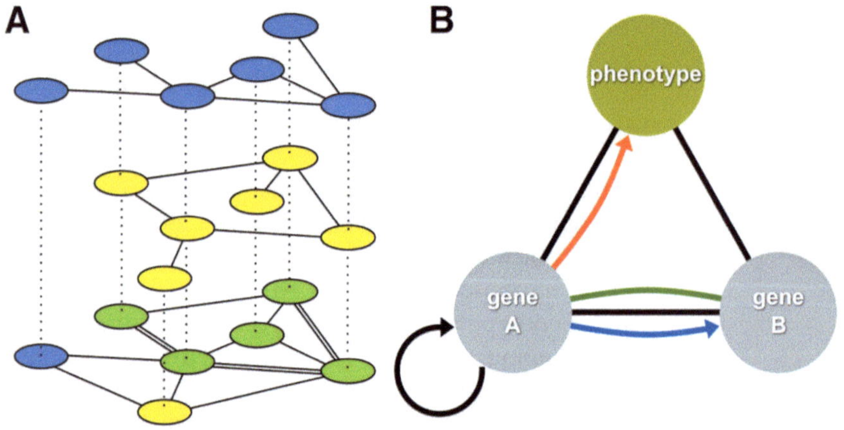

Fig. 1 The structure of a gene interaction network. (**a**) Schematic illustration the concepts of reproduction and completion. Node which are present in both the *blue* and *yellow layers* are shown in *green* in the final network (*at bottom*). Likewise, edges present in both layers are drawn as *double lines*. The final network is the union of all component layers. (**b**) Nodes (*circles*) are connected by edges (*lines*). Directed edges are drawn as an *arrow* pointing from source to target, whereas undirected edges are shown as *open-ended lines*. Different node types can be identified by color (here, *blue-gray* for genes/gene products and *green* for high-level phenotypes). Likewise, different edge types (i.e. belonging to different layers) can be visually differentiated (for example, *black* for transcript expression correlation, *green* for protein–protein binding, *blue* for transcription factor binding and *red* for gene perturbation). A *cis*-eQTL has been represented here as a self-loop (a directed edge with the same node as both source and target)

the edges—leading to possible false positives or false negatives. If the same interaction has been measured in other layers, then such reproduction provides a better argument for an edge in the final network. Secondly, any experiment can only address a subset of the total latent biological variation in a system—gene expression data sets are blind to effects from post-translational modification, for example. This means that certain edges or nodes will be visible to some layers and not others. The incorporation of data from diverse sources addressing the same biological process can help complete the final network model (*see* **Note 2**).

The aim of this chapter is to present a basic workflow to create a multilayer network from data sources typical for research with genetic reference populations. A methodological overview is provided, followed by a practical example using a phenotype from the field of adult neurogenesis (numbers of adult-born astrocytes in the hippocampus) augmented with publicly available transcript expression, genotype association, gene perturbation, and protein interaction data. A large variety of methods have been proposed for building, integrating, and analyzing biological networks. The constraints of space prevent a full discussion of all these other possible approaches, but the reader familiar with the basic workflow presented here should have little problem incorporating variations and adapting the methods to their specific needs.

2 Methods

2.1 Preparation

The methodology described below is merely one approach of many, but lends itself well to the type of data typically available to a researcher working with complex phenotypes in genetic reference populations. The example analysis presented here has been performed using the R software [2]. A number of alternatives are available, including various online workflows—but these may not always offer the flexibility required. The reader is encouraged to investigate other tools that best suit their specific data and question. Several helpful tools and resources are presented elsewhere in this book—including datasets which can be directly incorporated into the pipeline described below.

In the following steps, the individual network layers will be added—each from a particular data source—and these will then be joined together for analysis as a single, multilayer, unit.

2.1.1 The Phenotype of Interest

The analysis will be centered around a quantitative phenotype of interest (the "seed"). This phenotype (or trait—the terms here are used interchangeably) can be any measurable and heritable characteristic. The raw phenotype data measured by the researcher are first read in and the strain means calculated. At this point it is important to ensure that the strain designations are equivalent for both the phenotype and transcript data—evolving nomenclature and spelling errors/variations can result in common strains not being recognized—and thus result in a significant loss of statistical power.

2.1.2 Transcript Expression Correlation Network

The next step is to read in transcript expression data. This should be from a relevant tissue and can be custom-generated by the researcher or retrieved from a public resource. The overlap of strains between this and the phenotype data should be as large as possible to maximize power. Once read in, the data may need to be transposed to ensure that the variables (genes) are in the columns and observations (strains) in the rows (this is required for the R functions we will use). The phenotype is then correlated to each of the transcripts to identify those with similar expression patterns. Transcripts with a correlation above a certain threshold are retained, their data retrieved, and are then also correlated to each other. This two-step procedure yields a correlation matrix comprising pairwise links between the phenotype and all coexpressed genes.

2.1.3 QTL Mapping of Transcripts

Links from genome to transcript expression can be established through QTL mapping of transcript expression profiles. Such links also contain information on causality (i.e. they can be represented as directed edges)—this is because, while a genomic sequence variant can modulate gene expression, the reverse is not true. Thus, QTL data add powerful hypotheses about the directionality of information flow through the network. Expression data for all of the genes in the matrix generated in the previous step are mapped to the distribution of genomic markers for the population of interest. The presence of an expression QTL (eQTL) is evidence of a polymorphism affecting expression. If the differentially expressed gene itself is the source of the polymorphism (i.e. the gene and QTL are at the same place in the genome), then the association is known as a *cis*-eQTL. Such genes can be flagged as potential upstream drivers of the networks in which they are involved.

2.1.4 Supporting Gene–Gene Interaction Data

Many of the gene–gene interactions in the network so far may be spurious (arising from subtle measurement errors in the high-throughput expression dataset) or indirect. The presence of the same interactions in data from other sources can provide further support for these links. Also, because transcript expression does not predict protein expression well, some proteins may be involved but not be represented in the transcript expression network. This can be partly remedied by incorporating protein–protein binding interactions which may suggest additional members of functional pathways. The generation of such data is not within the reach of most laboratories—however many generic data sets are available. Here we will use a composite dataset consisting of protein interaction data, literature co-mention, and transcript coexpression and tissue coexpression which has been collated by the STRING project [3, 4].

2.1.5 Perturbation Data

Another source of causal information is the use of experimental perturbations. Such data are almost invariably generated ad hoc for a particular study and are thus an expensive addition to the analysis. They offer, however, the possibility to identify the direction of some gene–gene links and define new gene-phenotype associations and can thus be helpful in determining the causal structure of gene networks. Often, such data have already been gathered previously in the form of single-gene experiments and other links can be gleaned from a search of the published literature. In a few cases, a relevant database exists which can be used as a data source. Links of this type will all be directed edges from gene to phenotype (where the phenotype is the same as used as the seed).

2.1.6 Hub Analysis to Identify Candidate Pathways

Once the network layers have been created, as separate lists of interactions (*see* **Note 3**), these lists can be simply concatenated to form the multidimensional network. This method yields a network where each pair of nodes can be linked by multiple edges. It is also possible to merge such multiple edges by calculating a new weight combining the weights of the component edges to make the network more accessible to certain analysis methods. Here, however, we will retain the separate edges as an unweighted multi-edge network (Fig. 1b)—the weighted edges having been thresholded as described below. Now the network is ready for analysis and visualization. Depending on the goal of the analysis, different types of analysis are possible. Linking the phenotype to genes enriched in known pathways could suggest a function for a poorly characterized phenotype. Alternatively, the researcher might be interested in discovering novel genes potentially regulating a phenotype. A simple analysis of the latter type is to identify hub genes—those which have the highest number of connections to other genes—which is the approach we will take here.

2.2 Example Analysis

This section presents an example analysis following the steps described above. The data used are all publicly available and, as such, the analysis represents a methodology which can be applied to many questions to generate testable hypotheses—without any initial investment of experimental work. The example here addresses an aspect of adult neurogenesis—the survival of adult-born astrocytes in the murine hippocampus. In the dentate gyrus of most mammals, a pool of neural stem cells persists into adulthood and these cells can differentiate into mature granule cell neurons and astrocytes. Although the generation of new astrocytes is often regarded as being of incidental interest in the field, this phenotype will serve here well as a base for a novel analysis for demonstration.

1. The seed phenotype, the fraction of BrdU retaining cells in the dentate gyrus co-labeled with the astrocyte marker S100-β [5], is first retrieved from the GeneNetwork database (*http://www.genenetwork.org*) where it has the Record ID 10798. To read these data into R, we can utilize the GeneNetwork scriptable interface (*http://www.genenetwork.org/CGIDoc.html*). The following R command creates the appropriate URL, connects to the GeneNetwork server and reads the data into the R session.

```
astr <- read.table("http://robot.genenetwork.org/
webqtl/main.py?cmd=get&db=BXDPublish&probeset=10798&
format=col", sep="\t", header=TRUE)
```

2. Measurements of hippocampal mRNA expression in the BXD panel have been previously generated [6] and are available from GeneNetwork as dataset GN112. A data dump was obtained as a flat text file from *http://datafiles.genenetwork.org/download/GN112/GN112_MeanDataAnnotated_rev081815.txt* and read in to the R session. Note that there are several lines of header information in this file which need to be removed or skipped, and the data for control probes (prefixed with "AFFX") were also removed for this analysis. When using the R command read.table, the parameters sep="\t" and comment.char="" will also need to be set. There are data for several strains in this file—remove those which are not from the BXD panel. Columns with other information (except for the probeset and gene symbol) can also be removed—special characters in the long gene names cause trouble when reading in the file, and this information will not be used by the analysis.

3. Now the data from overlapping strains for the seed phenotype and BXD hippocampus expression can be correlated. Using the command cor with the parameter use="pairwise.complete.obs" allows missing data to be skipped properly. All correlates with a Pearson's *r* value of greater than 0.6 are retained.

4. For this analysis, we are fortunate to have a database of curated gene perturbation results available (MANGO; *http://mango.adult-neurogenesis.de*; [7]). There is also an application programming interface (API) available and a script to allow the database to be easily queried from R. We will retrieve all genes reported to positively affect the numbers of adult-born astrocytes or their differentiation. This is achieved with the following two lines of code:

```
source("http://mango.adult-neurogenesis.de/api/mango-query.R")
mango.astr.correlates <- mango.query(process="numbers,
differentiation", cellstage="astr", effect="positive",
expression="false")
```

5. Our first layer can now be constructed; consisting of the genes just retrieved from MANGO as source nodes and the astrocyte phenotype ("ASTR") as target node. We will use the MGI gene symbol as the node identifiers.

```
mango.network <- cbind("Layer"="MANGO", "Source"=
as.character(mango.astr.correlates$symbol), "Target"="ASTR")
```

6. The BXD transcript expression layer can now be generated in which all of the genes correlating to the seed phenotype are correlated to each other. In order to calculate gene–gene correlations between the MANGO correlates, the relevant data will need to be extracted (using the GeneID identifier) from the BXD expression dataset. Then these are concatenated with the data for the BXD correlates (from **step 3**). The resulting data (with genes as columns and strains as rows) is correlated (as in **step 3**). Just passing the one data matrix as argument means that this will be correlated to itself, thus yielding a complete gene–gene network.
7. To format the network, as in **step 5**, loop is required to traverse each unique gene pair and, if its correlation is above 0.6, to add it as an edge in the layer.
8. Using the core network genes from **step 6**, the corresponding genes are retrieved from another hippocampal expression dataset. We will use here expression data from the CXB cross which was generated on the same platform in parallel with the BXD dataset above. It can be downloaded at *http://datafiles.genenetwork.org/download/GN99/GN99_MeanDataAnnotated_rev081815.txt* and processed as in **steps 2**, **6**, and **7** to yield a separate gene–gene coexpression layer.
9. Because *cis*-eQTLs present a causal hypothesis liking genomic variant and gene expression, they are a powerful tool for identifying upstream genes in a network. Firstly, we need to search for all *cis*-QTLs in the BXD hippocampus dataset. From the main GeneNetwork page, select the appropriate database: *Mouse > BXD > Hippocampus mRNA > Hippocampus Consortium M430v2 (Jun06)* and enter into the "*Combined*" field the query: *cislrs=(0 1000 10) pvalue=(-1 0.05)*. This returns all 4926 probesets with a significant (genome-wide corrected *p*-value below 0.05) QTL within 10 Mb of the probeset position (this interval is influenced by linkage structure, *see* **Note 4**). This table can now be downloaded ("*Download table*" button), formatted as a tab-delimited text file and read into the R session.
10. The STRING database (*http://string-db.org*) can be accessed through a convenient R package; *STRINGdb*. Install this from BioConductor (*https://www.bioconductor.org*) and use the following code to identify interactions involving the genes in the

core set of nodes (from network.data, the data matrix created in **step 6**). We use an interaction score threshold of 400, the default used by the STRING web interface.

```
require(STRINGdb)
string_db <- STRINGdb$new(version="10", species=10090,
score_threshold=400, input_directory="")
mapped <- as.matrix(string_db$map(cbind("S
ymbol"=colnames(network.data)), "Symbol",
removeUnmappedRows=TRUE))
rownames(mapped) <- mapped[, "STRING_id"]
interactions <- string_db$get_
interactions(rownames(mapped))
string.network <- cbind("Layer"="STRING", "Source"=ma
pped[interactions$from, "Symbol"], "Target"=mapped[in
teractions$to, "Symbol"])
```

Note that the STRING database was not used here to introduce new nodes into the network. Because the data from STRING is heavily influenced by literature co-mention and other interactions from nonrandomly sampled data sources, the interactions carry some "bias" and run the risk of introducing a circular argument into the analysis. This is discussed further in **Note 2**.

11. Now all the layers have been generated, they can be merged into a single multilayer network. This is done simply by concatenating the layers;

```
network <- rbind(bxd.transcript.network, cxb.
transcript.network, mango.network, cis.qtl.network,
string.network)
```

12. Now, the network can be analyzed. For this example a simple hub detection will be performed. This approach calculates the number of connections linking to each node (the node degree) to identify the most highly connected gene, which is understood to reflect its importance in the network. This can be easily achieved by counting the number of times the gene appears in the list of source and target nodes.

```
nodes <- c(network[, "Source"], network[, "Target"])
degree <- as.matrix(table(nodes))
```

We can then sort the data from largest to smallest degree (removing the phenotype, which will obviously have the largest degree as it is linked to all other nodes by definition) and view the top hub genes.

```
degree <- degree[order(degree, decreasing=T), ]
degree <- degree[-which(names(degree)=="ASTR")]
#Remove phenotype from analysis
hubs <- degree[which(degree > 10)]
```

For this analysis, the top candidate hubs, and their degree, are;

Gene	*Miat*	*Kif3a*	*AU067697*	*Stat3*	*Ahi1*	*Notch1*	*Ylpm1*
Degree	17	13	12	11	10	10	10

The first three have no known role in astrocyte maturation, and are thus novel candidates for further study. The gene *Miat* (also known as Gomafu) encodes a long noncoding RNA which has already been associated with cell-type specification of neurons [8] and oligodendrocytes [9]. The protein encoded by *Kif3a* is a component of the primary cilium, which has been assigned a role in regulating radial astrocytes in the adult hippocampus [10]. The gene *AU067697* is an uncharacterized transcript which nevertheless may be of interest for further investigation as it also associated with a significant *cis*-eQTL. It is important to note that *Miat*, the top candidate from the above analysis, would not have been identified as such using traditional analyses relying on a single data source. It is only through its presence in a transcript correlation cluster and its association via STRING to two known modulators of astrocyte differentiation that it stands out here as a key potential regulatory gene.

13. Finally, it is useful to visualize the network we have created. To do this, the software Cytoscape (*http://www.cytoscape.org*) is used. The network is written to file:

```
write.table(network, file="Network.txt",
sep="\t", quote=FALSE, row.names=FALSE)
```

and then imported via the menu commands *File > Import > Network > File...* and assigning the correct columns for the interaction (layer), source, and target. The Style tab enables many options to be set so that, for example, edges from different layers are colored differently. The nodes can also be distributed for clarity and to aid in the identification of clusters. Many algorithms are available—a visually appealing option is found under *Layout > yFiles Layouts > Organic* which is what was used to generate the layout shown in Fig. 2.

3 Notes

1. Network structure and terminology. Networks are a way of representing systems comprised of interactions between elements (Fig. 1). In graph theory, the field of mathematics concerned with the study of networks, the interactions are termed edges and the elements they connect are known as vertices, or nodes as is more commonly seen in the biological literature. A set of nodes, joined by edges, is termed a network or, more formally, a graph. In systems genetics, most nodes will represent genes or phenotypes, although other biological entities can be used, such as proteins, genomic loci, or metabolites. The edges can likewise represent various interaction types; examples are coexpression, physical interaction or regulation.

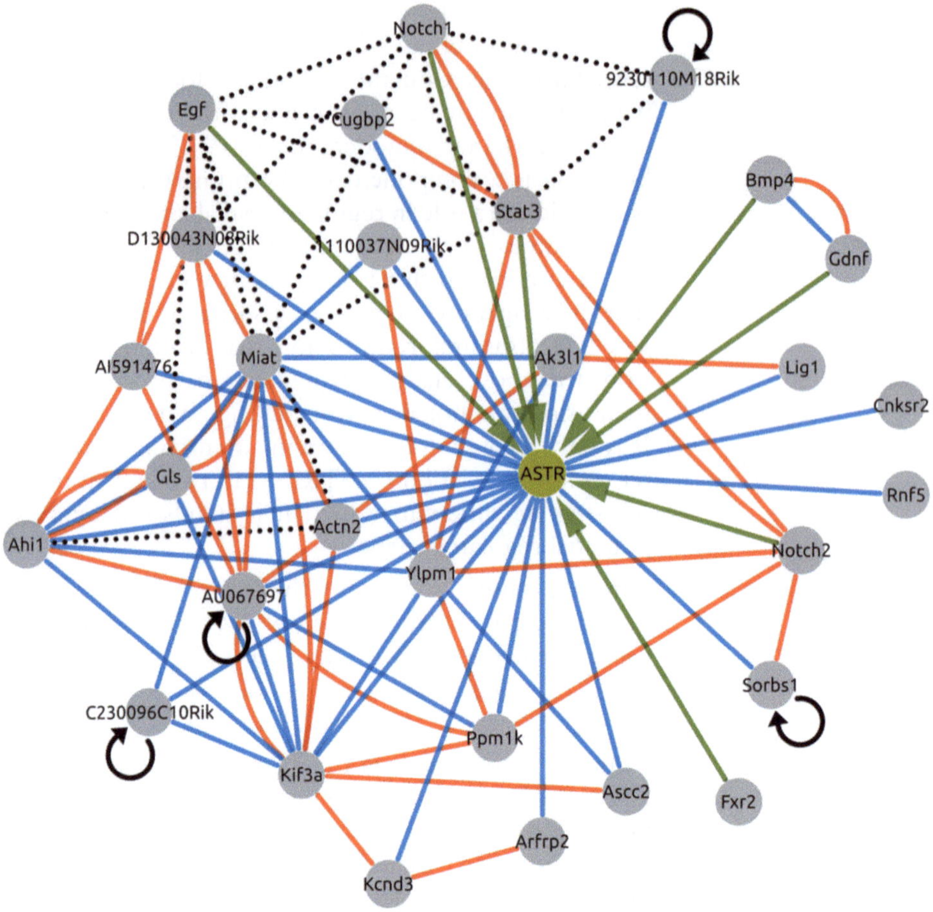

Fig. 2 Multidimensional network associated with astrocyte differentiation. A core set of genes (*gray nodes*) was selected based on their association to the adult-born astrocyte phenotype (ASTR; *green node*) either from BXD expression correlation (*blue edges*) or from curated literature links (*green arrows*). Additional interactions between the core genes were obtained from a second expression dataset (CXB; *red edges*) and the STRING database (*black dotted lines*). Genes which exhibit a cis-eQTL in the BXD dataset are marked by the presence of self-directed edges (*black arrows*)

In addition, edges can be either directed, where a causal relationship between source and target node is described, or undirected in which case no causal information is presented. Edges can also be weighted, where a quantitative value can be associated with the interaction, or unweighted where a meaningful value is not available. The sub-network comprising all of the edges of a single type is referred to here as a layer and the integration of several edge types yields a multilayer or multidimensional network.

2. Each separate network layer is essentially always incomplete in that not all of the possible interactions describing the biological system are present. This fact can lead to subtle sampling

bias that the researcher should be aware of. The reasons behind this incompleteness fall into two broad categories:

Firstly, each layer surveys only a specific type of interaction; such as transcript covariance, protein binding, or tissue co-expression. Thus, relationships between elements may not be visible in a particular layer. For example; two genes might be considered linked if their protein products both bind to a common target. That information, however, cannot ever be gleaned from knowledge of their transcript expression patterns. Such limitations are unavoidably inherent in the biological nature of the data. An extension of this issue when dealing with genetic populations is that the segregating polymorphisms are not uniformly distributed across all genes so that some genes will exhibit more expression variation than others. Genes that are not polymorphic within the panel studied thus constitute blind spots for QTL mapping and may exhibit less variation in expression leading to different correlation patterns. Both of these problems can be potentially overcome by incorporating additional data sources; either of differing layer types or derived from different genetic populations (the concept of "completion" referred to in the introduction).

The second, technical, factor influencing network completeness concerns the coverage of the available data. Some data sources, such as whole-genome transcript expression, cover an essentially complete set of all transcripts, but this is not the case for most other biological network types. Missing data, where certain interactions have simply not been measured, thus constitute additional blind spots. In such cases, it becomes important how nonpresent network edges are dealt with. An edge is evidence of an interaction between the nodes. Absence of an edge, however, can be interpreted in two ways; either as evidence of *no* interaction, or as the absence of evidence—two very different concepts. Missing data in an incompletely measured network layer is absence of evidence and it then becomes crucial to establish whether the interactions which *were* measured were selected randomly.

In the case of a whole-transcriptome expression dataset, presence of a strong interaction between two genes is more likely to be biologically important than a weak interaction. Such an assertion is not, however, necessarily true for a network based on literature co-mention. Such a network is created from a small subset of all possible gene–gene interactions and, more importantly, is dependent on the choice by the original author to present both genes in the same manuscript—or to study a particular candidate. Decisions like these are usually influenced by the existing literature and present a partly circular argument.

Thus, nonrandomly sampled ("biased") network layers need to be treated with caution to avoid inadvertently creating circular arguments.

3. Network data formats. Network data can be stored in many different formats, some of which allow additional information like node and edge metadata and layout parameters. The simplest form, which we will use here, is a tab-delimited text file consisting of at least three columns; the first contains the name of the layer, the next two columns contain the names of the source and target nodes respectively. The distinction between source and target node is important for directed layers, for undirected layers, these are interchangeable. Additional columns can hold information about the edge weight and distance.
4. Due to the genetic structure of recombinant inbred populations, there are often large linkage blocks which are inherited together. The transcript expression patterns of genes in these blocks can show similarities due to their inheriting the same alleles—rather than being due to functional interactions. This means that they will correlate more strongly than expected and therefore will tend to cluster in an expression correlation network. Because the patterns of linkage are determined by the population used, the only way to provide stability against such confounds is to look at the same interactions in different genetic backgrounds. The integration of data from multiple crosses into an analysis, as done in the example above, thus offers a practical solution to this problem.

References

1. Overall RW, Williams RW, Heimel JA (2015) Collaborative mining of public data resources in neuroinformatics. Front Neurosci 9:90. doi:10.3389/fnins.2015.00090
2. Team RC (2014) R: a language and environment for statistical computing
3. von Mering C, Huynen M, Jaeggi D et al (2003) STRING: a database of predicted functional associations between proteins. Nucleic Acids Res 31:258–261. doi:10.1093/nar/gkg034
4. Szklarczyk D, Franceschini A, Kuhn M et al (2011) The STRING database in 2011: functional interaction networks of proteins, globally integrated and scored. Nucleic Acids Res 39:D561–D568. doi:10.1093/nar/gkq973
5. Kempermann G, Chesler EJ, Lu L et al (2006) Natural variation and genetic covariance in adult hippocampal neurogenesis. Proc Natl Acad Sci 103:780–785. doi:10.1073/pnas.0510291103
6. Overall RW, Kempermann G, Peirce J et al (2009) Genetics of the hippocampal transcriptome in mouse: a systematic survey and online neurogenomics resource. Front Neurosci 3:55. doi:10.3389/neuro.15.003.2009
7. Overall RW, Paszkowski-Rogacz M, Kempermann G (2012) The mammalian adult neurogenesis gene ontology (MANGO) provides a structural framework for published information on genes regulating adult hippocampal neurogenesis. PLoS One 7:e48527. doi:10.1371/journal.pone.0048527
8. Sone M, Hayashi T, Tarui H et al (2007) The mRNA-like noncoding RNA Gomafu constitutes a novel nuclear domain in a subset of neurons. J Cell Sci 120:2498–2506. doi:10.1242/jcs.009357
9. Mercer TR, Qureshi IA, Gokhan S et al (2010) Long noncoding RNAs in neuronal-glial fate specification and oligodendrocyte lineage maturation. BMC Neurosci 11:14. doi:10.1186/1471-2202-11-14
10. Han Y-G, Spassky N, Romaguera-Ros M et al (2008) Hedgehog signaling and primary cilia are required for the formation of adult neural stem cells. Nat Neurosci 11:277–284. doi:10.1038/nn2059

Chapter 11

RNA-Seq in the Collaborative Cross

Richard Green, Courtney Wilkins, Martin T. Ferris, and Michael Gale Jr.

Abstract

The Collaborative Cross (CC) is a large panel of inbred mouse strains currently being developed for multiple areas of research. Scientists are taking integrated omics-style approaches to collecting data in order to obtain a deeper understanding of the biological mechanisms underlying a number of diverse disease phenotypes. As the cost of the next generation sequencing (NGS) decreases, RNA-sequencing (RNA-seq) has become the favored approach to transcriptomic analyses versus microarrays due to increases in sensitivity and resolution. This is particularly the case with newly defined genomes, where experimental annotation has not caught up to the new microarray platforms. Traditional RNA-seq approaches are not ideal when working with results from collaborative cross studies, as the genomes across individual strains differ considerably. In this chapter we will provide an overview of how to effectively perform RNA-seq analysis from data obtained from the CC mice.

Key words Analysis tools, Collaborative Cross, RNAseq

1 Introduction

In studies with Genetic Resource populations (GRPs) like the Collaborative Cross, standard RNAseq analysis approaches may not yield suitable results. To analyze this data accurately, custom genomic annotations and specialized tools need to be used. This chapter outlines the steps need to download, install, and run the necessary programs to generate RNAseq results with collaborative cross data.

2 Methods

2.1 Software Tools

All the work outlined in this chapter was done in Red Hat Enterprise Linux 7 (Fig. 1). These steps should be nearly identical for CentOS, although the tools and install procedures outlined below may vary for other Linux distributions (i.g. Debian, Ubuntu).

Klaus Schughart and Robert W. Williams (eds.), *Systems Genetics: Methods and Protocols*, Methods in Molecular Biology, vol. 1488, DOI 10.1007/978-1-4939-6427-7_11,

Software/Data	Description	Additional Information
Bowtie2, Tophat2, STAR	Alignment Tools	http://bowtie-bio.sourceforge.net/bowtie2/index.shtml http://tophat.cbcb.umd.edu https://code.google.com/p/rna-star/
FASTQC	Quality control for sequencing data	http://www.bioinformatics.babraham.ac.uk/projects/fastqc/
Git	Download tool	https://git-scm.com/downloads
Gunzip	Unpacking tool	
Ht-seq	Converted mapped reads into gene counts for statistical analysis	http://www-huber.embl.de/HTSeq/
Illumina igenomes	Illumina curated annotation for NGS	https://support.illumina.com/sequencing/sequencing_software/igenome.html
Lapels	Converts to alignments to	https://code.google.com/p/lapels/

Fig. 1 List of tools and corresponding references

	reference coordinates	
Pseudo Genomes	Genomic sequences genomic sequences for the Collaborative Cross (CC) mouse strains	http://www.csbio.unc.edu/CCstatus/index.py?run=Pseudo
Samtools	Tool to manipulate SAM and BAM files	http://www.htslib.org/
Suspenders	Merge lapel data files	https://github.com/holtjma/suspenders
Wget	Linux download tool	https://www.gnu.org/software/wget/

Fig. 1 (continued)

2.2 Working with Pseudogenomes

The first step in analyzing RNA-seq data for a collaborative cross (CC) experiment is to obtain the reference annotation for the genomes you want to map your raw sequences against. Since the CC mice are not traditional laboratory mouse strains (they are derived from 8 founder strains) we need to download their relevant CC genomes (also referred to as the pseudogenomes or in silico genomes). These genomes were created using data from DNA sequencing and genotyping. To determine which of genomes to use look at the UNC identifiers of the CC lines included in the experimental design. Each CC line was derived as a cross of two "engineered" inbred (RI) lines, and the UNC identifier denotes these intercrosses (RIX) lines as the cross of the parental RI lines (dam x sire). For example, a Collaborative Cross mouse with the UNC identifier of 8016x8034 is the result of crossing an 8016 mother with an 8034 father. The pseudogenome of each parental

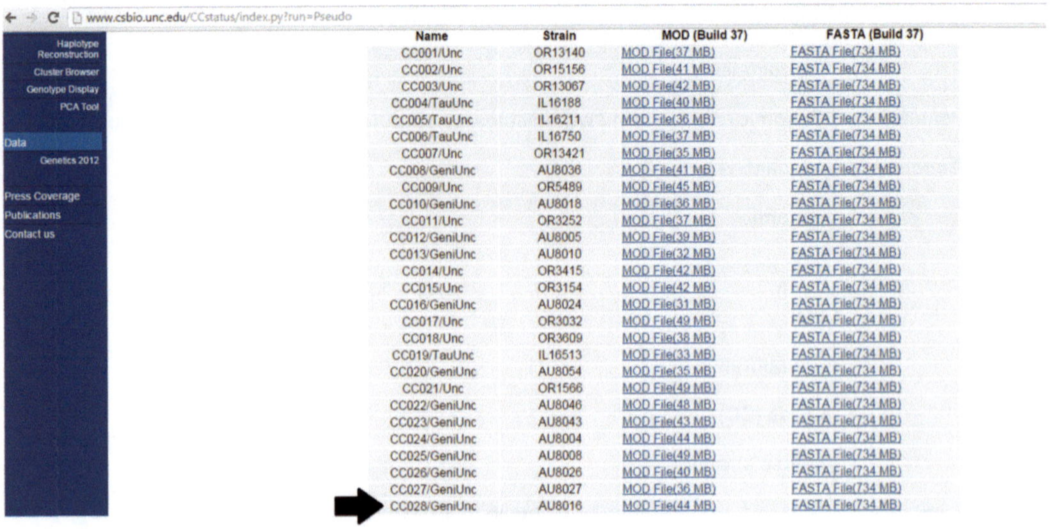

Fig. 2 Screenshot from CC website showing list of CC lines

line can be downloaded from the UNC systems genetics page: http://www.csbio.unc.edu/CCstatus (Fig. 2).

Look closely at the genome listings for additional information about each RI strain. The 8016 (black arrow) is listed as AU8016 because the strain originated in Australia. The genome also has an official UNC name of CC028/GeniUnc. Downloading these genomes and building indexes can take some time, so we recommend only downloading the CC genomes that relevant to your study. For each CC genome, there is a fasta and corresponding MOD file to be downloaded. The fasta file contains the raw genome sequence, and the MOD file is what will help us remap the FASTA sequences back to the reference mouse (mm10) coordinates/genes. The MOD file is similar to a VCF (Variant Call Format) for the pseudogenome. In this example, we will be mapping the CC line 16211x16557, so we will pull down the genomes for 16211 and 16557 using the linux tool wget. If you don't have wget installed, run the command:

```
sudo yum install wget
```

From there type wget and the paste the url previously identified for each selected genome to download.

We will download the CC pseudogenome and corresponding genomic MOD files, which we will use later.

```
wget http://www.csbio.unc.edu/CCstatus/pseudo2/IL16557.fa.gz
wget http://www.csbio.unc.edu/CCstatus/pseudo2/IL16557.mod
wget http://www.csbio.unc.edu/CCstatus/pseudo2/ IL16211.fa.gz
wget http://www.csbio.unc.edu/CCstatus/pseudo2/IL16211.mod
```

You will see the following messages:

```
--2015-12-02 09:40:37-- http://www.csbio.unc.edu/
CCstatus/pseudo2/IL16557.fa.gz
```

```
Resolving www.csbio.unc.edu (www.csbio.unc.edu)… 152.2.132.8
Connecting to www.csbio.unc.edu
(www.csbio.unc.edu)|152.2.132.8|:80… connected.
HTTP request sent, awaiting response… 200 OK
Length: 770349360 (735M) [application/x-gzip]
Saving to: 'IL16557.fa.gz'
1% [>                ] 7,807,318   2.75MB/s
```

The genome (fa) files are then unpacked with the following command:

```
$ gunzip *.gz
The files should now have a .fa extension and no longer a .gz
```

2.3 Install Bowtie2

To install bowtie2 go to the link (below) and download one of the pre-built binaries. There are binaries available for the Intel x86_64, which includes Linux, Mac, and Windows.

http://sourceforge.net/projects/bowtie-bio/files/bowtie2/

Once Bowtie2 is installed, we will build indexes for the pseudosequences using the bowtie2 command. Please note that this can take some time.

```
bowtie2-build <reference file> <output/index>
$ bowtie2-build -f IL16557.fa IL16557.cc
Settings:
Output files: "IL16557.cc.*.bt2"
Line rate: 6 (line is 64 bytes)
Lines per side: 1 (side is 64 bytes)
Offset rate: 4 (one in 16)
FTable chars: 10
Strings: unpacked
Max bucket size: default
Max bucket size, sqrt multiplier: default
Max bucket size, len divisor: 4
Difference-cover sample period: 1024
Endianness: little
Actual local endianness: little
Sanity checking: disabled
Assertions: disabled
Random seed: 0
Sizeofs: void*:8, int:4, long:8, size_t:8
Input files DNA, FASTA:
IL16557.fa
Building a SMALL index
Reading reference sizes
```

While the indexes of the pseudogenomes are building, we can proceed to obtaining quality scores for the raw sequencing reads. Raw sequencing data should come off the machine in the form of a fastq or fastq.gz (compressed fastq file).

2.4 FASTQC

FASTQC is quality control program that runs on raw sequencing data. To obtain the most recent version of this software, please go to: http://www.bioinformatics.babraham.ac.uk/projects/download.html and following the installation instructions. Once FASTQC is installed, the command below will check the quality of the raw sequencing data following the creation of a QC directory to store all FASTQC output.

```
mkdir ../QC
mkdir ../QC/FASTQC
fastqc -noextract -t 10 -o ../QC/FASTQC *.fastq.gz
--noextract do not compress the results after making them
-t number of fastq files that can be run at the same time (set to 10)
-o directory to write quality results

$fastqc --noextract -t 10 -o ../QC/FASTQC/NORRNA *.fastq.gz
```

This will produce a detailed html report about the raw sequencing data that is beyond the scope of this chapter. Depending on the output of the FASTQC reports, additional processing may be required, such as trimming bases off the beginning or the end of a sequence due to low quality scores. Once sequence quality is within acceptable limits, we can begin setting up the tools necessary to map CC data to the corresponding pseudogenomes. Additional information about FASTQC and interpreting quality scores can be found here: http://www.bioinformatics.babraham.ac.uk/projects/fastqc/.

2.5 Producing Results with Tophat2

To download Tophat2 go to: http://tophat.cbcb.umd.edu.

On the tophat website under releases you can download the version that is the best match to the linux version you are using. For Red Hat Enterprise 7, The prebuilt Linux x86_64 binary works. Once it is downloaded to your system, you can execute the following commands to unpack and install Tophat2.

```
tar xvfz tophat-<version>.Linux_x86_64.tar.gz
cd ~/bin
ln -s ~/tophat-<version>.Linux_x86_64/tophat2
```

Mapping samples to the pseudogenomes can take time. Before we begin, we should consider applying additional processing cores to the tophat2 run. Tophat2 is a multithreaded process (the program can run in parallel using multiple core processors to speed up the mapping against the genome index). First check how many cores are on your linux machine with the nproc command:

```
$ nproc
32
```

The computer used in this example has 32 cores. Now we can designate how many cores we would like to provide toward the mapping. Our computer has 32, but we will only designate 10 to allow for other processes to continue.

Since this is a F1 from two RI strains, you will need to map the data against both parental RI lines. This means you will need to map the sample once with the dam index 16211 and then with the sire index 16557.

To run tophat2 on your sample against IL16211, run this command:

```
tophat2 -p 10 -o matAlignmentDir16211 IL16211.cc IL16211_
D12_WNV.fastq.1.gz \ IL16211_D12_WNV.fastq.2.gz
```

p : number of core processors to dedicate to mapping
o: the directory to write mapped files

To run tophat2 on your sample against IL16557 run this command:

```
tophat2 -p 10 -o matAlignmentDir16557 IL16557.cc IL16557_
D12_WNV.fastq.1.gz \ IL16557_D12_WNV.fastq.2.gz
```

An example of the tophat2 mapping against IL16211 is shown below:

```
/usr/local/bin/tophat2 -p 10 -o matAlignmentDirIL16211
IL16211.cc IL16211_D12_WNV.fastq.1.gz IL16211_D12_WNV.
fastq.2.gz
[2015-12-08 09:33:07] Beginning TopHat run (v2.1.0)
-----------------------------------------------
[2015-12-08 09:33:07] Checking for Bowtie
                  Bowtie version:        2.2.5.0
[2015-12-08 09:33:07] Checking for Bowtie index files (genome)..
[2015-12-08 09:33:07] Checking for reference FASTA file
[2015-12-08 09:33:07] Generating SAM header for IL16211.cc
```

Once this is run against both pseudogenomes, it will produce results two separate directories: *matAlignmentDir16211and matAlignmentDir16557*. You may find that Tophat2 can be slow, and later in this chapter we discuss STAR, an alternate mapping tool which may produce faster results. For both tools, increasing the number of processors can speed up mapping slightly. Once the mappings are finished you will see a bam file(s) which you will process through lapels.

Obtain lapels by going to this website:

https://pypi.python.org/pypi/lapels

Next, we will get the lapels directory with compiled code:

```
mkdir lapels
cd lapels
wget https://pypi.python.org/packages/source/l/lapels/la-
pels- \ 1.0.6.tar.gz#md5=59f91f83269dae2425b17cf0e204916d
```

You will see the following messages:

```
--2015-12-05 10:05:26-- https://pypi.python.org/packages/
source/l/lapels/lapels-1.0.6.tar.gz
Resolving pypi.python.org (pypi.python.org)… 199.27.74.223
Connecting to pypi.python.org (pypi.python.
org)|199.27.74.223|:443… connected.
HTTP request sent, awaiting response… 200 OK
Length: 22141 (22K) [application/octet-stream]
Saving to: 'lapels-1.0.6.tar.gz.'
```

Then unpack the downloaded directory

```
tar -xzf lapels-1.0.6.tar.gz
```

Once lapels is downloaded, type easy_install lapels_<version>.gz . Note that you may need root access to perform this, so the command would be:

```
sudo easy_install lapels_<version>.gz
```

Now your pipeline and index have been assembled you can begin setting up lapels and suspenders.

2.6 Producing Results with Lapels

Mapping to the pseudogenomes will not produce the proper coordinates due to insertions and deletions that are unique to each pseudogenome. To make the data interpretable we need to run it through lapels. Lapels will take the alignments and convert them back to the standard mouse annotation (mm10).

Typing the command pylapels will confirm that the program was correctly installed and will display all its parameters:

```
$ pylapels
usage: pylapels [-h] [-V] [-q | -v] [-f] [-a alias.csv] [-t] [-n]
               [-p nProcesses] [-c chromList] [-o out.bam]
               in.mod in.bam
```

The important step is assigning the reference MOD file, the bam file you are working on, and the new annotated bam you are creating.

We will refer to our example directory ref used to store our MOD files, and use the command –n to create a new bam file with converted mm10 coordinates:

```
pylapels -n -o IL16557_D12_WNV_lapels.bam ref/IL16557.mod
matAlignmentDir16557/accepted_hits.bam
n: create a new file
p : number of core processors to dedicate to mapping
o: where to write a new, annotated bam file
sudo pylapels -n -o IL16557_D12_WNV_lapels.bam ref/
IL16557.mod matAlignmentDir/accepted_hits.bam
```

You will see the following messages:

```
[2015-12-11 10:17:29] root : INFO: input MOD file: ref/IL16557.
mod
[2015-12-11 10:17:29] root : INFO: input BAM file: matAlignment-
Dir/accepted_hits.bam
[2015-12-11 10:17:29] root : INFO: output BAM file: IL16557_D12_
WNV_lapels.bam
[2015-12-11 10:17:29] root : INFO: Using fallback alias settings.
[2015-12-11 10:17:31] root : INFO: use a single process
[2015-12-11 10:17:31] root : INFO: workder 0 processes chromosome 'chr1'
[2015-12-11 10:17:31] root : INFO: alias 'chr1' used for 'chr1' in MOD
[2015-12-11 10:17:33] root : INFO: alias 'chr1' used for 'chr1' in BAM
[2015-12-11 10:17:33] annotator : INFO: [chr1]: 6721163 read(s) found in BAM
[2015-12-11 10:17:38] annotator : INFO: [chr1]: 0%
[2015-12-11 10:17:45] annotator : INFO: [chr1]: 1%.....
```

2.7 Producing Results with Suspenders

Once lapels finishes on both pseudogenome files, it can be merged through suspenders.

Suspenders accepts two bam files that have been processed through lapels and merges them into a single file. Traditionally, these are two different bams produced from two separate in silico genomes which typically represent the mother's (Dam) and father's (Sire) genomes. There are additional uses for suspenders that are beyond the scope of this tutorial. Please refer to the literature for more information. The next step is to download and install suspenders.

To obtain suspenders, you will need to install git (if you haven't already done so) and pull down necessary code documentation. To install git, use the following command:

```
sudo yum install git
```

Next, create a directory where you will clone the code and documentation.

```
mkdir suspenders
cd suspenders
git clone https://github.com/holtjma/suspenders
```

Once the folder has been cloned from git, enter the command below to install suspenders. Please note that you should be in the directory that contains the suspenders folder, and depending on your system you may need to perform this install procedure as a super user (sudo).

```
sudo easy_install suspenders
```

Once suspenders is installed, the simplest command to run it is as follows:

```
pysuspenders -t ./merged.bam ./<DAM>.bam ./<SIRE>.bam
with our data we note the position of our bams (DAM and SIRE)
pysuspenders -t ./merged.bam ./IL16211_D12_WNV_lapels.bam \
./IL16557_D12_WNV_lapels.bam
```

For suspenders to work successfully it must contain:

1. A BAM file that mapped to the mother's in silico genome.
2. A BAM file that mapped to the father's in silico genome.
3. Both Bam files need to be preprocessed through Lapels.
4. Both Bam files need to be sorted by read name and not by coordinates (will produce a warning).

```
Below is an example of a successful run using Suspenders
$ pysuspenders ./merged.bam ./IL16211_D12_WNV_lapels.bam
./IL16557_D12_WNV_lapels.bam
[2015-12-10 14:02:43] INFO: Merge Type: Quality
[2015-12-10 14:02:43] INFO: Filter Type: Unique->Quality->Random
[2015-12-10 14:02:43] INFO: Input Type: Lapels Input
[2015-12-10 14:02:43] INFO: Inputs: ['./IL16211_D12_WNV_
lapels.bam', './IL16557_D12_WNV_lapels.bam']
[2015-12-10 14:02:43] INFO: Output: ./merged.bam
[2015-12-10 14:02:43] INFO: Number of processes: 1
```

As suspenders successfully merges the bam files it will prompt the user the following messages.

```
[2015-12-10 16:07:53] INFO: [Master] Processed 24500000 read
names…
[2015-12-10 16:08:23] INFO: [Master] Processed 24600000 read
names…
```

```
[2015-12-10 16:08:54] INFO: [Master] Processed 24700000 read
names...
[2015-12-10 16:09:24] INFO: [Master] Processed 24800000 read
names...
[2015-12-10 16:09:55] INFO: [Master] Processed 24900000 read
names...
[2015-12-10 16:10:26] INFO: [Master] Processed 25000000 read
names...
[2015-12-10 16:10:57] INFO: [Master] Processed 25100000 read
names...
[2015-12-10 16:11:27] INFO: [Master] Processed 25200000 read
names...
[2015-12-10 16:11:57] INFO: [Master] All reads scanned. Waiting on
workers to finish processing...
[2015-12-10 16:11:57] INFO: Merge complete!
```

For more information on Suspenders (and lapels), refer to "Read Annotation Pipeline for High-Throughput Sequencing Data" by James Holt, et al. [1]. Additional information about MOD files please refer to "Transforming genomes using mod files with applications" by Huang, et al. [2].

2.8 Generating Gene Count Data (Cufflinks and HT-Seq)

In order for us to obtain quantitative gene information from out merged bam file we will run our merged sample with cufflinks and ht-seq. These are two different tools that produce quantitative gene count data in different ways. Cufflinks will provide output in FPKMs (Fragments Per Kilobase of transcript per Million mapped reads), while ht-seq will generate gene counts in the form of integers. To run either of these tools you will need to apply an annotation in the form of a Gene Transfer Format (GTF) file. This is a file that contains information about the genomic structure. Igenomes (see table for url information) is a website maintained by the Illumina corporation which houses annotation (pre-build indexes and GTF files) for NGS analysis. To load annotation, create a directory and download the desired annotation data:

```
mkdir annotation
cd annotation
wget ftp://ussd-ftp.illumina.com/Mus_musculus/UCSC/mm10/ \
Mus_musculus_UCSC_mm10.tar.gz
gunzip Mus_musculus_UCSC_mm10.tar.gz
```

Cufflinks installation requires several dependencies that are beyond the scope of this tutorial. For complete installation instructions, please refer to the information found here:

https://github.com/cole-trapnell-lab/cufflinks

Once Cufflinks is installed, you can produce FPKMS by kicking off the following command:

```
cufflinks -p 8 -G annotation/genes.gtf merged.bam
G: location of annotation file (on the form of a gff/gtf)
p : number of core processors to dedicate to mapping
$ cufflinks -p 8 -G annotation/genes.gtf merged.bam
```

You will see the following messages:

```
[15:54:43] Loading reference annotation.
[15:54:51] Inspecting reads and determining fragment length distribution.
> Processed 25507 loci.
[*************************] 100%
> Map Properties:
>       Normalized Map Mass: 23410.34
>       Raw Map Mass: 23410.34
>       Fragment Length Distribution: Truncated Gaussian (default)
>                     Default Mean: 200
>                  Default Std Dev: 80
[15:55:01] Estimating transcript abundances.
[continues…]
```

Once completed it will generate a series of new files in the directory:

logs unmapped.bam skipped.gtf transcripts.gtf isoforms.fpkm_tracking genes.fpkm_tracking

Many of these files are table delimited files that contain information about enriched genes, transcripts, and isoforms in the data set. genes.fpkm_tracking specifically will tell you about expression levels of genes in a data set and can be loaded directly into other software (Excel, R, etc.).

2.9 Gene Counts Using HT-Seq

Now that we have results in cufflinks run quantitative gene expression with ht-seq. To install ht-seq you need to download the code from the link below. From there you will unpack following the install instructions.

https://pypi.python.org/pypi/HTSeq

Please note that ht-seq is a python program and runs in python version 2.5. At the time of this writing Python 3 is not yet supported. The bam may need to be additionally sorted by coordinates using samtools in order to be run through ht-seq. Running the ht-seq command of the merged bam (below) should produce the following output.

```
$ htseq-count merged.bam --format=bam annotation/genes.
gtf > merged_bam_genecounts.txt
```

You will see the following messages:

```
100000 GFF lines processed.
200000 GFF lines processed.
300000 GFF lines processed.
400000 GFF lines processed.
500000 GFF lines processed.
600000 GFF lines processed.
700000 GFF lines processed.
760316 GFF lines processed.
39490 SAM alignment pairs processed.
```

Opening the merged_bam_genecounts.txt file and produces the following output:

```
$ head merged_bam_genecounts.txt
0610005C13Rik 4
0610007P14Rik 349
0610009B22Rik 84
```

```
0610009L18Rik 4
0610009O20Rik 213
0610010B08Rik 0
0610010F05Rik 219
0610010K14Rik 186
0610011F06Rik 48
0610012G03Rik 142
```

The first column contains the gene identifier from the mouse (mm10) genes.gtf annotation. The next column holds the gene counts as integers that calculate each gene's abundance. These files can be concatenated together and run through differential expressed tools like EdgeR and DEseq2. To get more information and examples go here:

http://cgrlucb.wikispaces.com/file/view/edgeR_Tutorial.pdf

2.10 Mapping with STAR Instead of Tophat2

Both STAR and Bowtie2/Tophat will align unmapped reads to a pseudogenome. STAR is known to produce results faster than Tophat with similar accuracy. To run STAR, your computer must possess enough memory and processing cores. Please refer to the STAR website for hardware specifications. Bam files produced from the STAR aligner will not work natively with lapels and suspenders. Additional steps need to be performed, which we will review in the following sections.

2.10.1 Installing STAR

```
wget https://github.com/alexdobin/STAR/archive/STAR<version>.tar.gz
tar -xzf STAR<version>.tar.gz
cd STAR<version>
or use git
git clone https://github.com/alexdobin/STAR.git
cd STAR
make STAR
```

To build an index with STAR, run the following command:

```
$ STAR --runMode genomeGenerate --genomeDir /index/STAR/
IL16211 --genomeFastaFiles IL16557.fa --runThreadN 15
You will see the following messages:
Dec 10 13:40:30 ..... Started STAR run
Dec 10 13:40:30 … Starting to generate Genome files
```

2.10.2 Mapping to a Pseudogenome Using STAR

```
STAR --genomeDir /index/STAR/IL16211/ --readFilesCommand
zcat -outSAMtype \ BAM SortedByName --readFilesIn IL16211_
D12_WNV.fastq.1.gz \ IL16211_D12_WNV.fastq.2.gz --outFile-
NamePrefix IL16211_D12_WNV -runThreadN \ 15
```

Converting results to be interpreted by lapels/suspenders

```
samtools fillmd -b your_bam_file pseudogenome_fasta > bam_
file_after_fixed
```

For example:

```
samtools fillmd -b merged.bam /index/STAR/IL16211/IL16557.
fa > merged_fixed.bam
```

The remaining steps (lapels, suspenders, etc.) are the same as those previously shown.

3 Additional Information

For additional information about the CC NGS pipeline please refer to https://github.com/holtjma/suspenders/wiki/Lapels-and-Suspenders-Pipeline

Acknowledgements

Special thanks to James (Matt) Holt, Martin Ferris, Shunping Huang, Seth Greenstein, and Leonard Mcmillan at UNC for support. Thanks to UW Immunology and the Center for Innate Immunity and Immune Diseases for assistance (CIIID).

References

1. Holt J et al (2013) Read annotation pipeline for high-throughput sequencing data. In: Proceedings of the international conference on bioinformatics, computational biology and biomedical informatics, ACM, Washington, DC, USA, p 605–612
2. Huang S et al (2013) Transforming genomes using MOD files with applications. In: Proceedings of the international conference on bioinformatics, computational biology and biomedical informatics, ACM, Washington, DC, USA, p 595–604

Chapter 12

QTL Mapping and Identification of Candidate Genes in DO Mice: A Use Case Model Derived from a Benzene Toxicity Experiment

Dan Gatti, John E. French, and Klaus Schughart

Abstract

Diversity Outbred (DO) mice are a multiparental advanced generation intercross population derived from eight inbred strains which are genetically very diverse. They are maintained as an outbred population using a randomized mating design. Thus DO mice represent an ideal population to map phenotypic traits. Here, we provide a case study in which male DO mice were exposed to benzene and phenotyped for the number of micronucleated reticulocytes. We provide step-by-step R scripts for the analysis of phenotypes, genotypes, mapping of resistance gene loci and identification of candidate genes.

Key words Use case, DO mice, QTL mapping, Response to toxins

1 Introduction

Diversity Outbred (DO) mice are an advanced intercross derived from eight inbred strains, A/J, C57BL/6J, 129S1/SvImJ, NOD/ShiLtJ, NZO/HlLtJ, CAST/EiJ, PWK/PhJ, and WSB/EiJ. The genomes of the first five strains are largely derived from *M.m.domesticus* ancestry. The last three strains are derived from wild caught mice and contribute genomes from *M.m.castaneus*, *M.m.musculus*, and *M.m.domesticus*, respectively. Together, these strains contribute 36 million SNPs to the DO. These are the same eight strains there were used to create the recombinant inbred lines in the Collaborative Cross (CC). While mice in each CC line are inbred, DO mice are maintained as an outbred population using a random mating design with 175 breeding pairs. With each successive generation, DO mice accumulate more recombinations and the mapping resolution increases. As this book goes to press, the

Electronic supplementary material The online version of this chapter (doi:10.1007/978-1-4939-6427-7_12) contains supplementary material, which is available to authorized users.

Klaus Schughart and Robert W. Williams (eds.), *Systems Genetics: Methods and Protocols*, Methods in Molecular Biology, vol. 1488, DOI 10.1007/978-1-4939-6427-7_12,

DO mice are at generation 22 of outbreeding. Here, we provide a case study using the DO mice that will take the reader through the process of mapping a QTL and searching for candidate genes—starting with a set of phenotype data demonstrating a high degree of reproducibility and heritability. The phenotype data comes from a genotoxicity study in which male DO mice were exposed to benzene for 6 h a day, 5 days a week for 4 weeks [1] in inhalation chambers that simulate human occupational exposure. Mice were first phenotyped for the number of micronucleated reticulocytes (measure of chromosomal damage) and reticulocytes in the peripheral blood (before benzene exposure) as well as blood and bone marrow following exposure. The number of micronucleated reticulocytes is a continuous measure of DNA and chromosomal damage, in each DO mouse. The goal is to find gene(s) that influence the level of chromosomal DNA damage in the bone marrow following exposure to a genotoxicant.

2 Methods

2.1 Prepare the Environment and Load the Phenotype and Genotype Data

The use case builds on scripts written in R codes and we assume some familiarity with the R software. You should first create a folder in which you copy the data that we provide and where the results will be saved that will be generated by R (*see* **Note 1**). Copy the file containing the data sets named French_etal_MNRET_Data.txt and geno_301115.Rdata into your data folder (usually called data) and also the file containing the R codes (file: sysgenetbook) into your script folder (usually called script), *see* **Note 1**. The script is divided into four code blocks, corresponding to Chapters 2–5. Each block will run independently of the others. You also need to be connected to the internet to be able to download additional data and software packages (*see* **Note 2**).

Here, we use the data published by French et al. [1]. This data set can also be downloaded directly from the supplement materials of the publication as csv file. For convenience, we provide a .txt version of this file as French_etal_MNRET_Data.txt which can be easily uploaded into R.

Open R-Studio and set your current working directory to the directory containing the files for this demo. For example, if the files are in /home/you/DOQTL_demo, type setwd("/home/you/DOQTL_demo").

For the analysis itself, we need the R packages DOQTL and ggplot2 which can be downloaded from Bioconductor and CRAN and then installed into your R environment (*see* **Note 3**).

2.2 Import Data and Inspect It

Load the data, check its column names and dimensions (number of rows and columns) and cleanup the columns names for the subsequent analyses (code block 1). We now have the complete data set in R, consisting of all measurements for the different groups, doses, etc. We save it as "new_data_set.txt". We can get a first overview about the groups, e.g. a total of 598 mice were studied in two cohorts of 299 mice each (code block 1). For each treatment dose (0, 1, 10, 100 ppm benzene) 149, 150, 150, and 149 mice, respectively, were used (code block 1).

2.3 Inspect Data, Cleanup, and Save Phenotype Data for Further Analysis

For all following analyses, we first need to calculated and add columns that contain the values for the proportion of micronucleated reticulocytes with respect to the total number of reticulocytes counted, for each group (code block 2). Next, we obtain an overview of the phenotype data by creating boxplots for the different groups (pretreatment blood, post-treatment blood, post-treatment bone marrow) by treatment (Dose) (code block 2, Fig. 1). From the resulting figure it is clear that there is a dose-dependent increase

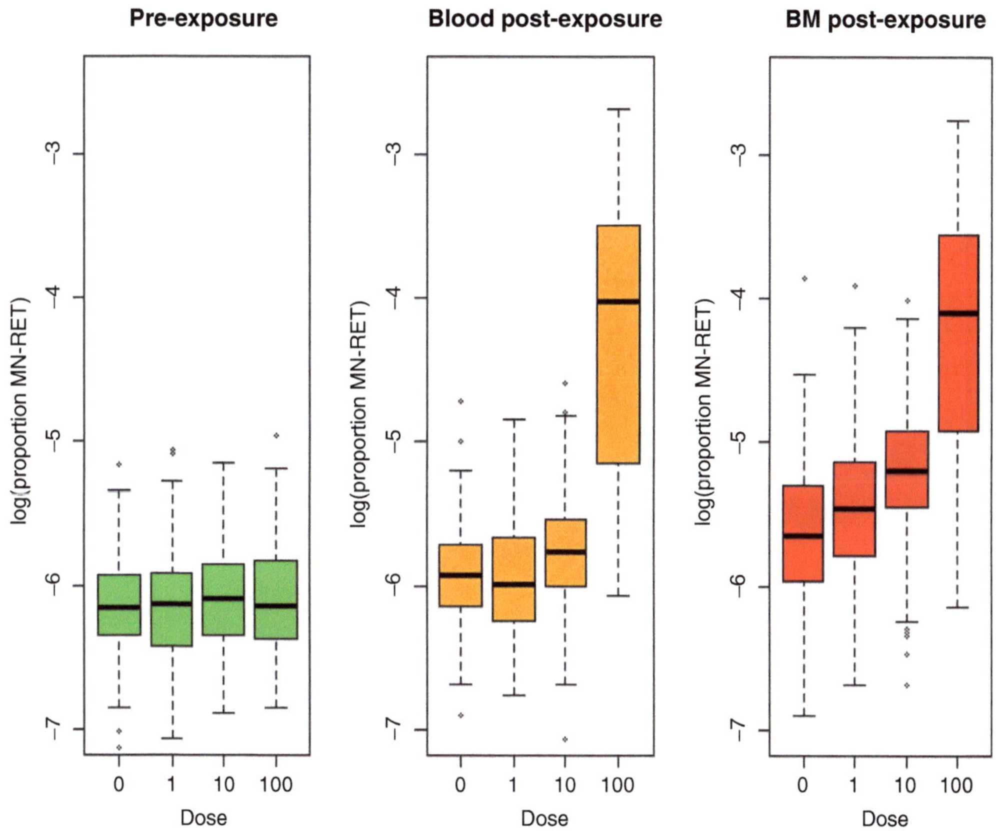

Fig. 1 Boxplots showing the proportion of micronucleated reticulocytes. Proportion of micronucleated reticulocytes per 1000 reticulocytes in each of the three different treatment groups (75 male DO mice): pretreatment blood (pre-exposure), post-treatment blood (blood post-exposure), and post-treatment bone marrow (BM post-exposure). The treatment groups exhibit an exposure-dependent response that is strongest at the 100 ppm benzene exposure level. Please note that for the pre-exposure group "Dose" simply refers to a group mice that has been used as untreated control for the different dose response groups

in micronucleated reticulocytes with increasing levels of benzene (maximal at 100 ppm). For the following analyses, we now add an additional column for sex (all male), remove all rows with NA values and save the data as "phen_dat_011215.txt" (code block 2).

3 ANOVA

Next, we perform an analysis of variance (ANOVA) to determine the factors and interactions that are significantly different between groups. We first need to re-format the data set (code block 3). Then, we view the effect sizes (mean values) for the factors dose and group (pre-blood, post-blood, post-bone marrow). It becomes clear that the treatment with the 100 ppm dose has the largest effect. The responses in the groups after treatment in blood and bone marrow have a similar effect size compared to each other and are clearly distinct from blood samples in untreated animals (Fig. 2). We then use ANOVA to investigate if these differences are statistically significant. The result of the ANOVA table reveals significant main effects (dose and group) and also a significant effect for their interaction (code block 3, Fig. 3). To identify significant pair-wise differences, we use the Tukey HSD post-hoc test with an adjusted *p*-value threshold of <0.05 (code block 3). Figure 4 shows that the post-treatment bone marrow measurements at dose 100 are significantly different to the pretreatment groups. In addition, the post-treatment bone marrow

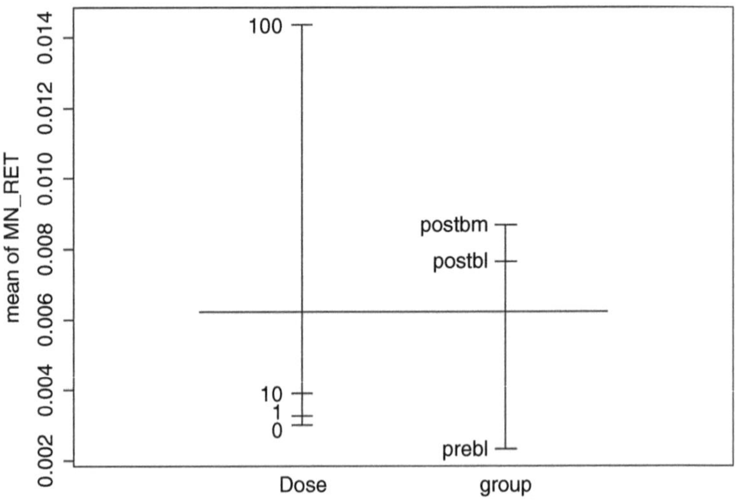

Fig. 2 Effect size for different experimental factors. The plot shows the effects for different treatment and dose on the mean number of micronucleated reticulocytes. The post-treatment effects in blood (post-bl) and bone marrow (postbm) are similar to each other but clearly distinguished from the effects in the blood in animals before treatment (prebl)

```
> summary(model)
               Df  Sum Sq  Mean Sq F value Pr(>F)
Dose            3 0.03903 0.013009  297.74 <2e-16 ***
group           2 0.01326 0.006629  151.73 <2e-16 ***
Dose:group      6 0.02013 0.003354   76.77 <2e-16 ***
Residuals    1687 0.07371 0.000044
---
Signif. codes:  0 '***' 0.001 '**' 0.01 '*' 0.05 '.' 0.1 ' ' 1
2 observations deleted due to missingness
```

Fig. 3 Result of ANOVA. The model (MN_RET ~ Dose*group) was used to perform ANOVA. Note significant effects of dose, group, and the interaction

```
> sel2 <- interact[grep("prebl", rownames(sel1),  value=T),];sel2
                          diff        lwr        upr        p adj
100:postbm-0:prebl   0.01726092 0.01470238 0.01981947 1.051159e-12
100:postbm-1:prebl   0.01720783 0.01462062 0.01979504 1.051159e-12
100:postbm-10:prebl  0.01712526 0.01458903 0.01966150 1.051159e-12
100:postbm-100:prebl 0.01711371 0.01457313 0.01965430 1.051159e-12
```

Fig. 4 Result of Tukey HSD. The table shows that the response in mice treated with 100 ppm was significantly different from the response in the respective control groups

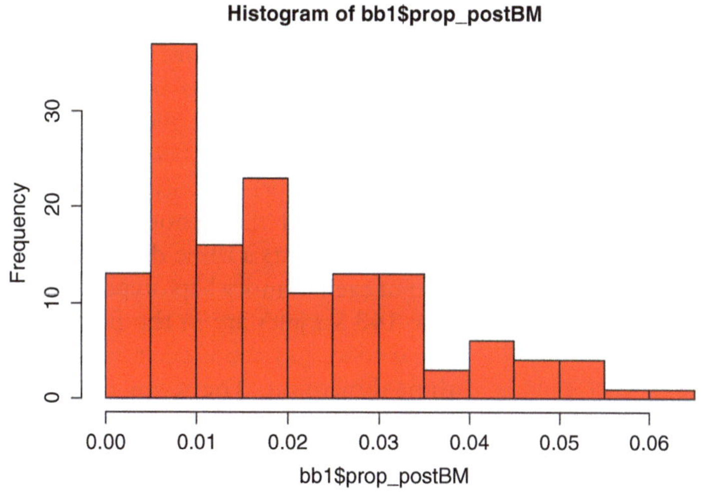

Fig. 5 Histogram of variance for bone marrow cells after treatment. The histogram shows the large variation in the response in bone marrow cells after treatment with a dose of 100 ppm benzene across the 149 DO mice treated with 100 ppm benzene

responses themselves exhibit a broad variation within the group (code block 3, Fig. 5). Thus, the post-treatment bone marrow response at a dose of 100 ppm in the individual DO mice represents a well suitable measurement for subsequent mapping of this trait to the respective individual DO genotypes.

4 Generate the Genotype Data

Each DO mouse is genetically unique and must be genotyped in order to perform genetic mapping. In general, DO mice are genotyped using the Mouse Universal Genotyping Array [2]. We reconstruct the DO genomes in terms of the founder haplotype blocks by estimating the probability that each mouse is in a given genotype state using a hidden Markov model. At the end of this process, we produce a matrix of haplotype probabilities with estimates that each mouse carries a haplotype from each founder at each marker. We provide the script to generate the genotype data from the MUGA results (code block 5). However, this process will take a very long time on a PC. You may skip this part and continue directly with the QTL mapping using the data set we provide (see below).

5 QTL Mapping

We first load the DOQTL (for QTL analysis) and ggplot2 (for drawing graphs) packages (code block 4). We then load the phenotype data table created above and select the data for blood post-treatment at dose 100 ppm (code block 4) containing sample ID, dose of benzene, sex, and proportion of bone marrow micronucleated reticulocytes (prop_postBM) as response variable (code block 4). We call this data set phen. We also need to write the sample IDs as rownames for phen, because this is used later to identify samples in the phenotype—genotype association analysis.

Next, load the genotype data, called probs (code block 4) which represents a three dimensional array containing the proportion of each founder haplotype at each marker for each DO sample. The 143 samples are in the first dimension, the eight founders in the second and the markers along the mouse genome are in the third dimension. The script will load a file that we provide here (geno_301115.Rdata). Alternatively, you may load the data set that you generated yourself with code block 5 (see above).

We have a look at the contents for the first 500 markers on Chr 1 of sample number 1 (code block 4, Fig. 6): Starting at the left, we see that this sample has genotype CD because both rows C and D are gray, indicating heterozygosity at this locus (values of 0.5 for each). Moving along the genome to the right, the genotype becomes DD where row D is black, then CD, AC, CH, CD, CH, etc.

We then need the locations of the markers on the genotyping array. The array is called the Mouse Universal Genotyping Array (MUGA) and contains 7856 SNP markers. Their locations are on The Jackson Laboratory's FTP site (ftp://ftp.jax.org/MUGA) and will be downloaded by the R script (code block 4). Next, we

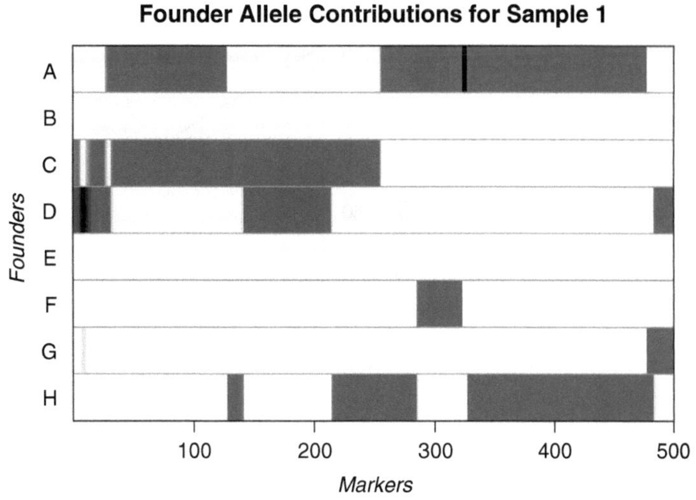

Fig. 6 Founder allele contribution. In this plot, the founder allele contributions, which range between 0 and 1, are colored from white (=0) to black (=1.0). A value of ~0.5 is *gray*. The markers are on the *X*-axis and the eight founders (denoted by the letters A through H) on the *Y*-axis. Starting at the *left*, we see that this sample has genotype CD because both rows C and D are *gray*, indicating values of 0.5 for each one. Moving along the genome to the *right*, the genotype becomes DD where row D is *black*, then CD, AC, CH, CD, CH, etc. The value at each marker sum to 1.0

need to create a matrix that accounts for the kinship relationships between the mice. We do this by looking at the correlation between the founder haplotypes for each sample at each SNP. For each chromosome, we create a kinship matrix using all markers except the ones on the current chromosome (Fig. 7). For example, to create the kinship matrix on Chr 1, we use only markers on Chr 2 through X. Simulations suggest that mapping using this approach increases the power to detect QTL (code block 4). The kinship values between pairs of samples range between 0 (no relationship) and 1.0 (completely identical).

Next, we introduce study cohort as a covariate for the subsequent mapping. For this, we must add the sample IDs to the rownames of the covariates because the "scanone" function will match up sample IDs in all of the data (code block 4). In order to map QTLs for prop_postBM, we use the scanone() function. To see the arguments for "scanone," type "help(scanone)" (code block 4). We can then plot the resulting QTL scan (Fig. 8) which shows a large peak on chromosome 10. Next, we want to assess the statistical significance of the QTL peak. This is most commonly done via permutation. We advise running at least 1000 permutations to obtain significance thresholds. In the interest of time, we only perform 100 permutations in this use case (code block 4). We then add thresholds to the previous QTL plot. We use significance

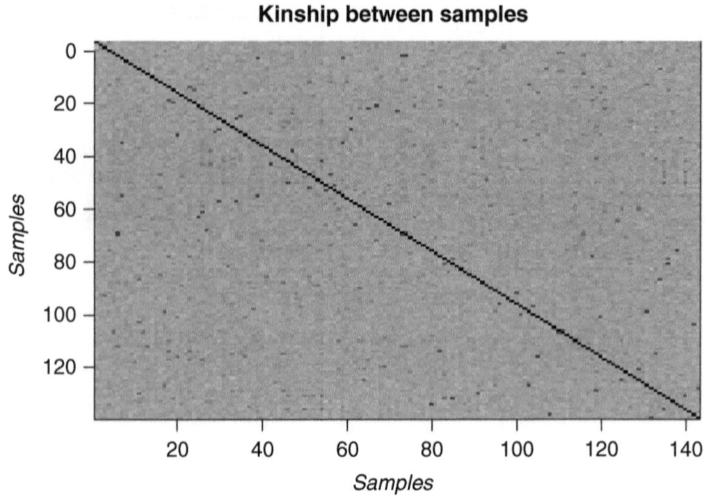

Fig. 7 Kinships. The figure shows the kinship between all pairs of samples. *White* (=0) indicates no kinship and *black* (=1) indicates full kinship. *Gray values* indicate varying levels of kinship between 0 and 1. The *black diagonal* of the matrix indicates that each sample is identical to itself. The *darker gray blocks off* of the diagonal may indicate siblings or cousins

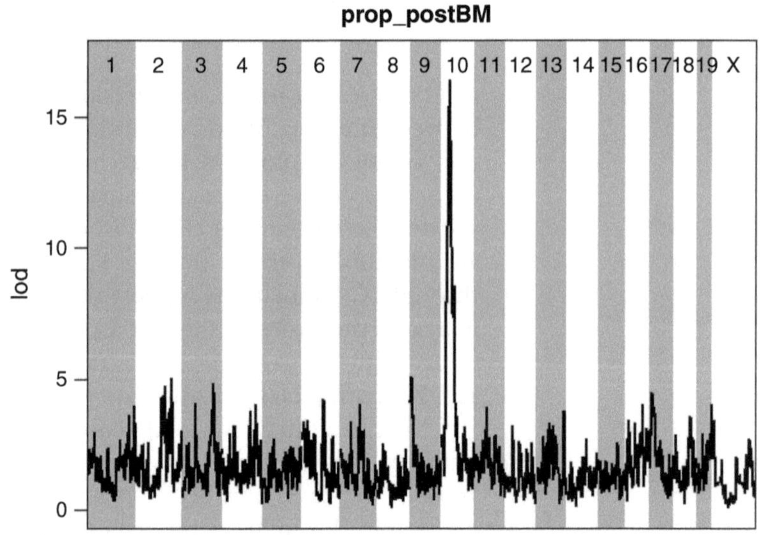

Fig. 8 QTL plot. The QTL plot shows the likelihood (Lod score) with which a particular marker is associated with the phenotype (proportion of micronucleated reticulocytes after exposure to 100 ppm benzene. A strong peak is seen on chromosome 10

thresholds at the $p<0.05$, 0.10, and 0.63 levels (red, orange, green) and add them to the QTL plot. This produces a figure (Fig. 9) similar to Fig. 3a of French et al. [1]. The signal on chromosome 10 is highly significant.

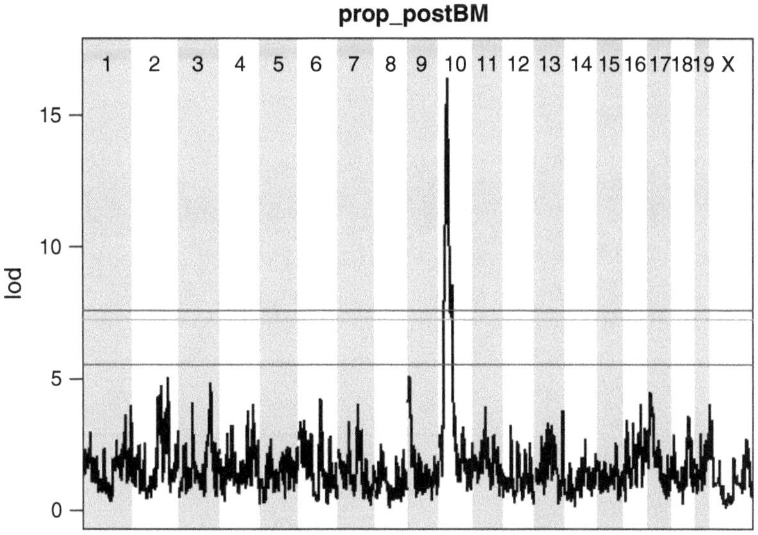

Fig. 9 QTL plot with thresholds. The same QTL plot as in Fig. 8 with significance thresholds at $p < 0.05$, 0.10, and 0.63 after performing 100 permutations. The QTL on chromosome 10 is highly significant. The figure has been reproduced from [1] with permission from the Journal Environmental Health Perspectives

We will now zoom in on the chromosome 10 QTL interval and look at the contribution of each of the eight founder alleles to the prop_postBM trait (code block 4). The mapping model fits a term for each of the eight DO founders. We can plot these coefficients across chromosome 10. The result is illustrated in Fig. 10 which is similar to Fig. 3b of French et al. [1] DO mice containing the CAST/EiJ allele around 32 Mb have lower levels of micronucleated reticulocytes. This means that the CAST/EIJ allele is associated with less DNA damage and has a protective allele. The figure also shows the LOD score, with the support interval for the peak shaded blue.

The support interval is then determined using the Bayesian Credible Interval (http://www.ncbi.nlm.nih.gov/pubmed/11560912). It represents the region most likely to contain the causative polymorphism(s). We can obtain this interval using the "bayesint" function (Fig. 11). Line 1 of the result table shows the Mb position (GRCm38) of the proximal end of the peak and line 3 shows the distal end. The maximum LOD score and its location are in line 2 of the table.

Next, we look at the distribution of prop_postBM across the 36 possible DO genotypes at the maximum peak location. Figure 12 which is similar to Fig. 3c from [1] shows the phenotype value on the *Y*-axis plotted against the 36 DO genotypes on the *X*-axis. Note that all of the samples with at least one CAST/EIJ allele (denoted by F) have low values. This suggests that the mode of inheritance may be dominant. Thus, the CAST/EIJ allele is driving the effect of the QTL. Also, note that some genotypes do not occur in this set of 143 samples.

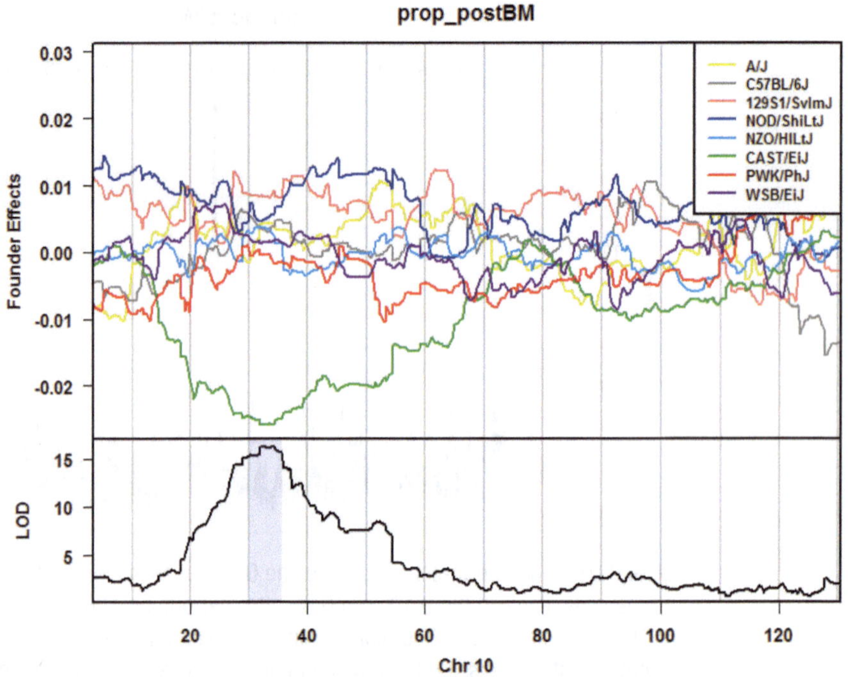

Fig. 10 Founder allele effect plot. The *top panel* shows the eight founder allele effects (or model coefficients) along Chr 10. DO mice containing the CAST/EIJ/EiJ allele around 32 Mb have lower levels of micronucleated reticulocytes. This means that the CAST/EIJ allele is associated with less DNA damage and has a protective allele. The *bottom panel* shows the LOD score, with the support interval for the peak *shaded blue*. The figure has been reproduced from [1] with permission from the Journal Environmental Health Perspectives

```
> interval
                SNP_ID Chr Mb_NCBI38       cM perc.var      lrs      lod                    p      neg.log10.p
1                 <NA>  10  29.98711 16.84414 38.90330 70.45913 15.30000                 <NA>               10
4262 backupUNC100621301  10  34.17711 18.84900 41.00691 75.46822 16.38772 1.15209550390118e-13 12.9385115182235
3                 <NA>  10  35.66405 19.11688 37.80354 67.95253 14.75570                 <NA>               10
```

Fig. 11 Bayesian Credible interval. Line 1 of the result table shows the Mb position (GRCm38) of the proximal end of the peak and line 3 shows the distal end. The maximum LOD score and its location are in line 2 of the table

6 Searching for Candidate Genes in QTL Interval

At this point, we have a 5.6 Mb wide support interval that contains polymorphism(s) that influence benzene-induced DNA damage. Next, we impute the DO founder sequences onto the DO genomes. The Sanger Mouse Genomes Project (http://www.sanger.ac.uk/resources/mouse/genomes/) has sequenced the eight DO founders and provides SNP, Indel and structural variant files for the strains [3]. We impute these SNPs onto the DO genomes and then perform association mapping. The function assoc.map() performs this analysis (code block 4). We plot the results of the association

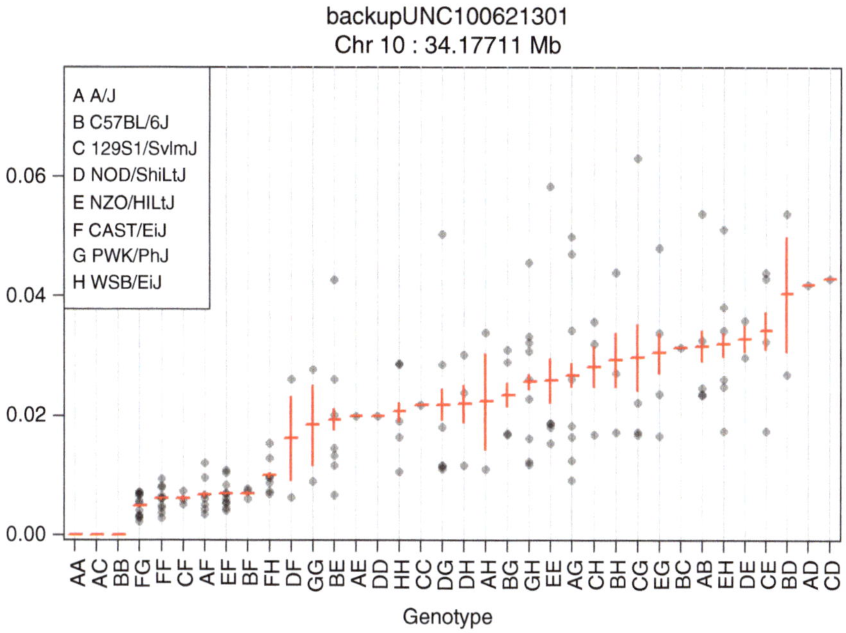

Fig. 12 Founder allele effects in QTL interval. The plot shows the phenotype value on the *Y*-axis plotted against the 36 DO genotypes on the *X*-axis. Note that all of the samples with at least one CAST/EIJ allele (denoted by F) have low values. The figure has been reproduced from [1] with permission from the Journal Environmental Health Perspectives

mapping using assoc.plot() with a threshold of 10 to highlight only SNPs with high LOD scores (Fig. 13) similar to Fig. 3d of French et al. [1]. There are about 60 genes (or noncoding RNAs) in the QTL interval.

At this point, the R scripts ends and further literature research and data base mining is employed to narrow down the list of candidate genes in the QTL interval.

One strategy for finding genes related to a phenotype is to search for genes with expression QTL (eQTL) in the same location. Ideally, one would have liver and bone marrow gene expression data in the DO mice from this experiment. Unfortunately, this data is not presently available. However, liver gene expression for a separate set of untreated DO mice is available. Thus, one can search for genes in the QTL interval that have an eQTL at the same location and look at the pattern of founder effects to see if CAST/EIJ stands out. The result of such a search is presented in [1] using gene expression data from liver, spleen, and kidney that were measured in 26 inbred strains, including the eight DO founders strain (http://cgd.jax.org/gem/strainsurvey26/v1). This analysis revealed that *Sult3a1* and its paralog *Gm4794* were the only genes with a different expression pattern in CAST/EIJ (Fig. 14). *Gm4794* is a paralog of *Sult3a1* and also contains a sulfotransferase domain [1]. Neither gene was expressed in the spleen. Both *Sult3a1* and

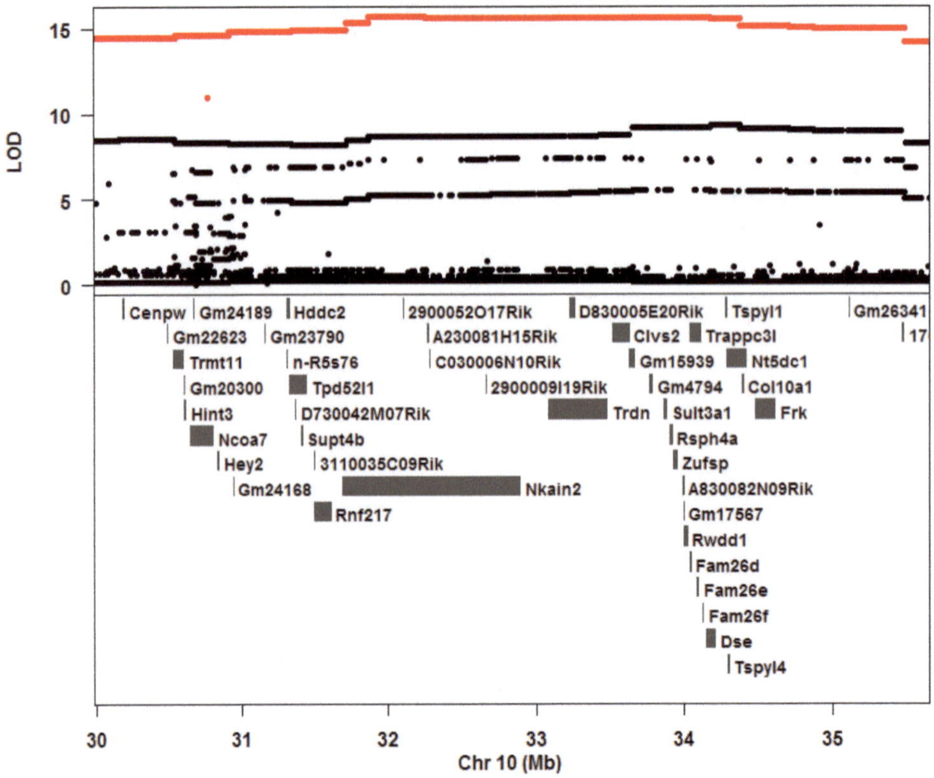

Fig. 13 Association mapping. Result of association mapping. There are about 60 genes (or noncoding RNAs) in the QTL interval. The figure has been reproduced from [1] with permission from the Journal Environmental Health Perspectives

Gm4794 have an eQTL in the same location on chromosome 10 (Fig. 15). In addition, linkage analysis revealed a suggested QTL for a susceptibility allele on Chr2 (LOD = 7.30) that spans ~8 Mb after the main effects of Chr10 QTL regressed out.

Next, we look at SNPs and structural variants that are specific for CAST/EIJ. For this, go to the Sanger Mouse Genomes. (http://www.sanger.ac.uk/sanger/Mouse_SnpViewer/rel-1505) website and enter *Sult3a1* into the gene box. Scroll down and check only the DO founders (129S1/SvImJ, A/J, CAST/EIJ/EiJ, NOD/ShiLtJ, NZO/HlLtJ, & WSB/EiJ) and then scroll up and press "Search." This will show the SNPs in *Sult3a1*. Select the "Structural Variants" tab and note the copy number gain in CAST/EIJ from 33,764,194 to 33,876,194 bp (Fig. 16). Click on the G (copy number gain) to see the location, copy this position (using this format: 10:33764194-33876194) and go to the Ensembl website (http://useast.ensembl.org/Mus_musculus/Info/Index). Enter the position into the search box and press "Go." You will see a figure similar to (Fig. 17) where one can zoom into see individual genes and transcripts. In order to visualize the size of the copy number gain, French et al. piled up the reads at each base [1]

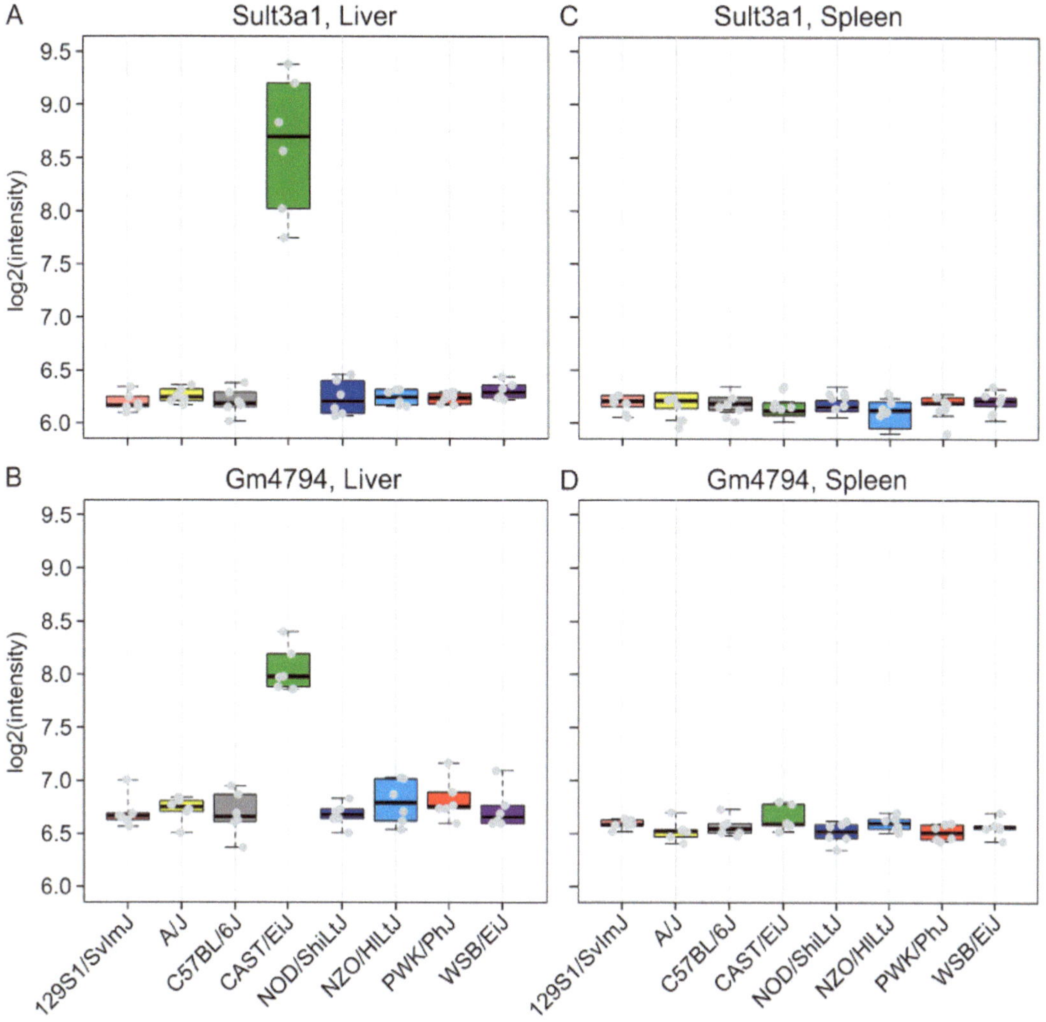

Fig. 14 Expression of *Sult3a* and *Gm4749* in CC founder strains. Expression of *Sult3a* and *Gm4749* is highest in the liver of CAST/EiJ mice compared to the other CC founder strains. The figure has been reproduced from [1] with permission from the Journal Environmental Health Perspectives. For details, *see* ref. 1

and they found a duplication in the CAST/EIJ founders that covers four genes (Fig. S3 in [1].: *Clvs2*, *Gm15939*, *Gm4794*, and *Sult3a1*. *Clvs2* is expressed in neurons and *Gm15939* is a predicted gene that may not produce a transcript. Although not orthologous genes, structural models comparing sulfotransfersase catalytic centers of mouse SULT3A1 and human SULT1A1 are similar. Further evidence of the potential importance of this finding to potential toxicity to humans is based on human CNV observed in different ethnic populations. Human SULT1A1 CNV range from 1 to 6 copies in Caucasian (5%-1 copy; 69%-2 copies, and 3 or more copies—26%) and African-American (38%-2 copies; 62%-3 or more copies) subpopulations [4, 5].

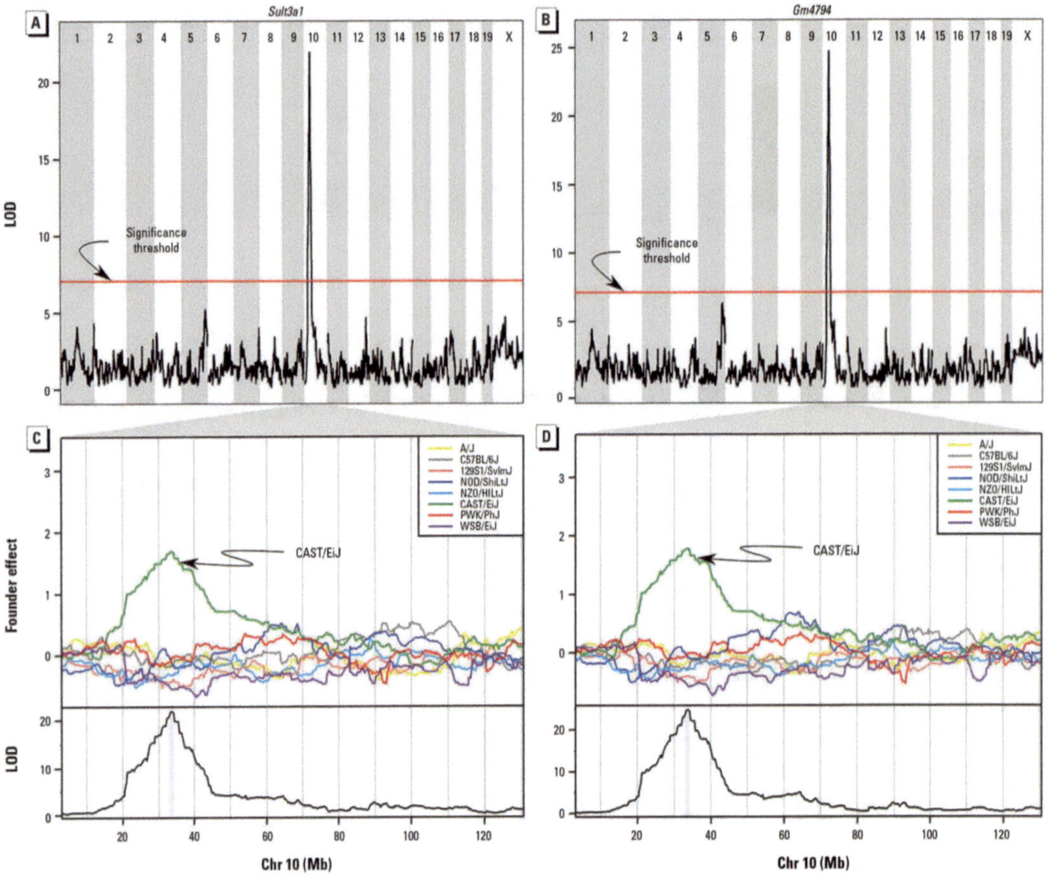

Fig. 15 Expression QTLs for *Sult3a* and *Gm4749*. Linkage analysis of expression levels for *Sult3a* and *Gm4749* revealed a cis-eQTL at the same chromosome 10 location to which benzene resistance mapped. The figure has been reproduced from [1] with permission from the Journal Environmental Health Perspectives. For details, *see* ref. 1

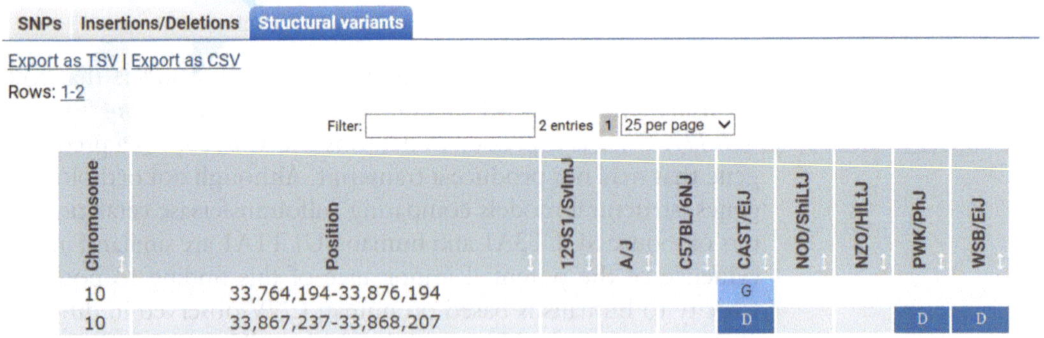
SNPs | Insertions/Deletions | Structural variants

Export as TSV | Export as CSV

Rows: 1-2

Filter: 2 entries 1 25 per page

Chromosome	Position	129S1/SvImJ	A/J	C57BL/6NJ	CAST/EiJ	NOD/ShiLtJ	NZO/HlLtJ	PWK/PhJ	WSB/EiJ
10	33,764,194-33,876,194				G				
10	33,867,237-33,868,207				D			D	D

Fig. 16 Structural variation of *Sult3a* locus. Structural Variants at the Sult3a locus. Image was taken from the http://www.sanger.ac.uk/sanger/Mouse_SnpViewer/rel-1505 website. Note the copy number gain in CAST/EIJ from 33,764,194 to 33,876,194 bp

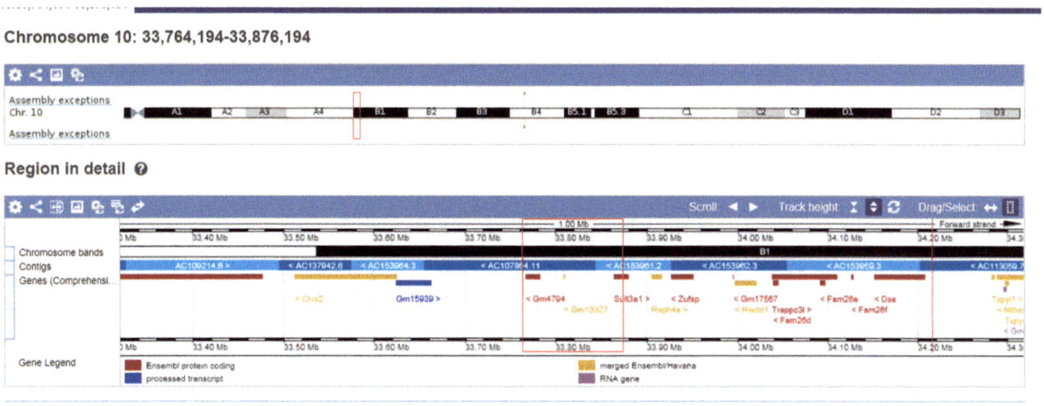

Fig. 17 Ensembl view *Sult3a* genomic region. The figure shows the genomic region around the *Sult3a* gene from the ENSEMBL website where additional information on the gene and its variants can be found. Image was taken from the http://useast.ensembl.org/Mus_musculus/Info/Index website

Thus, in summary we find two genes that meet the criteria for candidate genes: *Sult3a1* and *Gm4794*. Both represent sulfotransferases that may be involved in adding a sulfate group to phenol, one of the metabolites of benzene. Therefore, *Gm4794* and *Sult3a1* represent the strongest candidate genes explaining the high resistance to benzene in CAST/EIJ mice.

7 Notes

1. It is recommended to install R and R-Studio. R-Studio represents a graphical user interphase that allows easily writing and storing scripts as well as generating graphical outputs. Also, we recommend generating a folder R and two subfolders names data and scripts. You need to describe the paths to these folders as a first step in the R script. On a Windows PC, this path may look as follows: C:/Users/kls/Desktop/kls Daten/kls research/R/data/sysgenetbook, and for a Mac PC as follows:/Users/you/sysgenetbook.
2. Note that we wrote the scripts on a PC with European symbols. Thus, for example the symbols “ (quotation mark) may not work with R if you are using a PC with US special symbols and may need to be replaced by the symbols ‘ for the quotation mark.
3. To install the DOQTL package, perform the following R commands. For this, your computer must be connected to the Internet. Note that the installation of dependencies takes a while because the mouse and human genomes are downloaded.

```
> source("http://bioconductor.org/biocLite.R")
>biocLite("DOQTL")
```

To install the ggplot2 package, perform the following R commands. Your computer must be connected to the Internet.

```
>chooseCRANmirror() # select a mirror for the download from the proposed list
>install.packages("ggplot2")
```

8 Conclusions

The analysis of micronucleated reticulocytes after exposure to benzene in the DO mouse population has been quantitatively measured. Associating the differences for these phenotype measurements in individual DO mice with their respective genetic variation revealed a QTL that regulates resistance to benzene genotoxicity. Detailed analysis of genes in this QTL interval on chromosome 10 suggested several candidate genes regulating resistance to benzene. Additional evidence then narrowed down the candidate gene list to *Sult3a1* and *Gm4794*: Both genes have a liver eQTL in the same location on chromosome 10. Among the genes in QTL interval, only *Sult3a1* and *Gm4794* have differential expression of the CAST/EIJ allele in the liver. Finally, CAST/EIJ mice exhibit a copy number gain of these two genes in CAST/EIJ.

In conclusion, the analyses described in this use case have led to the following hypothesis [1]. Inhaled benzene is absorbed by the lungs into the blood stream and transported to the liver where it is metabolized. One class of genes that is involved in toxicant metabolism is sulfotransferases. *Sult3a1* is a phase II enzyme that conjugates compounds (such as phenol, which is a metabolite of benzene) with a sulfate group before transport into the bile. It is postulated that a high level of *Sult3a1* expression will remove benzene by this mechanism and thus be protective. The hypothesis is that the copy number gain in the CAST/EIJ mice increases liver gene expression of *Sult3a1* and *Gm4794*. High liver expression of these genes allows mice containing the CAST/EIJ allele to rapidly conjugate harmful benzene metabolites and excrete them from the body before they can reach the bone marrow and cause DNA damage. Further experimental validation is of course needed. However, this is a very likely plausible hypothesis.

In more general terms, this study also nicely demonstrates that DO mice are an excellent model system to discover new genes and loci by association mapping that are regulating variation in responses to environmental impacts as well as disease susceptibility. These findings are highly relevant to human health as they allow us to find genes and gene networks that are strongly correlated with disease phenotypes and which can then be tested in humans.

In many cases, studies in humans are often underpowered or very noisy because of not well-controlled environmental influences so that they do not allow easily the detection of significant and robust QTLs.

Acknowledgements

This work was supported by intra-mural grants from the Helmholtz-Association (Program Infection and Immunity) and a start-up grant from UTHSC awarded to KS and NIH grants GM076468 and GM070683 awarded to Gary Churchill of The Jackson Laboratory. The funders had no role in study design, data collection and analysis, decision to publish, or preparation of the manuscript.

References

1. French JE, Gatti DM, Morgan DL, Kissling GE, Shockley KR, Knudsen GA, Shepard KG, Price HC, King D, Witt KL, Pedersen LC, Munger SC, Svenson KL, Churchill GA (2015) Diversity outbred mice identify population-based exposure thresholds and genetic factors that influence benzene-induced genotoxicity. Environ Health Perspect 123(3):237–245. doi:10.1289/ehp.1408202
2. Morgan AP, Fu CP, Kao CY, Welsh CE, Didion JP, Yadgary L, Hyacinth L, Ferris MT, Bell TA, Miller DR, Giusti-Rodriguez P, Nonneman RJ, Cook KD, Whitmire JK, Gralinski LE, Keller M, Attie AD, Churchill GA, Petkov P, Sullivan PF, Brennan JR, McMillan L, Villena F (2015) The mouse universal genotyping array: from substrains to subspecies. G3 (Bethesda) 6:263–279. doi:10.1534/g3.115.022087
3. Baud A, Hermsen R, Guryev V, Stridh P, Graham D, McBride MW, Foroud T, Calderari S, Diez M, Ockinger J, Beyeen AD, Gillett A, Abdelmagid N, Guerreiro-Cacais AO, Jagodic M, Tuncel J, Norin U, Beattie E, Huynh N, Miller WH, Koller DL, Alam I, Falak S, Osborne-Pellegrin M, Martinez-Membrives E, Canete T, Blazquez G, Vicens-Costa E, Mont-Cardona C, Diaz-Moran S, Tobena A, Hummel O, Zelenika D, Saar K, Patone G, Bauerfeind A, Bihoreau MT, Heinig M, Lee YA, Rintisch C, Schulz H, Wheeler DA, Worley KC, Muzny DM, Gibbs RA, Lathrop M, Lansu N, Toonen P, Ruzius FP, de Bruijn E, Hauser H, Adams DJ, Keane T, Atanur SS, Aitman TJ, Flicek P, Malinauskas T, Jones EY, Ekman D, Lopez-Aumatell R, Dominiczak AF, Johannesson M, Holmdahl R, Olsson T, Gauguier D, Hubner N, Fernandez-Teruel A, Cuppen E, Mott R, Flint J (2013) Combined sequence-based and genetic mapping analysis of complex traits in outbred rats. Nat Genet 45(7):767–775. doi:10.1038/ng.2644
4. Gaedigk A, Twist GP, Leeder JS (2012) CYP2D6, SULT1A1 and UGT2B17 copy number variation: quantitative detection by multiplex PCR. Pharmacogenomics 13(1):91–111. doi:10.2217/pgs.11.135
5. Hebbring SJ, Adjei AA, Baer JL, Jenkins GD, Zhang J, Cunningham JM, Schaid DJ, Weinshilboum RM, Thibodeau SN (2007) Human SULT1A1 gene: copy number differences and functional implications. Hum Mol Genet 16(5):463–470. doi:10.1093/hmg/ddl468

Chapter 13

Visualization of Results from Systems Genetics Studies in Chromosomal Context

Karen Y. Oróstica and Ricardo A. Verdugo

Abstract

This chapter describes methods currently available for visualizing results from systems genetics experiments. Here, we abstract from the statistical methods used for genetic mapping, which are dependent on the specific resource being used, i.e. F2, RILs, or outbred populations among others. We use a public dataset with results from a mouse eQTL experiment for three examples of visualization: genome-wide dot plots of marker-by-gene association, karyotype-like plots, and circos plots. Dot plots give a first overview of the results from eQTL mapping, allowing detecting genome-wide patterns of *cis*- and *trans*-genetic association to transcription level. Karyotype-like plots provide chromosomal context and allow integrating multiple tracks of information in a single plot. Circos plots can, in addition, display long-range interactions to provide an overview of genetic connectivity at the genome level. All examples are developed and explained using R code, an open-source language with powerful statistical and graphical capabilities. The principles reviewed here, however, can be applied with other software options, organisms, and to any type of molecular phenotype that can be assigned to a genomic position.

Key words Tools, Systems Genetics, xQTL, Visualization, Chromosomal context

1 Introduction

The systems genetics approach can be a powerful tool for better modeling complex heritable traits [1]. By this method, a forward genetics approach is used both to mapping loci with genetic variation affecting organismal-level phenotypes as well as molecular phenotypes that may be relevant mechanistic intermediaries [2]. Current technology allows quantitatively surveying hundreds or thousands of biological molecules such as DNA sequence variations, epigenetic marks, and levels of transcripts, proteins, and metabolites. The underlying hypothesis is that by measuring molecular phenotypes that are under tighter genetic regulation than organism-level

Electronic supplementary material: The online version of this chapter (doi:10.1007/978-1-4939-6427-7_13) contains supplementary material, which is available to authorized users.

Klaus Schughart and Robert W. Williams (eds.), *Systems Genetics: Methods and Protocols*, Methods in Molecular Biology, vol. 1488, DOI 10.1007/978-1-4939-6427-7_13, © Springer Science+Business Media New York 2017

phenotypes, it may be possible to increase prediction accuracy for the target trait [3] as well as to produce testable hypothesis of causality that will suggest molecular mechanisms underlying phenotypes [4]. Different names have been given to QTLs associated to each of these types of molecular phenotypes. We can refer to them generically as xQTL. Because the most commonly profiled such phenotypes are gene expression levels measured by microarrays, we will develop examples for such expression QTL (eQTL).

Regardless of the genetic resource being used, the process can be partitioned in two mayor stages: (1) identification of loci associated to phenotypes, both molecular and physiological, and (2) modeling of networks of loci and phenotypes with the objective to discover association of the type *interaction* or *causal relation*. Here, we develop methods that are relevant for interpreting the results of the first part and that may suggest sensible models for the second part. Particularly, we will focus on displaying the results of a systems genetics experiment in a genomic layout. The aim is that by visualizing results in a chromosomal context, one may:

1. Identify errors in previous analysis steps.
2. Reveal relative proportion of cis- versus trans-genetic regulation.
3. Visualize hotspots of genetic variants regulating phenotypes.
4. Suggest genome-wide patterns of long-range genetic regulation or interactions.

2 Methods

2.1 Type of Data and Principles of Visualization

Any molecular phenotype with a genome position can be visualized in chromosomal context. The results from genetic mapping of these types of phenotypes can be summarized as a table containing loci as genomic regions or punctual positions and some sort of identification for the phenotypes. The data format will be similar to the one shown in Table 1.

In the examples that follow, we will use data from a study using Systems Genetics to genetically dissect hybrid sterility in male mice from a hybrid zone between two European subspecies of the house mouse, i.e. *M. musculus musculus* and *M. musculus domesticus* [5]. They investigated two phenotypes: relative testis weight (testis weight/body weight) and genome-wide testis gene expression pattern. Genetic mapping was done with the offspring of wild-caught mice by a GWAS approach using the Mouse Diversity Genotyping Arrays with 600K SNPs (Affymetrix, Santa Clara, CA). Abundance of 22K transcripts was profiled with the Whole Mouse Genome Microarray (Agilent, Santa Clara, CA). Genetic variation in gene expression was mapped both at the individual transcript level and for the first principal component of expression (PC1).

Table 1
BED format for data tracks

Probe.ID	Chr	Start	End	QTL.ID	QTL.Chr	QTL.Pos
A_52_P803348	19	18758768	18758827	JAX00339958	12	84287993
A_52_P486260	1	135806993	135806934	JAX00268096	1	137262920
A_52_P262997	11	115999665	115999606	JAX00322816	11	116044467
A_52_P333953	2	31658860	31658919	JAX00214519	19	21159325
A_52_P634829	7	111452887	111452828	JAX00232405	7	86614739
…						

Here we will use the following data obtained from the publication:

1. Significant phenotype-by-SNP associations.
2. Significant expression PC1-by-SNP associations.
3. Significant transcript-by-SNP associations.
4. Significant genetic interactions (SNP pairs) for testis expression PC1.

These data were obtained from the publication's material [5] or by personal communication. Probe-to-transcript mapping were obtained from the GEO website (http://www.ncbi.nlm.nih.gov/geo/query/acc.cgi?acc=GPL7202). All data are provided as electronic supplementary material accompanying this chapter.

2.2 Software Needed

The vizualization examples that we develop in this chapter are performed in the R programing language [6]. The software is open-source and is freely available from http://cran.r-project.org. Karyotype-like plots and circos plots demonstrated here require the chromPlot and OmicCircos R packages, respectively, which can be obtained from the Bioconductor repository (http://www.bioconductor.org). We assume basic knowledge of the R language and focus on the specifics for creating plots of genomic data. Readers unfamiliar with R are encouraged to consult online introductory material on installing, running, and using R (www.r-project.org). For brevity, we explain only the most important commands and concepts behind each visualization. Scripts with the full code and all necessary data files to replicate the graphs are provided in the accompanying website.

2.3 Genome-Wide Dot Plots

Dot plots give a quick overview of the distribution of molecular phenotypes and their respective genetic determinants of variation. The genomic position of the phenotype, transcripts in our example, is plotted according to its position (y-axis) and to the position of the QTL or most significant marker in a given locus (x-axis). This type of plot can be performed using R's built-in functionality, i.e. it does not require any extra R–package.

In order to display genomic positions across chromosomes on the same axis, we sort chromosomes by number, leaving sexual chromosomes last, and then we calculate cumulative positions. The units used for genomic position can vary depending on the application. Here we will use mega bases (one million basepairs) denoted by Mb, but one could use basepairs (bp) or centimorgans (cM) instead, for instance.

Dot plots are created with the plot() function in R as follows:

```
> plot(X, Y, axes=FALSE, frame.plot=TRUE,
pch=".")
```

where,

X	vector of positions of eQTL in Mb (start or mid position when segments)
Y	vector of positions for transcripts in Mb (molecular phenotype)
axes	we provide the logical value FALSE to the axes argument to omit axes
frame.plot	draw a square to delimit the plotting area
pch	sets the character used to represent data points. Use "." for dot plots.

The initial '>' symbol is not input but is used to indicate that the following text is R code. The X and Y are the names of R objects holding the vectors of cumulative positions in Mb for all SNPs and transcripts, respectively. This can be done with the cumsum() function in R. By default, plot()draws axes on the bottom and left of the plots, with automatic tickmarks indicating a scale of values in Mb. However, we omit this so that they can be drawn with chromosome names instead:

```
> axis(side=1, at=ticks, labels=labs, las=2)
> axis(side=2, at=ticks , labels=labs,
las=1)
```

Each line draws one axis,

side	axis placement, 1: bottom, 2: left
ticks	numeric vector with mid position in Mb of each chromosome
labs	character vector with chromosome names
las	argument to indicate text alignment of labels, 1: horizontal, 2: perpendicular

And lastly we can add vertical gray lines to delimit the borders of each chromosome:

```
> abline(v=seps, col="lightgray")
> abline(h=seps, col="lightgray")
```

With these two commands, we draw vertical and horizontal lines respectively, where, seps numeric vector with the start position for each chromosome

col argument to set the line's color

The resulting plot for the eQTLs reported by [5] is shown in Fig. 1a. Several aspects of the dataset are revealed by this plot. For comparison, we produced the same plot for a different dataset from an eQTL mapping in the BXD panel of recombinant strains reported by [7] (Fig. 1b). See the 01.Dotplot_outcross.R and 02.Dotplot_BXD.R scripts in the electronic supplementary materials for the full list of commands that created these plots. The amount of transcript-by-SNP associations that are detected depends on several factors, such as power, significance level, and mapping strategy. Here we are comparing two very different experiments. One is a GWAS on 185 mice generated from 63 mating pairs involving 37 unrelated females and 35 unrelated males and the other is a single-marker QTL mapping on 37 individuals from different recombinant strains from the BXD cross. Expression was measured in two different tissues, testis and liver respectively, but significance in both studies was determined by sample permutation. Therefore, although results from both studies are not directly comparable, it is useful to display side-by-side for illustration purposes.

As seen in Fig. 1, the GWAS revealed widespread genetic variation affecting transcript abundance, evidenced by a much denser plot than for BXD . This may be due to the large genetic variation present in the hybrid zone were the mice were sampled [5]. We

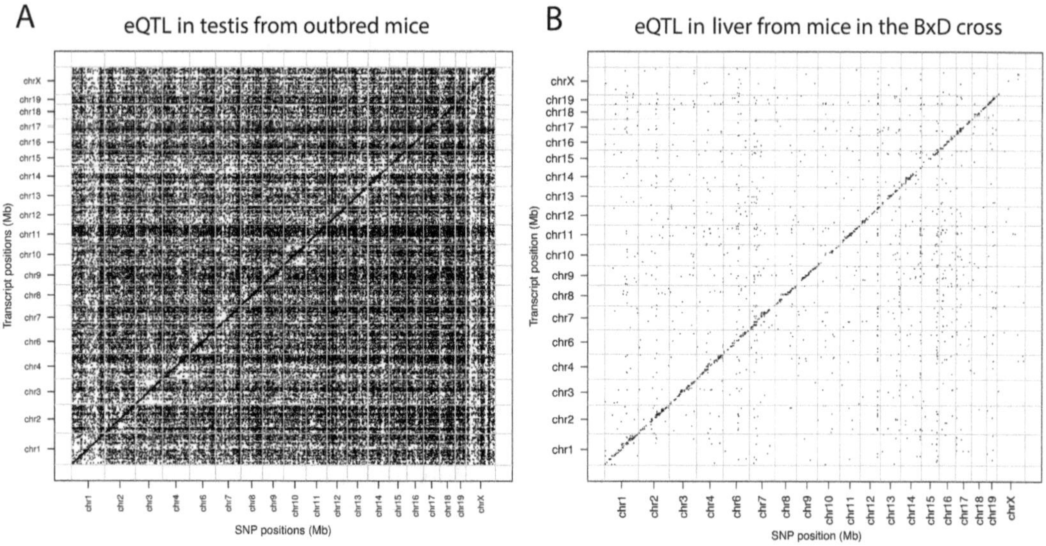

Fig. 1 Genome-wide scatter plots of eQTLs in system genetics studies. Genomic position of transcripts is shown by eQTL's SNP position in Mb. Only the most significant SNP per chromosome for each transcript is shown. (**a**) Results from eQTL mapping in the wild-derived male mice from [5]. (**b**) Results from eQTL mapping in the BxD recombinant inbred strains [7]

also evidence the presence of vertical and horizontal stripes of points in the GWAS, whereas BxD strains showed only vertical stripes. The latter are of more interest since they can represent regions of the genome with genetic variation affecting a large number of transcripts. Such loci may represent tans-eQTL shared among multiple transcripts and if colocalizing with QTL for the phenotype of interest, they may suggest a group of transcripts of interest to include in models to molecularly dissect such phenotype. By contrast, horizontal stripes may represent the effect of genome-wide differences between individuals, such as close relatedness or population structure from recent admixture. The study by [5] controlled for these effects when testing association for the organismal phenotype, relative testis weight, by including the kinship matrix as a random covariate. However, no such treatment was reported for molecular phenotypes, possibly because of computational restriction. The genome-wide dot plot reveals these effects and helps to orient further analyses by focusing on the regions and effects of most interest.

A second feature that is readily apparent from dot plots is the presence of a diagonal stripe of points, representing genetic loci that are in proximity to the gene they regulate. These are commonly referred to as cis-eQTL. By comparison, the GWAS experiment resulted in a lower proportion of cis- versus trans-eQTL than the BxD strains. cis-eQTL tend to have stronger effects and therefore are easier to detect, becoming dominant in experiments with low power of detection. Thus, detecting a diagonal line in genome-wide scatter plots provide good positive control; its absence should trigger checking for errors in data analyses or in the annotation of microarray probes for transcripts and SNPs.

Although a powerful tool, dot plots are limited in the number of data features they can show. Varying colors among data points can be used to add information about association significance (e.g. Fig. 3 in [7]) or some other property associated to transcripts and/or genetic loci. However, adding more layers of information can make the graph difficult to interpret. In addition, interaction effects among genetic loci or among molecular phenotypes cannot be represented in these plots. Therefore, other types of graphs are needed to inspect such properties of the data.

2.4 Karyotype-like Plots

Karyotypes are arrays of condensed metaphasic chromosomes used to detect large rearrangements, aneuploidies, polysomy, polyploidy, etc. One can use similar plots to display any type of genomic data along condensed chromosomes to provide location context. Location of data relative to important chromosomal structures such as centromere, telomeres, or heterochromatin is readily shown. The body of the chromosomes can be used to represent idiograms of G bands or some other type of data. Genomic data may also be represented on either side of the body. Depending on the type of data to be displayed, one could use histograms, points, connected lines, or rectangular segments, among others. These types of plots have the advantage of being able to display large and diverse data in a way that results familiar to most biologists.

Karyotype-like plots have been used in the context of systems genetics. In the past, we have used these plots to reveal preferential cis- over trans-regulation from QTL on gene expression levels in mouse congenic strains [8]. Here, following from the previous section, we use data from the paper by [5] to illustrate this type of visualization. We use the R-package chromPlot to display the results of eQTL mapping in chromosomal context. The chromPlot package can be obtained from Bioconductor (http://www.bioconductor.org) or from our website (http://genomed.med.uchile.cl/software).

In order to create a karyotype-like plot, one must first load the chromPlot package by:

```
> library(chromPlot)
```

We then load data necessary for drawing idiograms , which include human and mouse in the chromPlot package. For a mouse genome:

```
> data(mm10_gap)
> data(mm10_cytoBandIdeo)
```

These commands load data frames (tables) with the above names containing the genomic locations for centromeres and cytogenetic bands respectively. Then we must load the location of the genomic features that we want to display as histograms. Here, we will use:

refseq	RefSeq genes in the mouse genome (assembly mm10 from UCSC)
annot	RefSeq genes with probes in the microarray
eqtl_phenos	RefSeq genes with at least one eQTL from [5]
QTLs	QTLs for relative testis weight (testis/body weight) from [5]
eQTLs	Expression eQTLs from [5]

All of these tables contain the columns called Chrom, Start, and End with the name of the chromosome, and the start and end nucleotides for each genomic element. The annot and eqtl_phenos tables are created as subsets of the refseq table. This allows creating stacked barplots for number of genes falling in three categories: with an eQTL, tested but not associated to any SNP, and not tested (because not present in the array). Thus, the pattern of genes falling in these categories can be inspected genome-wide. The Probe-to-RefSeq mapping was taken from the GPL7202 table deposited at the GEO repository by the authors of [5].

The karyotype-like plot shown in Fig. 2 is created by:

```
> chromPlot(gaps = mm10_gap, bands=mm10_cy-
toBandIdeo, annot1= refseq, annot2=annot,
annot3=eqtl_phenos, segment=eQTLs_track,
segment2=QTLs, chr = c(13, 17, "X"), scale.
```

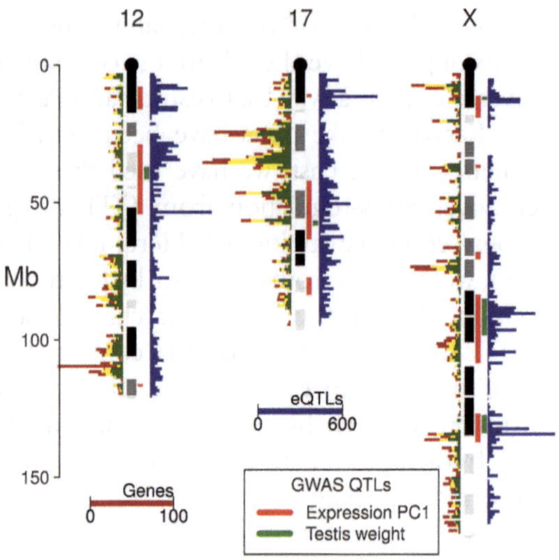

Fig. 2 Karyotype-like plot for three mouse chromosomes. Three types of data are shown. *Blue horizontal lines* on the body of chromosomes are typed SNP makers. The histogram on *left side* chromosomes represents number of genes in bins of 1 Mb. In *yellow* is the fraction of genes whose expression was tested by microarrays and *green* denotes genes whose expression was associated with at least one SNP in the GWAS from [5]. The color segments to the *right* of each chromosome represent GWAS regions for different phenotypes differentiated by color

```
title="Genes", segmentDesc="eQTL", segment-
2Desc=" GWAS QTLs", colSegments2="red",
cex=2, figCols=3)
```

This command tells the chromPlot function to use the mm10 coordinate system and creates histograms for genes to the left of each chromosome. The segment and segment2 arguments also create histograms if genomic features are smaller than the bin size (1 Mbp by default) or if too numerous to be plotted individually. Otherwise, they are plotted as colored lines to the right of chromosomes. In our example, eQTLs are plotted as histograms and testis weight QTLs as lines. Other arguments define text to be printed in the figure and change default colors or font size (cex). The figCols argument defines how many chromosomes are plotted per row (by default, chromPlot creates two rows).

The three chromosomes shown in Fig. 2 were selected because they contain a large number of eQTL and contain at least one QTL for relative testis weight, a fertility-related phenotype. A similar plot for all chromosomes is included in the electronic supplementary material (Figure S1.tif) and was created by the 03.Karyotype-like plots.R script. It becomes evident from Fig. 2 that the distribution of eQTLs along chromosomes is not uniform. Although, such conclusion would require a statistical test, this type of graph can suggest such pattern. One potential reason for this may be differences in gene density. By plotting a histogram of number of genes on the left side of chromosomes, we see some relation, especially on chromosome 17. However, this cannot

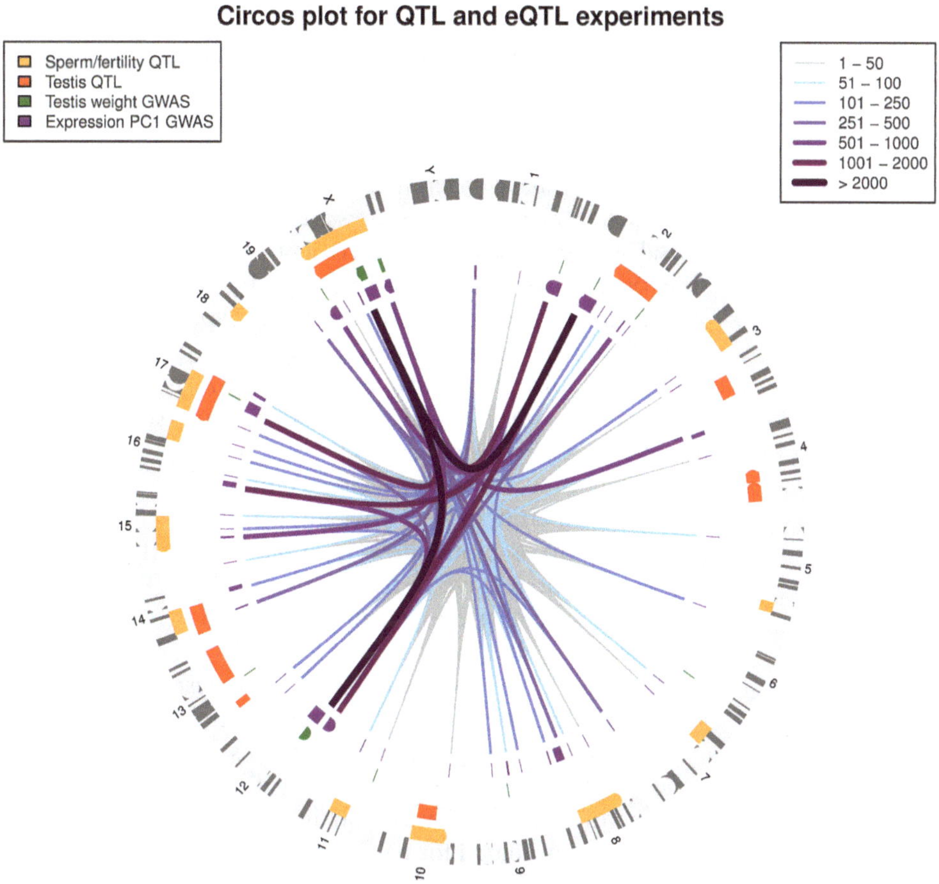

Fig. 3 Circos plot for QTL and eQTL experiments. *Yellow boxes* represent the QTLs for sterility phenotypes and *orange boxes* show QTL for relative testis weight as reported by [9, 10]. *Green boxes* indicate significant GWAS regions for relative testis weight and *purple boxes* denote significant GWAS regions for testis expression PC1 identified by [5]. The most inner track shows regions with significant genetic interactions reported in [5]. The color and line weight denote the number of significant pairwise interactions between SNPs for each region pair

explain completely the clustering in eQTLs. In fact, the largest number of eQTLs on chromosome 13 is present in a valley between two peaks of gene density. On chromosome X, the largest concentrations of eQTLs are around three QTL for testis weight. The region of chromosome X harboring the central QTL was the one explaining most overall variation in the testis transcriptome (PC1) as reported by [5]. This colocalization is highly unlikely by chance and is suggestive of presence of genetic factors on chromosome X that are driving low fertility, possibly by regulating the expression of a network of genes with eQTL in the region. Genes with cis-eQTL located in the testis QTL regions are obvious candidates for master regulators of such networks. Although beyond the scope of this chapter, formal testing of casual relations among genes should be applied in order to decompose the topology of the network and to identify regulator candidates [4].

2.5 Circos Plots

The visualization approaches revised so far are appropriate for inspecting global and local patterns of distribution of xQTL in the genome. However, they are not a good option for representing long-range interactions between genomic regions. By interaction, we mean any type of relationship of biological significance between elements that have different genomic position. For instance, gene-by-gene interactions may represent protein-protein interactions, a causal dependency of the expression of one gene on the expression of another gene, or a genotype-by-genotype effect from a statistical model on a quantitative phenotype. The type of interaction of interest will depend on the particular question under investigation. However, many times, visualizing such interactions genome-wide may reveal patterns that are suggestive of new hypotheses to be tested.

In Fig. 3, we have used a circos plot to summarize results from the eQTL experiment by Turner and Harr in [5]. Just as Karyotype-like plots, circos plots allow multiple tracks of information. In addition to the relative testis weight and PC1 QTLs, we included QTLs for several sperm fertility traits and relative testis weight taken from the literature as different tracks [9, 10]. Colocalization of phenotypic and expression QTLs suggest that both types of phenotypes are affected by the same genetic loci.

In addition to scanning for eQTLs, the authors tested for genetic interactions among SNPs in different chromosomes that significantly explained variation in gene expression. Interaction effects were evaluated for pairs of eQTL regions. The circos plot in Fig. 3 presents significant interactions as lines connecting loci. The color and width of the lines is proportional to the number of pairs of SNPs that interact in each region pair. By inspecting this plot, it becomes evident that loci don't act independently but that their effects are better modeled jointly and that gene expression may explain in part these interactions.

Here, we demonstrate how to create the circos plot in Fig. 3 by using the OmicCircos package in R. First, we load the OmicCircos functions:

```
> library(OmicCircos)
```

We then load the data tables as previously using the read.csv() function (not shown). The tracks in Fig. 3 consist of:

SpermFert QTLs for several sperm fertility traits from [9, 10]
testisWeight QTLs for relative testis weight from [9, 10]

TWgwas Regions spanning multiple SNPs associated at GWAS for testis weight in [5]
PC1gwas Regions spanning multiple SNPs associated at GWAS for Expression PC1 in [5]

RegionsPair Interaction effects between PC1gwas region pairs in [5]

The first four tracks have the BED format previously described. However, the RegionsPairs track is a table linking two QTL regions as shown in Table 2.

In order to draw the circos plot in Fig. 3, one must first create a system of coordinates on which each track of information will be plotted. This can be done as follows:

```
> plot(x=c(1, 800), y=c(1, 800), type = "n",
axes = FALSE, xlab = "", ylab = "", main =
"Circos plot for QTL and eQTL experiments");
```

The plot() function opens a graphics window and creates a new plotting area. Because, type="n", no plotting is actually performed and by axes=FALSE, the *x* and *y* axes are not drawn. However, this command has the effect of creating a plotting area between (1,1) at the bottom-left corner to (800,800) at the top-right corner. In this way, we obtain a square graphics window, which is necessary to draw a circle (instead of an ellipse). The range of values provided to *x* and *y* are arbitrary. However, the "radius" values later provided to circos() must be within this range (see bellow).

To add a track, one must decide at what distance from the center the data circle is to be plotted (the circle radius). This is set by the R argument of circos(). It is also important to tell circos() what system of genomic coordinates should be used, i.e. the exact UCSC name of the appropriate organism and genome assembly is needed.

First, let us plot the chromosomes:

```
> circos(xc=400, yc=400, R = 300, W = 15,
cir = "mm10", type = "chr", print.chr.lab =
TRUE, scale = FALSE);
```

The xc and yc parameters set the center of the circle used for the track in the *x* and *y* axes respectively. They are shown for completeness because 400 is the value by default and therefore could be omitted. This track will be drawn at a distance of 300 from the center of the plotting area ($R = 300$) and its width is set to 15

Table 2
Table liking pairs of QTL regions

seg1	po1	name1	seg2	po2	name2	freq	Colors
1	8010000	PC01	7	34970000	PC16	1	#BABABA
1	8010000	PC01	9	31940000	PC22	1	#BABABA
1	99030000	PC02	2	84120000	PC06	1	#BABABA
1	99030000	PC02	8	73660000	PC19	1	#BABABA
1	99030000	PC02	9	57230000	PC23	1	#BABABA
...							

The first six columns are required by the circos() function. The freq column contains the number of significant SNP pairs and the Colors column is used to plot a color gradient according to freq

(W= 15). We set to use the mm10 mouse genome by the cir parameter. This first track is special because it does not consist on actual data but it's used to plot chromosomes. This is done by setting type= "chr". We can ask to print chromosomes names by setting print.chr.lab=TRUE. Optionally, one can show a scale for the chromosome positions in Mb by setting scale=TRUE.

The following command is used to plot arcs for each QTL in the track:

```
> circos(R = 280, cir = "mm10", mapping =
SpermFert, type = "arc2", col = "#FCB14C",
print.chr.lab = FALSE, W = 10, scale =
FALSE, lwd = 10, B=FALSE);
```

The mapping parameter receives the data table in BED format. Since QTLs are large chromosomal segments, they can be plotted as arcs by setting type="arc2" as above. Arcs can be of two types: of variable ("arc") or fixed ("arc2") radius. The color is provided to the col argument as a quoted string with a color code supported by R. A gray background to a track can be set by setting B=TRUE. This can be useful to differentiate tracks more easily by alternating white and gray backgrounds. Commands similar to the one above must be used for each data track, adjusting the values of R and col so that arcs are placed at different distances from the center and in different colors. See also the 04.Circos plots.R script in the supplementary material for the full series of commands (Insert Table 3).

Finally, we can use circos() to plot lines connecting different genomic regions to represent the genetic interactions for PC1, reported by [5]. Data for this track was taken from the media-5 file in supplemental material of [5]. In the original data table, there was one line for each pair of interacting SNPs and the names of the two interacting regions is indicated. First, we summarize this by QTL region and determine the number of significant pairwise

Table 3
Colors by number of interacting SNPs used for connecting lines in Fig. 3

N° ranges	Ranges	Colors
1	1 < 50	"#BABABA"
2	51 < 100	"#A5C1DB"
3	101 < 250	"#828CC3"
4	251 < 500	"#8463AC"
5	501 < 1000	"#87459E"
6	1001 < 2000	"#81107C"
7	>2000	"#4A0247"

interactions (freq column in Table 2). We then define seven arbitrary ranges of SNP pair counts and assign a color to each range in order to give the impression of increasing intensity as counts increase [5]. Then, we must use the circos() function for to plot each range separately as in the following command line:

```
> circos(R = 210, cir = "mm10", W = 10, mapping = RegionsPair[(RegionsPair$freq >= 1 & RegionsPair$freq < 50),], type = "link", col = RegionsPair[(RegionsPair$freq >= 1 & RegionsPair$freq < 50),"Colors"], lwd=1);
```

For each range we use a different width and color by setting of lwd and col parameters. By setting type= "link", we tell circos to plot this track as connecting lines. The logical conditions that are placed within square brackets allow to subset the RegionsPairs track. In the above example, only the region pairs that meet the freq>=1 and freq<50 are provided to circos(). Note that in R, >= means "greater than or equal to". The legends are created with the R's builtin function legend(). See the 04.Circos plots.R script in the electronic supplementary material for the full list of commands used. All necessary data to produce these plots is also provided.

2.6 Summary of Results

In this chapter, we have demonstrated the use of three types of visualization for the inspection of results from Systems Genetics experiments in chromosomal context. Although simple in nature, a Dotplot revealed many properties of the data that were of significance for the following analyses. The appearance of vertical stripes locating trans-eQTL and horizontal stripes suggesting unaccounted population structure demonstrate that this global visualization is a must as one begins to inspect the results from Systems Genetics studies. Karyotype-like plots allowed incorporating new sources of information. Particularly, we were able to see colocalization of fertility and gene expression QTLs with peaks of eQTL frequency. This evidence suggested that loci harboring genetic variation affecting fertility traits may be mediating their effects through alteration in the expression of multiple genes. Whether those groups of genes form networks and whether those networks have a role in mediating QTL effects must be tested by alternative approaches. Finally, the Circos plot allowed inspecting the global pattern of gene-to-gene genetic interactions significantly affecting gene expression in testis. This visualization revealed that a few pairs of chromosomes account for most of the pairwise interactions affecting gene expression. In most cases, the interacting regions host phenotypic QTLs. However, some regions did not have QTLs identified from the GWAS nor from experimental crosses and the investigator may have missed them unless this type of analysis was performed.

3 Further Considerations and Limitations

We have presented three ways of visualizing genetic and genomic data using the R-programing language. This has the advantage of providing great flexibility for customizing the plot and has the potential to handle large amounts of data seemingly. However, this system does require some level of proficiency in a programing language. Although compared to others, this language is especially easy to learn, there is a learning curve for beginners. We hope that the examples and language used in this chapter make it easier for biologists and geneticists to get started and that we have provided the motivation to make the effort worthwhile.

Another limitation of the approach taken here is that the plots generated are not interactive, limiting the ability to explore the results at different levels of resolution or to direct users to further sources of information. There are some tools freely available that can provide this functionality. Of special mention, the Integrative Genomics Viewer can display genomic data in whole chromosomes or small genomic regions and allows simultaneous and diverse tracks of information in a similar manner than the tools demonstrated in this chapter [11]. By pointing, clicking, and dragging, the user can explore data at different resolutions and easily add public or private tracks of information. However, whole genome views as done here are no possible with this tool. The concepts provided in this chapter should prove useful when using any type of visualization tool, including IGV.

Acknowledgments

We acknowledge Dr. Bettina Harr at the Max-Planck-Institut fuer Evolutionsbiologie for making data available for this publication and for useful communications with the authors that helped improving the manuscript of this chapter. Development of the chromPlot package was funded by Grant FONDECYT 11121666.

References

1. Jansen RC, Nap JP (2001) Genetical genomics: the added value from segregation. Trends Genet 17(7):388–391
2. Civelek M, Lusis AJ (2014) Systems genetics approaches to understand complex traits. Nat Rev Genet 15(1):34–48. doi:10.1038/nrg3575
3. Jansen RC, Tesson BM, Fu J, Yang Y, McIntyre LM (2009) Defining gene and QTL networks. Curr Opin Plant Biol 12(2):241–246. doi:10.1016/j.pbi.2009.01.003
4. Schadt EE, Lamb J, Yang X, Zhu J, Edwards S, Guhathakurta D, Sieberts SK, Monks S, Reitman M, Zhang C, Lum PY, Leonardson A, Thieringer R, Metzger JM, Yang L, Castle J, Zhu H, Kash SF, Drake TA, Sachs A, Lusis AJ (2005) An integrative genomics approach to infer causal associations between gene expression and disease. Nat Genet 37(7):710–717. doi:10.1038/ng1589
5. Turner LM, Harr B (2014) Genome-wide mapping in a house mouse hybrid zone reveals

hybrid sterility loci and Dobzhansky-Muller interactions. eLife Sci 3:e02504. doi:10.7554/eLife.02504
6. Ihaka R, Gentleman R (1996) R: a language for data analysis and graphics. J Comput Graph Stat 5:299–314
7. Gatti DM, Zhao N, Chesler EJ, Bradford BU, Shabalin AA, Yordanova R, Lu L, Rusyn I (2010) Sex-specific gene expression in the BXD mouse liver. Physiol Genomics 42:456–468. doi:10.1152/physiolgenomics.00110.2009
8. Verdugo RA, Farber CR, Warden CH, Medrano JF (2010) Serious limitations of the QTL/microarray approach for QTL gene discovery. BMC Biol 8:96. doi:10.1186/1741-7007-8-96
9. Dzur-Gejdosova M, Simecek P, Gregorova S, Bhattacharyya T, Forejt J (2012) Dissecting the genetic architecture of F1 hybrid sterility in house mice. Evolution 66(11):3321–3335. doi:10.1111/j.1558-5646.2012.01684.x
10. White MA, Steffy B, Wiltshire T, Payseur BA (2011) Genetic dissection of a key reproductive barrier between nascent species of house mice. Genetics 189(1):289–U988. doi:10.1534/genetics.111.129171
11. Thorvaldsdottir H, Robinson JT, Mesirov JP (2013) Integrative Genomics Viewer (IGV): high-performance genomics data visualization and exploration. Brief Bioinform 14(2):178–192. doi:10.1093/bib/bbs017

Chapter 14

Using Baseline Transcriptional Connectomes in Rat to Identify Genetic Pathways Associated with Predisposition to Complex Traits

Laura Saba, Paula Hoffman, and Boris Tabakoff

Abstract

Although rat is a critical model organism in preclinical medications development, its use in systems genetics studies remains sparse. The PhenoGen database and website contain detailed information on the qualitative and quantitative aspects of the rat brain, liver, heart, and brown adipose transcriptome. This database has been generated using the HXB/BXH recombinant inbred panel and is being expanded to a hybrid rat diversity panel that includes many common inbred strains as well. By using such a panel, the PhenoGen project has created a renewable and cumulative resource for the rat genomics community. The database has been used to reconstruct the brain transcriptome identifying both annotated and unannotated transcribed elements that range in size from 20 nucleotides to over 30,000 nucleotides and elements that have a wide variety of roles in the cell including generation of proteins and regulation of the transcription and translation processes. In all 4 tissues, baseline transcriptional connectomes have been generated to model the relationships among transcripts. These connectomes can be used to identify genetic pathways associated with complex traits and to gain insight into biological function of individual transcripts. The PhenoGen website contains tools that allow the user to explore qualitative features of individual genes and to see how the gene relates to other genes within a tissue. The PhenoGen database and website continue to grow and to make use of the latest statistical methods for systems genetics creating a national resource for the rat genomics community.

Key words Systems genetics, Rat recombinant inbred panel, Weighted gene coexpression network analysis, RNA-Seq, Expression quantitative trait loci

1 Introduction

Although rats are an often-used resource in preclinical medications development including pharmacokinetics studies and toxicology studies, the use of rats in systems genetics studies is not common. Often, rats are preferred over mice for many basic physiological and behavioral studies due to their larger size and better-defined anatomy [1]. Their size is especially relevant to studies of the central nervous system due to the relatively large size of the brain,

Klaus Schughart and Robert W. Williams (eds.), *Systems Genetics: Methods and Protocols*, Methods in Molecular Biology, vol. 1488, DOI 10.1007/978-1-4939-6427-7_14, © Springer Science+Business Media New York 2017

which makes dissection and investigation of individual brain regions more tractable.

Many rat resources have been developed for genetically based investigations including selected lines [2], consomics [3], inbred strains [4], heterogeneous stock [5], and genetically modified rats [6]. In systems genetics research, one of the most valuable rat resources is recombinant inbred (RI) panels [7], which are composed of a number of strains that are isogenic within strain while each strain is a different mosaic of the parental strains. The inbred nature of the strains of these panels makes the accumulation of molecular, physiologic, and behavioral phenotypes possible and the controlled genetic relationship between strains ensures power to detect genetic effects at each locus (e.g., allele frequencies are approximately 50% at each locus). The relationship between rats from two different strains is similar to dizygotic twins or siblings, i.e., they share 50% of their genetics, and the relationship between rats within the same strain is similar to monozygotic twins, i.e., they are genetically identical.

Currently there are only a few recombinant inbred rat panels available. To our knowledge, the only panel that is currently being bred is the HXB/BXH RI panel originally derived from the gender-reciprocal crossing of the normotensive Brown Norway congenic strain (BN-Lx/Cub) carrying polydactyly-luxate syndrome and the hypertensive SHR/OlaIpcv inbred strain [8]. There are 30 RI strains available in this panel. This panel has been used to study a wide variety of traits ranging from metabolic syndrome [9] and pharmacogenetic effects of captopril [10] to alcohol preference [2]. Another popular panel, LEXF/FXLE RI panel, was originally derived from the gender-reciprocal crossing of the F344/Stm inbred strain and the LE/Stm strain [11] and is currently cryopreserved. This panel was originally designed for studies about chemical-induced tumors [12], but it has been used to study a wide variety of physiological and behavioral traits [11]. Both panels have been densely genotyped [4] and the parental strains have been fully sequenced [13, 14].

When using the systems genetics approach to studying complex traits, the RNA dimension becomes a crucial component and a natural starting point. RNA is the first *quantitative* link between DNA sequence and phenotype. Also transcription is the first step where environment can influence downstream outcomes. Our expanding knowledge of the role of RNA transcripts has led to the realization that RNAs do much more than simply code proteins [15]. Further exploring the RNA dimension can aid genome-wide association studies by linking biological function with the polymorphic loci in DNA that are associated with disease. Also, we now have many biologically motivated statistical tools (e.g., graph theory) that can model quantitative measures of RNA expression levels of individual transcripts to mathematically describe relationships among genes and can describe how these relationships may change after environmental perturbations.

In general, the goal of most systems genetics studies is to describe how gene products interact to produce a biological outcome. With RNA expression levels, complex hierarchical networks can be used to describe interactions among transcripts across time and space. For example, in brain, elements of a network are linked through both structure and function. Most complex pathologic traits can be conceptualized as systems disorders of failed regulation. These systems can be described through the application of graph theory and small world topology [16]. In this setting, small world topology indicates that the expression level of a single RNA transcript can be influenced by a variety of alternative pathways making the system both robust and efficient. This small world topology can be mathematically modeled using a scale-free network where most transcripts are only connected to a few other transcripts, but a small portion of transcripts ("hubs") are connected to many transcripts [17]. The PhenoGen project uses these types of methods to generate baseline transcriptional connectomes for individual tissues. These connectomes provide power for understanding the predisposition to disease, the etiology of pathology, and the response to medications or toxins.

In the transcriptome, the mathematical/statistical connection between two transcripts is based on their coexpression. The correlation of RNA expression levels over different environments indicates that the two transcripts are involved in similar biological processes [18]. With respect to developing the baseline connectome in PhenoGen, the different "environments" represent the different genetic backgrounds of the HXB/BXH RI panel. Weighted gene coexpression network analysis (WGCNA; [17]) has been used in the PhenoGen project to measure coexpression among transcripts and to form coexpression clusters. This method has several advantages over simply using a traditional correlation coefficient to measure the connectivity between two transcripts. In WGCNA, correlation coefficients are first mathematically transformed to mimic a biologically motivated scale-free network rather than a random network. Next these pairwise measures of direct connectivity are combined with an indirect measure of connectivity that quantifies the similarity of the other transcripts connected to the two transcripts, i.e., are their "friends" also "friends" with each other. This robust measure is used to form modules of transcripts with similar coexpression patterns.

These baseline transcriptional connectomes can be used in several different ways. The two most common types of analyses include a candidate gene-driven approach and a phenotype-driven approach (Fig. 1). In a candidate gene-driven approach, a researcher has a gene of interest, e.g., a known disease-related gene or a target of an effective therapeutic, but would like to know more about the

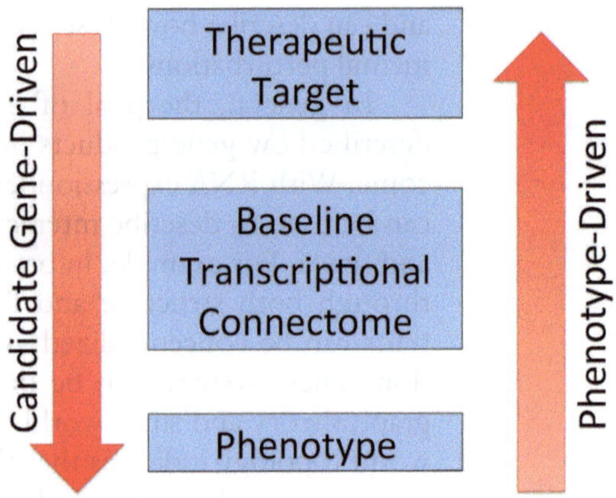

Fig. 1 Two primary approaches for utilizing the transcriptional connectome on the PhenoGen website (http://phenogen.ucdenver.edu). One of the goals of the PhenoGen project is to generate a "baseline" transcriptional connectome in several different tissues. The connectome can be used for two types of analysis: (1) candidate gene-driven analyses and (2) phenotype-driven analyses. In a candidate gene-driven analysis, a researcher would have a gene of interest (perhaps a known therapeutic target) and would want more information about the biological and genetic context within which this gene works and how the drug that targets this gene exerts its effects. In a phenotype-driven approach, the researcher has measured a phenotype on the RI rat population and wants to do an unbiased genome-wide search for candidate genetic (transcriptional) pathways for predisposition to the phenotype

biological/genetic context in which the gene works. The example below provides a step-by-step guide for this type of approach. Conversely, in a phenotype-driven approach, the researcher has measured a behavioral or physiological phenotype in the particular RI rat panel and wants to determine genetic pathways associated with the phenotype and to identify a potential therapeutic target. Recently, the derived baseline brain connectome of the HXB/BXH RI panel was used for studying genetic predisposition to alcohol consumption, an endophenotype of alcohol dependence [2].

2 Types of Data

The PhenoGen website (http://phenogen.ucdenver.edu) contains several well-curated RNA and DNA datasets for both mouse and rat. For this chapter, the focus is on the HXB/BXH RI rat panel. For the HXB/BXH panel, the site has: (1) full DNA sequence information on the two progenitor strains and detailed genotype information for the 30 HXB/BXH RI strains, (2) exon microarray data on RNA from 21 RI strains for brain, heart, liver, and brown adipose tissue, and (3) deep RNA-Seq data on the 30 RI strains for brain and liver and deep RNA-Seq data on the progenitor strains for heart and for female brain. This combination of high-quality data has allowed for the characterization of individual genes and isoforms based on exon inclusion, extensions/retractions of the untranslated regions, and general sequence characteristics of annotated and unannotated noncoding elements and for the quantification of RNA expression levels of all these transcripts. The expansion to more than one tissue allows not only for measuring differences in quantity between tissues but also differences in transcript characteristics. The use of a large panel rather than an individual strain allows for calculation of simple characteristics like heritability and more complex calculations of networks and eQTLs.

2.1 DNA Sequence

The progenitor strains of the HXB/BXH RI panel have been fully sequenced using DNA-Seq technology to generate over 1.7 billion reads [13]. In the RGSC 5.0/rn5 version of the genome, 51,329 SNPs and 66,470 small indels were identified between the BN-Lx/Cub strain and the BN reference genome and 3,578,145 SNPs and 1,089,050 small indels were identified between the SHR/OlaIpcv strain and the BN reference genome [2]. The HXB/BXH RI strains have been genotyped previously by the STAR Consortium at over 20,000 SNPs [4]. To convert their original genotype information to the rn5 version of the rat genome, the probe sequences from the genotyping array were aligned to the new version of the genome and only perfectly aligned probes were retained. SNPs were examined with respect to quality as outlined previously [19].

These data become important not only for the identification of quantitative trait loci (QTL) and for anchoring causal inference analyses within coexpression networks [20], but also become important when examining the integrity of microarray probes and when optimizing the quantification procedure in RNA-Seq data. The genotype data set used for mapping quantitative traits including eQTLs is available for download on PhenoGen along with the strain-specific genomes for BN-Lx and SHR for efficient aligning of RNA-Seq data.

2.2 RNA Expression: RNA-Seq

Initially, RNA was extracted from brain (male and female), liver (male), and heart (male) from the two progenitor strains for high-throughput sequencing on the HiSeq2000 ([2] for male brain data). RNA transcripts were first separated by size (<200 nucleotides and >200 nucleotides) and processed into libraries separately to optimize the detection of the entire transcriptome. The libraries generated from the small RNA fraction were sequenced using single-end 50 nucleotide reads. The long RNA fraction was first cleared of ribosomal RNA (rRNA) and then processed into libraries that were sequenced using 100 bp paired-end reads. Three to four samples per strain were analyzed and approximately 200 million reads (100 million read pairs) were generated for each sample from the long RNA fraction and 30 million single-end reads per sample for the small RNA fraction.

This deep RNA-Seq data on the parental strains allows for the reconstruction of tissue-specific transcriptomes. The reconstruction process identifies both novel genes and novel isoforms of annotated genes [21]. Algorithms for reconstruction of transcriptomes from RNA-Seq are still developing. For example, the precise end of a transcript remains ambiguous, but current algorithms do give high-quality information about novel exon-junctions in multi-exon transcripts and novel one-exon, possibly noncoding, transcripts. Analogously, in the small RNA fraction, the RNA-Seq data have been compared to annotated small RNA species (e.g., miRNA, snoRNA, tRNA) and novel transcripts have been identified via algorithms such as mirDeep2 [22] and snoSeeker [23]. On PhenoGen, the reconstructed transcriptome is visualized in a genome browser along with the RNA-Seq information on the read depth at individual nucleotides and the depth of coverage at exon-junctions. These reconstructed transcriptomes have also been used to improve the quality and interpretation of exon microarray data in different populations [2].

The current PhenoGen RNA-Seq database is being expanded to include the entire HXB/BXH RI panel. This database will continue to grow with a total of 4 samples per HXB/BXH RI strain and samples from 10 additional inbred strains to be completed by 2017 with even more strains added thereafter.

2.3 RNA Expression: Exon Arrays

RNA from four animals per strain (21 strains) was hybridized to separate arrays (i.e., one animal and one tissue per array). For each animal, RNA was extracted from brain, liver, heart (left ventricle), and brown adipose tissue, processed into cDNA, and hybridized onto arrays (*see* [2] for additional methodological details). The Affymetrix Rat Exon Array 1.0 ST contains over one million probe sets. These probe sets target transcripts whose annotation ranges from well studied and characterized genes that appear in all annotation databases to ab initio transcript predictions. Affymetrix provides guidance as to which probe sets to aggregate to obtain gene-level expression estimates, but the information that we have gathered from the RNA-Seq-derived reconstructed transcriptomes can be much more insightful and precise [2]. These expression databases have been well curated with extensive quality control procedures and batch effects adjustments. They have been used for heritability, eQTL, and gene coexpression network calculations.

3 Tools

The goals of the PhenoGen project are to generate high-quality, well-curated RNA expression data sets on genetically stable and renewable rat populations and to disseminate this information, including the transcriptional connectomes, to the research community. The PhenoGen website (http://phenogen.ucdenver.edu) includes the Genome/Transcriptome Data Browser that visualizes the data sets (raw, processed, and analyzed forms) and the 'Selected Feature' section that is the starting point for the candidate gene-driven approach discussed above. Much like the UCSC Genome Browser [24], the Genome/Transcriptome Data Brower displays the data as individual 'tracks' on the genome that can be customized to fit the researcher's interest. Unlike the UCSC Genome Browser, there are many additional visualization tools available in the 'Selected Feature' section when a feature (i.e., RNA transcript) has been chosen.

For the tracks, PhenoGen has integrated information from several public repositories: (1) Ensembl transcripts [25], (2) RefSeq transcripts [26], (3) behavioral and physiological QTL from the Rat Genome Database (RGD) [27], and (4) data from the UCSC RepeatMasker [28].

The DNA sequencing data set is included in the Browser to easily identify regions of the genome that harbor SNPs between either of the two parental strains of the HXB/BXH panel and the reference genome of the BN strain. When users hover over a particular sequence polymorphism, they will see what type of polymorphism is in that area (SNP, insertion, or deletion), the base pair change for the SNP/indel, and the precise location. In addition, this information is summarized in the 'Selected Feature' section below the browser as the number of exonic SNPs/indels by strain when a "gene" has been selected in the area.

The transcripts identified in the transcriptome reconstructions (both in the long RNA fraction and the small RNA fraction of the RNA-Seq data) can be included in the Browser to visually compare the alternative isoforms expressed in the different tissues. In the 'Selected Feature' table, a full description of the differences between the reconstructed transcript and similar annotated transcripts is given. The user can also visualize the read coverage at individual nucleotides across the genome in the region of the gene and the splice junctions captured in the RNA-Seq data.

The exon array data are also included in the Genome/Transcriptome Data Browser. The locations probed by the individual probe sets are displayed and can be color-coded in each tissue to represent percent of samples from the HXB/BXH RI panel with expression values above background, heritability of the expression values in the HXB/BXH RI panel, and the Affymetrix designation of annotation confidence (core, extended, or full). Locations of probe sets can be compared to the reconstructed and annotated transcriptomes.

From the Browser, users can click on individual transcripts to get further information in the "Selected Feature" section. The section currently has five different tabs: (1) Gene Details, (2) Gene eQTLs, (3) Probe Set Level Data, (4) miRNA Targeting Gene (multiMiR), and (5) WGCNA.

Gene Details contains general information about the gene, including links to other databases. It contains the information on the genomic variants and reconstructed transcripts mentioned previously. It also contains summary measures of probe set data from the exon array displayed in the corresponding track and a summary of the most significant eQTL for the transcript in the four different tissues surveyed.

Gene eQTLs contains a customizable Circos plot [29] that compares the eQTL profile for the transcript across the four different tissues (Fig. 2). The lines that connect one point in the circle to another point link the physical location of the transcript to the eQTL that surpass a specified *p*-value threshold. This graphic can be customized to only include a single tissue or subset of tissues or to only include specific chromosomes. The graphic is zoomable and downloadable.

Probe Set Level Data contains several different graphics that compare the expression of individual probe sets within the transcript and across tissues. Heat maps are used to compare the expression values between parental strains and between tissues and to visualize correlation patterns among probe sets. Bar charts are used to compare heritability of individual probe sets in the different tissues.

miRNA Targeting Gene (multiMiR) examines the predicted and validated microRNA binding sites on the transcript using the multiMiR package in R and its associated databases [30]. The individual microRNAs are listed and links are provided to mirBase [31] and to the detailed results of the prediction/validation.

WGCNA contains the information about the selected transcript within the weighted gene coexpression network analysis (WGCNA; [17]) performed on the transcript clusters on the array as defined by the transcriptome reconstruction in the parental strains (*see* [2] for further details about how modules are constructed). The initial view includes circles representing the coexpression modules that contain a transcript cluster related to the selected transcript. Clicking on one of the circles generates a coexpression module map where individual genes are represented by circles (or nodes) and coexpression between genes is represented as a line (or edge) (Fig. 3a). These maps are customizable and hovering over individual nodes gives the user additional detail about the transcript. The user can also change the view: (1) to see the Circos plot related to the module eigengene, a summary measure of the expression pattern among the transcripts within the module ([29]; Fig. 3b), (2) to see an interactive Sunburst plot ([32]; Fig. 3c) of the Gene Ontology terms [33] represented in the module, or (3) to see a graphical representation of microRNAs with binding sites on transcripts within the module ([30]; Fig. 3d).

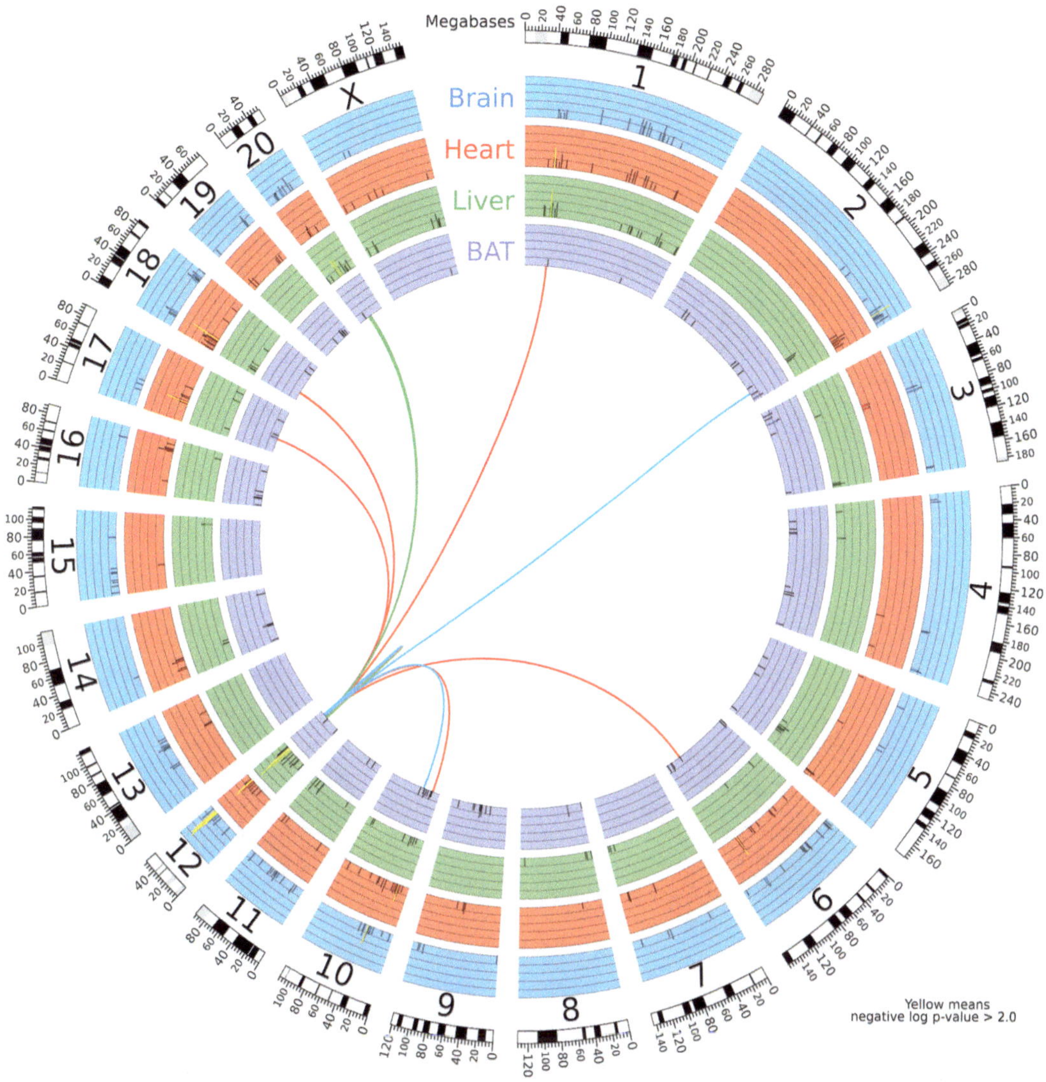

Fig. 2 Example Circos plot of eQTLs across tissues. The eQTLs in this graphic are for the purinergic receptor P2X, ligand-gated ion channel 4 (P2rx4) gene. The negative log base 2 transformation of locus-specific *p*-values are displayed as a bar chart where the position of the bar corresponds to the physical location of the associated SNP. Each of the four colored rings represent eQTLs calculated in different tissues of the HXB/BXH recombinant inbred panel using Affymetrix Rat Exon Arrays 1.0 ST. No bar is displayed if the p-value is greater than 0.2. SNPs with an association *p*-value less than 0.01 are highlighted in *yellow*. The *lines* that run through the center of the circle connect the physical location of the gene (e.g., Chr 12: 39.3 Mb) for P2rx4 to the eQTL with a *p*-value less than 0.01. The *color* of the line corresponds to the tissue used. This gene has a cis eQTL for brain, heart, and liver, but not brown adipose tissue (BAT). It also has several tissue-specific trans eQTL

4 Example (Step-by-Step)

The RNA expression data generated by the PhenoGen project and the related baseline transcriptional connectomes were primarily designed to be used in either (Fig. 1): (1) a candidate gene-driven

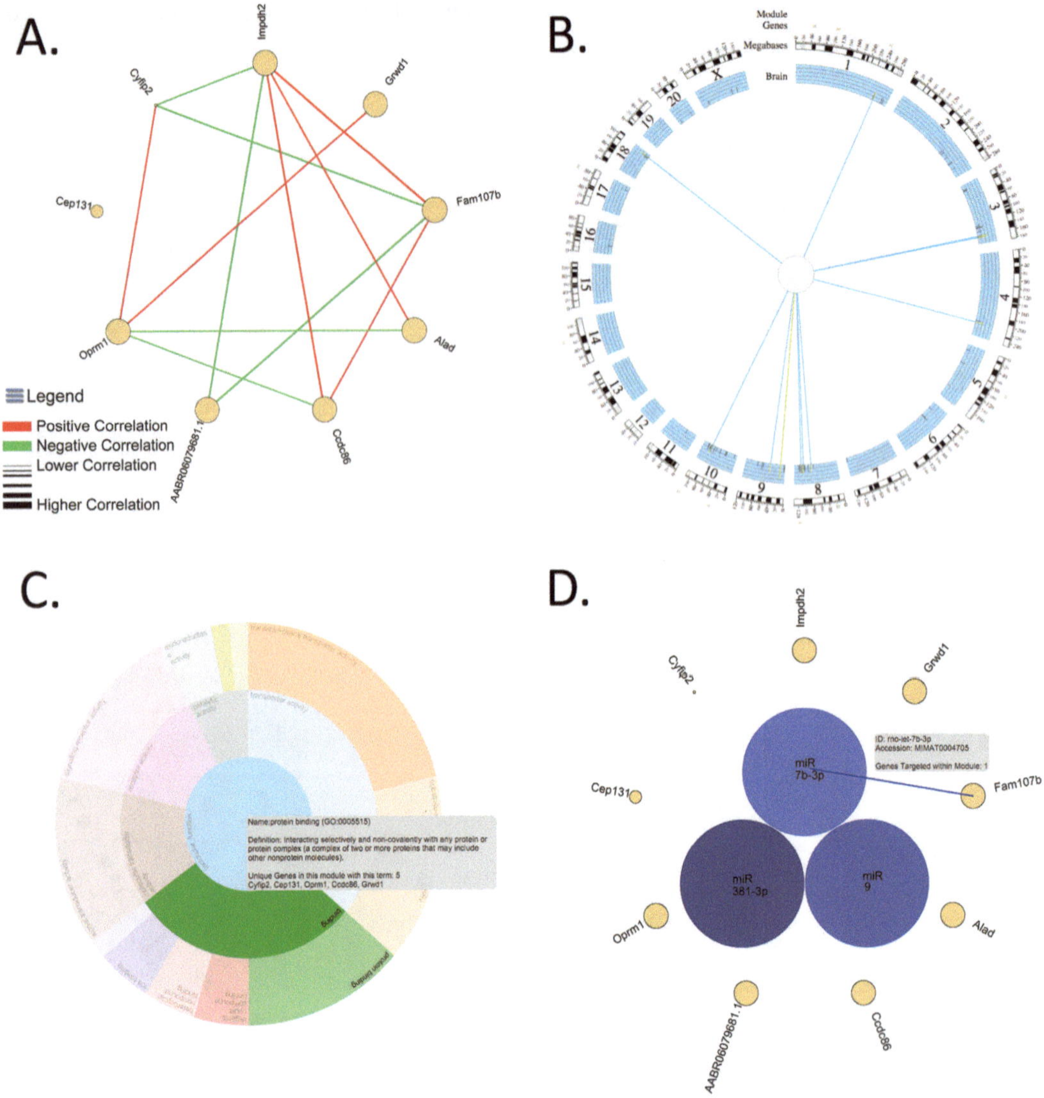

Fig. 3 Coexpression module characteristics. These four panels represent the detailed information displayed for coexpression modules on PhenoGen. The example given is the Azure module in brain that contains the opioid receptor, mu 1 (Oprm1) gene. (**a**) Node/edge plot of the expression module. Each of the nine transcripts in the module is displayed as a *circle*, where the size of the *circle* is representative of how connected the transcript is within the module. *Lines* between *circles* appear when the correlation coefficient is greater than 0.65 or less than −0.65. *Red lines* indicate a positive correlation between transcripts and a *green line* indicates a negative correlation between transcripts. (**b**) Circos plot of the module eigengene QTLs. The module eigengene (first principal component) was mapped to SNPs in the HXB/BXH genome. The log base 2 transformation of the locus-specific p-values are displayed in the *blue ring* by the position of the associated SNPs. eQTL with *p*-values less than 0.01 are highlighted in *yellow* and a *line* from the center of the circle is linked to the position. The eQTL with the minimum *p*-value has a *yellow line* from the center of the *circle* to its position. (**c**) Sunburst plot of gene ontology. A sunburst plot is used to display the hierarchical relationship among gene ontology terms related to genes in the module. When the user hovers over any section, a pop-up box appears with the name and definition of the ontology term and a list of genes associated with that term. The user can also click on any section and adjust the graphic to be centered at the term and show additional descendants of the term. (**d**) Common microRNA binding sites. This plot links microRNAs to individual genes with predicted or validated binding sites. The binding sites are determined using the multiMiR package in R

approach or (2) a phenotype-driven approach. In a recent publication, a phenotype-driven approach was used to identify a coexpression module in rat brain that was associated with alcohol consumption by the HXB/BXH RI panel and by several pairs of rat lines selectively bred for differences in alcohol preference [2]. By combining DNA-Seq, RNA-Seq, and microarray data from the RI panel and the selected lines, the systems genetics approach was able to identify a common pathway related to alcohol preference, rather than focusing on a common gene. Also, by incorporating the RNA-Seq data, previously unannotated transcripts and isoforms were quantified and included in the associated coexpression module. One unannotated transcript (possible lncRNA) was the most highly connected transcript (hub gene) within the module and follow-up studies are currently underway using CRISPR/CAS rat knockouts of this "gene."

The candidate gene-driven approach is outlined in a step-by-step guide below using the mu opioid receptor (Oprm1) as an example:

1. Navigate to the PhenoGen website (http://phenogen.ucdenver.edu).
2. Click on the *Genome/Transcriptome Data Browser* button at the top of the page.
3. Type Oprm1 into the *Gene Identifier or Region* box and change the *Species* to Rattus Norvegicus, then click *Go*. If multiple genes are returned (i.e., gene symbol is linked to multiple Ensembl Gene IDs), the user must pick one from a drop down list.
4. By default, the browser will open in the "Genome View" that contains tracks for annotated transcripts, BN-Lx and SHR SNPs and indels, and QTLs from RGD that overlap the region. The user can change the view by either adding or deleting individual tracks or they can choose one of the predefined views.
   (a) Select the predefined "Brain RNA-Seq" view from the *Initial View* box at the top of the page and click *Go*.
   (b) Add the Ensembl Protein-Coding Transcripts Gene track by clicking on the box with the three horizontal lines and the plus sign. In the window that opens, highlight the row labeled "Ensembl Protein-Coding Transcripts" and click the *Add Track* button.
   (c) To expand the Brain Illumina Total RNA (rRNA-depleted) Read Counts, click on the settings symbol to the right of the track title. Change the *View Density* to "Full" and click *Apply*. The Browser should now look like Fig. 4.

This image will give the user detailed information about the expression of this transcript in brain and highlight differences between the annotated version of the gene/isoform and the isoform detected in the RNA-Seq data. In the Oprm1 example, the reconstructed

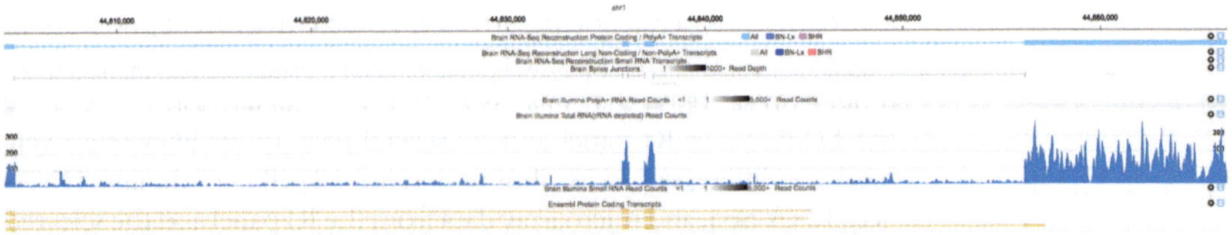

Fig. 4 Genome/Transcriptome Data Browser on PhenoGen (http://phenogen.ucdenver.edu). The first track in the browser image labeled "Brain RNA-Seq Reconstruction Protein-Coding/ PolyA+ Transcripts" contains the reconstructed version of the opioid receptor, mu 1 (Oprm1) gene in rat brain in *blue*. The *thicker bars* indicate the sections of the transcribed RNA retained in the mature RNA transcript and the *thin lines* in-between represent the intronic regions of the transcript. The arrows point from the 5′ end of the transcript to the 3′ end. The next two tracks "Brain RNA-Seq Reconstruction Long Non-Coding/Non-PolyA+ Transcripts" and "Brain RNA-Seq Reconstruction Small RNA Transcripts" do not have any information for this particular area of the genome because no *additional* transcripts were identified in the rRNA-depleted total RNA or in the small RNA fractions. The second track with information is labeled "Brain Splice Junctions." This track shows the RNA-Seq reads that span an exon junction. The *thicker vertical bars* indicate where the reads align and the *thin horizontal line* connecting the *thicker vertical bars* is the region not covered by the read. Users can hover over the illustration and the number of reads represented will be displayed. The next track "Brain Illumina PolyA+ RNA Read Counts" is a condensed form of the reads counts per nucleotide in the polyA+-selected RNA. The *blue* track labeled "Brain Illumina Total RNA (rRNA-depleted) Read Counts" has been expanded to show the read depth by nucleotide across the genomic region of Oprm1. This track visually confirms the extended 5′ untranslated region on the reconstructed Oprm1 transcript. In the final track in *yellow-labeled* "Ensembl Protein Coding Transcripts" are the annotated isoforms of the Oprm1

transcript is a close match to the ENSRNOT00000051837 version of Oprm1 with the exception of an extended 3′ untranslated region (UTR). This extension of the UTR is verified in the Brain Illumina Total RNA (rRNA-depleted) track that indicated a high level of read coverage in this region.

5. Click on the transcript for Oprm1 in the Brain RNA-Seq Reconstruction Long Non-Coding/Non-PolyA+ Transcripts track. This will activate the "Selected Feature" section below the Browser.

In this section, the user can browse summary information about this transcript with respect to HXB/BXH RI panel and related strains. For example, this section indicates that the SHR/OlaPrin strain has four exonic SNPs and seven exonic insertions/deletions. The user can add the SNP tracks to the Browser and find out exactly where those variants are located. Also in the transcript section, the difference between the reconstructed transcript and the Ensembl annotated transcript is summarized. For the novel Oprm1 transcript, both the 5′ UTR and the 3′ UTR are extended. This is helpful since the extension of the 5′ UTR was not easy to detect visually from the Browser. The probe set summaries indicate how many of the probe sets were detected above background in more than 1 % of the samples. Oprm1 has the most probe sets with expression values above background in brain and very few probe sets with expression values above background in liver, heart, or brown adipose. These details can be explored further by adding the microarray tracks to the Browser.

6. Click on the *Gene eQTL* tab of the “Selected Feature” section.

In this tab, a Circos plot will be generated. By default each of the four tissues are displayed as a separate ring of the circle and are color-coded. Each ring contains a bar chart where the bars are placed at the physical locations of the SNPs and the height of the bar indicates the negative log base 10 of the *p*-value. *p*-Values below the designated threshold are highlighted and a line is drawn from the physical location of Oprm1 on Chromosome 1 to the location of the highlighted *p*-value/SNP. The color of the line indicates the tissue in which the association was detected. Above the graphic is a set of controls that allow the user to change the appearance of the graphic including adjustment of the *p*-value threshold.

7. Click on the *miRNA Targeting Gene (multiMiR)* tab of the “Selected Feature” section.

The multiMiR algorithm queries several different databases for validated and predicted gene/miRNA pairs. In general, the more algorithms that predict a microRNA binding site, the higher the confidence that the binding site could be functional. With Oprm1, 28 microRNAs were predicted to have a binding site on Oprm1, but all 28 were predicted by only one database (low confidence).

8. Click on the *WGCNA* tab of the “Selected Features” section.

Two modules are shown on this tab because each contains a different transcript cluster associated with Oprm1. A transcript cluster is a group of probe sets from the exon array aggregated to derive one expression measure. For this analysis, transcript clusters were created based on the transcriptome reconstruction and on correlation patterns among individual probe sets that target the same gene. Coexpression modules are labeled with color names to distinguish one coexpression module from another and the size of the circles representing the module is associated with the number transcripts in the module. When the user hovers over either circle, the module name and the number of transcripts included in the module are displayed.

9. Click on the smaller circle, the azure module.

This will display the coexpression module as a series of circles (nodes) and lines (edges). The circles represent the genes contained within the coexpression module and the lines indicate the correlation between any two genes. The size of the circles indicates their connectedness within the coexpression module and the nodes are ordered in the circle by this trait. For the Oprm1 module, Impdh2 (inosine 5,-monophosphate dehydrogenase 2) is the gene with the highest connectivity and is referred to as the "hub gene" [34]. By adjusting the link correlation threshold values at the top of the graphic to 0.65 (i.e., only include lines when the correlation between two transcripts is greater than 0.65 or less than −0.65), the user can determine that Oprm1 is positively correlated with Grwd1 (Glutamate-rich WD repeat containing (1) and Cyfip2 (Cytoplasmic FMR1 interacting protein (2) and negatively correlated with Alad (Aminolevulinate dehydratase) and Ccdc86 (Coiled-coil domain containing 86).

10. Change the view in the WGCNA tab by clicking the radio button for Eigengene eQTL on the far right, above the coexpression module graphic.

In this Circos plot, the physical positions of the genes within the module are indicated on the outer parameter of the circle as "X"s. The *p*-values associated with the module eigengene QTL are shown as bars as described before. However, in this graphic the lines running through the middle of the graphic originate from the center of the circle and the line representing the QTL with the smallest *p*-value is highlighted in yellow. For the Oprm1 module, the minimum p-value is on chromosome 9, which does not contain any genes from the module. However there are suggestive QTL at the physical location of Ccdc86 on chromosome 1 and near Impdh2, the hub gene, on chromosome 8.

11. Change the view in the WGCNA tab by clicking the radio button for Gene Ontology on the far right.

This tab contains a Sunburst plot for the hierarchical Gene Ontology terms related to genes within the module. When the user hovers over any of the color-coded sections in this plot, a text box appears that contains the name and description of the gene ontology term and a list of genes from the coexpression module that are annotated to this term. The molecular function category of protein binding (light green section on outer ring) contains five genes from the coexpression module (Cyfip2, Cep131, Oprm1, Ccdc86, and Grwd1). Interestingly, not only is Oprm1 included in this group but also three of the four genes with the highest correlation with Oprm1.

12. Click on the section of the Sunburst plot labeled "protein binding."

By clicking on a GO term, the Sunburst plot adjusts to show more detail about the hierarchical GO terms that are descendants of "protein binding." Most of the descendants displayed are related to Oprm1 including G-protein alpha-subunit binding and cytoskeletal protein binding.

This quick exercise with Oprm1 gave several insights into its RNA expression. (1) Oprm1 is expressed in brain and not in liver, heart, or brown adipose tissue. (2) The variant expressed in brain has an extended 3′ UTR that is not currently indicated in any database. This could have implications for translational control. (3) There are four SNPs and seven indels in exonic regions of Oprm1 that could impact protein function. In humans, exonic SNPs in this gene have already been shown to affect function [34]. (4) There are several suggestive trans-eQTL for Oprm1 (minimum *p*-value located at chr9: 76.3 Mb). (5) There are no strong miRNA candidates for control of the expression of Oprm1. (6) The smaller coexpression module, which contains Oprm1, has nine transcripts including those most highly connected with Oprm1 (Cyfip2, Grwd1, Ccdc86, and Alad). (7) This module's hub gene is Impdh2, which is physically located on chromosome 8 near the module eigengene's QTL and the major eQTL for Oprm1.

5 Further Considerations and Limitations

Our knowledge about the RNA world [15] continues to expand. The databases, algorithms used, and tools developed as part of the PhenoGen project are dynamic and continue to evolve as methods and knowledge improve. This means that results presented on the PhenoGen website today will be improved a year from now. But this fact does highlight the resiliency of the data collected and the population used for building the database. The algorithm for identifying and quantifying transcripts in RNA-Seq data continues to evolve, but this evolution does not make RNA-Seq data captured previously obsolete. Instead the data can be reanalyzed with the latest tools and their relevancy will be retained. The nature of an inbred rodent panel is that data collected on the panel is cumulative and new/innovative technologies and molecular measures can be combined with historical data to get an even more detailed picture of the transcriptional networks within a tissue or organism.

Some perceive the RNA expression analysis of whole brain rather than brain parts as a limitation. Certainly the brain is a heterogeneous organ, and there are differences in brain gene expression across cell types and regions. The usual argument is that by measuring gene expression in whole brain, one will miss gene expression differences that only occur in a small brain region, or that a difference in one region will be counteracted by an opposite difference in another region, thereby confounding results. These objections, which may have had some justification in early brain transcriptome analyses, do not take into account the current state of technology, statistical and bioinformatic analysis of gene expression data. Important justification for investigation of whole brain gene expression data comes from the realization that the brain operates as a network of functionally linked cells and regions, as exemplified in functional neuroimaging (fMRI) studies [35]. One has to consider that if gene expression is measured in cells in an isolated brain region, the data cannot be analyzed in terms of functional events occurring in the regions to which particular neurons project, or in terms of the input that particular cells receive. However, the application of newer statistical and bioinformatics methods for analysis of gene coexpression data demonstrates that in many important ways, transcript levels can be related to the functional and structural connectivity in the brain (i.e., one can apply a systems approach to study whole brain gene expression) [36]. The goal of the PhenoGen project is to generate a resource that can be used by investigators to study a plethora of phenotypes. Limiting ourselves to a single brain region would limit the utility of the database, whereas, using whole brain will create the broadest opportunity for other researchers to use our transcriptome data to provide network-based insights into disease or other complex traits.

Currently the number of strains and the diversity of the genetic background of the HXB/BXH RI panel may be a limitation for complex systems genetics analyses. Public databases, such as GEO, can provide a broad set of transcriptional data when combined across experiments, but the trade-off lies in the quality of the data. With the PhenoGen database, environment, batch effects, and other technical factors are tightly controlled to avoid the confounding of biologically relevant results with differences of technique and environment.

6 Outlook

The PhenoGen databases, analysis techniques, and website continue to expand daily. The PhenoGen database will soon contain RNA-Seq data on RNA expression (both total RNA and small RNA) from liver and brain from 40 strains of our planned Hybrid Rat Diversity Panel (HRDP) and the goal is to expand to 96 strains. The HRDP, much like the mouse version [37], is a combination of recombinant inbred rat strains and classic rat strains. This combination gains power for genomic associations from the RI panels and precision for QTL analyses from the classic rat strains.

With more data comes more power to refine and improve current methods for transcript discovery and network modeling. With the attention to detail and the consistency of the methods for collecting data in the PhenoGen project, the data can be mined for a variety of tasks. The current methods for transcript discovery are in their infancy when considering the identification of precise transcription start and stop sites. This becomes increasingly important when examining the function of noncoding elements such as miRNA that can alter transcript levels and inhibit translation to protein. Tissue specific differences in the length of untranslated regions become another biological process to control gene expression levels and protein production. The RNA-Seq data generated as part of this project and knowledge about the transcription process, transcript stability, and protein translation can help to build better algorithms and models for the functional interpretation of our data.

The increase in sample size and the inclusion of many diverse genetic backgrounds can also be used to improve network discovery algorithms. With this breadth of data, our systems genetics approach can move beyond undirected associations among transcripts to directed models that can be used for causal inference and prediction [38]. These models allow for probability propagation that can be useful for bidirectional reasoning (e.g., given this genetic profile what is the probability of disease and given that the person has disease what is the probability of them having a disruption in this genetic pathway). Improving these types of models can have an enormous impact on a precision medicine approach to disease susceptibility and therapeutic efficacy.

Acknowledgement

This work was supported by NIAAA (AA013162).

References

1. Aitman TJ, Critser JK, Cuppen E, Dominiczak A, Fernandez-Suarez XM, Flint J, Gauguier D, Geurts AM, Gould M, Harris PC, Holmdahl R, Hubner N, Izsvak Z, Jacob HJ, Kuramoto T, Kwitek AE, Marrone A, Mashimo T, Moreno C, Mullins J, Mullins L, Olsson T, Pravenec M, Riley L, Saar K, Serikawa T, Shull JD, Szpirer C, Twigger SN, Voigt B, Worley K (2008) Progress and prospects in rat genetics: a community view. Nat Genet 40(5):516–522. doi:10.1038/ng.147
2. Saba LM, Flink SC, Vanderlinden LA, Israel Y, Tampier L, Colombo G, Kiianmaa K, Bell RL, Printz MP, Flodman P, Koob G, Richardson HN, Lombardo J, Hoffman PL, Tabakoff B (2015) The sequenced rat brain transcriptome – its use in identifying networks predisposing alcohol consumption. FEBS J 282(18):3556–3578. doi:10.1111/febs.13358
3. Cowley AW Jr, Liang M, Roman RJ, Greene AS, Jacob HJ (2004) Consomic rat model systems for physiological genomics. Acta Physiol Scand 181(4):585–592. doi:10.1111/j.1365-201X.2004.01334.x
4. Consortium S, Saar K, Beck A, Bihoreau MT, Birney E, Brocklebank D, Chen Y, Cuppen E, Demonchy S, Dopazo J, Flicek P, Foglio M, Fujiyama A, Gut IG, Gauguier D, Guigo R, Guryev V, Heinig M, Hummel O, Jahn N, Klages S, Kren V, Kube M, Kuhl H, Kuramoto T, Kuroki Y, Lechner D, Lee YA, Lopez-Bigas N, Lathrop GM, Mashimo T, Medina I, Mott R, Patone G, Perrier-Cornet JA, Platzer M, Pravenec M, Reinhardt R, Sakaki Y, Schilhabel M, Schulz H, Serikawa T, Shikhagaie M, Tatsumoto S, Taudien S, Toyoda A, Voigt B, Zelenika D, Zimdahl H, Hubner N (2008) SNP and haplotype mapping for genetic analysis in the rat. Nat Genet 40(5):560–566. doi:10.1038/ng.124
5. Solberg Woods LC (2014) QTL mapping in outbred populations: successes and challenges. Physiol Genomics 46(3):81–90. doi:10.1152/physiolgenomics.00127.2013
6. Li D, Qiu Z, Shao Y, Chen Y, Guan Y, Liu M, Li Y, Gao N, Wang L, Lu X, Zhao Y, Liu M (2013) Heritable gene targeting in the mouse and rat using a CRISPR-Cas system. Nat

Biotechnol 31(8):681–683. doi:10.1038/nbt.2661

7. Printz MP, Jirout M, Jaworski R, Alemayehu A, Kren V (2003) Genetic models in applied physiology. HXB/BXH rat recombinant inbred strain platform: a newly enhanced tool for cardiovascular, behavioral, and developmental genetics and genomics. J Appl Physiol (1985) 94(6):2510–2522. doi:10.1152/japplphysiol.00064.2003
8. Pravenec M, Klir P, Kren V, Zicha J, Kunes J (1989) An analysis of spontaneous hypertension in spontaneously hypertensive rats by means of new recombinant inbred strains. J Hypertens 7(3):217–221
9. Morrissey C, Grieve IC, Heinig M, Atanur S, Petretto E, Pravenec M, Hubner N, Aitman TJ (2011) Integrated genomic approaches to identification of candidate genes underlying metabolic and cardiovascular phenotypes in the spontaneously hypertensive rat. Physiol Genomics 43(21):1207–1218. doi:10.1152/physiolgenomics.00210.2010
10. Zicha J, Dobesova Z, Zidek V, Silhavy J, Simakova M, Mlejnek P, Vaneckova I, Kunes J, Pravenec M (2014) Pharmacogenetic analysis of captopril effects on blood pressure: possible role of the Ednrb (endothelin receptor type B) candidate gene. Physiol Res 63(2):263–265
11. Voigt B, Kuramoto T, Mashimo T, Tsurumi T, Sasaki Y, Hokao R, Serikawa T (2008) Evaluation of LEXF/FXLE rat recombinant inbred strains for genetic dissection of complex traits. Physiol Genomics 32(3):335–342. doi:10.1152/physiolgenomics.00158.2007
12. Shisa H, Lu L, Katoh H, Kawarai A, Tanuma J, Matsushima Y, Hiai H (1997) The LEXF: a new set of rat recombinant inbred strains between LE/Stm and F344. Mamm Genome 8(5):324–327
13. Hermsen R, de Ligt J, Spee W, Blokzijl F, Schafer S, Adami E, Boymans S, Flink S, van Boxtel R, van der Weide RH, Aitman T, Hubner N, Simonis M, Tabakoff B, Guryev V, Cuppen E (2015) Genomic landscape of rat strain and substrain variation. BMC Genomics 16:357. doi:10.1186/s12864-015-1594-1
14. Atanur SS, Birol I, Guryev V, Hirst M, Hummel O, Morrissey C, Behmoaras J, Fernandez-Suarez XM, Johnson MD, McLaren WM, Patone G, Petretto E, Plessy C, Rockland KS, Rockland C, Saar K, Zhao Y, Carninci P, Flicek P, Kurtz T, Cuppen E, Pravenec M, Hubner N, Jones SJ, Birney E, Aitman TJ (2010) The genome sequence of the spontaneously hypertensive rat: analysis and functional significance. Genome Res 20(6):791–803. doi:10.1101/gr.103499.109
15. Cech TR (2015) RNA World research-still evolving. RNA 21(4):474–475. doi:10.1261/rna.049965.115
16. Weiss JN, Karma A, MacLellan WR, Deng M, Rau CD, Rees CM, Wang J, Wisniewski N, Eskin E, Horvath S, Qu Z, Wang Y, Lusis AJ (2012) "Good enough solutions" and the genetics of complex diseases. Circ Res 111(4):493–504. doi:10.1161/CIRCRESAHA.112.269084
17. Zhang B, Horvath S (2005) A general framework for weighted gene co-expression network analysis. Stat Appl Genet Mol Biol 4:17. doi:10.2202/1544-6115.1128
18. Eisen MB, Spellman PT, Brown PO, Botstein D (1998) Cluster analysis and display of genome-wide expression patterns. Proc Natl Acad Sci U S A 95(25):14863–14868
19. Vanderlinden LA, Saba LM, Printz MP, Flodman P, Koob G, Richardson HN, Hoffman PL, Tabakoff B (2014) Is the alcohol deprivation effect genetically mediated? Studies with HXB/BXH recombinant inbred rat strains. Alcohol Clin Exp Res 38(7):2148–2157. doi:10.1111/acer.12471
20. Vansteelandt S, Lange C (2012) Causation and causal inference for genetic effects. Hum Genet 131(10):1665–1676. doi:10.1007/s00439-012-1208-9
21. Roberts A, Pimentel H, Trapnell C, Pachter L (2011) Identification of novel transcripts in annotated genomes using RNA-Seq. Bioinformatics 27(17):2325–2329. doi:10.1093/bioinformatics/btr355
22. Friedlander MR, Chen W, Adamidi C, Maaskola J, Einspanier R, Knespel S, Rajewsky N (2008) Discovering microRNAs from deep sequencing data using miRDeep. Nat Biotechnol 26(4):407–415. doi:10.1038/nbt1394
23. Yang JH, Zhang XC, Huang ZP, Zhou H, Huang MB, Zhang S, Chen YQ, Qu LH (2006) snoSeeker: an advanced computational package for screening of guide and orphan snoRNA genes in the human genome. Nucleic Acids Res 34(18):5112–5123. doi:10.1093/nar/gkl672
24. Meyer LR, Zweig AS, Hinrichs AS, Karolchik D, Kuhn RM, Wong M, Sloan CA, Rosenbloom KR, Roe G, Rhead B, Raney BJ, Pohl A, Malladi VS, Li CH, Lee BT, Learned K, Kirkup V, Hsu F, Heitner S, Harte RA, Haeussler M, Guruvadoo L, Goldman M, Giardine BM, Fujita PA, Dreszer TR, Diekhans M, Cline MS, Clawson H, Barber GP, Haussler D, Kent WJ (2013) The UCSC Genome Browser database: extensions and updates 2013. Nucleic Acids Res 41(Database issue):D64–D69. doi:10.1093/nar/gks1048

25. Cunningham F, Amode MR, Barrell D, Beal K, Billis K, Brent S, Carvalho-Silva D, Clapham P, Coates G, Fitzgerald S, Gil L, Giron CG, Gordon L, Hourlier T, Hunt SE, Janacek SH, Johnson N, Juettemann T, Kahari AK, Keenan S, Martin FJ, Maurel T, McLaren W, Murphy DN, Nag R, Overduin B, Parker A, Patricio M, Perry E, Pignatelli M, Riat HS, Sheppard D, Taylor K, Thormann A, Vullo A, Wilder SP, Zadissa A, Aken BL, Birney E, Harrow J, Kinsella R, Muffato M, Ruffier M, Searle SM, Spudich G, Trevanion SJ, Yates A, Zerbino DR, Flicek P (2015) Ensembl 2015. Nucleic Acids Res 43(Database issue):D662–D669. doi:10.1093/nar/gku1010
26. Pruitt KD, Brown GR, Hiatt SM, Thibaud-Nissen F, Astashyn A, Ermolaeva O, Farrell CM, Hart J, Landrum MJ, McGarvey KM, Murphy MR, O'Leary NA, Pujar S, Rajput B, Rangwala SH, Riddick LD, Shkeda A, Sun H, Tamez P, Tully RE, Wallin C, Webb D, Weber J, Wu W, DiCuccio M, Kitts P, Maglott DR, Murphy TD, Ostell JM (2014) RefSeq: an update on mammalian reference sequences. Nucleic Acids Res 42(Database issue):D756–D763. doi:10.1093/nar/gkt1114
27. Shimoyama M, De Pons J, Hayman GT, Laulederkind SJ, Liu W, Nigam R, Petri V, Smith JR, Tutaj M, Wang SJ, Worthey E, Dwinell M, Jacob H (2015) The Rat Genome Database 2015: genomic, phenotypic and environmental variations and disease. Nucleic Acids Res 43(Database issue):D743–D750. doi:10.1093/nar/gku1026
28. Rosenbloom KR, Armstrong J, Barber GP, Casper J, Clawson H, Diekhans M, Dreszer TR, Fujita PA, Guruvadoo L, Haeussler M, Harte RA, Heitner S, Hickey G, Hinrichs AS, Hubley R, Karolchik D, Learned K, Lee BT, Li CH, Miga KH, Nguyen N, Paten B, Raney BJ, Smit AF, Speir ML, Zweig AS, Haussler D, Kuhn RM, Kent WJ (2015) The UCSC Genome Browser database: 2015 update. Nucleic Acids Res 43(Database issue):D670–D681. doi:10.1093/nar/gku1177
29. Krzywinski M, Schein J, Birol I, Connors J, Gascoyne R, Horsman D, Jones SJ, Marra MA (2009) Circos: an information aesthetic for comparative genomics. Genome Res 19(9):1639–1645. doi:10.1101/gr.092759.109
30. Ru Y, Kechris KJ, Tabakoff B, Hoffman P, Radcliffe RA, Bowler R, Mahaffey S, Rossi S, Calin GA, Bemis L, Theodorescu D (2014) The multiMiR R package and database: integration of microRNA-target interactions along with their disease and drug associations. Nucleic Acids Res 42(17), e133. doi:10.1093/nar/gku631
31. Kozomara A, Griffiths-Jones S (2014) miRBase: annotating high confidence microRNAs using deep sequencing data. Nucleic Acids Res 42(Database issue):D68–D73. doi:10.1093/nar/gkt1181
32. Stasko J, Catrambone R, Guzdial M, McDonald K (2000) An evaluation of space-filling information visualizations for depicting hierarchical structures. Int J Hum-Comput St 53(5):663–694. doi:10.1006/ijhc.2000.0420
33. Gene Ontology C (2015) Gene ontology consortium: going forward. Nucleic Acids Res 43(Database issue):D1049–D1056. doi:10.1093/nar/gku1179
34. Langfelder P, Horvath S (2008) WGCNA: an R package for weighted correlation network analysis. BMC Bioinformatics 9:559. doi:10.1186/1471-2105-9-559
35. van den Heuvel MP, Hulshoff Pol HE (2010) Exploring the brain network: a review on resting-state fMRI functional connectivity. Eur Neuropsychopharmacol 20(8):519–534. doi:10.1016/j.euroneuro.2010.03.008
36. Wolf L, Goldberg C, Manor N, Sharan R, Ruppin E (2011) Gene expression in the rodent brain is associated with its regional connectivity. PLoS Comput Biol 7(5), e1002040. doi:10.1371/journal.pcbi.1002040
37. Bennett BJ, Farber CR, Orozco L, Kang HM, Ghazalpour A, Siemers N, Neubauer M, Neuhaus I, Yordanova R, Guan B, Truong A, Yang WP, He A, Kayne P, Gargalovic P, Kirchgessner T, Pan C, Castellani LW, Kostem E, Furlotte N, Drake TA, Eskin E, Lusis AJ (2010) A high-resolution association mapping panel for the dissection of complex traits in mice. Genome Res 20(2):281–290. doi:10.1101/gr.099234.109
38. Pearl J (2009) Causality. Cambridge University Press, Cambridge

Chapter 15

Precise Network Modeling of Systems Genetics Data Using the Bayesian Network Webserver

Jesse D. Ziebarth and Yan Cui

Abstract

The Bayesian Network Webserver (BNW, http://compbio.uthsc.edu/BNW) is an integrated platform for Bayesian network modeling of biological datasets. It provides a web-based network modeling environment that seamlessly integrates advanced algorithms for probabilistic causal modeling and reasoning with Bayesian networks. BNW is designed for precise modeling of relatively small networks that contain less than 20 nodes. The structure learning algorithms used by BNW guarantee the discovery of the best (most probable) network structure given the data. To facilitate network modeling across multiple biological levels, BNW provides a very flexible interface that allows users to assign network nodes into different tiers and define the relationships between and within the tiers. This function is particularly useful for modeling systems genetics datasets that often consist of multiscalar heterogeneous genotype-to-phenotype data. BNW enables users to, within seconds or minutes, go from having a simply formatted input file containing a dataset to using a network model to make predictions about the interactions between variables and the potential effects of experimental interventions. In this chapter, we will introduce the functions of BNW and show how to model systems genetics datasets with BNW.

Key words Causal network, Bayesian network modeling, Webserver, Probabilistic inference, Prediction

1 Introduction

Human individuals are different from each other genetically and phenotypically. Many phenotypic differences, such as susceptibility to diseases and response to medication, have significant health implications. Recently, the 1000 Genomes Project identified over 88 million DNA variants in the genomes of 2504 individuals from 26 populations around the world [1]. While most genetic variations are functionally neutral, a small portion (but still a large number) of genetic variations is responsible for the phenotypic diversity in the Human population. Many associations between genetic variations at the DNA level and phenotypic variations at the physiological and behavioral levels have been discovered in the past

Klaus Schughart and Robert W. Williams (eds.), *Systems Genetics: Methods and Protocols*, Methods in Molecular Biology, vol. 1488, DOI 10.1007/978-1-4939-6427-7_15,

decade, mainly by the genome-wide association studies (GWAS) [2, 3]. The NHGRI-EBI GWAS Catalog (http://www.ebi.ac.uk/gwas/) [4] contains about 22,000 genotype-phenotype associations with p-values $\leq 5.0 \times 10^{-8}$ (as of November 2015). More than 1500 human phenotypes, mostly disease-related, are involved in these associations.

While our knowledge of genotype-phenotype associations has grown rapidly, the molecular pathways through which genetic variants influence phenotypes remain largely unknown. There is a black box between a DNA variant and its associated phenotype. As the first attempt to glimpse into this black box, people studied the mRNA transcripts whose expression covaries with the genotype and phenotype of interest in the natural and experimental populations of Human, Mouse, and other model organisms [5–17]. The genetic loci regulating mRNA expression traits are called eQTLs (expression traits locus) [7]. Other molecular traits such as the abundance of proteins and metabolites can also be mapped as quantitative traits [18]. The new area of research that is intended to systematically identify the associations between the genotype, phenotypes, and molecular traits of a population is called Systems Genetics [19, 20] (or Genetical Genomics [21]). The Systems Genetics approach has been widely used to study development [22], behavior [23] and various diseases and disorders [24, 25] such as cancers [26–28], obesity [29, 30], cardiovascular diseases [31, 32], metabolic diseases [33, 34], substance dependence [35–37] and infectious diseases [38–40].

In this chapter, we introduce the Bayesian Network Webserver (BNW, http://compbio.uthsc.edu/BNW), a web-based causal network modeling platform [41]. One of our main inspirations when creating BNW was that, in many cases, several different tools would be required in order to comprehensively analyze a dataset using Bayesian networks. For example, different tools would be required to learn the structure of the network that explained the relationships observed in the data and to use this network to make predictions about the effect of experimental interventions. Therefore, Bayesian network modeling of a dataset often had both a steep learning curve for beginners and a significant time investment for experienced modelers. We created BNW with the aim of helping to overcome these bottlenecks by providing a single service that could be simply and quickly used to fully analyze a dataset with a Bayesian network. The general workflow for using BNW consists of the following steps: First, users upload a simply formatted text file containing their data and, if desired, specify constraints that should be used to limit searches during structure learning. BNW then automatically displays the network structure that best explains the data within a matter of seconds or minutes, allowing users to visualize the causal relationships between the variables in the data. Finally, users can immediately use the network to make testable predictions about the relationship between variables.

2 Bayesian Network Modeling of Systems Genetics Data

2.1 Systems Genetics Data

A typical systems genetics dataset consists of genotype data, molecular trait data, and phenotype data of a population (Fig. 1). Such datasets provide the opportunity to discover the correlative and regulatory relations across multiple biological scales—from DNA variants to disease phenotypes. A large number of systems genetics datasets are available from web databases such as GeneNetwork (http://genenetwork.org/) and GTEx Portal (http://www.gtexportal.org/). These websites also provide tools for searching, analyzing, and visualizing systems genetics data.

2.2 Causal Inference with Bayesian Networks

Many computational methods have been developed or adopted for analyzing systems genetics data. However, most of the methods can only capture the associative relations between the variables representing the genotype, phenotype, and OMICS traits. Bayesian networks are capable of discovering the causal relationships between the variables [42] and have been successfully used to model systems genetics data [13, 39, 43–45]. A Bayesian network is comprised of a set of random variables, a directed acyclic network graph representing the relations between the variables, and a set of parameters that define the distributions of the random variables. The network structure and the parameters can be learned from data.

The mathematical device that makes it possible to infer causal relations from observational data is called directed separation or d-separation [42], which can translate between the causal claims expressed in a directed graph and a statistical model of the data, as

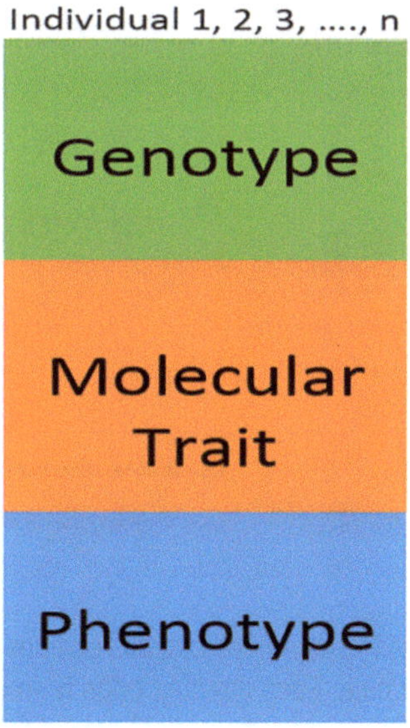

Fig. 1 Schematic representation of a systems genetics dataset (matrix)

shown in Fig. 2. The directed graph in Fig. 2 has four nodes that each represent a variable in the dataset connected by directed edges that indicate the causal relationships between the variables. Here A and B are the causes of C, and C is the cause of D. The causal relationships in the directed graph can be completely specified by a set of statements of conditional independence defined by d-separation, which gives the necessary and sufficient conditions for two sets of variables to be probabilistically independent, conditioned on some other variables [46]. For example, in Fig. 2, variables A and B are unconditionally independent: I (A, Φ, B), where Φ represents the empty set. If the causal model is correct, we should observe a corresponding probabilistic independence when inspecting the values of A and B in an observational dataset, that is, the joint probability of A and B should equal the probability of A times the probability of B: $P(\mathrm{A},\mathrm{B}) = P(\mathrm{A}) \times P(\mathrm{B})$. Another independence statement in the causal graph is that D is independent from A and B, conditioned on C: $\mathrm{C}: I(\mathrm{D}, \mathrm{C}\{\mathrm{A}, \mathrm{B}\})$. This is equivalent to a conditional probabilistic independence: $P(A,B,C,D) = P(D|\mathrm{C}) \times P(\mathrm{C}|\mathrm{A},\mathrm{B}) \times P(A,B)$. Thus, we can test the causal graph model by examining conditional probabilistic independence properties using the data.

The process of determining the network structure that appropriately represents the relationships between variables in a dataset is called structure learning. Structure learning is often the most computationally expensive part of using Bayesian networks to analyze data, as the number of possible network structures increases

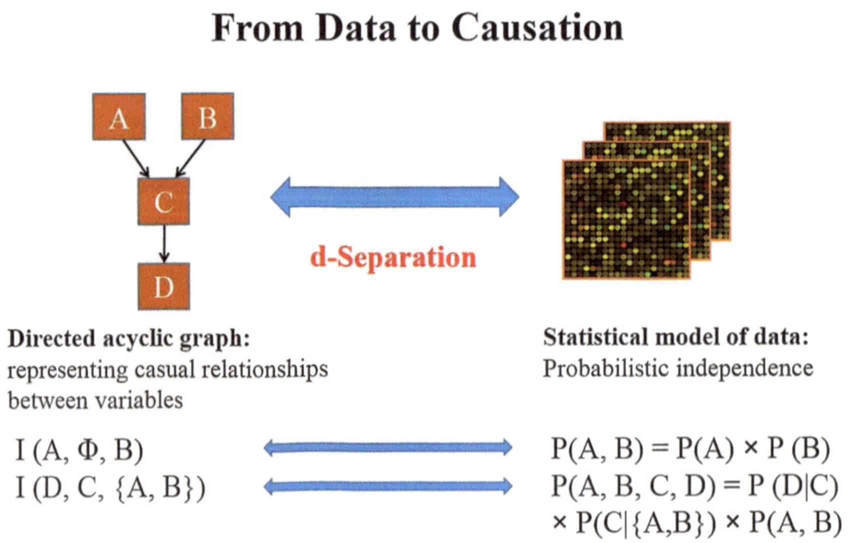

Fig. 2 Bayesian network for capturing causal relations from data. Each node in the directed acyclic graph represents a random variable. Each directed edge represents the cause and effect relationships between two variables. The d-Separation is a mathematical device that translates between the independence properties of the probabilistic distribution (*right*) and the causal relationships defined by the network structure (*left*)

super-exponentially with the number of variables. The most common method, and the one used in BNW, for structure learning from data involves searching and scoring possible network structures and selecting the highest scoring network. Each causal model that is generated during the search may include many d-separation statements, but we do not need to test them one by one. Instead, we can use a composite score to evaluate the graphical causal model against the data during the search. In Bayesian network modeling, we usually use a Bayesian score to evaluate the network $S(M:D) = \log P(M \mid D)$, where P is the posterior probability of a network M given dataset D. The Bayesian score can be written as: $S(M:D) = \sum_{i=1}^{N} S(X_i, P_i : D)$, where X_i represents the value of node (variable) i, P_i represents the values of the parent nodes of node i, and N is the total number of nodes in the network.

The structure learning method used in BNW can be broken down into three main steps: calculating local network scores, determining global structures that optimize network scores, and, if indicated by user settings, performing model averaging over high scoring structures.

2.2.1 Local Score Calculation

To begin structure learning in BNW, we calculate all possible local scores by performing an exhaustive search of local structures given the structural constraints specified by the user. Systems genetics datasets often contain both discrete (e.g., genotypes) and continuous (e.g., gene expression traits) variables. BNW allows for modeling such hybrid data by calculating local scores using conditional Gaussian distributions [47].

2.2.2 Model Averaging

Model averaging can be used to reduce the risk of over-fitting data to a single model. In BNW, model averaging is automatically performed over the *k*-best scoring structures [48], using a user selected value for *k*. To select features (i.e., directed edges between nodes), the posterior probability of each feature is calculated from a weighted average over the *k*-best networks where the weight is given by the score of the global network structure. This posterior probability, which ranges for 1 for features included in all high scoring networks to 0 for features in no high scoring networks, reflects confidence in the feature. Model averaging may be particularly advantageous when learning network models using small datasets. As the number of samples in a dataset increases, the differences between the scores of high scoring networks often also increase. Therefore, structure learning of small datasets may identify a group of structures with similar scores instead of a single network structure with a score significantly better than all other possible networks. Model averaging can be used to select features that are common to many high scoring networks.

2.3 Bayesian Network Webserver: Functions and User Interface

2.3.1 Structure Learning

The first step in Bayesian network modeling of a dataset is identifying the network structure. In BNW, users can either upload a known network structure or learn the network structure that best explains the data. Structure learning from a dataset identifies which directed edges between network variables (nodes) should be included in the network to represent the conditional independencies observed in the data. The structure learning method implemented in BNW can be used to learn the network structures of discrete, continuous, and hybrid (i.e., datasets containing both discrete and continuous variables) datasets. After uploading a text file containing a dataset, users can either immediately perform structure learning using default settings or add or modify structural constraints that can improve the performance of structure learning. By default, BNW limits the maximum number of parents for each node in the network to 4 and presents only the highest scoring network structure (i.e., no model averaging is performed).

BNW includes a structural constraint interface that provides users with options that can increase the speed of structure learning, aid in identifying robust network structures, and limit structure searches to biologically or physically meaningful networks by incorporating prior knowledge. The first section of the structural constraint interface allows users to set the following options that define global properties of the network structure search (Fig. 3).

Maximum Number of Parents

This option sets a limit on the number of immediate parents for every node in the network and can impact structure learning in two main ways. First, limiting the maximum number of parents can dramatically increase the speed of structure learning for larger networks. Second, this limit may also help in avoiding over-fitting a network model, as it prevents a variable from being directly influenced by a large number of the other variables in the network. By default, the maximum number of parents of a node in BNW is 4.

Number of Networks to Include in Model Averaging

This option specifies k, the number of the high scoring networks that will be included in model averaging. For $k=1$, only the highest scoring network is considered and no model averaging is performed. For other values of k, model averaging is performed over the k-best networks. Increasing the value of k will increase the time required to perform the structure learning search but may increase the performance of model averaging.

Model Averaging Selection Threshold

This option specifies the threshold that should be used to select directed edges to be included in the network given their posterior probabilities after model averaging. All directed edges with posterior probabilities greater than the threshold will be included in the network. It is ignored if $k=1$. By default, the threshold is set to 0.5.

Systems genetics datasets contain different types of variables (genotypes, molecular trains, and phenotypes) that will only

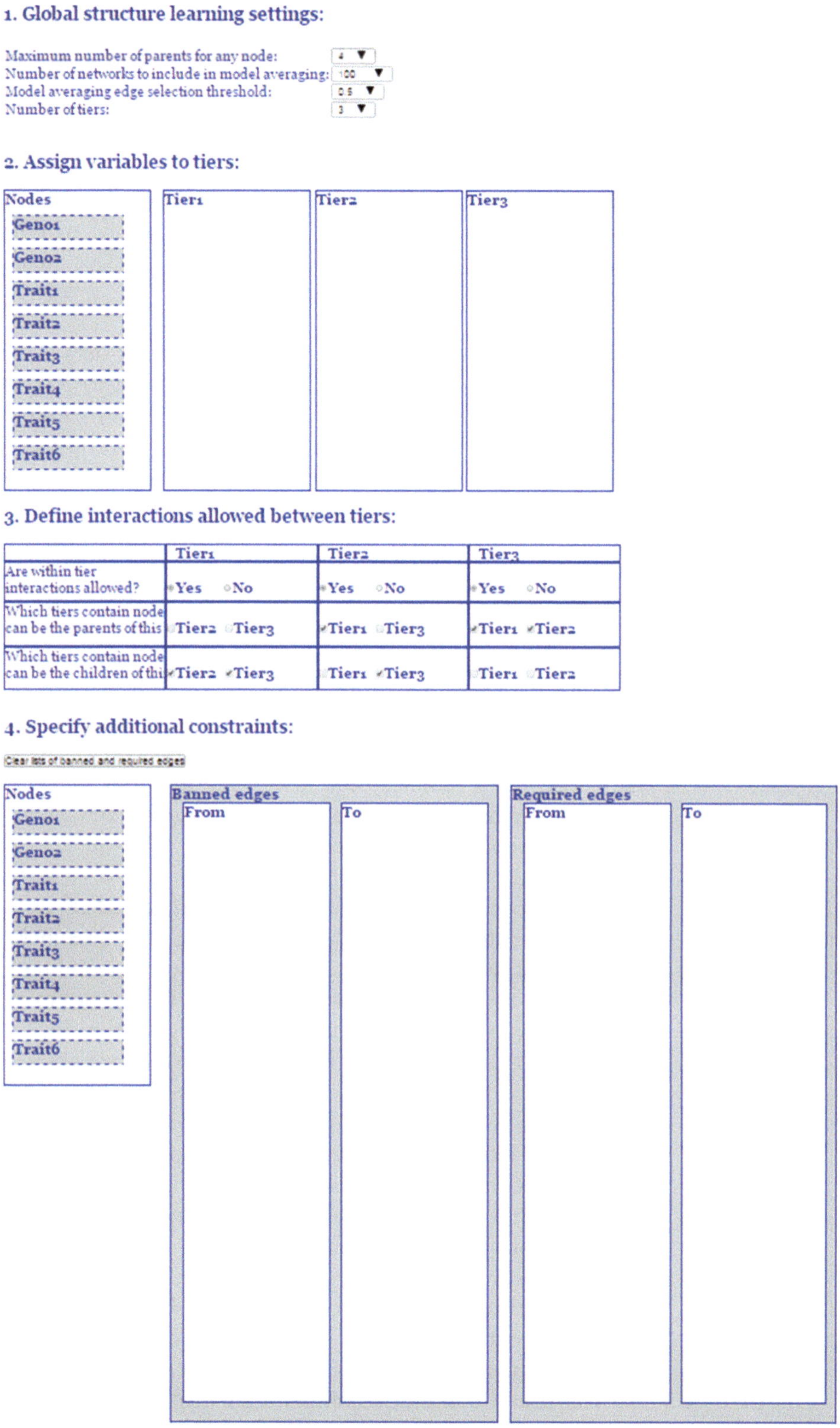

Fig. 3 Bayesian Network Webserver: Structural constraint interface

interact with each other through a limited set of casual relationships. For example, a phenotype variable should not be the parent or cause of a genotype, as the genotype predates the phenotype. In order to include this type of prior knowledge and focus structure learning searches on biologically relevant networks, the BNW structural constraint interface allows users to separate variables into different groups, or tiers, and specify the interactions that should be allowed between and within tiers. The specific workflow for using this section of the structural constraint interface includes the following steps:

Specifying the Number of Tiers and Assigning Variables to Tiers

Users of BNW can separate variables into as many tiers as is appropriate for their dataset. The number of tiers is set to 3 by default. Variables can then be assigned to appropriate tiers by simply dragging and dropping the variable name into the appropriate box. If the network variables are not assigned to tiers, this option will be ignored when structure learning is performed.

Tier Interactions

This section allows for the description of the interactions that are allowed within and between tiers. By default, edges are allowed within tiers for all tiers, and nodes within a tier are only allowed to be parents of nodes within lower ranking tiers. For example, nodes in Tier2 of a network with four tiers could be the parents of nodes in Tiers 3 and 4, but could not be the parents of nodes in Tier1. Users can modify the allowed interactions to fit the details of their networks. For example, they may want to prohibit interactions within a tier containing variables, such as genotypes, that do not causally depend on each other.

Specific Banned and Required Edges

Finally, users can enter lists of banned and required edges to identify specific interactions that should or should not be included in the network. Banned edges can be used if experimental testing has shown that a particular regulatory relationship does not exist, while required edges can specify known regulatory relationships.

After structure learning, it is possible that all variables will not be connected within a single network. Some variables may not be found to be associated with any other variables in the input dataset and may be left out of any network model, or there may be two or more distinct networks. In these cases, the highest scoring network model did not have all variables in a single network given the data and our scoring metric.

2.3.2 Parameter Learning

After structure learning is completed, BNW automatically performs parameter learning of the network model and displays the network model. Discrete variables in the network are displayed as bar charts and continuous variables are displayed as Gaussian distributions. The networks can be used to make predictions after clicking on a node and entering a value for that variable. Specifically, clicking on either the blue bar for a discrete node or the blue line for a continuous node brings up a pop-up box that can be used to enter a value for the node. After submitting a value for the variable, the distributions of the other nodes in the network will change, allowing for visualization of the impact of setting the variable to the given value. The distributions after the entered value is considered are shown in red, while the original distributions are shown in blue. The node for which data was entered is outlined in red.

2.3.3 Using Bayesian Network Models to Make Predictions

Two prediction modes are available in BNW: evidence and intervention. In the evidence mode, entered values will alter the distributions of the other variables in the network, but will not alter the network structure. In intervention mode, the intervention alters both the distributions of the network variables and the network structure. Specifically, the intervened variable becomes independent of its parents. Evidence mode is appropriate when making predictions of other network variables after the value of one variable in the network is observed, while intervention mode is appropriate for predictions after experimental interventions that alter the values of some variables in the network. To compare the evidence and intervention prediction modes, consider a genetic network model that was learned for a set of mouse strains:

Evidence

The evidence mode can be used to predict data generated for a new set of strains. For example, the model may predict that high expression of Gene1 results in high expression of Gene2. This prediction can be tested by measuring the expression levels of Gene1 and Gene2 in a new set of strains and comparing the predicted expression of Gene2 from the network model given the observed expression of Gene1 with the actual observed values of Gene2.

Intervention

In contrast, the intervention mode is appropriate for cases when you are not passively observing network interactions, but instead are experimentally perturbing the network. Again, consider a case in which the model predicts that high expression of Gene1 results in high expression of Gene2. To intervene on the network, a treatment that is known to cause high expression of Gene1 could be given. This treatment will cause all strains to have high expression of Gene1, and it will no longer be dependent on its parents in the original model. Also, the model will predict that Gene2 should have high expression in all strains. The expression levels of Gene2 in treated strains can be measured to test this predicted effect of intervention.

3 Modeling Systems Genetics Data with BNW: An Example

In this section, we will use an example to show how to create a causal network linking a genotype with gene expression traits and a phenotype. There are five variables in the dataset: Genotype, Gene1, Gene2, Gene3, and Phenotype. The input data file can be accessed by clicking on "Tutorials and example networks" on the BNW home page and then selecting Tutorial 2. Proper formatting of input files for use in BNW is fully described in the "Data formatting guidelines" section of the BNW FAQ page. Briefly, data files should be tab-delimited text files with the names of the variables in the first row of the file and the values of the variables for each sample or individual in the remaining rows. In this example, the dataset contains 500 measurements for each of the five variables in the network, and the input data file therefore contains five columns and 501 rows, with the first row containing the names of the variables. BNW automatically determines whether variables contain continuous or discrete data as long as a few simple formatting guidelines are followed. Here, we have one discrete variable (Genotype) which takes two integer values (1 and 2) and four continuous variables that are real numbers (Fig. 4).

3.1 Load Data and Set Parameters

To begin using BNW with this data file, select *Learn a network model from data* on the BNW homepage and load the data file, displaying what is shown in Fig. 4. We now have the option of either performing structure learning using the default BNW settings and no additional structural constraints or we can use modify settings and add structural constraints. Here, we will select *Go to structure learning settings and the BNW structural constraint interface* to select the latter option. The BNW structural constraint interface was previously described in Subheading 2.3.1. As shown in Fig. 5, the top of the page has several settings for global features of the structure search. We have kept the default settings, except we have changed the *Number of networks to include in model averaging* to 1000. As this is a small network with only five nodes, including many high scoring networks had little effect on the estimated time required to perform structure learning, and the estimated run time increased from 12 to 13 s. We also could increase the *Maximum number of parents for any node* to any value without

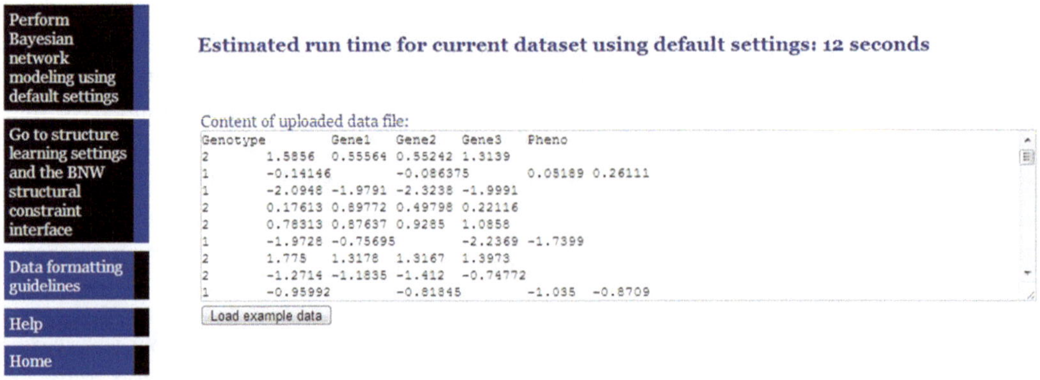

Fig. 4 Screenshot from BNW after uploading an input data file

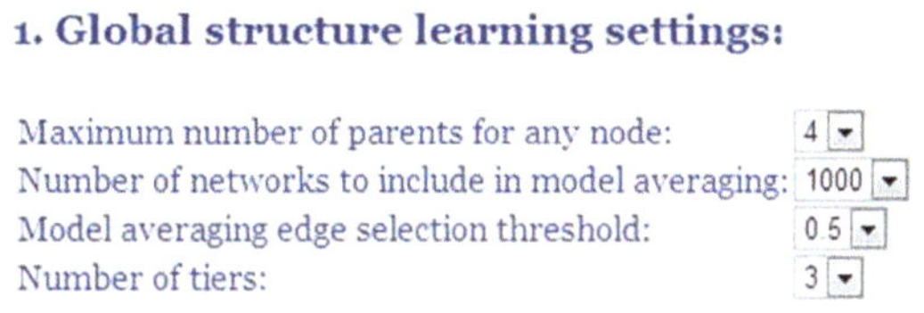

Fig. 5 Screenshot from the BNW structural constraint interface showing several setting that can be modified by users

significantly changing the run time for a network of this size. For larger networks with more than approximately ten nodes, changing these settings can have a major impact on estimated run times.

3.2 Assign Nodes to Tiers

In the next section of the structural constraint interface, we assign nodes to tiers which can be used to focus network searches on biologically meaningful networks. Our dataset contains a genotype, three gene expression traits, and a higher-order phenotype. Instead of considering all possible network models for this dataset, we may want to focus on models relevant to a question such as: How does variation in genotype and gene expression explain the variation observed in the phenotype? To address this question, we assign the network nodes to three tiers: Tier1 contains the genotype, Tier2 contains the gene expression traits, and Tier3 contains the phenotype (Fig. 6).

3.3 Specify Possible Interactions

The third section of the BNW structural constraint interface allows users to specify the interactions that are allowed within and between tiers (Fig. 7). By default, within tier interactions (i.e., nodes in TierX can be parents or children of other nodes in TierX) are allowed, but users may want to prevent within tier interactions for some cases. For example, it may be advantageous to prevent interactions between a tier that contained a set of some demographic variables (e.g., age, sex, and race), as these variables are not likely to be a causal factor that influences other variables within the tier. In this case, within tier interactions only apply to Tier2, and we do not have any prior knowledge that indicates that between gene interactions should not be allowed, so we will keep the default setting and allow within tier interactions.

The default settings for between tier interactions allow nodes within a tier to be the parents of all nodes in lower ranking tiers. Here, the Genotype node in Tier1 can be parents of the gene nodes in Tier2 and the Phenotype node in Tier3, the gene nodes in Tier2 can be the parents of the Tier3 Phenotype node, and the Tier3 Phenotype node cannot be the parents of nodes in any other tier. Therefore, by default, the Genotype node can be the direct parent of the Phenotype node. We may want to allow this interaction, as the genotype may influence the phenotype through genes or other factors that are not explicitly included as variables in the

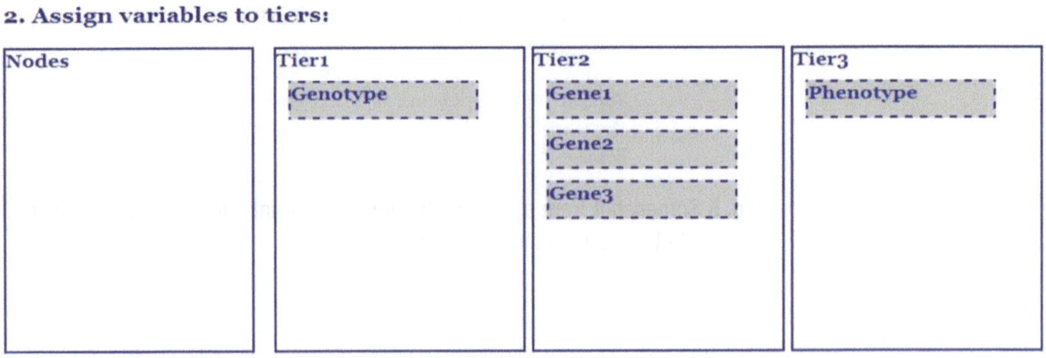

Fig. 6 Example of assigning network variables to tiers using the BNW structure learning interface

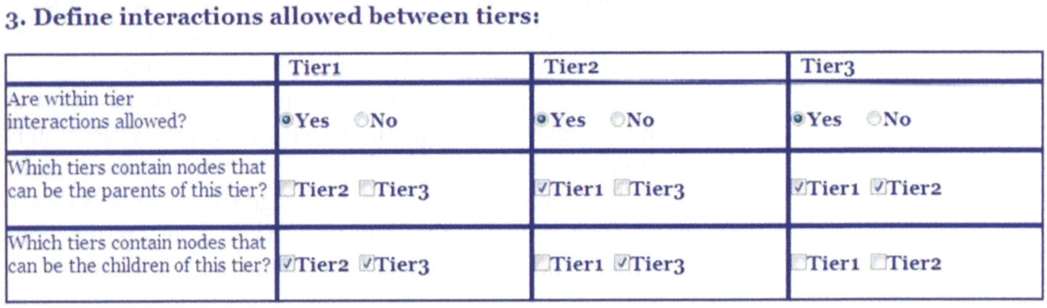

Fig. 7 Specifying the interactions that are allowed between and within tiers in the BNW structural constraint interface

network. If users do not want to allow this direct Genotype-Phenotype interaction, they can unclick the Tier3 box in the *Which tiers contain nodes that can be the children of this tier?* for Tier1. In this case, we have maintained the default settings (Fig. 7).

In this example, we will not specify any additional constraints in the fourth section of the structural constraint interface, and we can click *Perform Bayesian network modeling* on the upper left corner of the page.

3.4 Examine Network Structure

Figure 8 (left) shows the network structure after model averaging of the 1000 highest scoring networks; it can also be accessed by clicking "View network" under Tutorial 2 of the "Tutorials and example networks" page of the BNW website. Genotype directly influences two of the genes (Gene1 and Gene3), and two of the genes (Gene2 and Gene3) directly influence the Phenotype. In this case, although we did not prevent the Genotype from directly influencing the Phenotype, the highest scoring networks did not include this directed edge.

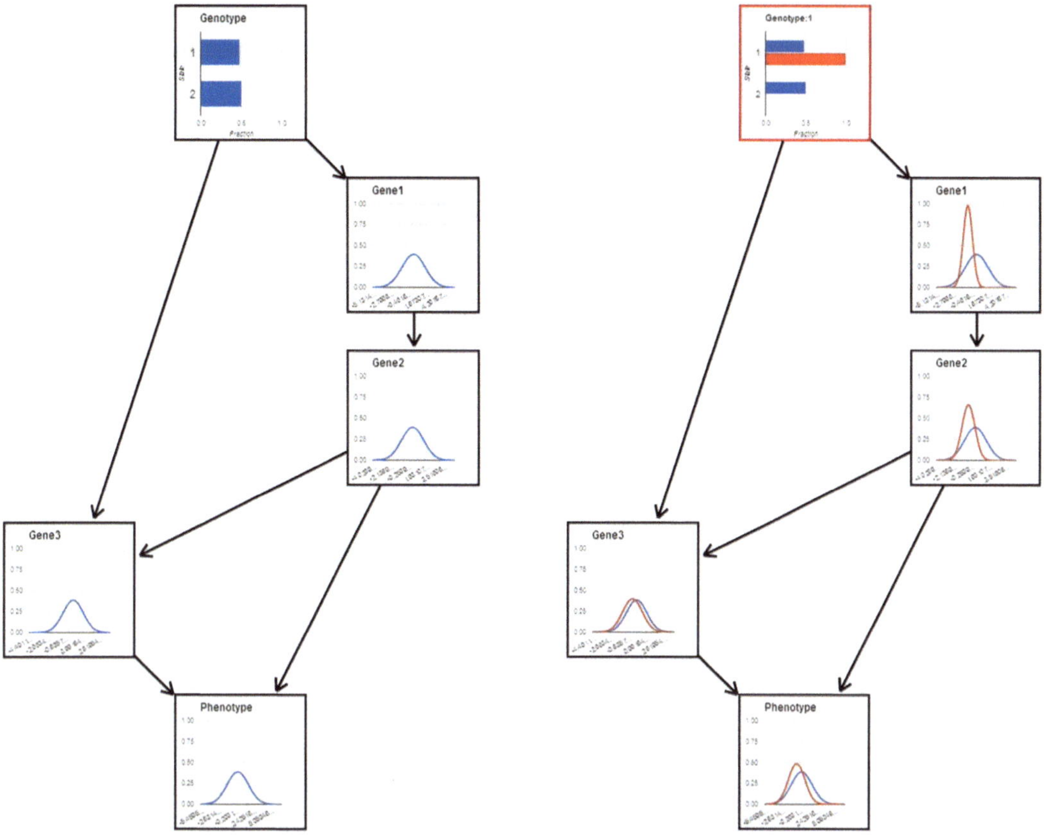

Fig. 8 The highest scoring causal network in BNW after modeling averaging for an example systems genetics dataset (*left*). Changes in the values of other network variables if the genotype is known to have a particular value (*right*)

3.5 Use the Network Model to Make Predictions

One exciting feature of BNW is that the causal network learned from the data can be immediately used to analyze the relationships between the variables in the network and make predictions about how experimental interventions may impact these predictions. For example, while the structure learned for this example dataset shows that the Genotype in the dataset directly impacts Gene1 and Gene3, the network structure alone is not able to fully describe this impact (e.g., Does having Genotype = 1 tend to increase or decrease the value of Gene1 and Gene3?). To more fully investigate the quantitative relationships between variables, users can click on a particular node of the network and enter a value for the variable as either evidence or an intervention (*see* Subheading 2.3.3). Here, we will consider the following experimental situation: The input data file used to learn the network structure and parameters contained data from a sample of individuals with a given disease with a severity that is quantified by the Phenotype variable in the network. Then, we encounter a new patient who is known to have

Genotype = 1 and would like to predict the severity of the disease in this patient. For this situation, we are not performing an experimental intervention to change the value of any of the variables in the network, and it is therefore appropriate to use the evidence mode to make predictions with the network. Specifically, we can use BNW for this situation by clicking on the Genotype node and entering "1" in the popup dialog box while in the evidence mode; the resulting network (Fig. 8 right) compares the predicted variable distributions when the Genotype is known to be 1 (shown in red) with the original distributions of the variables that correspond to case where nothing is known about the value of the Genotype (shown in blue). We see that if Genotype = 1 the network model predicts that the values of all of the other variables in the network are reduced. Even though an edge in the network does not directly connect Genotype with Gene2 and Phenotype, knowledge of the Genotype still influences the predicted values of these variables. For example, Genotype = 1 may cause a decrease in Gene1 and this decrease in Gene1 will subsequently cause a reduction in Gene2.

4 Discussion

Network modeling of biological datasets is often limited by the number of samples within a dataset, and the available data does not support the construction of precise and reliable large-scale networks in almost all cases. Furthermore, it has been found that modularity is a primary organizational principle of biological networks [49, 50]. Large-scale networks often consist of relatively independent network nodules (subnetworks). Effective computational methods have been developed to identify modules in various networks [51, 52]. Specifically, eQTL networks also have modular structures [53]. It is more practical to focus on the modeling of small networks and use these models to make testable predictions that are strongly supported by the data. Therefore, BNW is designed for precise modeling of relatively small networks.

The most computationally expensive step in BNW is structure learning, and BNW limits the size of datasets that can be used for structure learning. Currently, the maximum number of nodes when performing structure learning in BNW is 19, and the maximum number of samples is 10,000. We also estimate the time required for structure learning based on the input file size and the structure learning options provided by the user. Structure learning of networks on BNW should complete within approximately 10 min. For longer structure learning jobs, a structure learning package which is written in C is also available for download (http://compbio.uthsc.edu/BNW/downloads/BNW_src_files.tar).

BNW can also be integrated with other bioinformatic tools and databases to provide a web-based network modeling service. The GeneNetwork (GN, http://genenetwork.org), a multi-functional informatics platform for systems genetics studies, is using BNW as its network modeling module. Data pipelines have been set up between a mirror site of BNW (http://bnw.genenetwork.org) and GN. The users can use the various tools in GN, such as QTL mapping and correlation analysis, to select the variables to include in the subsequent network modeling. Then, users can, with a single mouse-click, send these variables to the BNW network building interface and start network modeling. The applications of BNW may go beyond systems genetics as it can be used as a general web-based engine for causal inference in various databases.

References

1. The Genomes Project, C (2015) A global reference for human genetic variation. Nature 526:68–74
2. Visscher PM, Brown MA, McCarthy MI, Yang J (2012) Five years of GWAS discovery. Am J Hum Genet 90:7–24
3. Hindorff LA, Sethupathy P, Junkins HA, Ramos EM, Mehta JP, Collins FS, Manolio TA (2009) Potential etiologic and functional implications of genome-wide association loci for human diseases and traits. Proc Natl Acad Sci 106:9362–9367
4. Welter D, MacArthur J, Morales J, Burdett T, Hall P, Junkins H, Klemm A, Flicek P, Manolio T, Hindorff L et al (2014) The NHGRI GWAS catalog, a curated resource of SNP-trait associations. Nucleic Acids Res 42:D1001–D1006
5. Brem RB, Yvert G, Clinton R, Kruglyak L (2002) Genetic dissection of transcriptional regulation in budding yeast. Science 296:752–755
6. Cheung VG, Spielman RS (2002) The genetics of variation in gene expression. Nat Genet 32(Suppl):522–525
7. Schadt EE, Monks SA, Drake TA, Lusis AJ, Che N, Colinayo V, Ruff TG, Milligan SB, Lamb JR, Cavet G et al (2003) Genetics of gene expression surveyed in maize, mouse and man. Nature 422:297–302
8. Bystrykh L, Weersing E, Dontje B, Sutton S, Pletcher MT, Wiltshire T, Su AI, Vellenga E, Wang J, Manly KF et al (2005) Uncovering regulatory pathways that affect hematopoietic stem cell function using 'genetical genomics'. Nat Genet 37:225–232
9. Cheung VG, Spielman RS, Ewens KG, Weber TM, Morley M, Burdick JT (2005) Mapping determinants of human gene expression by regional and genome-wide association. Nature 437:1365–1369
10. Hubner N, Wallace CA, Zimdahl H, Petretto E, Schulz H, Maciver F, Mueller M, Hummel O, Monti J, Zidek V et al (2005) Integrated transcriptional profiling and linkage analysis for identification of genes underlying disease. Nat Genet 37:243–253
11. Mehrabian M, Allayee H, Stockton J, Lum PY, Drake TA, Castellani LW, Suh M, Armour C, Edwards S, Lamb J et al (2005) Integrating genotypic and expression data in a segregating mouse population to identify 5-lipoxygenase as a susceptibility gene for obesity and bone traits. Nat Genet 37:1224–1233
12. Schadt EE, Lamb J, Yang X, Zhu J, Edwards S, Guhathakurta D, Sieberts SK, Monks S, Reitman M, Zhang C et al (2005) An integrative genomics approach to infer causal associations between gene expression and disease. Nat Genet 37:710–717
13. Li H, Lu L, Manly KF, Chesler EJ, Bao L, Wang J, Zhou M, Williams RW, Cui Y (2005) Inferring gene transcriptional modulatory relations: a genetical genomics approach. Hum Mol Genet 14:1119–1125
14. Bao L, Wei L, Peirce J, Homayouni R, Li H, Zhou M, Chen H, Lu L, Williams R, Pfeffer L et al (2006) Combining gene expression QTL mapping and phenotypic spectrum analysis to uncover gene regulatory relations. Mamm Genome 17:575–583
15. Li H, Chen H, Bao L, Manly KF, Chesler EJ, Lu L, Wang J, Zhou M, Williams RW, Cui Y (2006) Integrative genetic analysis of transcription modules: towards filling the gap between genetic loci and inherited traits. Hum Mol Genet 15:481–492

16. Bao L, Peirce JL, Zhou M, Li H, Goldowitz D, Williams RW, Lu L, Cui Y (2007) An integrative genomics strategy for systematic characterization of genetic loci modulating phenotypes. Hum Mol Genet 16:1381–1390
17. Alberts R, Lu L, Williams R, Schughart K (2011) Genome-wide analysis of the mouse lung transcriptome reveals novel molecular gene interaction networks and cell-specific expression signatures. Respir Res 12:61
18. MacLellan WR, Wang Y, Lusis AJ (2012) Systems-based approaches to cardiovascular disease. Nat Rev Cardiol 9:172–184
19. Kadarmideen HN, Von Rohr P, Janss LLG (2006) From genetical genomics to systems genetics: potential applications in quantitative genomics and animal breeding. Mamm Genome 17:548–564
20. Sieberts SK, Schadt EE (2007) Moving toward a system genetics view of disease. Mamm Genome 18:389–401
21. Jansen RC, Nap JP (2001) Genetical genomics: the added value from segregation. Trends Genet 17:388–391
22. Ha T, Swanson D, Larouche M, Glenn R, Weeden D, Zhang P, Hamre K, Langston M, Phillips C, Song M et al (2015) CbGRiTS: cerebellar gene regulation in time and space. Dev Biol 397:18–30
23. Mulligan MK, Williams RW (2015) Systems genetics of behavior: a prelude. Curr Opin Behav Sci 2:108–115
24. van der Sijde MR, Ng A, Fu J (2014) Systems genetics: from GWAS to disease pathways. Biochim Biophys Acta 1842:1903–1909
25. Civelek M, Lusis AJ (2014) Systems genetics approaches to understand complex traits. Nat Rev Genet 15:34–48
26. Li Q, Seo J-H, Stranger B, McKenna A, Pe'er I, LaFramboise T, Brown M, Tyekucheva S, Freedman ML (2013) Integrative eQTL-based analyses reveal the biology of breast cancer risk loci. Cell 152:633–641
27. Li Q, Stram A, Chen C, Kar S, Gayther S, Pharoah P, Haiman C, Stranger B, Kraft P, Freedman ML (2014) Expression QTL-based analyses reveal candidate causal genes and loci across five tumor types. Hum Mol Genet 23:5294–5302
28. Faraji F, Hu Y, Wu G, Goldberger NE, Walker RC, Zhang J, Hunter KW (2014) An integrated systems genetics screen reveals the transcriptional structure of inherited predisposition to metastatic disease. Genome Res 24:227–240
29. Kogelman LJA, Zhernakova DV, Westra HJ, Cirera S, Fredholm M, Franke L, Kadarmideen HN (2015) An integrative systems genetics approach reveals potential causal genes and pathways related to obesity. Genome Med 7:1–15
30. Dobrin R, Zhu J, Molony C, Argman C, Parrish M, Carlson S, Allan M, Pomp D, Schadt E (2009) Multi-tissue coexpression networks reveal unexpected subnetworks associated with disease. Genome Biol 10:R55
31. Ghosh S, Vivar J, Nelson CP, Willenborg C, Segrè AV, Mäkinen VP, Nikpay M, Erdmann J, Blankenberg S, O'Donnell C et al (2015) Systems genetics analysis of genome-wide association study reveals novel associations between key biological processes and coronary artery disease. Arterioscler Thromb Vasc Biol 35:1712–1722
32. Lusis AJ, Weiss JN (2010) Cardiovascular networks: systems-based approaches to cardiovascular disease. Circulation 121:157–170
33. Andreux PA, Williams EG, Koutnikova H, Houtkooper RH, Champy MF, Henry H, Schoonjans K, Williams RW, Auwerx J (2012) Systems genetics of metabolism: the use of the BXD murine reference panel for multiscalar integration of traits. Cell 150:1287–1299
34. Taneera J, Lang S, Sharma A, Fadista J, Zhou Y, Ahlqvist E, Jonsson A, Lyssenko V, Vikman P, Hansson O et al (2012) A systems genetics approach identifies genes and pathways for type 2 diabetes in human islets. Cell Metab 16:122–134
35. Ziebarth JD, Cook MN, Wang X, Williams RW, Lu L, Cui Y (2012) Treatment- and population-dependent activity patterns of behavioral and expression QTLs. PLoS One 7, e31805
36. Palmer RHC, McGeary JE, Francazio S, Raphael BJ, Lander AD, Heath AC, Knopik VS (2012) The genetics of alcohol dependence: advancing towards systems-based approaches. Drug Alcohol Depend 125:179–191
37. Ziebarth JD, Cook MN, Li B, Williams RW, Lu L, Cui Y (2010) Biomedical sciences and engineering conference (BSEC), 2010. IEEE 2010:1–4
38. Kollmus H, Wilk E, Schughart K (2014) Systems biology and systems genetics – novel innovative approaches to study host-pathogen interactions during influenza infection. Curr Opin Virol 6:47–54
39. Miyairi I, Ziebarth J, Laxton JD, Wang X, van Rooijen N, Williams RW, Lu L, Byrne GI, Cui Y (2012) Host genetics and chlamydia disease: prediction and validation of disease severity mechanisms. PLoS One 7, e33781
40. Emery FD, Parvathareddy J, Pandey AK, Cui Y, Williams RW, Miller MA (2014) Genetic control of weight loss during pneumonic Burkholderia pseudomallei infection. Pathog Dis 71:249–264

41. Ziebarth JD, Bhattacharya A, Cui Y (2013) Bayesian Network Webserver: a comprehensive tool for biological network modeling. Bioinformatics 29:2801–2803
42. Pearl J (2000) Causality: models, reasoning, and inference. Cambridge University Press, Cambridge
43. Cui Y (2007) In: Deng HW (ed) Current topics in human genetics: studies of complex diseases. World Scientific, Singapore, pp 433–448
44. Cui Y (2006) In: Shannon F, Rao S (eds) Microarrays and transcription networks. Landes Bioscience, Georgetown, KY, pp 114–126
45. Tasaki S, Sauerwine B, Hoff B, Toyoshiba H, Gaiteri C, Chaibub Neto E (2015) Bayesian network reconstruction using systems genetics data: comparison of MCMC methods., Genetics
46. Shipley B (2000) Cause and correlation in biology. Cambridge University Press, Cambridge
47. Bøttcher SG, Dethlefsen C (2003) Deal: a package for learning bayesian networks. J Stat Softw 8:1–19
48. Tian J, He R, Ram L (2010) Bayesian model averaging using the k-best Bayesian network structures. Proc Conf Uncertain Artif Intel 2010:589–597
49. Bolouri H, Davidson EH (2002) Modeling transcriptional regulatory networks. Bioessays 24:1118–1129
50. Davidson EH (2010) Emerging properties of animal gene regulatory networks. Nature 468:911–920
51. Mitra K, Carvunis A-R, Ramesh SK, Ideker T (2013) Integrative approaches for finding modular structure in biological networks. Nat Rev Genet 14:719–732
52. Aittokallio T, Schwikowski B (2006) Graph-based methods for analysing networks in cell biology. Brief Bioinform 7:243–255
53. Bao L, Xia X, Cui Y (2010) Expression QTL modules as functional components underlying higher-order phenotypes. PLoS One 5, e14313

Chapter 16

Systems Genetics as a Tool to Identify Master Genetic Regulators in Complex Disease

Aida Moreno-Moral, Francesco Pesce, Jacques Behmoaras, and Enrico Petretto

Abstract

Systems genetics stems from systems biology and similarly employs integrative modeling approaches to describe the perturbations and phenotypic effects observed in a complex system. However, in the case of systems genetics the main source of perturbation is naturally occurring genetic variation, which can be analyzed at the systems-level to explain the observed variation in phenotypic traits. In contrast with conventional single-variant association approaches, the success of systems genetics has been in the identification of gene networks and molecular pathways that underlie complex disease. In addition, systems genetics has proven useful in the discovery of master *trans*-acting genetic regulators of functional networks and pathways, which in many cases revealed unexpected gene targets for disease. Here we detail the central components of a fully integrated systems genetics approach to complex disease, starting from assessment of genetic and gene expression variation, linking DNA sequence variation to mRNA (expression QTL mapping), gene regulatory network analysis and mapping the genetic control of regulatory networks. By summarizing a few illustrative (and successful) examples, we highlight how different data-modeling strategies can be effectively integrated in a systems genetics study.

Key words Tools and strategies for systems genetics, Expression QTL (eQTL), Gene network, Molecular pathway, *trans*-regulators, Master genetic regulator

1 Introduction and Background

Systems genetics can be generally defined as an integrative data-modeling strategy aimed to identify the molecular determinants, pathways and gene networks that underlie human disease and complex traits. Given the intrinsic "holistic" nature of biological processes underlying complex disease, systems genetics lies within the broader confines of systems biology, in which the various components of the system under investigation are considered (and analyzed) together. From an operational point of view, the systems genetics approach leverages the flow of biological information from the genetic (i.e., DNA sequence variations) and epigenetic levels to

Klaus Schughart and Robert W. Williams (eds.), *Systems Genetics: Methods and Protocols*, Methods in Molecular Biology, vol. 1488, DOI 10.1007/978-1-4939-6427-7_16, © Springer Science+Business Media New York 2017

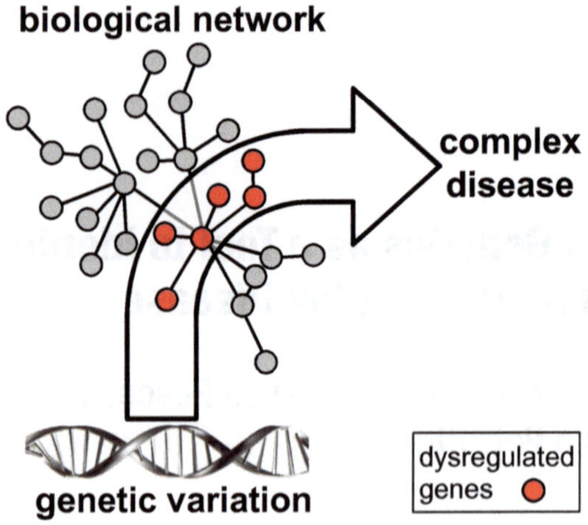

Fig. 1 The systems genetics workflow links sequentially three layers of biological information: genetic variation, biological networks, and complex disease phenotypes. In this, the systems genetics approach is aimed to identify DNA sequence variant(s) that regulate biological networks (or specific network-components, highlighted in *red*) that are dysfunctional in disease. For instance, a deleterious mutation in a transcription factor (TF) in turn alters the mRNA expression levels of the TF and TF-targets, which can result in the dysregulation of an essential transcriptional program in a disease relevant tissue. Depending on several context-specific factors, the mutation can exert its effect on specific components of the biological network, for instance acting on a subset of TF-targets that are expressed in a specific cell-type and developmental stage

the cellular (e.g., transcripts, proteins, metabolites) and whole-body phenotype levels (e.g., a clinical or physiological trait) to provide a systems-level description of disease traits. Therefore, the flow of biological information (also called the *central dogma of biology*) represents the working process of a typical systems genetics study design, as it is exemplified in Fig. 1. Assessing genetic variation, cellular level phenotypes (e.g., mRNA, protein abundance) and whole-body traits in the same system permits to link DNA sequence variants with specific components of the biological networks that are dysregulated in the disease process, Fig. 1. Depending on the analytical pipeline chosen, systems genetics studies can be carried out using "meta-dimensional" analyses, when the modeling is performed on all the data simultaneously, or "multi-staged" analyses when a stepwise, hierarchical approach is undertaken to reduce the search space progressively and identify primary genetic regulators of complex networks and pathways in disease [1].

In contrast with more traditional genetics strategies like genome-wide association studies (GWAS), the systems genetics approach can yield new insights into (1) the functional genetic and epigenetic control points of biological networks in disease, (2) the specific network components that are active (and dysregulated) during the disease process, (3) the potential point(s) of therapeutic intervention, and ultimately (4) the multifaceted molecular processes underpinning complex disease.

Here we detail the fundamental analytical steps required for a comprehensive systems genetics analysis of complex traits and disease. While this strategy can be generally applied to different cellular level phenotypes, here we focus on mRNA abundance (gene expression) and gene regulatory networks (or co-expression networks). In this, we set out to describe the following analytical Steps:

1. Assessing genetic and gene expression variation.
   - DNA sequence variation as a source of naturally occurring genetic perturbation.
   - The importance of measuring phenotypes at a cellular level.
2. Linking DNA sequence variation to mRNA: eQTL mapping.
   - *Cis*-eQTLs and *trans*-eQTLs.
   - Methods for eQTL mapping and applications.
3. Gene regulatory network analysis.
   - Main approaches for network inference.
   - Regulation of gene networks.
   - Associating networks to complex traits and disease susceptibility.
4. Mapping the genetic control of regulatory networks.
   - eQTL mapping of the network genes.
   - Genetic mapping of the network variability.
   - Genetic mapping of the network as multivariate response.

We will then review a few illustrative examples of *successful application of systems genetics approaches* to complex disease, in which different analytical strategies have been implemented in animal models and humans by analysis of:

- Gene network to annotate the function of GWAS-associated genes: Asxl2 [2].
- Transcription factors as trans-eQTLs in human GWAS: KLF14 [3].
- Genetic regulators of networks by TF-targets analysis: EBI2 [4].
- Genetic regulators of networks by genome-wide mapping of trans-eQTL clusters: Trem2 [5].
- Genetic regulators of networks by mapping their *trans*-acting genetic control: *SESN3* [6].

2 Methods

2.1 Assessing Genome-Wide Genetic and Gene Expression Variation

2.1.1 DNA Sequence Variation as a Source of Naturally Occurring Genetic Perturbation

Genetic variability is a distinctive feature of biological systems and is ultimately due to the changes in the sequence of bases in the nucleotides. These changes occur in the germ line or randomly during the replication of DNA and can also be introduced by environmental agents. Such variations characterize different alleles of genes in the gene pool both at a population and individual level. Regardless the way they are introduced, how they impact on the phenotype can be derived, at the most basic level, from the so called *central dogma of biology* [7]. In fact, this *dogma* states that the biological information (i.e., the amino acid sequence determining the genetic code) is sequentially transferred from DNA through RNA, affecting structural or functional activity of proteins and in turn more complex phenotypes. Hence, even a single nucleotide polymorphism (SNP) might have a major role in influencing the features of the main effectors of biological activity in the cell. In this regard, genetic variants can be seen as the most immediate and primary source of naturally occurring perturbation leading to more complex phenotypic traits. Genetic mapping strategies have been developed to identify variations in the DNA sequence associated with phenotypic traits. Traditionally genetic linkage and association studies, with frameworks for families or large cohorts of unrelated individuals, respectively, have been the major strategies implemented in this field. Recently, these have largely benefited from the advances in next generation sequencing (NGS) technologies gaining an unprecedented level of resolution and depth of DNA sequencing [8]. Different kinds of sequence variants can be used as genetic markers and tested in genetic linkage or association analyses, including microsatellite markers (i.e., repetitive DNA sequences of motifs), SNPs, small insertion/deletions and structural variants like inversions, translocations and copy number variants (CNV). These measurements of the DNA sequence variations have been extensively used to systematically probe the genome for variants associated (or genetically linked) with several complex phenotypes and disease traits. However, despite the high number of large-scale gene mapping efforts and the identification of major-effect genetic variants, a significant proportion of the genetic heritability of the trait (and disease) remains to be discovered—also called the "missing heritability" [9]. A possible explanation can be found in the fact that complex traits and many common diseases are determined by the interaction of small-effect genes that operate within complex molecular pathways [1]. Investigating these molecular pathways therefore requires the comprehensive and quantitative assessment of cellular level phenotypes (also called "intermediate phenotypes") in informative cellular and tissue systems.

2.1.2 The Importance of Measuring Cellular Level Phenotypes

Nowadays a relevant aid in dissecting complex traits and disease is offered by the comprehensive interrogation and quantification of intermediate cellular level phenotypes, including RNAs (*Transcriptomics*), proteins (*Proteomics*), and metabolites (*Metabolomics*). While NGS has become the technology of choice for *Transcriptomics* [10], mass spectrometry-based and nuclear magnetic resonance (NMR)-based methods are widely adopted in the rapidly evolving field of *Proteomics* [11] and *Metabolomics* [12, 13]. Moving on from single-feature analysis (e.g., single mRNA) to the investigation of genome-scale features is now possible because of the continuous improvement of high-throughput *-omics* technologies, which become cheaper and easier to use. In fact, this means narrowing down the quest for the determinants of a complex phenotype to the analysis of its underpinning "intermediate phenotypes" (e.g., proteins abundance). It is likely that the analysis of "intermediate phenotypes" might shed light on the missing heritability. For instance, the screening of "intermediate phenotypes" such as mRNAs or proteins in target cells can pinpoint which cellular components are dysregulated during the disease processes, and elucidate disease relevant interactions (e.g., protein complexes or protein–protein interactions). In contrasts with whole-body phenotypes, the quantification of cellular level phenotypes is usually more accurate, as it is possible to accumulate technical replicates and carry out orthogonal validations experiments (e.g., microarray-derived mRNA levels can be independently validated using quantitative RT-PCR). Perhaps most importantly, different "intermediate phenotypes" and data modalities (e.g., mRNA and protein abundances) can be measured within the same system where the genetic variability has been previously assessed, therefore allowing to discriminate the relative effects of DNA sequence variants on different cellular traits [14] and the consequences of regulatory variation from RNA to protein [15]. Any "intermediate phenotype" can then be investigated in relation to underlying genetic and epigenetic variations, to identify specific genomic loci of regulatory effect.

2.2 Linking DNA Sequence Variation to mRNA: eQTL Mapping

Expression quantitative trait loci or eQTLs are defined as regions of the genome that harbor sequence variants affecting the mRNA expression level of one or more genes [16]. After assessing DNA sequence variation at the genome level, each variant can be tested and statistically associated with variation in mRNA expression of a given gene. In other words, similarly to traditional quantitative trait loci (QTL), eQTLs also determine the observed variation of a trait, in this case the intermediate phenotype defined by gene expression [17]. In 2001, genome-wide eQTL mapping was initially proposed as a strategy for the identification of genes regulated by DNA sequence variants [18]. Since then, large maps of eQTLs have been generated in model organisms [19–23], human populations [24–26], tissues [27–32], specific cell-types [27, 33, 34], and in response to different stimuli [35–38].

2.2.1 Cis-eQTLs and Trans-eQTLs

Expression QTLs can be classified depending on whether the genetic regulatory effect is local (*cis*-acting) or distant (*trans*-acting) with respect of the regulated gene. *Cis*-eQTLs control the gene expression levels of a gene that is located nearby. Depending on the genetic system of study the definition of *cis*-eQTL can vary based on the recombination rate, marker density and size of the linkage disequilibrium (LD) blocks. For example, empirical simulation studies in the rat HXB/BXH recombinant inbred panel concluded with 10Mb either side from the original probe location as the optimal *cis*-eQTL window [20]. In humans, a *cis*-eQTL window is usually (and often operatively) defined as 1 Mb either side from the gene location [39]. Irrespective of the population or tissue analyzed, *cis*-eQTLs share some common features: (1) they tend to have a larger genetic effect size and are more heritable than *trans*-eQTLs [40], (2) they require less statistical power than *trans*-eQTLs to be detected at the genome-wide level, (3) they seem to display mainly additive effects [41] and (4) they are commonly located near to the transcription start site of the gene, possibly altering other *cis*-regulatory elements such as transcription factor binding sites [42]. In contrast, *trans*-eQTLs regulate the mRNA expression levels of genes that are located further away (even on a different chromosome) via an indirect effect such as, a long-range regulatory region or a protein encoded by a *trans*-acting eQTL (for instance, a transcription factor or RNA-binding protein). As in the case of *cis*-eQTLs, *trans*-eQTLs also present some communalities: (1) they usually have small effect sizes, making their genome-wide identification using small sample sizes challenging [43], (2) they appear to be more tissue-specific than *cis*-eQTLs [37] and (3) they have been observed to occur in clusters or *trans*-eQTL *hospots* [19, 44–46]. The number of reported *trans*-eQTLs so far is relatively low compared with the number of genes that have been shown to have a *cis*-eQTL (for example an eQTL study in whole blood with 922 subjects reported that around 80% of the expressed genes had a *cis*-eQTLs) [47]. Whether the different representation of *cis*- and *trans*-eQTLs reflects differences in the context specificity (e.g., cell-type [27]), the contribution of gene expression dynamics, statistical power, or a combination of these remains to be elucidated. In any case, both *cis*- and *trans*-acting eQTLs are increasingly recognized as important drivers or mediators of disease associated variants [48] and are now routinely studied in disease relevant tissues that are accessible for transcriptional analysis.

2.2.2 Methods for eQTL Mapping and Applications

Expression QTL mapping studies are carried out by measuring gene expression levels in a given tissue/cell-type using a population of samples where genetic variability is measured in parallel. Thus, two kinds of data modalities are required for eQTL analysis: genotype and gene expression [17]. Genotyping of common DNA variants is usually achieved by using SNP arrays; a technology that now allows the genotyping of about a million SNPs in an individual in a single run [49]. Initially, in eQTL studies genome-wide gene expression levels have been assessed by means of DNA microarray technology, which uses a predefined set of oligonucleotide probes (used to hybridize labeled cDNA or cRNA targets of known genes) that are immobilized on a solid support to assess intensity of expression that is measured as a fluorescent signal. However, this technology is now most frequently being replaced by whole-mRNA sequencing (RNA-seq). This is due to the fact that RNA-seq offers an improved dynamic range (more sensitivity for the detection of low expressed genes), higher resolution and true genome-wide representation (the mRNAs level quantification is not limited to the genes for which there are oligonucleotide probes available on the microarray) and also allows to map the relative ratios of different transcripts isoforms, which can be used for the detection of splicing-QTLs (sQTLs) [50, 51]. Despite these fundamental technical differences, good reproducibility has been reported between eQTLs detected by microarrays and RNA-seq analyses [52–54].

Similarly to traditional QTL mapping for clinical traits, eQTLs can be identified by either using association tests (in unrelated subjects at the population level) or linkage analysis (in the case of families or experimental crosses). Typically, a statistical test is performed for every gene mRNA and genotyped marker, depending on the study design and population (e.g., as implemented in the Matrix eQTL approach [55]). In this, for each SNP the hypothesis under testing is whether the genetic sequence variant shows any effect onto the expression level of the gene (e.g., whether gene expression is associated with a given allele). The identification of a significant eQTL suggest the presence of a regulatory variant at the locus, yet often, due to the lack of resolution of genetic variation and the potentially complex LD structure present between genetic markers, finding the actual "causal" variant remains challenging [56]. To overcome these difficulties, fully multivariate eQTL mapping methodologies that account for linkage disequilibrium have been developed (e.g., see [111]). Please refer to the chapter "*Expression QTL mapping and analysis: a Bayesian perspective*" for a detailed review of this class of multivariate eQTL mapping approaches.

In addition, there are a few approaches that can be pursued to disentangle eQTL effects in sets of SNPs, for instance by studying the local sequence conservation with the aim of finding out which regions are more prone to be regulatory and/or deleterious. Following this, eQTLs have been shown to be more prevalent in conserved regions and regulatory elements [57, 58]. Another approach is to use computational strategies to predict the consequences of DNA sequence variations on specific regulatory regions such as transcription factor binding sites [59], RNA-binding protein motifs [60] or DNA structure [61]. Mapping of eQTLs and identification of the regulatory sequence variant can aid ascertaining or prioritizing disease-causing variants. For example, integration of eQTLs with disease-associated sequence variants allows the identification of the genetic machinery that shapes a trait in a given tissue or cell-type or in response to specific stimuli [38]. Another major use of eQTLs is to pinpoint the causal variants among the set of SNPs identified by GWAS, often tagging regions encoding multiple variants in linkage disequilibrium, multiple genes or noncoding regions [17]. Several studies have now showed that GWAS hits are commonly enriched for eQTLs. These eQTLs that are co-localizing with GWAS hits are deemed as good candidates for the regulation of the complex trait, and they can give mechanistic insights into the pathogenesis of complex disease in a given tissue [62, 63].

2.3 Gene Regulatory Network Analysis

Gene network-based analyses are central to systems genetics and represent an important step in the analytical workflow (see Fig. 1) [64]. Amongst various types of biological networks, gene regulatory networks provide a representation of genes responsible for biochemical activities in the cell, where the network edges are indicating the causal relationships and type of interaction between genes [65]. A “simpler” snapshot-representation of regulatory networks is provided by gene co-expression networks. Most commonly, gene co-expression networks represent the functional relationships between genes in a given tissue and are traditionally represented as undirected graphs where genes are nodes and are connected through edges when they have a significant co-expression relationship [66]. Gene co-expression networks are (typically) undirected graphs because the causal relationships between nodes in the network are not specified, and the edges represent a correlation (association) between the genes’ expression profiles. However, the occurrence of a pattern (signature) of co-expressed genes can indicate the activation of a transcriptional program or a pathway, which can be specific to the disease process. For instance, the gene co-expression relationships can be associated with the disease state (i.e., by differential co-expression), therefore suggesting differential activation or the dysfunction of coordinated transcriptional programs in disease [67–70]. Methods to test for differential co-expression include DiffCoEx [71], or methods based on Higher-Order Generalized Singular Value Decomposition which are scalable to multiple conditions [72].

In general terms, the co-expression networks inferred from transcriptional analysis of a disease system can be *reactive*, in the context of the pathogenesis of the disease, or *causal* when they are directly involved in the etiology [73]. *Reactive* networks can reveal co-expressed genes that regulate (or are involved with) adaptive processes and might influence the progression of the disease. On the other hand, the opportunity offered by the analysis of *causal* co-expression networks is particularly interesting in the context of GWAS and whole-exome sequencing (WES) studies. In fact, co-expression networks can be used to leverage data-rich GWAS, in particular to either prioritize or annotate GWAS variants showing modest effects, and thus providing new insights into the underlying disease mechanisms. Gene co-expression network analysis has become a common strategy to annotate GWAS results, typically by testing for overrepresentation of GWAS-SNPs (including nonsignificant associations) in specific co-expression networks (see below for overview of the methods). This integrated GWAS-gene co-expression network strategy can be informative by providing the functional context for the associated variants with a given trait [74–76].

2.3.1 Main Approaches for Gene Network Inference

Starting from the quantification of gene expression levels across different samples in a given tissue system, the main aim of gene co-expression network analysis is to identify groups of genes sharing a similar expression profile. The tools used for this purpose mainly fall into two broad categories: latent factors methods [77–79] and reverse engineering approaches [71, 80]. The former are based on the assumption that the observed gene expression variability is influenced by a number of unmeasured covariates (latent factors) and the network is derived by modeling the gene expression profile as a linear combination of the latent factors estimated from the data. On the other hand, reverse engineering approaches typically model the structure of the gene co-expression networks as graphs (with nodes representing genes and edges representing co-expression relationships). Assuming a general framework where a gene expression matrix ($g \times s$) represent the expression levels of g genes in each sample s, as the first step of the network inference one can obtain a similarity matrix $g \times g$ (for instance calculating the pairwise correlations of the values of gene expression in the population). Then, by applying a threshold on the similarity matrix one can derive an adjacency matrix in which each value in $g \times g$ is either 0, when the correlation is below a given threshold, or 1, when the value in $g \times g$ is above the threshold, thus indicating that the pair of genes can be represented as two nodes connected by an edge in the network. In this process, one can use different measures of associations (e.g., Pearson's correlation, Spearman's rank correlation, Kendall, biweight correlation, mutual information, and partial correlation) and adopt experiment-specific thresholds to define statistically significant associations.

Nowadays, there is a plethora of freely available bioinformatics tools that can be used to infer gene co-expression networks from genome-wide transcriptional data. Here we describe two approaches (WGCNA and GeneNet) that are most commonly used in biomedical sciences and mention some alternative strategies. Weighted gene co-expression network analysis (WGCNA) [81] is a popular method for gene network analysis that is based on the following steps: (1) computation of similarity matrix; (2) soft-threshold selection; (3) computation of a Topological Overlap Matrix (TOM); and (4) gene clustering. While WGCNA is widely used and the software implementation (R package, http://labs.genetics.ucla.edu/horvath/CoexpressionNetwork/Rpackages/WGCNA) is supported by a helpful tutorial, the application of WGCNA in real-case scenarios requires the specification of few critical parameters. First, the choice of the type of correlation for the gene similarity measure is not trivial and might affect the network inference downstream. For instance, the Pearson correlation is not able to capture nonlinear dependencies and is not robust to outliers as it is the case for Spearman, Kendall, or biweight correlations, which are also more appropriate in case of small sample size [82]. For analysis of nonlinear relationships between gene expression profiles, mutual information-based approaches (such as the Mutual Information NETworks (MInet) implemented in a Bioconductor R package [83]) are considered more powerful approaches. The second step of WGCNA is to use a soft-thresholding procedure to assign a weight to the connection of each gene pair in the adjacency matrix [81]. The appropriate soft-threshold value has to be selected to fulfill the scale-free topology criterion, i.e., assuming a few highly connected nodes (hubs) and the majority of genes with fewer connections [84, 85]. A TOM is then computed from the adjacency matrix to capture the interconnectedness of any two genes in relation to their shared neighbors and so two genes will have high topological overlap if they share a large number of connections. Finally, a hierarchical clustering analysis of the TOM is performed and gene co-expression networks (or modules) are obtained by cutting the dendrogram at given height.

Alternative methods for gene network inference include, for example, Algorithm for the Reconstruction of Accurate Cellular Networks (ARACNe) [86], Context Likelihood of Relatedness (CLR) [87], Gaussian Graphical Modeling (GGM) and ELMM (Empirical Light Mutual Min) [88]. The ARACNe approach allows inferring gene networks by identifying triplets of genes and removing systematically the weakest associations in each triplet. Similar to ARACNe, CLR re-weights the edges in the network using an adaptive background correction to exclude (or limit the detection of) false gene-gene correlations [87].

Another commonly used approach for gene network analysis in transcriptomics is Gaussian Graphical Modeling, where the covariance of the expression for any pair of genes is assumed to follow a multivariate Gaussian distribution and modeled accordingly. In contrast with WGCNA, GGM detects gene co-expression by means of partial correlations that are detects via computationally–efficient regularization approaches [89, 90]. A practical implementation of this class of approaches can be found in the R package *GeneNet*, which is based on two main steps: (1) stable estimation of the partial correlation matrix from the gene expression dataset; (2) testing for significant partial correlations and identification of conditional independence graphs (representing the "essential" co-expression connections between genes). Since *GeneNet* is based on partial correlations, this kind of gene association metrics is not ideal to identify networks with dense regions of highly interconnected genes. To overcome this potential limitation, newer algorithms like ELMM have been developed to recover undirected conditional independence graphs by limiting the multiple testing correction issue in dense network regions [88].

2.3.2 Regulation of Gene Networks

Gene networks represent common patterns of gene co-expression, which can be explained by transcriptional co-regulation and might reflect shared molecular functions between the co-expressed genes. The transcriptional regulation exerted by transcription factors (TFs), which can also be part of the gene co-expression network, can be investigated by testing for enrichment of transcription factor binding site (TFBS) motifs of known TFs in the putative promoter sequences of the genes in the network. Several comprehensive and manually curated databases of TFBSs are now accessible for these analyses, including TRANSFAC [91] and JASPAR [92]. The investigation of TFBSs and other functional enrichments in gene networks has become routine, and these enrichment tests can be performed using several tools, including web platforms like "WEB-based GEne SeT AnaLysis Toolkit" (WebGestalt) [93], the ToppGene Suite [94], or Database for Annotation, Visualization and Integrated Discovery (DAVID) [95]. These web-based tools are also commonly used to explore the functional coherence and significance of the genes in co-expression networks using information from other public resources such as Gene Ontology (GO) and Kyoto Encyclopedia of Genes and Genomes (KEGG), micro-RNA (miRNA) targets, protein–protein interactions, disease and drug-associated genes. In addition to regulation by TF, miRNA or upstream signaling molecules, gene networks can be regulated by genetic (and epigenetic) variations. Instead of relying on prior knowledge (e.g., TFBS and miRNA-targets) to identify putative regulators of the gene co-expression networks, it is possible to scan the genome for genetic (or epigenetic) variants that explain the co-expression pattern. The search for novel genetic regulators is the last step of the systems genetics approach and is described in **Step 4** (see 2.4.1, 2.4.2 and 2.4.3).

2.3.3 Associating Gene Networks to Complex Traits and Disease Susceptibility

Once gene co-expression networks are constructed, one can ask whether these are *reactive* or *causal* (see above) and if are associated with variation in a whole-body (quantitative) phenotype. For example, one can test whether the pattern of expression or co-expression is quantitatively associated with the disease status or variation of trait of interest. This mRNA-quantitative phenotype association analysis has been initially defined as "Quantitative Trait Transcript" (QTT) analysis and it is based on straightforward correlation analysis between any gene profiled and the variation in the quantitative trait followed by multiple testing correction [96]. This approach has been successfully applied in model organisms to identify genes and *cis*-eQTLs associated with whole-body phenotypes (like cardiac mass [97]) as well as other cardiovascular traits [98].

Given that the genes in the networks are co-expressed, a QTT-based approach can start by capturing the variability of network genes using for instance principal component analysis (PCA), and then testing the correlation of the PC obtained for each sample with the quantitative (clinical) traits. An alternative, more flexible approach is provided by Gene Set Enrichment Analysis (GSEA) [99] computational framework. The advantage of this method is that it does not require specifying the hard threshold used to derive a fixed gene-list from a genome-wide analysis (e.g., differentially expressed genes). In fact, in some cases, applying a specific threshold cutoff for a *p*-value or fold-change could lead to a potential loss of biological information. Instead, GSEA tests for overrepresentation of a specific gene-set at the top or bottom of a pre-ranked list (with the null hypothesis being the genes in the gene set are randomly distributed across the ranked list). For instance, in the case of QTT-based approach for gene networks, this analysis can be performed in two steps: (1) identification of gene networks to define specific gene-sets, (2) genome-wide correlation of each gene expression level with the quantitative trait of interest to define the pre-ranked list; (3) GSEA using the pre-ranked list and the gene-sets to identify networks significantly enriched for QTT-associated genes. For each gene-set (network) GSEA returns an enrichment score and a normalized enrichment score (NES, adjusted on the gene set size) for the gene-set tested as a measure of the strength of enrichment, as well as *p*-value for significance that is corrected for multiple testing using resampling-based procedures [99].

Gene co-expression networks can be genetically associated with disease using genetic susceptibility data (e.g., GWAS and WES). For instance, in order to leverage the information from GWAS, one can test for enrichment of GWAS-associated variants (SNPs) within the genes in the co-expression networks. In fact, the coordinated activity of the genes in the co-expression networks might help to explain the combined activity of susceptibility variants with small effect, which might not be easily detected when applying a stringent significant *p*-value cutoff used in GWAS (typically $p < 10^{-8}$). There are several available tools to perform these GWAS-enrichment analyses, and these have different performance (for a comprehensive review of pathway-based approaches for GWAS, please refer to ref. [100]). Here we discuss and present the analysis pipeline for Meta-Analysis Gene-set Enrichment of variaNT Association (MAGENTA) [101] and Multi-marker Analysis of GenoMic Annotation (MAGMA) [102], two commonly used methods for GWAS-enrichment analysis. MAGENTA [https://www.broadinstitute.org/mpg/magenta/] is based on GSEA; the input consists of a table with the *p*-value of the associated variants from the GWAS and their genomic location (notably genotyping data are not necessary). The tool is based on the following steps: (1) variants (e.g., SNPs) from the GWAS are assigned to genes based on their genomic location; (2) each gene is assigned a score dependent on the *p*-value of the relative tagging SNP; (3) the score is corrected using a step-wise multiple linear regression model which takes into account a different confounding factors such as: (i) the length of the gene, (ii) LD per gene region and per Kb, and (iii) the number of recombination *hotspots* (to assess these factors data from HapMap are used as reference when genotype data are not available); (4) GSEA test is carried out to test for significant overrepresentation of GWAS associated hits in the gene set defined by the gene co-expression network. MAGMA [http://ctglab.nl/software/magma] is another method implemented with the following steps: (1) gene analysis using principal components analysis of the underlying genotype to take LD into account; the derived PCs are then used as predictors for the phenotype using a linear regression model; (2) gene-set test, MAGMA can run two different test: a "self-contained" and a "competitive" gene-set analysis. The former tests whether the gene set contains any association overall, whereas the latter tests whether the genes included the gene-set are more associated than genes outside the gene set [100]. The model implemented in the MAGMA approach is generalizable to more general gene-level linear regression models to allow for simultaneous analysis of multiple covariates and gene-sets. Both MAGENTA and MAGMA have features that make them easy to use, computationally efficient and applicable to analysis of summary GWAS statistics, i.e., they do not require providing raw genotype data and estimate important covariates (e.g., LD structure at the locus) for which they can use existing public resources. Beyond these (and many other) approaches for GWAS-enrichment analysis, we highlight that the available methods encompass a number of hypotheses about how genetic effects (in combination) contribute to disease susceptibility. These differences in the underlying assumptions are highly likely to influence the results obtained, and indeed reveal that there is no real consensus about the best way to perform GWAS-enrichment analyses (or more generally gene-set enrichment tests). A critical and useful review of these and other relevant aspects is provided by Mooney and colleagues [103]. However, although caution is required in mining and interpreting the GWAS-enrichment analysis of gene networks, these approaches provide a mean to investigate whether co-expression networks inferred from the data in an unsupervised manner, can account for molecular processes and pathways important for the genetic etiology of the disease. In the section below—*Successful application of systems genetics approaches*—we summarize a few relevant studies where GWAS-enrichment and GWAS data were effectively leveraged using network-based (and/or eQTL) approaches to identify new genes and pathways for common human disease.

2.4 Mapping the Genetic Control of Networks

2.4.1 eQTL Mapping of the Network Genes

The large number of common genetic variants found associated with diseases suggests that common disease etiology is due to perturbations in complex gene networks, rather than variations in single genes [63]. Furthermore, dysregulation of the transcriptional regulation of complex tissues have been shown to trigger modifications in complex traits [62]. Gene co-expression network analysis can be used to identify gene networks underlying disease and pinpoint major-effect genes, which can be identified as genes outside the network (for example, *trans*-acting *master genetic regulators* of the network, i.e., genes that carry sequence variants controlling the expression levels of the network) [73] or genes inside the network (for example, a dysregulated TF affecting expression of its primary gene targets). The identification of multiple overlapping *trans*-eQTLs might point to the presence of a *trans*-acting genetic regulator of complex traits, as shown in the case of *KLF14* in human adipose tissue associated with several metabolic phenotypes [3] or *Trem2* in rat macrophages associated with inflammatory disease and bone mass in vivo [5]. In both these cases (detailed below, *Successful application of systems genetics approaches*) the genome-wide mapping of *trans*-eQTLs yielded new, testable hypotheses on coordinated *trans*-regulation of gene expression, which led to the identification of the underlying genetic regulator (by definition, *trans*-acting). These studies exemplify how systematic detection (and experimental validation) of *trans*-eQTLs at the genome-wide level is possible in humans and animal models, and perhaps most importantly, how this approach can reveal unexpected *trans*-regulators of otherwise undetectable molecular processes underlying whole-body phenotypes. Other examples of integrating *trans*-eQTL analysis in a network-based framework include the identification of genes and networks driving cardiovascular and metabolic phenotypes in mice and humans [104] and of *trans*-regulated gene co-expression modules in human monocytes [105]. Several surveys of *trans*-eQTLs have highlighted the important role of these genetic regulators, including studies in twins showing that across multiple tissues a substantial proportion of gene expression heritability is *trans*-acting [106], and that these *trans*-acting factors may indirectly regulate multigene pathways [107] or might help explaining downstream effects of many trait-associated variants [48, 108].

2.4.2 Genetic Mapping of the Network Variability

To identify the genetic control points of a gene regulatory network, the variability present within the co-expression network (transcriptional level) can be mapped to the genome (DNA sequence level). This can be achieved by (1) summarizing the variability of the genes in the network by the network-eigengene, applying a dimensionality reduction approach, such as principal component analysis [6] or Bayesian factors analysis [109], followed by (2) conventional gene mapping approaches. This strategy is less computationally demanding than fully multivariate mapping of the co-expression network (see below), but, as it requires to summarize the variability within the network into a single variable (e.g., network-eigengene), this might result in a loss of information, making also harder the interpretability of the results.

2.4.3 Genetic Mapping of the Network as Multivariate Response

Alternatively, multivariate strategies that do not require dimensionality reduction of the network and instead borrow information across all the genes in the network have been developed. For instance, the mixture over markers (MOM) method by Kendziorski and colleagues [110] combines information across multiple responses (genes in the network), and achieve a better control of the false discovery rate (FDR) by thresholding the estimated posterior probabilities than traditional univariate testing of each possible gene-genetic marker pair, without sacrificing power. However, MOM is not fully multivariate in the predictor-space, as it does not account for multiple effects of several markers on each expression trait. A second competitive strategy is based on a regression analysis of large number of responses (genes in the network) and predictors (genome-wide SNPs), providing a mean to estimate the propensity of a genetic marker to influence several gene expression traits at the same time, based on a hierarchical formulation of related regressions [111]. The hierarchical regression model is implemented in a fully Bayesian framework, using a stochastic search algorithm that efficiently probes (sparse) subsets of genetic markers in a high-dimensional data matrix to identify *hotspots* for the regulation of all network genes (here considered as a complex multivariate trait). Among the methodologies developed so far for this (*see* Chapter 8) Bayesian variable selection strategies have shown increased power for identification of *hotspots* power for identification of hotspots for the regulation of the gene network [111, 112]. This class of approaches has been further developed to allow the mapping of genetic regulators of networks simultaneously across multiple tissues [113]. An application of these Bayesian modeling strategies based on (1) network-dimensionality reduction and genetic mapping followed by (2) fully multivariate Bayesian mapping of the genetic control points of networks is summarized below (*Successful application of systems genetics approaches—Genetic regulators of networks by mapping their trans-acting genetic control*).

The three major strategies for mapping the genetic control of co-expression networks and identify *master genetic regulators* are graphically summarized in Fig. 2.

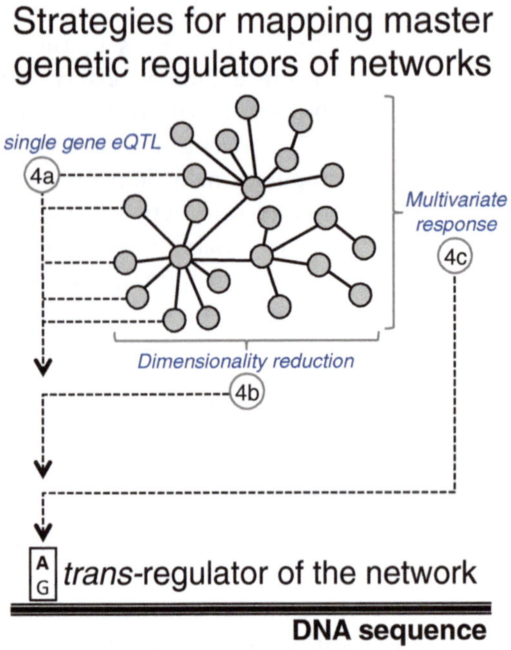

Fig. 2 Strategies for mapping *trans*-regulators of co-expression networks by means of: (4a) Detection of clusters of *trans*-eQTLs by mapping eQTL for each gene in the co-expression network (examples: *KLF14* [3], *Trem2* [5]); (4b) Summarization of the network co-expression by dimensionality reduction techniques and mapping the extracted feature (e.g., network eigengene) to the genome (example: *SESN3* [6]); (4c) Joint modeling of all network genes as a multivariate response followed by genetic mapping (example: *SESN3* [6])

3 Successful Application of Systems Genetics Approaches

For illustrative purposes, we focus on study designs where the systems genetics approach has been employed in model organisms and/or humans to identify "master genetic regulators," networks and new candidate genes for complex traits and disease. We prioritized studies where the findings have been translated (or were directly relevant) to complex human disease, and where different, often orthogonal, analytical strategies have been successfully implemented.

3.1 Gene Network to Annotate the Function of GWAS-Associated Variants

GWAS identified many gene variants associated with complex phenotypes; however, their functions often remain poorly understood. In their study, Farber and colleagues integrated GWAS and systems genetics in the Hybrid Mouse Diversity Panel (HMDP) to identify and functionally characterize novel genes for bone mineral density (BMD) [2]. The authors started by GWAS of BMD in the HMDP and then used (1) bone eQTL analysis of the GWAS-associated genes and (2) gene co-expression network analysis to pinpoint functional candidates, i.e., identify the genes that were the most likely causal for the association. Using functional annotation analysis of the associated variants (SNPs), the authors first prioritized additional sex combs like-2 (*Asxl2*) as the gene responsible for the BMD association detected on chromosome 12. This inference was further strengthened by the observation that *Asxl2* knockout mice had reduced BMD. Farber and colleagues went on and carried out gene co-expression analysis of cortical bone transcriptomic data by WGCNA [81], which identified a gene co-expression network that contained *Asxl2* along with 1334 other genes. Analysis of the network genes suggested that *Asxl2* was most closely connected with genes involved in myeloid cell differentiation and that *Asxl2* was involved in the differentiation of bone-resorbing osteoclasts. Additional RNA interference experiments showed that osteoclastogenesis was impaired in bone marrow macrophages in which *Asxl2* expression was reduced, therefore supporting the network-guided inference of the gene function. While in this case eQTL analysis did not provide the primary mean to prioritize new genes, these studies showed the power of network approaches to aid in the functional annotation of GWAS-associated genes of previously unknown function [2].

3.2 Transcription Factors as Trans-eQTLs in Human GWAS

Cis- and *trans*-eQTL analyses allowed the identification of Krüppel-like factor 14 (*KLF14*), a maternally imprinted gene, as a major regulator of multiple metabolic traits [3], thus proving the usefulness of the eQTL approach in dissecting complex phenotypes. In this study, Small and colleagues used gene expression data from the Multiple Tissue Human Expression Resource (MuTHER) consortium [28], derived from adipose tissue. The authors found that common genetic variants close to the *KLF14* gene, initially found to be associated with type 2 diabetes (T2D) [114] and with high-density lipoprotein (HDL) cholesterol levels [115] in a GWAS, were controlling the adipose tissue expression of the gene in *cis*. This suggested a causal role for *KLF14* at the locus and its implication in regulating the phenotypes tested in the GWAS. Since *KLF14* is a TF and controls the expression level of other genes in *trans*, the authors hypothesized that the GWAS SNPs were not only controlling the expression of *KLF14* but also the expression of known *KLF14*-target genes. In fact, they showed that a significant proportion of the *trans*-eQTL genes mapping to the *KLF14* locus were actually enriched for putative KLF-binding sites. Moreover, variants located nearby a subset of these *trans*-eQTLs showed also a direct association with several metabolic phenotypes. These variants, on the other hand, were highly correlated with the expression of the identified *trans*-eQTL genes for the locus in adipose tissue. This study showed that the implementation of *cis*- and *trans*-eQTL analyses to build on the GWAS is a powerful approach to provide additional functional information on the regulation of complex traits, which in this case has been used to identify a *KLF14*-associated regulatory network and reveal its involvement in regulating metabolic traits in human adipose tissue.

3.3 Genetic Regulators of Networks by TF-Targets Analysis

Analysis of co-expression networks by means of TF and TF-targets analysis led to the identification of Epstein–Barr virus induced gene 2 (*EBI2*) as a "master genetic regulator" of an antiviral gene network associated with type 1 diabetes (T1D) risk [4]. In this study, Heinig and colleagues pursued a cross-species systems genetics approach in which genome-wide co-expression network analysis in seven rat tissues yielded a network of co-expressed genes enriched for inflammatory genes and targets of the Interferon Regulatory Factor 7 (*IRF7*), a key transcription factor for type-I interferon-dependent immune responses [116]. The network was initially identified by integrated genome-wide analysis of eQTLs in TF and TFs-targets, which pinpointed to significant *trans*-regulation of *IRF7* gene expression and of many primary targets of *IRF7*. Then, by using Bayesian eQTL mapping approaches [117], the authors identified a common regulatory *hotspot* for the *IRF7*-driven network (iDIN) at the rat *Ebi2* gene, which was experimentally validated in primary rat macrophage cell culture experiments. Furthermore, the authors translated the findings to humans and reported that the human ortholog *EBI2* gene was co-localizing with a T1D-susceptibility locus identified by GWAS. By analyzing human monocytes from the Gutenberg Heart and Cardiogenics Study cohorts, they found that iDIN was also significantly conserved in human immune cells, and the iDIN genes in the ortholog human network were overrepresented for GWAS susceptibility variants for T1D. Taken together, the results reported in these studies showed how cross-species integration of eQTL/gene co-expression with transcription factors analysis can aid in the identification of novel gene networks and disease-susceptibility genes in human complex disease.

3.4 Genetic Regulators of Networks by Genome-Wide Mapping of Trans-eQTL Clusters

Using eQTL analysis in multinucleating macrophages, Kang and colleagues discovered a *trans*-regulated gene network associated with macrophage multinucleation and *Trem2* as its primary *trans*-acting regulator [5]. Cell multinucleation has been shown to be a distinctive feature of bone marrow derived macrophages in the Wistar-Kyoto rat [118]. Therefore, by taking advantage of rat strain-specific phenotypic differences in macrophage multinucleation, Kang and colleagues conducted genome-wide expression analysis in primary macrophages from 200 backcross rats showing a broad range of variability in their multinucleation [5]. Using multivariate Bayesian regression approaches for eQTL mapping [111], the authors identified 2357 eQTLs with the majority of transcripts being *cis*-eQTLs (67%). Importantly, this study identified a large *trans*-eQTL *hotspot* that mapped to the *Trem* family genes on rat chromosome 9, which regulated in *trans* the expression of 190 transcripts. These 190 transcripts formed a gene co-expression network enriched for genes regulating osteoclasts, which are multinucleating macrophages of the bone. The authors showed that at the *trans*-eQTL *hotspot* the *Trem2* gene regulated the macrophage multinucleation network (MMnet), which was consistent between rat and human macrophages. Further studies focused on the most significant *Trem2*-regulated *trans*-eQTL in the MMnet, *Kcnn4*, encoding an intermediate-conductance calcium-activated potassium channel, which the authors found to be implicated in macrophage multinucleation, bone homeostasis and inflammatory arthritis and glomerulonephritis [5]. Beyond the identification of *Trem2* as the master regulator of the *trans*-eQTL cluster, these studies reveal a complex gene network underlying macrophage multinucleation and pinpoint towards new regulators of this process (i.e., *Kcnn4* and other genes as part of the MMnet). The study provides new insights into the molecular pathways implicated in macrophage multinucleation, revealing a new potential therapeutic target for inhibition of bone resorption and chronic inflammation.

3.5 Genetic Regulators of Networks by Mapping Their Trans-Acting Genetic Control

Analysis of gene co-expression networks in the human hippocampus is limited by the lack of matched, good quality control tissue from healthy subjects. Starting from gene co-expression network analysis in 129 hippocampi form temporal lobe (TLE) patients, Johnson and colleagues [6] identified a large hippocampal gene co-expression network that was enriched for GWAS-susceptibility variants for epilepsy and for genes encoding proconvulsive cytokines and Toll-like receptor signaling. Since the Toll-like receptor pathway contributes to generating and perpetuating seizures in humans [119], the authors investigated the underlying genetic regulation of this pathway in the human hippocampus. Rather than mapping individual eQTLs for the network, genome-wide Bayesian mapping approaches were used to pinpoint the genetic regulation of the network as a whole, using a two steps strategy (see 2.4.2 and 2.4.3) [111]. By integrating PC analysis of the network (**step 1**) and fully multivariate Bayesian mapping (**step 2**), the authors therefore identified a single locus encoding the Sestrin-3 (*SESN3*) gene as a major positive regulator of the TLE-network in human hippocampus. As SESN3 is a member of the Sestrin family of proteins that have been shown to decrease intracellular reactive oxygen species and to confer resistance to oxidative stress [120], the authors hypothesized that *SESN3* might regulate neuro-inflammatory molecules (the TLE-network) through modulation of oxidative stress in the brain. To validate the role of SESN3 in the regulation of the epileptic-gene network and epileptic seizures the authors went to two distinct models: an epileptic mouse pilocarpine model [121] and a zebrafish model of convulsant-induced seizures [122]. The authors demonstrated that both the TLE-hippocampal network and its the positive regulation by *SENS3* were conserved and consistently reproduced in these experimental systems, therefore revealing an unexpected role for *SESN3* in regulating proinflammatory cytokines and their downstream effect on the central nervous system excitability and seizure susceptibility.

4 Further Considerations, Limitations and Outlook

One of the main successes of systems genetics has been the functional annotation of GWAS variants, which helped in predicting causal genetic pathways and moving away from the single-disease gene paradigm in complex disease [3, 4, 48, 62]. On the other side, the often critical constraint of accessing informative tissues (like in the case of brain disorders [6]) and issues related to cell-type specificity and/or time-dependency of the relevant regulatory processes occurring in disease, can make the application of the systems genetics strategy and interpretation of the results challenging. Another limitation is related to the accurate acquisition of high-dimensional "intermediate phenotype" data at a cost (and resolution) that allows carrying out an adequately powered study. It is not surprising that the first eQTL proposal [18] and pioneering fully integrated systems genetics [4] studies have been carried out in model organisms and translated to humans, since in these instances access to relevant target tissues was possible. In addition, in these cases systems genetics approaches were feasible by using a (relatively) small-size population in a model organism (e.g., mice and rats), being these studies also benefited from the fact that environmental components in model organisms can be better controlled and assessed than in humans.

As it happened over the past few years, the genomics technology constantly accelerates, and it is reasonable to expect that the coverage, resolution, and specificity of the -omics data generated will equally improve. Therefore, new concepts such as single-cell transcriptomics have started to emerge and change our understanding of the gene networks and complex regulatory mechanisms at the level of individual cells. For instance, in-depth and high-resolution single-cell analyses permit to disentangle the effects of gene variants and networks on cell-to-cell variability and temporal dynamics dependence in gene expression [123]. While these single-cells studies are still at their infancy (and to date have not be employed in full for systems genetics applications), single-cell network analyses are for example starting to reveal coordinated cell-to-cell processes underlying complex phenotypes and disease [14, 124–126]. Following the "mass production" of transcriptional [127] and genome sequencing profiling data at the single-cell resolution [128] in the same system, it is therefore possible to foresee the rise of a fully integrated systems genetics strategy at the single-cell level. This might allow the study the genetic processes and regulatory networks underlying disease by accounting for the variability due to heterogeneity, genomic diversity and stochastic gene expression across cells.

One important lesson coming out of the systems genetics approach is that the integration of different data-modalities (e.g., DNA-sequencing and transcriptomics) and analytical strategies (e.g., gene network and integrated GWAS analyses) can help to propose novel hypotheses on the complex regulation of patho-physiological traits, which, following experimental validation, will improve our understanding of the gene functions as well as the key genetic interactions occurring in disease. Perhaps more attracting is the application of systems genetics strategies to bridge the gap between the massive-scale genetics/genomics data generation [129] and efficient gene-target identification to improve or develop new drugs for the effective treatment.

Acknowledgments

We acknowledge funding from the British Heart Foundation (Ph.D. Studentship grant FS/11/25/28740; E.P. and A.M.M.), the European Union FP7 (ERG-239158, CardioNeT-ITN-289600, F.P.), Kidney Research UK (RP9/2013) (J.B.) and Medical Research Council Grant MR/M004716/1 (J.B. and E.P.) and Duke-NUS Graduate Medical School Singapore (E.P.).

References

1. Ritchie MD, Holzinger ER, Li R et al (2015) Methods of integrating data to uncover genotype–phenotype interactions. Nat Rev Genet 16:85–97. doi:10.1038/nrg3868
2. Farber CR, Bennett BJ, Orozco L et al (2011) Mouse genome-wide association and systems genetics identify Asxl2 as a regulator of bone mineral density and osteoclastogenesis. PLoS Genet 7:e1002038. doi:10.1371/journal.pgen.1002038
3. Small KS, Hedman AK, Grundberg E et al (2011) Identification of an imprinted master trans regulator at the KLF14 locus related to multiple metabolic phenotypes. Nat Genet 43:561–564. doi:10.1038/ng.833
4. Heinig M, Petretto E, Wallace C et al (2010) A trans-acting locus regulates an anti-viral expression network and type 1 diabetes risk. Nature 467:460–464. doi:10.1038/nature09386
5. Kang H, Kerloc'h A, Rotival M et al (2014) Kcnn4 is a regulator of macrophage multinucleation in bone homeostasis and inflammatory disease. Cell Rep 8:1210–1224. doi:10.1016/j.celrep.2014.07.032
6. Johnson MR, Behmoaras J, Bottolo L et al (2015) Systems genetics identifies Sestrin 3 as a regulator of a proconvulsant gene network in human epileptic hippocampus. Nat Commun 6:6031. doi:10.1038/ncomms7031
7. Crick F (1970) Central dogma of molecular biology. Nature 227:561–563
8. Ott J, Wang J, Leal SM (2015) Genetic linkage analysis in the age of whole-genome sequencing. Nat Rev Genet 16:275–284. doi:10.1038/nrg3908
9. Tenesa A, Haley CS (2013) The heritability of human disease: estimation, uses and abuses. Nat Rev Genet 14:139–149. doi:10.1038/nrg3377
10. Wang Z, Gerstein M, Snyder M (2009) RNA-Seq: a revolutionary tool for transcriptomics. Nat Rev Genet 10:57–63. doi:10.1038/nrg2484
11. Boersema PJ, Kahraman A, Picotti P (2015) Proteomics beyond large-scale protein expression analysis. Curr Opin Biotechnol 34C:162–170. doi:10.1016/j.copbio.2015.01.005
12. Fuhrer T, Zamboni N (2015) High-throughput discovery metabolomics. Curr Opin Biotechnol 31:73–78. doi:10.1016/j.copbio.2014.08.006
13. Ramautar R, Berger R, van der Greef J, Hankemeier T (2013) Human metabolomics: strategies to understand biology. Curr Opin Chem Biol 17:841–846. doi:10.1016/j.cbpa.2013.06.015

14. Albert FW, Treusch S, Shockley AH et al (2014) Genetics of single-cell protein abundance variation in large yeast populations. Nature 506:494–497. doi:10.1038/nature12904
15. Battle A, Khan Z, Wang SH et al (2014) Impact of regulatory variation from RNA to protein. Science 347:664–667. doi:10.1126/science.1260793
16. Bryois J, Buil A, Evans DM et al (2014) Cis and trans effects of human genomic variants on gene expression. PLoS Genet 10:e1004461. doi:10.1371/journal.pgen.1004461
17. Albert FW, Kruglyak L (2015) The role of regulatory variation in complex traits and disease. Nat Rev Genet 16:197–212. doi:10.1038/nrg3891
18. Jansen RC, Nap JP (2001) Genetical genomics: the added value from segregation. Trends Genet 17:388–391
19. Schadt EE, Monks SA, Drake TA et al (2003) Genetics of gene expression surveyed in maize, mouse and man. Nature 422:297–302. doi:10.1038/nature01434
20. Hubner N, Wallace CA, Zimdahl H et al (2005) Integrated transcriptional profiling and linkage analysis for identification of genes underlying disease. Nat Genet 37:243–253. doi: 10.1038/ng1522
21. King EG, Sanderson BJ, McNeil CL et al (2014) Genetic dissection of the Drosophila melanogaster female head transcriptome reveals widespread allelic heterogeneity. PLoS Genet 10:e1004322. doi:10.1371/journal.pgen.1004322
22. Fu J, Cheng Y, Linghu J et al (2013) RNA sequencing reveals the complex regulatory network in the maize kernel. Nat Commun 4:2832. doi:10.1038/ncomms3832
23. Rockman MV, Skrovanek SS, Kruglyak L (2010) Selection at linked sites shapes heritable phenotypic variation in C. elegans. Science 330:372–376. doi:10.1126/science.1194208
24. Stranger BE, Montgomery SB, Dimas AS et al (2012) Patterns of cis regulatory variation in diverse human populations. PLoS Genet 8:e1004322. doi:10.1371/journal.pgen.1002639
25. Spielman RS, Bastone LA, Burdick JT et al (2007) Common genetic variants account for differences in gene expression among ethnic groups. Nat Genet 39:226–231. doi:10.1038/ng1955
26. Storey JD, Madeoy J, Strout JL et al (2007) Gene-expression variation within and among human populations. Am J Hum Genet 80:502–509. doi:10.1086/512017
27. Dimas AS, Deutsch S, Stranger BE et al (2009) Common regulatory variation impacts gene expression in a cell type-dependent manner. Science 325:1246–1250. doi:10.1126/science.1174148
28. Nica AC, Parts L, Glass D et al (2011) The architecture of gene regulatory variation across multiple human tissues: the MuTHER study. PLoS Genet 7:e1002003. doi:10.1371/journal.pgen.1002003
29. Myers AJ, Gibbs JR, Webster JA et al (2007) A survey of genetic human cortical gene expression. Nat Genet 39:1494–1499. doi:10.1038/ng.2007.16
30. Emilsson V, Thorleifsson G, Zhang B et al (2008) Genetics of gene expression and its effect on disease. Nature 452:423–428. doi:10.1038/nature06758
31. Schadt EE, Molony C, Chudin E et al (2008) Mapping the genetic architecture of gene expression in human liver. PLoS Biol 6:e107. doi:10.1371/journal.pbio.0060107
32. Koopmann TT, Adriaens ME, Moerland PD et al (2014) Genome-wide identification of expression quantitative trait loci (eQTLs) in human heart. PLoS One 9:e97380. doi:10.1371/journal.pone.0097380
33. Lee MN, Ye C, Villani A-C et al (2014) Common genetic variants modulate pathogen-sensing responses in human dendritic cells. Science 343:1246980. doi:10.1126/science.1246980
34. Ye CJ, Feng T, Kwon H-K et al (2014) Intersection of population variation and autoimmunity genetics in human T cell activation. Science 345:1254665. doi:10.1126/science.1254665
35. Barreiro LB, Tailleux L, Pai AA et al (2012) Deciphering the genetic architecture of variation in the immune response to Mycobacterium tuberculosis infection. Proc Natl Acad Sci U S A 109:1204–1209. doi:10.1073/pnas.1115761109
36. Grundberg E, Adoue V, Kwan T et al (2011) Global analysis of the impact of environmental perturbation on cis-regulation of gene expression. PLoS Genet 7:e1001279. doi:10.1371/journal.pgen.1001279
37. Fairfax BP, Makino S, Radhakrishnan J et al (2012) Genetics of gene expression in primary immune cells identifies cell type-specific master regulators and roles of HLA alleles. Nat Genet 44:502–510. doi:10.1038/ng.2205
38. Fairfax BP, Humburg P, Makino S et al (2014) Innate immune activity conditions the effect of regulatory variants upon monocyte gene expression. Science 343:1246949. doi:10.1126/science.1246949

39. Stranger BE, Nica AC, Forrest MS et al (2007) Population genomics of human gene expression. Nat Genet 39:1217–1224. doi:10.1038/ng2142
40. Petretto E, Mangion J, Dickens NJ et al (2006) Heritability and tissue specificity of expression quantitative trait loci. PLoS Genet 2:e172. doi:10.1371/journal.pgen.0020172
41. Powell JE, Henders AK, McRae AF et al (2013) Congruence of additive and non-additive effects on gene expression estimated from pedigree and SNP data. PLoS Genet 9:e1003502. doi:10.1371/journal.pgen.1003502
42. Westra H-J, Franke L (2014) From genome to function by studying eQTLs. Biochim Biophys Acta 1842:1896–1902. doi:10.1016/j.bbadis.2014.04.024
43. Breitling R, Li Y, Tesson BM et al (2008) Genetical genomics: spotlight on QTL hotspots. PLoS Genet 4:e1000232. doi:10.1371/journal.pgen.1000232
44. Morley M, Molony CM, Weber TM et al (2004) Genetic analysis of genome-wide variation in human gene expression. Nature 430:743–747. doi:10.1038/nature02797
45. Brem RB, Yvert G, Clinton R, Kruglyak L (2002) Genetic dissection of transcriptional regulation in budding yeast. Science 296:752–755. doi:10.1126/science.1069516
46. Kirsten H, Al-Hasani H, Holdt L et al (2015) Dissecting the genetics of the human transcriptome identifies novel trait-related trans-eQTLs and corroborates the regulatory relevance of non-protein coding loci. Hum Mol Genet 24:4746–4763. doi:10.1093/hmg/ddv194
47. Battle A, Mostafavi S, Zhu X et al (2014) Characterizing the genetic basis of transcriptome diversity through RNA-sequencing of 922 individuals. Genome Res 24:14–24. doi:10.1101/gr.155192.113
48. Westra H-J, Peters MJ, Esko T et al (2013) Systematic identification of trans eQTLs as putative drivers of known disease associations. Nat Genet 45:1238–1243. doi:10.1038/ng.2756
49. LaFramboise T (2009) Single nucleotide polymorphism arrays: a decade of biological, computational and technological advances. Nucleic Acids Res 37:4181–4193. doi:10.1093/nar/gkp552
50. Chen L (2013) Statistical and computational methods for high-throughput sequencing data analysis of alternative splicing. Stat Biosci 5:138–155. doi:10.1007/s12561-012-9064-7
51. Monlong J, Calvo M, Ferreira PG, Guigó R (2014) Identification of genetic variants associated with alternative splicing using sQTLseekeR. Nat Commun 5:4698. doi:10.1038/ncomms5698
52. Pickrell JK, Marioni JC, Pai AA et al (2010) Understanding mechanisms underlying human gene expression variation with RNA sequencing. Nature 464:768–772. doi:10.1038/nature08872
53. Montgomery SB, Sammeth M, Gutierrez-Arcelus M et al (2010) Transcriptome genetics using second generation sequencing in a Caucasian population. Nature 464:773–777. doi:10.1038/nature08903
54. Ardlie KG, Deluca DS, Segre AV et al (2015) The genotype-tissue expression (GTEx) pilot analysis: multitissue gene regulation in humans. Science 348:648–660. doi:10.1126/science.1262110
55. Shabalin AA (2012) Matrix eQTL: ultra fast eQTL analysis via large matrix operations. Bioinformatics 28:1353–1358. doi:10.1093/bioinformatics/bts163
56. Battle A, Montgomery SB (2014) Determining causality and consequence of expression quantitative trait loci. Hum Genet 133:727–735. doi:10.1007/s00439-014-1446-0
57. Kudaravalli S, Veyrieras J-B, Stranger BE et al (2009) Gene expression levels are a target of recent natural selection in the human genome. Mol Biol Evol 26:649–658. doi:10.1093/molbev/msn289
58. Pennacchio LA, Ahituv N, Moses AM et al (2006) In vivo enhancer analysis of human conserved non-coding sequences. Nature 444:499–502. doi:10.1038/nature05295
59. Jolma A, Yan J, Whitington T et al (2013) DNA-binding specificities of human transcription factors. Cell 152:327–339. doi:10.1016/j.cell.2012.12.009
60. Li X, Quon G, Lipshitz HD, Morris Q (2010) Predicting in vivo binding sites of RNA-binding proteins using mRNA secondary structure. RNA 16:1096–1107. doi:10.1261/rna.2017210
61. Zhou T, Yang L, Lu Y et al (2013) DNAshape: a method for the high-throughput prediction of DNA structural features on a genomic scale. Nucleic Acids Res 41:W56–W62. doi:10.1093/nar/gkt437
62. Nicolae DL, Gamazon E, Zhang W et al (2010) Trait-associated SNPs are more likely to be eQTLs: annotation to enhance discovery from GWAS. PLoS Genet 6:e1000888. doi:10.1371/journal.pgen.1000888
63. Nica AC, Dermitzakis ET (2013) Expression quantitative trait loci: present and future. Philos Trans R Soc Lond B Biol Sci 368:20120362. doi:10.1098/rstb.2012.0362

64. Civelek M, Lusis AJ (2013) Systems genetics approaches to understand complex traits. Nat Rev Genet 15:34–48. doi:10.1038/nrg3575
65. De Smet R, Marchal K (2010) Advantages and limitations of current network inference methods. Nat Rev Microbiol 8:717–729. doi:10.1038/nrmicro2419
66. Stuart JM, Segal E, Koller D, Kim SK (2003) A gene-coexpression network for global discovery of conserved genetic modules. Science 302:249–255. doi:10.1126/science.1087447
67. Kim S, Hwang Y, Webster MJ, Lee D (2015) Differential activation of immune/inflammatory response-related co-expression modules in the hippocampus across the major psychiatric disorders. Mol Psychiatry. doi:10.1038/mp.2015.79
68. Wang K, Zhao L, Liu X et al (2014) Differential co-expression analysis of rheumatoid arthritis with microarray data. Mol Med Rep 10:2421–2426
69. Amar D, Safer H, Shamir R (2013) Dissection of regulatory networks that are altered in disease via differential co-expression. PLoS Comput Biol 9:e1002955. doi:10.1371/journal.pcbi.1002955
70. Min JL, Nicholson G, Halgrimsdottir I et al (2012) Coexpression network analysis in abdominal and gluteal adipose tissue reveals regulatory genetic loci for metabolic syndrome and related phenotypes. PLoS Genet 8:e1002505. doi:10.1371/journal.pgen.1002505
71. Tesson B, Breitling R, Jansen R (2010) DiffCoEx: a simple and sensitive method to find differentially coexpressed gene modules. BMC Bioinformatics 11:497. doi:10.1186/1471-2105-11-497
72. Xiao X, Moreno-Moral A, Rotival M et al (2014) Multi-tissue analysis of co-expression networks by higher-order generalized singular value decomposition identifies functionally coherent transcriptional modules. PLoS Genet 10:e1004006. doi:10.1371/journal.pgen.1004006
73. Rotival M, Petretto E (2014) Leveraging gene co-expression networks to pinpoint the regulation of complex traits and disease, with a focus on cardiovascular traits. Brief Funct Genomics 13:66–78. doi:10.1093/bfgp/elt030
74. Pérez-Palma E, Bustos BI, Villamán CF et al (2014) Overrepresentation of glutamate signaling in Alzheimer's disease: network-based pathway enrichment using meta-analysis of genome-wide association studies. PLoS One 9:e95413. doi:10.1371/journal.pone.0095413
75. Mercader JM, Puiggros M, Segrè AV et al (2012) Identification of novel type 2 diabetes candidate genes involved in the crosstalk between the mitochondrial and the insulin signaling systems. PLoS Genet 8:e1003046. doi:10.1371/journal.pgen.1003046
76. Voineagu I, Wang X, Johnston P et al (2011) Transcriptomic analysis of autistic brain reveals convergent molecular pathology. Nature 474:380–384. doi:10.1038/nature10110
77. Liebermeister W (2002) Linear modes of gene expression determined by independent component analysis. Bioinformatics 18:51–60
78. Lee DD, Seung HS (1999) Learning the parts of objects by non-negative matrix factorization. Nature 401:788–791. doi:10.1038/44565
79. Alter O, Brown PO, Botstein D (2000) Singular value decomposition for genome-wide expression data processing and modeling. Proc Natl Acad Sci U S A 97:10101–10106
80. Langfelder P, Horvath S (2008) WGCNA: an R package for weighted correlation network analysis. BMC Bioinformatics 9:559. doi:10.1186/1471-2105-9-559
81. Zhang B, Horvath S (2005) A general framework for weighted gene co-expression network analysis. Stat Appl Genet Mol Biol 4:Article17
82. Hardin J, Mitani A, Hicks L, VanKoten B (2007) A robust measure of correlation between two genes on a microarray. BMC Bioinformatics 8:220. doi:10.1186/1471-2105-8-220
83. Meyer PE, Lafitte F, Bontempi G (2008) minet: A R/bioconductor package for inferring large transcriptional networks using mutual information. BMC Bioinformatics 9:461. doi:10.1186/1471-2105-9-461
84. Barabási A-L, Gulbahce N, Loscalzo J (2011) Network medicine: a network-based approach to human disease. Nat Rev Genet 12:56–68. doi:10.1038/nrg2918
85. Carlson MRJ, Zhang B, Fang Z et al (2006) Gene connectivity, function, and sequence conservation: predictions from modular yeast co-expression networks. BMC Genomics 7:40. doi:10.1186/1471-2164-7-40
86. Margolin AA, Nemenman I, Basso K et al (2006) ARACNE: an algorithm for the reconstruction of gene regulatory networks in a mammalian cellular context. BMC Bioinformatics 7(Suppl 1):S7. doi:10.1186/1471-2105-7-S1-S7
87. Faith JJ, Hayete B, Thaden JT et al (2007) Large-scale mapping and validation of Escherichia coli transcriptional regulation from a compendium of expression profiles. PLoS Biol 5:e8. doi:10.1371/journal.pbio.0050008
88. Mahdi R, Madduri AS, Wang G et al (2012) Empirical Bayes conditional independence graphs for regulatory network recovery.

Bioinformatics 28:2029–2036. doi:10.1093/bioinformatics/bts312

89. Friedman J, Hastie T, Tibshirani R (2008) Sparse inverse covariance estimation with the graphical lasso. Biostatistics 9:432–441. doi:10.1093/biostatistics/kxm045
90. Schäfer J, Strimmer K (2005) An empirical Bayes approach to inferring large-scale gene association networks. Bioinformatics 21:754–764. doi:10.1093/bioinformatics/bti062
91. Matys V, Kel-Margoulis OV, Fricke E et al (2006) TRANSFAC and its module TRANSCompel: transcriptional gene regulation in eukaryotes. Nucleic Acids Res 34:D108–D110. doi:10.1093/nar/gkj143
92. Mathelier A, Zhao X, Zhang AW et al (2014) JASPAR 2014: an extensively expanded and updated open-access database of transcription factor binding profiles. Nucleic Acids Res 42:D142–D147. doi:10.1093/nar/gkt997
93. Wang J, Duncan D, Shi Z, Zhang B (2013) WEB-based GEne SeT AnaLysis Toolkit (WebGestalt): update 2013. Nucleic Acids Res 41:W77–W83. doi:10.1093/nar/gkt439
94. Chen J, Bardes EE, Aronow BJ, Jegga AG (2009) ToppGene Suite for gene list enrichment analysis and candidate gene prioritization. Nucleic Acids Res 37:W305–W311. doi:10.1093/nar/gkp427
95. Huang DW, Sherman BT, Lempicki RA (2009) Systematic and integrative analysis of large gene lists using DAVID bioinformatics resources. Nat Protoc 4:44–57. doi:10.1038/nprot.2008.211
96. Mackay TFC, Stone EA, Ayroles JF (2009) The genetics of quantitative traits: challenges and prospects. Nat Rev Genet 10:565–577. doi:10.1038/nrg2612
97. Petretto E, Sarwar R, Grieve I et al (2008) Integrated genomic approaches implicate osteoglycin (Ogn) in the regulation of left ventricular mass. Nat Genet 40:546–552. doi:10.1038/ng.134
98. Morrissey C, Grieve IC, Heinig M et al (2011) Integrated genomic approaches to identification of candidate genes underlying metabolic and cardiovascular phenotypes in the spontaneously hypertensive rat. Physiol Genomics 43:1207–1218
99. Subramanian A, Tamayo P, Mootha VK et al (2005) Gene set enrichment analysis: a knowledge-based approach for interpreting genome-wide expression profiles. Proc Natl Acad Sci U S A 102:15545–15550. doi:10.1073/pnas.0506580102
100. Wang K, Li M, Hakonarson H (2010) Analysing biological pathways in genome-wide association studies. Nat Rev Genet 11:843–854. doi:10.1038/nrg2884
101. Segrè AV, Groop L, Mootha VK et al (2010) Common inherited variation in mitochondrial genes is not enriched for associations with type 2 diabetes or related glycemic traits. PLoS Genet 6:e1001058. doi:10.1371/journal.pgen.1001058
102. De Leeuw CA, Mooij JM, Heskes T, Posthuma D (2015) MAGMA: generalized gene-set analysis of GWAS data. PLoS Comput Biol 11:e1004219. doi:10.1371/journal.pcbi.1004219
103. Mooney MA, Nigg JT, McWeeney SK, Wilmot B (2014) Functional and genomic context in pathway analysis of GWAS data. Trends Genet 30:390–400. doi:10.1016/j.tig.2014.07.004
104. Derry JMJ, Zhong H, Molony C et al (2010) Identification of genes and networks driving cardiovascular and metabolic phenotypes in a mouse F2 intercross. PLoS One 5:e14319. doi:10.1371/journal.pone.0014319
105. Rotival M, Zeller T, Wild PS et al (2011) Integrating genome-wide genetic variations and monocyte expression data reveals trans-regulated gene modules in humans. PLoS Genet 7:e1002367. doi:10.1371/journal.pgen.1002367
106. Grundberg E, Small KS, Hedman ÅK et al (2012) Mapping cis- and trans-regulatory effects across multiple tissues in twins. Nat Genet 44:1084–1089. doi:10.1038/ng.2394
107. Weiser M, Mukherjee S, Furey TS (2014) Novel distal eQTL analysis demonstrates effect of population genetic architecture on detecting and interpreting associations. Genetics 198:879–893. doi:10.1534/genetics.114.167791
108. Langley SR, Bottolo L, Kunes J et al (2013) Systems-level approaches reveal conservation of trans-regulated genes in the rat and genetic determinants of blood pressure in humans. Cardiovasc Res 97:653–665. doi:10.1093/cvr/cvs329
109. Stegle O, Parts L, Piipari M et al (2012) Using probabilistic estimation of expression residuals (PEER) to obtain increased power and interpretability of gene expression analyses. Nat Protoc 7:500–507. doi:10.1038/nprot.2011.457
110. Kendziorski CM, Chen M, Yuan M et al (2006) Statistical methods for expression quantitative trait loci (eQTL) mapping. Biometrics 62:19–27. doi:10.1111/j.1541-0420.2005.00437.x
111. Bottolo L, Petretto E, Blankenberg S et al (2011) Bayesian detection of expression quantitative trait loci hot spots. Genetics 189:1449–1459. doi:10.1534/genetics.111.131425

112. Scott-Boyer MP, Imholte GC, Tayeb A et al (2012) An integrated hierarchical Bayesian model for multivariate eQTL mapping. Stat Appl Genet Mol Biol 11:4. doi:10.1515/1544-6115.1760
113. Lewin A, Saadi H, Peters JE et al (2016) MT-HESS: an efficient Bayesian approach for simultaneous association detection in OMICS datasets, with application to eQTL mapping in multiple tissues. Bioinformatics 32(4):523–532
114. Voight BF, Scott LJ, Steinthorsdottir V et al (2010) Twelve type 2 diabetes susceptibility loci identified through large-scale association analysis. Nat Genet 42:579–589. doi:10.1038/ng.609
115. Teslovich TM, Musunuru K, Smith AV et al (2010) Biological, clinical and population relevance of 95 loci for blood lipids. Nature 466:707–713. doi:10.1038/nature09270
116. Honda K, Yanai H, Negishi H et al (2005) IRF-7 is the master regulator of type-I interferon-dependent immune responses. Nature 434:772–777. doi:10.1038/nature03464
117. Petretto E, Bottolo L, Langley SR et al (2010) New insights into the genetic control of gene expression using a Bayesian multi-tissue approach. PLoS Comput Biol 6:e1000737. doi:10.1371/journal.pcbi.1000737
118. Rotival M, Ko J-H, Srivastava PK et al (2015) Integrating phosphoproteome and transcriptome reveals new determinants of macrophage multinucleation. Mol Cell Proteomics 14:484–498. doi:10.1074/mcp.M114.043836
119. Maroso M, Balosso S, Ravizza T et al (2010) Toll-like receptor 4 and high-mobility group box-1 are involved in ictogenesis and can be targeted to reduce seizures. Nat Med 16:413–419. doi:10.1038/nm.2127
120. Budanov AV, Sablina AA, Feinstein E et al (2004) Regeneration of peroxiredoxins by p53-regulated sestrins, homologs of bacterial AhpD. Science 304:596–600. doi:10.1126/science.1095569
121. Mazzuferi M, Kumar G, Rospo C, Kaminski RM (2012) Rapid epileptogenesis in the mouse pilocarpine model: video-EEG, pharmacokinetic and histopathological characterization. Exp Neurol 238:156–167. doi:10.1016/j.expneurol.2012.08.022
122. Baxendale S, Holdsworth CJ, Meza Santoscoy PL et al (2012) Identification of compounds with anti-convulsant properties in a zebrafish model of epileptic seizures. Dis Model Mech 5:773–784. doi:10.1242/dmm.010090
123. Wills QF, Livak KJ, Tipping AJ et al (2013) Single-cell gene expression analysis reveals genetic associations masked in whole-tissue experiments. Nat Biotechnol 31:748–752. doi:10.1038/nbt.2642
124. Xue Q, Lu Y, Eisele MR et al (2015) Analysis of single-cell cytokine secretion reveals a role for paracrine signaling in coordinating macrophage responses to TLR4 stimulation. Sci Signal 8:59. doi:10.1126/scisignal.aaa2155
125. Pina C, Teles J, Fugazza C et al (2015) Single-cell network analysis identifies DDIT3 as a nodal lineage regulator in hematopoiesis. Cell Rep 11:1503–1510. doi:10.1016/j.celrep.2015.05.016
126. Moignard V, Woodhouse S, Haghverdi L et al (2015) Decoding the regulatory network of early blood development from single-cell gene expression measurements. Nat Biotechnol 33:269–276. doi:10.1038/nbt.3154
127. Macosko EZ, Basu A, Satija R et al (2015) Highly parallel genome-wide expression profiling of individual cells using nanoliter droplets. Cell 161:1202–1214. doi:10.1016/j.cell.2015.05.002
128. Wang J, Fan HC, Behr B, Quake SR (2012) Genome-wide single-cell analysis of recombination activity and de novo mutation rates in human sperm. Cell 150:402–412. doi:10.1016/j.cell.2012.06.030
129. Yang C, Li C, Wang Q et al (2015) Implications of pleiotropy: challenges and opportunities for mining Big Data in biomedicine. Front Genet 6:229. doi:10.3389/fgene.2015.00229

Part III

Systems Genetics Use Cases: Mapping and Combining Multiple Phenotypic Traits

Part III

Chapter 17

Genomic Control of Retinal Cell Number: Challenges, Protocol, and Results

Patrick W. Keeley, Irene E. Whitney, and Benjamin E. Reese

Abstract

This chapter considers some of the challenges in obtaining accurate and consistent estimates of neuronal population size in the mouse retina, in order to identify the genetic control of cell number through QTL mapping and candidate gene analysis. We first discuss a variety of best practices for analyzing large numbers of recombinant inbred strains of mice over the course of a year in order to amass a satisfactory dataset for QTL mapping. We then consider the relative merits of using average cell density versus estimated total cell number as the target trait to be assessed, and why estimates of heritability may differ for these two traits when studying the retina in whole-mount preparations. Using our dataset on cell number for 12 different retinal cell types across the AXB/BXA recombinant inbred strain set as an example, we briefly review the QTL identified and their relationship to one another. Finally, we discuss our strategies for parsing QTL in order to identify prospective candidate genes, and how those candidates may in turn be dissected to identify causal regulatory or coding variants. By identifying the genetic determinants of nerve cell number in this fashion, we can then explore their roles in modulating developmental processes that underlie the formation of the retinal architecture.

Key words Recombinant inbred strain, Haplotype, QTL, SNP, Variant, Neuron number

1 Introduction

The number of neuronal populations that make up the mammalian retina has increased over the past several years, as the five major classes of neuron have been progressively subdivided into their respective types based on morphological, physiological, and molecular characteristics [1]. The total number of retinal cell types, in the mouse, is now in the neighborhood of 85 [2], and there is some chance that more will become identified [3], including a few that disobey cardinal rules of retinal organization, for instance, that each cell type should provide a uniform and minimally complete coverage of the retinal surface [4]. Most of these populations are organized locally as retinal mosaics with varying degrees of regularity [5], with each cell type contributing to visual processing within its local territory [6]. Due to the diverse functions of these

Klaus Schughart and Robert W. Williams (eds.), *Systems Genetics: Methods and Protocols*, Methods in Molecular Biology, vol. 1488, DOI 10.1007/978-1-4939-6427-7_17, © Springer Science+Business Media New York 2017

different neuronal types, it is not surprising that the sizes of these populations vary conspicuously, from millions of rod photoreceptors to only hundreds of some amacrine cells [7, 8]. During development, many factors coalesce to achieve the final number of neurons within each population, including those regulating proliferation, fate determination, differentiation, and apoptosis. The total number of neurons within any given population, therefore, is a complex trait controlled by multiple genes underlying those various processes, subject to the influence of genetic variants.

Quantitative analysis of the total retinal ganglion cell population, which itself is composed of ~40 different types of ganglion cell [2], first demonstrated substantial heritable variation between different strains of mice [9]. Subsequent studies, using a panel of recombinant inbred strains, revealed that such variation could be mapped to genomic loci where causal genetic variants must reside [10]. Quantifying the total retinal ganglion cell population is reasonably straightforward by sampling multiple sites from a cross section of the optic nerve to determine an average axonal density, which when multiplied by the cross-sectional area of the nerve yields an estimated total number. Such sampling generates large differences in the estimates of the total population between different strains, in the presence of low variation amongst individuals of the same strain. While this might simply be a unique feature of the retinal ganglion population associated with its substantial and well-documented overproduction during development, a study of horizontal cell number based on analyzing immunostained retinas from seven different inbred laboratory strains, including the A/J and C57BL/6J strains [11], suggested that such interstrain variation might be present amongst other populations of retinal neurons as well. That study, in conjunction with related interests in the density dependency of mosaic order and dendritic outgrowth [12], prompted us to undertake a systematic analysis of multiple retinal cell types across 26 recombinant inbred (RI) strains derived from these same parental strains, the AXB/BXA strain set. That analysis revealed large variation to be present in every retinal cell type analyzed, approaching a twofold variation in number for some cell types, while minimal variation was found within each strain. For nearly every retinal cell type examined, we were able to map a considerable proportion of the variance observed across the strain set to genomic loci, yet rarely did that variation in cell number map to the same genomic locus for the different cell types [13]. Retinal cell number is therefore precisely specified within a strain, yet the presence of genetic variants renders large differences in total number between the strains, with their effects upon cell number disproportionately affecting select types of retinal neurons. Such variation in cell number is apparently well tolerated across different mouse strains, mitigated by dendritic plasticity during process outgrowth and synaptogenesis, ensuring uniformity in retinal coverage and connectivity with afferent populations, for some cell types [14].

The ability to identify genomic loci associated with such variation in quantitative traits (i.e., quantitative trait loci, or QTL) is dependent upon a reliable (and preferably accurate) determination of the trait of interest, in this case, the size of a neuronal population. One great attraction of working within the retina, unlike most other locations in the central nervous system, is that one can visualize the entire population of a given cell type in a whole-mount preparation, without the need for sectioning and the attendant correction for duplicate counting of sectioned cells, or for the application of dissector techniques [15]. Furthermore, by virtue of its unambiguous boundaries, one can precisely determine the areal size of the retina, and then multiply it by an average cellular density to estimate the total number of cells. Doing so consistently, over the course of 120 or so retinas might seem a challenging task, but with a streamlined approach and attention to detail, the exercise has proven fruitful, allowing us to detect QTL associated with 11 of the 12 cell types investigated to date [13], and to identify candidate genes at those loci controlling the variation in three of them so far, including the dopaminergic amacrine cells, the cone photoreceptors, and the horizontal cells [16–18]. We focus herein on our protocol and its rationale before briefly reviewing a portion of those results, paying particular attention to best practices we have found that ensure accuracy and reproducibility, permitting the successful mapping of the genomic control of retinal cell number.

2 Protocol

2.1 Coordinating Mouse Shipments

Mice of the AXB/BXA strain set (as well as other RI strain sets) are available from The Jackson Laboratory (JAX) as repository strains. Our routine has been to order all of the strains at the outset, but with the instruction to send only three strains at any given time, with each shipment arriving every 2–3 weeks. This latitude will enable JAX to send strains as they become available at the desired age, while providing enough time for tissue preparation, immunostaining, and analysis between shipments. Parental strains (in our case, the C57BL/6J and A/J strains) and their reciprocal F1 crosses (AB6F1/J and B6AF1/J) are readily available from JAX, and we routinely order and analyze them in advance. We use such parental and F1 strains to establish staining and sampling parameters when counting a new cell type, and for training naïve counters that are blinded to strain identity. Once reproducible results are obtained, the RI strains are ordered. We also order an additional shipment of the parental and F1 strains towards the end of the RI strain analysis, to confirm a lack of experimental drift in the analysis due to technical or human factors, such as a decrease in staining quality or a shift in the counter's criteria for identifying cells.

We routinely order four mice for each RI strain, seeking to ensure a minimal sample size of three mice per strain. If more than one mouse is excluded due to technical errors (poor perfusion, torn retina, nonuniform staining, etc.), additional mice of those strains are ordered near the end of the analysis to bolster the sample size. We have restricted our analysis to using only female mice, and we have the mice shipped between 6 and 10 weeks of age, well after retinal neurogenesis and differentiation have taken place.

2.2 Tissue Preparation

Animals are anesthetized with a lethal dose of sodium pentobarbital, and once they are unresponsive (e.g., no reaction to tail pinch), the thoracic cavity is opened and the heart is exposed. A 23 G butterfly needle is inserted into the left ventricle and a small incision is made to the right atrium. Two to three milliliters of saline is administered to rinse the blood from the vasculature (sufficient clearing of blood is indicated by a loss of color in the extremities and blood vessels of the ribcage), after which freshly prepared 4 % paraformaldehyde (PFA) in 0.1 M sodium phosphate buffer (PB, pH 7.4) is administered for 15 min to fix the tissues. We conduct perfusions on a ventilated down-draft table, and employ a gravity perfusion technique, in which the needle is attached via tubing to a large container of PFA approximately 1 m above the level of the mouse, yielding a flow rate of ~5 mL of fixative per minute (Fig. 1).

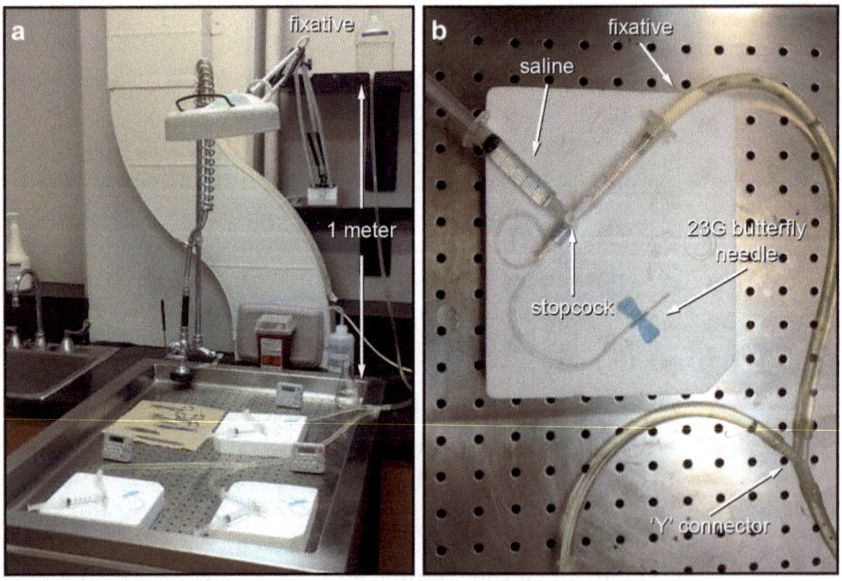

Fig. 1 Mice are perfused on a ventilated down-draft table. (**a**) A bottle of fixative (4 % paraformaldehyde) is positioned 1 m above the table. Plastic tubing connected to the bottle is split twice using "Y" connectors, allowing for three perfusions to be performed in parallel. (**b**) The flow of fixative and saline (administered via a syringe) is controlled independently for each mouse via a stopcock attached to a 23 G butterfly needle

Once the perfusion is complete, eyes are removed and immersed in fixative for an additional 15 min at room temperature and then transferred to PB, keeping left and right eyes separate. As a routine, we initiate a second and then third perfusion to be running simultaneously, staggering each perfusion to maintain a constant workflow. While such intracardial perfusion is our preferred method for preserving retinal tissue—as it allows for fixation of the eye before any mechanical stress associated with removal from the orbit, as well as clearing of blood from vessels to improve the quality of antibody staining—an alternative approach is to remove the eye and immediately immerse it into PFA for a total of 30 min.

Each eye is transferred to PB and maintained at 4 °C until the next day, when the retina is to be dissected. Each retina is then dissected from the eye in a petri dish using a binocular dissecting microscope, with care taken to remove the entirety of the tissue (Fig. 2a). The cornea and lens are first removed, and the retinal circumference is carefully separated from the ciliary body at the ora serrata. The optic nerve head is severed from the optic nerve at the posterior surface of the retina, freeing the entire retina from the cup, typically leaving the retinal pigment epithelium (RPE) within the cup. Using camel hair brushes, the retina is gently brushed open on the surface of a glass slide while being kept moist, and four relieving radial cuts are made, each extending half-way to the optic nerve head, to allow the retina to lie flat. Excess vitreous is carefully brushed free of the inner retinal surface, along with any remaining portions of iris and ciliary body. The "intactness" of the entire retina is then qualitatively assessed, taking note of any regions where cuts or tears have removed portions of the retina. Figure 2b contrasts a whole retina with one in which portions of the peripheral retina are missing (arrows). Since getting an accurate measurement of retinal area is essential for estimating the total number of cells, incomplete retinas are always excluded from further analysis.

Each retina is kept in a separate vial in PB until all retinas (typically 24, from the 12 mice in a shipment) have been dissected; our routine has been to have different individuals, working in parallel, dedicated to dissecting the left versus right retinas from the entire strain set. Each mouse is then randomly assigned a unique identifying number, and the identities of the mice are concealed until the entire RI strain set has been analyzed. All subsequent stages of retinal processing and analysis are conducted in numerical order, thereby ensuring that the investigators are blind to strain while also minimizing batch effects by this random intermingling of individuals within each set of three shipped strains.

2.3 Immunofluorescence

Whole retinas, labeled with their unique identifiers, are then transferred to phosphate-buffered saline (PBS) in the individual wells of a 96-well cell culture plate, with left and right retinas being segregated to the top versus bottom halves of the plate. Using the culture plate allows for each retina to be tracked throughout the

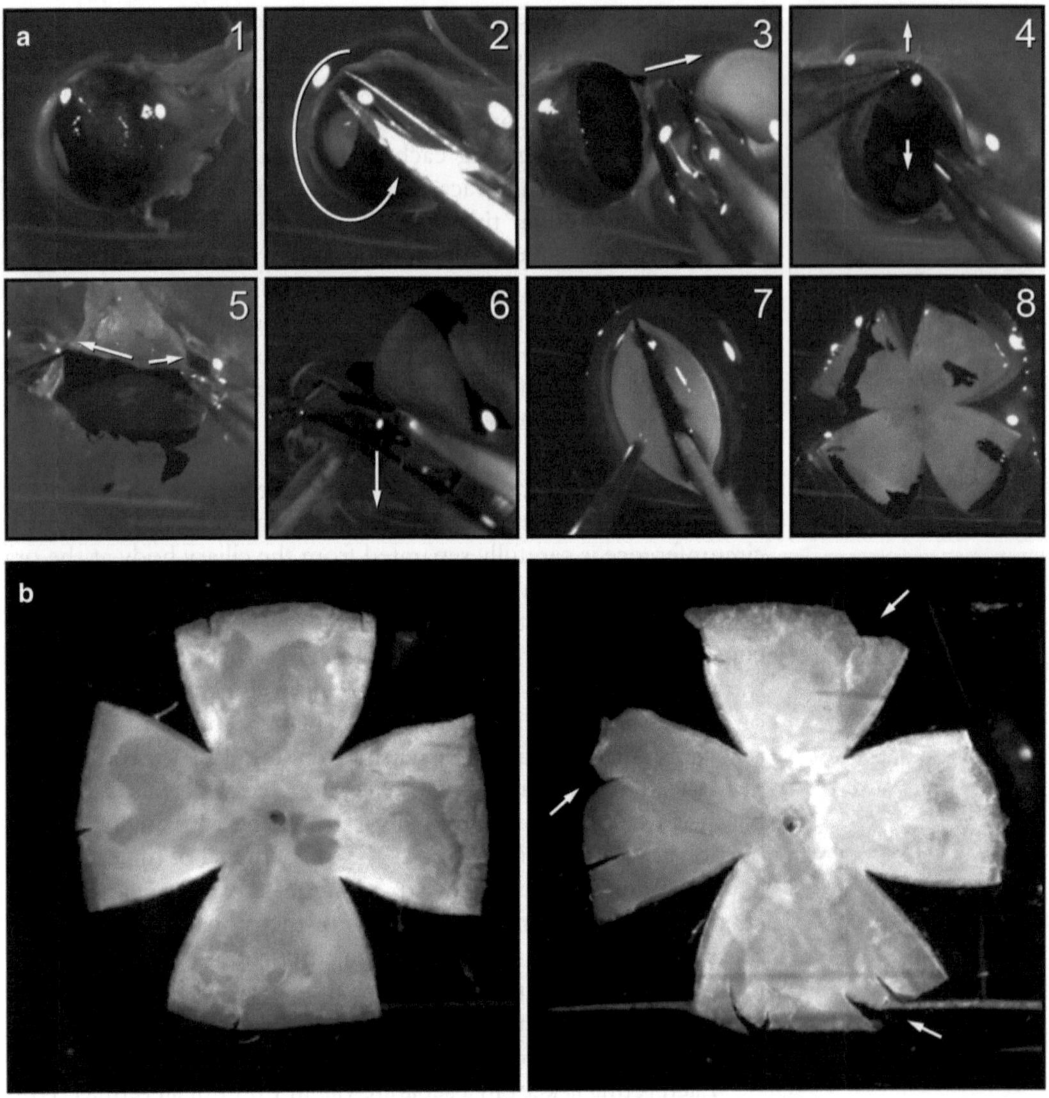

Fig. 2 (**a**) Example of a retinal dissection. (*1*) Eye is placed in a petri dish containing 0.1 M phosphate buffer. (*2*) A puncture is made at the margin of the cornea using a #11 scalpel blade and the cornea is removed by cutting along the perimeter with scissors. (*3*) The lens is gently removed from the eye. (*4*) The retina is detached from the ciliary body using forceps inserted beneath the retina, with each tine on either side of the remainder of the eye cup. (*5*) The remainder of the eye cup is gently torn at multiple locations (one such tear is shown), to allow access to the optic nerve head. (*6*) Fine forceps are then slowly inserted to sever the optic nerve at the posterior pole behind the retina. (*7*) The retina is transferred to a glass slide using camel hair brushes, with care taken to keep the retina moist with phosphate buffer. Using a #22 scalpel blade, four reliving cuts are made to allow the retina to lie flat. (*8*) The retina is cleaned of excess vitreous and remaining pieces of the iris and ciliary body. (**b**) Examples of retinal whole-mounts. The entire retina is present in the example on the *left*, while the example on the *right* has lost several small regions at the retinal perimeter (*arrows*) in the process of dissection and preparing the whole-mount

immunostaining protocol, while minimizing the amount of antibodies required, as each well needs only 150 μL of antibody solution per incubation. We immuno-label the left versus right retinas with different antibodies, and commonly double-label each retina with compatible primary antibodies in order to analyze a total of four cell types per mouse. All 12 left or right retinas in a run are stained at the same time using fresh solutions, made up in one vial, to minimize differences in staining quality across retinas. Since lot-to-lot variation in antibody quality is not uncommon, and can cause once reliable antibodies to exhibit unusual staining characteristics, we recommend acquiring a sufficient supply of a validated antibody to ensure the same lot can be used across the entire RI strain set, as well as for related follow-up projects, for instance, using chromosome substitution strains of mice to confirm the presence of an identified QTL on a particular chromosome. Most antibodies can be stored as small aliquots, to avoid repeated freeze-thaw cycles, and kept at −80 °C for long-term storage.

The following staining protocol is used as a general template, with durations occasionally being increased for certain antibodies. All steps are done at 4 °C with moderate agitation of the 96-well cell culture plate on an agitator (orbital shaker or rocker), and all solutions are exchanged using a micropipettor taking care not to damage the retinas. Retinas are first immersed in 5 % normal donkey serum (NDS), for 3 h. The NDS is diluted in PBS containing 1 % Triton-X (PBS-X). This solution is then replaced with PBS for 10 min, with this rinsing step being repeated two additional times. Retinas are then immersed in the primary antibodies at the required dilutions in PBS-X (with two different primary antibodies being prepared as a mixture, to permit the labeling of different cell types in each retina), for three nights on the agitator at 4 °C. Retinas are subsequently given three 10-min rinses in PBS, and then transferred to a mixture of appropriate donkey-anti-IgG secondary antibodies conjugated to distinguishable fluorochromes diluted in PBS-X, and left on the agitator at 4 °C overnight. The next morning, retinas are given two 10-min rinses in PBS, and finally, a rinse in PB.

2.4 Quantification

Each individual immunostained retina is carefully transferred to a glass slide and then brushed out flat, and a coverslip is overlaid using either PB (if the retina is to be recovered after the microscopy) or a fluorescence-preserving aqueous mounting media (e.g., FluoroGel), taking care to avoid putting excess pressure on the retina. Any remaining buffer on the surface of the slide is allowed to dehydrate, or is carefully blotted dry, after which the coverslip is affixed to the slide using nail varnish. Slides are then transferred to a fluorescence microscope for analysis. Counts can be made by hand using a tally counter, or made digitally using a computer receiving images via a video camera attached to the microscope. In the latter case, images can be captured on the computer to permit immediate counting while the shutter on the microscope is closed to avoid further bleaching of fluorescence. Alternatively, images can be taken across all retinas in one or two imaging sessions, thus ensuring the best quality labeling while allowing quantification to proceed according to counters' schedules. Another advantage to this approach is that each field is archived for post-hoc analyses (such as determining cell positioning or cell size), for evaluation of a counter's criteria, or for identifying drift in staining quality. Each cell type should be quantified completely (i.e., across every RI strain) by a single counter, to avoid any interindividual differences in criteria for identifying a cell to be counted.

Depending on the density of the cell type being analyzed, we take different approaches to sampling the retina in order to quantify cell number. For most cell types, either one mid-eccentric location or two (central versus peripheral) locations (being centered ~0.5 mm from the optic nerve head and from the peripheral margin of the retina, respectively) per retinal quadrant are analyzed (e.g., Fig. 3a), with sampling fields ranging from ~15,500 to 226,000 μm^2, depending on the density of the population of interest. Once a sampling field is identified using these criteria, and the relevant objective lens is brought into focus, the position of the field may be shifted by a field-width to avoid any small tears, large blood vessels or overlying debris that might preclude imaging the population. For small and denser cell types (e.g., rod and cone photoreceptors), we use smaller sampling fields (from ~225 to 14,000 μm^2, respectively); since each of these smaller fields represents a tiny fraction of the overall retinal area, we increase the number of fields sampled (~14–16 fields per retina) by sampling at repeated intervals from a 1 mm^2 grid (e.g., Fig. 3b). Conversely, with extremely sparse cell types (e.g., the dopaminergic amacrine cell), we count every cell by systematically moving the field-of-view across the entire retina (Fig. 3c). We use a microscope equipped with *X*-*Y* stage encoders and a digital camera, both interfaced to a computer running Bioquant software (Bioquant Image Analysis Corporation, Nashville, TN), in order to track each cell's precise location on the retina to ensure no single cell is counted twice. Of course, only the latter procedure determines the true total number of cells (assuming every cell is labeled); the former procedures are a compromise, influenced by many factors, to generate an estimate of that total number. Because cell density does not vary substantially across the mouse retina, however, these sampling protocols generate reasonable estimates of total cell number, being commensurate with estimates from other studies using different sampling protocols [7, 19]. Critically, they show large differences in estimated number between different strains of mice while generating conspicuously low differences in the number of cells present between mice of the same strain. Despite the large differences in the amount of retinal area directly sampled via these different

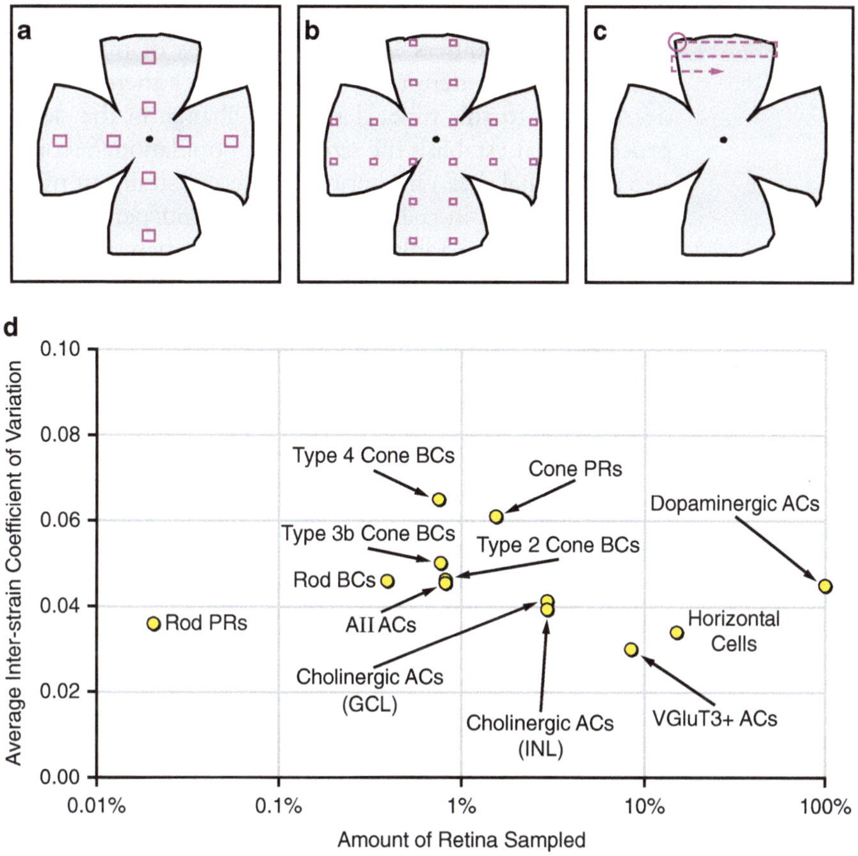

Fig. 3 Tracings of a retinal whole-mount illustrating three different sampling protocols. (**a**) Each quadrant of the retina is sampled at a central and a peripheral locus (e.g., cholinergic amacrine cells). (**b**) Multiple samples are collected at 1 mm intervals across the retinal surface (e.g., cone photoreceptors). In (**a**) and (**b**), total cell number is estimated by calculating an average density from these samples and multiplying by retinal area. (**c**) The entire surface of the retina is scanned, with the position of each cell being plotted to determine total cell number (e.g., dopaminergic amacrine cells). (**d**) The average coefficient of variation (CoV) across the strains is plotted, for each population of retinal neuron, as a function of the proportion of the retinal area sampled. *PRs* photoreceptors, *BCs* bipolar cells, *ACs* amacrine cells

procedures, from 0.02 to 100 %, the average coefficient of variation (CoV) for the size of a population, for each cell type, showed no obvious correlation with the amount of retina sampled (Fig. 3d).

2.5 Estimating Total Cell Number

With the exception of such sparse cell types, in which the total number of cells in the retina is directly quantified, an estimate of total number is determined by multiplying the average density by the retinal area. The reasons for calculating total number, rather than simply using average density measurements, for each mouse are numerous. First, even after the processes of neuronal production and programmed cell death are finished, yielding the final number of total neurons, the retina continues to expand, achieving its mature size into the third postnatal month, but exactly when this is achieved, and whether it differs for different strains of mice, has not been determined. The density of most cell types therefore continues to decrease up to this time, despite no change in the developmental processes that establish the size of the population. Second, both eye size and retinal area vary between different strains of mice [13, 20], but the variation in cell number is often independent of this variation in retinal size (Table 1). Indeed, even when there is a significant correlation between retinal area and total cell number, the correlation between cell density and total cell number is consistently greater (e.g., Fig. 4). Strain differences in cell density, therefore, translate into real differences in total number, yet it is difficult to discern such real density differences from others that can be an artifact of preparing the retina for analysis. In particular, differences in fixation, dissection, and whole-mounting can yield differences in the size of the retina when laid out upon the slide that in turn produce corresponding variations in cell density.

For instance, Table 2 shows the average cell densities from four retinas from the BXA4 strain that were immunostained to label the

Table 1
The Pearson correlation coefficient (*r*), and associated *p*-value (*p*), for the co-variation between total cell number and retinal area, or between total cell number and average cell density, across all of the RI and parental strains, for each of the 12 cell types analyzed

	Retinal area		Density	
Cell type	*r*	*p*	*r*	*p*
AII amacrine cells	0.60	**0.001**	0.93	**<0.001**
Cholinergic amacrine cells (INL)	0.39	**0.036**	0.85	**<0.001**
Cholinergic amacrine cells (GCL)	0.37	**0.050**	0.81	**<0.001**
Cone photoreceptors	0.25	0.201	0.94	**<0.001**
Dopaminergic amacrine cells	0.02	0.911	0.98	**<0.001**
Horizontal cells	0.26	0.179	0.94	**<0.001**
Rod bipolar cells	0.33	0.083	0.76	**<0.001**
Rod photoreceptors	0.53	**0.002**	0.71	**<0.001**
Type 2 cone bipolar cells	0.43	**0.016**	0.87	**<0.001**
Type 3b cone bipolar cells	0.32	0.091	0.89	**<0.001**
Type 4 cone bipolar cells	-0.01	0.942	0.96	**<0.001**
VGluT3+ amacrine cells	0.46	**0.011**	0.88	**<0.001**

Significant correlation coefficients are indicated in *bold*. Note that there is a consistently greater correlation between estimated total cell number and cell density than with retinal area

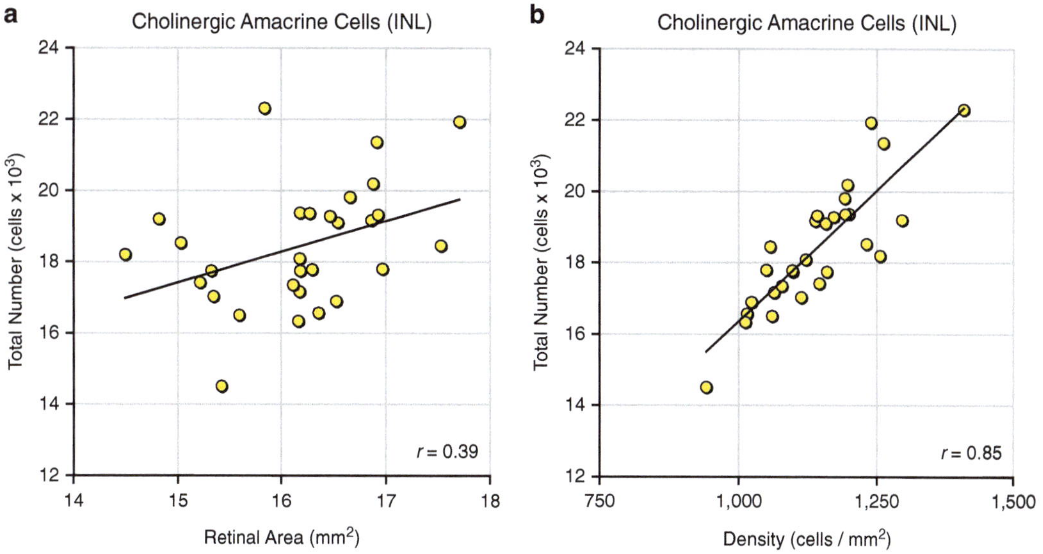

Fig. 4 (**a**) Estimated total number of cholinergic amacrine cells in the inner nuclear layer for each strain, as a function of retinal area. Note the large variation in total cell number, from ~14,000 to 22,000 cells, that is largely independent of the variation in the average size of the retinas for each strain ($p=0.036$). (**b**) Estimated total number is more strongly correlated with average cell density ($p<0.001$)

Table 2
Average density of cholinergic amacrine cells in the inner nuclear layer for four individual mice of the BXA4 strain, alongside their retinal area measures, and the estimated total number of cells per retina

Retina	Density (cells/mm²)	Retinal area (mm²)	Total number (cells)
1	1,216	14.565	17,751
2	1,111	16.308	18,112
3	1,039	16.439	17,078
4	1,020	17.836	18,190
CoV	0.081	0.081	0.029

Note the inverse relationship between average cell density and retinal area, yielding a reduction in the variation of estimated total cell number across individual retinas, as indicated by the coefficient of variation (CoV)

population of cholinergic amacrine cells in the inner nuclear layer (INL). There is conspicuous variation in average density between individuals, and while that variation could plausibly reflect real biological variation, the sizes of each of those retinas also vary, inversely with average cell density. Indeed, when multiplied together to estimate the total number of cholinergic amacrine cells in these four retinas, there is a substantial reduction in the variance in this measure relative to that for cell density. Within a strain, therefore, the variation across individuals in both total cell number and average cell density will reflect stochastic biological processes as well as some shared degree of technical error; the estimated total number, however, will correct for this one conspicuous and substantive contributor to that technical error, that introduced through the process of dissecting the retina and preparing the whole-mount.

Such estimates of total cell number generate tighter within-strain statistics than do average densities, while preserving the conspicuous between-strain statistics that reflect the actions of genetic variation across the strains. By maximizing the latter while minimizing the former, we should increase the "heritability" of our target traits [21], in turn heightening the likelihood we will detect QTL controlling them. Table 3 plots the heritability (h^2) for the estimated total cell number trait versus the average cell density trait for each of the 12 cell types, showing this fairly consistent increase when estimating total cell number. That differences in cell density can so readily arise for technical reasons means that measuring retinal area accurately is critical for eliminating this major source of error. We do so routinely by tracing the retinal perimeter using a 10× objective with Bioquant software that interfaces the *X*-*Y* positions traced on the monitor with those on the microscope stage. Doubtless one

Table 3
Heritability estimates (h^2; being the proportion of the variance of the trait across all individual mice that can be ascribed to an effect of strain) across the RI strains for average cell density and total cell number, for each of the 12 different cell types

Cell type	h^2 (density)	h^2 (total number)
AII amacrine cells	0.67	0.74
Cholinergic amacrine cells (INL)	0.49	0.68
Cholinergic amacrine cells (GCL)	0.44	0.63
Cone photoreceptors	0.60	0.67
Dopaminergic amacrine cells	0.80	0.87
Horizontal cells	0.66	0.89
Rod bipolar cells	0.37	0.44
Rod photoreceptors	0.53	0.48
Type 2 cone bipolar cells	0.50	0.58
Type 3b cone bipolar cells	0.57	0.64
Type 4 cone bipolar cells	0.70	0.69
VGluT3+ amacrine cells	0.65	0.82

could prepare the retina in such a manner as to greatly reduce the ease with which dissection or wholemounting might distort true retinal size, for instance, by increasing tissue fixation; yet we find a substantial reduction in the quality of immunostaining following overfixation, while also reducing the ease with which the retinal pigment epithelium can be dissected free from the retina.

The retina is therefore a pliant tissue, and this feature itself will vary depending upon the quality of the initial fixation achieved through perfusion. The subsequent act of dissecting the retina, like the final preparation of the wholemount following immunostaining, is consequently susceptible to the skills of the histologist. Figure 5 shows, for example, a collection of 106 retinas in which the size of the very same retinas had been measured twice: first at the time of the initial cell counts, and then a second time after the retinas had been returned to a vial and stored at 4 °C for 2–3 weeks before remounting by a second histologist, without any coaching to avoid undue stretch or compaction in the process of remounting. While a strong correlation exists between the two determinations of retinal area ($r = 0.75$), conspicuous differences were found for many individual retinas. The majority showed a larger area upon remounting and tracing, as might have been expected, but a number of others showed a reduction in area that could not be accounted for by missing portions. And although some of this variability might be eliminated with specific attention to achieving a

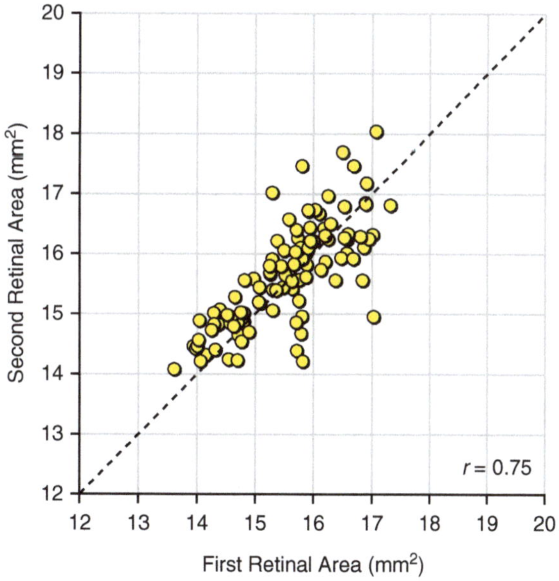

Fig. 5 Left retinal areal measurements for 106 individual RI, F1, and parental strain mice that were determined twice, the second time (plotted on the *Y* axis) after the retinas remained in a vial in PB at 4 °C for weeks following the initial determination

consistent degree of spread during the process of mounting the retina upon the slide, the exercise makes clear the critical need to measure retinal area at the time of determining cell density.

2.6 Timetable

The following schedule is a suggested course of action for a typical "run" in which animals are received and perfused, retinas are dissected and stained, and cell populations are quantified. After the initial run, all subsequent runs can begin on Day 15, so that the week of collecting a new set of retinas overlaps with the last week of analysis from the previous run, thus ensuring a constant turnover of retinas to be analyzed. By having some individuals (typically two) dedicated to perfusing mice and harvesting the left versus right retinas, and others (typically another two) dedicated to counting different populations of cells in the left versus right retinas, we have found a 2-week cycle to be generally sufficient. Using this timetable, the analysis of 26 RI strains, two parental strains (twice), and two F1 crosses can be completed in roughly 6 months, although in practice it will take about 9 months to complete a project, accommodating for some variability in counters' schedules, the occasionally limited availability of some RI strains, and the need to reorder some of the strains.

Day 1: 4% PFA solution is made in advance (50 mL per mouse).

Day 2: Three strains of mice (four mice per strain) are received in the morning. All 12 animals are perfused in the afternoon, and all eyes are immediately collected.

Days 3–4: Retinas are dissected from both left and right eyes. Animals are then coded to conceal strain identity in all subsequent steps.

Day 5: Retinas are placed into protein block, followed by primary antibodies.

Day 8: Retinas are placed into secondary antibodies.

Day 9: Retinas undergo final rinse step, and are ready for analysis.

Days 10–22: Cell densities and retinal areas are quantified and recorded.

3 Results

Table 4 shows the range of variation in the number of the 12 different retinal cell types examined. (The original strain data, for each cell type, are provided as a supplement in Keeley et al. [13].) Every cell type examined varied in number across the strain set, though the magnitude varied considerably. The rod bipolar cells, for instance, showed a 26% increase from the lowest strain to the highest strain, while the dopaminergic amacrine cells showed a 298% increase. For all but one of these cell types, the variation in

Table 4
Magnitude of the variation in cell number across the RI, F1, and parental strains, for each cell type

Cell type	Minimum		Maximum		Range	% Increase
AII amacrine cells	57,141	BXA26	86,557	BXA2	29,416	51%
Cholinergic amacrine cells (INL)	14,506	A/J	22,304	BXA26	7,798	54%
Cholinergic amacrine cells (GCL)	12,434	A/J	18,414	BXA12	5,980	48%
Cone photoreceptors	116,158	A/J	204,419	AXB23	88,261	76%
Dopaminergic amacrine cells	160	AXB12	637	BXA12	477	298%
Horizontal cells	9,884	A/J	18,942	BXA14	9,058	92%
Rod bipolar cells	202,415	AXB15	254,176	BXA7	51,761	26%
Rod photoreceptors	6,051,100	A/J	8,227,260	BXA12	2,176,160	36%
Type 2 cone bipolar cells	41,908	BXA4	60,570	B6AF1/J	18,662	45%
Type 3b cone bipolar cells	50,243	AXB1	80,469	BXA12	30,226	60%
Type 4 cone bipolar cells	29,351	BXA14	50,260	AXB2	20,909	71%
VGluT3+ amacrine cells	10,415	BXA25	14,857	BXA12	4,442	43%

The sizes of the respective populations in the strains with the lowest number of cells (minimum) and the highest number of cells (maximum) are presented, as is the range of this variation. That range is also expressed as a proportion of the strain with the lowest number of cells

cell number mapped to one or more genomic loci where the likelihood ratio statistic (LRS) crossed either the suggestive (<0.67) or significant (<0.05) threshold defined by permutation testing of the original strain data, for each cell type (Fig. 6). Table 5 plots the estimated additive effect attributed to the largest QTL present in Fig. 6 for each cell type. When considered relative to the range of variation across the different strains (Table 4), each one of these QTLs accounts for a conspicuous proportion of that variation, from a low of 23% to a high of 40%. While attention is drawn to those largest QTLs, we regard lesser suggestive QTL as also worthy of consideration, particularly when composite interval mapping, controlling for the effect of the largest QTL, drives a suggestive QTL to the significant threshold.

As mentioned in the Introduction, one striking feature about the collection of maps in Fig. 6 is that they show the variation in cell number, for each cell type, maps to distinct genomic loci. Variation in cell number, consequently, cannot be due to a variant gene or genes simply controlling the overall size of the retina, modulating, for instance, proliferation. This point has already been

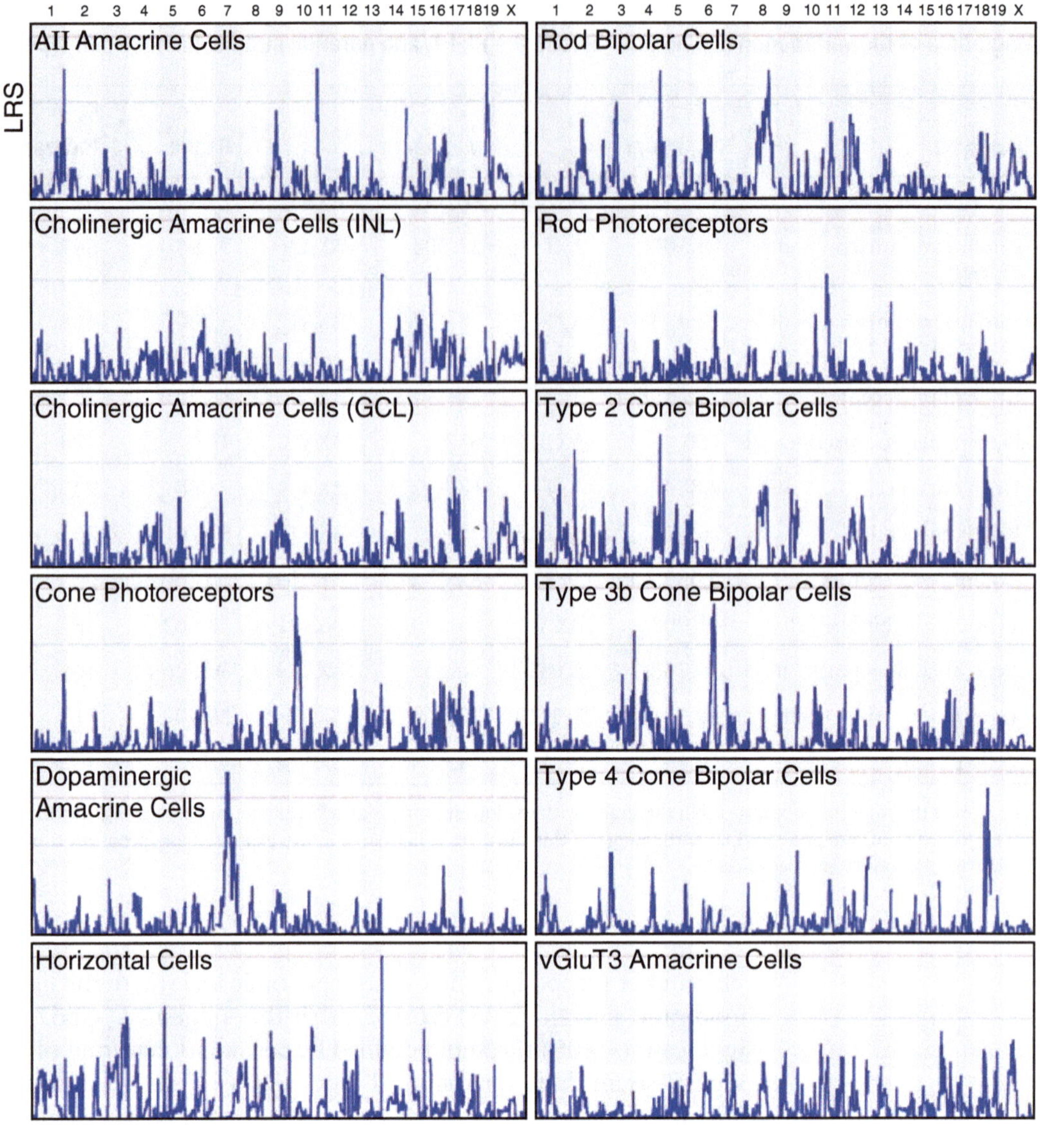

Fig. 6 Whole genome maps for the variation in cell number across the strain set, for each cell type. Each *blue* trace plots the likelihood ratio statistic (LRS) across the genome, with each chromosome represented on the *X*-axis (as indicated in the top panels). The *horizontal pink* and *gray lines* indicate the significant and suggestive thresholds, determined through permutation testing of the original strain data

made when considering the lack of correlation between retinal area and total cell number above (Fig. 4). But it is made that much more obvious by the fact that the strain distribution pattern for each cell type is unique, there being minimal co-variation between any pairs of cell types across the strain set (*see* [13], for the associated co-variation matrix and a fuller discussion of this point). While the parental B6/J strain invariably has more of every cell type than

Table 5
The location and peak LRS of the largest QTL identified for each of the 12 cell types

Cell type	Largest QTL	LRS	Additive effect (valence)	Magnitude of effect
AII amacrine cells	Chr 11	14.6	11,640 *(B)*	40%
Cholinergic amacrine cells (INL)	Chr 16	10.4	1,859 *(B)*	24%
Cholinergic amacrine cells (GCL)	-	-	-	-
Cone photoreceptors	Chr 10	14.7	21,946 *(B)*	25%
Dopaminergic amacrine cells	Chr 7	15.2	149 *(B)*	31%
Horizontal cells	Chr 13	18.1	2,490 *(B)*	28%
Rod bipolar cells	Chr 4	14.0	19,293 *(A)*	37%
Rod photoreceptors	Chr 11	12.0	494,913 *(B)*	23%
Type 2 cone bipolar cells	Chr 2	12.5	6,127 *(A)*	33%
Type 3b cone bipolar cells	Chr 6	14.4	8,607 *(A)*	29%
Type 4 cone bipolar cells	Chr 18	16.7	7,491 *(A)*	36%
VGluT3+ amacrine cells	Chr 5	14.7	1,591 *(A)*	36%

The additive effect of having two alleles of the indicated valence is given, as is the percentage of the total range in cell number across the strains that is accounted for by the QTL

does the A/J strain, the ratio between them varies considerably (Table 6), and the positioning of each recombinant inbred strain, relative to the parental strains, is unique for each cell type.

For example, Fig. 7a, d shows the strain distribution patterns for the Type 3b cone bipolar cells and for the dopaminergic amacrine cells, respectively. Note that the parental strains (in red and green) are near the extremes of the strain distribution for the latter cell type (Fig. 7d), while they are nearly identical for the former cell type (Fig. 7a). For these cone bipolar cells, there is still considerable variation outside of the range defined by the parental strains, indicating that there must be allelic variants distinguishing the parental strains that negate one another's additive effects upon total cell number. This is borne out in the whole genome map for the Type 3b cone bipolar cells (Fig. 7b), where multiple QTLs are detected on Chrs 3, 6, 13, and 17 (with that on 17 nearly reaching significance when controlling for the effect of the primary QTL on Chr 6—not shown). The valence of those QTL effects differs, with *A* alleles on Chrs 6 and 17 acting to increase trait values, while *A* alleles on Chrs 3 and 13 work in the opposite direction, decreasing trait values (i.e., loci where *B* alleles increase trait values). Consequently, different RI strains will show the additive effects of

Table 6
Total estimates for each cell type for the two parental strains, A/J and B6/J, and their ratio

Cell type	B6/J	A/J	Ratio (B/A)
AII amacrine cells	69,223	64,777	1.07
Cholinergic amacrine cells (INL)	19,539	14,506	1.35
Cholinergic amacrine cells (GCL)	16,680	12,434	1.34
Cone photoreceptors	196,897	116,158	1.70
Dopaminergic amacrine cells	617	219	2.81
Horizontal cells	18,471	9,884	1.87
Rod bipolar cells	213,873	209,009	1.02
Rod photoreceptors	7,624,090	6,051,100	1.26
Type 2 cone bipolar cells	55,275	47,878	1.15
Type 3b cone bipolar cells	58,722	57,365	1.02
Type 4 cone bipolar cells	36,940	31,772	1.16
VGluT3+ amacrine cells	12,264	10,644	1.15

Note that while the B6/J strain contains more of every type of cell sampled, the extent to which the size of each population in B6/J exceeds that for the A/J strain varies

these distinct loci to yield total numbers that may be either greater or less than the numbers displayed by either of the parental strains, based on their haplotypes at each of these four loci (Fig. 7c).

Even where only one QTL is detected (e.g., Fig. 7e), we know that more than just a single genetic variant should participate in defining total cell number, because the strain distribution pattern does not parcel the strains into only two phenotypic groupings, with some strains containing a high number of cells and others containing a low number. In some cases, evidence for other participating loci may have gone undetected due to insufficient fractionation of the genome within a particular strain set, preventing the participating variant to segregate with the haplotypes present. In other cases, however, the single genomic locus detected may conceal the presence of multiple independent genetic contributors. For instance, the variation in dopaminergic amacrine cell number across the strains is graded, from the lowest to the highest strain (Fig. 7d), yet it maps only a single, conspicuous, QTL on Chr 7 (Fig. 7e), for which *B* alleles are associated with an increase in cell number that is 31% of the range observed across the strains (Table 5). Use of various consomic, subconsomic, and congenic strains has allowed for a detailed dissection of this locus, revealing that it conceals at least three different participants affecting dopaminergic amacrine cell number [17].

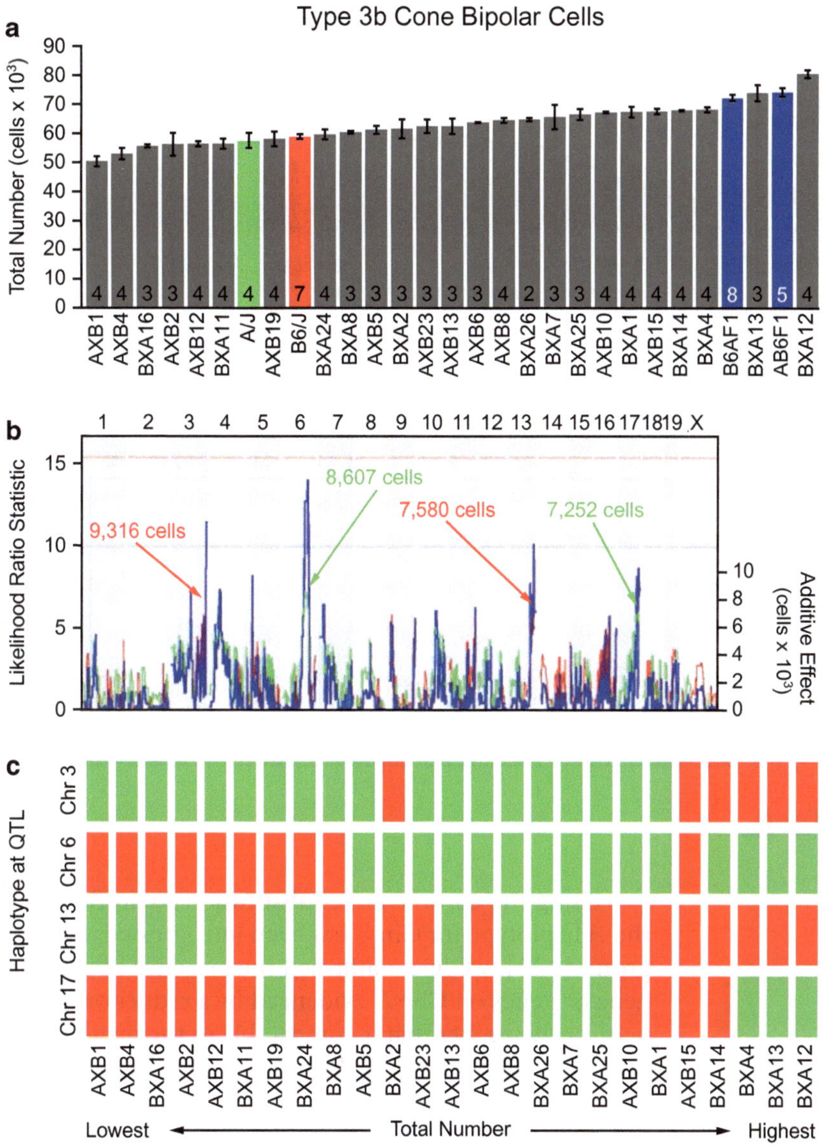

Fig. 7 (**a**) Average total number ± standard error of Type 3b cone bipolar cells for each of the RI, parental, and F1 strains, ranked from lowest to highest number of cells. *Numbers* at the base of each bar indicate sample size per strain. (**b**) Whole genome map for the variation in Type 3b cone bipolar cell number. *Blue* trace indicates the LRS, while the *pink* and *gray horizontal lines* indicate the significant and suggestive thresholds defined by permutation mapping, as in Fig. 6. Additionally, the red and green traces now show the additive effect of either *B* or *A* alleles across the genome. The additive effects of two alleles (*red* for *B*, *green* for *A*) at each of the four QTL are indicated. (**c**) Haplotype for each RI strain at each of the four loci for Type 3b cone bipolar cell number, with *green* indicating the presence of *A* alleles and red indicating the presence of *B* alleles at each locus. As in (**a**) above, the strains are ranked from lowest to highest number of Type 3b cone bipolar cells. (**d**) Average total number of dopaminergic amacrine cells for each of the RI, parental, and F1 strains, ranked from lowest to highest number, with all other conventions as in (**a**). (**e**) Whole genome map for the variation in dopaminergic amacrine cells, with the same conventions as in (**b**)

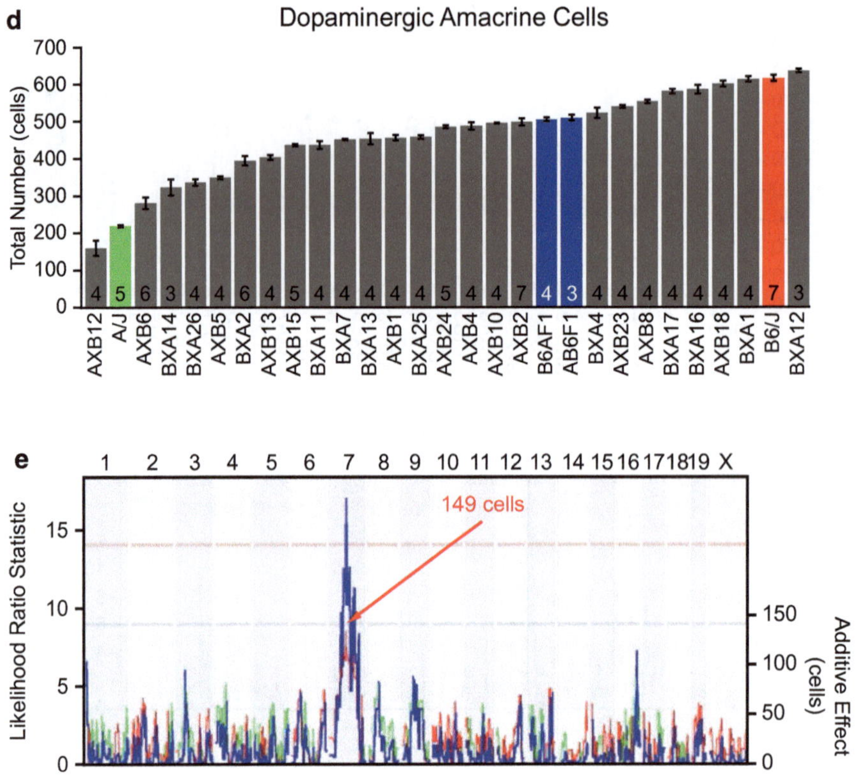

Fig. 7 (continued)

Using the chromosome substitution strain B6.A<7>, in which the *A* haplotype for Chr 7 has been introgressed on the B6/J background [22], we first confirmed the presence of allelic variants in a gene or genes on Chr 7, evidenced by a reduction in total dopaminergic cell number that is comparable to the magnitude ascribed to the QTL effect itself [17]. Within the genomic region identified, at 45.3 Mb, an SNP in the promoter of the proapoptotic *Bax* gene was noted, disrupting a p53 binding site in the B6/J strain that has been shown to modulate gene expression in a luciferase assay [23]. This variant may underlie the difference in *Bax* gene expression detected between the two parental strains during the postnatal period, precisely when naturally occurring cell death is most extensive within the developing mouse retina, with greater expression in the strain containing fewer dopaminergic amacrine cells (the A/J strain). Knocking out *Bax* gene function yielded a fourfold increase in dopaminergic cell number [17], confirming a role for this gene in the regulation of this trait.

This *Bax* gene variant cannot be the only contributor to the variation in dopaminergic amacrine cell number in this genomic region of interest: there is another gene (or genes), yet to be identified, present more distally on Chr 7, between 73.8 and 122.3 Mb, for which *B* alleles increase dopaminergic amacrine cell number.

This was determined by analyzing subconsomic mice in which the *B* haplotype for this 55 Mb portion of Chr 7 had been introgressed on an A/J genetic background. These mice show an increase in dopaminergic amacrine cell number (relative to A/J) of a magnitude roughly one-sixth of the range displayed across the strains. One participating gene within this region, the *Tyr* gene, at 87.3 Mb, was found to increase dopaminergic amacrine cell number when its function is disrupted, gleaned from assessing cell number in the c2J mouse (which carries a functional mutation in *Tyr* on the B6/J genetic background, rendering the mouse albino). This increase in cell number is roughly one-eighth of the range exhibited across the RI strains. Since A/J mice also contain a mutation in *Tyr* rendering it nonfunctional (and them albino), the increase seen in the subconsomic mice must be dampened by the presence of the functional *B* allele of *Tyr* lowering cell number. Therefore, the real effect of the other gene (or genes) in this 55 Mb region for which *B* alleles increase trait values must be approaching an effect with a magnitude nearer one-third of the overall range observed across the strains [17].

To identify such causal genes at a given QTL, it is necessary to prioritize them as candidates for further testing, particularly since the number of genes within a QTL can be quite large, sometimes exceeding 100 genes. Genes are evaluated using three major criteria: (1) identification of variants within or near the gene (such as SNPs, short INDELs, or structural variants) that differentiate the parental strains; (2) expression of the gene within the retina, with higher priority given to those genes known to be expressed during retinal development; and (3) a known role of the gene in developmental processes, with extra attention paid to those genes known to be involved in neuronal development. Information on each gene is gathered from several databases, such as the Sanger Institute Mouse Genomes Project for identifying parental variants (www.sanger.ac.uk/resources/mouse/genomes) [24]; the Mouse SAGE Retina database (cepko.med.harvard.edu) [25], as well as retinal cell type-specific expression data [3, 26, 27] for determining retinal expression at different developmental time-points; and PubMed and protein function databases, such as UniProt (www.uniprot.org) [28], for assessing functional roles for each gene. We also compare the expression level of a given gene from a microarray analysis of ocular mRNA derived from the same strain set [18], seeking to correlate any variation in expression with variation in cell number. Such co-variation may be indicative of a *cis*-regulatory variant controlling transcription, that in turn influences total cell number.

For instance, the QTL for horizontal cell number (Fig. 6) is situated at the distal tip of Chr 13, extending for 7 Mb and containing ~70 genes (Fig. 8a). Using the aforementioned criteria, two top candidate genes were identified in this interval: *Fst* and *Isl1*. *Fst*, at 115.24 Mb, was considered an attractive candidate, for previous

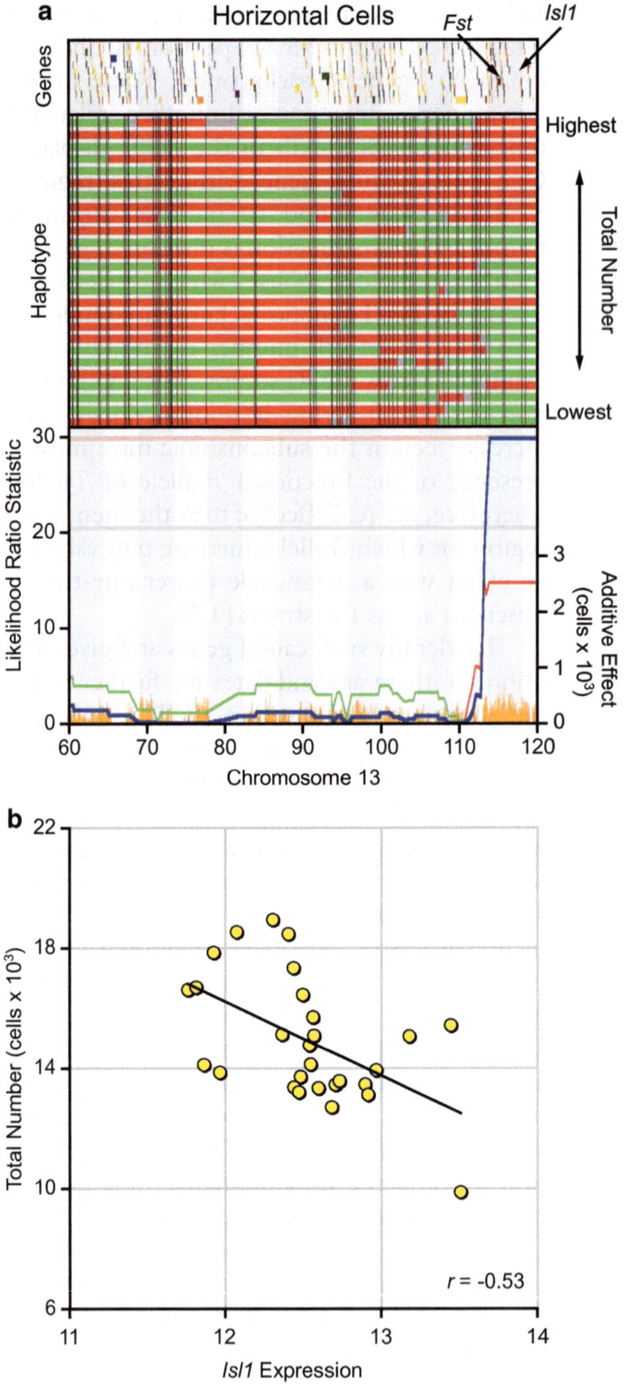

Fig. 8 (**a**) Map of Chr 13 showing the distally positioned QTL, with an LRS score of 30, just reaching the significant threshold defined by permutation mapping of the strain data. The Chr 13 haplotype for each RI strain (with *B* in *red*, *A* in *green*, heterozygous loci in *blue*, and unknown regions in *gray*) is shown above the map, while gene density is plotted across the very top. The positions of two candidate genes, *Fst* and *Isl1*, are indicated therein. SNP density across Chr 13 (shown in *gold*) is plotted at the bottom, while the number of horizontal cells for each strain is plotted to the right of each haplotype. (**b**) Total horizontal cell number across the RI and parental strains correlates negatively with *Isl1* expression levels derived from microarray analysis ($p = 0.004$)

studies in the chick retina had demonstrated the addition of follistatin during the neurogenetic period increased the number of horizontal cells [29]. Upon examining the retinas of mice in which *Fst* was conditionally eliminated from the retina, however, we found no difference in the number of horizontal cells. We turned our attention next to *Isl1*, at 117.08 Mb, particularly as its conditional loss reduces the numbers of ganglion cells, cholinergic amacrine cells, and bipolar cells in the mouse retina [30]. While all of those cell types express *Isl1* during early development, the horizontal cells do not [31], yet when we examined such retina-specific *Isl1* conditional knock-out mice, we found a conspicuous *increase* in the number of horizontal cells, elevating its candidacy as a causal gene at this locus.

Sequencing the two parental *Isl1* genes confirmed the presence of four previously identified synonymous SNPs, finding no additional SNPs in the coding region. Sequence analysis of putative regulatory regions, in turn, identified numerous SNPs and short INDELs. We first sought evidence for expression differences between the parental strains using qPCR, finding higher expression in the A/J strain (the strain having roughly half as many horizontal cells as the B6/J strain), during the period of horizontal cell genesis, as well as in maturity. That such expression differences might be meaningfully related to total horizontal cell number was in turn supported by a comparison of *Isl1* expression across the entire strain set, where a significant negative correlation was detected (Fig. 8b). We also confirmed that *Isl1* expression in the retina is *cis*-regulated, using an allele-specific expression assay measuring the relative abundance of *A* versus *B Isl1* transcripts in the F1 progeny of reciprocal crosses of the parental strains [16].

These various expression studies together support the hypothesis that a regulatory variant in *Isl1* modulates its expression, and that such variation in expression participates in the control of horizontal cell number. Narrowing the list of noncoding variants (i.e., those found in upstream, downstream, untranslated, and intronic regions) is an equally challenging task as is ranking candidate genes. First, such variants are parsed using the ECR Browser (ecrbrowser.dcode.org) [32] to identify those that are present in evolutionarily conserved regions (ECRs), since conservation of noncoding genomic loci implies the presence of important regulatory sequences (Fig. 9). These ECRs can then be analyzed for predicted transcription factor binding sites using programs such as rVista 2.0 (rvista.dcode.org) [33], which may be disrupted or created by the presence of the candidate variant. For instance, an SNP was identified in an ECR of the 5′ untranslated region of *Isl1* (indicated in Fig. 9), in which a T for G substitution in a hexanucleotide sequence creates an E-box in the B6/J strain, a binding site for the family of bHLH transcription factors, some of which are well-known transcriptional repressors [34]. Curiously, it is the A/J strain variant that is the more evolutionarily conserved of the two sequences, suggesting that this novel binding site in B6/J may act

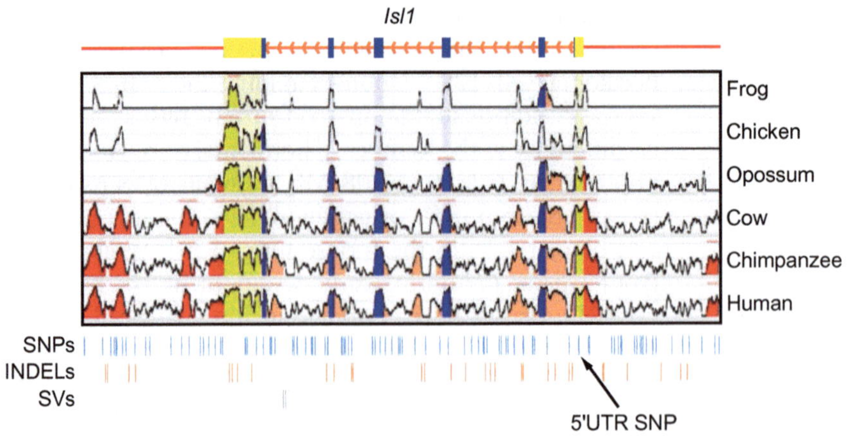

Fig. 9 Schematic showing evolutionarily conserved regions (ECRs) of the *Isl1* gene, between the mouse and six different species. Regions of at least 100 bp with 78 % homology to the mouse genome were defined as ECRs and are indicated where the traces are filled with color. The structure of the *Isl1* gene is indicated above the graph with coding regions (*blue*), untranslated regions (*yellow*), introns (*orange*), and upstream/downstream regions (*red*). Single nucleotide polymorphisms (SNPs), short insertions and deletions (INDELs), and structural variants (SVs) between the A/J and B6/J strains are indicated below, including the SNP creating the novel E-box in B6/J

to repress the expression of *Isl1* by interfering with a conserved enhancer region. Specifically, this *de novo* E-box may recruit transcription factors that act through passive repression, for example, by competing for DNA binding with, or masking the functional domains of, conserved transcriptional activators [35].

4 Conclusions

Understanding the neuronal composition of a structure—both the diversity of neuronal types and the number of each type—is a current challenge of the NIH BRAIN Initiative, as is establishing the wiring diagram interconnecting these various cell types. Studies on the retina make clear that the former concerns bear directly upon the latter one: the independent variation in those numbers has direct implications for neuronal connectivity between pre- and postsynaptic partners [13]. Variation in homotypic and afferent numbers shapes distinct aspects of dendritic differentiation and the resultant connectivity between such populations [14]. The pursuit of QTL associated with such variation in neuronal number provides us with a causal anchor to understand the molecular genetic control of these demographic traits, from which we can understand the developmental processes modulating the size of a population, ultimately shaping the connectivity of the nervous system, and providing a foothold for understanding disorders of the nervous system [36].

References

1. Masland RH (2011) Cell populations of the retina: the proctor lecture. Invest Ophthalmol Vis Sci 52(7):4581–4591. doi: 10.1167/iovs.10-7083
2. Sanes JR, Masland RH (2015) The types of retinal ganglion cells: current status and implications for neuronal classification. Annu Rev Neurosci 38:221–246. doi:10.1146/annurev-neuro-071714-034120
3. Macosko EZ, Basu A, Satija R, Nemesh J, Shekhar K, Goldman M, Tirosh I, Bialas AR, Kamitaki N, Martersteck EM, Trombetta JJ, Weitz DA, Sanes JR, Shalek AK, Regev A, McCarroll SA (2015) Highly parallel genome-wide expression profiling of individual cells using nanoliter droplets. Cell 161:1202–1214
4. Helmstaedter M, Briggman KL, Turaga SC, Jain V, Seung HS, Denk W (2013) Connectomic reconstruction of the inner plexiform layer in the mouse retina. Nature 500:168–174. doi:10.1038/nature12346
5. Reese BE, Keeley PW (2015) Design principles and developmental mechanisms underlying retinal mosaics. Biol Rev 90:854–876. doi:10.1111/brv.12139
6. Masland RH (2012) The neuronal organization of the retina. Neuron 76:266–280
7. Jeon C-J, Strettoi E, Masland RH (1998) The major cell populations of the mouse retina. J Neurosci 18:8936–8946
8. Masland RH, Rizzo JF, Sandell JH (1993) Developmental variation in the structure of the retina. J Neurosci 13:5194–5202
9. Williams RW, Strom RC, Rice DS, Goldowitz D (1996) Genetic and environmental control of variation in retinal ganglion cell number in mice. J Neurosci 16:7193–7205
10. Williams RW, Strom RC, Goldowitz D (1998) Natural variation in neuron number in mice is linked to a major quantitative trait locus on Chr 11. J Neurosci 18:138–146
11. Williams RW, Strom RC, Zhou G, Yan Z (1998) Genetic dissection of retinal development. Sem Cell Dev Biol 9:249–255
12. Reese BE, Raven MA, Stagg SB (2005) Afferents and homotypic neighbors regulate horizontal cell morphology, connectivity and retinal coverage. J Neurosci 25:2167–2175
13. Keeley PW, Whitney IE, Madsen NR, St. John AJ, Borhanian S, Leong SA, Williams RW, Reese BE (2014) Independent genomic control of neuronal number across retinal cell types. Dev Cell 30:103–109
14. Reese BE, Keeley PW, Lee SC, Whitney IE (2011) Developmental plasticity of dendritic morphology and the establishment of coverage and connectivity in the outer retina. Dev Neurobiol 71:1273–1285
15. Kaplan S, Odaci E, Canan S, Onger ME, Aslan H, Unal B (2012) The dissector counting technique. Neuroquantology 10:44–53
16. Whitney IE, Raven MA, Ciobanu DC, Poché RA, Ding Q, Elshatory Y, Gan L, Williams RW, Reese BE (2011) Genetic modulation of horizontal cell number in the mouse retina. Proc Natl Acad Sci U S A 108:9697–9702
17. Whitney IE, Raven MA, Ciobanu DC, Williams RW, Reese BE (2009) Multiple genes on chromosome 7 regulate dopaminergic amacrine cell number in the mouse retina. Invest Ophthalmol Vis Sci 50:1996–2003
18. Whitney IE, Raven MA, Lu L, Williams RW, Reese BE (2011) A QTL on chromosome 10 modulates cone photoreceptor number in the mouse retina. Invest Ophthalmol Vis Sci 52:3228–3236
19. Strettoi E, Volpini M (2002) Retinal organization in the bcl-2-overexpressing transgenic mouse. J Comp Neurol 446:1–10
20. Zhou G, Williams RW (1999) Eye1 and Eye2: gene loci that modulate eye size, lens weight, and retinal area in the mouse. Invest Ophthalmol Vis Sci 40:817–825
21. Hegmann JP, Possidente B (1981) Estimating genetic correlations from inbred strains. Behav Genet 11:103–114
22. Singer JB, Hill AE, Burrage LC, Olszens KR, Song J, Justice M, O'Brien WE, Conti DV, Witte JS, Lander ES, Nadeau JH (2004) Genetic dissection of complex traits with chromosome substitution strains of mice. Science 304:445–448
23. Semaan SJ, Nickells RW (2010) A single nucleotide polymorphism in the Bax gene promoter affects transcription and influences retinal ganglion cell death. ASN Neuro 2(2):00032. doi:10.1042/AN20100003
24. Keane TM, Goodstadt L, Danecek P, White MA, Wong K, Yalcin B, Heger A, Agam A, Slater G, Goodson M, Furlotte NA, Eskin E, Nellaker C, Whitley H, Cleak J, Janowitz D, Hernandez-Pliego P, Edwards A, Belgard TG, Oliver PL, McIntyre RE, Bhomra A, Nicod J, Gan X, Yuan W, van der Weyden L, Steward CA, Bala S, Stalker J, Mott R, Durbin R, Jackson IJ, Czechanski A, Guerra-Assuncao JA, Donahue LR, Reinholdt LG, Payseur BA, Ponting CP, Birney E, Flint J, Adams DJ (2011) Mouse genomic variation and its effect on phenotypes and gene regulation. Nature 477:289–294, doi:nature10413 [pii]

25. Blackshaw S, Harpavat S, Trimarchi J, Cai L, Huang H, Kuo WP, Weber G, Lee K, Fraioli RE, Cho SH, Yung R, Asch E, Ohno-Machado L, Wong WH, Cepko CL (2004) Genomic analysis of mouse retinal development. PLoS Biol 2(9), e247
26. Kay JN, Chu MW, Sanes JR (2012) MEGF10 and MEGF11 mediate homotypic interactions required for mosaic spacing of retinal neurons. Nature 483:465–469
27. Siegert S, Cabuy E, Scherf BG, Kohler H, Panda S, Le YZ, Fehling HJ, Gaidatzis D, Stadler MB, Roska B (2012) Transcriptional code and disease map for adult retinal cell types. Nat Neurosci 15:487–495. doi:10.1038/nn.3032
28. Bateman A, Martin MJ, O'Donovan C, Magrane M, Apweiler R, Alpi E, Antunes R, Ar-Ganiska J, Bely B, Bingley M, Bonilla C, Britto R, Bursteinas B, Chavali G, Cibrian-Uhalte E, Da Silva A, De Giorgi M, Dogan T, Fazzini F, Gane P, Cas-Tro LG, Garmiri P, Hatton-Ellis E, Hieta R, Huntley R, Legge D, Liu WD, Luo J, MacDougall A, Mutowo P, Nightin-Gale A, Orchard S, Pichler K, Poggioli D, Pundir S, Pureza L, Qi GY, Rosanoff S, Saidi R, Sawford T, Shypitsyna A, Turner E, Volynkin V, Wardell T, Watkins X, Watkins CA, Figueira L, Li WZ, McWilliam H, Lopez R, Xenarios I, Bougueleret L, Bridge A, Poux S, Redaschi N, Aimo L, Argoud-Puy G, Auchincloss A, Axelsen K, Bansal P, Baratin D, Blatter MC, Boeckmann B, Bolleman J, Boutet E, Breuza L, Casal-Casas C, De Castro E, Coudert E, Cuche B, Doche M, Dornevil D, Duvaud S, Estreicher A, Famiglietti L, Feuermann M, Gasteiger E, Gehant S, Gerritsen V, Gos A, Gruaz-Gumowski N, Hinz U, Hulo C, Jungo F, Keller G, Lara V, Lemercier P, Lieberherr D, Lombardot T, Martin X, Masson P, Morgat A, Neto T, Nouspikel N, Paesano S, Pedruzzi I, Pilbout S, Pozzato M, Pruess M, Rivoire C, Roechert B, Schneider M, Sigrist C, Sonesson K, Staehli S, Stutz A, Sundaram S, Tognolli M, Verbregue L, Veuthey AL, Wu CH, Arighi CN, Arminski L, Chen CM, Chen YX, Garavelli JS, Huang HZ, Laiho KT, McGarvey P, Natale DA, Suzek BE, Vinayaka CR, Wang QH, Wang YQ, Yeh LS, Yerramalla MS, Zhang J, Consortium U (2015) UniProt: a hub for protein information. Nucleic Acids Res 43:D204–D212. doi:10.1093/nar/gku989
29. Edqvist PH, Lek M, Boije H, Lindbäck SM, Hallböök F (2008) Axon-bearing and axon-less horizontal cell subtypes are generated consecutively during chick retinal development from progenitors that are sensitive to follistatin. BMC Dev Biol 8:46. doi:10.1186/1471-1213X-1188-1146
30. Elshatory Y, Everhart D, Deng M, Xie X, Barlow RB, Gan L (2007) Islet-1 controls the differentiation of retinal bipolar and cholinergic amacrine cells. J Neurosci 27:12707–12720
31. Elshatory Y, Deng M, Xie X, Gan L (2007) Expression of the LIM-homeodomain protein Isl1 in the developing and mature mouse retina. J Comp Neurol 503:182–187
32. Ovcharenko I, Nobrega MA, Loots GG, Stubbs L (2004) ECR Browser: a tool for visualizing and accessing data from comparisons of multiple vertebrate genomes. Nucleic Acids Res 32:W280–W286. doi:10.1093/nar/gkh355
33. Loots GG, Ovcharenko I (2004) rVISTA 2.0: evolutionary analysis of transcription factor binding sites. Nucleic Acids Res 32:W217–W221. doi:10.1093/nar/gkh383
34. Fischer A, Gessler M (2007) Delta–Notch—and then? Protein interactions and proposed modes of repression by Hes and Hey bHLH factors. Nucleic Acids Res 35:4583–4596
35. Thiel G, Lietz M, Hohl M (2004) How mammalian transcriptional repressors work. Eur J Biochem 271:2855–2862. doi:10.1111/j.1432-1033.2004.04174.x
36. Geschwind DH, Flint J (2015) Genetics and genomics of psychiatric disease. Science 349:1489–1494. doi:10.1126/science.aaa8954

Chapter 18

Systems Genetics Analysis to Identify the Genetic Modulation of a Glaucoma-Associated Gene

Sumana R. Chintalapudi and Monica M. Jablonski

Abstract

Loss of retinal ganglion cells (RGCs) is one of the hallmarks of retinal neurodegenerative diseases, glaucoma being one of the most common. Recently, γ-synuclein (SNCG) was shown to be highly expressed in the somas and axons of RGCs. In various mouse models of glaucoma, downregulation of *Sncg* gene expression correlates with RGC loss. To investigate the regulation of *Sncg* in RGCs, we used a systems genetics approach to identify a gene that modulates the expression of *Sncg*, followed by confirmatory studies in both healthy and diseased retinas. We found that chromosome 1 harbors an eQTL that modulates the expression of *Sncg* in the mouse retina and identified *Pfdn2* as the candidate upstream modulator of *Sncg* expression. Downregulation of *Pfdn2* in enriched RGCs causes a concomitant reduction in *Sncg*. In this chapter, we describe our strategy and methods for identifying and confirming a genetic modulation of a glaucoma-associated gene. A similar method can be applied to other genes expressed in other tissues.

Key words Use case, Eye, Retinal neurodegenerative diseases SNCG, PFDN2, Primary retinal ganglion cells, Systems genetics, eQTL, Flow cytometry, siRNA transfection

1 Introduction

Glaucoma is the world's leading retinal neurodegenerative disease that causes irreversible vision loss due to degeneration of retinal ganglion cell (RGC) somas and their axons [1]. Little is known about the molecular identity of RGCs and molecular changes occurring within degenerating RGCs in glaucoma. Synucleins are small proteins associated with neurodegenerative diseases and some forms of cancer. They are studied predominantly in the brain. In diseases such as Alzheimer's and Parkinson's, mutant synucleins are key components of pathological inclusions [2, 3]. In contrast, information about their presence and functions in ocular tissues is scarce. Among the three known isoforms (α, β, and γ), γ-synuclein (SNCG) is expressed in the retina and optic nerve head [4, 5]. *Sncg* has been suggested to be a marker for glaucoma due to the association between RGC degeneration and the loss of *Sncg* mRNA and

Klaus Schughart and Robert W. Williams (eds.), *Systems Genetics: Methods and Protocols*, Methods in Molecular Biology, vol. 1488,
DOI 10.1007/978-1-4939-6427-7_18,

protein expression in both human glaucoma patients and animal models of glaucoma [6]. SNCG is involved in cellular signaling and modulates the level of transcription of specific genes and therefore the reduction of SNCG in RGCs may have vital consequences for these cells. In neuroretinal cells, SNCG plays an essential regulatory role in resistance to stress, and neuroprotection [7]. Another report demonstrated that a reduction in SNCG protein levels initiates an apoptotic death cascade due to the dephosphorylation of BCL2-binding protein [8]. Collectively, these studies suggest that loss of SNCG is a putative marker of RGC degeneration, yet the molecular targets and biological relevance of aberrant SNCG expression remains largely unknown.

Most of the knowledge regarding RGC loss in glaucoma has been gathered using techniques that rely on studies in animal models and cell lines, as well as from the identification of disease-related genes [9]. In spite of extensive studies using these techniques for past several years, molecular mechanisms associated with RGC death and glaucoma are still under debate. Recent advances in technology now provide tools that are capable of tracking genome-wide expression changes occurring in progressive pathological processes and diseases such as glaucoma. It is becoming increasingly evident that the application of network-based genetics or systems genetics approaches can not only provide insights into the roles of individual genes or developmental pathways but also illuminate relationships between different levels of a biologic system, such as the genome, transcriptome, and phenome [10]. One such resource of systems genetics is the GeneNetwork website and resource (www.genenetwork.org) that provides access to a wide variety of data such as genotypes (e.g., SNPs), phenotypes that are obtained from groups of related individuals—including human families, experimental crosses of strains of mice and rats, and organisms as diverse as ***Drosophila melanogaster***, ***Arabidopsis thaliana***, and barley—messenger RNA (mRNA) expression levels from populations of different strains of mice and rats. The tools on the site provide a wide range of functions ranging from simple graphical displays of variation in gene expression or other phenotypes, scatter plots of covariation among traits (Pearson or rank order), simple and complex network graphs, analysis of principal components and synthetic traits, QTL mapping using marker regression, interval mapping, and pair scans for epistatic interactions.

The genetic architecture of the eye is complex and includes variation of expression at the mRNA level. In order to analyze complex traits under different conditions, such as interactions between different genes, response to environment, and effects of stochastic events, it is necessary to use a genetic reference population. Genetic reference populations allow for the control of these conditions while maintaining genetic complexity that models aspects of human populations. Gene expression quantitative trait locus (eQTL) mapping has been widely used to define genomic regions whose genotype is correlated with the RNA expression in a panel of genetically diverse individuals [11]. The goal of such an analysis is to identify clusters of co-regulated genes, to discover candidate genes that may regulate the expression of these clusters, to elucidate normal tissue-specific physiology, and to seek candidate genes underlying disease-related phenotypes. In the present study, we used the BXD-based retina expression dataset to determine which gene regulates *Sncg* expression in murine RGCs. The outcomes of our investigation may provide clues to understanding the molecular mechanisms that account for the degenerative changes in RGCs in glaucoma.

2 Methods

2.1 The Use of Animal Models for Expression Genetics of the Eye

The largest and best-characterized murine reference panel is the BXD family of recombinant inbred (RI) mouse lines [12] (Fig. 1). These strains are well suited for integrating data across multiple phenotype domains spanning molecular, morphological, physiological, and behavioral traits. BXDs have been used extensively in

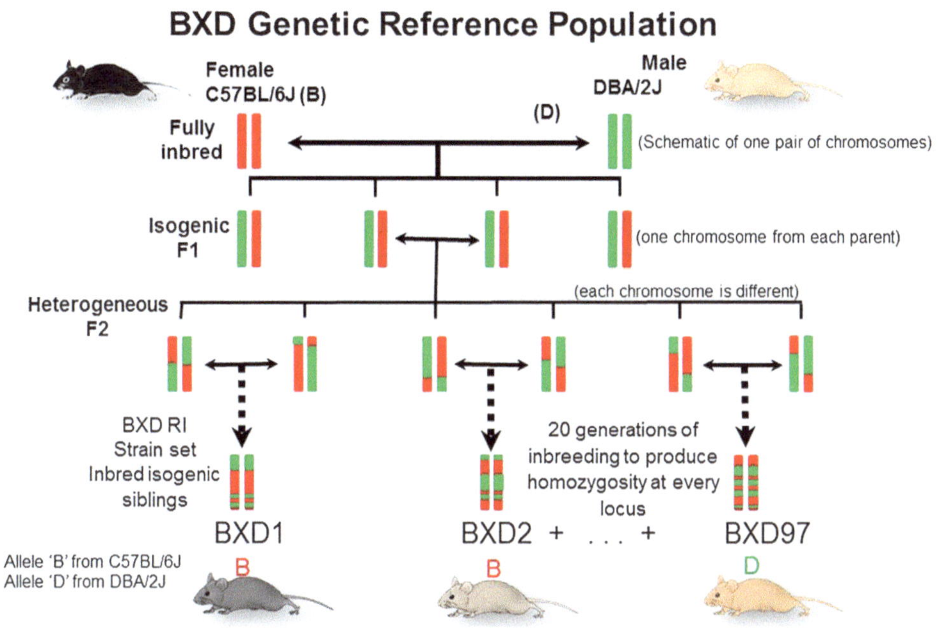

Fig. 1 Schematic representation of breeding strategy for creating a panel of RI mice from two parental strains—C57Bl/6J and DBA/2J. Following a cross between the parents, the F1 generation consists of individuals that inherited one chromosome from each parent. Intercrosses were then carried out between F1 individuals, generating recombination in the F2 generation. These F2 progeny were inbred until generation F20+, at which point the genome was isogenic and the strains are considered fully inbred. This genetic reference population has extensive variation between lines, and virtually no genetic variation between individuals of one line

genetic and genomic studies of the eye and central visual system [13–15]. The DBA/2J strain of mice is one of the parental strains of the BXD family. It closely mimics human hereditary pigmentary dispersion glaucoma and is therefore a widely used glaucoma model [16]. The greatest utility of this BXD family is that it can be used to map the chromosomal positions of sequence variants that cause differences in gene expression. Furthermore, reproducibility of eQTLs in the BXD family [17] and in an F2 cross between the same parental strains has been reported to be very high [18]. Over the last two decades, the BXD family has been exploited to study the genetics of immune function and infectious disease [19, 20]. It has also been used in behavioral and neuropharmacological research [21, 22]. The eyes and retinas of this BXD family have also been well studied for more than a decade, and we now possess extensive cytological and morphometric data on their eyes and retinas that can be studied with reference to differences in gene expression. Due to the fixed genotypes of each BXD strain and the availability of massive databases of expression data, it is possible to study correlations between two or more genes and identify upstream genetic regulators.

2.2 Strategy Used for This Study

In the present study, we used C57BL/6J and DBA/2J mice, as well as the BXD RI family as a collective genetic reference panel to determine which gene regulates *Sncg* expression in murine RGCs. We used two comprehensive and complementary expression datasets for the retina of the BXD mouse strains and aged DBA/2J mice. By examining the changes that occur in the gene expression profiles, we were able to identify an upstream modulator of *Sncg*. We also introduced partial correlation analysis to solidify the relationship between *Sncg* and its upstream modulator in the mouse retina. We have combined cutting-edge methodologies of systems genetics, meta-analyses with immunohistochemistry, FACS sorting, and gene knockdown studies to identify and validate the identity of a genetic modulator of *Sncg*, a gene that has been previously implicated in retinal ganglion cell death in glaucoma. The purpose of this chapter is to introduce a resource that binds together many datasets related to the genome, transcriptome, eye, and central visual system. Our goal is to improve the efficiency of making discoveries related to eye function and disease.

3 Methods

3.1 Mouse Strains

3.1.1 BXD Strains

To identify the genomic regions that modulate *Sncg* expression in the retina we used 2- to 4-month-old BXD strains for eQTL analysis. In this study retinas from a total of 80 BXD strains including 75 BXD RI strains ($n=307$), C57BL/6J ($n=4$), DBA/2J ($n=4$), the progeny of reciprocal F1 crosses of C57BL/6J ($n=4$), and DBA/2J ($n=4$) were used for generating microarray data.

Both sexes were equally represented. All animals used within this study were purchased from The Jackson Laboratory or were obtained from the breeding colonies of Drs. Robert Williams and Lu Lu at the University of Tennessee Health Science Center (UTHSC, Memphis, TN). Animals were housed under cyclic light (12 h on:12 h off) with 35 % humidity in a specific pathogen-free environment, with water and chow available ad libitum at UTHSC.

3.1.2 DBA/2J Mice

To investigate the relationship between *Sncg* and candidate gene in retinas of a mouse model of glaucoma—aged DBA/2J mice—we used GEO dataset generated by Howell et al. [23]. As noted by the authors, their DBA/2J mouse panel exhibited varying degrees of glaucoma-associated damage at 10.5 months including an increase in IOP and optic nerve head damage. Retinas from aged DBA/2J mice derived from the Howell et al. dataset had the following degrees of axonal damage: Severe (greater than 50 % axons lost); Moderate (10–50 % of axons lost); No/early 1 (no detectable damage); No/early 2 (no detectable damage); and Control (no glaucoma; 10.5-month-old DBA/2J-*Gpnmb+*/Sj mice). These eyes were selected because they encompassed a range of glaucoma severity where the RGC loss is prominent.

3.1.3 C57BL/6J Mice

A total of 80 mice at 8 weeks of age were used to isolate enriched retinal ganglion cells from the retinas and perform immunohistochemistry. Both sexes were equally represented.

3.1.4 BXD66 with High Intraocular Pressure

In our immunohistochemical analyses, a female BXD66 mouse aged ~12 months was used as a model of glaucoma due to the elevated average intraocular pressure (IOP) of this strain (18.50 ± 1.47 mmHg; unpublished observation but available on www.GeneNetwork.org). This particular strain has a high degree of optic nerve damage inclusive of many degenerating axons and gliotic scarring that are apparent on cross sections (unpublished data). All procedures involving mice were approved by the Animal Care and Use review board of UTHSC and followed the ARVO Statement for the Use of Animals in Ophthalmic and Vision Research in addition to the guidelines for laboratory animal experiments (Institute of Laboratory Animal Resources, Public Health Service Policy on Humane Care and Use of Laboratory Animals).

3.2 RNA Isolation, Microarray Hybridization, and Data Normalization

Retinal RNA isolation, microarray hybridization, and microarray data normalization were performed as reported earlier [13, 15, 24, 25]. Isolated RNAs were obtained from 80 mouse strains (total $n = 326$) [males ($n = 162$) and females ($n = 164$)]; independent retinal samples were hybridized to 346 Illumina Sentrix® Mouse Whole Genome-6 version 2.0 arrays (Illumina, San Diego, CA). Post-hybridization staining and washing were performed according to the manufacturer's protocols. The arrays were scanned and images were quantified. Two arrays were prepared and analyzed for each BXD strain using retinas (four each) pooled from both genders with an age range (48–118 days). The data was globally normalized with rank invariant and stabilization ($2z+8$). Detailed procedure for mouse genome arrays, annotation, and statistical analysis can be found in a published report [26].

The GEO Dataset generated by Dr. Simon John and colleagues at the Jackson Laboratory used a total of 110 Mouse Genome 430 version 2.0 GeneChip arrays (Affymetrix). 60 ONHs and 50 retinas were assessed in their study (NCBI accession number GSE26299). This GEO data was normalized in the identical manner and uploaded onto GeneNetwork. This dataset was derived from retinas of DBA/2J mice with varying degrees of glaucomatous optic nerve damage [23].

3.3 HEI Retina Database

The HEI Retina Database presents the retinal transcriptome profiles of 80 BXD strains including 75 BXD RI strains, the parental strains, the reciprocal crosses in a highly interactive website, GeneNetwork. The analytical tools within GeneNetwork allow for: (a) analysis of the dataset by identifying genetic variability across the BXD strains; (b) determining covariation among transcript expressions; (c) constructing genetic networks of the mouse retinal development; and (d) defining the genomic loci and causal models of linkage underlying complex traits in the retina.

3.4 Immunohistochemistry

To determine which retinal cell types express SNCG and PFDN2 immunohistochemical analysis was performed. Murine retinal sections embedded in low melting point agarose were prepared following our published methods [27]. Sections were permeabilized and blocked in blocking reagent (0.1% Triton X-100 and 0.5% Normal goat serum in PBS) for 1 h. Anti-prefoldin 2 (PFDN2; goat polyclonal IgG, Cat no: sc-19834, Santa Cruz Biotechnology) and anti-γ-synuclein (SNCG; rabbit polyclonal IgG, Cat no: GTX110483, GeneTex) were used as primary antibodies. Primary antibodies were diluted 1:50 in blocking reagent and incubated overnight at 4 °C. Alexa Fluor-tagged secondary antibodies specific for the antibody species and isotype of primary antibodies (1:200; Invitrogen) and TO-PRO3 iodide (1:4000; Invitrogen) were used to indicate the presence of the antigens of interest and nuclei, respectively. Image acquisition was performed using Nikon C1 confocal microscope within the Imaging Core Facility in the Hamilton Eye Institute. All microscope settings, including laser levels, were held constant to allow for relative comparisons of signal intensity within and between experiments. Images were minimally processed (signal intensities were not manipulated) using Nikon EZC1 confocal microscopy software.

3.5 Retinal Ganglion Cell Isolation

To validate our systems genetics outcome and determine if the candidate gene modulates SNCG expression in RGCs, we isolated enriched primary RGCs using fluorescence-activated cell sorting and cell surface antibodies.

3.5.1 Cell Suspension of Retinal Cells

C57BL/6J mice were sacrificed by cervical dislocation followed by enucleation. Collected retinas were dissociated without using any enzymatic digestion in PBS 1% FBS. The cell suspension were filtered through a Falcon 70 μM nylon strainer (BD Biosciences, San Jose, CA) and centrifuged at 1200 rpm for 7 min at RT. Cells were resuspended in PBS/1% FBS and kept on ice until ready to use.

3.5.2 Cell Labeling

To discriminate between live and dead cells a 1:100 solution of an amine-reactive dye, Zombie Aqua™ (BioLegend), diluted in PBS was used. Suspended retinal cells were treated with 5 μL of anti-CD16/32 (FcγR II/III block, BioLegend) to minimize nonspecific binding of antibodies. To determine the degree of autofluorescence, an aliquot of unlabeled retinal cells was used as negative control. The following isotype controls were used to confirm the specificity of primary and secondary antibodies: mouse IgG1 (clone MOPC-21, BioLegend) PE-Cy7, AF700, FITC PerCP-Cy5.5, and rat IgG2b (clone RTK4530, BioLegend). The AbC™ Anti-Mouse Bead Kit (Life Technologies, Carlsbad, CA) was used as a single-fluorochrome reference and instrument calibration. The following primary antibodies were used to label cells for 30 min on ice: anti-CD90.1 PerCP-Cy5.5 (Thy1.1, clone OX-7, BioLegend; exhibits no cross-reactivity with CD90.2) and anti-CD90.2 Alexa Fluor-700 (Thy1.2, clone 30-H12, BioLegend; exhibits no cross-reactivity with CD90.1) as a Pan-Thy1 marker; and anti-CD48 PE-Cy7 (clone HM48-1, BioLegend). We performed a positive selection for RGCs using a Thy1^{+} (Pan-Thy1) antibody and a negative selection using a CD48 antibody to remove monocytes, macrophages, and microglia. This positive and negative selection strategy allowed us to enrich for RGC with a Live Pan-Thy1^{+}CD48^{neg} phenotype. Cells were isolated by fluorescent activated cell sorting (FACS) using a BD Biosciences FACSAria Cell Sorter (BD Biosciences).

3.5.3 Sorting Strategy

Live retinal cells based on Zombie Aqua™ negativity were positively selected for Pan-Thy1 expression. Live PanThy1^{+} cells were selected based upon CD48 expression. The collected population of RGCs had the following characteristics: Live PanThy1^{+}CD48^{neg} cells.

3.6 Gene Knockdown Studies

To directly investigate if *Pfdn2* modulates *Sncg* expression in enriched RGCs, the levels of *Pfdn2* and *Sncg* mRNA transcripts and protein levels were measured using qPCR and flow cytometric analyses after transfecting the isolated RGCs with *Pfdn2* siRNA.

3.6.1 siRNA Transfection in Primary RGCs

Live RGCs were cultured in 96-well plates for 24 h in RGC culture media [28]. After 24 h, the RGC media was replaced with Accell delivery media (Dharmacon) containing SMARTpool siRNA targeting mouse *Pfdn2*, a pool of 4 different siRNAs targeting the gene to increase the potency (Target sequence 1: GCAAA GAACUGAACGAAUU, Target sequence 2: UGAUUAAAU GUUUUGGUCA, Target sequence 3: GAUUCCCACUUGU AAUUUC, Target sequence 4: GGACUGUCAAAGAAGUGCU; Cat no: E-062703-00-0005; Dharmacon), or a nontargeting GFP fluorescent siRNA (Cat no: D-001950-01-05; Dharmacon) at a final concentration of 1 μM according to the manufacturer's protocol.

3.6.2 RNA Isolation, cDNA Synthesis, and Quantitative Polymerase Chain Reaction

To investigate if *Pfdn2* modulates *Sncg* expression in enriched RGCs, the levels of *Pfdn2* and *Sncg* mRNA transcripts were measured by qPCR after transfecting the isolated RGCs with *Pfdn2* siRNA. RNA from 1.0×10^6 retinas cells was extracted using the Qiagen® miRNeasy Mini Kit (Qiagen, Valencia, CA) per manufacturer's protocol. Briefly, chloroform was added to the lysed cells to precipitate the RNA. The extract was passed through a spin column followed by on-column DNase digestion to increase purity and yield. RNA purity was assessed by analysis on NanoDrop Spectrophotometer. Genomic DNA elimination and First-Strand cDNA synthesis were performed using SuperScript® VILO™ cDNA Synthesis Kit (Life Technologies). The thermal cycling conditions included 25 °C for 10 min, 42 °C for 60 min, and 85 °C for 5 min. cDNA was pre-amplified for *Sncg* and *Pfdn2* using TaqMan® Gene Expression Assays and TaqMan® PreAmp Master Mix as per manufacturer's protocols (Life Technologies Inc., Applied Biosystems). 2.5 μL of diluted pre-amplified cDNA (1× Tris/EDTA at 1:5 ratio) was used in each plate for qRT-PCR. The ready-to-use TaqMan® Universal Master Mix II with UNG (Applied Biosystems) and following amplicons were used for the qRT-PCR reaction: *Sncg*, Mm00488345_ml; *Pfdn2*, Mm00448103_ml; and *Gapdh*, Mm99999915_gl (Life Technologies). Roche LightCycler® 480 Real-Time PCR system (Roche, Indianapolis, IN) was used for qPCR in 96-well plate format with a 10 μL final reaction volume. All reactions were performed in triplicates from three independent biological replicates. A relative quantification method (that is, the ΔΔCT method) was used to quantify difference in *Sncg* or *Pfdn2* with GAPDH used as the reference f in each sample. The relative quantification (R_q) was calculated using the following formula:

$$R_q = 2^{-\Delta\Delta C_T},$$

where $\Delta C_T = (C_T$ target gene$) - (C_T$ reference gene$)$ and $\Delta\Delta C_T = (\Delta C_T$ treatment sample$) - (\Delta C_T$ reference control sample$)$.

3.7 Flow Cytometry Analyses

To investigate if *Pfdn2* modulates *Sncg* expression in enriched RGCs, the levels of PFDN2 and SNCG protein levels were measured by FACS after transfecting the isolated RGCs with *Pfdn2* siRNA. Cells were fixed for 1 h at 4 °C using BD Cytofix/Cytoperm™ Buffer (BD Biosciences). Anti-gamma synuclein (GeneTex Inc.) and anti-prefoldin 2 (Santa Cruz Biotechnology, San Diego, CA) antibodies were diluted in permeabilization buffer and incubated with cells for 1 h at 4 °C. Alexa Fluor-568 goat anti-rabbit IgG or Alexa Fluor-488 donkey anti-goat IgG (Molecular Probes) were used as secondary antibodies. Cells were kept in PBS/1% FBS until ready for analysis. Data acquisition was performed on a BD LSRII Flow Cytometer (BD Biosciences) and analyses were performed using FlowJo vX10.0.6 (Tree Star, Inc., Ashland, OR).

4 Working Example

4.1 Variation in mRNA Expression as a Micro-trait

Variation in gene expression, often caused by natural genetic variation, is a major factor causing intra-species phenotypic differences. Hence, investigating the primary and causal sequence variants that modulate expression levels has been the focus of research in recent years due to its relevance to the differential disease risk among individuals. Transcript abundance can be used as a measure of the level of that gene's expression in each individual and can be analyzed as a quantitative trait or micro-trait. mRNA molecules are the product of a single gene with a specific well-defined single chromosomal locations. This makes it possible to identify the genetic variations that underlie inherited differences in gene transcription by expression-QTL (eQTL) analysis. eQTL refers explicitly to the mapped locus that influences the variable mRNA level and not the mRNA expression trait (the QT) itself. Hence, expression genetics can uncover the complex hierarchical networks that link genetic variation, through mRNA and protein levels, to clinical phenotypes that influence disease risk and progression. In this worked example we have used *Sncg* as a quantitative micro-trait to map its eQTL and identify the candidate gene modulating its expression.

4.2 Probe Annot ation, Variation in Transcript Expression, and Heritability

4.2.1 Variability Among Probe Sets for Single Genes

The goal of the gene expression microarray experiments is to obtain a list of genes that are upregulated or downregulated under particular conditions. In microarray, multiple probe sets assigned to the same gene detect cases of alternative splicing, use of alternative poly(A) sites, or errors. The expression patterns of the alternating splice isoforms and differences in their processing often vary among strains and among cell types. As a result, different probe sets for single genes can have different sets of QTLs.

On the Illumina arrays, *Sncg* is represented by two probe sets—ILMN_2939277 and ILMN_2598478 (Fig. 2a–c). ILMN_2939277 hybridized to Chr14: 35.184041 Mb within

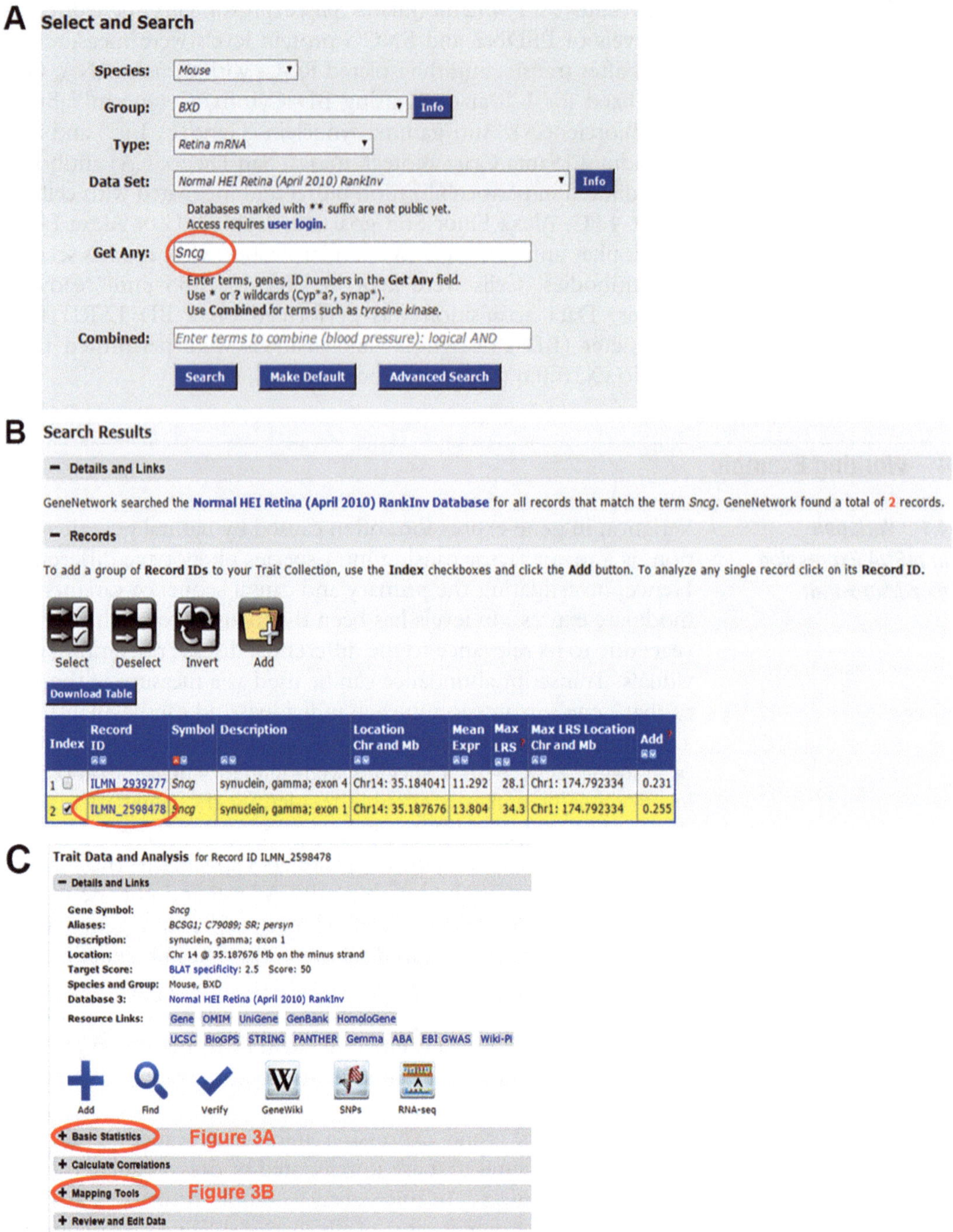

Fig. 2 Introduction to GeneNetwork. Step 1, Panel (**a**): Link to the GeneNetwork search page at www.genenetwork.org/. Choose Species = Mouse, Group = BXD, Type = Retina mRNA, Dataset = Normal HEI Retina (April 2010) RankInv. Enter search term "Sncg" into the search term field labeled "Get Any." Click on the Search button, your computer will send search term "Sncg" to GN, which then looks through all the records for matching terms. The search results will display a list of two data sets or Record IDs. Step 2, Panel (**b**): Both of the Record IDs are measurements of *Sncg* expression. Both measurements of mRNA levels target different parts of the mRNA molecule: #1 = Exon 4, or #2 = Exon 1. Which of the two should you pick? The best choice is usually the one which corresponds to coding sequence. In this case, both records correspond to coding region, but the highlighted Record ID ILMN_2598478 (#2) has higher LRS. Click the blue text ILMN_2598478 in the

exon 4 and ILMN_2598478 hybridized to Chr14: 35.187676 Mb within exon 1 of *Sncg*. ILMN_2939277 probe set had a mean expression level of 11.3 and likelihood ratio statistic (LRS) score of 28.1 versus ILMN_2598478 probe set that had a mean expression level of 13.8 and LRS score of 34.3. Both probe sets map to the same location within Chr14 at 35.187676 Mb. Using the UCSC Mouse Genome browser (http://ucscbrowserbeta.genenetwork.org/) both the probe sets were verified for presence or absence of SNPs within the binding site that could result in a false positive result. A single SNP (SNP I.D: rs30705163) was identified within the ILMN_2939277 probe binding site (data not shown). In contrast ILMN_2598478 probe binding site was SNP-free (data not shown). Because the ILMN_2598478 probe set was SNP-free, had higher average expression levels and higher LRS, it was selected to represent *Sncg* in our analyses.

On the Affymetrix panel used by Howell et al., *Sncg* is represented by one probe set, 1417788_at, which hybridizes to exon 1 of *Sncg* (Chr 14 at 35.183635). It is SNP-free, with a mean expression of 10.

4.2.2 Variation in Array Signal Across Strains

The expression level of a highly variable gene in an individual is considered as the "phenotype," which is possibly influenced by genetic determinants. Genetic analysis can therefore be used to map and to identify the genes and/or regulatory regions that control expression phenotypes. Genetic variations influencing gene expression may be within the regulatory sequences, such as promoters, enhancers, splice sites, or secondary structure motifs of the target gene, and so are genetically in *cis*, or they may be variations in the proteins and RNAs that interact with *cis*-regulatory sequences and so are genetically in *trans*.

The expression of *Sncg* in the retina among BXD mice varied (Fig. 3a) with BXD15 having the lowest *Sncg* gene expression of 12.65 ± 0.03 (expression $Log_2 \pm SEM$) and BXD61 having the highest at 14.47 ± 0.10. The average expression among all BXDs was 13.80 ± 0.33. The parental lines, C57BL/6J and DBA/2J, had *Sncg* expression levels of 13.75 ± 0.15 and 14.27 ± 0.09, respectively. These measurements are presented on a $\log_2$ scale and each unit represented a twofold difference in mRNA concentration in the retina.

4.2.3 Heritability of Variation in Expression

Heritability of gene expression data is the fraction of variation caused by genetic effects. In the retina expression dataset the

Fig. 2 (continued) "Search Results" window. Step 3, Panel (**c**): Clicking the blue text ILMN_2598478 will generate a new page called the "Trait Data and Analysis." The top of this page contains background information, including the database that was used, the mRNA or trait identifier, gene symbol and aliases, the chromosomal location and megabase position (Mb) of *Sncg* in the mouse genome. Other important links included in this page are: NCBI, OMIM, GenBank, BioGPS, STRING, PANTHER, Gemma, and the Allen Brain Atlas (ABA). Farther down the page there are four *grey horizontal bands* labeled Basic Statistics, Calculate Correlations, Mapping Tools, and Review and Edit Data that allow you to do more advanced analyses. These basic tools available on GN network were used to identify the candidate gene modulating *Sncg* expression in the retina

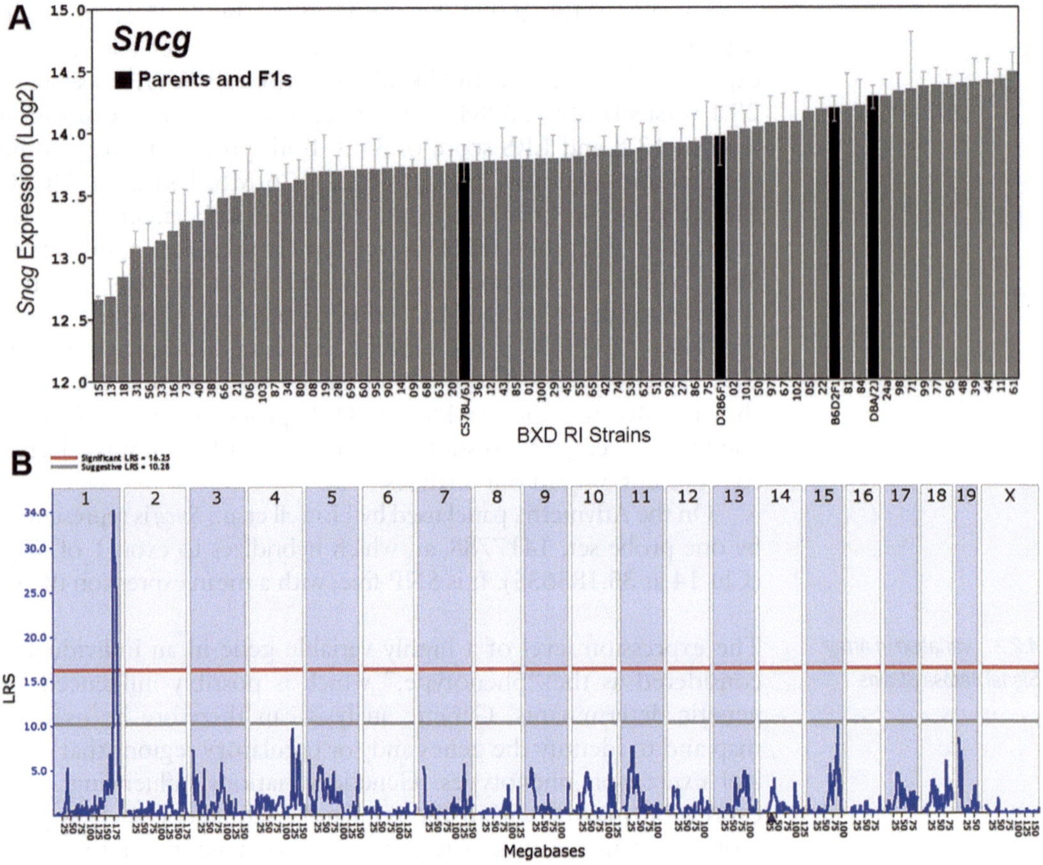

Fig. 3 Expression levels of *Sncg* across BXD strains and QTL mapping. Panel (**a**): Rank-ordered mean *Sncg* levels across the BXD recombinant inbred family. Values denote normalized relative expression levels on Log_2 scale (mean ± SEM). Panel (**b**): A significant *trans*-eQTL for *Sncg* is present on chromosome 1 between 167 and 190 Mb. *Red* and *gray horizontal lines* dictate significant and suggestive thresholds, respectively. The *purple triangle* indicates the location of *Sncg* within the genome

variation in expression was due to genetic differences between the genetic diversity of the strains, stability of environment, and technical error and confounds.

Heritability was calculated using the formula of Hegmann and Possidente [29]: $h^2 = 0.5\text{Vg}/(0.5\text{Vg} + \text{Ve})$, where h^2 is the heritability, Vg is the genetic variance, and Ve is the environmental variance. The heritability of the variation in *Sncg* expression was 0.57. This value indicates that 57% of the variation in the expression was due to genetic effects and the remaining 43% was due to environmental influences.

4.3 QTL Mapping

Expression quantitative trait loci (eQTLs) are regions of the genome containing DNA sequence variants that influence the expression level of one or more genes. Simple interval mapping was carried out using the WebQTL module on GeneNetwork to identify any significant eQTL(s) that modulate *Sncg* expression.

Genome-wide significance levels were estimated by performing 2000 permutations. Mapping results showed a significant trans-eQTL for *Sncg* on distal chromosome 1 with likelihood ratio statistics (LRSs) (Fig. 3b). The maximum LRS score of the locus was 34.1 Mb, which is equivalent to a logarithm of odds (LOD) ratio of 7.40. The confidence interval of this strong QTL extends from 171.5 to 183.5 Mb.

4.4 Genetic Correlations

4.4.1 Partial Correlation Analysis

To identify the candidate gene that modulates the expression of *Sncg* in the retina, partial correlation analysis was performed within GeneNetwork. A partial correlation reflects the level of association between a primary variable (i.e., *Sncg* expression level) and a target variable (i.e., upstream regulator of *Sncg* expression level) after controlling for one or more variables (i.e., the genetic variability of the trans-eQTL peak on Chr1) [30]. By holding constant the genetic variation of this region, any residual biological variation more accurately reflects the correlation between the expression of the *Sncg* and the candidate regulatory genes. In essence, application of this methodology removes genes with false positive correlations from the list of potential candidate modifiers. The correlation between the expression of *Sncg* (primary variable) and the expression of candidate regulatory gene within the retina database (target variables) was measured after mathematically controlling for the markers—rs8242766 (Chr1 at 172.981863) and rs4136041 (Chr1 at 177.366982)—that straddle the *trans*-eQTL (Fig. 4a–c).

4.4.2 Candidate Gene Identification in BXD Strains

Applying partial correlation analysis allowed us to identify a single gene candidate, *Prefoldin 2* (partial Pearson correlation value: $r=0.656$; $p=3.73\times10^{-13}$) (Figs. 4c and 5a). No other genes in that interval had significant expression levels or significant correlation values. This outcome solidified *Pfdn2* as the candidate upstream modulator of *Sncg*. A summary of both *Sncg* and *Pfdn2* is given in Table 1.

4.5 Candidate Gene Validation

4.5.1 Variation in Transcript Expression Across BXD Strains

On the Illumina array, *Pfdn2* is represented by one probe set—ILMN_129667—that hybridizes to exon 4 (Chr1 at 173.286888 Mb on the plus strand). The expression level of *Pfdn2* varied among the BXD strains from a low of 12.49 ± 0.10 in BXD13 to a high of 14.41 ± 0.2 in BXD39, with an average expression of 13.45 ± 0.08 (Fig. 5b).

4.5.2 Heritability of Variation in Expression in BXD Strains

The heritability of the variation in *Pfdn2* expression was 0.39, where 39 % variation in the expression was due to genetic effects and the remaining 61 % was due to environmental influences.

4.5.3 Genetic Correlations in BXD Strains

Genetic correlation is an estimate of the additive genetic effect that is shared between a pair of traits. The correlation coefficient is commonly used as a measure of the divergence of gene expression

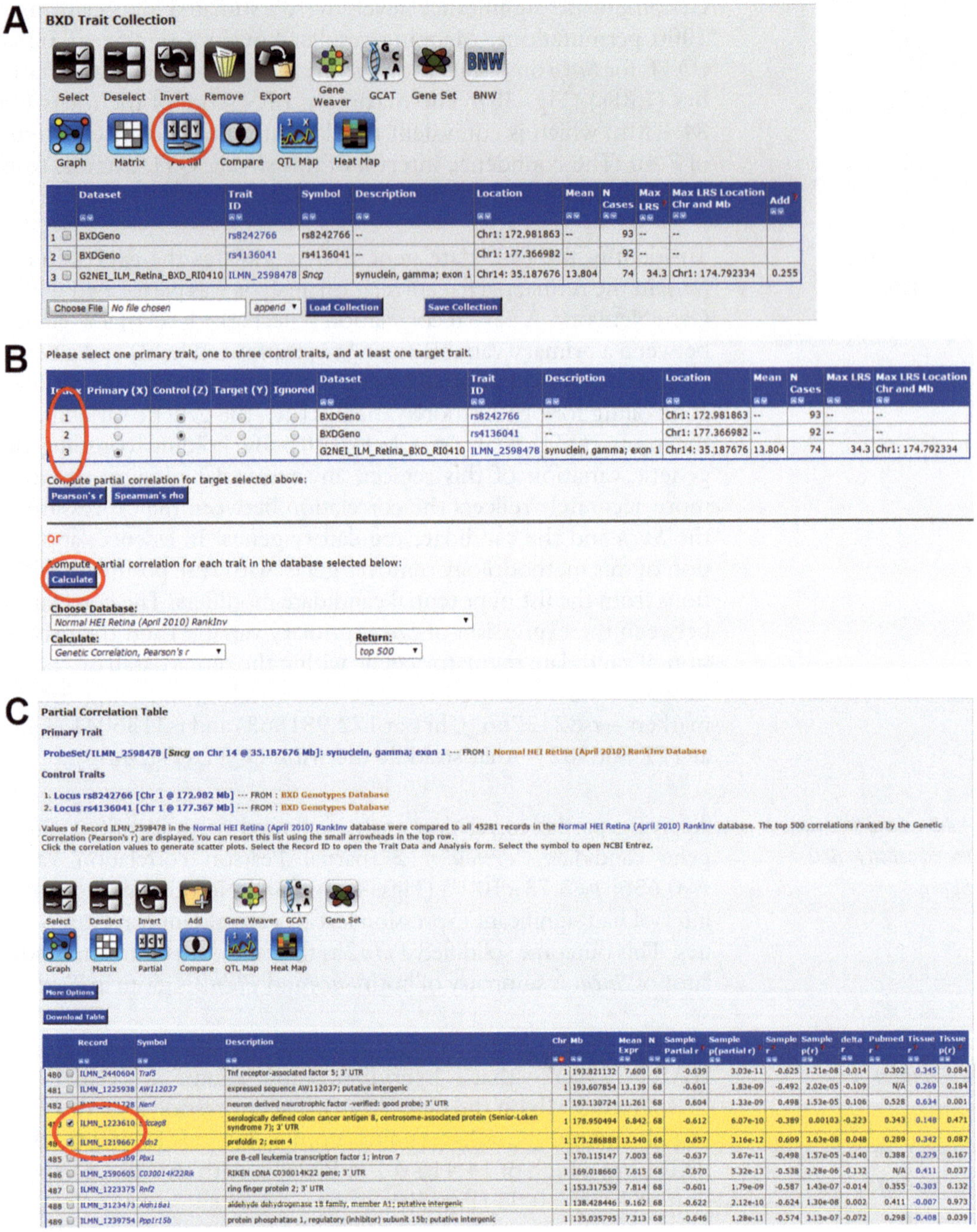

Fig. 4 Partial correlation analysis. Panel (**a**): The screenshot depicts the partial correlation analysis using the tools available on GN. Select and add *Sncg* into the BXD Trait Collection in addition to two SNP markers rs8242766 and rs4136041, which are located at the borders of the QTL on Chr 1 between 172 and 178 Mb (the location of the trans-eQTLs). The BXD collection will have one transcript and 2 SNPs in it. Select all transcript trait "Sncg" and the two SNP markers and then initiate a partial correlation analysis marked "Partial." Panel (**b**): Set the two SNP markers rs8242766 and rs4136041 as your CONTROL column. Select a single primary trait you will test "Sncg" probe set ILMN_2598478. Select the target traits in Normal HEI Retina (April 2010) RankInv. Panel (**c**): "IGNORE" all of the transcripts which are not with Chr 1 QTL peak to identify the candidate gene modulating *Sncg* expression

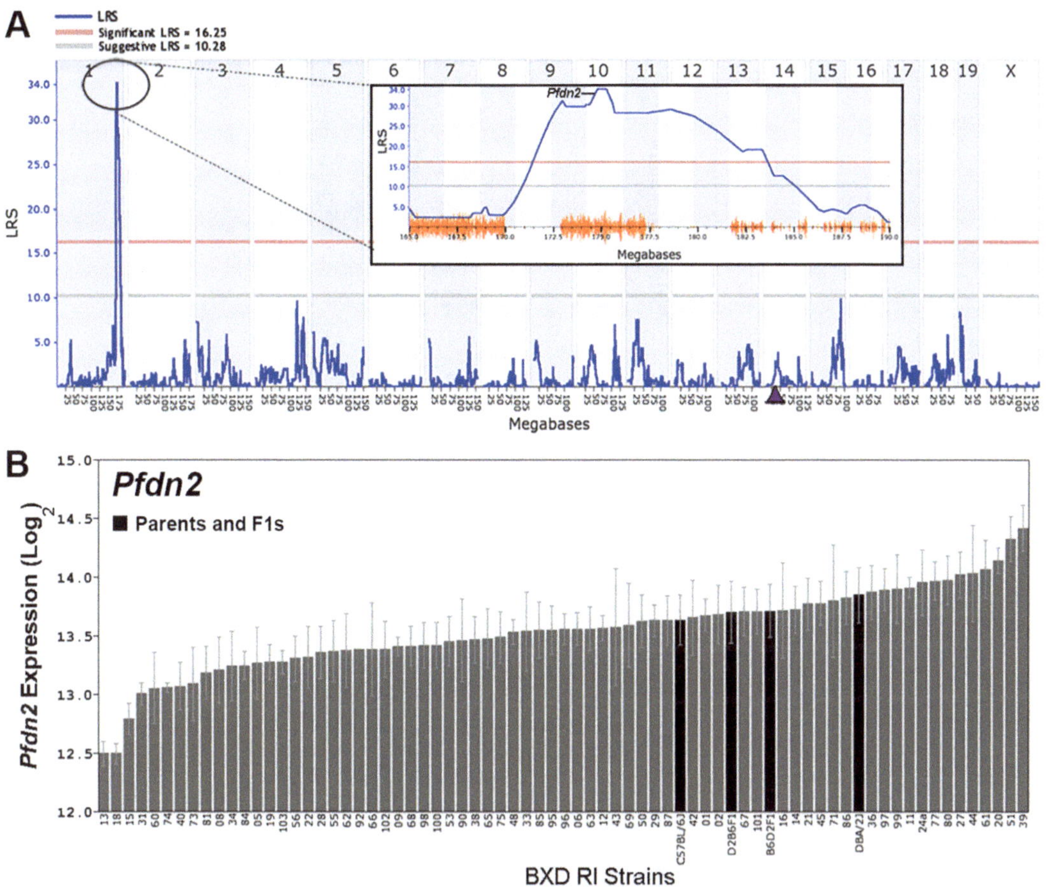

Fig. 5 Location of *Pfdn2* on Chr1 QTL peak and expression levels of *Pfdn2* across BXD strains. Panel (**a**): The *insert* depicts a zoomed-in view of the trans-eQTL on Chr1, which shows the location of *Pfdn2* within the LRS peak. Panel (**b**): Rank-ordered mean *Pfdn2* levels across the BXD recombinant inbred family. Values denote normalized relative expression levels on Log_2 scale (mean ± SEM)

profiles in a given population. The expression levels of *Sncg* and *Pfdn2* in the retinas of BXD mice aged 1-2 months were positively correlated (Pearson correlation value: $r = 0.609$; $p = 3.38 \times 10^{-09}$; Fig. 6a), which demonstrates that the two genes co-vary in the healthy retina.

4.5.4 Variation in Transcript Expression Across DBA/2J Strains in the Howell et al. Dataset

The average expression levels of *Sncg* and *Pfdn2* across the glaucomatous DBA/2J mice from the Howell dataset were 10.00 ± 0.85 and 11.00 ± 0.22, respectively (Fig. 6b). A significant variation in the expression of *Sncg* was noted among these mice, which ranged between a low of 8.08 in a mouse with a high degree of glaucomatous damage and a high of 10.72 in a mouse with a low degree of glaucomatous damage. *Pdfn2* expression varied in an inverse manner with an expression of 10.62 in a mouse with a low degree

Table 1
Summary of candidate genes

Gene ID	Location	Max LRS value and location	Probe set ID and BLAT specificity	Markers flanking the gene	Mean expression level	SNPs and indels
Sncg	Chr 14 at 35.187676 Mb on the minus strand	34.3; Chr1:174.792334	ILMN_2598478; 2.5	D14Mit45 at 35.170431 Mb rs13482135 at 36.200742 Mb	13.804	• Six intronic SNPs • Two deletions
Pfdn2	Chr 1 at 173.286888 Mb on the plus strand	12.8; Chr5:85.890222	ILMN_1219667; 3.1	NES13033186 at 173.149434 Mb rs3722740 at 173.679247 Mb	12.8	• 48 intronic SNPs • One 3'UTR SNP • One synonymous SNP in exon 4 (A → A, gcA → gcG) • Four insertions • Six deletions

Summary table showing candidate gene location, LRS values, expression level, and polymorphisms of glaucomatous damage to 11.54 in a mouse with a high degree of glaucomatous damage.

4.5.5 Genetic Correlations in DBA/2J Strains

The direct Pearson correlation between *Sncg* and *Pfdn2* across the glaucomatous D2 panel was significant ($r = -0.819$; $p = 1.09 \times 10^{-14}$; Fig. 6b).

4.5.6 Cellular Localization in Mouse Retina

To compile additional evidence to support our hypothesis that *Pfdn2* modulates *Sncg* expression, we determined the subcellular localization of both proteins in retinas from healthy mouse eyes using immunohistochemical analyses (Fig. 7). SNCG was strongly detected in GCL in the retinas of 3-month-old mice indicating that SNCG expression is limited to cells in the GCL (Fig 7a–c). PFDN2 immunofluorescence was ubiquitous throughout the retina in the NFL, GCL, IPL, OPL, and ONL (Fig. 7d–f).

4.5.7 Validation by RT-PCR and Flow Cytometry

To directly investigate if SNCG and PFDN2 are expressed in RGCs, enriched RGCs were isolated using a novel flow cytometry-based RGC isolation method (Fig. 8a) using our published protocol [31]; viable populations of murine RGCs were obtained using this method. To validate if *Pfdn2* modulates *Sncg* expression in enriched RGCs, the levels of *Pfdn2* and *Sncg* mRNA transcripts and protein levels were measured after transfecting the cells with *Pfdn2* siRNA.

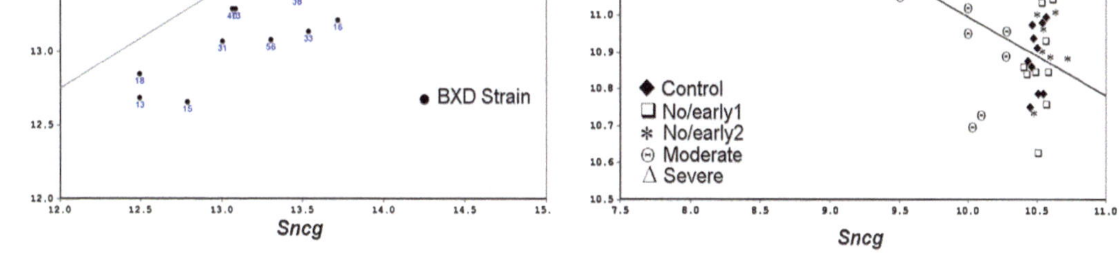

Fig. 6 Pearson correlation between *Sncg* and *Pfdn2* expression in retinas of BXD mice. Panel (**a**): The graph represents a strong positive correlation between expression levels of *Sncg* and *Pfdn2* ($p = 3.37 \times 10^{-09}$) in the retina of young mice with no glaucomatous damage. *Numbers* indicate BXD strains, and the parents. Panel (**b**): A strong negative correlation between *Sncg* and *Pfdn2* ($p < 10^{-14}$) is present in retinae from aged DBA/2J glaucoma mice. Optic nerves from aged DBA/2J mice derived from the Howell et al. data set had the following degrees of axonal damage: Severe (greater than 50 % axons lost, *triangle*); Moderate (10–50 % of axons lost, Greek capital theta); No/early 1 (no detectable damage, *asterisk*); No/early 2 (no detectable damage, *square*); and Control (no glaucoma; 10.5-month-old DBA/2J-*Gpnmb+*/Sj mice, *diamond*)

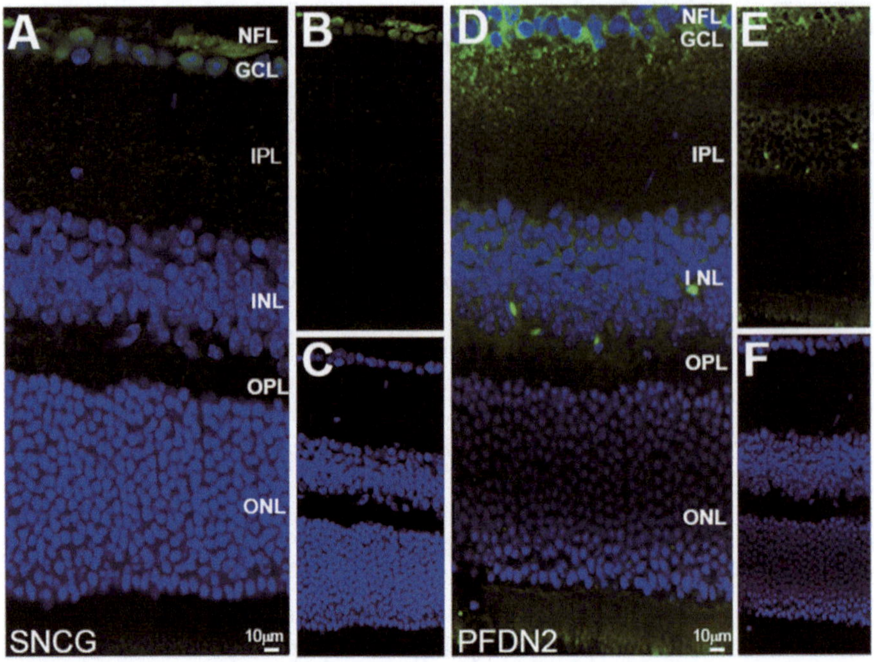

Fig. 7 Detection of SNCG and PFDN2 in healthy mouse retina. Panels **a–c**: In the retina from a healthy control mouse SNCG (*green*) immunoreactivity was detected in the GCL. Panels **d–f**: PFDN2 (*green*) was ubiquitously expressed in all retinal cells. To-PRO-3 iodide (*blue*) counterstained the nuclei. Merged images are shown in Panels **a** and **d**. *NFL* nerve fiber layer, *GCL* ganglion cell layer, *IPL* inner plexiform layer, *INL* inner nuclear layer, *OPL* outer plexiform layer, *ONL* outer nuclear layer. Scale bars: 10 µm

Quantitative transcriptional analyses demonstrate that both *Sncg* and *Pfdn2* are expressed in enriched primary RGCs (Fig. 8b). A significant knockdown of *Pfdn2* expression (61 % reduction; $p \leq 0.001$) was observed in the cells transfected by *Pfdn2* siRNA compared to cells exposed to control siRNA. Furthermore, knockdown of *Pfdn2* expression using siRNA resulted in a significant reduction in the expression of *Sncg* mRNA compared to samples treated with control siRNA (57 % reduction; $p \leq 0.001$) and samples without any siRNA treatment (89 % reduction; $p \leq 0.001$) (Fig. 8b), demonstrating that decreasing *Pfdn2* expression levels caused a significant reduction in *Sncg* expression in primary enriched murine RGCs. These results strongly support our hypothesis that *Pfdn2* modulates *Sncg* in RGCs.

Quantitative protein analyses demonstrate that both SNCG and PFDN2 are expressed in enriched primary RGCs. The transcriptional downregulation of *Pfdn2* reduced both PFDN2 and SNCG protein levels. Flow cytometry analyses (Fig. 8c) after cell transfection resulted in a significant reduction in the number of PFDN2^{+} cells (18.8 ± 8.2 % when transfected with *Pfdn2* siRNA

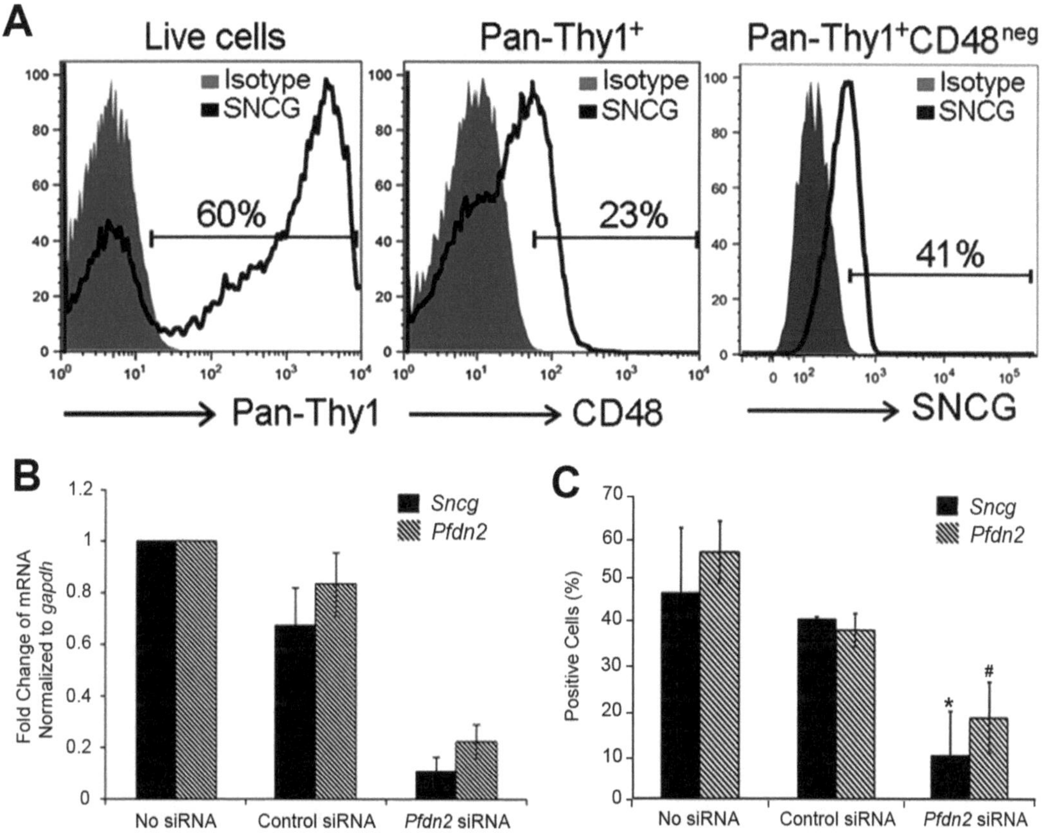

Fig. 8 Panel (**a**) Our cell sorting strategy to enrich for RGCs. Cells were positively selected based on Pan-Thy1 staining. The Pan-Thy1$^+$ population was negatively selected using a CD48 antibody to have a final phenotype of Live Pan-Thy1$^+$CD48^{neg} cells. An aliquot of the sorted cells was evaluated for intracellular SNCG labeling, which exhibit 41 % positivity. Panel (**b**) Transcriptional analyses for *Sncg* and *Pfdn2* in primary mouse RGCs demonstrate a significant downregulation of *Sncg* mRNA after *Pfdn2* siRNA treatment ($^*p < 0.001$). There is a significant difference between the level of *Pfdn2* expression in the control siRNA treatment group and the *Pfdn2* siRNA treatment group ($^\#p < 0.001$). There is also a significant difference between the level of *Sncg* expression in the control siRNA treatment group and the *Pfdn2* siRNA treatment group ($^*p < 0.001$). Results are presented as a fold change after normalizing to the level of *Gapdh* mRNA. Panel (**c**): Protein analyses by flow cytometry show a significant reduction in the percentage of enriched RGCs that were immunopositive for SNCG and PFDN2 after *Pfdn2* siRNA treatment. There is a significant difference between the number of cells that are immunopositive for PFDN2 ($^\#p < 0.05$) and SNCG ($^*p < 0.05$) after transfection with *Pfdn2* siRNA. Specific isotype controls were used to distinguish between positive and false positive cells. Results are shown as means ± SEM from three independent biological replicates performed in triplicate

compared to 39.1 ± 3.8 % when transfected with Control siRNA, $p = 0.049$). Under this same experimental condition, we also measured a statistically significant reduction in the percentage of enriched RGCs that were immunopositive for SNCG (10.2 ± 10.1 % when transfected with *Pfdn2* siRNA compared to 41.7 ± 0.6 % when transfected with Control siRNA, $p = 0.047$).

4.6 Molecular Pathway Identification

4.6.1 Correlation Comparison

To identify additional gene variants and pathways that are correlated with both *Sncg* and *Pfdn2* we performed correlation comparison. As a first step in this process, the transcript levels of our genes of interest—*Sncg* and *Pfdn2*—were compared using Pearson correlation with all 45,281 probe sets present on the Illumina V6.2 array. A list of top 500 genetically correlated genes in the retina database was generated using the "Correlation" tool. To produce a set of shared correlated transcripts, we selected all the common transcripts of *Sncg* and *Pfdn2* within the list of the top 500 correlates of both genes. The gene transcripts with expression levels less than seven were eliminated from the list; the remaining list of 163 shared correlates was analyzed by Gene Ontology (GO) enrichment analysis WebGestalt (http://bioinfo.vanderbilt.edu/webgestalt), as described previously [13, 25] (Fig. 9).

4.6.2 Gene Ontology Analysis

To identify genes that were co-regulated with both *Sncg* and *Pfdn2* and shared a functional relationship, we performed correlation comparison and Gene Ontology (GO) enrichment analysis. GO enrichment analysis allows users to query, browse, and visualize ontologies and gene product annotation data. We can use this to list highly correlated genes in terms of their function, location etc. This tool identifies GO terms that are significantly associated with the input gene lists, and visualizes the enriched GO terms in a directed acyclic graph (DAG).

Ten of 32 (31%) categories in the graph were statistically overrepresented in the GO tree generated from the 163 shared correlates. Of these overrepresented categories five (62%) clusters contained genes related to mitochondrial function: "Oxidoreductase activity" (11 genes); "Hydrogen ion transmembrane transporter activity" (four genes); and "Structural constituent of ribosome" (four genes); "NADH dehydrogenase activity" (two genes); and

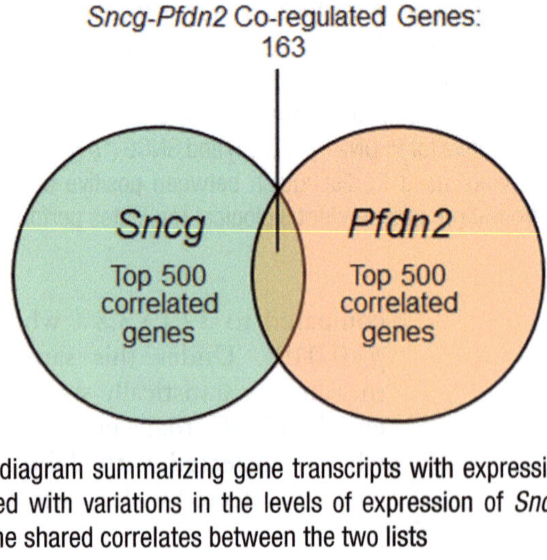

Fig. 9 Venn diagram summarizing gene transcripts with expression levels that are correlated with variations in the levels of expression of *Sncg* and *Pfdn2*, along with the shared correlates between the two lists

"Cytochrome-c oxidase activity" (two genes) (Fig. 10). The gene products encoded by these transcripts are present in mitochondria and are involved in maintaining the potential difference across the inner mitochondrial membrane and maintaining normal physiological function (Table 2).

5 Conclusions

The findings presented in this study provide strong support for the value of systems genetics, especially GeneNetwork, in discovering new upstream regulators of genes that can be confirmed by molecular analysis.

We exploited partial correlation analysis to solidify the relationship between *Sncg* and *Pfdn2* in the mouse retina. The statistically significant partial correlation between *Sncg* and *Pfdn2* reflects gene–gene interactions, as well as a regulatory relationship between the genes.

In healthy retinas, the protein levels of both genes are significantly correlated in a positive manner. In contrast, the relationship between the genes appears to have changed to an inverse relationship with reduced *Sncg* levels in glaucoma as the disease progressed.

Pfdn2 is located in the QTL rich region (*Qrr1*) on chromosome 1, which is a genomic region of unusually high gene density and contains major regulatory QTLs for various behavioral, metabolic, physiologic, and immunological processes [32–35] including diverse epileptic traits [36, 37]. Knocking down *Pfdn2* expression in primary murine RGCs significantly reduced the expression of *Sncg*, confirming that *Pfdn2* regulates *Sncg* expression in murine RGCs.

6 Further Considerations and Limitations

The use of BXD mice limits the number of genomes investigated to only two common inbred strains (C57BL/6J and DBA/2J). A possible weakness in nominating genes that act through gene expression is that there is a probability of missing gene variants that cause null or poorly trafficked proteins which may not be detected in the microarray.

7 Outlook

Systems genetics can be used to study various induced and inherited models of glaucoma, thereby identifying molecules and pathways for further mechanistic evaluation. The high level of strain variation in gene expression is a powerful tool that can be used to identify candi-

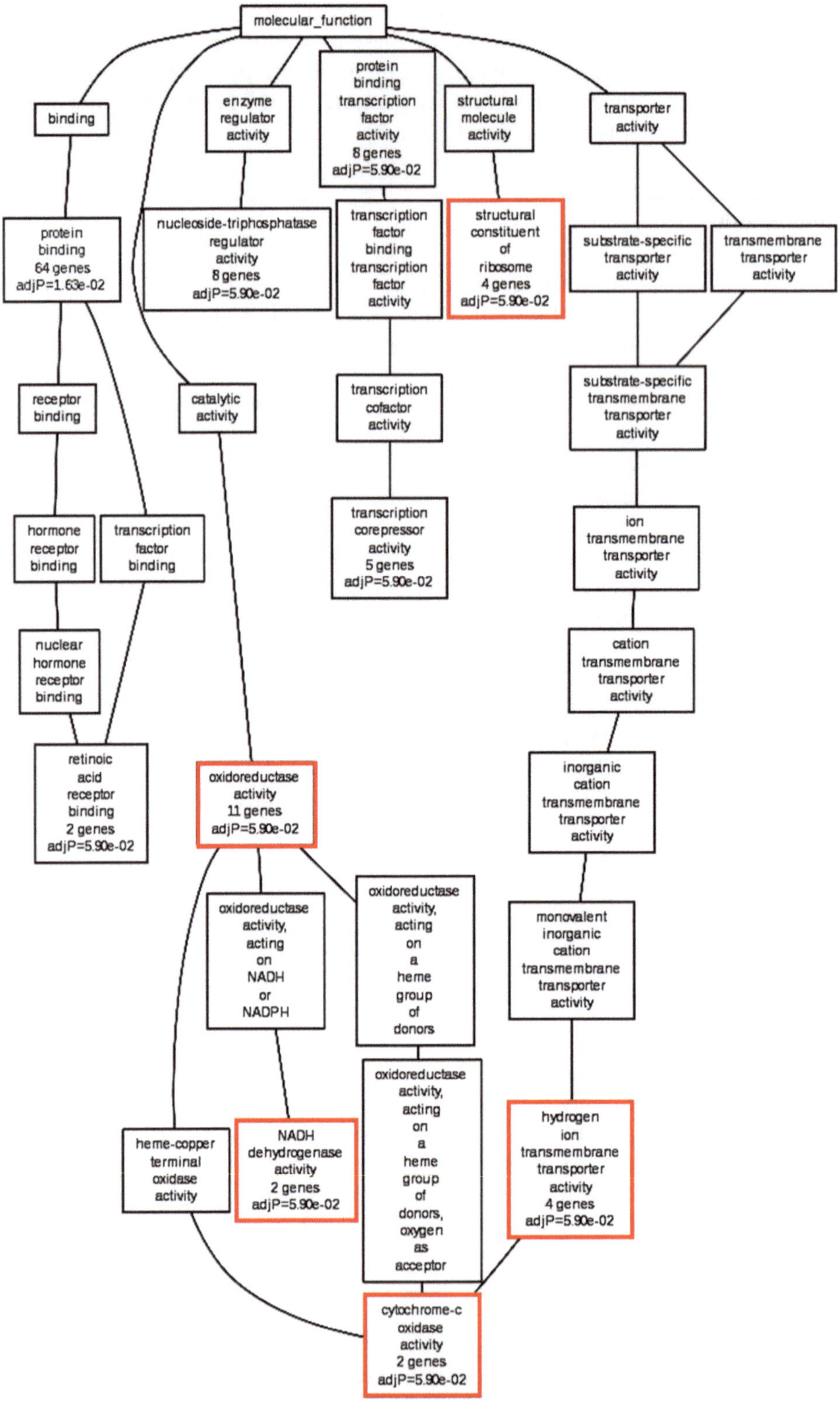

Fig. 10 Gene ontology analysis of the shared correlates of *Sncg* and *Pfdn2* is presented. Categories that are statistically overrepresented are presented in (adjusted $p = 5.9 \times 10^{-2}$ for all groups). Categories indicated with *red boxes* indicate mitochondria-associated genes

Table 2
Gene ontology

Database: molecular function name: Cytochromec oxidase activity ID:GO:0004129					
C = 16; O = 2; E = 0.12; R = 17.06; rawP = 0.0060; adjP = 0.0590					
Index	**User ID**	**Gene symbol**	**Gene names**	**Entrez gene**	**Ensemble**
1	ILMN_2657141	*Surf1*	Surfeit gene 1	20930	ENSMUSG00000015790
2	ILMN_1254971	*Cox6b1*	Cytochrome c oxidase, subunit VIb polypeptide	110323	ENSMUSG00000036751

Database: molecular function Name: NADH dehydrogenase activity ID:GO:0003954					
C = 18; O = 2; E = 0.13; R = 15.17; rawP = 0.0075; adjP = 0.0590					
Index	**User ID**	**Gene symbol**	**Gene names**	**Entrez gene**	**Ensemble**
1	ILMN_1220362	*Ndufa12*	NADH dehydrogenase (ubiquinone) 1 alpha subcomplex, 12	66414	ENSMUSG00000020022
2	ILMN_2985053	*Ndufv2*	NADH dehydrogenase (ubiquinone) flavoprotein 2	72900	ENSMUSG00000024099

Database: molecular function Name: oxidoreductase activity ID:GO:0016491					
C = 663; O = 11; E = 4.86; R = 2.26; rawP = 0.0095; adjP = 0.0590					
Index	**User ID**	**Gene symbol**	**Gene names**	**Entrez gene**	**Ensemble**
1	ILMN_2752552	*Sod1*	Superoxide dismutase 1, soluble	20655	ENSMUSG00000022982
2	ILMN_2610531	*Spr*	Sepiapterin reductase	20751	ENSMUSG00000033735
3	ILMN_1220362	*Ndufa12*	NADH dehydrogenase (ubiquinone) 1 alpha subcomplex, 12	66414	ENSMUSG00000020022
4	ILMN_2657141	*Surf1*	Surfeit gene 1	20930	ENSMUSG00000015790
5	ILMN_1258815	*Jmjd1c*	Jumonji domain containing 1C	108829	ENSMUSG00000037876
6	ILMN_2985053	*Ndufv2*	NADH dehydrogenase (ubiquinone) flavoprotein 2	72900	ENSMUSG00000024099
7	ILMN_2512849	*Uqcrh*	Ubiquinolcytochrome c reductase hinge protein	66576	ENSMUSG00000063882
8	ILMN_1220100	*Ywhae*	Tyrosine 3monooxygenase/ Tryptophan 5monooxygenase activation protein, epsilon polypeptide	22627	ENSMUSG00000020849

(continued)

Table 2
(continued)

Database: molecular function Name: oxidoreductase activity ID:GO:0016491					
C = 663; O = 11; E = 4.86; R = 2.26; rawP = 0.0095; adjP = 0.0590					
Index	**User ID**	**Gene symbol**	**Gene names**	**Entrez gene**	**Ensemble**
9	ILMN_2594149	*Sdhc*	Succinate dehydrogenase complex, subunit C, integral membrane protein	66052	ENSMUSG00000058076
10	ILMN_1254971	*Cox6b1*	Cytochrome c oxidase, subunit VIb polypeptide 1	110323	ENSMUSG00000036751

Database: molecular function Name: hydrogen ion transmembrane transporter activity ID:GO:0015078					
C = 71; O = 4; E = 0.52; R = 7.69; rawP = 0.0018; adjP = 0.0590					
Index	**User ID**	**Gene symbol**	**Gene names**	**Entrez gene**	**Ensemble**
1	ILMN_2657141	*Surf1*	Surfeit gene 1	20930	ENSMUSG00000015790
2	ILMN_1254971	*Cox6b1*	Cytochrome c oxidase, subunit VIb polypeptide 1	110323	ENSMUSG00000036751
3	ILMN_1249783	*Atp5c1*	ATP synthase, H+ transporting, mitochondrial F1 complex, gamma polypeptide 1	11949	ENSMUSG00000025781
4	ILMN_2512849	*Uqcrh*	Ubiquinolcytochrome c reductase hinge protein	66576	ENSMUSG00000063882

Database: molecular function Name: structural constituent of ribosome ID:GO:0003735					
C = 81; O = 4; E = 0.59; R = 6.74; rawP = 0.0030; adjP = 0.0590					
Index	**User ID**	**Gene symbol**	**Gene names**	**Entrez gene**	**Ensemble**
1	ILMN_1229110	*Mrpl33*	Mitochondrial ribosomal protein L33	66845	ENSMUSG00000029142
2	ILMN_2646163	*Rpl7*	Ribosomal protein L7	19989	ENSMUSG00000043716
3	ILMN_2883147	*Mrpl12*	Mitochondrial ribosomal protein L12	56282	ENSMUSG00000039640
4	ILMN_1236904	*Mrpl23*	Mitochondrial ribosomal protein L23	19935	ENSMUSG00000037772

date genes for human ocular disease, assemble genetic networks regulating tissue-specific gene expression, and identify complex interactions among gene variants that generate variation in eye structure and function. When coupled with genomic data, systems genetics analyses can validate these mechanistic insights on disease occurrence and can lead to the development of future therapies.

Acknowledgments

We would like to thank Dr. Robert Williams (UTHSC) and Dr. Vanessa Morales-Tirado (UTHSC) for contributing to the research design and execution of our original study that is published in *The FEBS Journal*. We would like to thank Dr. Dan Rosson (UTHSC) for technical assistance with the cell sorting. We thank Dr. Michael Whitt (UTHSC), Dr. Tony Reiner (UTHSC), and Dr. R.K. Rao (UTHSC) for providing C57BL/6J mice eyes. We also thank Dr. Lu Lu (UTHSC) for his assistance in generating the BXD microarray datasets that were used in these analyses. We also thank Dr. Eldon Geisert (Emory University) and Mr. Bill Orr (UTHSC) for formatting the dataset of Howell et al. (NCBI accession number GSE26299) so that it could be mined within GeneNetwork. Funding provided by Juvenile Diabetes Research Foundation Grant (VMT), Research to Prevent Blindness Award (PI: James C. Fleming), National Eye Institute Vision Core Grant (PI: Dianna Johnson), National Eye Institute EY021200 (MMJ).

References

1. Lee DA, Higginbotham EJ (2005) Glaucoma and its treatment: a review. Am J Health Syst Pharm 62(7):691–699
2. Kruger R, Schols L, Muller T, Kuhn W, Woitalla D, Przuntek H, Epplen JT, Riess O (2001) Evaluation of the gamma-synuclein gene in German Parkinson's disease patients. Neurosci Lett 310(2-3):191–193
3. Rockenstein E, Hansen LA, Mallory M, Trojanowski JQ, Galasko D, Masliah E (2001) Altered expression of the synuclein family mRNA in Lewy body and Alzheimer's disease. Brain Res 914(1-2):48–56
4. Lavedan C, Leroy E, Dehejia A, Buchholtz S, Dutra A, Nussbaum RL, Polymeropoulos MH (1998) Identification, localization and characterization of the human gamma-synuclein gene. Hum Genet 103(1):106–112
5. Surguchov A, McMahan B, Masliah E, Surgucheva I (2001) Synucleins in ocular tissues. J Neurosci Res 65(1):68–77
6. Soto I, Oglesby E, Buckingham BP, Son JL, Roberson ED, Steele MR, Inman DM, Vetter ML, Horner PJ, Marsh-Armstrong N (2008) Retinal ganglion cells downregulate gene expression and lose their axons within the optic nerve head in a mouse glaucoma model. J Neurosci 28(2):548–561. doi:10.1523/jneurosci.3714-07.2008
7. Surgucheva I, Shestopalov VI, Surguchov A (2008) Effect of gamma-synuclein silencing on apoptotic pathways in retinal ganglion cells. J Biol Chem 283(52):36377–36385. doi:10.1074/jbc.M806660200
8. Wilding C, Bell K, Beck S, Funke S, Pfeiffer N, Grus FH (2014) Gamma-Synuclein antibodies have neuroprotective potential on neuroretinal cells via proteins of the mitochondrial apoptosis pathway. PLoS One 9(3):90737. doi:10.1371/journal.pone.0090737
9. Nickells RW (2012) The cell and molecular biology of glaucoma: mechanisms of retinal ganglion cell death. Invest Ophthalmol Vis Sci 53(5):2476–2481. doi:10.1167/iovs.12-9483h
10. Civelek M, Lusis AJ (2014) Systems genetics approaches to understand complex traits. Nat Rev Genet 15(1):34–48. doi:10.1038/nrg3575

11. Gibson G, Powell JE, Marigorta UM (2015) Expression quantitative trait locus analysis for translational medicine. Genome Med 7(1):60. doi:10.1186/s13073-015-0186-7
12. Peirce JL, Lu L, Gu J, Silver LM, Williams RW (2004) A new set of BXD recombinant inbred lines from advanced intercross populations in mice. BMC Genet 5:7. doi:10.1186/1471-2156-5-7
13. Lu H, Wang X, Pullen M, Guan H, Chen H, Sahu S, Zhang B, Chen H, Williams RW, Geisert EE, Lu L, Jablonski MM (2011) Genetic dissection of the Gpnmb network in the eye. Invest Ophthalmol Vis Sci 52(7):4132–4142. doi:10.1167/iovs.10-6493
14. Jablonski MM, Freeman NE, Orr WE, Templeton JP, Lu L, Williams RW, Geisert EE (2011) Genetic pathways regulating glutamate levels in retinal Muller cells. Neurochem Res 36(4):594–603. doi:10.1007/s11064-010-0277-1
15. Geisert EE, Lu L, Freeman-Anderson NE, Templeton JP, Nassr M, Wang X, Gu W, Jiao Y, Williams RW (2009) Gene expression in the mouse eye: an online resource for genetics using 103 strains of mice. Mol Vis 15:1730–1763
16. Williams PA, Howell GR, Barbay JM, Braine CE, Sousa GL, John SW, Morgan JE (2013) Retinal ganglion cell dendritic atrophy in DBA/2J glaucoma. PLoS One 8(8), e72282. doi:10.1371/journal.pone.0072282
17. Burgess-Herbert SL, Cox A, Tsaih S-W, Paigen B (2008) Practical applications of the bioinformatics toolbox for narrowing quantitative trait loci. Genetics 180(4):2227–2235. doi:10.1534/genetics.108.090175
18. Peirce JL, Li H, Wang J, Manly KF, Hitzemann RJ, Belknap JK, Rosen GD, Goodwin S, Sutter TR, Williams RW, Lu L (2006) How replicable are mRNA expression QTL? Mamm Genome 17(6):643–656. doi:10.1007/s00335-005-0187-8
19. Bystrykh L, Weersing E, Dontje B, Sutton S, Pletcher MT, Wiltshire T, Su AI, Vellenga E, Wang J, Manly KF, Lu L, Chesler EJ, Alberts R, Jansen RC, Williams RW, Cooke MP, de Haan G (2005) Uncovering regulatory pathways that affect hematopoietic stem cell function using 'genetical genomics'. Nat Genet 37(3):225–232. doi:10.1038/ng1497
20. Miyairi I, Tatireddigari VR, Mahdi OS, Rose LA, Belland RJ, Lu L, Williams RW, Byrne GI (2007) The p47 GTPases Iigp2 and Irgb10 regulate innate immunity and inflammation to murine Chlamydia psittaci infection. J Immunol 179(3):1814–1824
21. Chesler EJ, Lu L, Shou S, Qu Y, Gu J, Wang J, Hsu HC, Mountz JD, Baldwin NE, Langston MA, Threadgill DW, Manly KF, Williams RW (2005) Complex trait analysis of gene expression uncovers polygenic and pleiotropic networks that modulate nervous system function. Nat Genet 37(3):233–242. doi:10.1038/ng1518
22. Gaglani SM, Lu L, Williams RW, Rosen GD (2009) The genetic control of neocortex volume and covariation with neocortical gene expression in mice. BMC Neurosci 10:44. doi:10.1186/1471-2202-10-44
23. Howell GR, Macalinao DG, Sousa GL, Walden M, Soto I, Kneeland SC, Barbay JM, King BL, Marchant JK, Hibbs M, Stevens B, Barres BA, Clark AF, Libby RT, John SW (2011) Molecular clustering identifies complement and endothelin induction as early events in a mouse model of glaucoma. J Clin Invest 121(4):1429–1444. doi:10.1172/jci44646
24. Templeton JP, Freeman NE, Nickerson JM, Jablonski MM, Rex TS, Williams RW, Geisert EE (2013) Innate immune network in the retina activated by optic nerve crush. Invest Ophthalmol Vis Sci 54(4):2599–2606. doi:10.1167/iovs.12-11175
25. Lu H, Li L, Watson ER, Williams RW, Geisert EE, Jablonski MM, Lu L (2011) Complex interactions of Tyrp1 in the eye. Mol Vis 17:2455–2468
26. Freeman NE, Templeton JP, Orr WE, Lu L, Williams RW, Geisert EE (2011) Genetic networks in the mouse retina: growth associated protein 43 and phosphatase tensin homolog network. Mol Vis 17:1355–1372
27. Nookala S, Gandrakota R, Wohabrebbi A, Wang X, Howell D, Giorgianni F, Beranova-Giorgianni S, Desiderio DM, Jablonski MM (2010) In search of the identity of the XAP-1 antigen: a protein localized to cone outer segments. Invest Ophthalmol Vis Sci 51(5):2736–2743. doi:10.1167/iovs.09-4286
28. Winzeler A, Wang JT (2013) Purification and culture of retinal ganglion cells from rodents. Cold Spring Harb Protoc 2013(7):643–652. doi:10.1101/pdb.prot074906
29. Hegmann JP, Possidente B (1981) Estimating genetic correlations from inbred strains. Behav Genet 11(2):103–114
30. Mulligan MK, Wang X, Adler AL, Mozhui K, Lu L, Williams RW (2012) Complex control of GABA(A) receptor subunit mRNA expression: variation, covariation, and genetic regulation. PLoS One 7(4), e34586. doi:10.1371/journal.pone.0034586
31. Chintalapudi SR, Morales-Tirado VM, Williams RW, Jablonski MM (2015) Multipronged approach to identify and validate a novel upstream regulator of Sncg in mouse retinal ganglion cells. FEBS J 283(4):678–693
32. Mozhui K, Ciobanu DC, Schikorski T, Wang X, Lu L, Williams RW (2008) Dissection

of a QTL hotspot on mouse distal chromosome 1 that modulates neurobehavioral phenotypes and gene expression. PLoS Genet 4(11), e1000260. doi:10.1371/journal.pgen.1000260

33. Ding Q, Cecarini V, Keller JN (2007) Interplay between protein synthesis and degradation in the CNS: physiological and pathological implications. Trends Neurosci 30(1):31–36. doi:10.1016/j.tins.2006.11.003
34. Koonin EV, Wolf YI, Aravind L (2001) Prediction of the archaeal exosome and its connections with the proteasome and the translation and transcription machineries by a comparative-genomic approach. Genome Res 11(2):240–252. doi:10.1101/gr.162001
35. Lehner B, Sanderson CM (2004) A protein interaction framework for human mRNA degradation. Genome Res 14(7):1315–1323. doi:10.1101/gr.2122004
36. Yalcin B, Willis-Owen SA, Fullerton J, Meesaq A, Deacon RM, Rawlins JN, Copley RR, Morris AP, Flint J, Mott R (2004) Genetic dissection of a behavioral quantitative trait locus shows that Rgs2 modulates anxiety in mice. Nat Genet 36(11):1197–1202. doi:10.1038/ng1450
37. Leygraf A, Hohoff C, Freitag C, Willis-Owen SA, Krakowitzky P, Fritze J, Franke P, Bandelow B, Fimmers R, Flint J, Deckert J (2006) Rgs 2 gene polymorphisms as modulators of anxiety in humans? J Neural Trans (Vienna, Austria: 1996) 113(12):1921–1925. doi:10.1007/s00702-006-0484-8

Chapter 19

Genetic Dissection of Variation in Hippocampal Intra- and Infrapyramidal Mossy Fibers in the Mouse

Anna Delprato and Wim E. Crusio

Abstract

This chapter describes the genetic analysis of a morphometric neuroanatomic trait. We used the extended BXD family of recombinant inbred mouse strains with the intent to analyze the genetic bases of heritable differences in hippocampal neurocircuitry and to identify Quantitative Trait Loci that underlie these variations. A detailed description of a GeneNetwork analysis is provided using data for the intra- and infrapyramidal mossy fiber (IIPMF) terminal fields which are strongly correlated with spatial navigation/radial maze learning.

Key words Complex traits, Hippocampal morphometry, Intra- and infrapyramidal mossy fibers, Quantitative trait loci, Recombinant inbred mice, Systems genetics

1 Introduction

Brain disorders, including neurological and psychiatric disorders, are among the most serious and, at the same time, intractable diseases, causing immeasurable human suffering. Understanding the genetics of variation in the brain should therefore be a priority. However, before proceeding with the genetic analysis of a neuronal characteristic, there are a number of criteria that should be satisfied. First, the neuronal phenotype should ideally be known to be important in the regulation and/or modulation of behavior. Second, an important practical consideration is that the phenotype can be measured in an efficient way. Third, a prerequisite for this type of research obviously is the presence of heritable variation in the character of interest.

The mammalian hippocampus is known to be affected in a number of important disorders, such as Alzheimer's Disease [1] or schizophrenia [2]. Notably, it plays an important role in learning and memory, especially spatial navigation [3, 4]. Inbred strains of mice show large differences in their ability to learn spatial navigation tasks [5, 6] and this has been shown to covary with variations

Klaus Schughart and Robert W. Williams (eds.), *Systems Genetics: Methods and Protocols*, Methods in Molecular Biology, vol. 1488, DOI 10.1007/978-1-4939-6427-7_19,

in the sizes of their hippocampal IIPMF [5, 7]. The IIPMF constitute one of the most important afferent pathways of the hippocampus [3], so it could be expected that variations in the size of this projection would form a bottleneck in the flow of information about the surrounding environment into the hippocampus. Taken together, the IIPMF satisfy the first criterion.

The hippocampus has a laminar structure [8] and the IIPMF projection is rather discrete (Fig. 1a). In addition, this projection can easily be stained specifically by means of Timm's stain [9], *see* Fig. 1b. A morphometric analysis of the IIPMF is therefore relatively easy to perform, so that the IIPMF also satisfy the second criterion.

It has long been known that inbred strains of mice show large, heritable differences in the structure of their brains [10, 11]. In particular, the sizes of their hippocampal intra- and infrapyramidal mossy fiber (IIPMF) projections are very variable [12, 13]: expressed as a percentage of the total size of the CA3 and CA4 (hilus) fields combined, sizes go from about 0.8% in NZB/B1NJ up to around 4% in C3H/HeJ mice [14]. It has been shown that these variations have a high heritability. In a diallel-cross study, comprising the inbred strains BA, C57BL/6J, C57BR/cdJ, BALB/cJ, and DBA/2J, we could show that about 53% of the variation observed is attributable to genetic differences between strains [15]. Wahlsten et al. [16] found a heritability of about 35% for the IIPMF in a cross between the inbred strains C57BL/6J and C57BL/6J, whereas Lassalle et al. [17] estimated heritability in a sample of 26 BXD RI strains as 65%. Even though estimates

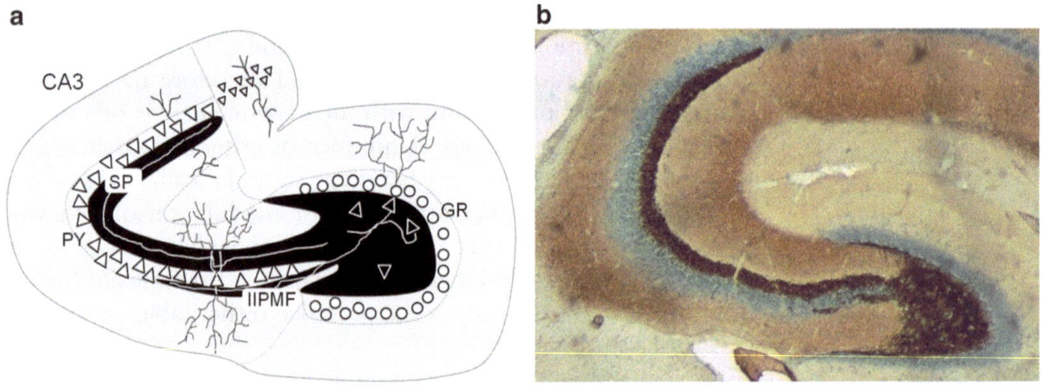

Fig. 1 (**a**) Schematic representation of the hippocampus, showing the localization of the intra- and infrapyramidal mossy fiber (IIPMF) terminal fields. The mossy fibers are the axons of the dentate granular cells (GR) that form Zn^{2+}-containing synapses on the dendrites of the pyramidal cells (PY) in the CA3 region. They form two major projections: the suprapyramidal mossy fibers (SP) forming synaptic contacts with the apical dendrites of the pyramidal cells and the IIPMF, mostly forming synaptic contacts with the basal dendrites of the pyramidal cells. (**b**) Timm-stained hippocampal section, counterstained with methylene blue for visual contrast (*note*: no counterstaining was employed for sections that were analyzed in this study). Areas containing zinc are marked by *dark silver* deposits, making the mossy fiber projections appear in *dark brown*

vary somewhat between studies, it should be noted that the only estimate below 50 % was obtained from a cross between two inbred strains that do not differ very much in the sizes of their IIPMF (C57BL/6J and BALB/cJ). All other estimates agree in a heritability of at least 50 % for the IIPMF. In short, the IIPMF satisfy the third criterion, too.

The BXD family of recombinant inbred strains was used for this genetic dissection because the parent strains of this set (C57BL/6J and DBA/2J) are at opposite extremes of the distribution of the IIPMF in different mouse strains. In addition, the parents have been fully genotyped and much phenotype and genetic information is available on the parental and RI strains, making it feasible to identify candidate genes within QTL intervals with more precision.

2 Materials

2.1 Subjects

Breeding pairs of 53 BXD strains were acquired from the University of Tennessee Health Center (Memphis, TN, USA) and the Center for Neurogenomics and Cognitive Research (Free University, Amsterdam, Netherlands). Breeding pairs of the parental strains (C57BL/6J and DBA/2J) were obtained from Charles River (L'Arbresle, France). Because of logistical problems (for example, some strains not breeding well), we did not achieve this goal for all strains. All animals used were housed and bred in the SPF mouse facility of the University of Bordeaux (Pessac) in a climate-controlled breeding room (temperature: 21 ± 1 °C, humidity: 55 ± 10%, 12 h light–dark cycle with lights on at 7 a.m.). Food (Safe, type 113, sterilized) and water (softened, sterilized) were available ad libitum. Animals were housed 2–4 in clear plastic cages (162 × 406 × 176 mm, Tecniplast) filled with poplar wood shavings (Souralit). Data were collected on 442 mice; a roughly balanced number of males and females were analyzed.

2.2 Sample Size

The additive genetic correlation between a phenotype and a genetic marker approaches the correlation between strain means and the genetic marker with increasing *n* per strain [18]. We aimed for a power such that the correlation between RI strain means and a given genetic marker would reach 95 % of the additive genetic correlation. The sample sizes necessary to obtain this goal were determined according to a method published earlier [19]. Using 0.50 for the heritability of the size of the IIPMF, the equations given by Crusio [19] indicate that sample sizes of five animals per strain would be needed to obtain the desired statistical power.

The precision with which QTLs can be localized does not only depend on a high additive-genetic correlation between the phenotype and genetic markers (i.e., a high heritability), but also on the number of RI strains used. In general, the more strains used, the narrower the chromosomal intervals in which QTL are localized will be. We therefore aimed to characterize as many BXD RI strains as possible, as well as the C57BL/6J and DBA/2J parentals. At the end of the project, data on 53 RIS were available with most strains (but not all) having the desired sample size of five males and five females.

2.3 Hippocampal Histology and Morphometry

For hippocampal histology and the morphometric analyses, male and female BXD and parental mice were perfused intra-cardially with sodium sulfide followed by glutaraldehyde. Brains were removed for histology, weighed, and placed for 24 h in a postfixative solution of 3% glutaraldehyde and 20% sucrose at room temperature. Next, 40 μm horizontal cryostat sections mounted on a glass slide were processed for Timm's silver sulfide stain for heavy metals, which specifically stains zinc [9] and allows the visualization via light microscopy of the terminal fields of the hippocampal projections in the form of colored bands and patches (for photomicrographs *see*, e.g. [15]). Starting at the midseptotemporal level, immediately after the disappearance of the septal pole, we measured every second section for a total of five sections for both the left and right hippocampi. Micrographs were made with a Leica DM6000 B microscope using ×10 objective lens and an automated procedure to create composite photographs. The morphometrical analysis was performed using ImageJ software (NIH v.1.48).

Initially, a subset of sections was analyzed independently by three persons. Discrepancies were discussed and measurements repeated until the three persons achieved an inter-observer correlation of 0.95 or better for measurements obtained from single sections. Final measurements of all animals were made by only one of these three persons.

3 Methods

3.1 Strain and Sex Differences

Data were obtained from 207 female (51 strains) and 215 male mice (53 strains), as well as for five males and five females for each of the parental strains. Statistical analyses to assess strain and sex effects were performed with SAS 9.3 [20]. A repeated-measures ANOVA indicated that left-right differences did not interact with either strain or sex, so all subsequent analyses were performed on left-right mean values. To determine strain and sex effects, IIPMF morphometry data were subjected to 2-way ANOVA with sex and strain as main factors. Heritabilities (h^2) were estimated according to the method of Hegmann and Possidente [21]. Briefly, h^2 was defined as the ratio of the variance between strains divided by the sum of the within-strain and between-strain variances. These variance components were derived from the expected mean squares obtained with the SAS procedure GLM. This resulted in an estimated heritability of 0.46, which was very close to our original estimate of 0.50. Accordingly, we found significant strain differences ($F_{52,342} = 8.61$, $p < 0.001$). There were no significant sex differences ($F_{52,342} = 0.06$) or sex by strain interactions ($F_{52,342} = 0.50$).

3.2 GeneNetwork Analysis

The genetic analyses were primarily done in GeneNetwork (http://genenetwork.org/), which is an open source bioinformatics resource for systems genetics. GeneNetwork is not only a repository for genetic, genomic, and phenotypic data related to recombinant inbred mice (mostly, but not exclusively, BXD), but also has a suite of statistical programs for data analysis that includes mapping and analyzing QTLs, examining phenotype/genotype correlations, and building interaction networks. A second generation of GeneNetwork, GeneNetwork2 (http://gn2.genenetwork.org/), is in the making with an expanded repertoire of analysis tools.

3.2.1 Trait Data Analysis and Basic Statistics

BXD and parental strain means and their standard errors were uploaded to the GeneNetwork database (trait IDs 16307 and 17476 for males and females, respectively), according to its guidelines for file formats (for detailed instructions and example files *see* http://www.genenetwork.org/webqtl/main.py?FormID=batSubmit). An illustrative follow along example for a systems genetics analysis using GeneNetwork can be found in Williams and Mulligan [22]. For our purposes, the initial step of the analysis involved using the basic statistics and graphing functions in GeneNetwork. These options can be found by selecting the trait ID and/or by selecting the data and adding it to one's "collection" using the select and add buttons. An icon menu with the different analyses tools will now appear. Particularly useful are the bar graph options which quickly rank order the strain means (from low to high). Data can also be ordered by strain labels. Further output options are a table of general statistics (means, variances, min, max, etc.) as well as probability and box plots.

3.2.2 Data Preparation

The next step is a check of the data for extreme high or low values (outliers), which are automatically highlighted in yellow. Before the mapping process, there are options to either mask outliers or to winsorize them. The latter is a simple statistical transformation of extreme values. An explanation of the procedure can be found at http://www.genenetwork.org/glossary.html#W. Because QTL mapping is sensitive to outliers, rather than exclude outlier strains, we winsorize the extreme values so that the data can still be used in the analysis. Keep in mind that if you decide to winsorize your data, you should always do so in the same manner for consistency of results. In the present case, data for the males were winsorized as follows: BXD69 (3.774 changed to 3.218), BXD61 (3.816 was changed to 3.219), and BXD62 (4.201 changed to 3.220).

3.2.3 Interval Mapping

In the GeneNetwork mapping tools, the interval mapping function calculates linkage maps for the entire genome. The significance levels associated with a particular linkage score are determined using a permutation test, whereas the confidence limits of a QTL interval are estimated using bootstrap sampling.

We began the mapping by generating an initial map for the whole genome to view the likelihood ratio statistic (LRS) on all chromosomes. The resulting data from the mapping study are downloadable in text format or can be copied and pasted to a spreadsheet. For the IIPMF trait data we used the default parameters (5000 permutations). The parental strains are not included in the analyses.

When determining the interval most likely containing the QTL, it is best to be conservative and select chromosomal intervals/areas that include both significant peaks and shoulders. In the IIPMF data for males, a significant peak ($p=0.029$) was observed at 48.1 Mb on the X chromosome, explaining 26% of the variance between strain means. The interval considered for candidate gene analysis ranged from 45 to 60 Mb. In females a suggestive peak ($0.05<p<0.10$) occurred at about the same chromosomal location. Another (suggestive) QTL was observed on chromosome 11 at position 42.6 Mb for both males (LRS = 15.2) and females (LRS = 15.9), explaining 28 and 25% of the variance between strain means, respectively. The interval, including the peak and shoulders, ranged from 34 to 43 Mb (Fig. 2). Levels of variance explained were estimated by using the squared correlation coefficient between the genetic marker at the peak location and the IIPMF strain means.

3.2.4 Identification of Candidate Genes

Next, the areas under the QTL peaks and shoulders that cross the suggestive threshold were screened for possible candidate genes using QTLminer [23]. QTLminer integrates information for all the genes present within a specific genomic region such as functional annotation, gene expression data, sequence polymorphisms, and KEGG pathway associations. It also includes information on whether the expression of a gene is *cis*-regulated, that is, when an expression QTL or "eQTL", is at the same chromosomal location as the gene in question. Candidate genes were selected using criteria used in similar studies: (1) a gene should be expressed in the relevant tissue (for our purpose, the hippocampus), (2) SNPs should be present within the candidate gene, (3) missense or non-synonymous mutations or other DNA variants should be present, and (4) the gene is *cis*-regulated [23].

QTLminer can be accessed from the GeneNetwork landing page under the search tab. To use QTLminer, specify the chromosome of interest and the relevant QTL interval and select the parental mouse strains. There is an option to use three datasets for inclusion of expression and *cis*-activity data. For our purposes, we used the three hippocampal gene expression datasets. Genes are

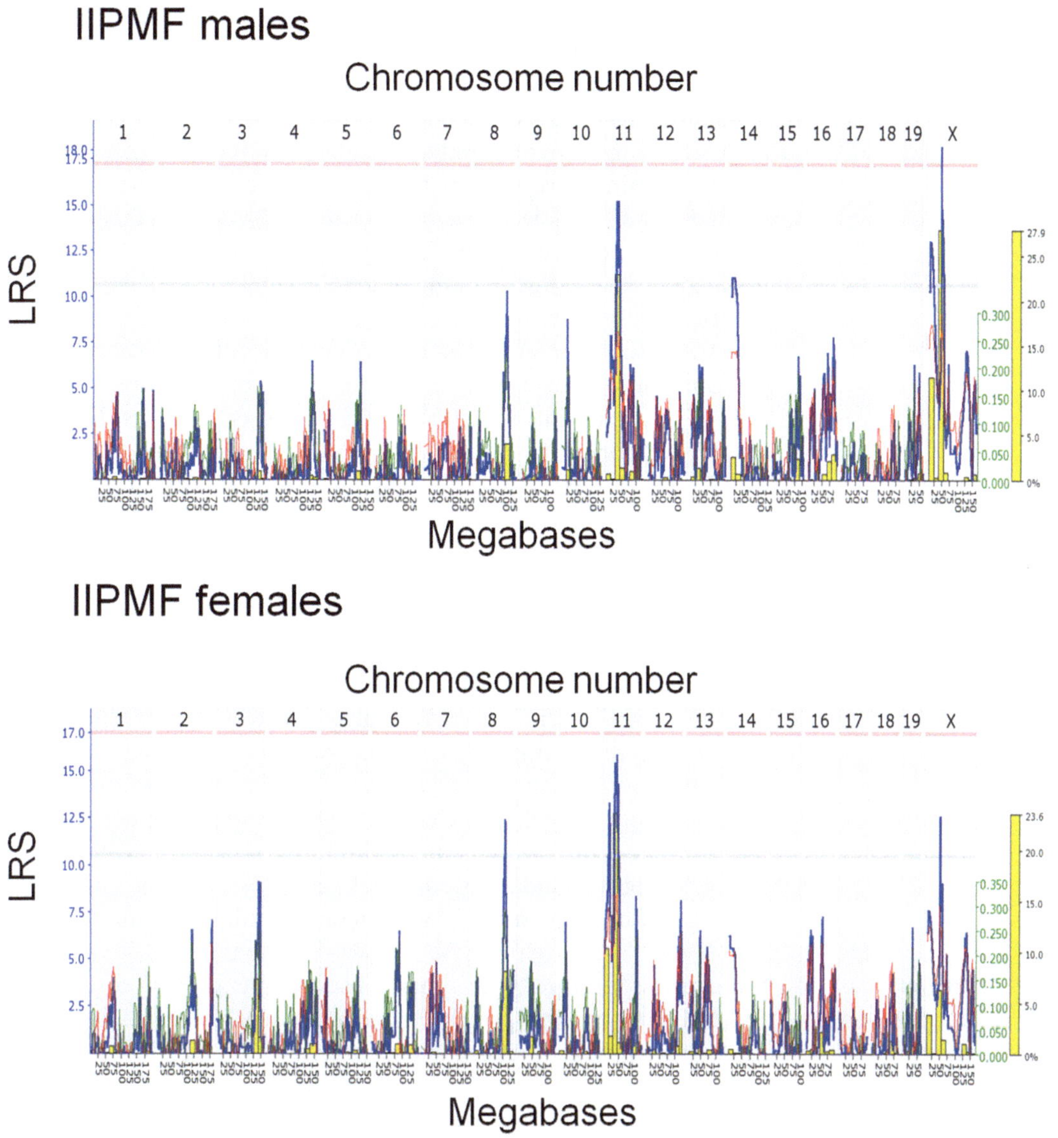

Fig. 2 Whole-genome scan for the IIPMF in males (*top*) and females (*bottom*). The *x*-axis represents chromosome number and megabase position and the *y*-axis represents the likelihood ratio statistic (LRS) of linkage. *Blue lines* represent LRS across the genome. The *pink* and *gray horizontal lines* are approximate threshold values which are used to assess whether a peak is significant ($p < 0.05$) or suggestive ($p < 0.63$), respectively. *Red* and *green lines* represent the additive genetic contribution; *red lines* indicate negative values (C57BL/6J alleles increasing trait values) and *green lines* indicate positive values (DBA/2J alleles increasing trait values). *Gray lines* are shown when the parental strain is unknown. The *yellow bars* indicate the relative frequency of peak LRS at a given location from 2000 bootstrap resamples

scored from 1 to 4 (one point for each criterion that is met). From this information, the candidate gene list can be narrowed down considerably. For example, if a gene is not expressed in the hippocampus or does not differ between the parental strains (i.e., does not possess SNPs or other DNA variants), it reasonably cannot be expected to be causatively involved in the observed phenotypic difference. A word of caution here: QTLminer streamlines the process of identifying candidate genes but it is not a definitive final step. Any candidate gene must subsequently be verified experimentally.

To obtain more information on the DNA variants from the refined gene candidate list, select the SNP/Variant browser tab from the GeneNetwork landing page. Here you can either search using an individual gene name or a SNP ID. One can also use this function by specifying a chromosome and an interval of interest. The strains considered in the analyses can be edited with the "cut" key until only the parental strains remain. This will limit the output of nonrelevant information. The other options here are self-explanatory.

For the IIPMF data, the peak LRS occurred in a gene sparse region on the X chromosome and, from an initial list of 107 genes, only one candidate gene remained based on the four selection criteria. For the highly suggestive peak on chromosome 11, an initial list of 25 genes was narrowed down to two candidates. However, for this case, we have additional data that support one candidate over the other (for details, *see* [24]). Both candidate genes for chromosomes X and 11 encode glycoproteins that interact with cell surface and extracellular matrix proteins. Both genes are also associated with neuronal processes. Functional associations and gene ontology for selected candidate genes can be further assessed using Gene (http://www.ncbi.nlm.nih.gov/gene), WebGestalt (http://bioinfo.vanderbilt.edu/webgestalt/) and/or literature mining in PubMed (http://www.ncbi.nlm.nih.gov/pubmed).

3.2.5 Other Mapping Functions in GeneNetwork

Sometimes, a large QTL may obscure the effects of another QTL. To test for this possibility, GeneNetwork offers a composite interval mapping function. First we determine which marker has the highest LRS, using a marker regression analysis (under mapping tools in GeneNetwork). To minimize the output, there is an option to display LRS only above a specified value. For our study we used a value of 12 which included both suggestive and significant LRS. Subsequently, we run a composite interval mapping analysis, which is a marker regression analysis controlling for the marker with the highest LRS. For the IIPMF data, the top ranking markers were rs13483748 on the X chromosome and rs6307831 on chromosome 11 and we ran composite interval mapping analyses for both of them. This procedure did not reveal any additional significant QTLs, although suggestive QTLs on chromosomes 14 and 7, respectively, were indicated.

3.2.6 Testing for Epistasis

Epistatic effects are nonadditive interactions between two or more loci. In this situation the combined effect is either greater or smaller than the summed effects of each gene alone. In GeneNetwork there is a "pair scan" option designed to detect such epistatic effects. The algorithm searches the genome for all possible pairs of chromosomal regions that are involved in two-locus interactions and analyzes the LRS scores. Again, as with the other QTL mapping functions, this is sensitive to outliers and data should be winsorized or left out of the analyses to prevent false positives. To run the pair scan function, select the "pair scan" option under mapping tools. The output is a two-dimensional plot.

The lower right half of the plot gives a summary of LRS values of the full model, representing cumulative effects of both loci and their possible interaction, corresponding to the column labeled "LRS Full" in the table beneath the plot. Only if the full LRS for a particular combination is significant, should we proceed with the analysis of possible epistatic effects. The upper left side of the plot indicates the LRS values for the presence of epistatic interactions, corresponding to the column labeled "LRS Interact" in the table beneath the plot.

In the present case of the IIPMF we do not detect any robust epistatic effects, because none of the full models reach statistical significance, so that subsequent analyses are not warranted. For an example where two loci have a significant epistatic effect, without any direct effect from either one of them, *see* Fig. 5 in Overall et al. [25].

3.2.7 Correlation and Network Analyses

A correlation analysis in GeneNetwork is a powerful tool which allows one to determine how different traits covary. There are two databases available for a correlation analysis: one containing mRNA expression datasets representing many tissues and brain regions and another containing a multitude of BXD phenotype data that has been amassed from many years of experimentation in many different laboratories. For our purposes, we considered covariates in the hippocampus mRNA expression datasets and the phenotypes database and chose to have the top 200 results for each returned. Only the top 30 or so correlates were significant and used in subsequent analyses. Because of the presence of outliers in the data we used Spearman rank correlations rather than Pearson's correlation, the former being insensitive to outlier effects. We used a stringent criterion and selected correlations with a p-value ≤ 0.02 and at least ten common strains.

Next, select these correlations and use the "Graph" option to generate a network view which will show connected entries. Positive correlations are shown as red lines and negative ones in blue. Line width corresponds to the strength of the correlation. There are many other parameters that can be adjusted to obtain an optimal graph view for the data as well. In addition, the data can be exported for further multivariate analyses using standard statistics packages.

4 Potential Pitfalls and Problems

It will be obvious from the foregoing, that any project in this field poses challenges with regard to required effort and cost: These experiments, encompassing many hundreds of animals, are much larger than generally is the case in neuroscientific research. It is therefore of prime importance that all potential problems and pitfalls that one may encounter during the execution of this work are carefully considered beforehand. In addition, one should realize that despite all the effort put into a project of this size, unequivocal identification of QTLs and identification of strong candidate genes are not a guaranteed outcome. Clearly, such high-risk long-term studies are not something to be undertaken lightly and it would be a distinct disservice to assign, for example, graduate students or postdocs to such a long-term project that will not render any publishable results before completion.

Given the expected duration of projects of this kind (generally several years), cohort effects are a particular concern. Such effects may cause variation in testing scores due to environmental factors (e.g., seasonal effects) or personnel-related factors (e.g., change of laboratory technician mid-way an experiment). To counter this possible problem, two things are essential. First, within and between-observer reliability has to be carefully evaluated by performing measurements twice. Training of personnel should continue until measurement reliability is satisfactory (e.g., an inter-observer correlation of 0.95 or better). Second, care should be taken that testing of strains is randomized as much as possible to ensure that animals from one particular strain will not all be observed at more or less the same time. This will allow for at least some statistical control of cohort effects, should these occur.

Another important possible confounding factor can be litter effects. Not all RI strains are good breeders and some strains produce larger litters than others, possibly leading to differences in pups not related to direct genetic effects, but to indirect genetic effects through maternal influences (e.g., quantity of milk available per pup). To reduce these effects, one can cull all newborn litters to a certain maximum number of pups (say, six), with at least two animals from each sex (ideally 3–3).

Acknowledgements

The data used in this book chapter as an example of the analysis of a neuroanatomical trait were part of a larger study that was originally published in an earlier article [24]. This work was supported by a grant from NIMH (R01MH072920) to W.E.C. Dr. Rob Williams (University of Tennessee Health Sciences Center, supported by grants from NIAAA-U01 AA016662 and U01 AA013499) and Drs Sabine Spijker and August B. Smit (Free University of Amsterdam) from the Neuro-BSIK Mouse Phenomics Consortium (BSIK03053) generously provided breeders of several BXD strains (for the origin of the latter strains, *see* [26]). We thank Raphael Pineau and Laetitia Medan for expert animal care and Marie-Paule Algéo, Brice Bonheur, and Alexis Cornuez for carrying out the histology and morphometry.

References

1. Mineur YS, McLoughlin D, Crusio WE, Sluyter F (2005) Genetic mouse models of Alzheimer's disease. Neural Plast 12(4):299–310
2. Gray JA, Feldon J, Rawlins JN, Hemsley DR, Smith AD (1991) The neuropsychology of schizophrenia. Behav Brain Sci 14(1):1–20
3. O'Keefe J, Nadel L (1978) The Hippocampus as a cognitive map. Clarendon, Oxford
4. Eichenbaum H, Dudchenko P, Wood E, Shapiro M, Tanila H (1999) The hippocampus, memory, and place cells: is it spatial memory or a memory space? Neuron 23(2):209–226
5. Crusio WE, Schwegler H (2005) Learning spatial orientation tasks in the radial-maze and structural variation in the hippocampus in inbred mice. Behav Brain Funct 1(1):3
6. Crusio WE (2013) Radial maze. In: Crusio WE, Sluyter F, Gerlai RT, Pietropaolo S (eds) Behavioral genetics of the mouse: genetics of behavioral phenotypes, vol 1, Cambridge handbooks in behavioral genetics. Cambridge University Press, Cambridge, pp 299–303
7. Schwegler H, Crusio WE (1995) Correlations between radial-maze learning and structural variations of septum and hippocampus in rodents. Behav Brain Res 67(1):29–41
8. Stanfield BB, Cowan WM (1979) The morphology of the hippocampus and dentate gyrus in normal and reeler mice. J Comp Neurol 185:393–422
9. Danscher G, Zimmer J (1978) An improved Timm sulphide silver method for light and electron microscopic localization of heavy metals in biological tissues. Histochemistry 55(1): 27–40
10. Wimer RE, Wimer CC, Roderick TH (1969) Genetic variability in forebrain structures between inbred strains of mice. Brain Res 16(1):257–264
11. Roderick TH, Wimer RE, Wimer CC, Schwartzkroin PA (1973) Genetic and phenotypic variation in weight of brain and spinal cord between inbred strains of mice. Brain Res 64:345–353
12. Barber RP, Vaughn JE, Wimer RE, Wimer CC (1974) Genetically-associated variations in the distribution of dentate granule cell synapses upon the pyramidal cell dendrites in mouse hippocampus. J Comp Neurol 156(4):417–434
13. Vaughn JE, Matthews DA, Barber RP, Wimer CC, Wimer RE (1977) Genetically-associated variations in the development of hippocampal pyramidal neurons may produce differences in mossy fiber connectivity. J Comp Neurol 173(1):41–52
14. Schwegler H, Crusio WE, Brust I (1990) Hippocampal mossy fibers and radial-maze learning in the mouse: a correlation with spatial working memory but not with non-spatial reference memory. Neuroscience 34(2):293–298
15. Crusio WE, Genthner-Grimm G, Schwegler H (1986) A quantitative-genetic analysis of hippocampal variation in the mouse. J Neurogenet 3(4):203–214
16. Wahlsten D, Lassalle J-M, Bulman-Fleming B (1991) Hybrid vigour and maternal environment in mice. III. Hippocampal mossy fibres and behaviour. Behav Processes 23:47–57
17. Lassalle J-M, Halley H, Milhaud J-M, Roullet P (1999) Genetic architecture of the hippocampal mossy fiber subfields in the BXD RI mouse strain series: A preliminary QTL analysis. Behav Genet 29(4):273–282
18. Crusio WE (2007) An introduction to quantitative genetics. In: Jones BC, Mormède P (eds) Neurobehavioral genetics: methods and applications, 2nd edn. CRC Press, Boca Raton, FL, pp 37–54
19. Crusio WE (2004) A note on the effect of within-strain sample sizes on QTL mapping in recombinant inbred strain studies. Genes Brain Behav 3(4):249–251
20. SAS Institute Inc (1987) SAS/STAT guide for personal computers, version 6 edition. Sas Institute, Cary, NC
21. Hegmann JP, Possidente B (1981) Estimating genetic correlations from inbred strains. Behav Genet 11(2):103–114

22. Williams RW, Mulligan MK (2012) Genetic and molecular network analysis of behavior. Int Rev Neurobiol 104:135–157
23. Alberts R, Schughart K (2010) QTLminer: identifying genes regulating quantitative traits. BMC Bioinformatics11:516.doi:10.1186/1471-2105-11-516, 1471-2105-11-516 [pii]
24. Delprato A, Bonheur B, Algéo MP, Rosay P, Lu L, Williams RW, Crusio WE (2015) Systems genetic analysis of hippocampal neuroanatomy and spatial learning in mice. Genes Brain Behav 14(8):591–606
25. Overall RW, Kempermann G, Peirce J, Lu L, Goldowitz D, Gage FH, Goodwin S, Smit AB, Airey DC, Rosen GD, Schalkwyk LC, Sutter TR, Nowakowski RS, Whatley S, Williams RW (2009) Genetics of the hippocampal transcriptome in mouse: A systematic survey and online neurogenomics resource. Front Neurosci 3:55
26. Loos M, Mueller T, Gouwenberg Y, Wijnands R, van der Loo RJ, Birchmeier C, Smit AB, Spijker S (2014) Neuregulin-3 in the mouse medial prefrontal cortex regulates impulsive action. Biol Psychiatry 76(8):648–655

Chapter 20

Complex Genetics of Cardiovascular Traits in Mice: F2-Mapping of QTLs and Their Underlying Genes

Svitlana Podliesna, Connie R. Bezzina, and Elisabeth M. Lodder

Abstract

In this chapter, we will use the example of the identification of *Tnni3k* as a modulator of cardiac conduction to introduce you to the use of a murine F2-generation intercross as a powerful method for the identification of novel genes relevant for cardiovascular traits. Murine F2-progeny is a genetically diverse panel of mice with differences in phenotype manifestations, e.g. cardiovascular traits such as cardiomyopathy and ECG parameters. This chapter discusses the best strategies for using F2-mice for genetic mapping. Moreover, we provide an example of the feasibility of identification of new genes modulating cardiac function utilizing the technique of mapping quantitative trait loci (QTLs) and a systems genetics integration of available genetic, gene expression, and phenotypic data.

Key words Use case, Cardiovascular traits mapping

1 Introduction

Cardiovascular disease and its sequelae are a major burden on health and consequently on the health care systems in the western world [1]. As a result of the aging population and the increase of comorbidities such as diabetes and obesity, the incidence of cardiovascular disease is expected to rise in the coming years [2]. Sudden cardiac death caused by ventricular fibrillation is responsible for 20% of all natural deaths and most often occurs in the context of other cardiovascular problems such as myocardial infarction as a result of atherosclerotic disease [3]. Atrial fibrillation is the most common cardiac arrhythmia affecting up to 10% of the population and associated with increased risk of heart failure and sudden death [4].

Most cardiovascular disorders are clearly multifactorial with strong influences of both environmental (e.g. diet, exercise, smoking) and heritable factors interacting determining each individual's risk of disease [5]. As a result of this complexity, the identification of the underlying genetic factors in humans has remained largely limited to clear familial disease and genome wide association studies

Klaus Schughart and Robert W. Williams (eds.), *Systems Genetics: Methods and Protocols*, Methods in Molecular Biology, vol. 1488, DOI 10.1007/978-1-4939-6427-7_20, © Springer Science+Business Media New York 2017

[5]. These studies have identified major groups of genes/proteins affected in different types of disease (i.e. ion channels in cardiac arrhythmias [6], sarcomeric proteins in cardiomyopathy [7] and the LDL receptor pathway in hypercholesteremia and coronary artery disease [8]. Despite these results a large part of the heritability in cardiovascular disease remains unexplained [5]. In an attempt to overcome the technical difficulties of identifying the remaining genetic factors determining cardiovascular disease in human the research community has turned to mouse genetic studies.

The genetic makeup of the standard laboratory strains of mice is fundamentally different from that of the human population. As a result of consistent inbreeding, through sibling-sibling and child-parent mating, most laboratory mouse strains have become completely homozygous in their entire genome [9]. This unique genetic makeup provides the opportunity to tease out the effects of specific genetic variants without the experimental noise introduced by diversity in the genetic background, through specific breeding schemes between these lines.

Thus far genetic mapping approaches using a backcross or intercross breeding approach (N2 and filial generation 2 (F2) respectively) and in (recombinant) inbred lines (*see* Fig. 1 for an explanation of the breeding schemes) has been used successfully in multiple studies to identify genetic variants and genes underlying modulation of cardiovascular disease in mice i.e. in cardiomyopathy [10–14] and cardiac structure [15–21], blood pressure [18,

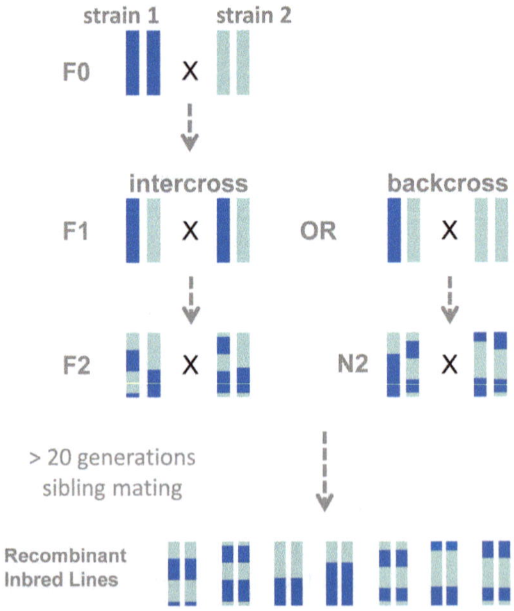

Fig. 1 Breeding scheme for obtaining an F2- or N2-generation and recombinant inbred lines from two inbred mouse strains. The *two colors* denote the genetic background of the different strains

Table 1
Overview of all murine cardiac QTLs

Trait	Phenotype	Study	Mouse cross	Detailed phenotype	Locus	Chr	Peak (cM)	Peak (Mb)	Borders	LOD score	Causal gene	Overlap
Cardiac structure	Cardiac weight and size	Sugiyama et al. [18]	BALB/cJ × CBA/CaJ (F2)	Heart weight	Hwq1	14	D14Mit67 38 (cM)		20–43 (cM)	2.9		
		Rocha et al. [17]	M16i × L6 (F2)	Cardiac weight	Hrtq1	2	52 (cM)		38–70 (cM)	6.4		Hfm1 (Htrfm1)
					Hrtq2	2	80 (cM)		76–84 (cM)	5.7		Hrtfm5; Hrq1
					Hrtq3	10	27 (cM)		14–45 (cM)	4.2		Cmn2
				Heart weight/ body weight (10 weeks)	Hrt%q1	3	30 (cM)		16–53 (cM)	3.5		Vms1
		Derry et al. [15]	C57BL/6J × A/J (F2)	Heart weight/ body weight (16 weeks)		6	30 (cM)			7.7		
						8	49 (cM)			8.1		Lodder; chr8 collagen
				Left ventricular weight/body weight (16 weeks)		8	34 (cM)			6.8		
		Hersch et al. [19]	Inbred strains (23 lines)	Atrial weight	AW ate (1)	5		91				
					AW ctr (2)	5		98–99				
				Atrial weight/ body weight	AW/BWS iso 1 (3)	15		96–97				
					VW/AW iso 10 (4)	6		78				
				Ventricular weight/atrial weight	VW/AW ctr (5)	7		129				
					VW/AW iso 10 (6)	9		115				Andreux, systolic blood pressure

(continued)

Table 1 (continued)

Trait	Phenotype	Study	Mouse cross	Detailed phenotype	Locus	Chr	Peak (cM)	Peak (Mb)	Borders	LOD score	Causal gene	Overlap
	Cardiac collagen deposition	Lodder et al. [20]	FVB/NJ –*Scn5a* 1798insD/+ × 129/P2–*Scn5a* 1798insD/+ (F2)	Cardiac collagen deposition	Chr8 dominant QTL	8	50 (cM)	61	42–63 (cM)	4.1	*Pdlim3*	Derry hw/bw locus
					Chr2 additive QTL (with Ch18 covariate)	2	0 (cM)	3	tel-22 (cM)	3.0	*Gpr158*	
					Chr18 covariate region	18	32 (cM)	63	tel-52 (cM)	2.5	*Fgf1*	Hrtfm4; Hrtfm7
Cardiac function	Cardiac function	Derry et al. [15]	C57BL/6J × A/J (F2)	Velocity Time Integral (10 weeks)		2	53 (cM)			5.4		
				Ejection fraction (10 weeks)		17	13 (cM)			4.5		
				Aorta diameter (10 weeks)		6	42 (cM)			5.3		

Heart disease	Atrial septal abnormalities	Kirk et al. [21]	QSi5 × 129T2/SvEms (F2)	Flap valve length (FVL)		8	60 (cM)	114	108-tel (Mb)	5.5		
						13	10 (cM)	37	6.5–48 (Mb)	4.6		
						19	9 (cM)	13	cen-46 (Mb)	6.0		
				Foramen ovale width (FOW)		1	34 (cM)	68	43–102 (Mb)	5.9		
						2	63 (cM)	137	115–154 (Mb)	4.4		
						4	24 (cM)	49	26–80 (Mb)	9.0		
				Crescent width (CRW)		7	43 (cM)	113	108–tel (Mb)	4.6		
	Progression of heart failure	Suzuki et al. [10]	*CalsequestrinTg* DBA/2J × C57BL6 F1 backcross (N2) to DBA/2J or to C57BL6	Survival	Hfm1 (Htrfm1)	2	D2Mit327 (41 cM)	69	34–48 (cM)	7.8		Hrtq1
					Hfm2 (Htrfm2)	3	D3Mit86 (72 cM)	147	72–87 (cM)	5.7	*Tnni3K*	Vms1
				Left ventricle end-diastolic dimension	Hfm2 (Htrfm2)	3	D3Mit86 (72 cM)	147	70–85 (cM)	9.3	*Tnni3K*	Vms1
		Le Corvoiser et al. [12]	*CalsequestrinTg* DBA2/J × AKR/J backcross to AKR/J (N2)	Survival	Hrtfm3	4	D4Mit236 (20 cM)	39	8–18 (cM)	4.3		Blizard-bp-chr4
					Hrtfm4	18	D18Mit28 (34 cM)	60	17–38 (cM)	5.1		Lodder chr18 collagen; Hrtm7
					Hrtfm4	18	D18Mit28 (34 cM)	60	32–38 (cM)	3.8		Lodder chr18 collagen; Hrtm7
				Fractional shortening	Hrtfm5	2	D2Mit138 (69 cM)	140	67–89 (cM)	4.8		
					Hrtfm6	13	D13Mit213 (60 cM)	108	59–75 (cM)	3.3		
				Left ventricular end-diastolic diameter	Hrtfm5	2	D2Mit138 (69 cM)	140	67–98 (cM)	5.1		Hrtq2; Hrq1
					Hrtfm6	13	D13Mit213 (60 cM)	108	59–75 (cM)	4.1		

(continued)

Table 1
(continued)

Trait	Phenotype	Study	Mouse cross	Detailed phenotype	Locus	Chr	Peak (cM)	Peak (Mb)	Borders	LOD score	Causal gene	Overlap
		Wheeler et al. [11]	*CalsequestrinTg* DBA/2J × AKR/J N2 to DBA/2J	Survival	Hrtfm2	3	D3Mit220 (79 cM)	154	74–79 (cM)	9.5	*Tnni3K*	Vms1
				Left ventricular end-diastolic diameter	Hrtfm2	3	D3Mit220 (79 cM)	154	74–79 (cM)	5.2	*Tnni3K*	Vms1
					Hrtfm7	18	D18Mit22 (13 cM)	25	10–18 (cM)	3.5		Lodder chr18 collagen; Hrtm4
	Dilated cardiomyopathy	Maddatu et al. [13]	(B6.CAST-MnmC/MnmC Ighmbp2nmd-2J/b × CAST/EiJ) F1 × (B6.CAST-MnmC/MnmC Ighmbp2nmd-2J/b) (N2)	Dilated cardiomyopathy	Cmn1	9	D9Mit17 (70 cM)		60–70 (cM)	5.6		
					Cmn2	10	D10Mit42 (34 cM)		28–44 (cM)	8.0		Hrtq3
					Cmn3	16	D16Mit64 (44 cM)		42–48 (cM)	4.3		
	Myocarditis	Wiltshire et al. [29]	A/J × B10.A (F2), C57BL/6J-Chr3A/NaJ (N2)	Myocarditis	Vms1 (Vms1.1a and Vms1.1b)	3	D3Mit19		D3Mit291-D3Mcg1; Vms1.1a: D3Mit116-D3Mcg1 (78.7–83.29 cM); Vms1.1b: D3Mit128-D3Mcg1(76.7–84 cM)	3.7	*Fpgt*, *H28*, and *Tnni3k*	Hrtfm2

Blood pressure	Blood pressure	Sugiyama et al. [18]	BALB/cJ × CBA/CaJ (F2)	Systolic blood pressure	Bpq6	15	16 (cM)		tel-25 (cM)	4.9		Hrq2
					Bpq7	7	42 (cM)		35–50 (cM)	6.1		
		Blizard et al. [22]	C57BL/6J × DBA/2J F2 and BXD recombinant inbred strains (22 lines)	Systolic blood pressure		4	9.9 (cM)		tel-37 (cM)	5.8		Hrtfm3
		Derry et al. [15]	C57BL/6J × A/J (F2)	Systolic blood pressure at 10 weeks		1	56 (cM)			4.8		
		Andreux et al. [23]	BXD recombinant inbred strains (43 lines)	Systolic blood pressure		9		113		4.7	*Ubp1* (known candidate; Koutnikova et al 2009, pmid 19662162)	VW/AW iso 10 (6)
		Hersch et al. [19]	inbred strains (23 lines)	Systolic blood pressure	Systolic blood pressure iso1 (17)	1		40				

(continued)

Table 1 (continued)

Trait	Phenotype	Study	Mouse cross	Detailed phenotype	Locus	Chr	Peak (cM)	Peak (Mb)	Borders	LOD score	Causal gene	Overlap
ECG parameters	Heart rate	Smolock et al. [24]	C3HeB × SJL (N2)	Heart rate		7	D7Mit350.1 (41cM)		D7Mit21.1-D7Mit1 01.1 (0.5–66 cM)	6.7		GABA receptor genes A
		Sugiyama et al. [18]	BALB/cJ × CBA/CaJ (F2)	Heart rate	Hrq1	2	72 (cM)		60–80 (cM)	4.0		Hrtfm5; Hrtq2
					Hrq2	15	26 (cM)		20–35 (cM)	3.1		Bpq6
		Howden et al. [25]	inbred strains (30 lines) and recombinant inbred strains (AXB/BXA) (29 lines)	Heart rate	HR1	6		54	52–56 (Mb)	3.8		
				HR variability in high frequency	Hrvhf1	5		54	46–56 (Mb)	3.1		
		Blizard et al. [22]	C57BL/6J × DBA/2J F2 and BXD recombinant inbred strains (22 lines)	Heart rate (female)		1	72 (cM)		48–86 (cM)	7.9		
						5	54 (cM)		45–64 (cM)	8.5		
		Derry et al. [15]	C57BL/6J × A/J (F2)	Heart rate at 10 weeks		1	49 (cM)			4.5		
		Scicluna et al. [27]	FVB/NJ–Scn5a1798insD/+ × 129/P2 Scn5a1798insD/+ (F2)	Baseline heart rate		4			136–151 (Mb)	4.5		Scicluna HR post flec
				Post-flecainide heart rate		4			136–151 (Mb)	4.2		Scicluna HR baseline
		Andreux et al. [23]	BXD recombinant inbred strains (43 lines)	Heart rate		19		33		4.5	*Pten* (known candidate; Zu et al 2011; pmid: 21421815)	
		Hersch et al. [19]	Inbred strains (23 lines)	Heart rate	Heart rate ate	17			33 (Mb)			

ECG parameters	Scicluna BP et al. [27]; Lodder et al. [28]	FVB/NJ–Scn5a1798insD/+×129/P2 Scn5a1798insD/+ (F2)	Baseline PR interval		3		135–159 (Mb)	5.0	*Tnni3K*	Vms1
			Baseline QRS interval		7		77–124 (Mb)	4.7		
			Post-flecainide QRS interval		3		88–159 (Mb)	3.9		
			Post-flecainide venticular tachycardia		4		103–151 (Mb)	4.8		Scicluna HR loci
	Hersch et al. [19]	Inbred strains (23 lines)	QRS interval	QRS ctr (9)	6	14.4				
			QTc interval	QTc iso10 (10)	4	123				
				QTc ate	14	85				

cM centimorgan, *Mb* megabase, *LOD* logarithm (base 10) of odds, *AW* atrial weight, *VW* ventricular weight, *ctr* baseline, *iso1* low dose isoproterenol, *iso10* high-dose isoproterenol, *ate* atenolol, *LRS* likelihood-ratio score, *hw/bw* heart weight/body weight, *HR* heart rate

19, 22, 23], heart rate [18, 19, 22–25], cardiac electrographical (ECG) parameters [19, 26–28] and in myocarditis [29]. A concise summary of the results of these studies is given in Table 1.

All these studies employ quantitative trait mapping to uncover quantitative trait loci (QTL) associated with the trait of interest. The principle is based on the underlying assumption that if a particular genetic locus drives the phenotypic difference between the two founder strains, this locus should segregate with this particular phenotype [30].

In this chapter, we will use the example of the F2-generation intercross to explain the process of genetic mapping in cardiovascular disease. The results of this example are described in detail in Scicluna et al. JMCC 2011 and in Lodder et al. PLoS Genetics 2012 [27, 28].

2 Methods

2.1 Determining Phenotypic Differences Between Two Inbred Strains

Starting at the beginning: in order to be able to determine the underlying genetic variants affecting your phenotype of interest, differences in this phenotype need to exist between the two founder strains. In many cases these differences are first observed while backcrossing a genetic variant of interest (e.g. a gene knockout, mutation, or overexpression transgene) into a different mouse strain to obtain a pure genetic background after the line was generated. In such cases it is often observed that the phenotype, as a result of the introduced mutation, is different in mice with different genetic backgrounds. If this difference is both large enough and of particular interest, it may be the starting point for genetic mapping studies to identify the underlying genetic cause. Furthermore, for many phenotypes, spanning the entire spectrum of disease, metabolic traits, and body composition, large screens have now been performed in multiple inbred strains. Most of this data is publicly available in online repositories such as the mouse phenome database (http://phenome.jax.org/). This reference data provides a wealth of information that allows choosing the right strains to start the F2-study. It is worthwhile to inspect this publicly available data before starting, as the ideal strains for mapping are not by definition the same strains that triggered the original interest. Other strains might have even more extreme phenotypic differences for example.

Before embarking on an F2-screen, thoroughly phenotype the two founder lines that will be used in the screen. Determine the variability within the strains (i.e. the environmental and technical variability of your phenotype) and the variation between the two strains (i.e. the variation resulting from the genetic differences between the two strains). Based on these results perform a power calculation to determine the number of F2-mice that will be needed to be able to detect any genetic loci driving the difference.

2.2 Crossing Your F2-Population

Once you have determined which strains to use and the number of F2-mice needed, plan your breedings to obtain these mice. Keep in mind the following points while planning:

1. The planning should be made in such a way that the phenotypic analysis of the F2-mice will be feasible. Think about the number of mice that can be phenotyped per week and what age range the F2-mice may be for the phenotypic analysis.
2. Calculate the number of cages that will be present simultaneously for these experiments, will this fit within the animal facility?
3. Make sure to leave room in the planning for absence of personnel due to holidays, conferences etc.
4. Not all mouse inbred lines give the same number of offspring in their litters; the F1 generation producing the F2's may differ from the two founder strains in this respect. Perform some test breedings to assess the number of pups expected per litter in your F1 breeding.
5. Keep track of the breeding; it is important at a later stage to know the pedigree of any given F2-mouse. For mapping of loci on the X-chromosome it is essential to know the strain of the parental grandmother (pgm), as the F1 father will transmit this X-chromosome in its entirety, without crossing overs, to his female offspring.
6. For most phenotypes differences exist between the sexes, plan accordingly. Either decide on one sex only (which will double the number of breedings needed to obtain the required number of mice), or take the variation between the sexes along in the power calculations and include sex as a covariate in the QTL analyses.

2.3 Obtaining Phenotypic Data in Your Population

Phenotype the mice in a consistent and thorough manner; keep in mind the following considerations:

1. Each mouse is effectively an $N=1$ experiment, no other mice in the F2-population will have the exact same genetic makeup. Therefore it is important to minimize the technical variability as much as possible.
2. Label all tubes and forms in advance with a unique code per F2-mouse to prevent sample swaps.
3. Think about the measurements to be performed, what other phenotypes can be taken along with a relatively small investment of time and money (e.g. body weight, tibia length etc.).
4. Measure all phenotypes that may affect the trait of interest (e.g. heart weight for ECG parameters).
5. Keep samples of the tissue of interest for RNA expression analysis and validation of results.
6. Don't forget to take samples for genotyping.

2.4 Obtaining Genotypic Data of Your F2-Population

As each of the mice represents a unique reshuffling of the genetic variations of the two founder strains, all F2-mice will need to be genotyped. Considering the number of mice (typically at least 300) it is worthwhile to invest time to determine the right balance between cost and mapping density:

1. It is only useful to genotype genetic loci that differ between the two founder strains. The most common inbred lines have now been completely sequenced (http://www.sanger.ac.uk/science/data/mouse-genomes-project) [31, 32].
2. Most other lines have been genotyped with SNP-arrays for which the data is available online (http://phenome.jax.org/).
3. Multiple SNP-arrays covering the mouse genome are available from different suppliers. As the distance between the genetic breakpoints of the recombinations in the F2-generation will generally be large (a typical haploblock will be several Mbs), it is not necessary to type with high density as most SNPs will be in LD.
4. Make sure that enough informative SNPs are included to generate a genetic map of the F2-animals, i.e. there should be no large gaps between the informative SNPs.
5. Type the parental strains on the same array; although the data of your parental strains is most likely available, invest some arrays on typing the parental strains to exclude any potential identity mistakes.
6. Perform thorough quality control (QC) of the obtained array data (see below), exclude any results where the quality of the genotype calling is unsure. Unreliable genotypic data will only dilute any effects you want to measure, or worse, introduce spurious correlations.

2.5 Obtaining Expression Data of the Tissue of Interest

To facilitate identification of the causal genes underlying the detected QTLs at a later stage, we need to detect transcripts that are regulated differentially between the two founder strains. In order to obtain this information for the tissue of interest, transcript abundance data needs to be obtained of the F2-mice. After completion of the phenotyping of the F2-mice, the mice will be sacrificed. At this stage harvest any tissue of interest for gene expression analysis.

2.6 (Multiple) QTL Mapping

2.6.1 Data QC; Genotypic Data

Standard QC for genotypic data obtained from SNP-arrays needs to be performed. As each genotype will be unique, stringent QC is necessary, at least the following criteria should be followed:

1. Genotype calling rate: remove samples with a calling rate <95 % as this is an indication of poor hybridisation.
2. SNPs with a call rate <95 % should be removed, as should SNPs with a MAF <45 % (all SNPs are expected at a 50/50 ratio based on random recombination from both founder strains).

3. Sex matching: SNP status on the X chromosome should match the sex of the mouse, all mismatches should be removed and investigated to identify the source of the mismatch.
4. Overall array intensity, depending on the type of array criteria, concerning the overall hybridisation intensity should be within the ranges published by the producer.
5. Check the genotyping for systematic errors using LOD scores [33].
6. Check the map of the informative SNPs for a regular distribution of the markers; while it is not possible to change this distribution at this stage, you need to be aware of potential gaps in your genetic map.

2.6.2 Data QC; Phenotypic Data

QTL mapping of your phenotype of interest will depend on the quality of and variation within the phenotypic data obtained. Like in any experiment, you want to minimize the technical variability to be able to distinguish the biological variability. This however, is even more important in an F2-study where, as mentioned before, effectively each mouse is an experiment of $N = 1$ and technical variability cannot be averaged out over multiple measurements. Inspect the data for outliers and investigate the potential underlying problems, do not remove any data points without good reasons to do so! Based on the recombination of the different alleles of the two founder strains, it is to be expected that some of the F2-offspring will have more extreme phenotypes than either of the parental lines. The exact QC procedures will of course differ depending on the type of measurements performed. For the RNA expression data, if measured by expression array, make sure to follow general QC guidelines for expression array analysis:

1. Remove samples with overall low expression rates.
2. Remove badly performing duplicate probes.
3. Normalize expression levels within and between arrays.

2.6.3 QTL Mapping

After obtaining high-quality genotypic and phenotypic data it is time for the association analysis to obtain the loci of interest. QTL mapping can be performed using multiple tools, the principle is based on the underlying assumption that if a particular genetic locus drives the phenotypic difference between the two founder strains, this locus should segregate with this particular phenotype in the resulting F2-offspring. Therefore, the basic statistic tests whether a particular marker is correlated to the phenotype of interest; the resulting score gives the logarithm of the odds (LOD-score).

The QTL package in R (https://www.r-project.org/) [34] contains all necessary tools for QC, mapping, and plotting the results. The steps to take are described below; the code gives the basic commands needed in R to perform these steps, for details please refer to the tutorials accompanying the R/QTL package (http://www.rqtl.org/tutorials/).

Data Preparation

1. Organize the data in a cross file, in order to be able to relate the phenotypes to the genotypes; this data needs to be linked based on a unique identifier per mouse; the exact file format is described in detail in the vignette belonging to the QTL package. An example file is given in Table 2.
2. Read in the data file into a cross object:

```
> setwd("path to your data directory")
> library(qtl)
> library(sfsmisc)
> library(stringr)
> options(strings.as.factors=F)
> mydata <- read.cross(format = "csvr", file = "data.
csvr",as.is=T,    genotypes=c("AA","AB","BB"))
```

3. Move markers at identical positions slightly for mapping purposes:

```
> mydata <- jittermap(mydata)
```

4. Calculate genetic distances based on data:

```
> mydata <- est.rf(mydata)
> mydata <- calc.genoprob(mydata)
```

Table 2
Example csvr table, column two contains the chromosome number, column three the position in centimorgans; note that these two columns are empty in the phenotype rows

ID			Mouse_1001	Mouse_1002	Mouse_1003
Sex			1	0	0
body_weight			27.7	27.8	28
heart_weight			0.1	0.2	0.1
tibia_length			17.9	17.7	17.5
PR_Interval_(s)_Baseline_ch1			0.03227	0.03397	0.03908
P_Duration_(s)_Baseline_ch1			0.01668	0.01721	0.01456
QRS_Interval_(s)_Baseline_ch1			0.00873	0.0075	0.00777
QT_Interval_(s)_Baseline_ch1			0.04	0.04309	0.03302
QTc_(s)_Baseline_ch1			0.03487	0.03413	0.02721
pgm			0	0	1
rs13475701	1	0	AB	AB	AB
rs13475706	1	1.00*E*-06	AB	AB	AB
rs3716083	1	1.883054	AB	AB	AB
rs13475745	1	4.374489	AB	AB	AB
rs6173215	1	7.371609	BB	AB	AB
mCV24784983	1	8.768047	BB	AB	AB

5. Inspect the recombination map visually for gaps and other aberrations:

```
> plotRF(mydata)
```

FDR Calculations

In order to correct for multiple testing, false discovery rates (FDR) for the dataset need to be calculated by random permutations of phenotypic and genotypic data. Determine the FDR rate separately for the phenotypic traits, and for a random selection of expressed transcripts (to reduce computing power needed).

1. Calculate LOD thresholds by permutation; test first with a limited set of permutations as this step is time-consuming depending on your processor power and the number of animals:

```
> perm <- scanone(mydata, pheno.col =
c(2:ncol(mydata)), n.perm=10.000)
```

2. Summarize the results and define the thresholds to be used:

```
> summary(perm, alpha=0.1) -> thresh_suggest
> summary(perm, alpha=0.05/(length(mydata$pheno)-1))
-> thresh_sign
```

QTL Analysis per Trait

After these preparations it is time to run the QTL analysis itself for each trait of interest (or transcript of interest when running the expression QTLs). For an example of a significant QTL result *see* Fig. 2.

1. Run single QTL analyses for each phenotypic trait, repeat similarly for each expressed transcript. For the eQTL data it is useful to write the results per transcripts into a separate small .Rdata object and remove the result from your working space to reduce the amount of memory needed:

```
> for (i in c(2:(length(mydata$pheno)-1))){
> name <- paste("scanone", colnames(mydata$pheno[i]),
sep = ".")
> res <- scanone(mydata, pheno.col = i)
> eval(parse(text = paste(name, "<-", "res")))
> ####make merged names object for separate printing
and comparisons
```

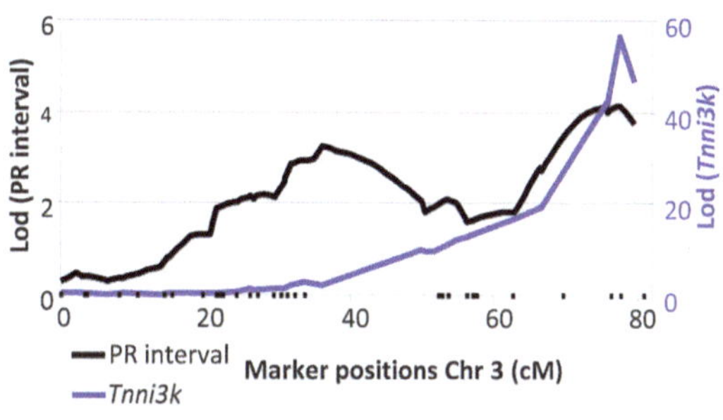

Fig. 2 LOD plot of PR-interval (*black*, *left y*-axis) and *Tnni3k* (*purple right y*-axis) on chromosome 3, adapted with permission from Lodder et al. PLoS Genetics 2012 [28]

```
> if (i < 3) {names.merged <-c()}
> names.merged <- c(names.merged, name)
}
```

2. Plot the results of your QTL analysis per phenotypic trait and inspect for significant results:

```
>
> mult.fig(length(names.merged), main="QTL results my
data", mfrow=c(4, 1))
> for (i in (2:length(names.merged)){
> eval(parse(text =
> paste('plot(', names.merged[i], ', main = "re-
sult', names.merged[i], '", col =    "black", lwd = 2,
bandcol="gray90", ylim = c(0,',ylim,'))', sep ="")))
> abline(h= thresh_suggest[grep(phenos[i],
colnames(thresh_suggest))],
> col= "gray70", lty = 2)
> abline(h= thresh_sign[grep(phenos[i],
colnames(thresh_sign))], col = "black")
}
```

2.7 Towards Identification of the Causal Gene

Obviously, finding genetic loci associated with the trait of interest is relevant but never the final end point of a study. To understand the biology driving the association we want to know which gene or preferably even which genetic variant is responsible. Ultimately, we want to understand the molecular process causing the association. The step from QTL to causative gene is generally not trivial however. In an F2-screen the resulting resolution of the obtained QTLs is low, i.e. the identified loci are large, typically more than 20 Mb in size [30] spanning dozens of genes. So how do we move from the obtained QTL towards candidate genes for the effect? Generally there are three main options for how a locus can affect the phenotype:

1. a protein changing mutation exists in one of the founder strains;
2. a mutation in an enhancer or other *cis*-acting regulatory element affects expression levels of an important gene in one of the founder strains;
3. a mutation in a trans-acting regulatory element affects expression levels of an important gene in one of the founder strains.

2.7.1 Identifying Protein Changing Mutations in the Founder Strains

For option (1), as most mouse inbred lines have been completely sequenced and this data is publicly available [32, 35, 36], it is easy to identify any protein changing variants between the founder strains. Any coding variants within the LOD interval could be causing the observed segregation of the locus with the phenotype in the F2-mice. To easily identify these coding variants you can use the mouse phenome browser (http://phenome.jax.org/).

2.7.2 Overlaying Expression QTL and Phenotypic QTL Data

Options (2) + (3) can be identified by overlapping eQTL data with our region of interest. In 2.6 we determined not only which loci affect the phenotypes of interest but also which loci affect the expression levels of which transcript. We can now overlay the eQTL data with the QTL of interest.

1. To do so, we first determine the borders of the QTL:

```
> lodint(scanone.pheno, chr=3, expandtomarkers=T,
drop=1.5)
```

2. This will yield a small table providing the upper and lower border of the QTL of your phenotype (pheno).

3. Now retrieve the relevant rows containing the LOD scores within the locus for each transcript and keep track of the transcripts for which the max LOD in this interval exceeds the significance threshold as determined in the FDR calculations above:

```
>
> upper <- "name of upper border marker"
> lower <-" name of lower border marker "
> chrom <- 3 ### enter your chromosome of interest here
> rows <- c(grep(paste(upper), rownames(scanone.phe-
no)): grep(paste(lower), rownames(scanone.pheno)))
> ### the object rows should now contain the row num-
bers of interest, as all sets have the same genotypic
markers this is identical for all QTL objects you gen-
erated, but do check this!!
> ### read all eQTL files one by one and write to the
temporary res object:
> for (eqtl in (2:length(colnames(eQTLdata$pheno)))){
> eval(parse(text = (paste(
> 'load("eQTLfilesC/scanone.', colnames(eQTLdata$pheno)
[eqtl], '_C.Rdata")', sep =""))))
> eval(parse(text = (paste('
> res <- scanone.', colnames(eQTLdata$pheno)[eqtl],
sep = ""))))
> ### prepare your results table
> eqtllist <- matrix(nrow=length(rows), ncol = 3)
> colnames(eqtllist) <- c("marker","chr", "pos")
> rownames(eqtllist) <- rep("x", nrow(eqtllist))
> for (i in (1:length(rows))){
> rownames(eqtllist)[i] <- paste(rows[i],"_LOD", sep="")
> r <- rows[i]
> eqtllist[i,1:3] <- c(rownames(scanone.pheno)[r],
scanone.pheno [r,"chr"], scanone.pheno [r, "pos"])
}
> ### read each line of your scanone object at specific
rows to identify possible eQTLs:
> if (max(res[rows,"lod"]) > treshsuggest){
> pos <- res[rows,]
> eqtllist <- cbind(eqtllist, pos[,"lod"])
> colnames(eqtllist)[length(colnames(eqtllist))] <-
paste(colnames(eQTLdata$pheno)[eqtl],"_C", sep="")
}
> ### remove the eQTL object again for space reasons:
> eval(parse(text = (paste(
> 'rm(scanone.', colnames(eQTLdata$pheno)[eqtl],
')',sep =""))))

}
```

4. Check your results table and save it, this table contains the data on all transcripts having a significant eQTL overlapping with the QTL of your phenotype.

2.8 Candidate Prioritization, Haplotype Analysis, and Other Follow-Up Studies

2.8.1 Prioritization

After 2.7 we now have:

1. The borders of the significant QTL for the phenotype of interest.
2. The transcripts with a significant eQTL overlapping the phenotype QTL.
3. Coding genomic changes: as the entire genomic sequence of the founder strains is known for most strains, inspect the QTL for any coding variations that may affect protein function.

We however still miss necessary information for prioritizing candidate genes for follow-up studies:

1. A list of all genes underlying the locus of interest, including annotation. Due to the limited power of eQTL studies (especially to detect trans eQTLs), eQTLs may remain undetected. Furthermore, unknown structural genomic variants may be present in one of the founder strains. It is therefore a good idea to manually inspect all the genes located within the locus to identify potentially interesting candidate genes.
2. Relative expression levels in the tissue of interest of these genes can be extracted from the expression data obtained in 2.5 and 2.6. Furthermore, public information available at GEO (http://www.ncbi.nlm.nih.gov/sites/GDSbrowser?acc=GDS3142 for example) and the mouse phenome database (http://phenome.jax.org/) provides additional information on the expression of the genes, not only in the heart, but other tissues as well and may provide insight in the tissue specificity of the candidate genes.

Integration of all the information above will yield a list of candidate genes that could potentially underlie the observed QTL effect. These genes merit validation and further investigation.

2.8.2 Haplotype Analysis

Phenotyping several inbred strains that differ at the locus of interest may provide a quick validation of the QTL effect (*see* Fig. 3). The mouse phylogeny viewer (http://msub.csbio.unc.edu/) [37] provides detailed analysis possibilities of genetic differences between the different inbred mouse lines and their phylogenetic origins. The tool offers an elegant browser in which the strains can be compared and colored by haplotype based on identity by descent. By comparing the haplotypes of different inbred mouse lines (including the F2-founders) at the position of the QTL, it is possible to quickly select inbred strains of interest for validation studies. It is worthwhile to check for any datasets including these mouse strains in the mouse phenome database [35] that could be related to the trait of interest (other studies may already have provided the data your need, or at least have performed related measurements that can be of interest).

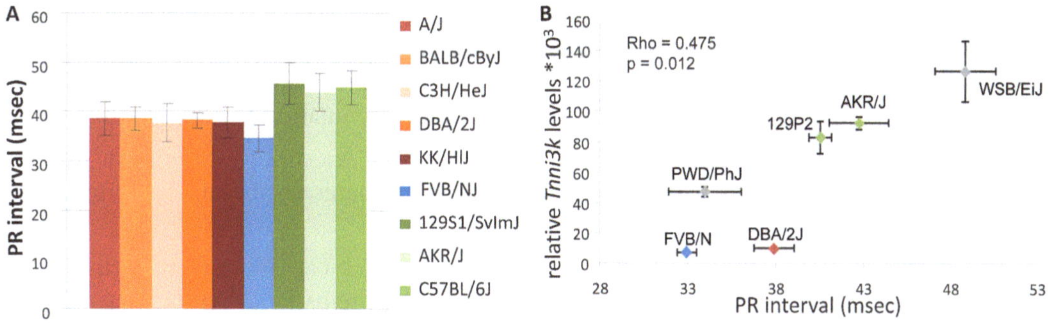

Fig. 3 In silico and in vivo correlation of *Tnni3k* levels and PR-interval duration. (**a**) In silico analysis: mouse inbred lines are colored based on the haplotypes in the minimal region of overlap; PR-interval duration obtained from the mouse phenome database (http://phenome.jax.org/) FVB/NJ (*blue*) (low expression) and *green* (high *Tnni3k* expression), inbred lines with the *reddish colors* carry rs49812611 (associated with nonsense mediated decay). (**b**) In vivo analysis of *Tnni3k* expression levels (*y*-axis) and PR-interval duration (*x*-axis) in six inbred mouse lines, *colors* again denote the haplotype (*gray* is unknown haplotype). Adapted with permission from Lodder et al. PLoS Genetics 2012 [28]

2.8.3 Follow-Up Studies

To ultimately prove the role of the newly identified gene in the F2-study it is essential to isolate the effect of the locus from the other genetic differences between the inbred strains tested. This can be achieved through different genetic mouse models:

1. Congenic lines: back breeding of one founder strain to the other while selecting for the locus of interest from line 1. This will result in the presence of the haplotype of line 1 in the genetic background of line 2. If the QTL effect is true, you would expect to phenotypically recapitulate the phenotype of line 1 in the resulting congenic line. This elegant approach ensures that physiological expression levels of the gene of interest are maintained. It is however very time-consuming.
2. Genetic targeting of the gene of interest: for many genes (conditional) targeted mouse models are now available through different consortia (e.g. EMMA, KOMP, ENU targeted alleles). This reduces the time needed to test the phenotypes in a model that specifically affects the gene of interest. When no model has yet been generated, CRISPR/Cas9 technology [38] now drastically reduces the time needed for the generation of targeted deletions and/or specific insertions of mutations in a mouse model.

A wide array of in vivo, ex vivo, and in vitro phenotyping options are available for the validation and more detailed analysis of the cardiac phenotype of interest in mice, depending on the phenotype:

1. Measurements of electrocardiographs (ECGs) can be performed acutely and for prolonged periods with telemetry.
2. Echocardiography provides insight in the structure and function of the intact heart.

3. Langendorff perfusion can be used for detailed electrical and optical mapping of the conduction and repolarisation of the heart ex vivo. These measurements can be combined with perfusion of drugs and electrical stimulation to challenge the heart.
4. Electrophysiology of isolated cardiomyocytes can provide insight in the underlying mechanisms affecting channel function and availability.
5. Molecular biological techniques can identify the interaction between different cardiac proteins, localisation, expression, and phosphorylation levels of proteins of interest.

A combination of these techniques in the appropriate mouse models can provide evidence of the effect of the QTL and prove the causality of the underlying genes.

3 Results

The results of this example are described in detail in Scicluna et al. JMCC 2011 and in Lodder et al. PLoS Genetics 2012 [27, 28].

Employing QTL mapping in a murine FVB/NJ-129P2-F2-intercross, our group has identified *Tnni3k* as a genetic modifier of cardiac conduction, in particular the PR-interval [28]. The PR-interval indicates the time between the onset of atrial depolarization and the onset of ventricular depolarization. Atrioventricular conduction disease is characterized by the increase of PR-interval duration on the surface electrocardiogram (ECG). Moreover, PR-interval prolongation is a strong predicting factor for atrial fibrillation (AF), the most common sustained arrhythmia [4].

We performed our study in a F2-hybrid mouse population generated by crossing of two different inbred strains: FVB/NJ and 129P2 that were sensitized for conduction disease [27]. Both inbred strains carried the *Scn5a*-1798insD mutation [39], homologous to the human *SCN5A*-1795insD mutation, identified in a Dutch family manifesting an "overlap syndrome" with bradycardia, conduction disease, and prolongation of QT-interval [40]. *SCN5A* encodes the cardiac sodium channel which is the main determinant of cardiac conduction. Importantly, two parental inbred strains (129P2-*Scn5a*$^{1798insD/+}$ and FVB/NJ-*Scn5a*$^{1798insD/+}$ mice) exhibited large differences in conduction disease severity [39]. Not surprisingly, murine F2-progeny generated by crossing of these parental strains is a powerful resource for mapping of novel genetic traits underlying cardiac conduction.

By integrating ECG and genome-wide genotypic data in 502 F2-progeny from 129P2-*Scn5a*$^{1798insD/+}$ and FVB/NJ-*Scn5a*$^{1798insD/+}$ mice intercross, we uncovered a locus modulating PR-interval on mouse chromosome 3 [27]. Further, we combined genome-wide gene expression measured in cardiac tissue with genotypic data in

109 F2-progeny to identify cardiac expression QTLs (eQTL) overlapping the PR-QTL on chromosome 3 (Fig. 2). Subsequently, we tested whether the identified eQTLs correlated with the PR-interval duration in the F2-mice. Four transcripts showed a significant correlation with PR-interval, of these only *Tnni3K* was abundantly and specifically expressed in the heart which made it the main candidate for the PR-interval modulating effect of the Chr3 PR-QTL [28].

In silico haplotype analysis (using the mouse phylogeny viewer (http://msub.csbio.unc.edu/) in a panel of inbred mouse strains identified three independent haplotypes in the *Tnni3K* genomic locus, which were associated with different levels of *Tnni3K* expression in these inbred strains. Phenotypic data in mouse phenome database (http://phenome.jax.org/) indicated a significantly longer PR-interval in strains with high *Tnni3K* expression (129S1/SvImJ, C57BL/6J, and AKR/J) compared to strains with low *Tnni3K* expression (A/J, BALB/cByJ, C3H/HeJ, KK/HlJ, DBA/2J, and FVB/NJ) (Fig. 3a). These findings were further validated in six inbred mouse strains

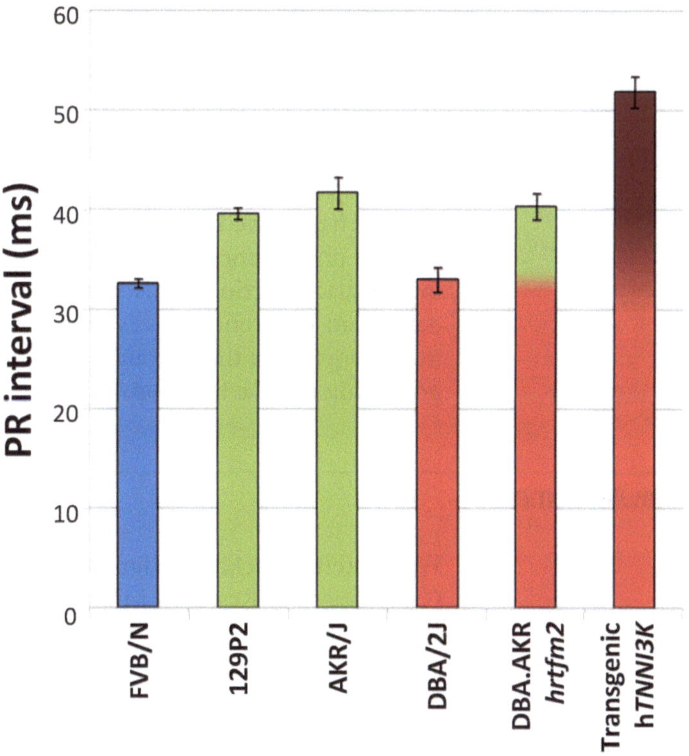

Fig. 4 *Tnni3k* prolongs the PR-interval. Congenic mice carrying the AKR/J green haplotype in a DBA/2J genetic background display the green haplotype PR-interval duration. Overexpression of tagged *hTNNI3k* in a DBA/2J background significantly prolongs the PR-interval. *Colors* show the haplotype of each strain at the *Tnni3k* locus, error bars indicate standard deviations. Adapted with permission from Lodder et al. PLoS Genetics 2012 [28]

harboring at least four independent haplotypes at *Tnni3K* eQTL region (Fig. 3b); *Tnni3K* expression levels significantly correlated to PR-interval duration in these inbred mouse strains [28].

Finally, we demonstrated that h*TNNI3K* overexpression in DBA/2J-*hTNNI3K* transgenic mice induced a pronounced PR-interval prolongation when compared to PR-interval of wild-type DBA/2J mice (with low *Tnni3K* expression). Importantly, DBA/2J.AKR/J congenic mice (with physiological levels of *Tnni3k* as a result of the presence of the AKR/J-derived *Tnni3k* allele in the DBA/2J genetic background) had PR-intervals comparable to the AKR/J mice (i.e. significantly longer than DBA/2J mice which lack *Tnni3k*, Fig. 4) [28].

Interestingly, two independent studies conducted in cardiomyopathy-sensitized mice (due to the cardiac-specific overexpression of the Ca^{2+} binding protein *calsequestrin*) generated by two different mouse N2 backcrosses indicated a causal link between *Tnni3K* expression and progression of cardiomyopathy [10, 14, 41]. Furthermore, by QTL mapping, *Tnni3K* was identified as a possible modulator of viral myocarditis [29]. A recent overview of the functions of *Tnni3k* including recent evidence of a potential role in human disease is given in a review by Milano et al. J Mol Cell Cardiol [42].

4 Conclusions and Summary

In this chapter we have given an overview of the methods used to identify novel genetic loci and the underlying causal genes affecting cardiac function. This systems genetics approach integrating data from different sources provides a powerful tool to identify novel targets for therapy and insight into the molecular processes governing cardiac function.

Acknowledgements

We gratefully acknowledge the support from the Netherlands CardioVascular Research Initiative (CVON-PREDICT project) to E.M.L. and C.R.B. and of the AMC foundation (Ph.D. scholarship) to S.P.

References

1. Fuster V (2014) Global burden of cardiovascular disease: time to implement feasible strategies and to monitor results. J Am Coll Cardiol 64:520–522
2. Jouven X, Desnos M, Guerot C, Ducimetiere P (1999) Predicting sudden death in the population: the Paris Prospective Study I. Circulation 99:1978–1983
3. Zipes DP, Wellens HJ (1998) Sudden cardiac death. Circulation 98:2334–2351
4. Heeringa J, Van Der Kuip DAM, Hofman A et al. (2006) Prevalence, incidence and lifetime

risk of atrial fibrillation: the Rotterdam study. Eur Heart J 27:949–953

5. Marsman RF, Tan HL, Bezzina CR (2014) Genetics of sudden cardiac death caused by ventricular arrhythmias. Nat Rev Cardiol 11:96–111
6. George AL Jr. (2013) Molecular and genetic basis of sudden cardiac death. J Clin Invest 123:75–83
7. Maron BJ, Maron MS, Semsarian C (2012) Genetics of hypertrophic cardiomyopathy after 20 years: clinical perspectives. J Am Coll Cardiol 60:705–715
8. Schunkert H, König IR, Kathiresan S et al. (2011) Large-scale association analysis identifies 13 new susceptibility loci for coronary artery disease. Nat Genet 43:333–338
9. Beck JA, Lloyd S, Hafezparast M et al. (2000) Genealogies of mouse inbred strains. Nat Genet 24:23–25
10. Suzuki M, Carlson KM, Marchuk DA, Rockman HA (2002) Genetic modifier loci affecting survival and cardiac function in murine dilated cardiomyopathy. Circulation 105:1824–1829
11. Wheeler FC, Fernandez L, Carlson KM, Wolf MJ, Rockman HA, Marchuk DA (2005) QTL mapping in a mouse model of cardiomyopathy reveals an ancestral modifier allele affecting heart function and survival. Mamm Genome 16:414–423
12. Le Corvoisier P, Park HY, Carlson KM, Marchuk DA, Rockman HA (2003) Multiple quantitative trait loci modify the heart failure phenotype in murine cardiomyopathy. Hum Mol Genet 12:3097–3107
13. Maddatu TP, Garvey SM, Schroeder DG et al. (2005) Dilated cardiomyopathy in the nmd mouse: transgenic rescue and QTLs that improve cardiac function and survival. Hum Mol Genet 14:3179–3189
14. Wheeler FC, Tang H, Marks OA et al. (2009) Tnni3k modifies disease progression in murine models of cardiomyopathy. PLoS Genet 5, e1000647
15. Derry JM, Zhong H, Molony C et al. (2010) Identification of genes and networks driving cardiovascular and metabolic phenotypes in a mouse F2 intercross. PLoS One 5, e14319
16. Rocha JL, Eisen EJ, Van Vleck LD, Pomp D (2004) A large-sample QTL study in mice: I. Growth. Mamm Genome 15:83–99
17. Rocha JL, Eisen EJ, Dale Van Vleck L, Pomp D (2004) A large-sample QTL study in mice: II. Body composition. Mamm Genome 15:100–113
18. Sugiyama F, Churchill G, Li R et al. (2002) QTL associated with blood pressure, heart rate, and heart weight in CBA/CaJ and BALB/cJ mice. Physiol Genomics 10(1):5–12
19. Hersch M, Peter B, Kang HM et al (2012) Mapping genetic variants associated with beta-adrenergic responses in inbred mice. PLoS One 7, e41032
20. Lodder EM, Scicluna BP, Beekman L et al. (2014) An integrative genomic approach identifies multiple genes involved in cardiac collagen deposition. Circ Cardiovasc Genet 7:790–798
21. Kirk EP, Hyun C, Thomson PC et al. (2006) Quantitative trait loci modifying cardiac atrial septal morphology and risk of patent foramen ovale in the mouse. Circ Res 98:651–658
22. Blizard DA, Lionikas A, Vandenbergh DJ et al. (2009) Blood pressure and heart rate QTL in mice of the B6/D2 lineage: sex differences and environmental influences. Physiol Genomics 36:158–166
23. Andreux PA, Williams EG, Koutnikova H et al. (2012) Systems genetics of metabolism: the use of the BXD murine reference panel for multiscalar integration of traits. Cell 150:1287–1299
24. Smolock EM, Ilyushkina IA, Ghazalpour A et al. (2012) Genetic locus on mouse chromosome 7 controls elevated heart rate. Physiol Genomics 44:689–698
25. Howden R, Liu E, Miller-DeGraff L et al. (2008) The genetic contribution to heart rate and heart rate variability in quiescent mice. Am J Physiol Heart Circ Physiol 295:59–68
26. Berthonneche C, Peter B, Schupfer F et al. (2009) Cardiovascular response to beta-adrenergic blockade or activation in 23 inbred mouse strains. PLoS One 4, e6610
27. Scicluna BP, Tanck MWT, Remme CA et al. (2011) Quantitative trait loci for electrocardiographic parameters and arrhythmia in the mouse. J Mol Cell Cardiol 50:380–389
28. Lodder EM, Scicluna BP, Milano A et al. (2012) Dissection of a quantitative trait locus for pr interval duration identifies Tnni3k as a novel modulator of cardiac conduction. PLoS Genet 8, e1003113
29. Wiltshire SA, Leiva-Torres GA, Vidal SM (2011) Quantitative trait locus analysis, pathway analysis, and consomic mapping show genetic variants of Tnni3k, Fpgt, or H28 control susceptibility to viral myocarditis. J Immunol 186:6398–6405
30. Flint J, Valdar W, Shifman S, Mott R (2005) Strategies for mapping and cloning quantitative trait genes in rodents. Nat Rev Genet 6:271–286
31. Keane TM, Goodstadt L, Danecek P et al. (2011) Mouse genomic variation and its effect on phenotypes and gene regulation. Nature 477:289–294

32. Wong K, Bumpstead S, Van Der Weyden L et al. (2012) Sequencing and characterization of the FVB/NJ mouse genome. Genome Biol 13:R72
33. Lincoln SE, Lander ES (1992) Systematic detection of errors in genetic linkage data. Genomics 14:604–610
34. Broman KW, Wu H, Sen S, Churchill GA (2003) R/qtl: QTL mapping in experimental crosses. Bioinformatics 19:889–890
35. Bogue MA, Grubb SC (2004) The Mouse Phenome Project. Genetica 122:71–74
36. Yalcin B, Adams DJ, Flint J, Keane TM (2012) Next-generation sequencing of experimental mouse strains. Mamm Genome 23:490–498
37. Wang JR, de Villena FP, McMillan L (2012) Comparative analysis and visualization of multiple collinear genomes. BMC Bioinformatics 13(Suppl 3):13
38. Sander JD, Joung JK (2014) CRISPR-Cas systems for editing, regulating and targeting genomes. Nat Biotechnol 32:347–355
39. Remme CA, Verkerk AO, Nuyens D et al. (2006) Overlap syndrome of cardiac sodium channel disease in mice carrying the equivalent mutation of human SCN5A-1795insD. Circulation 114:2584–2594
40. Bezzina C, Veldkamp MW, van Den Berg MP et al. (1999) A single Na(+) channel mutation causing both long-QT and Brugada syndromes. Circ Res 85:1206–1213
41. Tang H, Xiao K, Mao L, Rockman HA, Marchuk DA (2013) Overexpression of TNNI3K, a cardiac-specific MAPKKK, promotes cardiac dysfunction. J Mol Cell Cardiol 54:101–111
42. Milano A, Lodder EM, Bezzina CR (2015) TNNI3K in cardiovascular disease and prospects for therapy. J Mol Cell Cardiol 82:167–173

Chapter 21

Systems Genetics of Liver Fibrosis

Rabea A. Hall and Frank Lammert

Abstract

This systems genetics analysis comprises quantitative measurements of hepatic fibrogenesis in mouse models and mapping of quantitative traits in mouse genetic reference populations. It is part of a large mapping project of fibrogenic genes including the analyses of experimental crosses from different inbred mouse strains. Extensive quantitative trait loci (QTL) mapping of fibrosis phenotypes and liver expression profiling in combination with in silico mapping facilitated the identification of QTL regions and underlying candidate genes that confer fibrosis susceptibility also in humans. Moreover, the approach led to the identification of interacting QTLs and gene networks in liver fibrosis, providing a key experimental platform for the development of novel, more precise therapeutic interventions. Here, we provide a use case for the application of different analysis tools and the integration of multiple datasets determined in F2 intercrosses and BXD recombinant inbred lines to identify, finemap and affirm fibrosis susceptibility loci.

Key words GeneNetwork, Genetic reference population, Hepatic fibrosis, Quantitative trait locus mapping

1 Introduction

The liver plays a central role in metabolism, energy supply and storage as well as detoxification. Moreover, it is equipped with a unique wound healing and repair machinery that allows total regeneration of liver after removing up to 70% of the original organ mass [1]. During acute liver injury the wound healing response is activated and injured tissue is replaced by new cells or extracellular matrix (ECM) proteins forming a fibrotic scar. Once the damaging stimulus is eliminated, the repair structures resolve. In diseases that cause chronic injury the regenerative mechanisms turn pathological and an imbalance of fibrotic progression and regression occurs, resulting in the accumulation of scar tissue [2]. Heptic stellate cells (HSC) and also portal fibroblasts represent the main cellular sources of ECM proteins. In their quiescent form HSC store vitamin A. They are localized in the space of Disse between sinusoidal endothelial cells and hepatocytes. The HSC are activated by hepatocyte apoptosis and profibrogenic cytokines,

Klaus Schughart and Robert W. Williams (eds.), *Systems Genetics: Methods and Protocols*, Methods in Molecular Biology, vol. 1488, DOI 10.1007/978-1-4939-6427-7_21, © Springer Science+Business Media New York 2017

chemokines, and growth factors (in particular transforming growth factor-β), which are released by infiltrating immune cells and HSC themselves. They transdifferentiate into myofibroblasts, which implies changes in gene expression and cellular behavior, such as loss of vitamin A, cell proliferation, chemotaxis, contractility, and increased ECM production [3]. The major ECM constituent is collagen, and during fibrogenesis the typical ECM composition changes from type IV collagen, heparan sulfate proteoglycans, and laminin to fibrillary collagens type I and III [4]. The increasing scar disrupts liver structure and function and eventually hepatic fibrosis progresses to end-stage cirrhosis and liver cancer, which cause further complications and mortality. Although at the earlier stages of cirrhosis regression is still possible, this requires time, which the patient's deteriorating state and frequent infections do often not allow for [5]. Globally over two million deaths were estimated to be related to chronic liver diseases in 2013, which corresponds to an increased death rate of 44 % since 1990 [6]. Though new therapies for chronic hepatitis B and C virus infections have improved survival rates, the prevalence of other etiologies such as alcoholic liver diseases and nonalcoholic fatty liver diseases (NAFLD) is increasing [7].

Irrespective of etiology, age, or sex, fibrosis susceptibility varies among patients with chronic liver diseases. This variation in the progression rate distinguishes "slow" from "rapid fibrosers" [8] and indicates the relevance of genetic factors. In contrast to rare monogenic liver diseases, such as hereditary hemochromatosis or Wilson disease, most common liver diseases are complex and are influenced by multiple genes and gene–gene interactions. The identification of these genes is complicated due to low effect sizes, epistatic effects, and environmental factors [9]. Twin studies are a valuable source to study the heritability of diseases. Twins grow up under similiar environmental conditions; monozygotic twins share an identical genetic background, whereas dizygotic twins share 50 % of their genes. The identification of traits shared in monozygotic but not dizygotic twins implies genetic effects. Recently Loomba et al. [10] studied the concordance of NAFLD and fibrotic NAFLD. They observed a robust correlation of both diseases in monozygotic but not in dizygotic twins. Hepatic fibrosis showed an estimated heritability of 50 % ($h^2 = 0.5$, $p = 6.1 \times 10^{-11}$). The identification and characterization of inherited mediators of hepatic fibrosis provides additional information for individualized treatment. Human association studies have identified multiple fibrosis-related genes and several genetic factors are being tested in replication studies and clinical trials to assess the prediction of fibrosis predisposition [9, 11, 12]. However, only few studies have considered testing for associations of multiple polygenic risk factors simultaneously [13, 14]. Both studies were performed in patients with chronic viral hepatitis. Hence, more studies are needed to detect multigenic associations and core pathways underlying susceptibility to liver diseases. The early identification of fibrotic mediators is required to allow close monitoring and, if applicable, preventive measures and specific treatment. To date, fibrosis often remains undetected before progressing to end-stage liver disease. Hence, noninvasive markers and imaging techniques are needed to ensure early identification of injury [15, 16]. A deeper knowledge of the individual genetic constitution will open new avenues to target fibrogenic pathways and to stop progression or even promote resolution of fibrosis.

2 Methods

2.1 Models of Liver Fibrosis

Human tissue samples of diseased livers are normally derived from biopsies or surgical resection. Therefore, research samples are limited and might not be representative. There are several ways to model liver diseases experimentally. Mice provide a suitable model system as they share more than 95 % of the human genome. They reproduce rapidly, environmental influences can be systematically controlled, they are genetically modifiable, and their genetic diversity can be reduced by inbreeding.

There are different ways to induce hepatic fibrosis in mice, and standard operating procedures have been published recently to provide transferablility of experimental results among different research groups [17]. Cholestatic fibrosis can be induced by surgical intervention (bile duct ligation) or develops spontaneously in genetically modified mice, e.g. after knockout of the multidrug resistance gene 2 (*Mdr2 or Abcb4*), which functions as a phosphatidylcholine floppase in the hepatocanalicular membrane [18]. Another possibility of fibrosis induction is by application of hepatotoxic drugs such as acetaminophen [19] or carcinogenic challenge with diethylnitrosamine [20]. The most established model for fibrosis induction is intraperitoneal injection of carbon tetrachloride (CCl_4) [21]. The chronic model involves repeated CCl_4 injections for a period of several weeks, e.g. injections twice a week for 6–12 weeks or three injections for 4 weeks, depending on strain susceptibility. Dosages also vary from 0.5 to 0.7 ml CCl_4/kg mouse weight dissolved in mineral or corn oil. The metabolism of CCl_4 by cytochrome P450 leads to trichloromethyl radicals (CCl_3^*), which initiate the formation of toxic metabolites such as mutated nucleic acids, peroxidation of lipids, and hypomethylation of proteins. This results in severe oxidative stress, hepatocellular damage, and necrosis, followed by chronic inflammation and hepatic fibrosis.

The in vivo models reflect the complex mechanisms during hepatic injury and interactions of the various resident hepatic cell types as well as infiltrating immune cells. Cell culture experiments of primary cells allow a controlled observation of different mediators and their effects in single-cell culture or in cocultured cells. Hepatocytes are the most common resident cells (~80 %), the collagen producing HSC represent approximately 10%, and the remaining cells are Kupfer cells (macrophages) and endothelial cells. The isolation of HSC requires 5–6 mice (6-month-old) to obtain the appropriate number of cells, whereas one mouse is adequate for the isolation of primary hepatocytes. Experimental observations in these model systems need to be compared to human samples.

2.2 Mapping Quantitative Trait Loci (QTL) for Hepatic Fibrosis

The identification of disease-related loci by genetic linkage is possible by QTL mapping. The first systematic analysis of hepatic fibrosis in mice was performed in seven inbred strains (A/J, AKR/J, BALB/cJ [BALB], C57BL/6J [BL/6], C3H/HeJ, DBA/2J [DBA], and FVB/NJ [FVB]) [22]. After the hepatotoxic challenge with CCl_4 for 6 weeks slow (A/J, AKR, FVB), intermediate (BL/6, DBA) and rapid fibrosers (BALB, C3H) were identified (Fig. 1). These differences in fibrogenesis were contributed to genetic background [22].

2.3 F2 Intercross: [A/J × BALB] F_1 × [A/J × BALB] F_1

To identify the underlying loci in a subsequent mapping analysis, the fibrosis susceptible strain BALB and the resistant strain A/J were crossed to obtain 385 F_2 littermates. Microsatellite markers were genotyped genome-wide by PCR using genomic DNA derived from spleen. Fibrosis progression was characterized by determining the collagen-specific amino acid hydroxyproline in liver hydrolysates (Fig. 1) and quantification of liver histologies using a semiquantitative fibrosis score. For linkage analysis, phenotypes and genotypes were correlated using MapManager. For this intercross, two QTLs were localized on chromosomes 2 and 15 and significantly associated to both hydroxyproline concentrations and fibrosis stages and designated as *Hfib2* and *Hfib1*, respectively (Table 1).

For the identification of the underlying quantitative trait genes in silico gene mapping was integrated to narrow down QTL intervals [23]. A 140K single nucleotide polymorphism (SNP) map was used to identify similar haplotypes that correspond to the

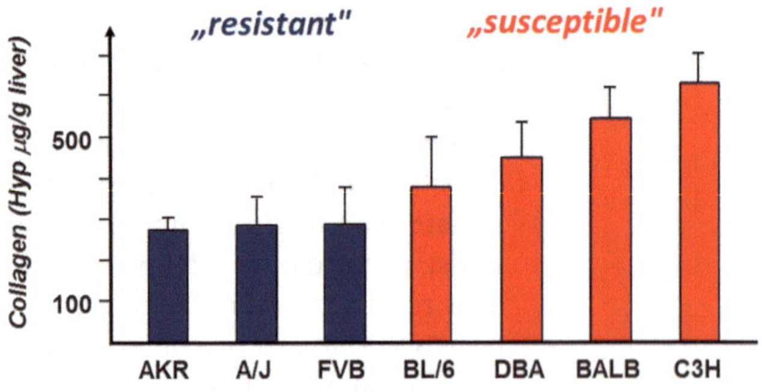

Fig. 1 Hepatic collagen concentrations (determined as µg hydroxyproline/g liver) in inbred mouse strains after CCl_4 intoxication for 6 weeks. Lines were stratified into fibrosis-resistant (low collagen levels) and fibrosis-susceptible strains displaying pronounced hepatic collagen accumulation (Figure adapted from Hillebrandt et al. 2002 [22])

Table 1
QTLs associated with hepatic fibrogenesis (designated *Hfib*) in chronological order of discovery

Locus	Chr	Mb	Cross	Gene
Hfib1	15	55.2–64.4	AxB	*A1bg*, *Mtss1* [22]
Hfib2	2	20.9–65.7	AxB/BxF	*Hc* [23, 27]
Hfib3	5	90.2–94.9	AxB	*Cxcl9* [26]
Hfib4	11	83.5–89.5	BxD	*Expi*, *Msi2* [33]
Hfib5	1	163.6–171.8	BxF	*Fasl* [27]

Locus: name of QTL; Chr: chromosomal location; Mb: Megabase position on chromosome; Cross:experimental crosses used for mapping analysis; AxB: A/J and BALB; BxF: BALB and FVB; BxD: BL/6 and DBA; Gene: potential candidate gene found in QTL region

phenotypic characteristics of the seven inbred strains listed above [24] (Fig. 2). The analysis verified the results of the experimental QTL analysis by identifying the strongest associations on chromosomes 2 and 15 (Fig. 3). Within these refined QTL regions, hepatic gene expression data was used to define candidate genes that showed at least threefold increased hepatic expression (Fig. 2). Among the nine top candidates were complement factor C5 (a.k.a. hemolytic complement, *Hc*) on chromosome 2 and α-1-B glycoprotein (*A1bg*) and metastasis suppressor 1 (*Mtss1*) on chromosome 15.

Interestingly, some mouse strains inherited a 2-bp deletion in the *Hc* gene leading to C5 deficiency, which was associated with lower fibrosis susceptibility. The causal relationship of the *Hc* gene was validated using congenic *Hc* knockout mice and *Hc* transgenic mice [23]. The results taken together demonstrated the profibrogenic effect of C5. In addition, inhibition of the C5a receptor in vivo ameliorated hepatic fibrosis. The potential relevance of *C5* in humans is supported by the association of *C5* haplotype tagging polymorphisms and C5 serum concentrations [23, 25].

The same cross was used for a reverse genetics approach specifically focusing on the analysis of chemokines located in a small *Cxc* chemokine cluster on mouse chromosome 5 (*Hfib3*) (Table 1). Three tagging SNPs positioned distal, in the middle and the proximal section of the gene cluster were introduced in a regression analysis of the F_2 mice. The distal and middle tagging SNP were significant associated to the histologic stages of liver fibrosis. Here, in particular *Cxcl9*, a gene in close proximity to the distal SNP, was differentially expressed in the parental strains, which pointed towards its antifibrotic function. These findings were confirmed in CXCL9 receptor (*Cxcr3*) knockout mice and could also be validated in a human cohort of patients with chronic hepatitis C virus infection by haplotype analysis. Carriers of the identified risk allele also showed reduced CXCL9 serum levels [26].

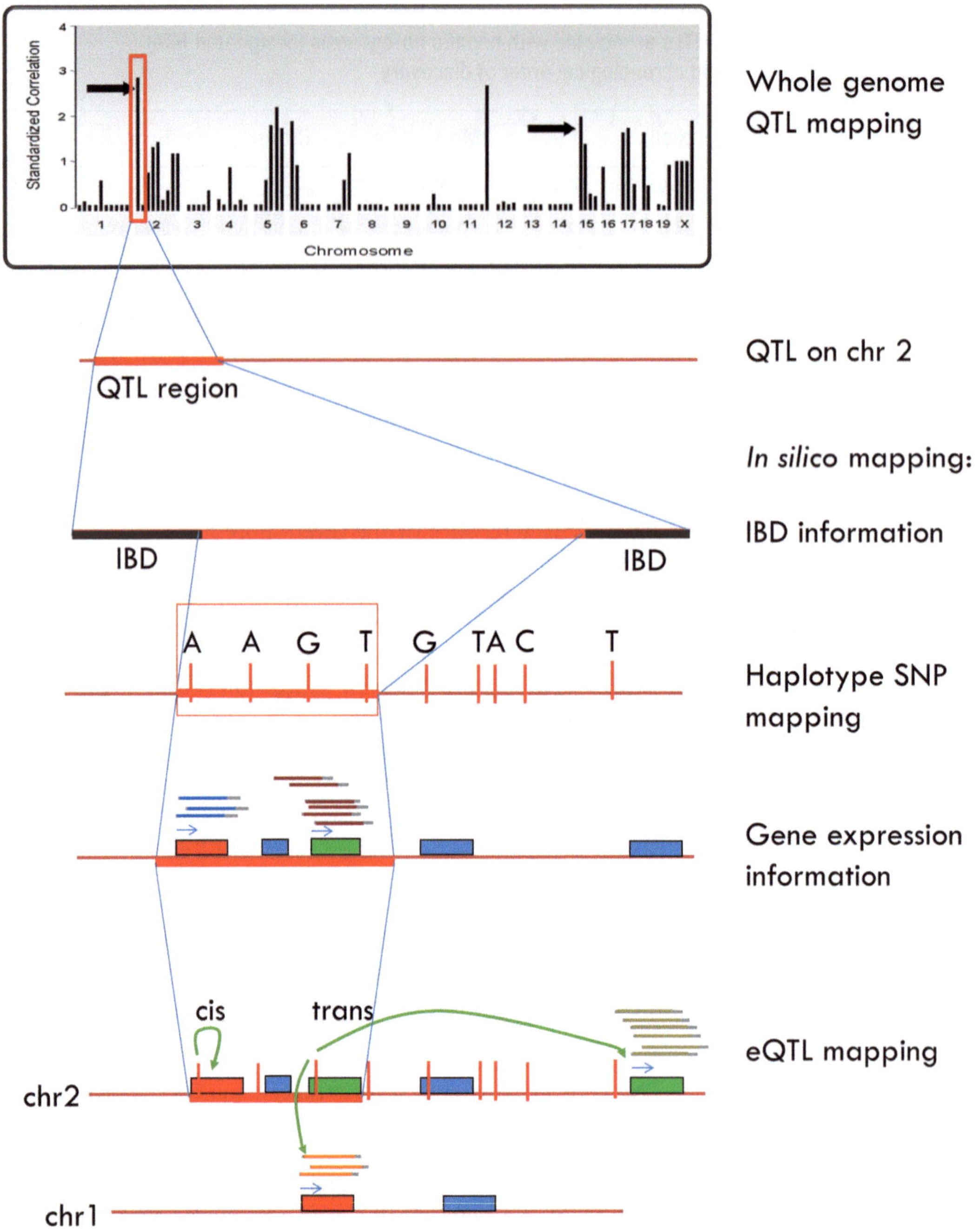

Fig. 2 Steps to refine QTL regions by in silico analyses: A QTL is identified by experimental QTL mapping. In silico mapping allows to narrow QTL regions identical by decent (IBD) and identifies haplotypes correlating with phenotypic expression. The identification of differentially expressed genes (*green*: overexpressed; *red*: downregulated) in QTL regions. eQTL analyses identify regulatory loci in QTL regions (*cis*: local regulation of genes, e.g. variants in promoter regions; *trans*: regulation of distant genes, e.g. variants in transcription factors). (Adapted from Hillebrandt et al. 2002 [22] and Cervino et al. 2007 [24])

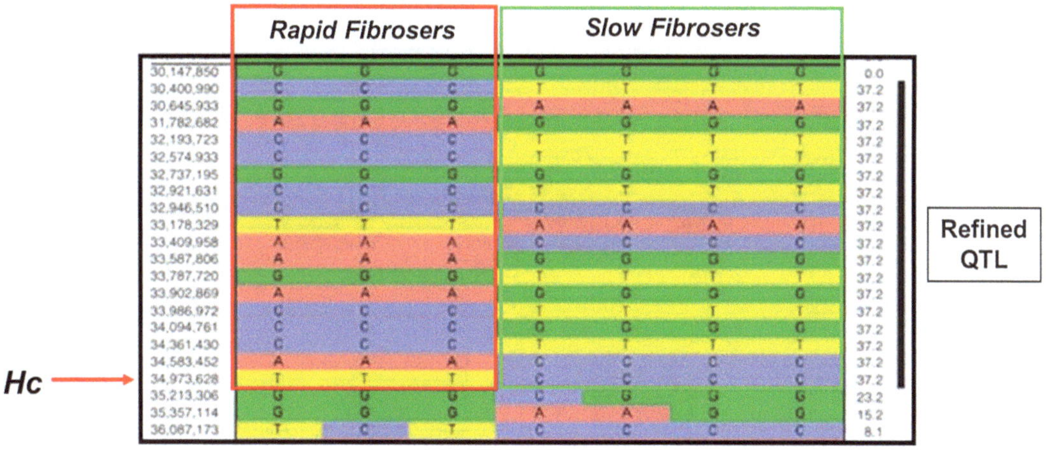

Fig. 3 Refined QTL identified by in silico haplotype mapping. Distinct haplotypes for rapid and slow fibrosing strains on chromosome 2 were identified by sliding three-SNP windows and calculating *F*-statistics by pairing genotype and phenotype at each haplotype. Underlying this region is the *Hc* gene encoding complement factor C5 (Figure adapted from Hillebrandt et al. 2005 [23])

2.4 F2 Intercross: [BALB × FVB] F_1 × [BALB × FVB] F_1

For the second intercross study, the fibrosis-susceptible strain BALB was crossed with another resistant strain (FVB) to identify additional fibrosis QTLs [27]. In this approach the effect of multiple QTLs was mapped using the "fitqtl" function in R/qtl to identify fibrogenic networks [28]. A network of 11 interacting QTLs associated to the histological fibrosis stage or hydroxyproline concentrations was constructed with a single overlapping locus on chromosome 1. Overall, the effects of the single loci were very small (≤4 %), which renders them challenging to detect. Therefore, composite interval mapping was applied to reduce the residual variation of the stronger loci and to allow the identification of weaker loci or epistatic interactions. Single loci were jointly analyzed in multiple QTL models. These models explained up to 25 % of the phenotypic variance. This second intercross verified the profibrogenic QTL *Hfib2* on chromosome 2 as well as the underlying quantitative trait gene *Hc/C5*, with the deletion of the gene resulting in marked reduction of fibrosis. The cross identified two QTLs on chromosome 1 associated to fibrosis stage and hydroxyproline concentrations with overlapping QTL regions. This overlapping QTL on chromosome 1 was designated as fibrosis QTL *Hfib5*. The most promising candidate gene Fas ligand (*Fasl*) is functionally involved in hepatic fibrogenesis [29]. The analysis of a nonsynonymous *Fasl* variant identified a significant allele effect on hepatic hydroxyproline concentrations. Furthermore, all interacting loci covered regions with genes that have been linked to fibrosis in other studies (Fig. 4). This study indicates that the effects of disease-related genes for a polygenic trait depend on genetic interactions and that it is advisable to screen mouse strains with different genetic backgrounds.

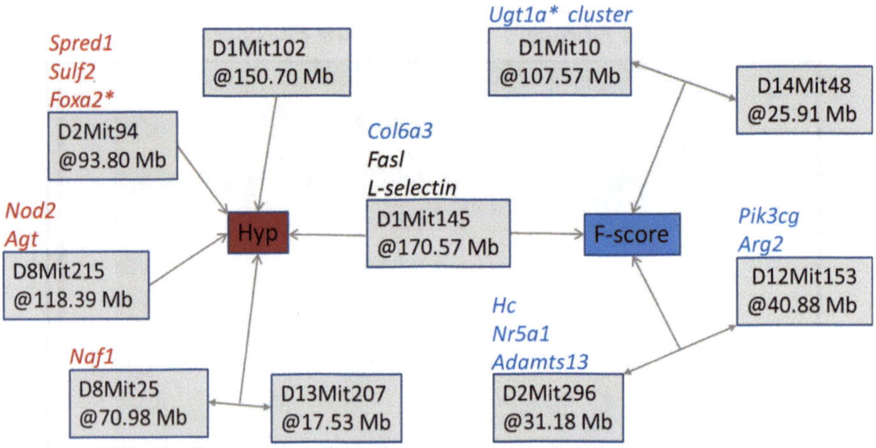

Fig. 4 Multiple QTL network determined in the (BALB × FVB) F_2 intercross mice. Each node represents a locus; edges resemble associations either to phenotypes or other loci. Indicated above the nodes are potential candidate genes that might underlie each locus and are functionally associated to fibrosis (Figure adapted from Hall et al. 2015 [27], with permission of Springer)

2.5 Systems Genetics in BXD Recombinant Inbred Lines

To address the complexity of the genetic regulation of fibrosis, systems genetics allows an untargeted approach by integration of large-scale genomics into genetic association studies. As experimental framework, systems genetics avails of a genetic reference population (GRP). A GRP allows the integration of multiple quantitative traits on the defined genetic backgrounds of homozygous inbred lines that were derived by crossing two or more inbred strains. Genotypes of the GRP lines show the typical mosaic structure of the recombined parental chromosomes. The BXD recombinant inbred mouse lines are a GRP derived from a cross of BL/6 and DBA mice [30]. A major advantage of the panel is that the lines can be analyzed using the web-based GeneNetwork database [31]. In GeneNetwork, the genotype data of the BXD lines are deposited together with various trait data that have been studied in the BXD by different research groups (http://genenetwork.org). The database can be used to perform QTL mapping of own datasets or to calculate correlations with other open access traits.

The identification of fibrosis QTL was performed in two different BXD studies: (1) in vivo fibrosis was induced by long-term CCl_4 application as described above [32]; and (2) in vitro hepatocyte injury was studied [33]. In vivo data can be accessed at GeneNetwork, searching for Group: BXD → Type: Phenotypes → IDs: 14355–14396; or expression data of these mice: Type: Liver mRNA → Data Set: *SUH BXD Liver CCl4-treated Affy Mouse Gene 1.0.ST*. The in vitro data are stored at IDs: 16299–16240 searching the BXD phenotypes.

2.5.1 In Vivo

BL/6 and DBA resemble inbred strains of intermediate fibrosis susceptibility (Fig. 1). However, due to transgressive segregation, the BXD lines exceed the parental phenotypes. Fibrosis heritability estimated in the BXD panel corresponds with the values estimated in the human twin studies ($h^2 \sim 0.5$). QTL mapping of the phenotypic traits (hydroxyproline concentration and fibrosis stage) identified nine fibrosis-associated QTLs. For the discovery of the quantitative trait genes and regulatory mechanisms hepatic expression profiles of the fibrotic BXD lines were obtained by whole-genome microarray analysis. The RNA expression levels were implemented as quantitative traits in an expression QTL (eQTL) analysis. By eQTL mapping *cis*- and *trans*-acting loci affecting hepatic gene expression were identified (Fig. 2). Colocalizing regions of phenotype- and expression-associated QTLs (mainly *cis*QTLs) suggest a common association of regulatory mechanisms for gene expression and phenotypic variation. Further selection criteria for the identification of candidate genes were defined to allow explorative in silico analyses in GeneNetwork: (1) *cis* quantitative trait genes that correlate to phenotypes across the BXD panel were identified, pointing to causal relationships of gene regulation and phenotypic expression; (2) GeneNetwork variant browser was used to identify coding nsSNP variants that differ between the parental strains; and (3) hepatic expression profiles were compared to identify differential gene regulation. The results of the eQTL analysis in CCl_4-treated mice were compared to eQTL data of saline-treated controls. Differentially regulated genes in diseased and normal mice also represent potential candidates (Fig. 5). The candidate genes with potentially causal relationships to fibrosis were then screened to identify coexpression clusters, which were combined in a regulatory fibrosis network [32].

2.5.2 In Vitro

The aim of the in vitro study was to reduce the complexity of in vivo experiments by focusing on a central fibrogenic signaling pathway in a primary cell culture experiment in hepatocytes [33]. The study showed that TGF-β-induced cell damage is genetically controlled. We mapped a prominent locus on chromosome 11, and a combination of experimental, genetics, and bioinformatics approaches was applied to identify a critical network of genes underlying this locus. The overlap of the in vitro locus with a locus identified in vivo [33] confirms that TGF-β-induced hepatocellular injury resembles a suitable substitute for liver fibrogenesis. These two studies show that the systems genetics approach and the integration of different datasets in a combined analysis allow narrowing large loci and validation of mapping results.

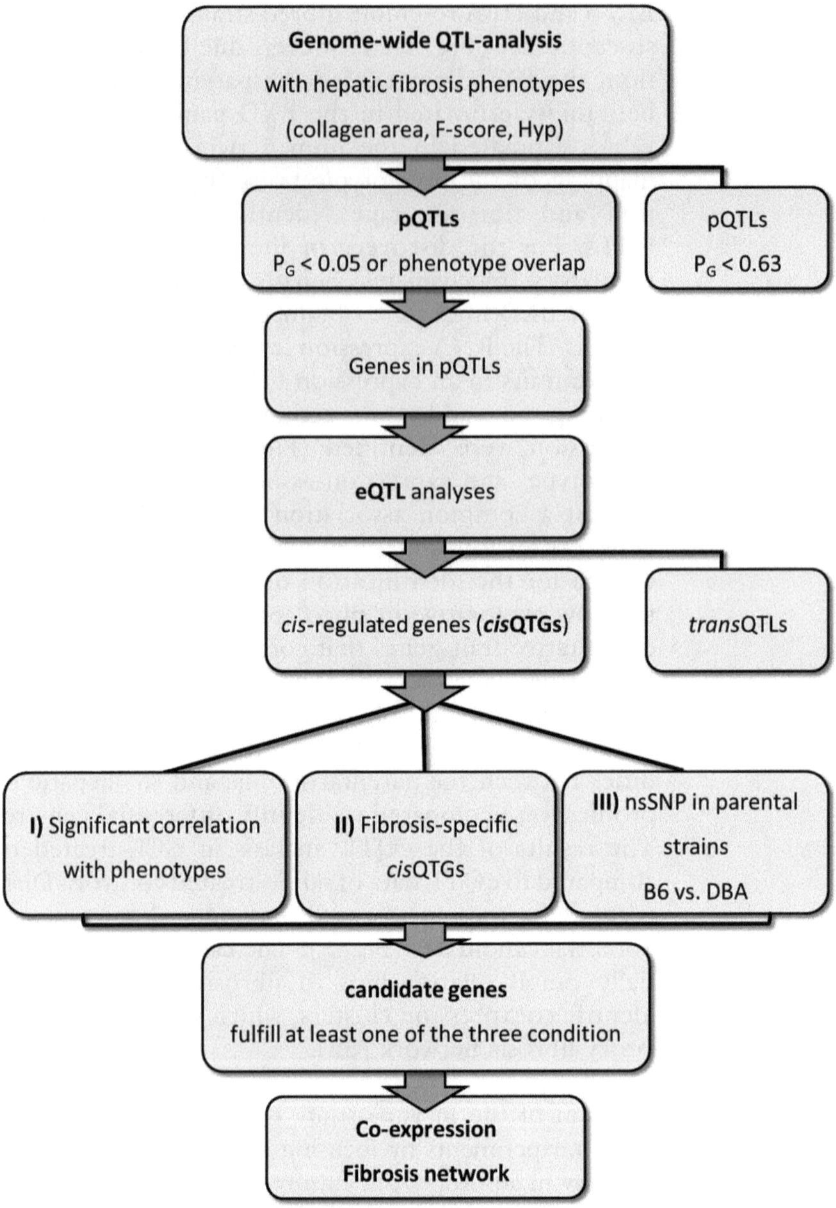

Fig. 5 Workflow of the explorative in silico analysis following QTL mapping. Different selection criteria were used to eliminate genes incoherent with the phenotype and identify genes with creedal associations (Figure taken from Hall et al. 2014 [32])

3 Outlook

Hepatic fibrosis is a complex trait that is influenced by a large number of common variants with low effect sizes. This makes it difficult to dissect underlying susceptibility genes. The hepatic fibrosis studies were part of a large mapping project that included the analysis of multiple experimental crosses derived from different mouse strains.

The mapping studies in the [A/J × BALB]F1 × [A/J × BALB]F_1 and [BALB × FVB]F_1 × [BALB × FVB]F_1 intercrosses have shown that the identification of unknown risk factors (i.e. *C5*) as well as the translation of experimental findings from mouse to human is possible. The two-way cross of inbred strains limits genetic information to regions that are polymorphic, therefore it is recommended to introduce higher diversity by screening different strains and cross combinations. Data upload into open access databases (such as GeneNetwork) allows an integrated analysis of different resources, phenotypic, genomic, and proteomic traits as well as data derived from different tissues and diseases. Fibrogenesis is a common pathway that is induced by injury in almost every tissue [34], which implies core fibrogenic pathways [35] and common systemic modifiers [2], but also tissue-specific mechanisms [36]. Hence, data integration enhances the chance to detect genuine modifiers across organs. GeneNetwork is a valuable platform that can be used by researchers without advanced skills of bioinformatics to perform systems genetics analyses. The next step would be to establish software tools that allow researchers to combine datasets from multiple resources and mapping analyses in different crosses and species (e.g. intercross, recombinant inbred lines, and human data).

References

1. Taub R (2004) Liver regeneration: from myth to mechanism. Nat Rev Mol Cell Biol 5(10):836–847
2. Wynn TA (2007) Common and unique mechanisms regulate fibrosis in various fibroproliferative diseases. J Clin Invest 117(3):524–529
3. Friedman SL (2008) Hepatic fibrosis – overview. Toxicology 254(3):120–129
4. Friedman SL (2008) Mechanisms of hepatic fibrogenesis. Gastroenterology 134(6):1655–1669
5. Marcellin P, Gane E, Buti M, Afdhal N, Sievert W, Jacobson IM, Washington MK, Germanidis G, Flaherty JF, Schall RA, Bornstein JD, Kitrinos KM, Subramanian GM, McHutchison JG, Heathcote EJ (2013) Regression of cirrhosis during treatment with tenofovir disoproxil fumarate for chronic hepatitis B: a 5-year open-label follow-up study. Lancet 381(9865):468–475
6. Jalan R (2015) Emerging trends in hepatology: 30 years of the Journal of Hepatology and 50 years of EASL. J Hepatol 62(1 Suppl):S1–S3
7. Byrne CD, Targher G (2015) NAFLD: a multisystem disease. J Hepatol 62(1 Suppl):S47–S64
8. Poynard T, Ratziu V, Benmanov Y, Di Martino V, Bedossa P, Opolon P (2000) Fibrosis in patients with chronic hepatitis C: detection and significance. Semin Liver Dis 20(1):47–55
9. Krawczyk M, Müllenbach R, Weber SN, Zimmer V, Lammert F (2010) Genome-wide association studies and genetic risk assessment of liver diseases. Nat Rev Gastroenterol Hepatol 7(12):669–681
10. Loomba R, Schork N, Chen CH, Bettencourt R, Bhatt A, Ang B, Nguyen P, Hernandez C, Richards L, Salotti J, Lin S, Seki E, Nelson KE, Sirlin CB, Brenner D (2015) Heritability of hepatic fibrosis and steatosis based on a prospective twin study. Gastroenterology 149(7):1784–1793
11. Weber S, Gressner OA, Hall R, Grünhage F, Lammert F (2008) Genetic determinants in hepatic fibrosis: from experimental models to fibrogenic gene signatures in humans. Clin Liver Dis 12(4):747–757, vii
12. Karlsen TH, Lammert F, Thompson RJ (2015) Genetics of liver disease: From pathophysiology to clinical practice. J Hepatol 62(1 Suppl):S6–S14
13. Huang H, Shiffman ML, Friedman S, Venkatesh R, Bzowej N, Abar OT, Rowland CM, Catanese JJ, Leong DU, Sninsky JJ, Layden TJ, Wright TL, White T, Cheung RC (2007) A 7 gene signature identifies the risk of developing cirrhosis in patients with chronic hepatitis C. Hepatology 46(2):297–306

14. Richardson MM, Powell EE, Barrie HD, Clouston AD, Purdie DM, Jonsson JR (2005) A combination of genetic polymorphisms increases the risk of progressive disease in chronic hepatitis C. J Med Genet 42(7), e45
15. Krawczyk M, Grünhage F, Lammert F (2013) Identification of combined genetic determinants of liver stiffness within the SREBP1c-PNPLA3 pathway. Int J Mol Sci 14(10): 21153–21166
16. Arslanow A, Stokes CS, Weber SN, Grünhage F, Lammert F, Krawczyk M (2015) The common PNPLA3 variant p.I148M is associated with liver fat contents as quantified by controlled attenuation parameter (CAP). Liver Int 36(3):418–426
17. Omary MB, Cohen DE, El-Omar EM, Jalan R, Low MJ, Nathanson MH, Peek RM, Jr., Turner JR (2016) Not all mice are the same: Standardization of animal research data presentation. Hepatology 63(6):1752–1754
18. Fickert P, Fuchsbichler A, Wagner M, Zollner G, Kaser A, Tilg H, Krause R, Lammert F, Langner C, Zatloukal K, Marschall HU, Denk H, Trauner M (2004) Regurgitation of bile acids from leaky bile ducts causes sclerosing cholangitis in Mdr2 (Abcb4) knockout mice. Gastroenterology 127(1):261–274
19. Mossanen JC, Tacke F (2015) Acetaminophen-induced acute liver injury in mice. Lab Anim 49(1 Suppl):30–36
20. Tolba R, Kraus T, Liedtke C, Schwarz M, Weiskirchen R (2015) Diethylnitrosamine (DEN)-induced carcinogenic liver injury in mice. Lab Anim 49(1 Suppl):59–69
21. Scholten D, Trebicka J, Liedtke C, Weiskirchen R (2015) The carbon tetrachloride model in mice. Lab Anim 49(1 Suppl):4–11
22. Hillebrandt S, Goos C, Matern S, Lammert F (2002) Genome-wide analysis of hepatic fibrosis in inbred mice identifies the susceptibility locus *Hfib1* on chromosome 15. Gastroenterology 123(6):2041–2051
23. Hillebrandt S, Wasmuth HE, Weiskirchen R, Hellerbrand C, Keppeler H, Werth A, Schirin-Sokhan R, Wilkens G, Geier A, Lorenzen J, Kohl J, Gressner AM, Matern S, Lammert F (2005) Complement factor 5 is a quantitative trait gene that modifies liver fibrogenesis in mice and humans. Nat Genet 37(8):835–843
24. Cervino AC, Darvasi A, Fallahi M, Mader CC, Tsinoremas NF (2007) An integrated in silico gene mapping strategy in inbred mice. Genetics 175(1):321–333
25. Sendler M, Beyer G, Mahajan UM, Kauschke V, Maertin S, Schurmann C, Homuth G, Volker U, Volzke H, Halangk W, Wartmann T, Weiss FU, Hegyi P, Lerch MM, Mayerle J (2015) Complement Component 5 Mediates Development of Fibrosis, via Activation of Stellate Cells, in 2 Mouse Models of Chronic Pancreatitis. Gastroenterology 149(3):765–776 e710
26. Wasmuth HE, Lammert F, Zaldivar MM, Weiskirchen R, Hellerbrand C, Scholten D, Berres ML, Zimmermann H, Streetz KL, Tacke F, Hillebrandt S, Schmitz P, Keppeler H, Berg T, Dahl E, Gassler N, Friedman SL, Trautwein C (2009) Antifibrotic effects of CXCL9 and its receptor CXCR3 in livers of mice and humans. Gastroenterology 137(1): 309–319, e301–303
27. Hall RA, Hillebrandt S, Lammert F (2015) Exploring multiple quantitative trait loci models of hepatic fibrosis in a mouse intercross. Mamm Genome 27(1-2):70–80
28. Broman KW, Wu H, Sen S, Churchill GA (2003) R/qtl: QTL mapping in experimental crosses. Bioinformatics 19(7):889–890
29. Hammam O, Mahmoud O, Zahran M, Aly S, Hosny K, Helmy A, Anas A (2012) The role of fas/fas ligand system in the pathogenesis of liver cirrhosis and hepatocellular carcinoma. Hepat Mon 12(11), e6132
30. Peirce JL, Lu L, Gu J, Silver LM, Williams RW (2004) A new set of BXD recombinant inbred lines from advanced intercross populations in mice. BMC Genet 5:7
31. Wang J, Williams RW, Manly KF (2003) WebQTL: web-based complex trait analysis. Neuroinformatics 1(4):299–308
32. Hall RA, Liebe R, Hochrath K, Kazakov A, Alberts R, Laufs U, Böhm M, Fischer HP, Williams RW, Schughart K, Weber SN, Lammert F (2014) Systems genetics of liver fibrosis: identification of fibrogenic and expression quantitative trait loci in the BXD murine reference population. PLoS One 9(2), e89279
33. Liebe R, Hall RA, Williams RW, Dooley S, Lammert F (2013) Systems genetics of hepatocellular damage in vivo and in vitro: identification of a critical network on chromosome 11 in mouse. Physiol Genomics 45(20):931–939
34. Rockey DC, Bell PD, Hill JA (2015) Fibrosis—a common pathway to organ injury and failure. N Engl J Med 373(1):96
35. Mehal WZ, Iredale J, Friedman SL (2011) Scraping fibrosis: expressway to the core of fibrosis. Nat Med 17(5):552–553
36. Walkin L, Herrick SE, Summers A, Brenchley PE, Hoff CM, Korstanje R, Margetts PJ (2013) The role of mouse strain differences in the susceptibility to fibrosis: a systematic review. Fibrogenesis Tissue Repair 6(1):18

Chapter 22

Systems Genetics Analysis of Iron and Its Regulation in Brain and Periphery

Byron C. Jones and Leslie C. Jellen

Abstract

In this contribution, we demonstrate the utility of the systems genetics—systems biology approach to the study of iron regulation while employing a comprehensive database. We describe our work in iron regulation in the brain and periphery under normal iron and iron-restricted dietary conditions in the BXD family of recombinant inbred mouse strains. Using multiple measures, we showed wide variation among the strains in the effect of being fed an iron-restricted diet for 100 days in every measure from brain and from the periphery. All data were entered into GeneNetwork (www.genenetwork.org), a database that contains genotypic, phenotypic, and gene expression data (Rosen et al., Methods Mol Biol 401:287–303, 2007). Using this resource, we were able to ask the following four questions concerning possible candidate genes underlying our measures: (1) what is the range of response for each of the measures? (2) Does the pattern of variability show continuous (additive genetic) or discrete (Mendelian) distribution across strains? (3) Are there genetic markers that are associated with the variability in the measures? (4) Are there genes in near the markers that contain associated allelic differences, and whose expression is related to the variability in the measures? Other questions that we could address include: (5) what is the association among the measures between the sexes? (6) What is the association among the measures, e.g., is liver iron status under the diets related to brain iron? (7) What is the relationship between our measures and other phenotypic parameters—i.e., is there an association between our brain iron measures and neurochemical phenotypes extant in the database? And finally, (8) are there gene networks that underlie single or combined measures?

Key words Use case, Quantitative trait loci analysis, Genetic mapping, Genetic correlation, Gene network

1 Introduction

Iron has many important biological roles, ranging from participating in energy production and regulation to serving as cofactor in production of important molecules. Iron is active in these processes when in its ferrous (+2 charge) state. In this state, however, iron can produce free radicals via the Fenton reaction and can cause serious damage to all kinds of cells, including neurons [2].

Klaus Schughart and Robert W. Williams (eds.), *Systems Genetics: Methods and Protocols*, Methods in Molecular Biology, vol. 1488, DOI 10.1007/978-1-4939-6427-7_22, © Springer Science+Business Media New York 2017

Fortunately, there are numerous proteins that regulate iron, its transport and availability and even its oxidative state. Ferritin contains intracellular iron and can oxidize ferrous iron to ferric, which is nearly inert biologically. When work is needed, ferritin can reduce ferric iron to the ferrous state. The aim of this chapter is to present the many roles of iron and how we can use systems genetics and systems biology to elucidate how iron is regulated in multiple organ systems.

Iron deficiency is a major problem worldwide, but especially in so-called developing countries. In fact, it has remained on the World Health Organization's "hit list" for several years. In the most severe condition, iron deficiency can lead to anemia; however, less-severe iron deficiency can lead to other problems, including neurological, affective, cognitive, and motor difficulties [3, 4]. Iron deficiency in infancy or early childhood can lead to cognitive developmental difficulties that can last into adolescence and adulthood [5–7]. While it is not always clear to what extent these trace back to iron deficiency or poor socioeconomic background [8], experimental evidence from animal models is corroborative, with iron deficiency causing irreversible biochemical and behavioral changes [9–11]. Iron deficiency in childhood is also associated with attention deficit disorder [12], and the attention deficits may carry over into adulthood. In adulthood, the effects of iron deficiency are reversible and include attentional, cognitive, mood, and performance deficits [13–16] as well as restless legs syndrome [17, 18]. In our work, we are interested in the role of iron in the ventral midbrain and striatum. The ventral midbrain contains the cell bodies of dopamine neurons and iron has profound effects on dopamine neuron development [19–21], dopamine function in adulthood [4, 11] and in neurodegeneration in aging [22].

Iron overload can produce serious cellular damage via the Fenton reaction to produce free oxygen radicals. Hereditary hemochromatosis, African iron overload, thalassemia, and sickle cell disease are examples of genetic-based consequences. In fact, iron overload can lead to major organ failure. In the nervous system, iron overload or dysregulation is a suspected risk factor for Parkinson's disease [23, 24] and can contribute to other neurodegenerative diseases such as Alzheimer's disease. Of particular importance, in both iron deficiency and iron overload, there are large inter-individual differences in response, including compensatory actions.

There are numerous proteins involved in the regulation of iron. The most commonly referred to proteins are:

Ferritin. Ferritin is an intracellular protein that is found in two forms, light and heavy. Both forms store iron and the heavy form limits iron toxicity by oxidizing the highly active and toxic ferrous (+2) to the less reactive ferric (+3) iron. The light form is believed to participate in electron transport. Low concentrations of ferritin indicate low iron status or anemia and high concentrations can indicate iron overload or hemochromatosis. This protein is highly conserved across species.

Transferrin and receptor. Transferrin is a glycoprotein that binds to and transports iron in body fluids. Cells have transferrin receptors that bind to iron-carrying transferrin. Transferrin is pulled into the cell, is acidified and releases its iron and is then transported back out of the cell to gather more iron.

Hepcidin and ferroportin. Ferroportin is the protein most responsible for moving iron out of the cytoplasm. Hepcidin is a small protein that binds to, and inhibits ferroportin, thus inhibiting iron release from cells. Hepcidin also has antibacterial actions.

IRP1 and IRP2. Iron regulatory proteins 1 and 2 monitor iron status and during times of deficiency or overload, regulate the expression of ferritin, transferrin, and other proteins involved in iron homeostasis.

Ceruloplasmin. While considered to be the primary protein in copper transport, ceruloplasmin also reduces iron toxicity by converting Fe^{2+} to Fe^{3+}.

This list of proteins involved in iron regulation is not exhaustive and one of the advantages of our systems genetics analysis using recombinant inbred mice is the possible discovery of hitherto unknown regulatory proteins. Also, we can ask the question, which of the regulatory protein genes underlie individual differences in iron concentrations in the various tissues and which respond differentially, by expression, to iron deficiency or overload?

Iron regulation in humans shows wide individual differences for almost any measure and across tissues. By using genetic reference populations of rodents, we can study probable genetic bases to answer basic questions concerning iron regulation and what is causing the individual differences.

All examples shown in this contribution are from www.GeneNetwork.org [1]

2 Methods

1. What is the range of iron concentration in the brain and what is the range of iron concentration in the liver and for hemoglobin?

   To answer this question, we fed male and female mice from 22 of the BXD recombinant inbred mouse strains together with the parental C57BL/6J (B6) and DBA/2J (D2) inbred strains one of two diets differing in iron concentration. The adequate iron diet (AD) contained 240 ppm Fe and the iron-deficient diet (ID) contained 4 ppm Fe. The diets were administered at weaning (postnatal day 21) for a period of 100 days. The data presented below show iron content for the AD and ID diets in the ventral midbrain, dorsal striatum (caudate-putamen) (Fig. 1), hemoglobin (Fig. 2), and liver (Fig. 3). The brain iron data and peripheral iron data were published separately [25, 26]. To answer the first question, all figures show wide, genetic variation in basal values and in response to having been fed an iron-deficient diet.

2. What is the nature of distribution of differences across the strains?

   Again, inspection of the figures shows continuous variation in basal and treatment conditions. This indicates that each of the traits is under additive genetic influence, i.e., under the influence of several genes.

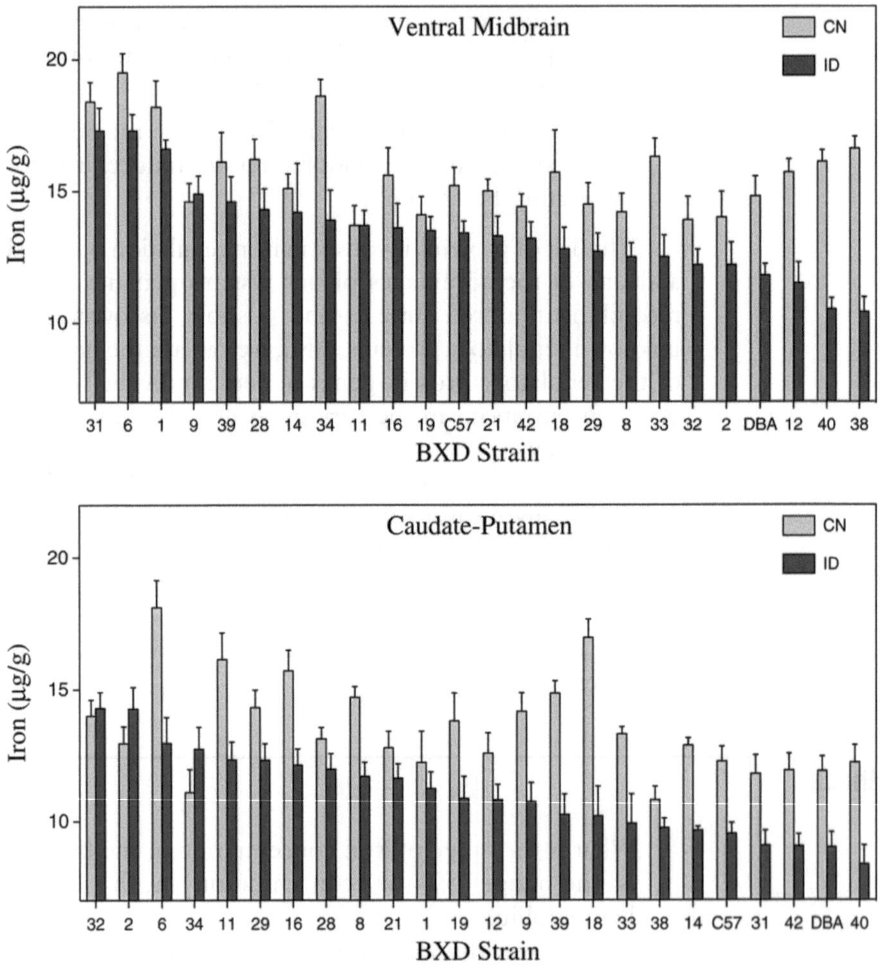

Fig. 1 Effect of different iron containing diets on iron concentration in the ventral midbrain (*top panel*) and caudate-putamen (*bottom panel*) in BXD mouse strains (sexes combined). The control diet (CN) contained 240 ppm Fe in a standard rodent formulation (AIN-93) and the iron-deficient diet (ID) contained 4 ppm Fe in an otherwise identical formulation. The diets were administered from weaning at 21 days until 121 days of age

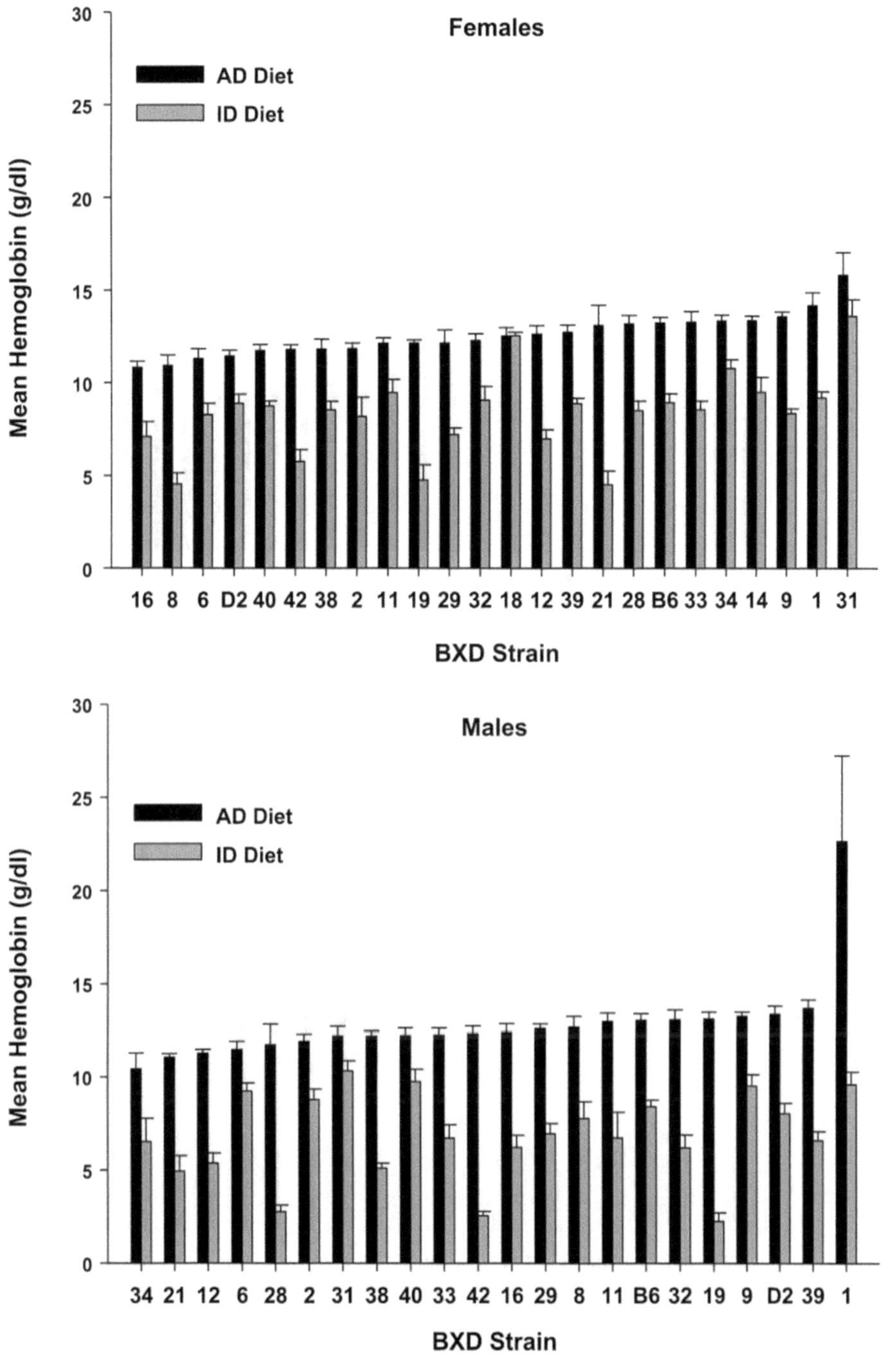

Fig. 2 Effect of different iron containing diets on hemoglobin concentration in female (*top panel*) and male (*bottom panel*) BXD mice. The control diet (AD) contained 240 ppm Fe in a standard rodent formulation (AIN-93) and the iron-deficient diet (ID) contained 4 ppm Fe in an otherwise identical formulation. The diets were administered from weaning at 21 days until 121 days of age

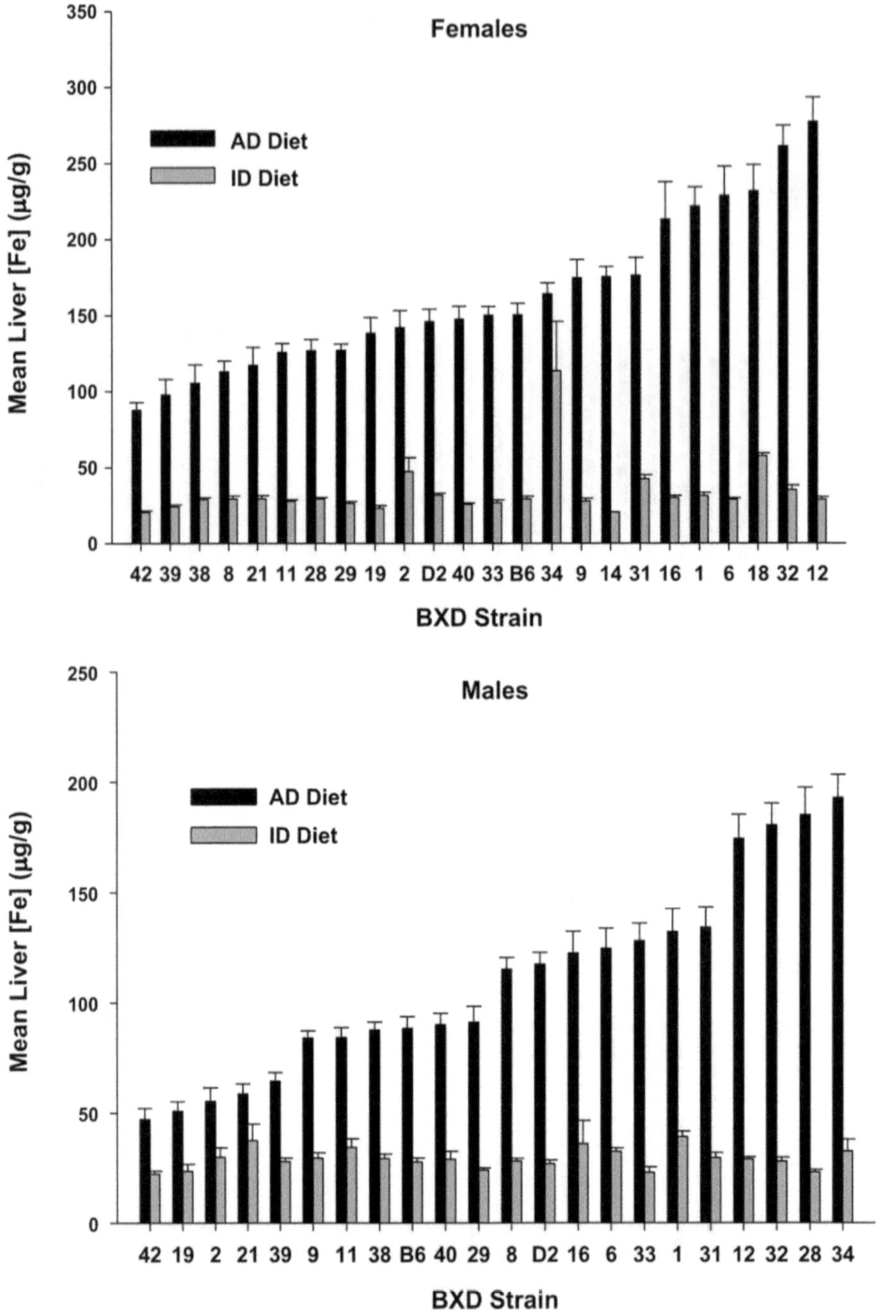

Fig. 3 Effect of different iron containing diets on liver iron concentration in female (*top panel*) and male (*bottom panel*) BXD mice. The control diet (AD) contained 240 ppm Fe in a standard rodent formulation (AIN-93) and the iron-deficient diet (ID) contained 4 ppm Fe in an otherwise identical formulation. The diets were administered from weaning at 21 days until 121 days of age

3. Are there genetic markers associated with any of the traits?

   For this question, we focus on iron concentration in the ventral midbrain (the original major focus of the study) for the diets. To answer the question, we performed quantitative trait loci (QTL) analysis on a derived iron regulation-related eigentrait. This eigentrait was derived by performing principal component analysis on ventral midbrain iron concentration. We combined ventral midbrain iron values for males and females and for both dietary conditions. This technique is useful when several related individual values produce weak but suggestive signals at the same location. Figure 4 illustrates the mapping of the iron regulatory eigentrait using GeneNetwork.org. The process is a point-biseral correlation between the markers (originating in the C57BL/6 or DBA/2 strain) and the continuously variable trait.

4. What are possible candidate genes located near the marker associated with the eigentrait?

   To answer this question, we need to satisfy two criteria. First, for the candidate gene, is its expression (transcript abundance) correlated with the trait, and second, is the gene *cis*-regulated? A search for genes near the QTL ~102 Mb produced two possibilities. The first, *Cd44*, codes the CD44 antigen, its expression correlated −0.574 with the iron regulatory eigentrait and is *cis*-regulated. The second, *Slc1a3*, the high-affinity glial glutamate transporter, is also correlated, 0.672 and is also *cis*-regulated (Fig. 5). We believe that the latter is the more likely candidate based on biology—amino acid transporters can possibly transport metals; however, the nomination is provisional until verification can be performed by manipulation of the genes, i.e., knockout, amplification, etc.

5. To what degree are the measures correlated between the sexes and for each condition?

   Figure 6 shows moderate Pearson *r* correlations between the sexes in each condition.

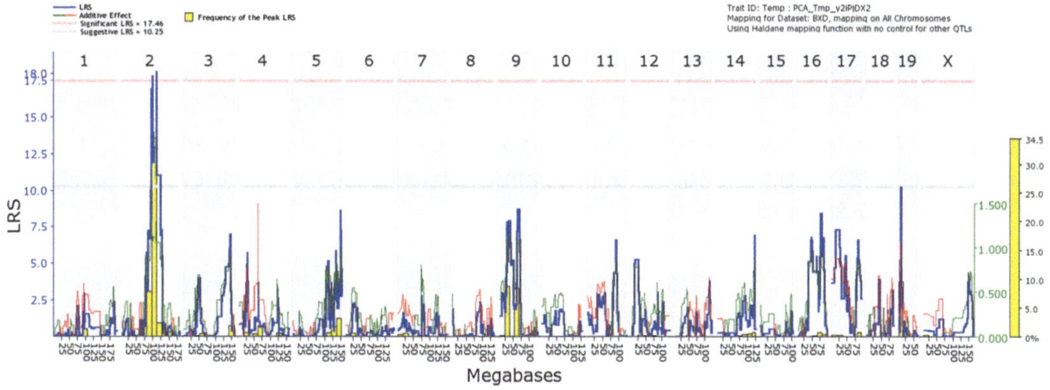

Fig. 4 Genetic mapping of iron regulation eigentrait vs. polymorphic genomic markers. We note a significant association between one or more markers on chromosome two and our eigentrait

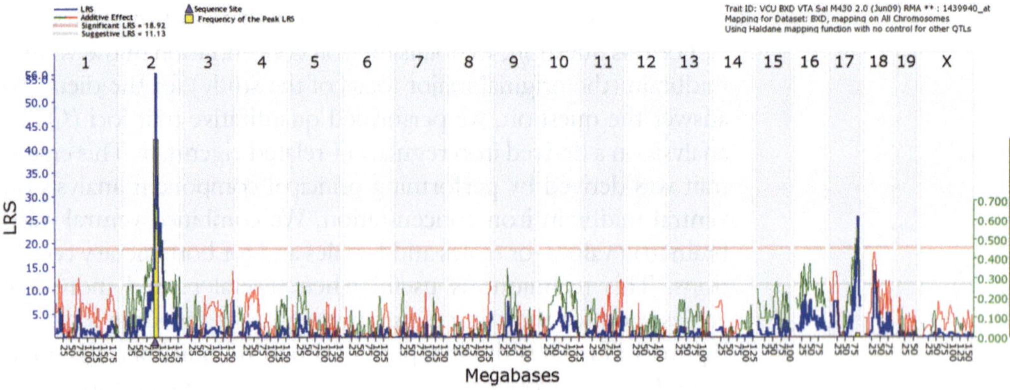

Fig. 5 Genetic mapping of the expression of a possible candidate gene near the marker associated with the iron regulatory eigentrait. The blue triangle located on the *X*-axis, chromosome 2 at 102 Mb is the coding region for the gene that produces, *Slc1a2*, the high-affinity glial glutamate transporter

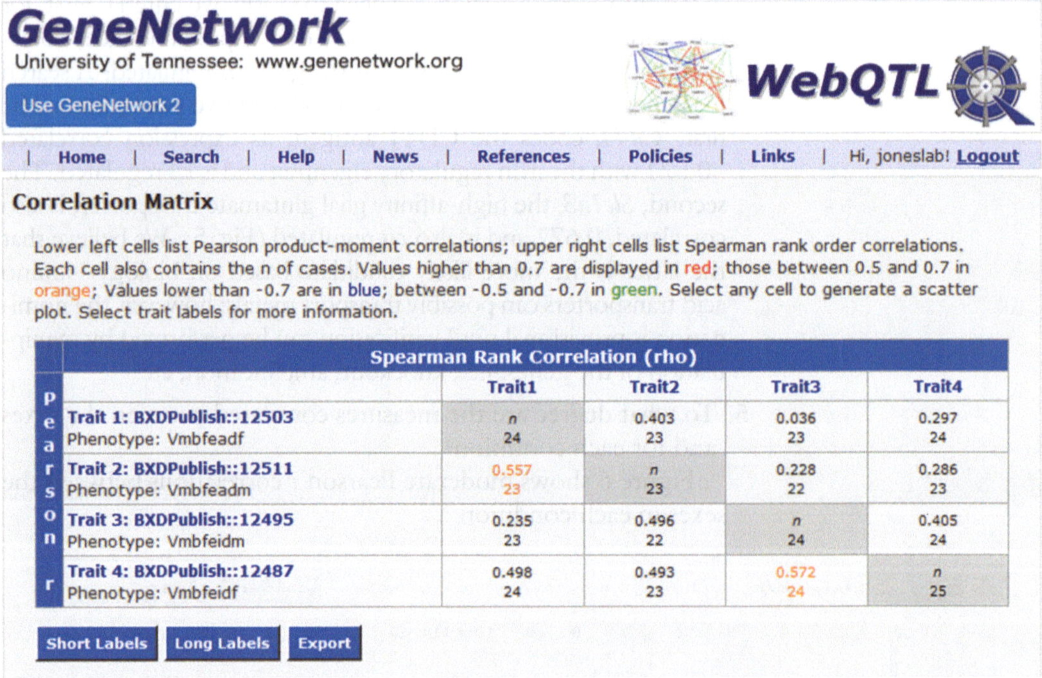

Correlation Matrix

Lower left cells list Pearson product-moment correlations; upper right cells list Spearman rank order correlations. Each cell also contains the n of cases. Values higher than 0.7 are displayed in red; those between 0.5 and 0.7 in orange; Values lower than -0.7 are in blue; between -0.5 and -0.7 in green. Select any cell to generate a scatter plot. Select trait labels for more information.

Pearson r / Spearman Rank Correlation (rho)	Trait1	Trait2	Trait3	Trait4
Trait 1: BXDPublish::12503 Phenotype: Vmbfeadf	*n* 24	0.403 23	0.036 23	0.297 24
Trait 2: BXDPublish::12511 Phenotype: Vmbfeadm	0.557 23	*n* 23	0.228 22	0.286 23
Trait 3: BXDPublish::12495 Phenotype: Vmbfeidm	0.235 23	0.496 22	*n* 24	0.405 24
Trait 4: BXDPublish::12487 Phenotype: Vmbfeidf	0.498 24	0.493 23	0.572 24	*n* 25

Short Labels | Long Labels | Export

Fig. 6 Matrix of genetic correlations by sex and by dietary treatment. The correlations are for strain means. Pearson *r* correlations are shown on the *left* and Spearman rank-order correlations, on the *right*. For Pearson 4, values at 0.41 are significant at $p < 0.05$

6. What is the correlation across the systems for iron concentration?

   The correlation matrix for liver and ventral midbrain iron and hemoglobin collapsed across sex is presented in Fig. 7. The mostly weak correlations indicate to us that iron regulation is tissue specific.

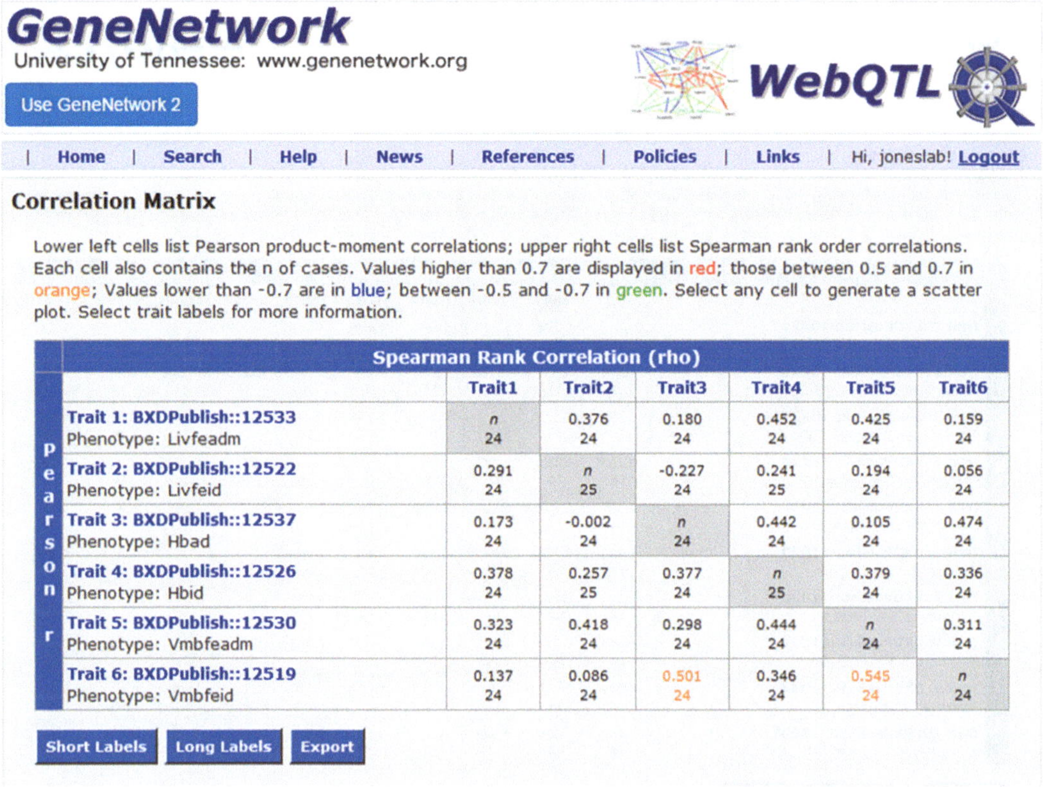

GeneNetwork
University of Tennessee: www.genenetwork.org

Use GeneNetwork 2

WebQTL

| Home | Search | Help | News | References | Policies | Links | Hi, joneslab! Logout

Correlation Matrix

Lower left cells list Pearson product-moment correlations; upper right cells list Spearman rank order correlations. Each cell also contains the n of cases. Values higher than 0.7 are displayed in red; those between 0.5 and 0.7 in orange; Values lower than -0.7 are in blue; between -0.5 and -0.7 in green. Select any cell to generate a scatter plot. Select trait labels for more information.

Spearman Rank Correlation (rho)	Trait1	Trait2	Trait3	Trait4	Trait5	Trait6
Trait 1: BXDPublish::12533 Phenotype: Livfeadm	*n* 24	0.376 24	0.180 24	0.452 24	0.425 24	0.159 24
Trait 2: BXDPublish::12522 Phenotype: Livfeid	0.291 24	*n* 25	-0.227 24	0.241 25	0.194 24	0.056 24
Trait 3: BXDPublish::12537 Phenotype: Hbad	0.173 24	-0.002 24	*n* 24	0.442 24	0.105 24	0.474 24
Trait 4: BXDPublish::12526 Phenotype: Hbid	0.378 24	0.257 25	0.377 24	*n* 25	0.379 24	0.336 24
Trait 5: BXDPublish::12530 Phenotype: Vmbfeadm	0.323 24	0.418 24	0.298 24	0.444 24	*n* 24	0.311 24
Trait 6: BXDPublish::12519 Phenotype: Vmbfeid	0.137 24	0.086 24	0.501 24	0.346 24	0.545 24	*n* 24

Pearson r

Short Labels | Long Labels | Export

Fig. 7 Genetic correlation matrix between sexes and across liver iron concentration, hemoglobin, and ventral midbrain iron concentration. The correlations (Pearson *r* on the *left* and Spearman rank order on the *right*) are based on strain means and significance at $p < 0.05$ with 22 df is $r = 0.41$

7. Do our measures correlate with similar or other phenotypes that are listed in the database?

What about associations across tissues? Figure 8 is a correlation matrix for iron, copper, and zinc in ventral midbrain and hippocampus. Here, we show almost no association among the metals in the ventral midbrain, but high intercorrelations among the metals in the hippocampus. This finding lends further evidence for tissue-specific regulation of iron and other metals. Because the metals tended to be highly inter-correlated in the hippocampus, we performed principal component analysis and derived one eigentrait. When we performed QTL mapping on the trait, we observed two suggestive QTLs, one on chromosome 9 and the other on chromosome 14 (Fig. 9). The best candidate gene on chromosome is RIKEN 1700063D05, a theoretical gene for which the function is unknown at present. The correlation with the eigentrait is $r = 0.645$ and Fig. 10 shows that the gene is *cis*-regulated. On chromosome 14, there are three possible candidates (Fig. 11), *Peli2*, *Samd4, and Cnih. Cnih*

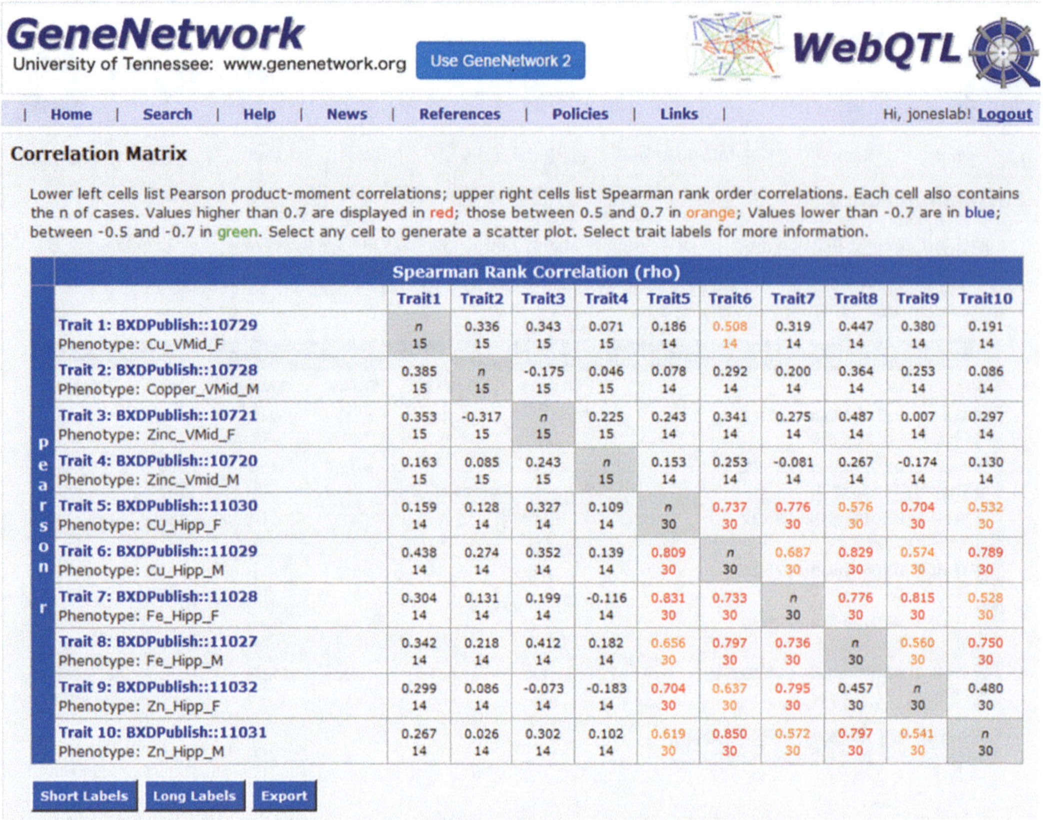

GeneNetwork
University of Tennessee: www.genenetwork.org
Use GeneNetwork 2
WebQTL

| Home | Search | Help | News | References | Policies | Links | Hi, joneslab! Logout

Correlation Matrix

Lower left cells list Pearson product-moment correlations; upper right cells list Spearman rank order correlations. Each cell also contains the n of cases. Values higher than 0.7 are displayed in red; those between 0.5 and 0.7 in orange; Values lower than -0.7 are in blue; between -0.5 and -0.7 in green. Select any cell to generate a scatter plot. Select trait labels for more information.

Spearman Rank Correlation (rho) (upper right); Pearson r (lower left)

	Trait1	Trait2	Trait3	Trait4	Trait5	Trait6	Trait7	Trait8	Trait9	Trait10
Trait 1: BXDPublish::10729 Phenotype: Cu_VMid_F	n 15	0.336 15	0.343 15	0.071 15	0.186 14	0.508 14	0.319 14	0.447 14	0.380 14	0.191 14
Trait 2: BXDPublish::10728 Phenotype: Copper_VMid_M	0.385 15	n 15	-0.175 15	0.046 15	0.078 14	0.292 14	0.200 14	0.364 14	0.253 14	0.086 14
Trait 3: BXDPublish::10721 Phenotype: Zinc_VMid_F	0.353 15	-0.317 15	n 15	0.225 15	0.243 14	0.341 14	0.275 14	0.487 14	0.007 14	0.297 14
Trait 4: BXDPublish::10720 Phenotype: Zinc_Vmid_M	0.163 15	0.085 15	0.243 15	n 15	0.153 14	0.253 14	-0.081 14	0.267 14	-0.174 14	0.130 14
Trait 5: BXDPublish::11030 Phenotype: CU_Hipp_F	0.159 14	0.128 14	0.327 14	0.109 14	n 30	0.737 30	0.776 30	0.576 30	0.704 30	0.532 30
Trait 6: BXDPublish::11029 Phenotype: Cu_Hipp_M	0.438 14	0.274 14	0.352 14	0.139 14	0.809 30	n 30	0.687 30	0.829 30	0.574 30	0.789 30
Trait 7: BXDPublish::11028 Phenotype: Fe_Hipp_F	0.304 14	0.131 14	0.199 14	-0.116 14	0.831 30	0.733 30	n 30	0.776 30	0.815 30	0.528 30
Trait 8: BXDPublish::11027 Phenotype: Fe_Hipp_M	0.342 14	0.218 14	0.412 14	0.182 14	0.656 30	0.797 30	0.736 30	n 30	0.560 30	0.750 30
Trait 9: BXDPublish::11032 Phenotype: Zn_Hipp_F	0.299 14	0.086 14	-0.073 14	-0.183 14	0.704 30	0.637 30	0.795 30	0.457 30	n 30	0.480 30
Trait 10: BXDPublish::11031 Phenotype: Zn_Hipp_M	0.267 14	0.026 14	0.302 14	0.102 14	0.619 30	0.850 30	0.572 30	0.797 30	0.541 30	n 30

Short Labels | Long Labels | Export

Fig. 8 Multiple metal regulation between the ventral midbrain and hippocampus. We correlated the concentrations of iron, zinc, and copper in the ventral midbrain and hippocampus for males and females

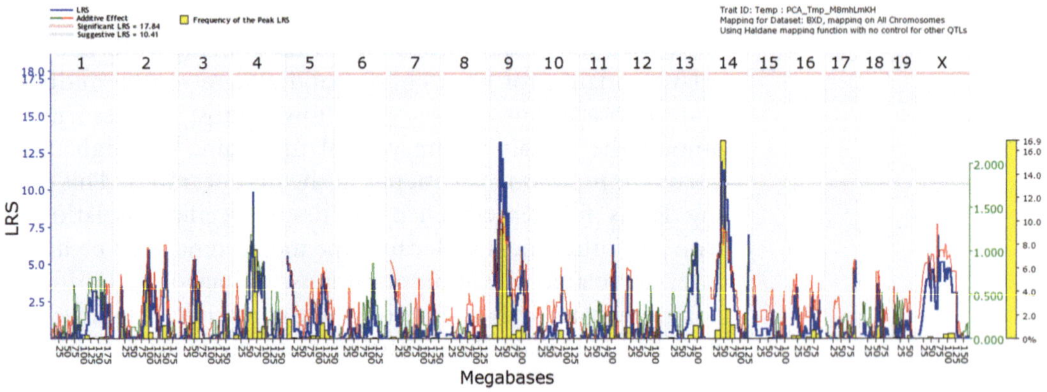

Fig. 9 QTL mapping of the principal component derived from all of iron, copper, and zinc measures in the hippocampus

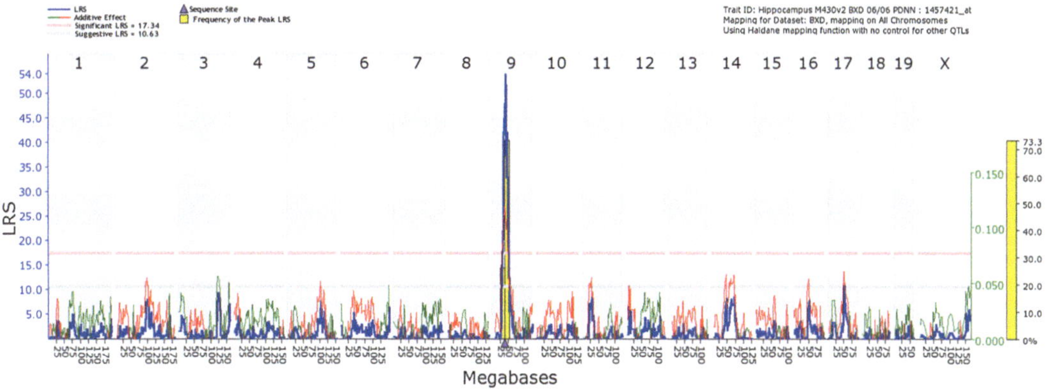

Fig. 10 QTL map for expression of candidate gene RIKEN 1700063D05 gene

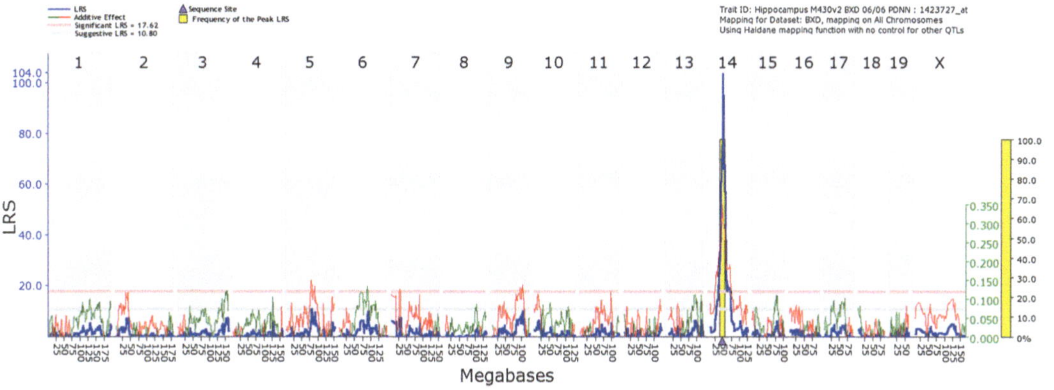

Fig. 11 QTL map for expression of candidate gene *Cnih*

is the most likely of the three, with a correlation of $r = 0.527$ and being highly *cis*-regulated. More importantly, this gene has high biological relevance, as it is associated with abnormal AMPA glutamatergic neurotransmission.

8. Are there gene and phenotypic networks that are interrelated to the target measure(s)?

   Figure 12 presents a gene-phenotypic network related to the iron-copper-zinc eigentrait. This network was constructed using GeneNetwork.org and following on the correlated gene expression above. We then used a feature, Compare Correlates, to query the database about related phenotypes. We present just a few for illustration. We found measures related to hippocampal function, i.e., swimming speed during Morris water maze acquisition [27] and neurochemistry, i.e., dopamine D_2 receptor density in the ventral midbrain [28].

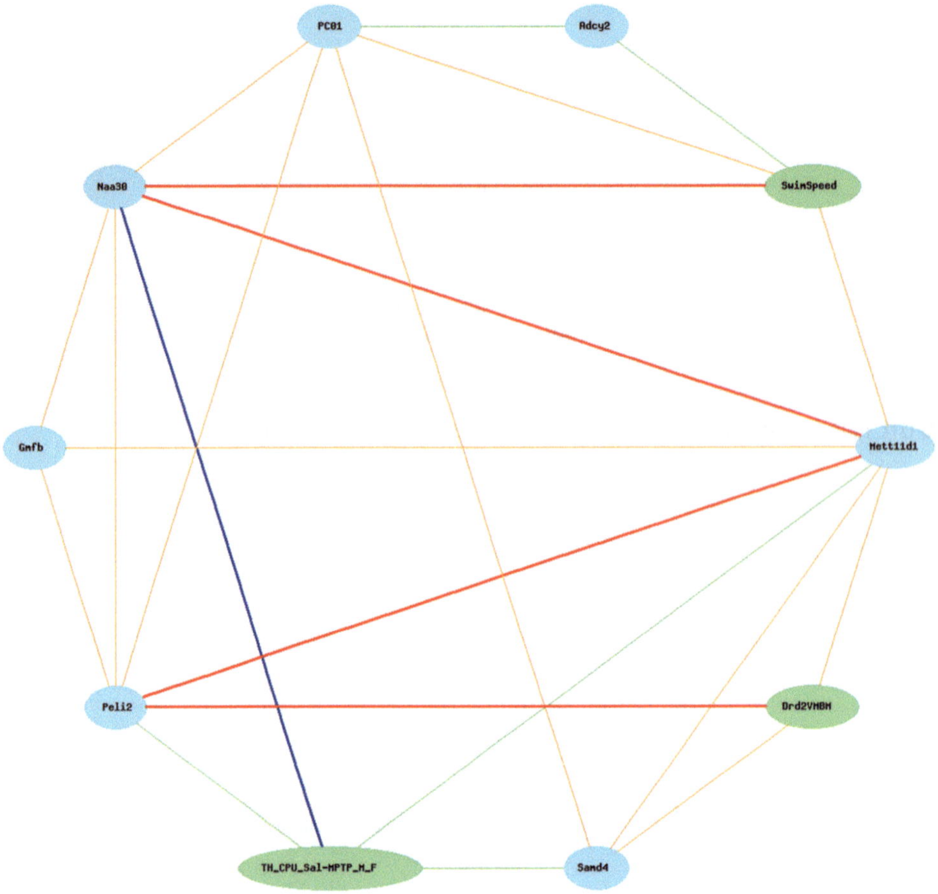

Fig. 12 Gene and phenotype network related to iron, copper, and zinc eigentrait. The eigentrait is designated by PC01. The phenotypes (*green nodes*) are dopamine D2 receptor densities in the ventral midbrain [28], swim speed during acquisition phase of Morris water maze acquisition [27] and tyrosine hydroxylase response to the neurotoxin, MPTP (1-methyl-4-phenyl-1,2,3,6-tetrahydropyridine) treatment [29]. Gene transcript abundances (*blue nodes*) are sterile alpha domain containing 4 (*Samd4*), cornichon homolog (*Cnih*), adenylate cyclase 2 (*Adcy2*), glia maturation factor, beta (*Gmfb*), methyltransferase 11 domain containing 1 (*Mett11d1*), and N(alpha) acetyltransferase 30, NatC catalytic subunit (*Naa30*)

3 Conclusions

In this short chapter, we have illustrated the value of systems genetics approach to complex traits analysis. We have shown how to assess the variability among phenotypes, using genetic reference populations of mice to model individual differences. We have also shown how to search for genes that underlie these individual traits and indeed, this approach can be used to discover genes previously not known that influence the phenotypes. Systems genetics can also be used effectively to monitor what changes occur concomitantly with changes in the targeted phenotype. Finally, systems genetics/biology methods can elucidate gene networks that operate in concert, influence other related phenotypes and can serve as a powerful means to identify mechanisms and also to generate hypotheses.

References

1. Rosen GD, Chesler EJ, Manley KF, Williams RW (2007) An informatics approach to systems neurogenetics. Methods Mol Biol 401:287–303
2. Halliwell B, Gutteridge JM (1986) Oxygen free radicals and iron in relation to biology and medicine: some problems and concepts. Arch Biochem Biophys 246(2):501–514
3. Beard JL (2001) Iron biology in immune function, muscle metabolism and neuronal functioning. J Nutr 131(2):568S–580S
4. Beard J (2003) Iron deficiency alters brain development and functioning. J Nutr 133(5): 1468S–1472S
5. Lozoff B, Jimenez E, Hagen J, Mollen E, Wolf AW (2000) Poorer behavioral and developmental outcome more than 10 years after treatment for iron deficiency in infancy. Pediatrics 105(4):e51
6. Beard JL (2008) Why iron deficiency is important in infant development. J Nutr 138(12): 2534–2536
7. Lukowski AF, Koss M, Burden MJ, Jonides J, Nelson CA, Kaciroti N, Jimenez E, Lozoff B (2010) Iron deficiency in infancy and neurocognitive functioning at 19 years: evidence of long-term deficits in executive function and recognition memory. Nutr Neurosci 13(2):54–70
8. Grantham-McGregor S, Ani C (2001) A review of studies on the effect of iron deficiency on cognitive development in children. J Nutr 131(2):649S–668S
9. Ben-Shachar D, Ashkenazi R, Youdim MB (1986) Long-term consequence of early iron-deficiency on dopaminergic neurotransmission in rats. Int J Dev Neurosci 4(1):81–88
10. Beard J, Erikson KM, Jones BC (2003) Neonatal iron deficiency results in irreversible changes in dopamine function in rats. J Nutr 133(4):1174–1179
11. Felt BT et al (2006) Persistent neurochemical and behavioral abnormalities in adulthood despite early iron supplementation for perinatal iron deficiency anemia in rats. Behav Brain Res 171(2):261–270
12. Konofal E, Lecendreux M, Arnulf I, Mouren MC (2004) Iron deficiency in children with attention-deficit/hyperactivity disorder. Arch Pediatr Adolesc Med 158(12):1113–1115
13. Haas JD, Brownlie T (2001) Iron deficiency and reduced work capacity: a critical review of the research to determine a causal relationship. J Nutr 131(2):676S–690S
14. Murray-Kolb LE, Beard JL (2007) Iron treatment normalizes cognitive functioning in young women. Am J Clin Nutr 85(3):778–787
15. Shariatpanaahi MV, Shariatpanaahi ZV, Moshtaaghi M, Shahbaazi SH, Abadi A (2007) The relationship between depression and serum ferritin level. Eur J Clin Nutr 61(4): 532–535
16. McClung JP, Karl JP, Cable SJ, Williams KW, Nindl BC, Young AJ, Lieberman HR (2009) Randomized, double-blind, placebo-controlled trial of iron supplementation in female soldiers during military training: effects on iron status, physical performance, and mood. Am J Clin Nutr 90(1):124–131
17. Earley CJ, Allen RP, Beard JL, Connor JR (2000) Insight into the pathophysiology of restless legs syndrome. J Neurosci Res 62(5):623–628
18. Connor JR, Boyer PJ, Menzies SL, Dellinger B, Allen RP, Ondo WG, Earley CJ (2003) Neuropathological examination suggests impaired brain iron acquisition in restless legs syndrome. Neurology 61(3):304–309
19. Youdim MB, Ben-Shachar D, Yehuda S (1989) Putative biological mechanisms of the effect of iron deficiency on brain biochemistry and behavior. Am J Clin Nutr 50(3):607–617
20. Erikson KM, Jones BC, Beard JL (2000) Iron deficiency alters dopamine transporter functioning in rat striatum. J Nutr 130(11):2831–2837
21. Lozoff B, Georgieff MK (2006) Iron deficiency and brain development. In: Hayflick SJ (ed) Seminars in pediatric neurology, Vol 13, No 3. WB Saunders, Philadelphia, PA, pp 158–165
22. Youdim MB, Stephenson G, Shachar DB (2004) Ironing iron out in Parkinson's disease and other neurodegenerative diseases with iron chelators: a lesson from 6-hydroxydopamine and iron chelators, desferal and VK-28. Ann N Y Acad Sci 1012(1):306–325
23. Dexter DT, Wells FR, Lee AJ, Agid F, Agid Y, Jenner P, Marsden CD (1989) Increased nigral iron content and alterations in other metal ions occurring in brain in Parkinson's disease. J Neurochem 52(6):1830–1836

24. Zucca FA, Segura-Aguilar J, Ferrari E, Muñoz P, Paris I, Sulzer D, Sarna T, Casella L, Zecca L (2015) Interactions of iron, dopamine and neuromelanin pathways in brain aging and Parkinson's disease. Prog Neurobiol pii:S0301-0082(15)00101-X
25. Jellen LC, Unger EL, Lu L, Williams RW, Rousseau S, Wang X et al (2012) Systems genetic analysis of the effects of iron deficiency in mouse brain. Neurogenetics 13(2):147–157
26. Yin L, Unger EL, Jellen LC, Earley CJ, Allen RP, Tomaszewicz A et al (2012) Systems genetic analysis of multivariate response to iron deficiency in mice. Am J Physiol Regul Integr Comp Physiol 302(11):R1282–R1296, http://doi.org/10.1152/ajpregu.00634.2011
27. Kempermann G, Gage FH (2002) Genetic determinants of adult hippocampal neurogenesis correlate with acquisition, but not probe trial performance, in the water maze task. Eur J Neurosci 16(1):129–136
28. Jones BC, Tarantino LM, Rodriguez LA, Reed CL, McClearn GE, Plomin R, Erwin VG (1999) Quantitative-trait loci analysis of cocaine-related behaviours and neurochemistry. Pharmacogenet Genomics 9(5): 607–618
29. Jones BC, Miller DB, O'Callaghan JP, Lu L, Unger EL, Alam G, Williams RW (2013) Systems analysis of genetic variation in MPTP neurotoxicity in mice. NeuroToxicology 37:26–34

Chapter 23

Systems Genetics of Obesity

Gudrun A. Brockmann, Danny Arends, Sebastian Heise, and Ayca Dogan

Abstract

Obesity is a complex trait, determined by many genes and influenced by environmental factors. Mapping genomic loci contributing to obesity helps to identify gene variants responsible for differences in the phenotype. However, measuring fat content alone is often not sufficient to identify the underlying gene or genes. Besides in-depth phenotyping, well-designed genetic populations and the combined analysis of data of different origins are necessary to detect one of several genetic determinants. Structured mouse populations and linking information from different experiments help to simplify the complexity in the search for direct genetic effects or factors that are hidden in the genome. In this chapter we present an example of how the physicochemical characterization of adipose tissue in BXD recombinant inbred lines contributes to enlighten the obese phenotype of mice. We describe the search for gene(s) contributing to collagen content in adipose tissue of BXD strains using the GeneNetwork platform.

Key words Use case, Recombinant inbred strains (RIS), Collagen content, Adipose tissue, GeneNetwork

1 Introduction

Obesity is one factor of the metabolic syndrome predisposing for later onset of severe diseases as type 2 diabetes, cardiovascular complications, or fatty liver syndrome. Obesity is a complex trait that is determined by many genes of different effect size and influenced by environmental conditions such as diet composition and physical exercise. To quantify obesity, often body weight and total fat or adipose tissue mass are measured and impairments on glucose clearance or insulin sensitivity are tested. Additional endogenous phenotypes such as hormone availability, gene expression, or metabolites would be desirable to better understand mechanisms leading to obesity and potential consequences. In this chapter we dissect the determinants for excessive fat accumulation in adipose tissues of mice. While human populations are characterized by a high level of genetic and consequently also physiological diversity, mouse model populations are used to simplify the genetics determining obesity on a more homogenous genetic background.

Klaus Schughart and Robert W. Williams (eds.), *Systems Genetics: Methods and Protocols*, Methods in Molecular Biology, vol. 1488, DOI 10.1007/978-1-4939-6427-7_23,

In particular, structured populations such as crosses between inbred strains and recombinant inbred strains generated from crossbred or outbred populations resemble valuable resources for the search of causal genetic factors and modifiers.

In the past, linkage mapping of regions contributing to obesity, so-called quantitative trait loci (QTL), were often carried out in crossbred populations. The F_2 individuals originate from an intercross of two inbred strains and subsequent random mating of F_1 individuals. It is beneficial to choose two strains with extremely different phenotypes for fat accumulation for the experiment. Due to Mendelian laws, the F_1 generation is supposed to be genetically homogeneous showing only one phenotype. If our characteristic phenotype is not present in the F_1 generation, it is very likely that the underlying variation is recessive. In contrast, if all F_1 individuals are obese, most probably a major gene variant is dominant. In the case of obesity, the F_2 individuals likely show a continuous distribution of fat deposition. The reason for the distribution is chromosomal recombination between both parental strains during gametogenesis in F_1 individuals and random combination of germ cells during mating. As a result of QTL mapping in a F_2 population large chromosomal regions comprising a third or even half of a chromosome are identified containing the gene variant(s) causing obesity, but also many other genes. Finemapping of the QTL in an appropriate population is necessary to reduce the confidence interval of the QTL and thereby the number of positional candidate genes. For this purpose advanced intercross lines (AIL) can be generated from the F_2 mapping population by random mating in each subsequent generation. Due to many recombination events, the AIL has a sufficiently high level of genetic resolution to fine map the whole QTL interval into smaller regions.

Successful QTL mapping and subsequent finemapping can also be carried out in backcrosses of crossbreeding experiments. A very conclusive example is mapping of the *Nob1* QTL for diet-induced obesity in an F1(NZOxSJL) × NZO backcross and the identification of a 7 bp deletion in the gene *Tbc1d1* as a suppressor of obesity and hyperglycemia [1]. The *Nob1* QTL was finemapped using recombinant congenic strains (RCS). Such RCS carry chromosomal intervals of the QTL region of different length of the donor strain on the genetic background of the recipient strain. The comparison of the phenotypes of the distinct RCS carrying different chromosomal intervals of the QTL shows which interval is responsible for obesity.

Another possibility for the identification of causal gene variations by direct finemapping is to apply a set of recombinant inbred strains (RIS). The most commonly used system of RIS resembles the BXD strains [2]. Here, the two parental inbred strains

C57BL/6J (B) and DBA/2J (D) were initially crossed to generate inbred strains from randomly chosen mating pairs in generation F_2. By repeated inbred mating of the offspring of particular F_2 parents for at least 20 generations distinct inbred strains are generated. During the course of repeated mating, many recombination events are accumulated leading to highly mixed genomes of both parental strains. Each strain resembles a distinct combination of the parental genomes, which facilitates finemapping in a panel of many RIS. Since all individuals of a RIS are genetically identical, only one individual of each strain has to be genotyped, while phenotypic measurements can be performed for several individuals within each strain to have a balanced phenotype. Beneficial is the circumstance that B and D are fully sequenced mouse strains and all data is available in public databases. Currently 101 BXD strains are available in three batches of 32, 9, and 60 strains which were generated in 1975, 2001, and 2005, respectively [3–5]. Ideally 25 and more inbred strains should be used for mapping.

Another interesting set of RIS, which also facilitates direct finemapping, constitutes the so-called Collaborative Cross (CC) [6]. Based on eight founder mouse strains, five of them referred to as being classical inbred (A/J, C57BL/6J, 129S1/SvImJ, NOD/ShiLtJ, and NZO/HILtJ) and three as wildtype-derived inbred strains (CAST/EiJ, PWK/PhJ, and WSB/EiJ), it is estimated that this panel is capturing the majority of the genetic and phenotypic diversity harbored in laboratory mouse stocks [7]. To date, all founder strains are sequenced which additionally facilitates the identification of causal gene variants responsible for phenotypical differences among the CC strains. Interestingly, in addition to the set of CC strains, a diversity outbred (DO) population was established from the same eight founder strains [8]. The DO population mimics a high level of heterozygosity that is found in natural populations and allows testing for epistatic interactions.

In this chapter, we describe the protocol for using BXD strains for identifying candidate genes associated with the fat and collagen content in reproductive adipose tissue. The decision for using this resource was driven by

- Low number of mice needed to phenotype.
- Availability of genotypes.
- Availability of the genome sequence of parental strains.
- Access to phenotypes of other researchers for the same strains.

For data analysis, we used the GeneNetwork platform, available at www.genenetwork.org/. This platform provides analysis tools, access to diverse data of BXD and other mouse resources and links to other species.

2 Methods

For the identification of most likely causal genes, following steps have to be performed:

- Phenotyping.
- Genotyping.
- QTL mapping via linkage analysis.
- Prioritizing positional candidate genes.
- Functional tests of candidate genes.

2.1 Phenotyping

In our study, we used 29 BXD RIS on a high fat diet to carry out linkage mapping towards the collagen content in epidydimal adipose tissue, liver, and skeletal muscle. We examined two to five males, since males are more prone to diabetes, for which obesity predisposes. Two or more, ideally 6–12, individuals per strain and sex should be phenotyped under the same environmental conditions to characterize a specific BXD strain. In case of obesity, you first should find at which age, under which diet and in which sex the gain of fat mass is highest. We suggest to measure body weight and body composition between weaning at about 3 weeks and adulthood at about 20 weeks. At the age and sex of fastest weight gain additional phenotyping should be performed which provides potential insight into physiological or metabolic changes associated with the obese phenotype.

Phenotypes that nobody has measured before are of highest interest since these could add new knowledge to the identification of mechanisms leading to obesity or conditions predisposing the onset of severe diseases. Such phenotypes include for example ATR-FTIR (Attenuated Total Reflectance Fourier Transform Infrared Spectroscopy) measurements of the composition of adipose tissues. An ATR spectrum carries information of a thin sample layer close to an ATR crystal. Such spectra provide information on molecular structure, composition, and content of biomolecules such as lipids, proteins, nucleic acids, and carbohydrates in tissues and cells simultaneously [9, 10].

All phenotypes have to undergo a rigorous quality check (*see* **Note 1**). Data have to be corrected for systematic environmental factors such as season, litter size, number of offspring per cage, or person who performed the measurement, for example. Outliers having three standard deviations higher or lower than the mean for some traits have to be omitted from analyses. In our ATR-FTIR measurements of the epididymal adipose tissue we detected two BXD strains as outliers and excluded them from further analyses [9]. Furthermore, for linkage analysis, phenotypic data should follow a normal distribution. If data are not normally distributed, they should be e.g. logarithmically transformed to improve the distribution.

BXD phenotype data are uploaded *to the GeneNetwork database* as follows:

Strain means of each normalized phenotypic measurement of each strain can be done either as temporary or as permanent data. Here we will upload two phenotypes temporarily:

1. Under the **home** button, select: **Batch Submission**.
2. Select from the **dropdown box** the cross or recombinant inbred set on which you have measured phenotype data; in our example we will choose *BXD*.
3. **Download**, **read, and follow** "*The guide for naming your traits.*" This guide will help you come up with valid and descriptive names for your phenotypes and makes it easier for others to find your phenotype data. Not adhering to this guide will cause the upload of your data to fail (*see* **Note 2**).
4. Prepare your phenotype input file with a professional text editor (not MS Word). The structure of the input file is explained in the GeneNetwork example files *Sample* and *Sample2*. The only difference between these two files is that *Sample* is column oriented (phenotypes are in columns) and *Sample2* is row oriented. The first line in your input file is a **@format line** defining if phenotypes are in *columns* or *rows*. In our example, phenotypes are in columns; strains are automatically in rows. Add to the **@format line** phenotype names of each trait followed by the heads of the first column containing the mean value of trait 1, the second column for the standard error (**SE**) of trait 1, and the third column for the sample size (**N**) (*see* **Note 3**). The same information will be given for trait 2. The columns containing SE and N are optional, however, if you use them, they have to be named **SE** and **N** and need to follow the phenotype they belong to; otherwise they will not be interpreted correctly, which can lead to erroneous results. Make sure to add a line brake at the end of this @format line. In the example below, replace <tab> by a real tab character, *<newline>* by a real newline (line brake), and replace "*Trait1*" and "*Trait2*" by the phenotype names (see bullet 3):

```
@format=column<tab>Trait1<tab>Standard er-
ror Trait 1[SE]<tab>Sample size trait 1
[N]<tab>Trait 2<tab> Standard error Trait
2[SE]<tab>Sample size trait 2[N] <newline>
```

Here we add data on *BXD* 14, 15, and 16 to our phenotype input file:

```
@format=column<tab>Metabolism, ATR-
FTIR, 20 weeks: collagen [mg]<tab>SE
<tab>N<tab>Metabolism, ATR-FTIR, 20 weeks:
Leptin [mg]<tab>SE<tab>N <newline>
```

```
BXD14<tab>421.972<tab>3.03449<tab>97<tab>10.9<tab>1.0<
tab>97<newline>
BXD15<tab>455.929<tab>4.26894<tab>31<tab>41.7<tab>4.4<
tab>97<newline>
BXD16<tab>448.976<tab>4.52902<tab>49<tab>21.7<tab>5.7<
tab>97<newline>
```

5. Click the **upload** button, select the file you created, and press the **next** button. When uploading is finished, a new window will open showing the uploaded phenotypes as a GeneNetwork collection.

Now you can check your uploaded phenotypes. Clicking Basic Statistics, a menu opens up that lists basic statistics such as sample size, mean, median, and standard deviation of our phenotype. Outliers will be recognized and highlighted in yellow. Look at the **Probability plot** this shows you how close your phenotype distribution is to a normal distribution. A normally distributed phenotype will show a straight line at $x=y$; samples that significantly deviate from the expected normal distribution are highlighted on the plot.

2.2 Retrieving BXD Genotype Data

The advantage of the BXD strains is that they are inbred and have dense genotype information available in GeneNetwork (*see* **Notes 4–6**). Genotyping of the first 32 *BXD* strains was initially performed in the late 1970s by Benjamin Taylor, who in the 1990s produced and genotyped an additional 11 BXD strains. BXD strains 43 through 100 were created by Lu Lu and Robert Williams (UTHSC), and by Jeremy Peirce and Lee Silver (Princeton University) (*see* **Notes 7–9**).

Collaboration with members of the CTC consortium resulted in a combined effort to genotype all BXD strains using microsatellite and SNP markers leading to a map consisting of about 7500 informative markers. In the future, new GigaMuga microarrays will be used to create an even denser SNP map of the BXD lines [Personal communication R. Williams]. The parental strains C57BL/6J (B) and DBA/2J (D) were also deeply genotyped using the mouse diversity array [11], and additionally sequenced (datasets are available at Ensembl). Genetic markers in GeneNetwork are linked to their corresponding Ensembl data. All informative sequence variants are used during QTL mapping.

Marker genotypes for one or multiple BXD individuals can be found as follows:

- Use the **Search** option.
- Set the Species field to *Mouse*.
- Set the Group field to *BXD*.
- Set the Type field to *Genotypes*.
- In the **Get Any** field type: *POSITION=(chr1 10 15)*

– Clicking **Search** will display all markers located on chromosome 1 between 10 and 15 Mb.

2.3 QTL Mapping of Your Phenotype via Linkage Analysis

For linkage analyses, you need phenotypes and genotypes of your examined BXD strains. Now you can analyze your own phenotypes, such as collagen content in different tissues, or you might also be interested in phenotypes that are stored from other researchers in the database.

1. For calling the phenotype of interest, return to the **Search** page at the main menu at genetwork.org. Use the Get Any option and search the whole database for the keyword: *Obesity* (at the time of writing this returns slightly more than 400 search results). Since we cannot do much with 400+ results, we have to narrow down our search, e.g. by combining obesity with collagen. Return to the **Search** page, set Species to *Mouse* and Group *to BXD*, use the Combined Search *option and* search for: *Obesity collagen*. In this case, we will receive three search results: An experiment in 2011 measured '*FTIR_liver_collagen, males*', '*FTIR_muscle_collagen, males*' and the '*FTIR_AdiposeTissue_collagen, males*' phenotype on BXD mice, stored under IDs 15096, 15104, and 15089. Click the record **15089** of the '*FTIR_AdiposeTissue_collagen, males*' search result. A new window will open up; this is the **Trait Overview Page** where the following categories are listed:
   - Details and Links—Shows details about the experimental data such as owner of the data or title of the paper. Being consistent in naming phenotypes and linking to the article describing the data will allow others to quickly and easily find back your data. When adding your own data to GeneNetwork, be precise when describing sample collection and data (pre)processing to make sure people can correctly interpret and reuse your experimental data.
   - Mapping tools—Performs QTL mapping of the selected trait.
   - Review and edit data—Allows the user to edit data before using any of the above-mentioned tools. These changes are not saved in the database, but persist during the user session.
2. For performing a linkage analysis to find a genomic region of that controls the collagen content in adipose tissue, click the **mapping tools** button. The **mapping tools** page provides four different types of QTL mapping:
   - Interval mapping—performs standard interval mapping using the Haldane mapping function. This method allows controlling for the effect of one other genetic marker when performing a QTL scan for a single or multiple

chromosomes. The result will give information about additive and/or dominance effects for suitable datasets. The method determines significance by permutation and bootstrapping to find confidence intervals.

- Marker regression—computes and displays Likelihood Ratio Statistics (LRS) for individual markers in a way comparable to genome wide association (GWA) analysis. This method also lists additive effects (phenotype units per allele) and dominance deviations for suitable datasets. The method determines significance by permutation.
- Composite interval mapping—performs interval mapping while allowing to control for one or multiple markers as cofactor. To find a suitable 'control marker' marker regression is often used iteratively. This allows us to use forwards selection to build a more complex multiple QTL models:
- Pairscan—searches for pairs of chromosomal regions that are involved in a two-locus epistatic interaction.

In our example, we select **interval mapping**, accept the default parameters and click the **compute** button. A new window will open, showing a progress bar. When the calculation is complete the screen shows a QTL profile similar to Fig. 1a. Zooming to a single chromosome to see the chromosome in detail, click the **name of chromosome** with the highest LRS score (chromosome 12, Fig. 1a).

3. Define a QTL confidence region by performing a bootstrap test and plotting the results on top of the QTL profile. Click the bootstrap test button and press the **remap** button. GeneNetwork performs 2000 bootstraps to estimate the confidence interval. As a result, we see several yellow bars under the major QTL peek. These bars tell us how much confidence there is that the

Fig. 1 (continued) markers across all chromosomes (*blue*). Additive effects are shown as *colored lines* in *green* indicating that D alleles increase the phenotypic values and *red* indicating that the B allele increases the phenotypic values. (**b**) Interval mapping and bootstrapping a confidence interval on chromosome 12 of the ATR-FTIR spectra for collagen in adipose tissue of 20 weeks old BXD mice using GeneNetwork. The *x*-axis shows genetic locations in megabases. The *y*-axis shows the LRS score of the association between the collagen measurements and the genetic markers across chromosome 12 (*blue*). In *yellow* we see the confidence interval around the top marker, leading to a confidence interval between 27 and 30 Mb. (**c**) Haplotype analysis of chromosome 12 in BXD mice using GeneNetwork. The *x*-axis shows the genetic markers across chromosome 12. The *y*-axis shows the different BXD strains. *Colors* represent the origin of the genetic marker (Paternal, Maternal, Heterozygous, and Unknown). (**d**) Heat map of all FTIR traits across the three different tissues (Figure 2 from Dogan et al. 2013, reprinted with permission from BMC Genomics). Heat map showing the QTL regions for different traits obtained by ATR-FTIR spectroscopy. Intensity of colors shows chromosomal regions with high linkage statistics (LRS) and the color encodes the allelic effect

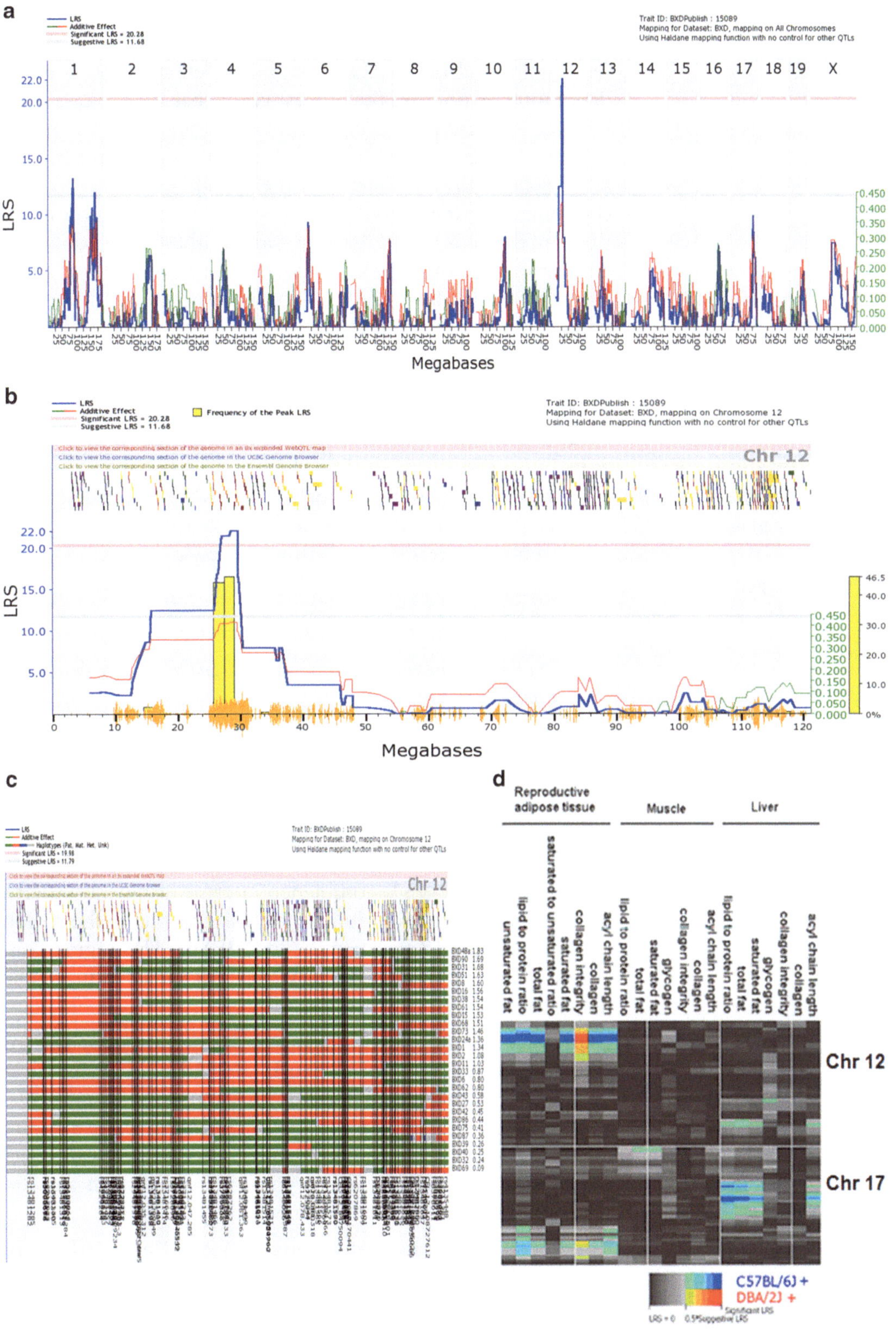

Fig. 1 (**a**) Interval mapping of the ATR-FTIR spectra for collagen in adipose tissue of 20 weeks old BXD mice using GeneNetwork. The *x*-axis shows genetic locations in megabases, chromosome numbers are listed on *top*. The *y*-axis shows the LRS score of the association between the collagen measurements and the genetic

real variant will be in this interval. To define the 95% confidence interval go from the highest yellow bar (in our case 49%, Fig. 1b) and add bars to both sides until the sum of the bars is higher than 95%. Check the box for **haplotype analysis**, look at the yellow bars and determine the confidence interval. Zoom (by using the **Remap** button) to only show the region inside the confidence interval. After remapping the QTL, bootstraps and performing the haplotype analysis a new window will open. In this window above the QTL profile we now see the genotypes of the individual BXD strains in Green (paternal marker) and Red (maternal marker), showing the recombination breakpoints in each individual strain (Fig. 1c). Bellow the plot there is a table which lists all genes located in this interval allowing us to do follow up research on these genes.

2.4 Prioritizing Positional Candidate Genes in the QTL Region

Prioritizing positional and potential functional genes underlying a QTL effect is an important step for the identification of causal genes or even mutations. Prioritization comprises

- *Localization of the positional candidate genes in the QTL region.*
- *Expression of the positional candidate genes in the target tissues.*
- *Density of nonsynonymous (ns) SNP and InDels in the positional candidate genes.*
- *cis-Regulated expression of positional candidate genes.*
- *Known or inferred gene function of positional candidate genes.*
- *Correlation of examined trait with other phenotypes.*

1. For obtaining a list of **positional candidate genes**, zoom to the region of the 95% confidence interval using the options panel above the plot. In the case of collagen, this is the region on chromosome 12 between 25 and 30 Mb. This region contains 36 protein coding genes. Genes will be listed in a table under the QTL plot.
2. **Expression of the positional candidate genes** can be investigated by searching GeneNetwork. Expression of genes was measured in multiple tissues during multiple experiments. Search GeneNetwork for the gene name and tissue you are interested in. If any matching dataset is found you can select the experiment and phenotype you are interested in and get information about the gene expression.
3. The **density of nonsynonymous (ns) SNP and InDels in positional candidate genes** is displayed in the table under the QTL profile. GeneNetwork calculates the density of sequence variants per gene and allows sorting of data by density. A general rule of thumb: Higher nsSNP density means more likely to be considered a candidate gene.

4. Information on cis-**regulated expression of positional candidate genes** can be obtained from a wealth of microarray data. Use the **search** page to find other experiments that have measured these genes in an expression QTL (eQTL) study. Use the '*Map QTLs on GN data*' guide to find out if genes in the confidence interval show the character of any *cis*-eQTL. This would indicate that sequence variants in the gene or in close proximity regulate the transcript amount of this gene. Genes that show a *cis*-eQTL are considered as candidate genes. There are five suitable data sets available for looking up eQTL information on adipose tissue: Adipose tissue (without further differentiation) mRNA was measured in the F_2 cross BH/HB and in the BH (Apoe0) dataset. White adipose tissue mRNA was measured in the F_2 crosses CastB6/B6Cast and C3H/JxC57BL6/J, and brown adipose tissue in BXD RIS. The closer the available expression data to the local action of the causal gene, the more informative are eQTL data.
5. **Known or inferred gene function of positional candidate genes**. Gene name and MGI description are listed in a table under the QTL profile. This table lists all genes in the visible area of the plot. Make sure to zoom to just the confidence interval. Besides gene name and MGI description additional information is available as a link-out when clicking on **individual gene names**. This will redirect you to the corresponding NCBI gene information page.
6. **Correlation of the QTL phenotype with other phenotypes** measured in the same population and colocation of QTL of the target phenotype with QTL of the correlated phenotypes are a hint for potential pleiotropic effects of the priorized gene or genetic linkage of causal genes (Fig. 1d). A special case is the correlation of the target phenotype with the expression of the priorized gene(s) (RNA or protein amounts). This refers to colocalization of the QTL of the target phenotype with the eQTL position. Correlation can also be examined between the target QTL phenotype and expression of all genes in the QTL interval. If the gene expression strongly correlates with the QTL phenotype, this further strengthens the assumption that this gene might be causal (*see* **Note 12**).

   For performing a correlation analysis:

- Go to the **Trait Overview Page**, as described in **step 3**, point 1. From the Trait Overview Page select **Calculate Correlation**.
- Calculate correlation for the phenotype **FTIR_AdiposeTissue_collagen, males** and return the 500 highest correlated gene expression phenotypes.

- Select Pearson correlation, which assumes a normal distribution, or Spearman correlation, which uses nonparametric testing.
- Next we have to choose which dataset to calculate correlation towards. Since we are interested in genes expressed in adipose tissue, underlying our QTL we change the **database** field to: *EPFL/LISP BXD CD Brown Adipose Affy Mouse Gene 2.0 ST Exon Level (Oct13) RMA*.
- Results will be sorted based on their correlation and the table allows to sort by correlation coefficients or significance of correlation, you can choose the number of phenotypes/results you want have returned.

3 Results

The results of the example are described in detail in Dogan et al. [9]. The 29 examined BXD strains showed a wide range of obesity under a high fat diet. The phenotypic distribution of all phenotypes was more extreme among BXD strains than between the parental strains B and D. The epididymal adipose tissue and liver weights ranged from 0.37 ± 0.08 to 3.82 ± 0.07 g and 1.11 ± 0.08 to 3.17 ± 0.13 g, respectively. The strains also varied widely in the macromolecular composition of these two tissues, which was measured by ATR-FTIR spectroscopy. Differences between strains were not significant in muscle. These phenotypic observations let us expect to find QTL for adipose tissue and liver, but less likely for muscle.

Indeed, the QTL analysis revealed the most significant QTL on chromosome 12 between 26 and 30 Mb. We will focus on this QTL in this section. The identified region significantly affected the total fat content (LRS = 21.5), saturated (LRS = 21.5), and unsaturated fat (LRS = 21.2), the relative content of collagen (LRS = 22.1) and collagen integrity (LRS = 17.6) in adipose tissue. Under the specific experimental conditions of feeding high fat diet, the B allele was elevating all traits, except for collagen integrity, where the allele of strain D was increasing. For example, BXD strains carrying the chromosome 12 QTL allele of strain B versus strains carrying allele D had about 2.5-fold more relative contents of total fat, saturated and unsaturated fats, and collagen content, but also 5.5-fold lower collagen integrity (*see* **Note 10**).

Twenty four genes are located in the chromosome 12 region of 4 Mb. Based on the search for sequence variations between the parental strains B and D and their predicted functional consequences, the expression of genes in adipose tissue as well as in liver, and QTL for gene expression patterns in segregating populations (eQTL), we suggested *Rsad2* (viperin) and *Colec11* (collectin-11) as potential quantitative trait candidate genes for the locus on chromosome 12. *Rsad2* encodes the radical S-adenosyl methionine domain containing two protein, which might modulate the lipid droplet contents and lipid biosynthesis. *Colec11* encodes the protein collectin subfamily member 11, which might play a role in apoptotic cell clearance and maintenance of adipose tissue.

Two and one nonsynonymous (ns) SNP between the parental strains B and D were found in *Rsad2* and *Colec11*, respectively. However, potential effects of those nsSNPs on the protein function are unknown. Gene expression data accessible at BioGps (*see* **Note 11**) [12] showed higher transcript amounts of *Rsad2* and *Colec11* (2.78 and 1.93 times, respectively) in adipose tissue of lean compared to obese mice. This suggests a *cis*-acting gene variant leading to differential gene activation (*see* **Note 13**). Further evidence for *cis*-regulation of the two genes comes from genetic variation between B and D in the gene regions and information on expression QTLs (eQTL). Expressed QTLs (eQTL) for *Rsad2* and *Colec11* were described for adipose tissues in the F_2 cross CastXC57BL6/J (CastB6/B6Cast F2), and for *Colec11* additionally in the F_2 cross C3H/JxC57BL6/J (BH/HB F2). Moreover, four out of ten synonymous SNPs in the coding region of *Rsad2* that exist between B and D also occur between Cast/J and B mice of the eQTL experiment. Therefore, it is very likely that the genetic variation in regulatory sequence motifs in this gene could be responsible for expression or functional differences in both reference populations. The measurement of transcript amounts in adipose tissue of BXD strains carrying either the B or D allele of *Rsad2* and *Colec11* revealed 1.4 times higher mRNA amounts of *Rsad2* of D carriers ($p=0.03$), while no expression difference was found for *Colec11*. Therefore, we suggest *cis*-acting genetic variation that interacts with environmental changes on the cellular level leading to differential gene activation of *Rsad2* on a high fat diet (*see* **Note 14**).

Finally, we detected pleiotropic effects of the chromosome 12 locus (*see* **Note 15**). The colocalization of QTL affecting the total fat content, saturated and unsaturated fat as well as the relative content of collagen and collagen integrity in adipose tissue is consistent with the high phenotypic correlation between these traits. Collagen itself was positively correlated with all other traits in adipose tissue, except collagen integrity. This suggests that a single gene could regulate the complex of cellular changes. *Rsad2* could potentially act as such gene. The Rsad2 protein is located on the cytosolic side of the endoplasmic reticulum (ER) where it is involved in protein-protein interactions. It may affect the proper folding of proteins or anchoring of proteins to the membrane [13, 14]. As such protein, it could have a function during endoplasmic reticulum stress. *Rsad2* colocalizes also with the adipocyte differentiation-related protein on the outer layer of lipid droplets and interacts with an enzyme on the endoplasmic reticulum which is required for the generation of cholesterol [14, 15]. Rsad2 may also alter the lipid content and quantity of lipid droplets through interaction of the lipid droplet with the endoplasmic reticulum [15]. Furthermore, Rsad2 is likely also required for the T-cell receptor mediated activation of NFkB and AP-1, which are important regulators of inflammatory cytokine production in white adipose tissue [16, 17].

The colocalization of *Rsad2* on lipid droplets as lipid storage organelles, the association of lower gene expression with higher fat deposition and our genetic mapping results suggests that *Rsad2* might control the formation of lipid droplets while impairment of *Rsad2* likely enhances fat accumulation. However, the final prove of function of *Rsad2* by cell-based experiments or genetically modified organisms has to be provided (*see* **Note 16**).

4 Notes

1. Phenotypic data should be quality checked and preprocessed before being uploaded to GeneNetwork. This includes normalization of data, removal of outliers or windsorization, eventually transformation of data to obtain normal distribution.
2. When uploading data to GeneNetwork for permanent and public storage, make sure to follow the GeneNetwork naming guide for phenotypes.
3. When uploading your own data make sure that for any phenotype that has SE and N values the SE and N columns are located directly after the phenotype mean they belong to in the @format line.
4. An F_2 population is informative for complex traits such as obesity since all loci contributing to the phenotype occur in either homozygous or the heterozygous state. This allows for the identification of additive, dominant, and recessive effects and the analysis of epistatic interaction.
5. In F2 individuals, different genetic backgrounds potentially lead to different fat mass, which has to be taken into consideration for the analysis and correct interpretation of the data.
6. For the generation of a RCS an individual carrying the QTL interval of the donor strain is crossed back to the recipient strain. Afterwards, only offspring carrying the donor QTL interval or recombinations in this region are chosen for further mating to the recipient strain. After ten generations of backcrossing, homozygous RCS with different chromosomal intervals of the QTL are obtained.
7. The different batches of BX RIS should be considered in an experiment since confounding effects due to hidden private or

de novo mutations among mice within an inbred strain might account to the phenotypic variation.

8. In GeneNetwork, temporary phenotype data is available only to you so that tools can access your data. Uploading allows you to make data public or keep it private until for example publication in a journal. To add permanent data you need to have a GeneNetwork user account.
9. Despite extensive genotyping of the BXD strains, it is always possible that hidden mutations could segregate or have happened during the course of generating or maintaining the BXD animals, which are unknown and therefore are not in the database. Nevertheless, such unknown mutations could contribute to the phenotypical differences between strains.
10. BXD strains are not suited to detect dominance effects since all strains are inbred and heterozygous genotypes do not occur.
11. If you check expression data of your genes in expression databases, e.g. BioGPS, it is necessary to normalize expression levels with housekeeping genes (we used *Actc*, *Gapdh*, *Rps29*, *B2m*, *Ppia*, *Gusb*, *Tbp*). Take care for the specific environmental conditions under which the gene expression study was performed.
12. The correlation analysis requires that gene expression and classical phenotypes were measured in the same population in the current or previous studies.
13. For providing evidence for *cis*-regulated expression of positional candidate genes, ideally, expression data are needed for the population we perform the QTL study in and for the tissue we expect the gene acts on. If such data are not available, it might be helpful to examine available eQTL data from other resource populations. However, the closer the available expression data to the local action of the causal gene with respect to strain and tissue, the more informative are eQTL data.
14. Using different expression databases allows us to expand the number of possible candidate genes. For example, when looking into the brown adipose tissue mRNA database for BXD strains we only find *Kidins220* as a possible candidate gene showing a very significant eQTL (LRS = 54.4) on chromosome 12 at 15.7 Mb. In addition, the expression of this gene in brown adipose tissue is highly correlated with the collagen content in white adipose tissue (0.83). Therefore, *Kidins220* should be considered as a putative candidate gene although it is 10 Mb away of the peak QTL position and slightly outside of the confidence interval of our example trait collagen.
15. The observation that a particular genomic region affects several traits could be indicative for pleiotropic effects. The under-

lying genetic effect could be caused by a single gene or mutation or by two or several genes or mutations in close linkage.

16. The final prove of principle for a suggested candidate gene is the gene effect in a cell based system or on the whole organism.

Acknowledgements

We acknowledge permanent support by the German Research Foundation (DFG), the German Ministry of Education and Research (BMBF), the GeNeSys Network and the COST action SYSGENET BM 0901.

References

1. Chadt A, Leicht K, Deshmukh A, Jiang LQ, Scherneck S, Bernhardt U, Dreja T, Vogel H, Schmolz K, Kluge R, Zierath JR, Hultschig C, Hoeben RC, Schurmann A, Joost HG, Al-Hasani H (2008) Tbc1d1 mutation in lean mouse strain confers leanness and protects from diet-induced obesity. Nat Genet 40(11):1354–1359. doi:10.1038/ng.244
2. Crabbe JC, Belknap JK (1993) Behavior genetic analyses of drug withdrawal. Alcohol Alcohol Suppl 2:477–482
3. Taylor BA, Rowe L (1984) Genes for serum amyloid A proteins map to chromosome 7 in the mouse. Mol Gen Genet 195(3):491–499
4. Groves MG, Rosenstreich DL, Taylor BA, Osterman JV (1980) Host defenses in experimental scrub typhus: mapping the gene that controls natural resistance in mice. J Immunol 125(3):1395–1399
5. Peirce JL, Lu L, Gu J, Silver LM, Williams RW (2004) A new set of BXD recombinant inbred lines from advanced intercross populations in mice. BMC Genet 5:7. doi:10.1186/1471-2156-5-7
6. Churchill GA, Airey DC, Allayee H, Angel JM, Attie AD, Beatty J, Beavis WD, Belknap JK, Bennett B, Berrettini W, Bleich A, Bogue M, Broman KW, Buck KJ, Buckler E, Burmeister M, Chesler EJ, Cheverud JM, Clapcote S, Cook MN, Cox RD, Crabbe JC, Crusio WE, Darvasi A, Deschepper CF, Doerge RW, Farber CR, Forejt J, Gaile D, Garlow SJ, Geiger H, Gershenfeld H, Gordon T, Gu J, Gu W, de Haan G, Hayes NL, Heller C, Himmelbauer H, Hitzemann R, Hunter K, Hsu HC, Iraqi FA, Ivandic B, Jacob HJ, Jansen RC, Jepsen KJ, Johnson DK, Johnson TE, Kempermann G, Kendziorski C, Kotb M, Kooy RF, Llamas B, Lammert F, Lassalle JM, Lowenstein PR, Lu L, Lusis A, Manly KF, Marcucio R, Matthews D, Medrano JF, Miller DR, Mittleman G, Mock BA, Mogil JS, Montagutelli X, Morahan G, Morris DG, Mott R, Nadeau JH, Nagase H, Nowakowski RS, O'Hara BF, Osadchuk AV, Page GP, Paigen B, Paigen K, Palmer AA, Pan HJ, Peltonen-Palotie L, Peirce J, Pomp D, Pravenec M, Prows DR, Qi Z, Reeves RH, Roder J, Rosen GD, Schadt EE, Schalkwyk LC, Seltzer Z, Shimomura K, Shou S, Sillanpaa MJ, Siracusa LD, Snoeck HW, Spearow JL, Svenson K, Tarantino LM, Threadgill D, Toth LA, Valdar W, de Villena FP, Warden C, Whatley S, Williams RW, Wiltshire T, Yi N, Zhang D, Zhang M, Zou F, Complex Trait C (2004) The Collaborative Cross, a community resource for the genetic analysis of complex traits. Nat Genet 36(11):1133–1137. doi:10.1038/ng1104-1133
7. Threadgill DW, Miller DR, Churchill GA, de Villena FP (2011) The collaborative cross: a recombinant inbred mouse population for the systems genetic era. ILAR J 52(1):24–31
8. Svenson KL, Gatti DM, Valdar W, Welsh CE, Cheng R, Chesler EJ, Palmer AA, McMillan L, Churchill GA (2012) High-resolution genetic mapping using the mouse diversity outbred population. Genetics 190(2):437–447. doi:10.1534/genetics.111.132597
9. Dogan A, Lasch P, Neuschl C, Millrose MK, Alberts R, Schughart K, Naumann D, Brockmann GA (2013) ATR-FTIR spectroscopy reveals genomic loci regulating the tissue response in high fat diet fed BXD recombinant inbred mouse strains. BMC Genomics 14:386. doi:10.1186/1471-2164-14-386, 1471-2164-14-386 [pii]

10. Sen I, Bozkurt O, Aras E, Heise S, Brockmann GA, Severcan F (2015) Lipid profiles of adipose and muscle tissues in mouse models of juvenile onset of obesity without high fat diet induction: a fourier transform infrared (FT-IR) spectroscopic study. Appl Spectrosc 69(6): 679–688. doi:10.1366/14-07443
11. Yang H, Ding Y, Hutchins LN, Szatkiewicz J, Bell TA, Paigen BJ, Graber JH, de Villena FP, Churchill GA (2009) A customized and versatile high-density genotyping array for the mouse. Nat Methods 6(9):663–666, doi:nmeth.1359 [pii]10.1038/nmeth.1359
12. Wu C, Delano DL, Mitro N, Su SV, Janes J, McClurg P, Batalov S, Welch GL, Zhang J, Orth AP, Walker JR, Glynne RJ, Cooke MP, Takahashi JS, Shimomura K, Kohsaka A, Bass J, Saez E, Wiltshire T, Su AI (2008) Gene set enrichment in eQTL data identifies novel annotations and pathway regulators. PLoS Genet 4(5), e1000070. doi:10.1371/journal.pgen.1000070
13. Chin KC, Cresswell P (2001) Viperin (cig5), an IFN-inducible antiviral protein directly induced by human cytomegalovirus. Proc Natl Acad Sci U S A 98(26):15125–15130. doi:10.1073/pnas.011593298
14. Hinson ER, Cresswell P (2009) The N-terminal amphipathic alpha-helix of viperin mediates localization to the cytosolic face of the endoplasmic reticulum and inhibits protein secretion. J Biol Chem 284(7):4705–4712. doi:10.1074/jbc.M807261200
15. Wang X, Hinson ER, Cresswell P (2007) The interferon-inducible protein viperin inhibits influenza virus release by perturbing lipid rafts. Cell Host Microbe 2(2):96–105. doi:10.1016/j.chom.2007.06.009
16. Lu L, Wei L, Peirce JL, Wang X, Zhou J, Homayouni R, Williams RW, Airey DC (2008) Using gene expression databases for classical trait QTL candidate gene discovery in the BXD recombinant inbred genetic reference population: mouse forebrain weight. BMC Genomics 9:444. doi:10.1186/1471-2164-9-444
17. Gatti D, Maki A, Chesler EJ, Kirova R, Kosyk O, Lu L, Manly KF, Williams RW, Perkins A, Langston MA, Threadgill DW, Rusyn I (2007) Genome-level analysis of genetic regulation of liver gene expression networks. Hepatology 46(2):548–557. doi:10.1002/hep.21682

Chapter 24

Social Interactions and Indirect Genetic Effects on Complex Juvenile and Adult Traits

David G. Ashbrook and Reinmar Hager

Abstract

Most animal species are social in one form or another, yet many studies in rodent model systems use either individually housed animals or ignore potential confounds caused by group housing. While such social interaction effects on developmental and behavioral traits are well established, the genetic basis of social interactions has not been researched in as much detail. Specifically, the effects of genetic variation in social partners on the phenotype of a focal individual have mostly been studied at the phenotypic level. Such indirect genetic effects (IGEs), where the genotype of one individual influences the phenotype of a second individual, can have important evolutionary and medically relevant consequences. In this chapter, we give a brief outline of social interaction effects, and how systems genetics approaches using recombinant inbred populations can be used to investigate indirect genetic effects specifically, including maternal genetic effects. We discuss experimental designs for the study of IGEs and show how indirect genetic loci can be identified that underlie social interaction effects, their mechanisms, and consequences for trait variation in focal individuals.

Key words Social interactions, Indirect genetic effects, Maternal genetic effects, Parental care, Systems genetics, BXD, Cross-fostering

1 Introduction

1.1 Social Interactions and Indirect Genetic Effects

Social interactions can influence a wide variety of behavioral, developmental, and disease-related traits [1–4], including courtship [5–7], play behavior [8, 9], aggression [10, 11], and parental care [12–15]. Traits expressed during social interactions are referred to as interacting phenotypes [16], and are predicted to follow a different evolutionary trajectory from nonsocial traits [16–18]. This has important implications for a diversity of disciplines, ranging from behavioral ecology [17, 19] to quantitative genetics [20–22]. One way by which social interactions can affect the fitness of an organism is "social selection" [18, 23, 24]. This occurs when a phenotype expressed in one individual directly alters fitness in a second individual; however, the phenotype of the second individual

Klaus Schughart and Robert W. Williams (eds.), *Systems Genetics: Methods and Protocols*, Methods in Molecular Biology, vol. 1488, DOI 10.1007/978-1-4939-6427-7_24,

is not altered. By contrast, a phenotype expressed in one individual may alter the phenotype in a second individual. Therefore, the phenotype of a focal individual depends not only on its own genotype, but also, in part, on the genotype of interacting individuals (Fig. 1). At a population level, if genetic variation underlies trait variation in social interactants (e.g. offspring, parents, siblings, cage, or litter mates) indirect genetic effects (IGEs) occur [11, 16, 20, 24–30]. The most obvious IGE is the effect of a mother's genotype on her offspring's phenotype, independent of any shared genetic material [31–33]. These maternal genetic effects (MGEs) have been identified in the 1940s [34] and were explored in various animal models, primarily in relation to domestic livestock [34–39]. Falconer famously referred to maternal effects as a "frequent, and often troublesome, source of environmental resemblance, particularly with mammals" ([40], p. 156) because they cause phenotypic similarities among siblings in a litter or a brood due to shared maternal environment, thus confounding estimates of how much variation among litter mates is due to genetic differences.

IGEs are considered an important source of phenotypic variation and are particularly relevant to development in social animals. In mammals, for example, the social environment is provided by parents and siblings during early development, and it is during this early developmental period that IGEs have the most significant impact [42, 43]. IGEs are also important from an evolutionary perspective because they have a genetic basis and can thus respond to selection and alter the genotype—phenotype relationship compared to what would be predicted from just considering direct effects [16]. Therefore, IGEs represent an environment which can respond to selective pressures [16, 17, 44, 45]. Further, since interactions are two-way, IGE-related genes can evolve in both individuals, causing feedback between the individuals [16].

Commonly, studies focus on identifying direct genetic effects (DGEs) , i.e. effects of own genotype on trait variation, e.g. [46].

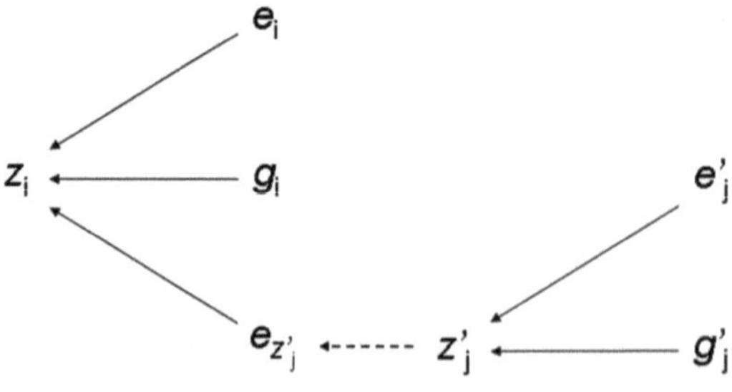

Fig. 1 Sources of variation for a hypothetical trait z_i showing indirect genetic effects. Variation in trait i is affected by genetic variation g_i (direct genetic effect) and e_i, the random environmental component. The indirect genetic effect $e_z{'}_j$ caused by the phenotype of individual j, z'_j also contributes to trait variation in the focal phenotype. From Harris and Hager 2009 [41]

While the importance of IGEs for understanding evolutionary dynamics and causes of variation has been established in theoretical work [24, 27, 45, 47], empirical research has predominantly adopted a phenotypic approach in that the effects of a given trait on a partner's trait were assessed (e.g. [11, 13, 29, 30, 42, 48–51]). Far fewer studies have attempted to investigate the genetics underlying IGEs, i.e. directly identifying genetic variants in social partners that affect trait variation in a focal individual [52]. For IGEs to occur two requirements need to be met. First, the focal trait must be phenotypically plastic in response to the interacting individual's phenotype [53], e.g. a mother's suckling behavior may increase due to offspring solicitation. Second, the indirect effect has to have a genetic basis, otherwise it is simply an environmental effect, rather than in IGE [20, 31, 50].

To illustrate the concept of IGEs we can look at how variance components are partitioned (Fig. 1). The expression of a phenotype in a focal individual (z_i), is dependent on the genotype of that focal individual (g_i) and the environment (e_i) [$z_i = g_i \times e_i$]. Many genetic mapping studies, however, treat the phenotype as being dependent on the genotype [$z_i = g_i$] assuming controlled environmental conditions in the laboratory. For phenotypes influenced by IGEs their expression is also dependent on the genotype of the interacting individual (g'_j; $z_i = g_i \times g'_j \times e_i \times e'_j$; Fig. 1). We will show below that mapping of IGEs is possible by controlling for environmental variance components (i.e. all animals are kept under the same conditions) and g_i (as all animals are of the same genotype). Therefore, the equation can be simplified to $z_i = g'_j$: differences in the phenotype of the observed individual are due to the genotype of the interacting individual (mediated by the interacting individual's phenotype; z'_j). This means that standard mapping methods, such as interval mapping [54], can be used, replacing the genotype of the focal individual with the genotype of the interacting individual.

1.2 The Prevalence of Indirect Genetic Effects

IGEs are most important in affecting trait variation in early life, which is evinced by the fact that up to ~50 % of the variance in preweaning weight can be caused by maternal effects [48], and that ~40 % of the variance in maternal performance is heritable [21, 48]. This is perhaps unsurprising since offspring are dependent on mothers for both warmth and nutrition.

However, the consequences of IGEs may extend well into later life affecting the development of adult traits, including disease traits. The most likely traits to be affected significantly (i.e. a large degree of variation in these traits can be attributed to IGEs, and especially MGEs) are developmental and behavioral traits. Here, IGEs have been demonstrated in many species including domestic farm animals (e.g. [39]), *Arabidopsis thaliana* [30], red squirrels (*Tamiasciurus hudsonicus*) [55], sheep (*Ovis aries*) [56], pigeons (*Columba livia*) [57], European starlings (*Sturnus vulgaris*) [58], dung beetles (*Onthophagus taurus*) [13] and burying beetles, *Nicrophorus pustulatus* [59, 60] and *N. vespilloides* [61]. Developmental traits have been particularly well studied in laboratory mice [48, 62–69], with multiple maternal genetic effect loci identified in mice [33, 49, 70, 71].

Among behavioral traits influenced by IGEs are social dominance, which is relevant to many species [17, 72], with more dominant individuals having greater access to resources, such as food [73], mates [74] and territories [75], and thus significant effects on fitness [76]. Indeed, IGEs on social dominance have been identified in cattle (*Bos taurus*) [77, 78], red deer (*Cervus elaphus*) [76] and cockroaches (*Nauphoeta cinerea*) [17]. Wilson et al. [11] showed that aggressive behavior in deer mice (*Peromyscus maniculatus*) is influenced by IGEs, however, specific loci were not mapped in this study. It may be possible to investigate the degree to which social dominance and aggression is mediated by IGEs using recombinant inbred systems in a design as described below (Fig. 2). Aggressive behavior of genetically uniform individuals raised by genetically variable mothers could be investigated to find aggression-related loci caused by MGEs. Similarly, courtship displays have been shown to be influenced by IGEs [50]. Although demonstrated in the fruit fly (*Drosophila serrata*), mice and other rodents also show courtship behavior [5–7] and given DGEs have been identified [79, 80], it is likely that also IGEs may influence courtship. IGEs have been demonstrated in other systems e.g. chemical displays in *Drosophila melanogaster* [81] antipredator behavior in guppies (*Poecilia reticulata*) [82], elaborate secondary sexual characteristics [83], laying date in red-billed gulls (*Larus novaehollandiae scopulinus*) [28] and mortality due to pecking behavior in chickens (*Gallus gallus*) [84].

1.3 Maternal Genetic Effects

MGEs are the best known example of IGEs and have been widely studied. In many species, and particularly mammals, the mother provides a large part of the early life environment, and therefore changes in the mother's phenotype, due to her genotype, will alter the phenotype in offspring. In most mammals, offspring are dependent on their mother for warmth, nutrition, and protection during the post-natal period. Indeed, differences in maternal behavior between different strains of mice significantly correlate with offspring survival [85]. Two of these post-natal maternal behaviors, food provisioning (lactation) and nestbuilding [86], can easily be quantified in a laboratory situation.

There are considerable differences in maternal care between different strains of rodents, suggesting genetic differences between the strains underlie differences in maternal behavior [85, 87–90], and indeed, several DGE loci have been mapped for maternal care in mice [91, 92]. Further, it has been shown in both humans and rodents that differences in maternal behavior can alter offspring

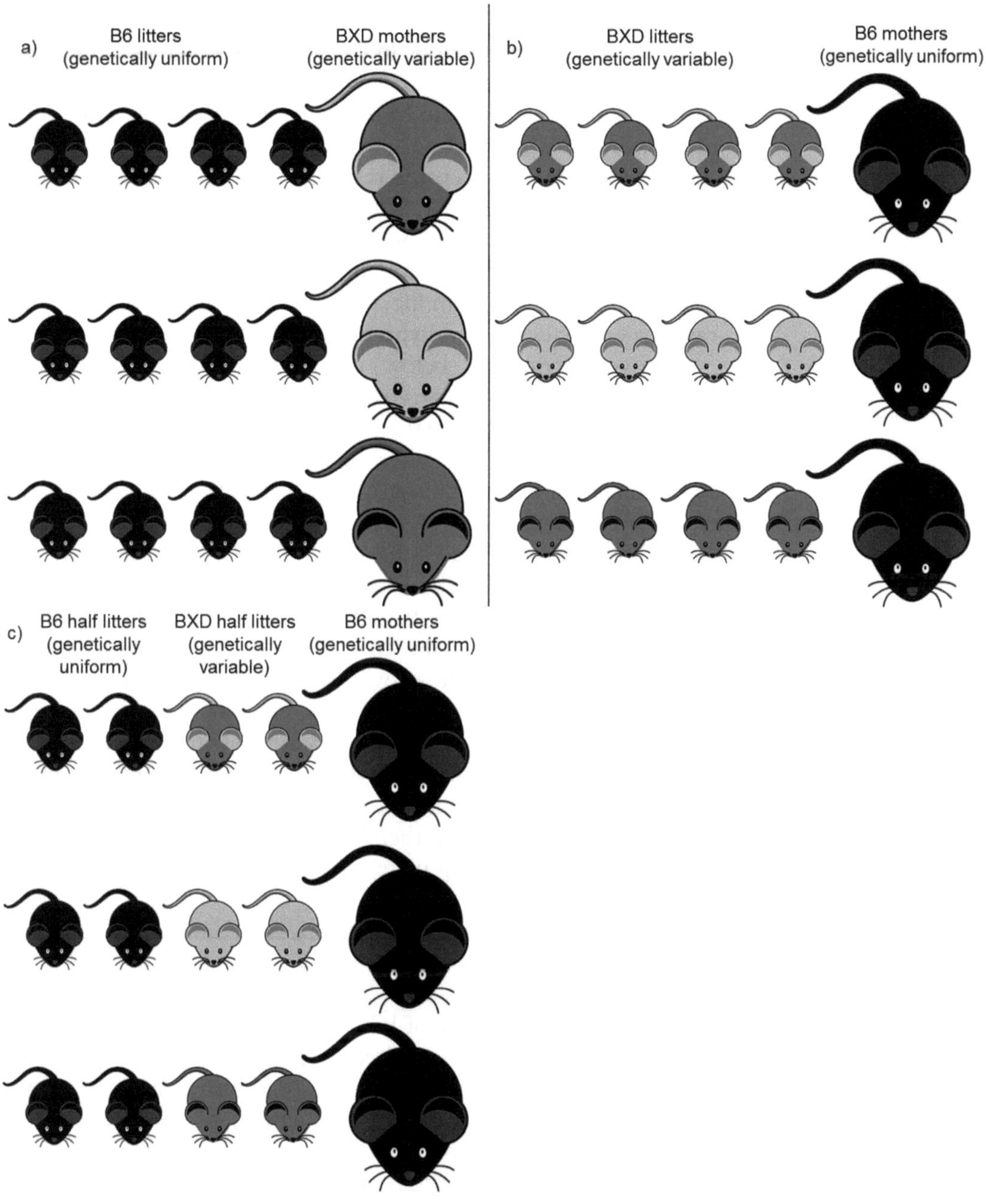

Fig. 2 Cross-fostering experimental designs. Here we use the BXD recombinant inbred panel as an example of genetically variable lines (*shades of gray*), and the C57BL/6J (B6) as an example of a genetically uniform inbred line (*black*). In each case the phenotype of the B6 members of a family will be mapped as a function of the genotype of the BXD family members. Panels **a** and **b** represent full-litter cross-fostering between BXD and B6 strains. In **a**, differences between the phenotypes of the genetically uniform B6 litters are due to the genotype of their genetically variable BXD mothers (maternal genetic effect; MGE). In **b**, the differences between the phenotypes of the genetically uniform B6 mothers are hypothesized to be due to the genotype of their genetically variable BXD foster-litters (offspring genotype having an indirect genetic effect; IGE). In a half-litter cross-fostering design, shown in panel **c**, differences in phenotype of the genetically uniform B6 litters or B6 mother are hypothesized to be due to the genotype of the BXD half-litter (IGE of nest mate and offspring respectively)

phenotypes as adults, particularly social behavior, including disease-related traits such as anxiety and stress traits which are influenced by early life environment [93–95], including MGEs [96–98]. These changes are associated with expression of a number of genes in offspring [99], including glucocorticoid [100] $GABA_A$ [101], and oxytocin [102] receptors resulting in altered behavioral phenotypes later in life [33, 97, 99, 100, 102–115]. These MGEs can also have a significant effect on metabolic disease [116], potentially mediated by altering birth weight [117], which in turn can have health implications later in life, predisposing to coronary heart disease, type 2 diabetes, and hypertension [118].

Cross-fostering (see below) can be used to separate the influence of offspring and maternal genome on offspring traits and therefore to explore how natural variation in maternal care (i.e. by different genotypes) can alter offspring phenotype [119–124]. These maternal effects can have significant effects on the phenotype. of offspring, such as maternal nursing on offspring adult blood pressure [125] and behavior [126, 127]. Further, maternal effects may also affect the rate and direction of evolution [31, 128]. However, there have been few studies to identify the genetic loci influencing these traits [33, 49, 70, 71, 129]. Moreover, MGEs can easily be confused with maternal environmental effects and therefore experiments need to be carefully designed to separate these influences [31]. Finally, it is not only the maternal genotype which can alter offspring phenotype (for example maternal care shaping offspring central nervous system function [130]), but offspring can alter their mothers' phenotypes (e.g. increased solicitation behavior increasing maternal provisioning behavior [22]).

2 Methods

2.1 Cross-Fostering

A critical step in the study of IGEs. is to ensure that one can partition phenotypic variance in direct and indirect components, notwithstanding experimental limitations (e.g. distinguishing prenatal from postnatal maternal effects, see below). Maternal effects can account for a large proportion of phenotypic variance, especially during early life, and for some traits explain more variation than direct genetic effects [33, 97, 99, 100, 102–115]. However, maternal and offspring genotype are correlated (i.e. half their genes are shared), and in inbred lines they are fully confounded, thus separating the effects of their respective genotypes is difficult. To remove this confounding effect cross-fostering has been used, both in the laboratory and in the field [119, 131]. It has been carried out across taxa, including insects, mammals, and birds [3, 15, 55, 58, 59, 132] and across disciplines, from behavioral ecology studying, for example, kin recognition [133, 134] to medical sciences, e.g. for models of metabolic [135, 136] or psychiatric [137] diseases.

Cross-fostering designs can also be used to analyze epigenetic effects, for instance finding epigenetic markers in offspring genotype caused by differences in maternal care, e.g. [107]. Although originally intended to investigate the effect of maternal genotype on offspring phenotype, simple changes enable us to examine effects in the opposite direction, of offspring genotype on maternal phenotype (Fig. 2). It is important to note, however, that cross-fostering does not control for pre-natal maternal effects [69], and in the case of inbred lines offspring and biological mother genotype will broadly be identical. Embryo-transfer could be used to examine these pre-natal maternal effects, but these are challenging techniques and not really suitable for genetics experiments due to the low sample size generated [69]. Of course, there is also the question about maternal effects occurring prior to embryo transfer.

In the following, we will give a brief outline of the cross-fostering technique established in our group [138, 139] to highlight a number of points which may be considered when carrying out this procedure. For any experiment we would use nulliparous females because previous litters may themselves effect IGEs on the mother's phenotype. For example, if a previous litter showed increased levels of solicitation, the mother may have a reduced residual ability to nurse a subsequent litter, while at the same time greater maternal experience may also be a confound. This may affect her ability to provide for the next litter, resulting in an IGE of the first litter on the second litter, mediated by the first litter's IGE on the mother (an example of inter-brood conflict [140, 141]).

Females should be placed in individual cages before parturition. If the sire or other females are present when the pups are born they then may have IGEs on the pups which may confound the apportioning of IGEs. Further, leaving the sire may increase the risk of infanticide (although the level of infanticide depends on the male's genotype. [142, 143]), and leaving females may increase the amount of care the pups receive, due to communal nursing behaviors [144–147]. Finally, to increase acceptance of a foster litter, cross-fostered pups should be exposed to the nesting material in the foster mother's cage so they adopt a scent familiar to the foster mother [148, 149].

2.2 Studying IGEs in Family Interactions

As an example of empirical research in IGEs, in our work we focus on studying the interaction between parent and offspring, and between siblings, largely during the lactation period, i.e. the first three postnatal weeks as during this period offspring depend on maternal provisioning and care, and sibling competition, e.g. over teats and nest position, is highest [3, 139, 150]. The aim of these studies is to determine IGEs on developmental and behavioral traits in experimental populations of recombinant inbred BXD mice.

To obtain quantifiable parameters of behavioral traits and maternal provisioning a detailed and tested assay is used measuring maternal and offspring behaviors and maternal provisioning. Maternal behaviors are subdivided into behaviors directed at offspring, including nursing, suckling, retrieval, and those not directed to young. Maternal provisioning may be measured on 3 days during lactation (for details see e.g. [151]). Behavioral recordings follow standard procedures (e.g. [152]) and are either taken by a human observer or on video. Specifically, we use the following protocol in mice to study parent offspring interactions:

1. Separate mothers from their litters 4 h prior to the planned behavioral observation. Animals may be weighed at this stage. This way, we standardize the motivation by mother so show maternal behavior and offspring motivation to solicit behavior. Otherwise, some mothers may have nursed just prior to an observation, while others nursed last several hours ago. This will cause differences between females in their behavior that is not due to treatment or genotype differences.
2. Mothers stay in the home cage with water and food provided, offspring are transferred to a new cage, which is placed on a heat mat to provide warmth.
3. After 4 h, mothers and pups are weighed and rejoined in their original home cage, and behavioral recording commences.
4. We record behavior in both mothers and offspring over 15 min in 20 s intervals. For example, maternal behaviors included suckling and nestbuilding, offspring behavior includes solicitation and sibling competition.
5. Maternal provisioning is then measured as maternal or offspring weight change over the 2 h period following the 4 h separation.

We further note, that for adult phenotypes it is important that no IGEs occur between group housing during lactation and testing later. For example, animals from the same half litter in Fig. 2c would have to be housed together post-weaning but separate from the other half litter, lest there is the continued possibility of IGEs.

2.3 Analysis and QTL Mapping

We are interested in identifying particular loci that have IGEs on the traits measured, rather than merely demonstrating the existence of IGEs generally (which is possible by partitioning variance components, e.g. [29]). Generally, quantitative trait loci (QTL) are segments of the genome containing sequence variants that affect a particular phenotype [40], and QTL mapping relies principally on measuring correlations between genetic markers and phenotypic traits in a population. Individuals are scored for their phenotype for a particular trait, and their genotype at a marker. If there is a difference in mean phenotype between those individuals with one genotype at a particular locus compared with the other, than we can infer that there is a QTL linked to that marker [40, 153].

Prior to mapping we need to ensure that covariates are appropriately controlled for because animals may differ in a focal trait not because of differences in genotype but because of differences in covariates such as body weight, sex, litter size, epoch, batch etc. By using recombinant inbred lines, the same trait can be mapped repeatedly in the same genotype and we obtain line averages for a particular trait. It is straightforward to obtain residuals from univariate linear models where the covariates have been controlled for, systematically removing nonsignificant covariates from the model, and then map the residuals (e.g. [46]).

As illustrated in Fig. 2, the phenotype of genetically uniform individuals can be mapped as a function of genetically variable social partners, whereas the phenotype of the genetically variable individual can be mapped as a function of its own genotype. This means that any standard QTL mapping tools can be used, for example the r/QTL package [154], or GeneNetwork [155], as described elsewhere in this book. Once loci are identified, potential candidate genes may be suggested using gene annotations (e.g. Gene Ontology annotations [156]) and correlation with other experimental data contained within GeneNetwork (e.g. [157–160].

2.4 Results: IGE Mapping Using Recombinant Inbred Populations

To illustrate the concept of IGE mapping, here we map traits in genetically uniform B6 offspring as a function of genetically variable BXD mothers in a cross-fostering design (Fig. 2a). Specifically, we map body weight of B6 litters cross-fostered to BXD mothers on postnatal day 14.

1. Using 40 BXD genotypes, we detected a significant IGE QTL locus on the X chromosome, from 46.067–55.908 Mbp (the locations of the first markers either side of the peak with a LOD score >1.5 lower than the peak LOD), where mothers carrying the B6 allele increase the weight of the foster-litter by ~2.5 g on day 14 (Fig. 3). The mapped trait has been calculated from the residuals of a GLM to adjust for pup number, initial pup weight, and the foster-mother's weight, all of which have a significant effect on the raw phenotype. The confidence interval contains 67 genes, 25 of which have nsSNPs or insertions/deletions (indels). This shows that our method is able not only to find MGEs, but to identify QTL for them which contain a relatively small number of candidate genes.
2. Conversely, we can look at the effects of genetic variation in offspring on maternal traits. Here, genetically uniform B6 mothers were nursing BXD litters (Fig. 2b), and maternal body weight was measured on post-natal day 10. Using 37 BXD genotypes, we mapped B6 maternal traits and obtained the residuals of a GLM, with the initial maternal weight, initial number and weight of BXD pups. , number and weight of BXD pups on day 10 and the proportion of males in the litter

added as covariates, removing, in a stepwise process the least significant predictors to find the minimal model with BXD litter weight on day 10 and B6 female body weight. As above, this maps a suggestive QTL ($p=0.128$) on chromosome 9, from 47.862 to 51.758 Mbp (>1.5 LOD drop threshold; Fig. 4), where the B6 allele increases maternal weight by 0.44 g. Again, this shows that genetic variation, and indeed specific IGE loci, in the social environment can modify traits in a focal individual. In this example, we need to remember that there is no genetic variation among B6 females and, once adjusted for nongenetic sources of variation such as initial weight and BXD litter weight, any variation is predicted to be due to variation in offspring genotype. One possible explanation is that BXD genotypes differ in the level of solicitation of maternal resources, which may affect maternal bodyweight.

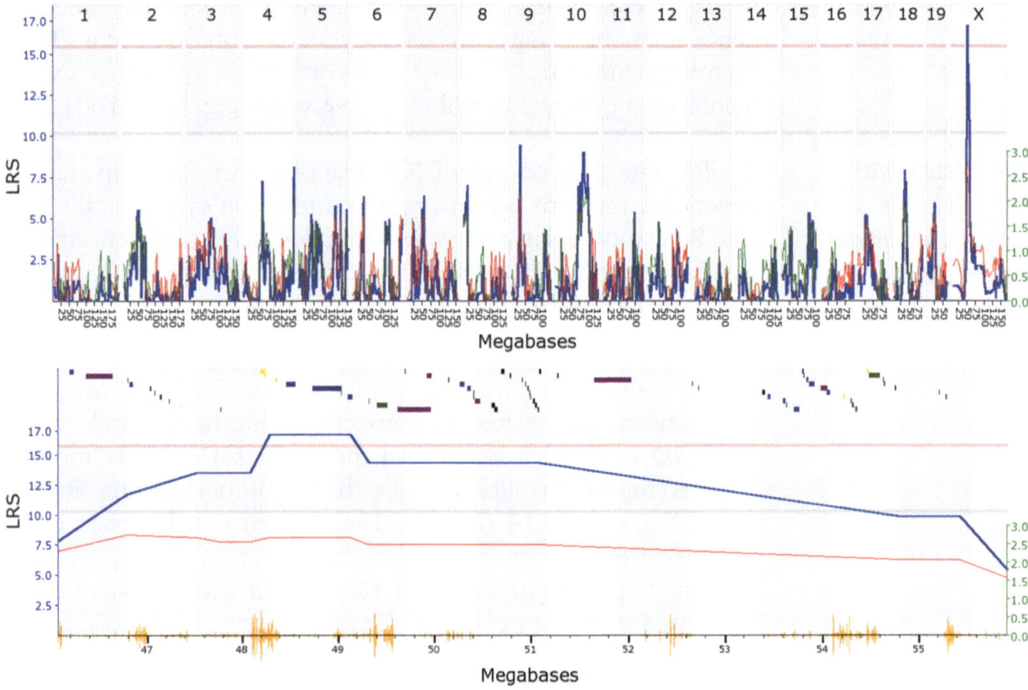

Fig. 3 Maternal indirect genetic effect modifying offspring weight on day 14. Genome scan of B6 litter weight on post-natal day 14 as a function of the BXD genotype of their adoptive mothers. The *upper panel* shows the whole-genome scan, the *lower panel* the enlarged QTL location. The *blue line* represents the genome scan, showing the likelihood ratio statistic (LRS, where LOD scores are LRS/4.61) associated with each marker across the 19 autosomal and the X chromosome. The *top, pink, line* marks genome-wide significance ($p=0.05$), the *lower, gray, line* marks the suggestive significance threshold ($p=0.63$). The *green* or *red line* shows the additive coefficient, with *green* showing that the DBA/2J allele increases trait values, and *red* that the C57BL/6J allele increases trait values. The *green* axis on the *right* shows by how much the respective alleles increase trait values. In this case, we can see that the C57BL/6J allele at the QTL position increases the trait value by ~2.5 g. On the *lower panel*, colored blocks at the *top* of the panel show the location of genes. QTL were mapped using interval mapping as implemented in GeneNetwork.org

3. One important caveat, which applies equally to mapping direct genetic effects, is to ensure that the trait to be mapped is corrected for variation in appropriate covariates. In principle, the aim is to control for variation in the mapping trait that is not due to differences in the genotype as otherwise spurious association and loci may appear. To illustrate this point, in the above same experiment as above, we also analyzed day 14 body weight of B6 mothers nursing BXD litters. The mapped trait was calculated from the residuals of a GLM, with the initial number and weight of BXD pups, the number and weight of BXD pups on day 14 and the proportion of males in the litter added as covariates, removing, in a stepwise process again, the least

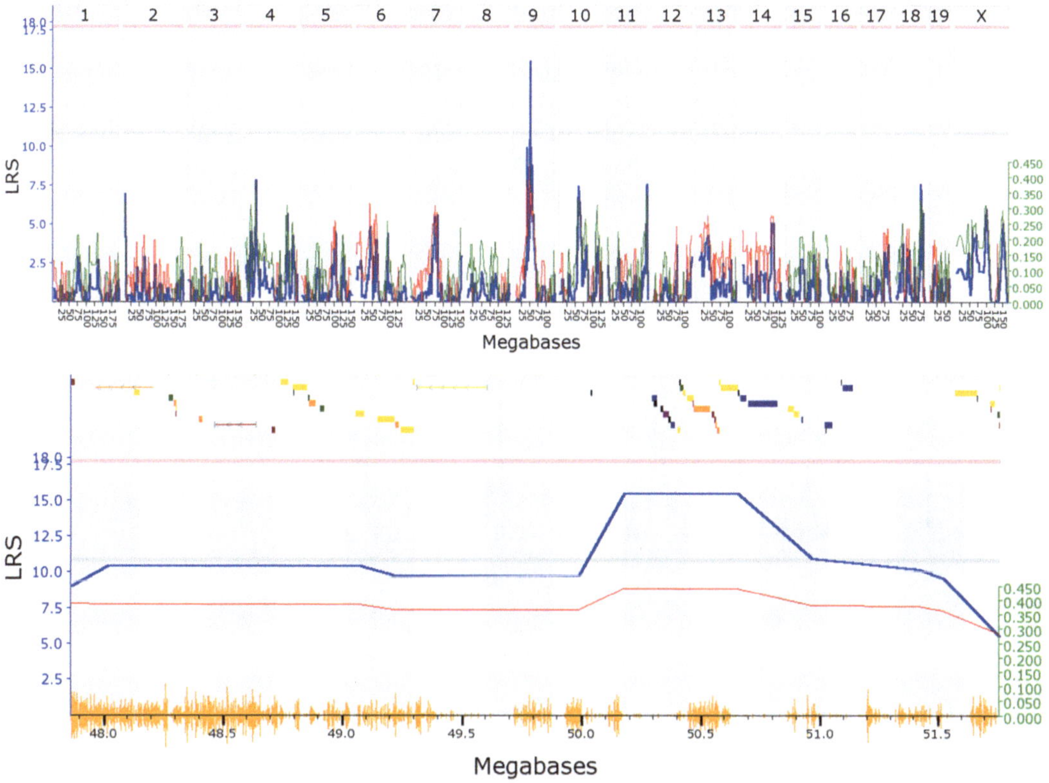

Fig. 4 Offspring indirect genetic effect modifying maternal weight on day 10. Genome scan of B6 maternal weight on post-natal day 10 as a function of the BXD genotype of their adoptive litters. The *upper panel* shows the whole-genome scan, the *lower panel* the enlarged QTL location. The *blue line* represents the genome scan, showing the likelihood ratio statistic (LRS) associated with each marker across the 19 autosomal and the X chromosome. The *top, pink, line* marks genome-wide significance ($p = 0.05$), the *lower, gray, line* marks the suggestive significance threshold ($p = 0.63$). The *green* or *red line* shows the additive coefficient, with *green* showing that the DBA/2J allele increases trait values and *red* that the C57BL/6J allele increases trait values. The *green* axis on the *right* shows by how much the respective alleles increase trait values. In this case, we can see that the C57BL/6J allele at the QTL position increases the trait value by ~0.44 g. On the *lower panel*, colored blocks at the top of the panel show the location of genes. QTL were mapped using interval mapping as implemented in GeneNetwork.org

significant predictors to find the minimal model with only BXD litter weight as a significant predictor. This produced a significant IGE QTL on chromosome 6, from 95.446 to 113.996 Mbp (>1.5 LOD drop threshold; Fig. 5, upper panel). However, when the same model was run, but now including B6 mothers' initial weight as a significant additional predictor, the QTL disappeared, even at the suggestive level, when mapping the residuals (Fig. 5, lower panel). This is an obvious example as it would be difficult to explain how the B6 allele increased maternal weight by almost 10 g when nursing a BXD litter.

2.5 Mechanisms of IGEs and Further Considerations

Using a cross-fostering design with recombinant inbred and inbred strains is one way of mapping IGEs and has the advantage that we are able to control and define genetic variation effects, but other methods may also be suitable, e.g. F2 mapping populations [33, 123]. Once an IGE has been detected the interesting question is how does genetic variation in the social environment affect a focal trait? To answer this question, relevant behavioral and physiological traits have to be measured to enable investigating any

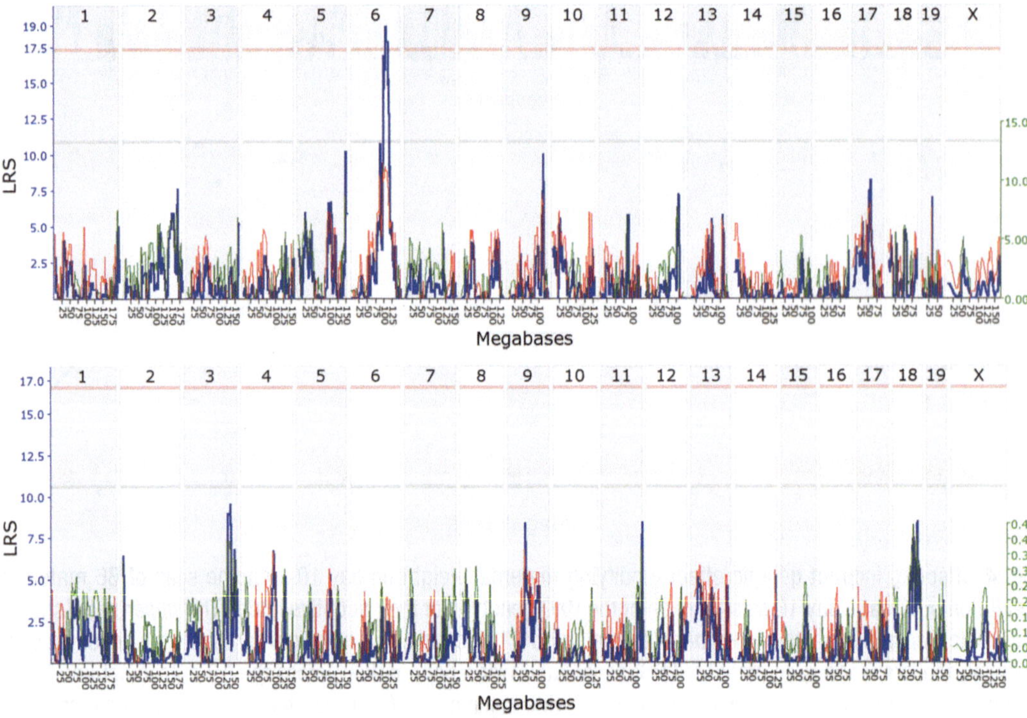

Fig. 5 Offspring indirect genetic effect modifying foster-mother weight on day 14. Genome scan of residual B6 foster-mother weight on post-natal day 14 as a function of BXD offspring genotype. The *upper panel* shows the whole-genome QTL scan, when maternal initial weight was not considered in the GLM, while the *lower panel* shows the QTL scan when maternal weight was included. The *blue line* represents the genome scan, showing the likelihood ratio statistic (LRS) associated with each marker across the 19 autosomal and the X chromosome

correlation between traits of animals in the social environment and the focal individual based on predictions about the nature of these interactions. . For example, in parent offspring interactions, offspring solicitation is predicted to influence maternal provisioning or proxy measures but not necessarily maternal cognitive traits. Ideally, one would want to demonstrate a direct genetic effect on a given trait that significantly correlates with a trait in the focal individual, and the genotype in the social environment shows an IGE on that same trait. This may well be extended to expression profiles as it is often difficult to determine the exact effect of a set of candidates on higher-level complex traits, such as anxiety affected by differences in maternal care [96, 97].

Acknowledgements

We would like to thank Beatrice Gini and Sophie Lyst for help with data collection and colony management. This research is supported by NERC grants NE/I001395/1 and NE/F013418/1.

References

1. Arndt SS, Laarakker MC, van Lith HA et al (2009) Individual housing of mice—impact on behaviour and stress responses. Physiol Behav 97:385–393. doi:10.1016/j.physbeh.2009.03.008
2. Mashoodh R, Franks B, Curley JP, Champagne FA (2012) Paternal social enrichment effects on maternal behavior and offspring growth. Proc Natl Acad Sci U S A 109(Suppl):17232–17238. doi:10.1073/pnas.1121083109
3. Hager R, Johnstone RA (2006) The influence of phenotypic and genetic effects on maternal provisioning and offspring weight gain in mice. Biol Lett 2:81–84. doi:10.1098/rsbl.2005.0403
4. Frank SA (2007) All of life is social. Curr Biol 17:R648–R650. doi:10.1016/j.cub.2007.06.005
5. Whitney G, Coble JR, Stockton MD, Tilson EF (1973) Ultrasonic emissions: do they facilitate courtship of mice? J Comp Physiol Psychol 84:445–452. doi:10.1037/h0034899
6. Doty RL (1974) A cry for the liberation of the female rodent: courtship and copulation in rodentia. Psychol Bull 81:159–172. doi:10.1037/h0035971
7. Neunuebel JP, Taylor AL, Arthur BJ, Egnor SR (2015) Female mice ultrasonically interact with males during courtship displays. Elife doi:10.7554/eLife.06203
8. Cox KH, Rissman EF (2011) Sex differences in juvenile mouse social behavior are influenced by sex chromosomes and social context. Genes Brain Behav 10:465–472. doi:10.1111/j.1601-183X.2011.00688.x
9. Lukas M, Wöhr M (2015) Endogenous vasopressin, innate anxiety, and the emission of pro-social 50-kHz ultrasonic vocalizations during social play behavior in juvenile rats. Psychoneuroendocrinology 56:35–44. doi:10.1016/j.psyneuen.2015.03.005
10. Anholt RRH, Mackay TFC (2012) Genetics of aggression. Annu Rev Genet 46:145–164. doi:10.1146/annurev-genet-110711-155514
11. Wilson AJ, Gelin U, Perron M-C, Reale D (2009) Indirect genetic effects and the evolution of aggression in a vertebrate system. Proc R Soc B Biol Sci 276:533–541. doi:10.1098/rspb.2008.1193
12. Smiseth PT, Kölliker M, Royle NJ (2012) What is parental care? In: Royle NJ, Smiseth PT, Kölliker M (eds) Evolution of parent care, 1st edn. Oxford University Press, Oxford, pp 1–18
13. Hunt J, Simmons LW (2002) The genetics of maternal care: direct and indirect genetic effects on phenotype in the dung beetle Onthophagus taurus. Proc Natl Acad Sci U S A 99:6828–6832. doi:10.1073/pnas.092676199
14. Kölliker M, Richner H (2001) Parent–offspring conflict and the genetics of offspring

solicitation and parental response. Anim Behav 62:395–407. doi:10.1006/anbe.2001.1792
15. Kölliker M, Brinkhof MWG, Heeb P et al (2000) The quantitative genetic basis of offspring solicitation and parental response in a passerine bird with biparental care. Proc R Soc B Biol Sci 267:2127–2132. doi:10.1098/rspb.2000.1259
16. Moore AJ, Brodie ED III, Wolf JB (1997) Interacting phenotypes and the evolutionary process: I. Direct and indirect genetic effects of social interactions. Evolution 51:1352–1362. doi:10.2307/2411187
17. Moore AJ, Haynes KF, Preziosi RF, Moore PJ (2002) The evolution of interacting phenotypes: genetics and evolution of social dominance. Am Nat 160(Suppl):S186–S197. doi:10.1086/342899
18. Wolf JB, Brodie ED III, Moore AJ (1999) Interacting phenotypes and the evolutionary process. II. Selection resulting from social interactions. Am Nat 153:254–266. doi:10.1086/303168
19. Trivers RL (1974) Parent-offspring conflict. Am Zool 14:249–264. doi:10.1093/icb/14.1.249
20. Wolf JB, Brodie ED III, Cheverud JM et al (1998) Evolutionary consequences of indirect genetic effects. Trends Ecol Evol 13:64–69. doi:10.1016/S0169-5347(97)01233-0
21. Cheverud JM (2003) Evolution in a genetically heritable social environment. Proc Natl Acad Sci 100:4357–4359. doi:10.1073/pnas.0931311100
22. Kölliker M, Brodie ED III, Moore AJ (2005) The coadaptation of parental supply and offspring demand. Am Nat 166:506–516. doi:10.1086/491687
23. West-Eberhard MJ (1983) Sexual selection, social competition, and speciation. Q Rev Biol 58:155–183. doi:10.2307/2828804
24. Trubenová B, Hager R (2014) Social selection and indirect genetic effects in structured populations. Evol Biol 41:123–133. doi:10.1007/s11692-013-9252-5
25. Agrawal AF, Brodie ED III, Wade MJ (2001) On indirect genetic effects in structured populations. Am Nat 158:308–323. doi:10.1086/321324
26. McGlothlin JW, Brodie ED III (2009) How to measure indirect genetic effects: the congruence of trait-based and variance-partitioning approaches. Evolution 63:1785–1795. doi:10.1111/j.1558-5646.2009.00676.x
27. Bijma P (2010) Multilevel selection 4: modeling the relationship of indirect genetic effects and group size. Genetics 186:1029–1031. doi:10.1534/genetics.110.120485
28. Teplitsky C, Mills JA, Yarrall JW, Merilä J (2010) Indirect genetic effects in a sex-limited trait: the case of breeding time in red-billed gulls. J Evol Biol 23:935–944. doi:10.1111/j.1420-9101.2010.01959.x
29. Wolf JB (2003) Genetic architecture and evolutionary constraint when the environment contains genes. Proc Natl Acad Sci U S A 100:4655–4660. doi:10.1073/pnas.0635741100
30. Mutic JJ, Wolf JB (2007) Indirect genetic effects from ecological interactions in *Arabidopsis thaliana*. Mol Ecol 16:2371–2381. doi:10.1111/j.1365-294X.2007.03259.x
31. Kirkpatrick M, Lande R (1989) The evolution of maternal characters. Evolution 43:485. doi:10.2307/2409054
32. Wolf JB, Wade MJ (2009) What are maternal effects (and what are they not)? Philos Trans R Soc B Biol Sci 364:1107–1115. doi:10.1098/rstb.2008.0238
33. Wolf J, Cheverud JM (2012) Detecting maternal-effect loci by statistical cross-fostering. Genetics 191:261–277. doi:10.1534/genetics.111.136440
34. Dickerson GE (1947) Composition of hog carcasses as influenced by heritable differences in rate and economy of gain. Res Bull Iowa Agric Exp Stn 354:489–524
35. Willham RL (1963) The covariance between relatives for characters composed of components contributed by related individuals. Biometrics 19:18–27. doi:10.2307/2527570
36. Willham RL (1972) The role of maternal effects in animal breeding. 3. Biometrical aspects of maternal effects in animals. J Anim Sci 35:1288–1293
37. Hanrahan JP, Eisen EJ (1973) Sexual dimorphism and direct and maternal genetic effects on body weight in mice. Theor Appl Genet 43:39–45. doi:10.1007/BF00277832
38. Hanrahan JP (1976) Maternal effects and selection response with an application to sheep data. Anim Prod 22:359–369. doi:10.1017/S0003356100035637
39. Ellen ED, Rodenburg TB, Albers GAA et al (2014) The prospects of selection for social genetic effects to improve welfare and productivity in livestock. Front Genet 5:377. doi:10.3389/fgene.2014.00377
40. Falconer DS, Mackay TFC (1996) Introduction to quantitative genetics, 4th edn. Longman, Burnt Mill
41. Harris WE, Hager R (2009) On the evolution of reproductive skew: a genetical view. In: Hager R, Jones CB (eds) Reproductive skew in vertebrates. Cambridge University Press, Cambridge, pp 467–479

42. Cheverud JM, Moore AJ (1994) Quantitative genetics and the role of the environment provided by relatives in behavioral evolution. In: Boake CRB (ed) Quantitative genetic studies of behavioral evolution. University of Chicago Press, Chicago, IL, pp 67–100
43. Wolf JB, Brodie ED III (1998) The coadaptation of parental and offspring characters. Evolution 52:299–308. doi:10.2307/2411068
44. Bijma P, Wade MJ (2008) The joint effects of kin, multilevel selection and indirect genetic effects on response to genetic selection. J Evol Biol 21:1175–1188. doi:10.1111/j.1420-9101.2008.01550.x
45. McGlothlin JW, Moore AJ, Wolf JB, Brodie ED III (2010) Interacting phenotypes and the evolutionary process. III. Social evolution. Evolution 64:2558–2574. doi:10.1111/j.1558-5646.2010.01012.x
46. Hager R, Lu L, Rosen GD, Williams RW (2012) Genetic architecture supports mosaic brain evolution and independent brain-body size regulation. Nat Commun 3:1079. doi:10.1038/ncomms2086
47. Trubenová B, Hager R (2012) Phenotypic and evolutionary consequences of social behaviours: interactions among individuals affect direct genetic effects. PLoS One 7:e46273. doi:10.1371/journal.pone.0046273
48. Cheverud JM (1984) Evolution by kin selection: a quantitative genetic model illustrated by maternal performance in mice. Evolution (NY) 38:766. doi:10.2307/2408388
49. Wolf JB, Vaughn TT, Pletscher LS, Cheverud JM (2002) Contribution of maternal effect QTL to genetic architecture of early growth in mice. Heredity (Edinb) 89:300–310. doi:10.1038/sj.hdy.6800140
50. Petfield D, Chenoweth SF, Rundle HD, Blows MW (2005) Genetic variance in female condition predicts indirect genetic variance in male sexual display traits. Proc Natl Acad Sci U S A 102:6045–6050. doi:10.1073/pnas.0409378102
51. Camerlink I, Turner SP, Bijma P, Bolhuis JE (2013) Indirect genetic effects and housing conditions in relation to aggressive behaviour in pigs. PLoS One 8:e65136. doi:10.1371/journal.pone.0065136
52. Bailey NW, Hoskins JL (2014) Detecting cryptic indirect genetic effects. Evolution 68:1871–1882. doi:10.1111/evo.12401
53. Donohue K (2003) Setting the stage: phenotypic plasticity as habitat selection. Int J Plant Sci 164:S79–S92. doi:10.1086/368397
54. Haley CS, Knott SA (1992) A simple regression method for mapping quantitative trait loci in line crosses using flanking markers. Heredity (Edinb) 69:315–324. doi:10.1038/hdy.1992.131
55. McAdam AG, Boutin S, Réale D, Berteaux D (2002) Maternal effects and the potential for evolution in a natural population of animals. Evolution 56:846–851
56. Neser FWC, Erasmus GJ, van Wyk JB (2001) Genetic parameter estimates for pre-weaning weight traits in Dorper sheep. Small Rumin Res 40:197–202. doi:10.1016/S0921-4488(01)00172-9
57. Aggrey SE, Cheng KM (1993) Genetic and posthatch parental influences on growth in pigeon squabs. J Hered 84:184–187
58. Smith HG, Wettermark KJ (1995) Heritability of nestling growth in cross-fostered European Starlings Sturnus vulgaris. Genetics 141:657–665
59. Rauter CM, Moore AJ (2002) Evolutionary importance of parental care performance, food resources, and direct and indirect genetic effects in a burying beetle. J Evol Biol 15:407–417.doi:10.1046/j.1420-9101.2002.00412.x
60. Rauter CM, Moore AJ (2002) Quantitative genetics of growth and development time in the burying beetle Nicrophorus pustulatus in the presence and absence of post-hatching parental care. Evolution 56:96–110. doi:10.1554/0014-3820(2002)056[0096:QGOGAD]2.0.CO;2
61. Head ML, Berry LK, Royle NJ, Moore AJ (2012) Paternal care: direct and indirect genetic effects of fathers on offspring performance. Evolution 66:3570–3581. doi:10.1111/j.1558-5646.2012.01699.x
62. Cheverud JM, Leamy LJ, Atchley WR, Rutledge JJ (1983) Quantitative genetics and the evolution of ontogeny: I. Ontogenetic changes in quantitative genetic variance components in randombred mice. Genet Res 42:65. doi:10.1017/S0016672300021492
63. Riska B, Rutledge JJ, Atchley WR (1985) Covariance between direct and maternal genetic effects in mice, with a model of persistent environmental influences. Genet Res 45:287–297. doi:10.1017/S0016672300022278
64. Nagai J, Bakker H, Eisen EJ (1976) Partitioning average and heterotic components of direct and maternal genetic effects on growth in mice using crossfostering techniques. Genetics 84:113–124
65. Williams WR, Eisen EJ, Nagai J, Bakker H (1978) Direct and maternal genetic effects on body weight maturing patterns in mice. Theor Appl Genet 51:249–260. doi:10.1007/BF00273772
66. Cheverud JM, Leamy LJ (1985) Quantitative genetics and the evolution of ontogeny. III. Ontogenetic changes in correlation structure among live-body traits in randombred mice. Genet Res 46:325–335. doi:10.1017/S0016672300022813

67. Cheverud JM, Routman EJ, Duarte FA et al (1996) Quantitative trait loci for murine growth. Genetics 142:1305–1319
68. Vaughn TT, Pletscher LS, Peripato A et al (1999) Mapping quantitative trait loci for murine growth: a closer look at genetic architecture. Genet Res 74:313–322
69. Cowley DE, Pomp D, Atchley WR et al (1989) The impact of maternal uterine genotype on postnatal growth and adult body size in mice. Genetics 122:193–203
70. Casellas J, Farber CR, Gularte RJ et al (2009) Evidence of maternal QTL affecting growth and obesity in adult mice. Mamm Genome 20:269–280. doi:10.1007/s00335-009-9182-9
71. Wolf JB, Leamy LJ, Roseman CC, Cheverud JM (2011) Disentangling prenatal and postnatal maternal genetic effects reveals persistent prenatal effects on offspring growth in mice. Genetics 189:1069–1082. doi:10.1534/genetics.111.130591
72. Drews C (1993) The concept and definition of dominance in animal behaviour. Behaviour 125:283–313. doi:10.1163/156853993X00290
73. Herberholz J, McCurdy C, Edwards DH (2007) Direct benefits of social dominance in juvenile crayfish. Biol Bull 213:21–27
74. Willisch CS, Neuhaus P (2010) Social dominance and conflict reduction in rutting male Alpine ibex, Capra ibex. Behav Ecol 21:372–380. doi:10.1093/beheco/arp200
75. Franck D, Ribowski A (1993) Dominance hierarchies of male green swordtails (*Xiphophorus helleri*) in nature. J Fish Biol 43:497–499. doi:10.1111/j.1095-8649.1993.tb00586.x
76. Wilson AJ, Morrissey MB, Adams MJ et al (2011) Indirect genetics effects and evolutionary constraint: an analysis of social dominance in red deer, *Cervus elaphus*. J Evol Biol 24:772–783. doi:10.1111/j.1420-9101.2010.02212.x
77. Sartori C, Mantovani R (2013) Indirect genetic effects and the genetic bases of social dominance: evidence from cattle. Heredity (Edinb) 110:3–9. doi:10.1038/hdy.2012.56
78. Moore AJ (2013) Genetic influences on social dominance: cow wars. Heredity (Edinb) 110:1–2. doi:10.1038/hdy.2012.85
79. Bechstein P, Rehbach N-J, Yuhasingham G et al (2014) The clock gene *Period1* regulates innate routine behaviour in mice. Proc Biol Sci 281:20140034. doi:10.1098/rspb.2014.0034
80. Zakany J, Duboule D (2012) A genetic basis for altered sexual behavior in mutant female mice. Curr Biol 22:1676–1680. doi:10.1016/j.cub.2012.06.067
81. Kent C, Azanchi R, Smith B et al (2008) Social context influences chemical communication in D. melanogaster males. Curr Biol 18:1384–1389. doi:10.1016/j.cub.2008.07.088
82. Bleakley BH, Brodie ED III (2009) Indirect genetic effects influence antipredator behavior in guppies: estimates of the coefficient of interaction psi and the inheritance of reciprocity. Evolution 63:1796–1806. doi:10.1111/j.1558-5646.2009.00672.x
83. Miller CW, Moore AJ (2007) A potential resolution to the lek paradox through indirect genetic effects. Proc R Soc B Biol Sci 274:1279–1286. doi:10.1098/rspb.2006.0413
84. Bijma P, Muir WM, Ellen ED et al (2007) Multilevel selection 2: estimating the genetic parameters determining inheritance and response to selection. Genetics 175:289–299. doi:10.1534/genetics.106.062729
85. Carlier M, Roubertoux P, Cohen-Salmon C (1982) Differences in patterns of pup care in *Mus musculus domesticus* I—comparisons between eleven inbred strains. Behav Neural Biol 35:205–210. doi:10.1016/S0163-1047(82)91213-4
86. Klug H, Bonsall MB (2014) What are the benefits of parental care? The importance of parental effects on developmental rate. Ecol Evol 4:2330–2351. doi:10.1002/ece3.1083
87. McIver AH, Jeffrey WE (1967) Strain differences in maternal behavior in rats. Behaviour 28:210–216. doi:10.1163/156853967X00244
88. Myers MM, Brunelli SA, Shair HN et al (1989) Relationships between maternal behavior of SHR and WKY dams and adult blood pressures of cross-fostered F1 pups. Dev Psychobiol 22:55–67. doi:10.1002/dev.420220105
89. Champagne FA, Curley JP, Keverne EB, Bateson PPG (2007) Natural variations in postpartum maternal care in inbred and outbred mice. Physiol Behav 91:325–334. doi:10.1016/j.physbeh.2007.03.014
90. Chourbaji S, Hoyer C, Richter SH et al (2011) Differences in mouse maternal care behavior – is there a genetic impact of the glucocorticoid receptor? PLoS One 6:e19218. doi:10.1371/journal.pone.0019218
91. Peripato AC, Cheverud JM (2002) Genetic influences on maternal care. Am Nat 160(Suppl):S173–S185. doi:10.1086/342900
92. Peripato AC, De Brito RA, Vaughn TT et al (2002) Quantitative trait loci for maternal performance for offspring survival in mice. Genetics 162:1341–1353
93. Anisman H, Zaharia MD, Meaney MJ, Merali Z (1998) Do early-life events permanently alter behavioral and hormonal responses to stressors? Int J Dev Neurosci 16:149–164. doi:10.1016/S0736-5748(98)00025-2
94. Akers KG, Nakazawa M, Romeo RD et al (2006) Early life modulators and predictors of

adult synaptic plasticity. Eur J Neurosci 24:547–554. doi:10.1111/j.1460-9568.2006.04921.x
95. Branchi I, Cirulli F (2014) Early experiences: building up the tools to face the challenges of adult life. Dev Psychobiol 56:1661–1674. doi:10.1002/dev.21235
96. Lerch S, Brandwein C, Dormann C et al (2014) What makes a good mother? Implication of inter-, and intrastrain strain "cross fostering" for emotional changes in mouse offspring. Behav Brain Res 274:270–281. doi:10.1016/j.bbr.2014.08.021
97. Priebe K, Brake WG, Romeo RD et al (2005) Maternal influences on adult stress and anxiety-like behavior in C57BL/6J and BALB/cJ mice: a cross-fostering study. Dev Psychobiol 47:398–407. doi:10.1002/dev.20098
98. Holmes A, le Guisquet AM, Vogel E et al (2005) Early life genetic, epigenetic and environmental factors shaping emotionality in rodents. Neurosci Biobehav Rev 29:1335–1346. doi:10.1016/j.neubiorev.2005.04.012
99. Francis DD, Champagne FA, Liu D, Meaney MJ (1999) Maternal care, gene expression, and the development of individual differences in stress reactivity. Ann N Y Acad Sci 896:66–84. doi:10.1111/j.1749-6632.1999.tb08106.x
100. Liu D, Diorio J, Tannenbaum B et al (1997) Maternal care, hippocampal glucocorticoid receptors, and hypothalamic-pituitary-adrenal responses to stress. Science 277:1659–1662. doi:10.1126/science.277.5332.1659
101. Caldji C, Diorio J, Anisman H, Meaney MJ (2004) Maternal behavior regulates benzodiazepine/GABAA receptor subunit expression in brain regions associated with fear in BALB/c andC57BL/6mice. Neuropsychopharmacology 29:1344–1352. doi:10.1038/sj.npp.1300436
102. Francis DD, Champagne FC, Meaney MJ (2000) Variations in maternal behaviour are associated with differences in oxytocin receptor levels in the rat. J Neuroendocrinol 12:1145–1148. doi:10.1046/j.1365-2826.2000.00599.x
103. Boccia ML, Pedersen CA (2001) Brief vs. long maternal separations in infancy: contrasting relationships with adult maternal behavior and lactation levels of aggression and anxiety. Psychoneuroendocrinology 26:657–672. doi:10.1016/S0306-4530(01)00019-1
104. Champagne FA, Weaver ICG, Diorio J et al (2003) Natural variations in maternal care are associated with estrogen receptor alpha expression and estrogen sensitivity in the medial preoptic area. Endocrinology 144:4720–4724. doi:10.1210/en.2003-0564
105. Champagne FA, Francis DD, Mar A, Meaney MJ (2003) Variations in maternal care in the rat as a mediating influence for the effects of environment on development. Physiol Behav 79:359–371. doi:10.1016/S0031-9384(03)00149-5
106. Weaver ICG, Diorio J, Seckl JR et al (2004) Early environmental regulation of hippocampal glucocorticoid receptor gene expression: characterization of intracellular mediators and potential genomic target sites. Ann N Y Acad Sci 1024:182–212. doi:10.1196/annals.1321.099
107. Weaver ICG, Cervoni N, Champagne FA et al (2004) Epigenetic programming by maternal behavior. Nat Neurosci 7:847–854. doi:10.1038/nn1276
108. Veenema AH, Blume A, Niederle D et al (2006) Effects of early life stress on adult male aggression and hypothalamic vasopressin and serotonin. Eur J Neurosci 24:1711–1720. doi:10.1111/j.1460-9568.2006.05045.x
109. Veenema AH, Bredewold R, Neumann ID (2007) Opposite effects of maternal separation on intermale and maternal aggression in C57BL/6 mice: link to hypothalamic vasopressin and oxytocin immunoreactivity. Psychoneuroendocrinology 32:437–450. doi:10.1016/j.psyneuen.2007.02.008
110. Murgatroyd C, Patchev AV, Wu Y et al (2010) Dynamic DNA methylation programs persistent adverse effects of early-life stress. Nat Neurosci 13:649. doi:10.1038/nn0510-649e
111. Curley JP, Rock V, Moynihan AM et al (2010) Developmental shifts in the behavioral phenotypes of inbred mice: the role of postnatal and juvenile social experiences. Behav Genet 40:220–232. doi:10.1007/s10519-010-9334-4
112. Gudsnuk KMA, Champagne FA (2011) Epigenetic effects of early developmental experiences. Clin Perinatol 38:703–717. doi:10.1016/j.clp.2011.08.005
113. Curley JP, Jensen CL, Mashoodh R, Champagne FA (2011) Social influences on neurobiology and behavior: epigenetic effects during development. Psychoneuroendocrinology 36:352–371. doi:10.1016/j.psyneuen.2010.06.005
114. Veenema AH (2012) Toward understanding how early-life social experiences alter oxytocin- and vasopressin-regulated social behaviors. Horm Behav 61:304–312. doi:10.1016/j.yhbeh.2011.12.002
115. Franks B, Champagne FA, Curley JP (2015) Postnatal maternal care predicts divergent weaning strategies and the development of social behavior. Dev Psychobiol 57(7):809–817. doi:10.1002/dev.21326
116. Wells JCK (2007) The thrifty phenotype as an adaptive maternal effect. Biol Rev 82:143–172. doi:10.1111/j.1469-185X.2006.00007.x
117. Lunde A, Melve KK, Gjessing HK et al (2007) Genetic and environmental influences on birth weight, birth length, head circumfer-

ence, and gestational age by use of population-based parent-offspring data. Am J Epidemiol 165:734–741. doi:10.1093/aje/kwk107
118. Barker DJP, Eriksson JG, Forsén T, Osmond C (2002) Fetal origins of adult disease: strength of effects and biological basis. Int J Epidemiol 31:1235–1239. doi:10.1093/ije/31.6.1235
119. Ressler RH (1962) Parental handling in two strains of mice reared by foster parents. Science 137:129–130. doi:10.1126/science.137.3524.129
120. Bester-Meredith JK, Marler CA (2001) Vasopressin and aggression in cross-fostered California Mice (*Peromyscus californicus*) and White-Footed Mice (*Peromyscus leucopus*). Horm Behav 40:51–64. doi:10.1006/hbeh.2001.1666
121. Champagne FA, Weaver ICG, Diorio J et al (2006) Maternal care associated with methylation of the estrogen receptor-α1b promoter and estrogen receptor-α expression in the medial preoptic area of female offspring. Endocrinology 147:2909–2915. doi:10.1210/en.2005-1119
122. Hager R, Cheverud JM, Wolf JB (2009) Change in maternal environment induced by cross-fostering alters genetic and epigenetic effects on complex traits in mice. Proc Biol Sci 276:2949–2954. doi:10.1098/rspb.2009.0515
123. Cox KH, So NLT, Rissman EF (2013) Foster dams rear fighters: strain-specific effects of within-strain fostering on aggressive behavior in male mice. PLoS One 8:e75037. doi:10.1371/journal.pone.0075037
124. Peña CJ, Neugut YD, Champagne FA (2013) Developmental timing of the effects of maternal care on gene expression and epigenetic regulation of hormone receptor levels in female rats. Endocrinology 154:4340–4351. doi:10.1210/en.2013-1595
125. Gouldsborough I, Black V, Johnson IT, Ashton N (1998) Maternal nursing behaviour and the delivery of milk to the neonatal spontaneously hypertensive rat. Acta Physiol Scand 162:107–114. doi:10.1046/j.1365-201X.1998.0273f.x
126. Caldji C, Diorio J, Meaney MJ (2000) Variations in maternal care in infancy regulate the development of stress reactivity. Biol Psychiatry 48:1164–1174. doi:10.1016/S0006-3223(00)01084-2
127. Cameron NM, Fish EW, Meaney MJ (2008) Maternal influences on the sexual behavior and reproductive success of the female rat. Horm Behav 54:178–184. doi:10.1016/j.yhbeh.2008.02.013
128. Lande R, Kirkpatrick M (1990) Selection response in traits with maternal inheritance. Genet Res 55:189–197
129. Wilson AJ, Coltman DW, Pemberton JM et al (2004) Maternal genetic effects set the potential for evolution in a free-living vertebrate population. J Evol Biol 18:405–414. doi:10.1111/j.1420-9101.2004.00824.x
130. Nephew B, Murgatroyd C (2013) The role of maternal care in shaping CNS function. Neuropeptides 47:371–378. doi:10.1016/j.npep.2013.10.013
131. Kruuk LEB, Hadfield JD (2007) How to separate genetic and environmental causes of similarity between relatives. J Evol Biol 20:1890–1903. doi:10.1111/j.1420-9101.2007.01377.x
132. Lock JE, Smiseth PT, Moore AJ (2004) Selection, inheritance, and the evolution of parent-offspring interactions. Am Nat 164:13–24. doi:10.1086/421444
133. Aldhous P (1989) The effects of individual cross-fostering on the development of intrasexual kin discrimination in male laboratory mice, *Mus musculus.* Anim Behav 37:741–750. doi:10.1016/0003-3472(89)90060-2
134. Penn D, Potts W (1998) MHC-disassortative mating preferences reversed by cross-fostering. Proc Biol Sci 265:1299–1306. doi:10.1098/rspb.1998.0433
135. Matthews PA, Samuelsson A-M, Seed P et al (2011) Fostering in mice induces cardiovascular and metabolic dysfunction in adulthood. J Physiol 589:3969–3981. doi:10.1113/jphysiol.2011.212324
136. Wattez J-S, Delahaye F, Barella LF et al (2014) Short- and long-term effects of maternal perinatal undernutrition are lowered by cross-fostering during lactation in the male rat. J Dev Orig Health Dis 5:109–120. doi:10.1017/S2040174413000548
137. van Vugt RWM, Meyer F, van Hulten JA et al (2014) Maternal care affects the phenotype of a rat model for schizophrenia. Front Behav Neurosci 8:268. doi:10.3389/fnbeh.2014.00268
138. Hager R, Johnstone RA (2003) The genetic basis of family conflict resolution in mice. Nature 421:533–535. doi:10.1038/nature01239
139. Hager R, Johnstone RA (2005) Differential growth of own and alien young in mixed litters of mice: a role for genomic imprinting? Ethology 111:705–714. doi:10.1111/j.1439-0310.2005.01097.x
140. Macnair MR, Parker GA (1979) Models of parent-offspring conflict. III. Intra-brood conflict. Anim Behav 27:1202–1209. doi:10.1016/0003-3472(79)90067-8
141. Kilner RM, Hinde CA (2012) Parent-offspring conflict. In: Smiseth PT, Kölliker M, Royle N (eds) Evolution of parent care. Oxford University Press, Oxford, pp 119–132
142. Svare B, Kinsley CH, Mann MA, Broida J (1984) Infanticide: accounting for genetic variation in mice. Physiol Behav 33:137–152

143. Perrigo G, Belvin L, Quindry P et al (1993) Genetic mediation of infanticide and parental behavior in male and female domestic and wild stock house mice. Behav Genet 23:525–531
144. Sayler A, Salmon M (1969) Communal nursing in mice: influence of multiple mothers on the growth of the young. Science 164:1309–1310. doi:10.1126/science.164.3885.1309
145. König B (1997) Cooperative care of young in mammals. Naturwissenschaften 84:95–104. doi:10.1007/s001140050356
146. Weidt A, Lindholm AK, König B (2014) Communal nursing in wild house mice is not a by-product of group living: females choose. Naturwissenschaften 101:73–76. doi:10.1007/s00114-013-1130-6
147. Heiderstadt KM, Vandenbergh DJ, Gyekis JP, Blizard DA (2014) Communal nesting increases pup growth but has limited effects on adult behavior and neurophysiology in inbred mice. J Am Assoc Lab Anim Sci 53:152–160
148. Calamandrei G, Wilkinson LS, Keverne EB (1992) Olfactory recognition of infants in laboratory mice: role of noradrenergic mechanisms. Physiol Behav 52:901–907
149. Malenfant SA, Barry M, Fleming AS (1991) Effects of cycloheximide on the retention of olfactory learning and maternal experience effects in postpartum rats. Physiol Behav 49:289–294
150. Mock DW, Parker GA (1997) The evolution of sibling rivalry. Oxford University Press, Oxford
151. Hager R, Johnstone RA (2007) Maternal and offspring effects influence provisioning to mixed litters of own and alien young in mice. Anim Behav 74:1039–1045. doi:10.1016/j.anbehav.2007.01.021
152. Martin P, Bateson PPG (2007) Measuring behaviour: an introductory guide, 3rd edn. Cambridge University Press, Cambridge
153. Miles CM, Wayne M (2008) Quantitative trait locus (QTL) analysis. Nat Educ 1:208
154. Broman KW, Wu H, Sen S, Churchill GA (2003) R/qtl: QTL mapping in experimental crosses. Bioinformatics 19:889–890. doi:10.1093/bioinformatics/btg112
155. Wang J, Williams RW, Manly KF (2003) WebQTL: web-based complex trait analysis. Neuroinformatics 1:299–308. doi:10.1385/NI:1:4:299
156. Gene Ontology Consortium (2015) Gene Ontology Consortium: going forward. Nucleic Acids Res 43:D1049–D1056. doi:10.1093/nar/gku1179
157. Mulligan MK, Dubose C, Yue J et al (2013) Expression, covariation, and genetic regulation of miRNA biogenesis genes in brain supports their role in addiction, psychiatric disorders, and disease. Front Genet 4:126. doi:10.3389/fgene.2013.00126
158. Andreux PA, Williams EG, Koutnikova H et al (2012) Systems genetics of metabolism: the use of the BXD murine reference panel for multiscalar integration of traits. Cell 150:1287–1299. doi:10.1016/j.cell.2012.08.012
159. Alberts R, Schughart K (2010) QTLminer: identifying genes regulating quantitative traits. BMC Bioinformatics 11:516. doi:10.1186/1471-2105-11-516
160. Ashbrook DG, Delprato A, Grellmann C et al (2014) Transcript co-variance with Nestin in two mouse genetic reference populations identifies Lef1 as a novel candidate regulator of neural precursor cell proliferation in the adult hippocampus. Front Neurosci 8:418. doi:10.3389/fnins.2014.00418

Chapter 25

Complex Genetics of Behavior: BXDs in the Automated Home-Cage

Maarten Loos, Matthijs Verhage, Sabine Spijker, and August B. Smit

Abstract

This chapter describes a use case for the genetic dissection and automated analysis of complex behavioral traits using the genetically diverse panel of BXD mouse recombinant inbred strains. Strains of the BXD resource differ widely in terms of gene and protein expression in the brain, as well as in their behavioral repertoire. A large mouse resource opens the possibility for gene finding studies underlying distinct behavioral phenotypes, however, such a resource poses a challenge in behavioral phenotyping. To address the specifics of large-scale screening we describe how to investigate: (1) how to assess mouse behavior systematically in addressing a large genetic cohort, (2) how to dissect automation-derived longitudinal mouse behavior into quantitative parameters, and (3) how to map these quantitative traits to the genome, deriving loci underlying aspects of behavior.

Key words Inbred mouse strains, Spontaneous behavior, Sheltering, Automated home-cage, PhenoTyper, Systematic behavioral profile, Reference database, Genetic effect size, Mouse phenomics, High-throughput

1 Introduction

Recombinant inbred (RI) strains, including the panel of BXD strains, are a valuable resource to reveal the genetic contribution to particular phenotypes, including behavior [1]. Dedicated online tools that analyze the data derived from these resources, such as WebQTL [2], can be used to map genetic loci once phenotyping is completed. Given the complex multigene encoding of behavioral traits, behavioral phenotyping requires relatively large cohorts of RI strains. Traditionally, these have been characterized using batteries of standardized behavioral tests, which when used together, can measure a large part of the behavioral spectrum. This approach has been used successfully for RI strains (e.g., [3, 4]). However, given

Klaus Schughart and Robert W. Williams (eds.), *Systems Genetics: Methods and Protocols*, Methods in Molecular Biology, vol. 1488, DOI 10.1007/978-1-4939-6427-7_25, © Springer Science+Business Media New York 2017

that a behavioral phenotype is the result of a gene–environment interaction, the power in dissecting the genetic contribution easily decreases by confounding environmental factors. Standard test batteries are extremely labor intensive and the individual tests require ample human interference, thereby introducing many environmental cues, which are well known to interact with behavioral phenotypes [5]. Hence, large-scale screening of RI strains for genetic mapping studies would benefit from fully automated approaches devoid of human interference.

Several fully automated systems have been developed that measure a wide variety of behaviors ranging from spontaneous activity to more complex cognitive tests [6–9]. Standard short-lived tests typically take a few minutes and almost instantaneously produce behavioral summary scores that can be used readily for genetic mapping studies. In contrast, home-cage assessments provide longitudinal data and require parsing of the continuous data stream into discrete behavioral parameters amenable for genetic mapping. In this study, we recorded the spontaneous behavior of 43 BXD strains, 366 mice in total, for three consecutive days in a home-cage system by 24/7 overhead video tracking (PhenoTyper). Mice were housed individually to allow for the analysis of their spontaneous behavior. A set of algorithms and data analysis methods is presented that can be used to dissect complex home-cage activity profiles into discrete behavioral parameters.

Both for standard tests as well as automated home-cage tests, obtained behavioral parameters are related in terms of time or location in the apparatus, questioning the independence of these parameters. In this use case, we demonstrate how to examine to what extent these parameters detect unique genetic variation. Hereto we show how to use correlation analyses on the behavioral data to critically assess the complexity of the dataset. By using the online resource WebQTL, the reader can explore the relation between the newly measured behavioral parameters and data previously deposited in WebQTL, acquired in other behavioral tests, each aimed at measuring a different aspect of behavior. Finally, WebQTL can be used to identify loci harboring genetic variation contributing to each of the behavioral parameters assessed in the phenotypic screen.

2 Methods

2.1 Mice

As part of a larger screening project, breeding pairs of BXD lines and their parental lines (C57BL/6J and DBA/2J) were received from The Jackson Laboratory (Bar Harbor, Maine, USA), or from Oak Ridge National Laboratory (Oak Ridge, Tennessee, USA) in case they were not available from The Jackson Laboratory at the time (BXD43, BXD51, BXD61, BXD62, BXD65, BXD68, BXD69, BXD73, BXD75, BXD87, BXD90), and were bred in the facility of the Neuro-Bsik consortium of the VU University Amsterdam (Amsterdam, The Netherlands). After arrival in the screening facility at the age of 7–11 weeks mice were individually housed on sawdust in standard Makrolon type II cages enriched with cardboard nesting material (7:00 lights on, 19:00 lights off). A total of 366 BXD mice from 43 different strains (n=3–20 per strain) were tested. In addition, C57BL/6J (n=105) and DBA2/J (n=39) mice were screened as part of a larger project assessing common inbred mice [10]. Experiments were carried out in accordance with the European Communities Council Directive of 24 November 1986 (86/609/EEC), and with approval of the local animal care and use committee of the VU University.

2.2 Home-Cage Phenotyping

Individual mice were housed in a home cage environment (PhenoTyper model 3000, Noldus Information Technology, Wageningen, The Netherlands) for 7 consecutive days, of which the first 3 days were used to analyze spontaneous behavior (Maroteaux et al. 2012). Mice were introduced in the cage in the second half of the subjective light phase (14:00–16:00 h), and video tracking started at the onset of the first subjective dark phase (19:00 h). The cages (30×30×35 cm; $L \times W \times H$) were made of transparent Perspex walls with an opaque Perspex floor covered with bedding based on cellulose. A feeding station and a water bottle were attached on to two adjacent walls. A triangular shaped shelter compartment (height: 10 cm; nontransparent material) with two entrances was fixed in the corner of the opposite two walls. The top unit of each cage contained an array of infrared LEDs and an infrared-sensitive video camera used for video-tracking. The X-Y coordinates of the center of gravity of mice, sampled at a resolution of 15 coordinates per second were acquired and smoothed using EthoVision software (EthoVision HTP 2.1.2.0, based on EthoVision XT 4.1, Noldus Information Technology, Wageningen, The Netherlands). At occasions in which mice were not detected by the video tracking system, due to processor overload (missing frames) or in particular areas of the cage where image contrast was low (not found frames; particularly when climbing into the feeding station), linear interpolation (interpolated frames) between last observed position and newly detected position was implemented. Resulting digital tracks were processed to generate behavioral parameters using AHCODA™ analysis software (Sylics, Synaptologics BV, Amsterdam, The Netherlands), as described in detail previously [10].

2.3 Data Analysis

Behavioral data obtained in this study were uploaded to a publically accessible database designed for storage of longitudinal behavioral data (http://public.sylics.com/) and available online tools were used to calculate and plot strain means as well as standard error of the means. In addition, data of this study was uploaded into WebQTL (www.genenetwork.org), for correlation analyses (Spearman's Rho) as well as linkage mapping (Haley-Knott regression method for calculating significance).

3 Results

3.1 Description of the Acquired Behavioral Data

After introduction into an automated PhenoTyper home-cage, spontaneous behavior of individually housed BXD mice was video-tracked continuously for 3 consecutive days. The presence of an enclosed nontransparent shelter in the home-cage, in which mice created their nest, facilitated a distinction between active movements and moments of sheltering in their nest. Vast activity differences were observed between the BXD strains, which transgressed beyond the phenotypes of the parental strains (Fig. 1; data available at http://public.sylics.com/).

3.2 Studying Aspects of Longitudinal Behavior

The longitudinal behavior data was studied at four different time scales; i.e., a comparison for (1) habituation over 3 days, (2) distribution of activity in the dark vs. the light period, (3) transitions of the dark to the light phase, and the light to the dark phase, (4) characteristics of movement bouts. First, in the course of 3 days, mice adapted to their new home-cage, during which the majority of strains decreased their activity. In some strains this habituation effect was stronger (e.g., BXD55; Fig. 1, green line; decrease in the peak of activity at the onset of the dark phase by day 3 in BXD55) compared to others (e.g., BXD43; Fig. 1, blue line). Second, there

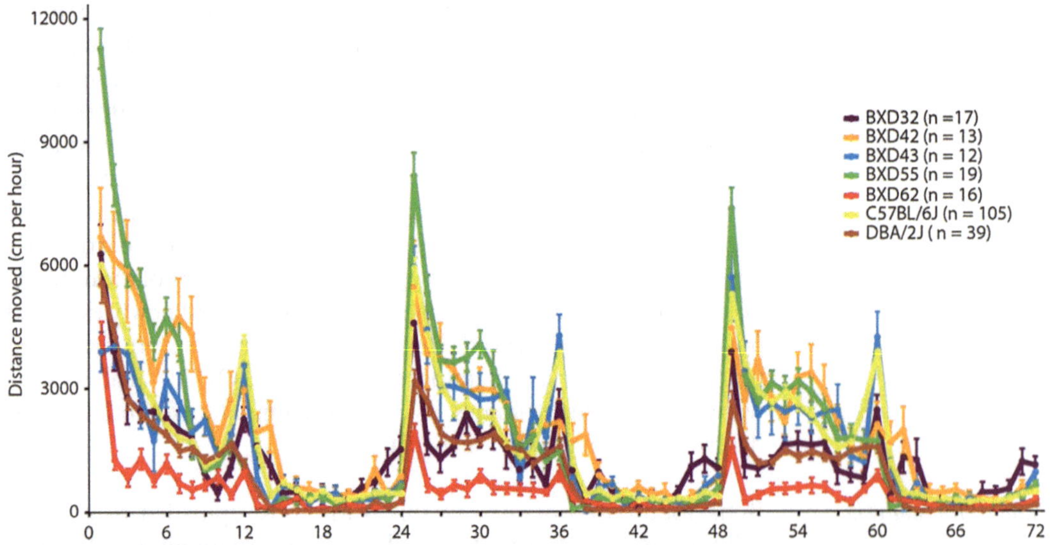

Fig. 1 Distance-moved plot exemplifying the differences in activity between BXD strains for a period of 3 days. Five BXD strains as well as the parental strains were selected to illustrate the variety of strain differences that can be observed in a longitudinal 3-day measurement of activity. Data available at http://public.sylics.com

was a clear strain difference in the distribution of activity between dark and light phases, with some strains displaying relatively high activity during the light phase (e.g., BXD32; Fig. 1, purple line) compared with others. Third, during the hours preceding or following dark/light phase transitions notable strain differences were observed. C57BL/6J are known to display a characteristic increase in activity towards the end of the dark phase (Fig. 1, yellow line; [10]), which was shared by some BXD strains (e.g., BXD43; Fig. 1, blue line), but clearly not by others (e.g., BXD42; Fig. 1, orange). The fourth time scale, not present in Fig. 1, described the characteristics of individual movement bouts that are in the order of seconds to minutes using Gaussian mixture model fitting to separate differently sized move-, arrest-, and shelter-segments according to our previously published methods [10]. Together, the analysis provided 115 parameters to study the genetic architecture of spontaneous behavior in the PhenoTyper home-cage.

3.3 Analysis of Interdependence of Behavioral Parameters

Given that data on all parameters were derived in the same cage and partly at the same moment in time, it is conceivable that some parameters measure similar genetic effects. To assess the dimensionality of the dataset, strain means were calculated for all 115 parameters and the degree of correlation between each of the parameters was calculated. As shown in the network graph (Fig. 2) several groups of parameters indeed are highly correlated, however, many parameters do not strongly correlate with any other parameter (i.e., $|r| < 0.5$). This analysis indicates that the set of 115 parameters is multidimensional, probing multiple dimensions of strains differences.

3.4 Analysis of Phenotypic Home-Cage Data of the BXD Strains in WebQTL

Using the phenotypic traits available in WebQTL, a number of in silico experiments can be performed. In particular, one can upload a set of traits and compute the correlation of these traits with those obtained on the same BXD strains previously. This will allow researchers to generate and test hypothesis on possible relation among traits. Given the vast amount of traits available in the WebQTL database and the associated multiple testing issue when correlating new traits against thousands of other traits, these analyses should be considered exploratory. First, similar traits may have been measured previously, which allows evaluating the reproducibility or robustness of the observations in the present study. Basic activity counts in the home-cage were obtained in a previous study [11]. Those data indeed correlated highly significantly with the measure of activity during the dark phase in the PhenoTyper (Fig. 3a; Spearman Rho 0.45, $P < 0.05$), showing that measuring home-cage activity is reproducible across studies and methods.

Second, correlation analyses with existing reference traits may contribute to the annotation of the new parameters. Our analysis of sheltering behavior distinguishes a class of long shelter visits

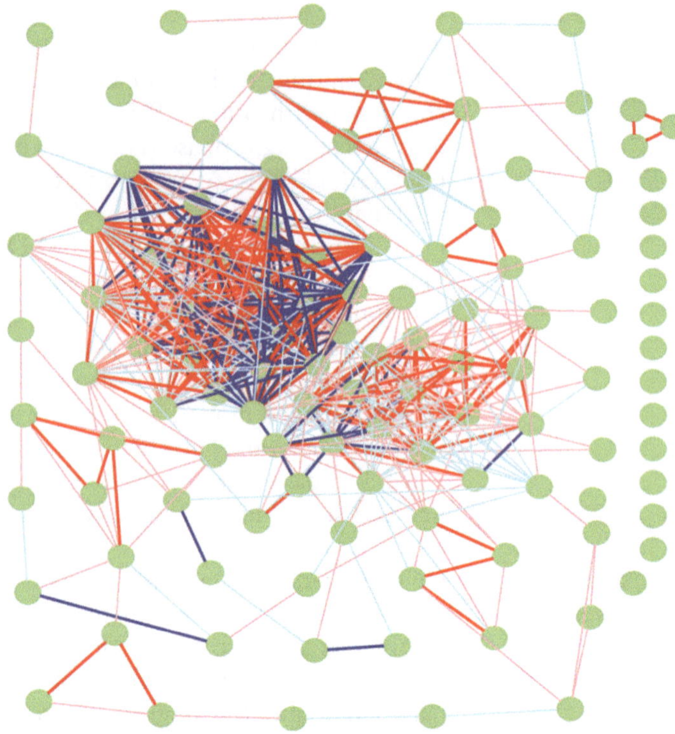

Fig. 2 Network graph of 115 parameters of spontaneous home-cage behavior, indicating the connectivity of values of these parameters in the dataset. Green circles (nodes) represent the 115 parameters. Lines (edges) represent significant Pearson correlation coefficients ($P<0.001$; $N=43$) between respective parameters ($r>0.7$ thick, intense red; $0.5>r>0.7$ thin, light red; $-0.5>r>-0.7$ thin, light blue; $-0.7>r$ thick, intense blue). This graph was generated with GeneNetwork using the GraphViz visualization toolkit from AT&T Research

(typically longer than 30 min), which may be sleeping episodes. In an attempt to confirm this hypothesis, we correlated this new measure with the only available measure of sleep need in WebQTL [12], and found no significant correlation (Fig. 3b; Spearman Rho 0.14, $P=0.63$). While awaiting other measures of sleep in the WebQTL database, the currently available data could not support our hypothesis that the duration of long shelter visits is related to sleep.

Third, some behavioral parameters in the PhenoTyper might be measuring similar behavior as that assessed in classical tests used for decades. For instance, we hypothesized that reduced climbing on top of the Shelter during the light phase might be an index of activity/anxiety-related behavior as seen in the number of visits to anxiogenic areas in conventional tests. When computing the correlation between this parameter and all other traits in the phenotype database in WebQTL, we found a strong correlation with the number of transitions in a light–dark box [3], a classical activity/anxiety test, conform the idea that OnShelter visits during the light

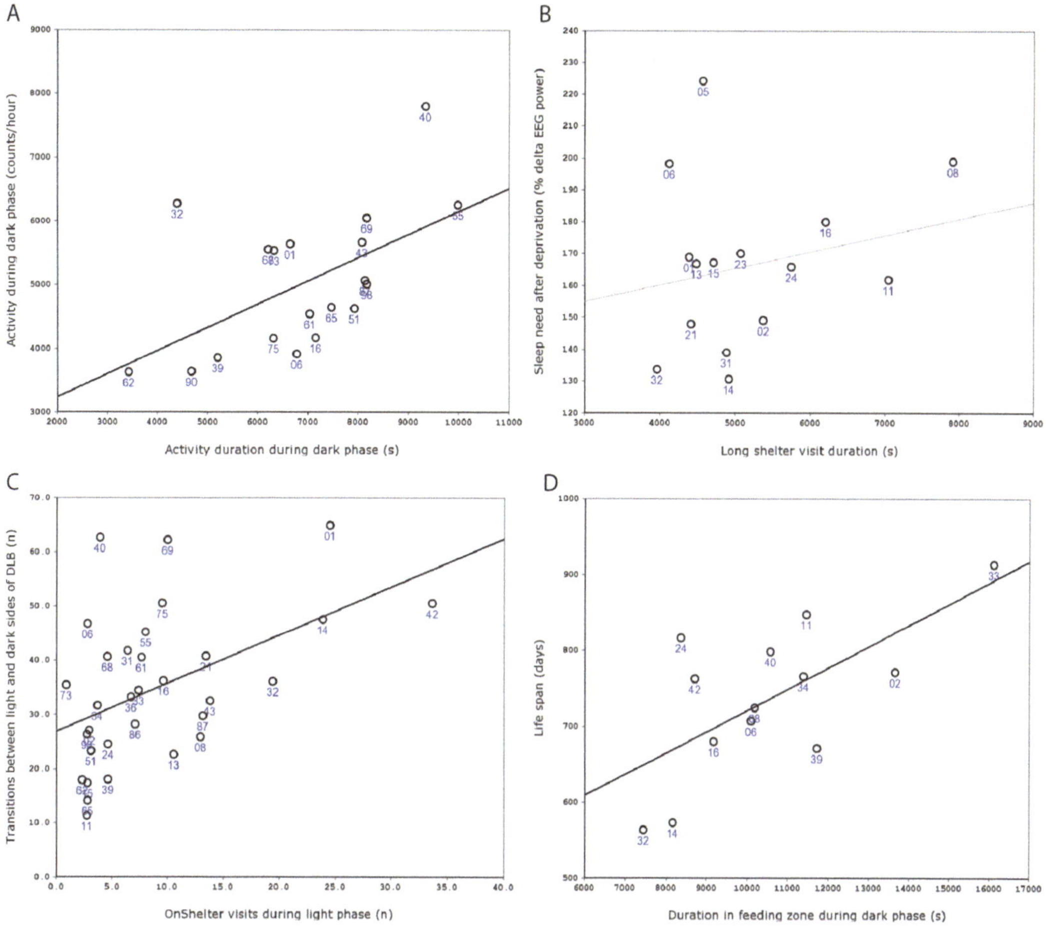

Fig. 3 Correlations between newly generated traits and the traits in the WebQTL data repository. (**a**) Activity (movement) during dark phase (night), in standard housing cage at 23 weeks of age (WebQTL RecordID 17717) was highly correlated with the activity cumulative activity duration during the dark phase in the PhenoTyper (WebQTL RecordID 18384). (**b**) Sleep need, indicated by the percentage of slow wave sleep delta EEG power after 6 h sleep deprivation (WebQTL RecordID 10143) was not correlated with the mean duration of a long shelter visit (WebQTL RecordID 18315). (**c**) Transitions between light and dark sides of a light–dark box (WebQTL RecordID 11391) were highly correlated with the number of OnShelter visits in the PhenoTyper during the light phase (WebQTL RecordID 18390). (**d**) Life span, as index of longevity (WebQTL RecordID 12563) correlated significantly with the cumulative duration in the feeding zone during the dark phase (WebQTL RecordID 18385). Graphs were generated using GeneNetwork

phase can be used as predictive index of (reduced) activity/anxiety-related behavior in conventional tests (Fig. 3a; Spearman Rho 0.48, $P<0.01$).

In addition to behavioral phenotypes, the WebQTL database contains many other traits that might be correlated to one of the 115 parameters of spontaneous behavior. For instance, longevity has been measured in BXD mice [13], and, after scanning the correlation of all 115 parameters with longevity, it shows that the cumulative duration in the feeding zone during the dark phase is a positive predictor of longevity (Fig. 3d; Spearman Rho 0.57, $P<0.05$).

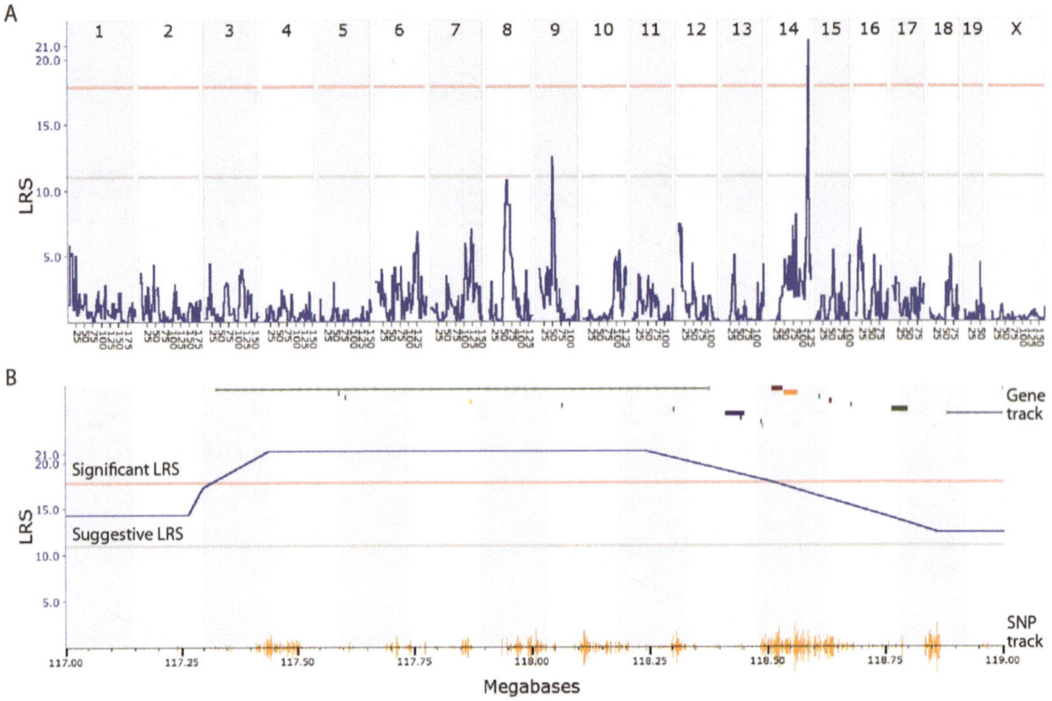

Fig. 4 QTL maps of habituation of the number of activity bouts during the light phase. The mean habituation ratio was used for whole genome mapping (outliers removed), and the LRS score (y-axis) quantifies the relation between genomic markers (x-axis) and the trait. The threshold for significance (genome-wide P = 0.05) and suggestive significance (genome-wide P = 0.63) is indicated. (**a**) Genome-wide significance was reached at a locus on chromosome 14. The magnified view (**b**) displays candidate genes potentially influencing the trait (Gene track). The height of the orange markers placed on the x-axis (SNP track) displays the number of SNPs between C57BL/6J and DBA/2J at each locus as presented in WebQTL. Graphs were generated with GeneNetwork

Taken together, the published traits in WebQTL can be used to understand more about the mechanisms underlying newly measured behaviors. Obviously, many more cross correlations might be made to test specific hypotheses.

3.5 Using Large-Scale Behavioral Data for QTL Mapping

The interval mapping tool of WebQTL computes linkage maps for the entire genome, used for mapping QTLs underlying particular behavioral traits. One example of a significant QTL on chromosome 14 is displayed in Fig. 4a for the parameter that describes the change in number of activity bouts during the light phase from day 1 to day 3, calculated as a habituation ratio (parameter value of day 3/parameter value of day 1). To delineate the QTL region, a 1 LOD drop-off (4.6 LRS units) was used, limiting the QTL region to approximately the 117–119 Mb region on this chromosome. While zooming in on the QTL map and plotting the gene track, one can appreciate that this is a rather extraordinary QTL with relatively few genes residing in the QTL region (Fig. 4b; *see* Table 1 for gene names).

Available gene expression data in WebQTL can be used to identify candidate genes in the QTL region, for prioritization in follow up experiments. Although not all genes are represented in some of the gene expression databases, it is clear that one of the transcripts, i.e., *Gpc6*, correlates significantly with of the habituation of the number of activity bouts during the light phase (Table 1), providing an attractive candidate gene for follow up experiments (Fig. 5; Spearman Rho 0.55, $P < 0.05$).

4 Further Considerations and Limitations

In general, the variation between mice of the same inbred strain (within-strain variation) appears to be larger for behavioral phenotypes compared to molecular, cellular, and anatomical phenotypes. Hence, larger sample sizes may be required for an accurate assessment of strain means. Relatively accurate assessment of strain means is essential for subsequent correlation and genetic mapping studies. Besides increasing the within-strain sample size, increasing the number of strains in the present study could have increased the number of identified QTL, in particular those of small effect size. When aiming for correlation analyses with traits obtained versus those in WebQTL, it is imperative to choose those RI strains for which relevant data is available in WebQTL (behavioral data set typically do not cover all strains). Given the large number of traits in WebQTL available for correlation analyses, exploratory correlation analyses will most likely detect false positive results due to multiple testing. Besides correcting for multiple testing (e.g. using Bonferroni) or setting the number of false positives at a predefined level (using FDR approaches) prior selection of traits in a particular domain (e.g. anxiety) may be a fruitful approach to increase confidence in the observed correlations.

5 Outlook

Home-cage phenotyping is an unbiased and efficient approach for screening large genetic mouse resources. As shown here, longitudinal data on spontaneous mouse activity can yield insights in aspects of behavior, and new connections of parameters with existing behaviors that can be used to map QTLs relevant for the specific trait. Certainly protocols are required that more specifically assess specific behavioral domains, such as motor function, aspects of cognitive performance or anxiety. The development of such protocols in instrumented home-cages such as IntelliCages and PhenoTypers [14–16] has only recently begun, and promises to deliver a highly versatile set of high-throughput protocols in the near future.

Table 1
Genes in the QTL region of the habituation of the number of activity bouts during the light phase

Gene	Description	Spearman correlation (Rho; n = 14) and respective transcript	
Gpc6	glypican 6	0.552*	NM_001079844
A830021K08Rik	RIKEN cDNA A830021K08		
8430413D17Rik	RIKEN cDNA 8430413D17		
4932442G11Rik	RIKEN cDNA 4932442G11		
3110035F07Rik	RIKEN cDNA 3110035F07		
5730405N03Rik	RIKEN cDNA 5730405N03		
Dct	Dopachrome tautomerase	−0.421	NM_010024
1700008A07Rik	RIKEN cDNA 1700008A07		
2700005E23Rik	RIKEN cDNA 2700005E23		
Tgds	TDP-glucose 4,6-dehydratase	0.481	NM_029578
Gpr180	G Protein-coupled receptor 180	0.029	NM_021434
4933431J24Rik	RIKEN cDNA 4933431J24		
Sox21	SRY-box containing 21	−0.099	NM_177753
1700027M17Rik	RIKEN cDNA 1700027M17		
1700044C05Rik	RIKEN cDNA 1700044C05		
Abcc4	ATP-binding cassette, sub-family C (CFTR/MRP), member 4	0.402	NM_001163675

*$P < 0.05$

Acknowledgements

We thank Rolinka van der Loo for operating the PhenoTyper systems and Ruud Wijnands for assistance, Noldus Information Technology for supplying software and Ben Loke, Cecilia Herrera, Raymond de Heer and Willem van der Veer for development of hardware, software and test scripts. The Neuro-BSIK Mouse Phenomics consortium: A.B. Brussaard [a], J.G.G. Borst [b], Y. Elgersma [b], N. Galjart [c], G.T. van der Horst [c], C.N. Levelt [d], C.M. Pennartz [e], A.B. Smit [f], B.M. Spruijt [g], M. Verhage [h] and C.I. de Zeeuw† [b], and the companies Noldus Information Technology (www.noldus.com) and Sylics (Synaptologics BV; www.sylics.com). [a] Department of Integrative Neurophysiology, Center for Neurogenomics and Cognitive Research, Neuroscience Campus Amsterdam, VU University Amsterdam, Amsterdam, The Netherlands. [b] Department of Neuroscience, Erasmus MC,

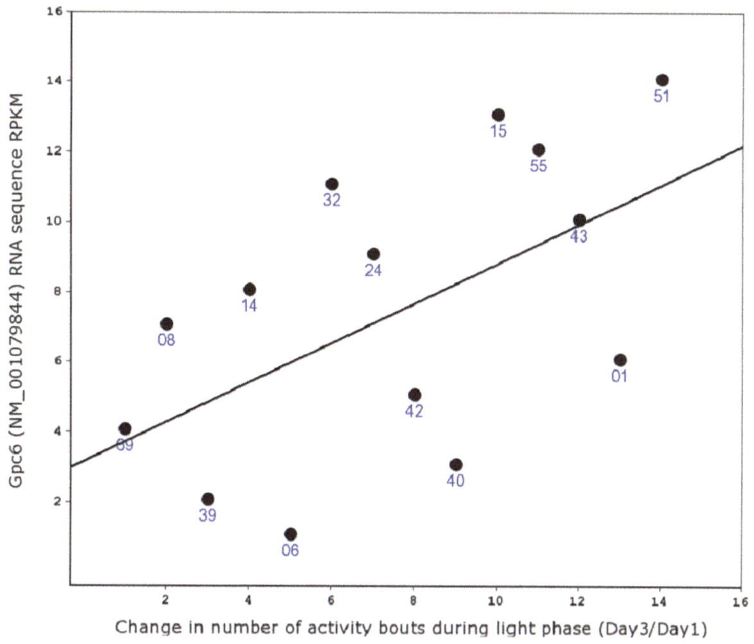

Fig. 5 Spearman correlation between of the habituation of the number of activity bouts during the light phase and the whole brain gene expression of *Gpc6*. The graph was generated with GeneNetwork

University Medical Center Rotterdam, Rotterdam, The Netherlands [c] Department of Cell Biology, Erasmus MC, University Medical Center Rotterdam, Rotterdam, The Netherlands [d] Netherlands Institute for Neuroscience, Amsterdam, The Netherlands [e] Swammerdam Institute for Life Sciences–Center for Neuroscience, University of Amsterdam, Amsterdam, the Netherlands [f] Department of Molecular and Cellular Neurobiology, Center for Neurogenomics and Cognitive Research, Neuroscience Campus Amsterdam, VU University Amsterdam, Amsterdam, The Netherlands [g] Department of Biology, University of Utrecht, Utrecht, The Netherlands [h] Department of Functional Genomics, Center for Neurogenomics and Cognitive Research, Neuroscience Campus Amsterdam, VU University Amsterdam, Amsterdam, The Netherlands. †Lead-author of the Neuro-BSIK Mouse Phenomics consortium, address correspondence to c.dezeeuw@erasmusmc.nl.

Conflict of interest: The authors declare no conflict of interest. M.L. is full time employee of Sylics (Synaptologics BV), a private, VU University spin-off company that offers mouse phenotyping services using AHCODA™. A.B.S. and M.V. participate in a holding that owns Sylics shares and have received consulting fees from Sylics.

References

1. Peirce JL, Lu L, Gu J et al (2004) A new set of BXD recombinant inbred lines from advanced intercross populations in mice. BMC Genet 5:7
2. Chesler EJ, Lu L, Shou S et al (2005) Complex trait analysis of gene expression uncovers polygenic and pleiotropic networks that modulate nervous system function. Nat Genet 37:233–242
3. Philip VM, Duvvuru S, Gomero B et al (2010) High-throughput behavioral phenotyping in the expanded panel of BXD recombinant inbred strains. Genes Brain Behav 9:129–159. doi:10.1111/j.1601-183X.2009.00540.x
4. Loos M, Staal J, Pattij T et al (2012) Independent genetic loci for sensorimotor gating and attentional performance in BXD recombinant inbred strains. Genes Brain Behav 11:147–156. doi:10.1111/j.1601-183X.2011.00754.x
5. Chesler EJ, Wilson SG, Lariviere WR et al (2002) Influences of laboratory environment on behavior. Nat Neurosci 5:1101–1102. doi:10.1038/nn1102-1101nn1102-1101 [pii]
6. de Visser L, van den Bos R, Kuurman WW et al (2006) Novel approach to the behavioural characterization of inbred mice: automated home cage observations. Genes Brain Behav 5:458–466
7. Vannoni E, Voikar V, Colacicco G et al (2014) Spontaneous behavior in the social homecage discriminates strains, lesions and mutations in mice. J Neurosci Methods 234:26–37. doi:10.1016/j.jneumeth.2014.04.026
8. Goulding EH, Schenk AK, Juneja P et al (2008) A robust automated system elucidates mouse home cage behavioral structure. Proc Natl Acad Sci U S A 105:20575–20582. doi:10.1073/pnas.0809053106
9. Voikar V, Colacicco G, Gruber O et al (2010) Conditioned response suppression in the IntelliCage: assessment of mouse strain differences and effects of hippocampal and striatal lesions on acquisition and retention of memory. Behav Brain Res 213:304–312. doi:10.1016/j.bbr.2010.05.019
10. Loos M, Koopmans B, Aarts E et al (2014) Sheltering behavior and locomotor activity in 11 genetically diverse common inbred mouse strains using home-cage monitoring. PLoS One 9, e108563. doi:10.1371/journal.pone.0108563
11. Williams EG, Mouchiroud L, Frochaux M et al (2014) An evolutionarily conserved role for the aryl hydrocarbon receptor in the regulation of movement. PLoS Genet 10, e1004673. doi:10.1371/journal.pgen.1004673
12. Franken P, Malafosse A, Tafti M (1999) Genetic determinants of sleep regulation in inbred mice. Sleep 22:155–169
13. Lang DH, Gerhard GS, Griffith JW et al (2010) Quantitative trait loci (QTL) analysis of longevity in C57BL/6J by DBA/2J (BXD) recombinant inbred mice. Aging Clin Exp Res 22:8–19
14. Endo T, Maekawa F, Voikar V et al (2011) Automated test of behavioral flexibility in mice using a behavioral sequencing task in IntelliCage. Behav Brain Res 221:172–181. doi:10.1016/j.bbr.2011.02.037
15. Vannoni E, Voikar V, Wolfer DP (2010) Higher motor impulsivity in DBA/2 than in C57BL/6 mice revealed by a novel simple reaction time task in intellicage. Present. Soc, Neurosci
16. Remmelink E, Loos M, Koopmans B et al (2015) A 1-night operant learning task without food-restriction differentiates among mouse strains in an automated home-cage environment. Behav Brain Res 283:53–60. doi:10.1016/j.bbr.2015.01.020

Chapter 26

Integrative Analysis of Genetic, Genomic, and Phenotypic Data for Ethanol Behaviors: A Network-Based Pipeline for Identifying Mechanisms and Potential Drug Targets

James W. Bogenpohl, Kristin M. Mignogna, Maren L. Smith, and Michael F. Miles

Abstract

Complex behavioral traits, such as alcohol abuse, are caused by an interplay of genetic and environmental factors, producing deleterious functional adaptations in the central nervous system. The long-term behavioral consequences of such changes are of substantial cost to both the individual and society. Substantial progress has been made in the last two decades in understanding elements of brain mechanisms underlying responses to ethanol in animal models and risk factors for alcohol use disorder (AUD) in humans. However, treatments for AUD remain largely ineffective and few medications for this disease state have been licensed. Genome-wide genetic polymorphism analysis (GWAS) in humans, behavioral genetic studies in animal models and brain gene expression studies produced by microarrays or RNA-seq have the potential to produce nonbiased and novel insight into the underlying neurobiology of AUD. However, the complexity of such information, both statistical and informational, has slowed progress toward identifying new targets for intervention in AUD. This chapter describes one approach for integrating behavioral, genetic, and genomic information across animal model and human studies. The goal of this approach is to identify networks of genes functioning in the brain that are most relevant to the underlying mechanisms of a complex disease such as AUD. We illustrate an example of how genomic studies in animal models can be used to produce robust gene networks that have functional implications, and to integrate such animal model genomic data with human genetic studies such as GWAS for AUD. We describe several useful analysis tools for such studies: ComBAT, WGCNA, and EW_dmGWAS. The end result of this analysis is a ranking of gene networks and identification of their cognate hub genes, which might provide eventual targets for future therapeutic development. Furthermore, this combined approach may also improve our understanding of basic mechanisms underlying gene x environmental interactions affecting brain functioning in health and disease.

Key words Use case, Genomics, Ethanol, Alcoholism, Genetics, Mouse, Brain, Gene networks, Bioinformatics

1 Introduction

Complex trait analysis describes the study of the vast majority of common diseases affecting humans. The most prevalent human diseases, such as heart disease, hypertension, cancer, or addiction,

Klaus Schughart and Robert W. Williams (eds.), *Systems Genetics: Methods and Protocols*, Methods in Molecular Biology, vol. 1488, DOI 10.1007/978-1-4939-6427-7_26,

are not caused by single genes or environmental events. Tackling the understanding of such complex diseases has been illusive given the interplay between multiple causal genetic and environmental factors. Traditional hypothesis-based research, such as exploring the role of a single molecule or signaling event, has largely failed to produce breakthrough discoveries for the understanding of complex traits.

The advent of high-throughput genotyping and gene expression studies over the last 15 years has produced a deluge of data regarding complex traits. It was the hope at the advent of such technology that answers to seemingly unsolvable complex diseases would shortly be forthcoming [1]. Novel insights and non-biased hypotheses have indeed been generated though high-throughput genomic and genetic analyses. The recent discovery of over 100 loci significantly linked to the risk for schizophrenia is perhaps one of the most striking examples of such success [2]. However, even in this case, the step from genetically associated loci to causal mechanisms leading to new therapeutics is still in the distant future. Paradoxically, particularly in regard to genomic studies such as microarrays and RNA-Seq, the insights or causal relationships generated from the voluminous and complex data seem slow to come and extremely tedious to derive. Gene expression studies, particularly in exceedingly heterogeneous tissues such as the brain, may produce hundreds or thousands of genes with "significant" changes in their expression (e.g. *see* [3]). Approaches to derive meaning or actionable conclusions from such information have been varied but generally of low productivity in regard to the amount of effort expended. In part, this may be from an often one-dimensional approach to the analysis of complex multivariate data. Producing long lists of significant genes and then doing a simple over-representation analysis for functional biological categories has been commonly used and does not, by itself, identify potential key mechanistic points or potential drug targets. Recent application of gene network analysis, when combined with phenotypic, genetic, and bioinformatics studies, has improved the identification of testable "hub genes" but expression correlations do not directly equate to causality associations.

Here we outline one approach to potentially identify causal relationships between gene networks derived from animal model studies and complex traits seen in human diseases such as alcohol use disorder. We describe the integration of genomic, genetic, and behavioral data across species to identify actionable targets underlying complex ethanol-related behaviors. While not presenting results on a single example project, we outline an approach and review existing literature for analysis of a complex phenotype such as behavioral responses to ethanol, including alcohol use disorders in humans.

2 Materials and Methods

The approach outlined below is somewhat hypothetical in terms of applicability to other experimental systems or diseases. We describe our actual procedures as performed to identify gene networks and candidate hub genes contributing to ethanol behavioral traits in animal model systems. These studies are designed to increase our understanding of the complex neurobiology underlying the disease of alcoholism in humans (now referred to as "alcohol use disorder") and to identify potential new targets for therapeutic development. The protocol is described for studies with microarrays but RNA-seq results could be substituted with minimal change in the overall approach. We describe key elements in deriving gene networks from microarray genomic data and give several examples of how bioinformatics tools can be used to superimpose external database information onto derived networks so as to increase the likely yield of biologically informative networks. We do not discuss details of RNA isolation and generating probes for microarray analysis as these are described extensively elsewhere in the literature.

2.1 Experimental Design

The approach discussed below assumes that an appropriate experimental design has been employed such that ethanol behavioral data and microarrays have been gathered on the same individual animals. Alternative designs are possible, but the gene network analysis by WGCNA (see below) requires sufficient numbers of samples (e.g. $n \geq 35$) to drive the formation of robust correlations across the network [4]. Thus, this method requires either sufficient numbers of animals from a single strain but across multiple doses/times (e.g. *see* [5]) or behavioral and genomic data across a genetic resource such at the BXD recombinant inbred panel derived from C57BL/6J (B6) and DBA2/J progenitors (e.g. *see* [3]). The goals with such designs are to ensure adequate variance in the data is derived from the biological phenotypes of interest and to directly correlate gene expression and behavior from the same animals. Extensive examples of ethanol behavioral studies (e.g. locomotor activity, anxiolytic-like behaviors, loss-of-righting reflex, withdrawal-induced seizures, or ethanol consumption) across BXD strains are found in GeneNetwork (www.genenetwork.org). Since analysis of behavioral genetic data within GeneNetwork is described elsewhere within this volume, we do not discuss it further here.

Deriving genomic and behavioral data across a large number of subjects is technically demanding and requires extensive planning in terms of experimental design so as to avoid uncontrolled or unknown environmental sources of variance in the data [6]. Such strategies are discussed further within Subheading 4. The keys to avoiding such "batch effects" are to avoid fluctuations in procedures (e.g. different staff doing behavioral or molecular analyses) and to perform a supervised randomization of the handling and treatment of animals and the processing of samples for RNA isolation and genomic analysis as much as possible. However, even with such measures, the possible presence of batch effects must be assessed and eliminated as described below.

2.2 Quality Control and Removal of Batch Effects from Microarray Data Using ComBat

As mentioned above, one common challenge in analyzing genome-wide expression data (or any other large set of biological data) is dealing with batch effects. Nonbiological or uncontrolled experimental variation is often introduced by the serial processing of batches of biological samples, or it can also be an issue when analyzing data collected across multiple experiments. In the case of large microarray experiments, factors could include sacrifice of animals at different times, sequential isolation of RNA in batches, or the hybridization and processing of arrays in batches. Even when the same experimenter has conducted experiment batches with the same equipment, subtle factors such as the age of reagents or the temperature of the laboratory can unknowingly introduce nonbiological variation into the genomic data collected. Despite extreme care, even expert experimentalists can introduce nonbiological variance into whole-genome expression data without their knowledge. For this reason, it is necessary to always assess for batch effects whenever completely parallel processing of samples is not possible.

The first step in eliminating batch effects is identifying them. Principle component analysis is an excellent way to do this and can be easily performed using the ggplot2 package for R (https://cran.r-project.org/web/packages/ggplot2/index.html). One must first identify all factors at which sample processing or data collection was possibly performed in a nonparallel manner. Once such factors have been identified, a simple and effective way to eliminate potential batch effects from microarray data is through use of the ComBat script for R (R Project for Statistical Computing http://www.r-project.org/). The ComBat method has been described in detail in a published report [7] and the software can be downloaded for free at http://www.bu.edu/jlab/wp-assets/ComBat/Download.html as a stand-alone or as part of the SVA package in the Bioconductor project (http://www.bioconductor.org/packages/release/bioc/html/sva.html) within the R statistical framework. The practical use of this method as a stand-alone is described below (*see* also **Notes 1** and **2**).

2.3 PCA Analysis and ComBat

1. Create a spreadsheet in which the first column is an identifier for each microarray, and each subsequent column contains information grouping the arrays into the experimental batches that were used in all nonparallel steps. You can also add columns containing other information, such as the binned age or weight of the animals used, cage location, or perhaps their sex or treatment

group. Each column should have a header describing the information within. Save this file as a tab delimited text file.

2. A second spreadsheet will contain the microarray expression data. The first column should contain the probe/gene IDs, and each subsequent column should contain the expression data for each microarray. Save this file as a tab delimited text file. Generally speaking, this array data has already been assessed for routine individual technical outliers (as may be caused by a degraded RNA sample) by use of standard low level microarray analysis metrics.

3. Run the following commands in R to perform principle component analysis. Lines marked with # indicate annotation lines rather than commands. This will create a .tiff image file containing a scatter plot of the first two principle components of your expression data, with each data point representing one array, colored by its batch. The below block of script is for coloring the data points by array batch, and in the hypothetical batches spreadsheet, the column header containing that information would be called "array.batch." Repeat this block of code for each column in your batches spreadsheet, replacing "array.batch" with the relevant column header. The plots created will look similar to Fig. 1a.

```
>library(ggplot2)
>expressiondata=read.
csv("expressionspreadsheet.txt", sep="\t")
>rownames(expressiondata)=expressionda
ta[,1]
>expressiondata=expressiondata[,-1]
```

- This calls the ggplot2 package and loads your expression data. Bold italic text will be changed to the file name of your particular expression data spreadsheet.

```
>pca=prcomp(t(expressiondata), scale=T)
arraydata=read.csv("batchesspreadsheet.
txt",sep="\t")
```

- This runs the principle component analysis of your expression data and loads your batch information spreadsheet, the name of which will replace the bold italic text.

```
>array.batch=arraydata$array.batch
>scores=data.frame(array.batch,
pca$x[,1:3])
>tiff("PCA by array batch.tiff",
width=750, height=750)
```

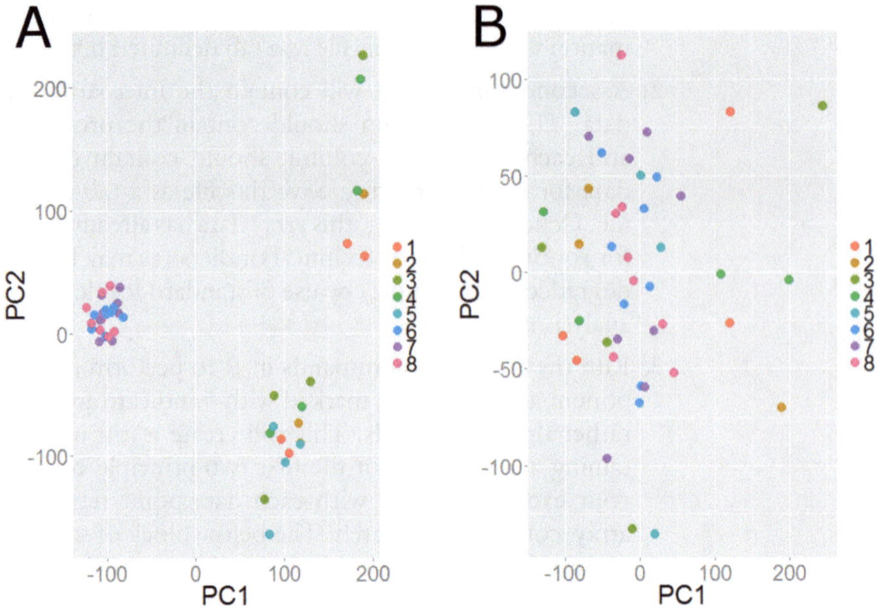

Fig. 1 Principle component plots showing a batch effect of array processing batch (**a**), which is corrected following ComBat (**b**)

```
>p=qplot(x=PC1, y=PC2, data=scores,
color=factor(array.batch), main="PCA plot
by array batch")
>p + theme(text = element_text(size = 20))
+ geom_point(size = 5)
>dev.off()
```

4. Look at your principle component plots and determine which batch effects need to be corrected by examining the data structure. Ideal data structure looks like a homogenous, somewhat circular cloud of data points, as shown in panel B of the figure. Any grouping of data points apart from the whole should be addressed. Look for the batch category that best identifies each segregating group of data points by color.
5. Once it has been determined which batch factor needs to be addressed, create a sample information spreadsheet where the first column is headed "Array name" and contains the identifiers of your microarrays, and a second column headed "Batch" which contains the relevant batch groupings. Do not put any extraneous information in this file. Save this spreadsheet as a tab delimited text file.
6. Run the following script to use ComBat for correcting an identified batch effect in your data. Make sure that the ComBat.R file you downloaded is in your working directory.

```
>Source("ComBat.R")
>ComBat("expressionspreadsheet.txt",
"sampleinfospreadsheet.txt")
```

The bold italic text should be replaced with the filenames of your expression data spreadsheet from **step 2**, and your sample information spreadsheet from **step 5**. A new file containing your adjusted data will automatically be written to your working directory. Instructions for more advanced usage of ComBat, as well as troubleshooting information can be found at http://www.bu.edu/jlab/wp-assets/ComBat/Usage.html.

7. Run principle component analysis again, repeating **step 3** but substituting the adjusted data file name for the original expression data file name. Also, change the name of the output .tiff file so that your original plot is not overwritten. Ensure that the data structure is homogeneous, as shown in Fig. 1b.

2.4 Network Identification with Weighted Gene Correlation Network Analysis

While multiple methods exist for clustering or network analysis of genomic data, the Weighted Gene Correlation Network Analysis algorithm (WGCNA) is a very widely used R software package to identify groups, known as modules, of correlated genes within microarray or other suitable data (Zhang and Horvath 2005). WGCNA is based on scale-free network topology, a model system that assumes a small number of highly connected nodes within a network. For transcriptomic data these nodes are referred to as 'hub genes'. Due to their high connectivity, hub genes represent potential therapeutic targets to affect ethanol responsive gene expression in the brain and, potentially, ethanol behaviors. WGCNA involves multiple analysis steps that are outlined in detail in a series of R tutorials produced by the Horvath laboratory. These extensive tutorials and associated papers applying WGCNA are a major reason for the popularity of this approach. Instructions for obtaining the WGCNA R package and dependencies can be found at: https://labs.genetics.ucla.edu/horvath/CoexpressionNetwork/Rpackages/WGCNA/. A brief overview of a sample WGCNA analysis of microarray data from ethanol-treated mouse brain follows but the actual commands and options for the analysis can be found within the tutorials from the Horvath web site. Quite often, simple copy/pasting of those commands with changing file names or parameters is sufficient for carrying out an initial WGCNA analysis.

2.5 Protocol Overview for WGNCA

1. An appropriate dataset must be chosen for WGCNA analysis. As with any clustering technique, it is essential to have substantial biological variation across samples and to have enough samples such that correlation networks have sufficient statistical power. Although firm thresholds are difficult to define, Iancu and colleagues, using a large microarray dataset from mouse striatum, determined that a $n \geq 35$ appeared optimal for defining network topology [4].

2. In order to identify networks with the most meaningful correlations, a variance filter is often first applied to e.g. a ComBat-corrected microarray dataset (*see* also **Note 3**). This variance filter eliminates genes showing minimal variation in expression across samples. The median absolute deviation (MAD) is our preferred method for variance quantification:

$$\mathrm{MAD} = \mathrm{median}_{\mathrm{i}}\left(X_{\mathrm{i}} - \mathrm{median}_{\mathrm{j}}\left(X_{\mathrm{j}}\right)\right).$$

   A histogram of MAD values is plotted to identify the lower tail of variance. The number of genes included in network analysis may be limited by computer power. However, ideally, the proportion of variance should be calculated at regular intervals to determine a MAD threshold at which the majority of variation within the total data is included. All data below this threshold may be assumed to be noise, and excluded from further analysis.

3. The next step in WGCNA involves uploading gene expression data, and modeling to determine a soft-thresholding power at which the data structure best fits scale-free topology. This is done using the pickSoftThreshold() function as part of the WGCNA package. This function will output a table of scale-free fit metrics at various soft-thresholding powers (Table 1). We use the scale-free fit index (SFT R^2) as our primary measure of scale-free fit. A scale-free of 0.9 or greater is ideal, however, a scale-free fit of 0.75 or greater can be acceptable.

Table 1

Scale-free metrics from WGCNA analysis

Power	SFT.R.sq	Slope	Truncated.R.sq	Mean.k.	Median.k.	Max.k.
1	0.028	0.345	0.456	746.978	761.733	1206.429
2	0.126	-0.597	0.843	254.498	250.843	573.545
3	0.340	-1.030	0.972	111.005	101.736	324.154
4	0.506	-1.422	0.973	56.536	47.248	202.307
5	0.681	-1.716	0.940	32.154	25.089	134.175
6	0.902	-1.497	0.962	19.906	14.498	94.774
7	0.921	-1.667	0.917	13.192	8.678	84.103
8	0.904	-1.724	0.876	9.249	5.394	76.315
9	0.859	-1.703	0.836	6.795	3.556	70.532
10	0.833	-1.663	0.831	5.194	2.382	65.752

Scale free metrics resulting from the function *pickSoftThreshold()* within WGCNA. Results show that a power of ≥5 results in acceptable scale-free fit index values (SFT.R.sq)

4. Network construction is the next step in WGCNA. Both manual and automatic network construction and module identification are outlined, in detail, in the WGCNA tutorials. During module identification, gene expression data is organized by topological overlap distance. This data can be visualized using a cluster dendrogram. One particular variable to pay attention to during module construction is the "deep-split." Deep-split is used to fine-tune the sensitivity of module detection by adjusting the branch cutting threshold within the dendrogram. Multi-dimensional scale plots using first and second principal components as the *x* and *y*-axes are another useful way to visualize modules in order to identify optimal deep-split value. We consider an optimal deep-split value to be one where there is minimal spatial overlap between modules.
5. Both modules and individual genes can then be correlated to phenotype data (*see* **Note 4**). Individual gene correlations are performed using gene expression measures such as RMA values. Modules are correlated to phenotype based on their first principal component, known in WGCNA as the module eigengene. The module eigengene is a value that explains the majority of gene expression variance within each identified module. Phenotype data can include many variables from behavioral data to technical variables such as RNA quality index of each microarray sample. This network correlation analysis is one of the most powerful features of WGCNA. Ideally, phenotypic and genomic data are derived from the same individual animals. Due to the limited sample size often seen in microarray studies, we recommend using Spearman Rank rather than Pearson correlation in order to minimize the influence of outliers. An example of network correlations to phenotypic data is shown in Fig. 2.
6. The identification of hub genes within modules is one of the final steps in WGCNA. Connectivity is usually the primary variable by which we identify hub genes. WGCNA produces several connectivity metrics with the command function: intramodularConnectivity(). For each gene, this command outputs its total connectivity within the network (kTotal), its connectivity within its assigned module (kWithin), the difference between kTotal and kWithin (kOut), and the difference between kWithin and kOut (kDiff). The intramodularConnectivity() function features an option to scale within module connectivity (kWithin) based on module size. We recommend scaling, as this eases identification of the most highly connected genes within their respective modules independent of module size.
7. Modules identified by WGCNA can be further interrogated for biological function enrichment using a wide variety of standard bioinformatics tools such as those for Gene Ontology or

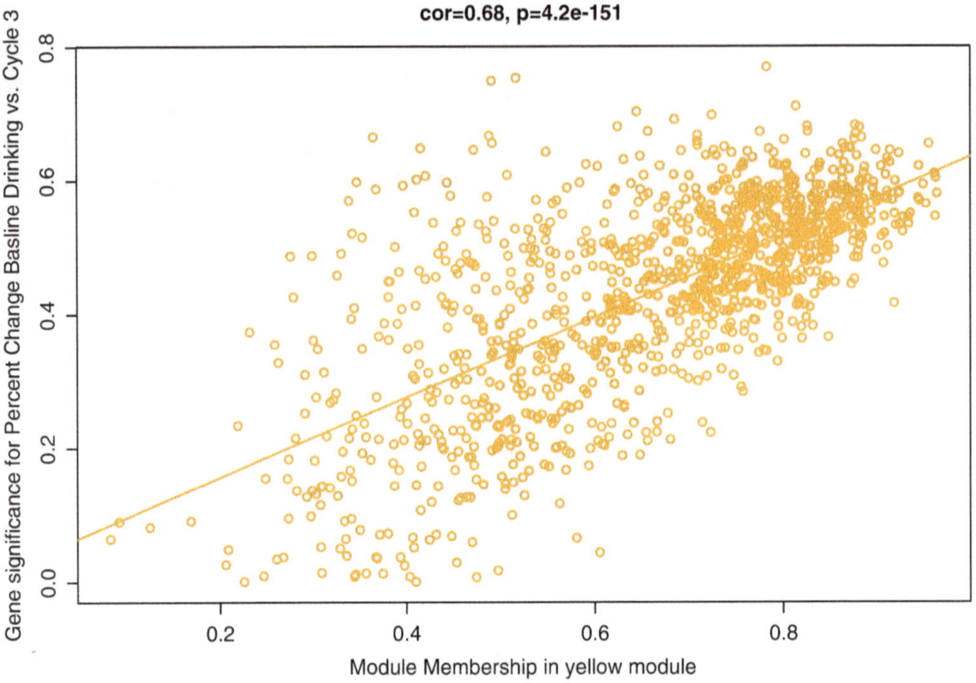

Fig. 2 Module membership correlation with gene significance. WGCNA analysis plot of member genes from one module. *X*-Axis is module membership scoring where higher values represent genes with greater connectivity. *Y*-Axis shows gene significance in terms of correlation of expression values versus a trait of interest. Genes with expression more highly correlated with trait of interest and showing higher connectivity (*top right corner*) are high value "hub genes." High correlation of the module membership with gene significance strongly suggests this module is involved in biological mechanisms of the trait

pathway analysis. Our laboratory frequently uses free web-based resources such as DAVID (http://david.abcc.ncifcrf.gov), ToppGene (https://toppgene.cchmc.org), WebGestalt (http://bioinfo.vanderbilt.edu/webgestalt/) or REVIGO (http://revigo.irb.hr) to identify and display over-represented functional categories. Additionally, module gene lists can be submitted to resources that construct networks based upon other biological information such as protein–protein interactions, published biological interaction, or transcription factor binding site analyses. Such tools include GeneMANIA (http://genemania.org) and the subscription-based Ingenuity Pathway Analysis (http://www.ingenuity.com). These tools can, in effect, validate the network structure of WGCNA-derived expression correlation networks.

8. WGCNA networks can be further validated and ranked based upon quantifying their overlap with other user defined gene lists obtained from differing biological contexts or from expression genetics datasets. For example, we frequently interrogate

WGCNA modules for overlap with public or our own gene sets within the GeneWeaver web-based application (www.geneweaver.org). Finally, WGCNA module genes can be interrogated for correlations in other expression datasets, phenotypic correlations, and conserved genetic regulators by the rich resources available within GeneNetwork (www.genenetwork.org).

2.6 Combined Expression, Protein–Protein Interaction, and Genetic Networks Using EW_dmGWAS

An ideal bioinformatics methodology for interrogating the significance and biological function of genomic expression data would be to superimpose other types of gene–gene interaction data or biological significance scores onto expression correlation networks. WGCNA, as described above, only produces expression correlation networks and phenotype/expression correlation. Several tools have been developed, however, to directly superimpose multiple types of biological interactions. These include GeneMANIA, mentioned above, where gene lists can be constructed into network via a combined scoring algorithm entailing mining multiple databases such as protein–protein interactions, other microarray expression correlations, and literature association. An additional resource whereby gene expression correlations can be combined with protein–protein interaction networks and genetic significance scores (such as *p*-values from human GWAS data), is the "edge-weighted dense-module searching of Genome Wide Association Studies" algorithm, or EW_dmGWAS, developed by Jia and colleagues at Vanderbilt University [8, 9].

EW_dmGWAS (referred to henceforth as dmGWAS) is a method that allows integration of gene expression data, genotypic data, and protein–protein interaction data to find gene networks that are associated with complex traits. The tool seeks to solve the two simultaneous problems generated by under-powered and complex genomic or genetic datasets. First, is there an underlying biological structure to GWAS results, including those not reaching genome-wide significance levels? Second, are expression correlation networks (particularly those from model organisms) relevant to genetic loading in complex traits? Using gene pairs from a predefined protein–protein interaction (PPI) background network, the program forms edge weights between two genes, by comparing the correlation in expression between the two genes in the control versus the treated sample. Correlation changes between the two conditions result in increased weighting, thus informing the structure of the network for the biological trait of interest. dmGWAS then uses these values, in conjunction with GWAS *p*-values, to determine whether or not genes are part of the same network. The importance attributed to GWAS *p*-values, relative to that attributed to expression correlation differences, is determined by the parameter *lambda*, which can be user defined or estimated by the program. The estimation algorithm chooses a value that will attribute equal importance to each set of values. The program also allows the user to define the stringency of the inclusion choice (as determined by parameter *r*).

The program starts with a seed gene as a module, and expands each module by adding genes that are associated with the current module in the PPI network and increase the module score by a factor of *r*, one at a time. Each time a gene is queried, a new module score is calculated from its node weight (GWAS *p*-value) and the weight (correlation difference) of each of its edges shared with genes already included in the module, using *lambda* to determine the relative reliance on node-weights and edge-weights. The expansion process is terminated when it reaches the point at which no additional genes increase the module score by a factor of *r*. The final module score is then standardized via *z*-score transformation with respect to a distribution of scores of 10,000 random modules of the same size. These scores can then be used to calculate the significance of each module. The resulting modules are networks of genes (nodes) whose protein products interact, have correlations (edges) differing between treatment and control groups, and

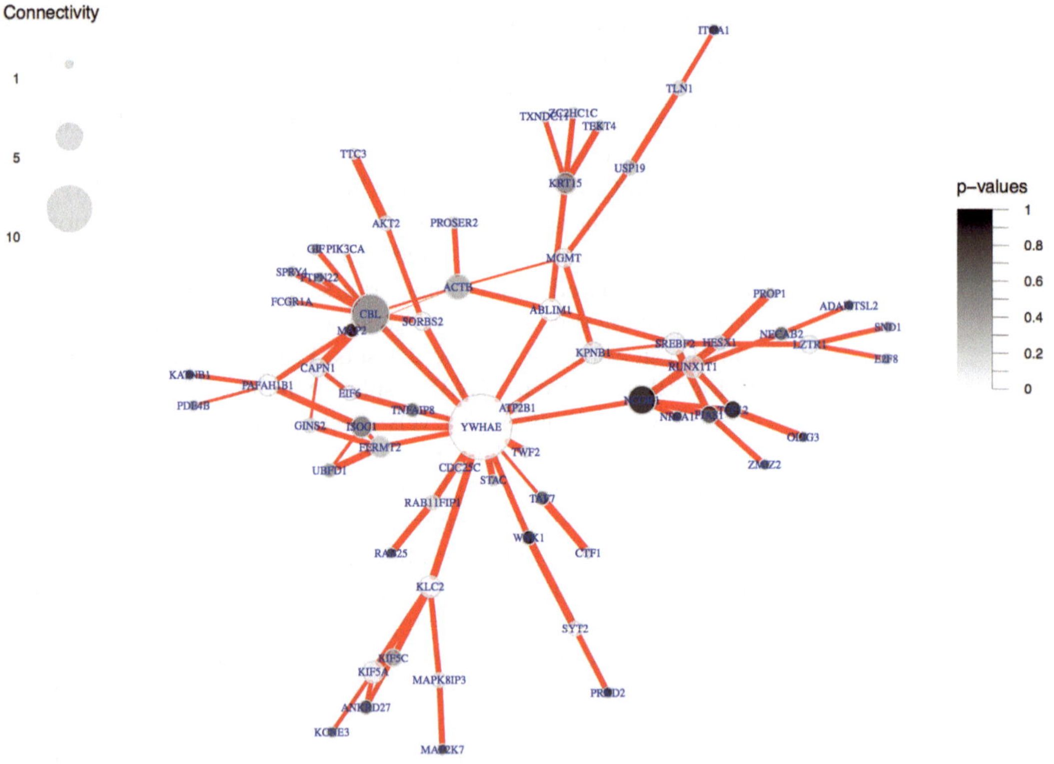

Fig. 3 ggPlot output of EW_dmGWAS results. Figure shows a module of genes from a dmGWAS analysis where protein–protein interactions form the background matrix of the plot but edges between genes are weighted by changes in correlations between the pair of genes as a function of treatment group. The density of gene (nodes) coloration is inversely proportional to p-values of GWAS results and the size of nodes correlates with the connectivity

have nodes enriched for genes with relatively low GWAS *p*-values. An example of results is shown in Fig. 3.

2.7 Protocol Overview for dmGWAS Analysis

1. *Formatting genome wide association study (GWAS) data.* For this step, a human GWAS results file is needed that contains SNP names and raw *p*-values for the association of each SNP with a trait of interest. Because the nodes of the dmGWAS network will represent genes, as opposed to SNPs, gene-wise *p*-values need to be calculated from the raw SNP *p*-values. This can be accomplished by using programs like VEGAS2 (Versatile Gene-Based Association Study) [10] or KGG (Knowledge-based mining system for Genome-wide Genetic studies) [11]. A data frame should result that contains one column of human gene symbols and one column for gene-wise *p*-values (*see* **Note 5**).
2. *Formatting microarray data.* Data from of microarray results are needed for calculating edge weights between nodes (genes) in the dmGWAS networks. We typically employ RMA (Robust Multichip Average) [12] or similar expression values, obtained from tissue with a treatment and control group, and gene symbols associated with each probeset ID. The data used for the example network displayed in Fig. 3 was obtained from a genomic study of acute ethanol (1.8 g/kg × 4 h) expression responses in prefrontal cortex of BXD mouse strains [3]. Because the GWAS data contains human gene symbols, expression data gene symbols (mouse) were converted to human using resources such as SwissProt (http://www.ebi.ac.uk/uniprot). The mouse gene symbols and their human homologues are then merged with the microarray data frame by probeset ID. The resulting data frame should contain a column for each of the following: human gene symbols, mouse gene symbols, probeset IDs, and RMA valucs.
3. *Filtering microarray data.* In forming edge weights between two genes, dmGWAS compares the correlation in expression between the two genes in the control versus the treated (ethanol) sample. Some gene symbols in the microarray dataset will likely have multiple RMA value data values due to redundancies in probeset design. We thus reduce this complexity by using the probeset for a given gene having the largest average expression value across all microarray samples. Low expression value probesets/genes are not removed from the array dataframe since they will not contribute appreciably to edge weighting although the cognate nodes may be vital to the formed dmGWAS network if they have high node values (low GWAS *p*-values). The filtered expression data frame should contain a column for each of the following: probesetIDs, mouse gene symbol, human gene symbol, and RMA expres-

sion values for each sample. Separate data frames will be needed for the control group and treatment group.

4. *Formatting protein–protein Interaction network data.* A predefined protein–protein interaction (PPI) network is used as a background network for the dmGWAS analysis. Although multiple such resources exist, the Protein Interaction Network Analysis (PINA) platform contains multiple curated public databases and interactions for human and five model organisms [13]. When using human GWAS data, a human PPI network is utilized. The data frame should contain two columns of protein IDs, with each row representing a pair-wise interaction. Protein IDs must be converted to human gene symbols to be matched with data in the gene expression and GWAS datasets. Such gene symbol cross-annotation data can be obtained from SwissProt (http://www.ebi.ac.uk/uniprot). The resulting final PPI data frame should contain two columns: one column for the gene symbol of the first member of an interaction pair, and a second column for gene symbols associated with the second member of each PPI pair.
5. *Checking data format for analysis.* Before you can run a dmGWAS analysis, it is critical for the data to be formatted in a specific manner, in order for the algorithms in the package to work. Four data frames are utilized by the dmGWAS command dms(), which will run the dense module search and output the result objects: GWAS, the two expression datasets, and the PPI data frame.
   - The GWAS data frame should contain one column of human gene symbols and one column of gene-wise GWAS *p*-values, in that order. This data frame will be entered in the "geneweight" argument of the dmGWAS command dms().
   - The "exp1" and "exp2" arguments will take the treatment and the control expression data frames, respectively. The final data frames must contain one column of human gene symbols followed by *n* (number of samples) columns of RMA values associated with each gene.
   - The PPI network data frame must have two columns, with each row containing the human gene symbols associated with each protein in a pair-wise interaction. The number of rows should equal the number of pair-wise interactions in your dataset. This data frame will be entered in the "network" argument.
   - The program will discard information on any genes that are not present in all four data frames. Therefore, a gene must be present in all four datasets in order to be included in the analysis.

6. *Run the dms() command, insert the aforementioned data frames in their respective arguments, and define r and lambda.* If you want to use the default *r* value (0.1), do not include the "r" argument in the command. If you want the program to estimate the lambda parameter, use *lambda="default."* Store the run as an object. This will save all of the output objects as a list. An example the script to run a dmGWAS analysis and save output is thus:

```
#Run dense module search and store results as dataframe
>library(igraph)
>library(dmGWAS)
>library(tcltk)
>res.list<-dms(network, geneweight, expr1, expr2, r=0.1, lambda="default")
>#res.list<-dms(network, geneweight, expr1, expr2, r=0.1, d=2, lambda="default") #default is d=1
>names(res.list)
>save(res.list, file="Results_dmGWAS_filename.Rdata")
#GWPI: igraph class object, node and edge-weighted PPI network
#genesets.clear: list containing all valid modules; name of each record=seed gene
#genesets.length.null.dis: list containing the randomization data for each size of module
#genesets.length.null.stat: list containing the statistical values of randomization data for each size of module
#module.score.matrix: matrix containing data for each module; data = gene (seed gene), Sm (module score), Sn (normalized module score)
#ordered.module.score.matrix: ordered matrix of module.score.matrix based on Sn
#select the top (100x"top")% of modules with respect to significance, and store as a data-frame
>selected<-moduleChoose(res.list, top=0.005, plot=T)
>Top0.5<-selected
>save(Top0.5, file="TopPoint5Percent.Rdata")
```

7. *Reading the results.* The results list contains an object for each of the following: igraph object of your input, the resulting module names (names of the seed genes) and their raw and standardized scores, and the raw scores of the permuted random modules. If you use the dmGWAS_3.0 package, the

lambda value will be stored in the name of a separate object that will be saved to your working directory automatically. This package will also output an object of module names and scores that is ordered by the standardized scores. The modules may overlap significantly. Identical modules are removed by the package, but no modules are merged based on overlap. It is up to the user to merge overlapping modules, if desired. Because the permuted module scores compose an approximately normal distribution, the standardized module scores can be treated as *Z*-scores, and therefore may be used to calculate *p*-values that represent module significance. If you choose to merge any modules, you must recalculate a standardized score based on a permutation of modules of the new merged module's size, in order to calculate its significance.

3 Results

The elements of a data analysis pipeline that are outlined in Subheading 2 will produce well normalized data suitable for network analysis by WGCNA and downstream bioinformatics studies, such as those produced by dmGWAS. The intended result of such studies is to identify networks of genes that can be ranked by their biological functions and perhaps by over-representation with other relevant genomic data or human GWAS results. The latter could greatly increase the biological value for both a given network and cognate "hub genes," producing better and fewer targets for complicated and time-consuming validation in animal models (e.g. by gene targeting or pharmacological interventions)—with the eventual goal of defining candidates for new therapeutic measures in humans.

The importance of experimental design and eliminating batch effects is clearly demonstrated in Fig. 1. Prior to ComBat normalization, the principle component analysis indicates at least three groupings of arrays within the data that do not correspond to known desired dependent variables (Fig. 1a). In this case, these were produced by differing batches of animals in an experiment conducted over an extended period of time. Following ComBat correction for such a batch effect, the microarray data is much more evenly distributed with little apparent subdivision by any known dependent variable (Fig. 1b).

The generation of scale-free networks by WGCNA analysis is a powerful tool for defining functional and regulatory groupings within complex sets of genomic data. The multi-step process that is well described by protocols from the Horvath laboratory produces modules that are logical choices for downstream interrogation in bioinformatics and biological validation experiments. Figure 2 demonstrates the power of this approach by displaying a module with remarkable correlation between the relative "connectivity" of genes within the module (x-axis) and the correlation of the expression of these genes with a theoretical quantitative behavioral phenotype of interest (y-axis). When such a striking correlation between module membership and trait correlation is seen, it serves to validate the module as intimately involved in the phenotype of interest. More importantly, this analysis also portrays robust candidate genes for future studies, as being those in the top right corner of the Scattergram in Fig. 2—with the strongest intramodular correlations and correlation to the phenotype.

Finally, our laboratory has used genomic studies in mouse models for nearly 20 years as an effort to increase our mechanistic understanding of ethanol-responsive behaviors relevant to the development of alcohol use disorder in humans. The defining of ethanol-responsive gene networks in mouse models, however, does not necessarily directly translate to a role in human disease. Thus, the importance of methodologies such as dmGWAS is that they can directly superimpose animal model genomic data with human genetic results, within a known framework of protein interactions. As shown in Fig. 3, such studies may therefore highlight particular aspects of networks and hub genes that have predictive validity for human disease.

4 Notes

1. The description on use of ComBat for correction of possible batch effects is brief and does not include the full details of the R module. Readers are referred to the source paper and the R module information for further information on the use of covariates and other factors.
2. ComBat should be used with great care and only after extensive characterization of quality control measures across arrays or RNA-seq samples. RNA samples should have uniformly high quality measures such as RNA integrity numbers (RIN) on bioanalyzer studies prior to genomic analysis. Outliers can greatly modify the results of ComBat corrections and should be eliminated from the analysis if possible. Correction for an independent variable that seems unrelated to the dependent variable of interest may serve to eliminate expression patterns actually related to the trait of interest. For example, ComBat correction for age differences amongst experimental (male) animals might eliminate brain gene expression patterns related to testosterone levels.
3. WGCNA analysis can be greatly affected by inclusion of genes having marginal expression levels or no variance across the experimental samples. The trimming protocols that are suggested are thus critical to downstream analysis. Furthermore,

efforts to decrease the number of genes included in the WGCNA analysis (for increasing processing speed) should be done with care since use of biased filters (such as statistical cutoffs) could significantly disrupt intramodular connectivity in some cases. A highly "responsive" network might actually disappear due to the lack of a few highly connected, but "unresponsive," hub genes.

4. WGCNA module eigengene x phenotype correlations should only be done for phenotypes with sufficient numbers of subjects and data distribution. Correlation of continuous module eigengenes with discontinuous or ordinal traits can produce markedly significant correlations of unclear significance.
5. In dmGWAS analysis, do not filter out genes based on GWAS p-values. dmGWAS is designed to incorporate the p-value of each gene in the decision to include or exclude the gene from a module. It is possible for a gene to play an important role in relevant networks, although genetic variation, specifically, may not contribute to the gene's association with the network. Protein–protein interactions and expression correlation changes might be more important drivers for inclusion of such a gene in a given network. Furthermore, the genetic variation in other genes associated with the same pathway may confer the relevance of the overall network. If such a gene with a poor GWAS p-value is central to the overall connectivity of a network, the module may be excluded from ranked results if the gene is excluded from the analysis.

Acknowledgements

This work was supported by grants from the National Institute on Alcohol Abuse and Alcoholism to Michael F. Miles (U01AA016667, P50AA022537) and Maren L. Smith (F31AA023134). Kristin M. Mignogna was partially supported by the VCU C. Kenneth and Dianne Wright Center for Clinical and Translational Research and a grant from the National Center for Advancing Translational Sciences (UL1TR000058). James W. Bogenpohl was supported by an Institutional Training Grant in Digestive and Liver Diseases from the National Institute of Diabetes and Digestive and Kidney Diseases (5T32DK007150).

References

1. Hughes TR, Marton MJ, Jones AR, Roberts CJ, Stoughton R, Armour CD, Bennett HA, Coffey E, Dai H, He YD, Kidd MJ, King AM, Meyer MR, Slade D, Lum PY, Stepaniants SB, Shoemaker DD, Gachotte D, Chakraburtty K, Simon J, Bard M, Friend SH (2000) Functional discovery via a compendium of expression profiles. Cell 102(1):109–126

2. Loh PR, Bhatia G, Gusev A, Finucane HK, Bulik-Sullivan BK, Pollack SJ, Schizophrenia Working Group of the Psychiatric Genomics C, de Candia TR, Lee SH, Wray NR, Kendler KS, O'Donovan MC, Neale BM, Patterson N, Price AL (2015) Contrasting genetic architectures of schizophrenia and other complex diseases using fast variance-components analysis. Nat Genet 47(12):1385–1392. doi:10.1038/ng.3431
3. Wolen AR, Phillips CA, Langston MA, Putman AH, Vorster PJ, Bruce NA, York TP, Williams RW, Miles MF (2012) Genetic dissection of acute ethanol responsive gene networks in prefrontal cortex: functional and mechanistic implications. PLoS One 7(4), e33575. doi:10.1371/journal.pone.0033575
4. Iancu OD, Darakjian P, Malmanger B, Walter NA, McWeeney S, Hitzemann R (2012) Gene networks and haloperidol-induced catalepsy. Genes Brain Behav 11(1):29–37. doi:10.1111/j.1601-183X.2011.00736.x
5. Smith ML, Lopez MF, Archer KJ, Wolen AR, Becker HC, Miles MF (2016) Time-course analysis of brain regional expression network responses to chronic intermittent ethanol and withdrawal: implications for mechanisms underlying excessive ethanol consumption. PLoS One 11(1), e0146257. doi:10.1371/journal.pone.0146257
6. Chesler EJ, Wilson SG, Lariviere WR, Rodriguez-Zas SL, Mogil JS (2002) Identification and ranking of genetic and laboratory environment factors influencing a behavioral trait, thermal nociception, via computational analysis of a large data archive. Neurosci Biobehav Rev 26(8):907–923, doi:S0149763402001033 [pii]
7. Johnson WE, Li C, Rabinovic A (2007) Adjusting batch effects in microarray expression data using empirical Bayes methods. Biostatistics 8(1):118–127. doi:10.1093/biostatistics/kxj037
8. Jia P, Zheng S, Long J, Zheng W, Zhao Z (2011) dmGWAS: dense module searching for genome-wide association studies in protein-protein interaction networks. Bioinformatics 27(1):95–102. doi:10.1093/bioinformatics/btq615
9. Wang Q, Yu H, Zhao Z, Jia P (2015) EW_dmGWAS: edge-weighted dense module search for genome-wide association studies and gene expression profiles. Bioinformatics 31(15):2591–2594. doi:10.1093/bioinformatics/btv150
10. Mishra A, Macgregor S (2015) VEGAS2: software for more flexible gene-based testing. Twin Res Hum Genet 18(1):86–91. doi:10.1017/thg.2014.79
11. Gui H, Li M, Sham PC, Cherny SS (2011) Comparisons of seven algorithms for pathway analysis using the WTCCC Crohn's disease dataset. BMC Res Notes 4:386. doi:10.1186/1756-0500-4-386
12. Irizarry RA, Bolstad BM, Collin F, Cope LM, Hobbs B, Speed TP (2003) Summaries of affymetrix GeneChip probe level data. Nucleic Acids Res 31(4), e15
13. Cowley MJ, Pinese M, Kassahn KS, Waddell N, Pearson JV, Grimmond SM, Biankin AV, Hautaniemi S, Wu J (2012) PINA v2.0: mining interactome modules. Nucleic Acids Res 40(Database issue):D862–D865. doi:10.1093/nar/gkr967
14. Langfelder, P., & Horvath, S. (2008). WGCNA: an R package for weighted correlation network analysis. BMC Bioinformatics, 9(1), 559. http://doi.org/10.1186/1471-2105-9-559

Chapter 27

Dissection of Host Susceptibility to Bacterial Infections and Its Toxins

Aysar Nashef, Mahmoud Agbaria, Ariel Shusterman, Nicola Ivan Lorè, Alessandra Bragonzi, Ervin Wiess, Yael Houri-Haddad, and Fuad A. Iraqi

Abstract

Infection is one of the leading causes of human mortality and morbidity. Exposure to microbial agents is obviously required. However, also non-microbial environmental and host factors play a key role in the onset, development and outcome of infectious disease, resulting in large of clinical variability between individuals in a population infected with the same microbe. Controlled and standardized investigations of the genetics of susceptibility to infectious disease are almost impossible to perform in humans whereas mouse models allow application of powerful genomic techniques to identify and validate causative genes underlying human diseases with complex etiologies. Most of current animal models used in complex traits diseases genetic mapping have limited genetic diversity. This limitation impedes the ability to create incorporated network using genetic interactions, epigenetics, environmental factors, microbiota, and other phenotypes. A novel mouse genetic reference population for high-resolution mapping and subsequently identifying genes underlying the QTL, namely the Collaborative Cross (CC) mouse genetic reference population (GRP) was recently developed. In this chapter, we discuss a variety of approaches using CC mice for mapping genes underlying quantitative trait loci (QTL) to dissect the host response to polygenic traits, including infectious disease caused by bacterial agents and its toxins.

Key words Bacterial infections, Host response, Collaborative Cross mice, Recombinant inbred lines, QTL and fine mapping, Heritability

1 Introduction

Infection is one of the leading causes of human mortality and morbidity. Exposure to microbial agents is obviously required, but sometimes not sufficient, for development of an infectious disease. It is now evident that non-microbial environmental and host factors, whether genetic or nongenetic, also play a key role in the onset, development and outcome of infectious disease, resulting in an astounding level of clinical variability between individuals in a population infected with the same microbe. Moreover, Louis

Klaus Schughart and Robert W. Williams (eds.), *Systems Genetics: Methods and Protocols*, Methods in Molecular Biology, vol. 1488, DOI 10.1007/978-1-4939-6427-7_27, © Springer Science+Business Media New York 2017

Pasteur himself, the founder of the microbial theory emphasized the importance of non-microbial factors, including host hereditary constitution, in the susceptibility to infection [1].

Investigations of the effect of natural variability in the development of infectious diseases were significantly boosted between 1911 and 1917 by Charles Nicolle's discovery of the coexistence of symptomatic and asymptomatic infections in human populations [2]. Various theories have been proposed to account for this heterogeneity. Since the 1930s, different genetic epidemiological studies, including observations of intra familial clustering of cases and interindividual phenotypic variability among infected people, as well as more sophisticated studies of concordance rates between monozygotic and dizygotic twins, have implicated human genetics as a central factor in susceptibility to disease [3–5]. This was first shown clearly in 1952 by the identification of children with a Mendelian (monogenic traits) primary immunodeficiency, X-linked recessive Bruton's agammaglobulinemia, showing severe infections and a lack of peripheral B cells and serum immunoglobulins [6, 7]. Two years later, the first non-Mendelian genetic factor was identified, when the heterozygous sickle-cell trait was shown to protect against *Plasmodium falciparum* malaria [8].

Until recently, medical genetics was mainly restricted to the study of relatively rare familial diseases that are controlled by a single major gene. However, more recently, increasing efforts have been directed towards defining the genetic basis of common diseases, which have a much greater impact on human health. Studies of animal models and epidemiological studies in humans have shown that many apparently nonhereditary diseases, including infectious diseases [9], develop predominantly in genetically predisposed individuals, and that this predisposition is caused by multiple genes [10]. Identification of these low-penetrance genes would allow the identification of individuals at high risk of disease, increase our understanding of the molecular mechanisms that underlie disease, and help to identify therapeutic targets.

It is obvious that controlled and standardized investigations of the genetics of susceptibility to infectious disease are almost impossible to perform in humans, due to the difficulty of controlling challenge, and because susceptibility is controlled by the cumulative effect and interactions of numerous genetic loci and environmental factors. Mouse models allow application of powerful genomic techniques to identify and validate causative genes underlying human diseases with complex etiologies. In previous studies, host genetic loci controlling a wide variety of complex traits were discovered by using QTL mapping studies in mice [11–21], and other species [22, 23]. However, the genes underlying these mapped loci remain unknown.

A huge drawback in traditional linkage analysis is its low mapping resolution that rarely leads to gene discovery. A novel and promising mouse genetic reference population for high-resolution mapping and subsequently identifying genes underlying the QTL, namely the Collaborative Cross (CC) mouse genetic reference population (GRP) was recently suggested and developed [24–27]. Several genetic reference panels of mice already exist; however, many have constraints.

Most of current animal models used in complex traits diseases genetic mapping, as well as in intestinal tumorgenesis, have limited genetic diversity. This limitation impedes the ability of system biologists to create incorporated network using genetic interactions, epigenetics, environmental factors, microbiota, and other phenotypes. The Collaborative Cross represents a genetically high diversity resource that has been, specifically designed for analysis of complex trait diseases.

The CC mice model was pronounced for the first time in the year 2002 at Complex Trait Consortium (CTC), aimed specifically for genetically complex traits research [24, 25, 28]. This unique genetic resource comprises a set of ~300 Recombinant Inbred Lines (RILs) that were created by full reciprocal matings between eight different mice strains, further called the CC founders. These eight founder strains are genetically diverse, while five common laboratory strains: A/J, C57BL/6J, 129S1/SvImJ, NOD/LtJ, NZO/HiLtJ and three founders are wild-derived strains: CAST/Ei, PWK/PhJ, and WSB/EiJ. As shown in Fig. 1, five laboratory strains and WSB are relatively close to each other in the phylogenetic tree, while two additional wild-derived strains are phylogenetically far from the others. This phylogenetic difference contributes to the high genetic diversity of final CC mice population, which does not exist in other mouse models.

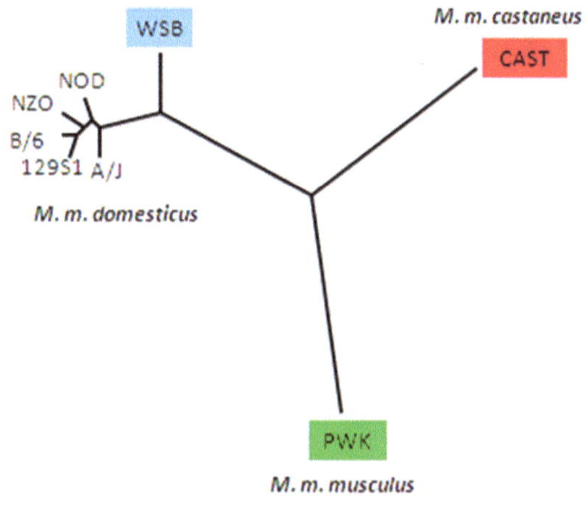

Fig. 1 Phylogenetic tree of the eight founder strains. The five inbred strains and WSB (wild-derived strain) belong to *Mus musculus domesticus* subspecies. Two other wild-derived strains, CAST and PWK, belong to *Mus musculus castaneus* and *Mus musculus musculus*, respectively [24]

The actual inbreeding process of CC lines was initiated in the year 2004. The first step was outbreed matings between different founder strains, resulting in generation of F1, *see* Fig. 2, which was performed at Jackson laboratory in the USA. Later on, F1 population was split into three CC cohorts, geographically located in different continents: (1) America—*University of North Carolina* (UNC), (2) Australia—*Western Australia* by Geniad (GND) , (3) Africa/Asia—*International Livestock Research Institute* (ILRI) in Kenya, which was transferred to *Tel Aviv University* (TAU) at the year 2007 [25]. A fully details and assessing of the CC cohort of Tel Aviv University are provided in recent publications [29–36]. Description about two other cohorts is represented in refs. 26, 27.

The genomes of all CC founder strains are introduced in a single CC line by well-planned breeding scheme, as presented in Fig. 2. In the first phase, there are three generation of outbreed matings, which produce litters with all possible permutations of genomes. The purpose of this phase is to accumulate recombinant

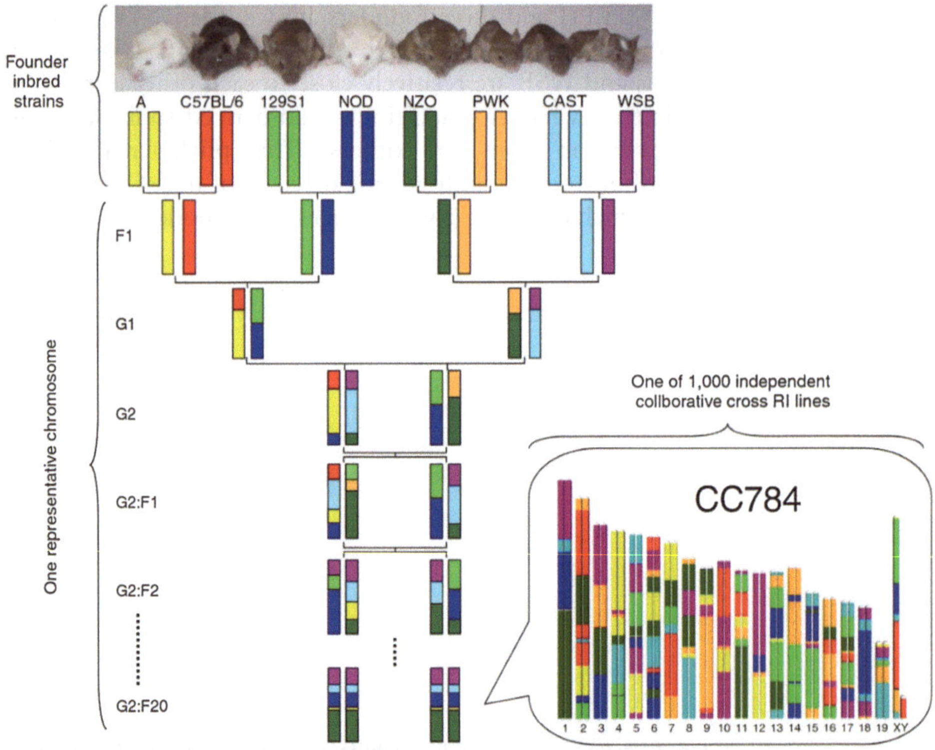

Fig. 2 The eight parental founder strains capture 90% of all genetic variation in CC mice, a funnel breeding scheme is used to randomize variation. A single breeding funnel results in one immortal CC recombinant inbred line that is a mosaic combination of the eight founder genomes

events, rearrange founder's genomes and to represent haplotypes of all eight founders in new generated CC lines. The next phase is inbreeding, brother–sister matings, for more than 20 generations, in order to reach about 99% of homozygosity with a roughly equivalent contribution of the eight founder strains [24]. Litters of 20 and higher of inbreeding generations will be syngeneic (consists of same genetic composition), which will be useful in follower studies. By changing order of founder strains that participate in phase of outbreed mating, we can generate completely different genetic mosaic of a new CC line. Consequently, the genomic component for each CC line will be unique. By using this breeding scheme (Fig. 2) we can generate a Genetic Reference Population (GRP), which are defined as sets of syngeneic individuals with fixed and known genomes. This new GRP would reflect much of the genetic variation present in natural populations.

The advantage of using the CC mice population for modifiers mapping is the numerous genetic variants segregating in the population (there are over 36 million SNPs), and the relatively high level of recombination events compared to other mouse RILs sets (4.4 million SNPS segregate between the founders [37]. The three wild-derived founders of the CC are representing different mice subspecies, *M.m castaneus, M.m musculus*, and *M.m. domesticus* , and contribute many sequence variants not segregating among classical laboratory strains descended from *M.m. domesticus* [38, 39]. Interestingly, based on the recent publications it was shown that the mapped QTLs in CC mice tends to map contrast alleles between the wild-derived strains and laboratory strains [29, 30, 40],and stimulation of QTL analysis in CC population showed that resolution of mapping will be less than 1 Mb [41].

Using this unique mouse resource population we were able to high resolution mapping of genetic factors underlying host susceptibility to very complex infectious diseases including, Aspergillosis (fungi infection) [30] and Klebsiella pneumonia (bacterial infection) [29]. Initially, all CC mice were genotyped with the mouse diversity array (MDA), which consists 620,000 SNP markers. After about six generations of inbreeding, all the CC lines were re-genotyped with 7500 SNPs by Mouse Universal Genotype Array (MUGA), and finally with MegaMUGA SNP, which is consists 75,000 SNPs [33], array to confirm their genotype status. Recently, we have completed an update of a merge genotype file for all the CC lines developed at Tel-Aviv University. Full genome data of the CC lines is available on http://csbio.unc.edu/CCstatus/index.py. Figure 3 represents genomic reconstruction of two CC lines, IL-18 and IL-507.

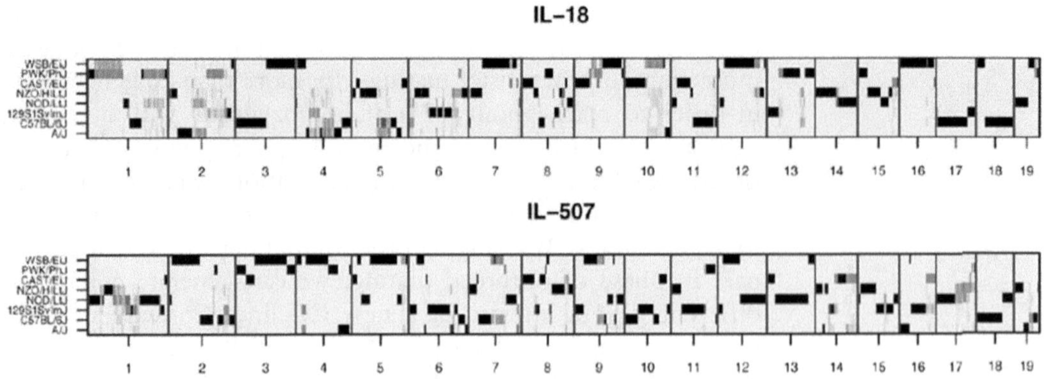

Fig. 3 Reconstructions of the genomes of representative CC lines IL-18 and IL-507 lines. The *x*-axis shows the 19 autosomes, the *y*-axis shows eight founder strains. *Black horizontal bands* represents fixed locus, contributed only by one founder, *gray bands* represent loci with residual heterozygosity [30]

2 Experimental Approaches

2.1 Variety of Genetic Approaches for Mapping Host Response to Infectious Diseases

Mice offer a powerful tool for elucidating the genetic architecture of behavioral and physiological traits, and are complementary to human studies. In crosses between genetically defined strains of mice, chromosomal regions responsible for the genetic variance of complex traits can be mapped as quantitative trait loci (QTL) in experimental populations available for precise study under defined conditions [14, 15]. Importantly, comparative mapping shows that the majority of murine genes have known homologues in the human genome emphasizing the relevance of QTL analysis and gene identification in the mouse model for understanding complex diseases in humans. Once QTL or the corresponding genes have been identified in the mouse, genetic analysis and cloning of the orthologue genes can then be extended successfully to humans. QTL mapping in rodents has been a successful strategy for identifying regions of the genome that play a role in many diseases and traits. Classical approaches of QTL studies involve using of F2 intercross or backcrosses populations, in which two parental strains that differ phenotypically and genetically are bred for two generations. With this strategy, F2 progeny are phenotyped for one or more interested quantitative traits and genotyped at polymorphic markers across the genome. Such data can be then analyzed by exploring the evidence for associations between the phenotype and the genotype at each marker.

These approaches have been used to identify hundreds of QTLs for a wide variety of phenotypes including susceptibility to several infectious diseases (Table 1). However, these study designs have seldom led to the identification of individual genes [42, 48, 49]. This is because the identified QTL regions are very large and typically contain hundreds of genes. Traditional approaches to the

Table 1
Summarization of number of QTL achieved using various strategies for QTL mapping

Approach	Disease	Trait	Chromosome/ QTL	Intervals	Reference
F2	Trypanosomiasis (*T. congolense*)	Mean survival (days)	Chr1 (Tir 3) Chr5 (Tir 2) Chr17 (Tir1)	10–40 cM	[11] [42]
F2	Periodontitis (*P. gingivalis*, *Fusobacterium nucleatum*)	Residual alveolar bone volume	Chr5 (Perio 1) Chr3 (perio2) Chr1 (perio3)	22.96 Mb (95%) 10.04 Mb (95%) 0.24 Mb (50%)	[13]
F2	Gastrointestinal nematode infections (*H. polygyrus*)	Fecal egg counts Total worm count	Cr1 Chr2 Chr8 Chr13 Chr17 Chr19	20–30 cM	[21]
Congenic	West Nile (WN) viruses (strain IS-98 ST1)	Mean survival (d)	(Chr5)Wnv	0.4 cM	[43]
AIL	Trypanosomiasis (*T. congolense* infection)	Mean survival (d)	Chr1 (Tir 3) Chr5 (Tir 2) Chr17 (Tir1)	1–4 cM	[14]
AIL	Malaria (*P.c. chabaudi*)		CHh5 Chr8 Chr17	13–16 cM	[18]
AIL	Gastrointestinal nematode	Fecal egg counts Immunological traits Blood packed cell volume	Chr1 Chr17	20–32 cM 17.9–18.4 cM	[16]
AIL	Pulmonary adenoma	Tumor count	Chr6 (Pas1)	1 cM	[20]
AIL	Trypanosomiasis (*T. congolense* infection)	Mean Survival (d)	Chr1 (Tir 3a and 3b) Chr5 (Tir 2)	1–7 cM	[44]
AIL	Malaria (*Plasmodium chabaudi*)	Parasitaemia	Chr11 (Char8)	23 cM	[19]

(continued)

Table 1 (continued)

Approach	Disease	Trait	Chromosome/ QTL	Intervals	Reference
AIL	Gastrointestinal nematode (*H. bakeri*)	Fecal egg counts Total worm count Immunological traits	Chr5 Chr8 Chr11		[17]
RIL (BXD)	H5N1	Survival time (d)	Chr2 (Qiver2) Chr7 (Qiver7) Chr11 (Qiver11) Chr15 (Qiver11) Chr17 (Qivr 17)	19 Mb (p< 0.001) 14 Mb (p< 0.001) 6 Mb (p< 0.05) 6 Mb (p< 0.05) 16 Mb (0.001)	[45]
RIL (BXD)	Chlamydia psittaci infection		Chr11	1.5 Mb	[46]
RIL (BXD)	Streptococcal sepsis	Survival time Bacteremia Tissue dissemination to spleen	Chr2 Chr2 Chr2 Chr2	12–25 Mb	[47]
CC	KLEB	Survival time (d)	CHh4 (kpr/1) Chr8 (kpr/2) Chr18 (kpr/3)	7.03 Mb (95% CI) 5.44 Mb (95% CI) 18.06 Mb (95% CI)	[29]
CC	*Aspergillus fumigatus*	Survival time (d)	Chr2 (aspr7) Chr3 (Aspr/6) Chr8 (Aspr/1) Chr10 (Aspr/2) Chr10 (Aspr/4) Chr15 (Aspr/3) Chr18 (Aspr/)	8.82 Mb (95% CI) 17.06 Mb (95% CI) 16.67 Mb (95% CI) 5.93 Mb (95% CI) 14.60 Mb (95% CI) 9.21 Mb (95% CI) 10.06 Mb (95% CI)	[30]
CC	Inflammatory bowel disease	7 Components of disease	Chr 12 (Ccc1) Chr 14 (Ccc2) Chr1 (Ccc3) Chr8 (Ccc4)		[28]

analysis of quantitative traits in mice evolved in an environment where genotyping was the most difficult and expensive step. Thus, for many years, crosses like F2s were used to map QTLs because they limited the amount of genotyping that was required. Only one or two recombinations per chromosome occur in these populations, so only a handful of markers per chromosome are needed. Unfortunately, this lack of recombination prevented these studies from identifying smaller QTLs.

Due to these limitations, alternative fine mapping approaches for QTL were developed including congenic strains and advanced intercross lines (AIL) [15, 17–20, 43, 44, 50] (Table 1).

Congenic strains are created by introgressing a small interval from one of the two inbred strains used to map the QTL onto the other strain by repeated rounds of backcrossing, in conjunction with maker-based selection for the putative QTL region. This approach is extremely time, animal, and labor intensive, and has only occasionally been brought to fruition [51–57]. Furthermore, it is now clear that traditional approaches are based on an unreliable premise; namely, that each locus identified using an F2 cross is caused by a single polymorphism. In fact, it is often true that apparently large QTLs detected in F2 crosses are due to multiple smaller QTLs that happen to be clustered in a single region [58, 59]. In other cases, F2 crosses may fail to identify true QTLs. For example, if two or more QTLs with opposite effects on the phenotype are closely linked, they may cancel out one another's effects on the phenotype. It is also suggested that gene-gene interactions are sometimes required such that when a QTL region is broken into smaller pieces, none of them will affect the phenotype individually [53]. Due to all of these limitations, F2 populations are being supplanted by other approaches.

Another alternative approach for QTL study is to perform linkage disequilibrium mapping in highly recombinant or outbred populations. In contrast to studies conducted in F2 intercross or backcrosses, in which there is 40–60 Mb between recombination events, distance between historical recombination events in outbred models is generally <5 Mb. With large sample sizes, the distance between recombination events within the population (as opposed to within an individual animal) is even smaller, leading to fine mapping of QTL and rapidly narrowing the number of potential candidate genes. The advanced intercrosses lines (AIL), which were first proposed by Darvasi and Soller [60], is the simplest outbred model. Those strains are created by breeding two inbred strains for many generations, resulting in the accumulation of many historical recombinations. Although AIL populations have been a successful model for fine-mapping multiple traits including obesity [61], metabolic syndrome [62], methamphetamine sensitivity [63], and others [15, 18, 20, 44, 64–67] (Table 1), it is still limited by its ability to identify only those QTL that segregate within the two starting populations.

In attempt to overcome this limitation, a similar model with some design modifications called heterogeneous stock (HS) animals has been suggested. Unlike the AIL, the HS strains are created by combining eight inbred strains and then outbreeding in a way that minimizes inbreeding. After 50 generations of outbreeding, the genetic make-up of the resulting progeny represents a random mosaic of the founding animals, with the average distance between recombination events approaching a single centimorgan [68], enabling the fine-mapping of QTL to only a few megabases. In contrast to AI, the underlying genetic architecture is more complex and the underlying ancestral haplotypes need to be determined prior to analysis. Determining this underlying structure provides increased information from what is obtained from the genotypes themselves and lends to improved genetic mapping [68]. This is done by determining ancestral probabilities with a dynamic programming algorithm initially developed by Mott and colleagues [68].

Despite the clear successes of using HS mice and rats for genetic fine-mapping, one of the disadvantages of the HS strategy is that, as with an F2 intercross, each animal is genetically and phenotypically distinct. Because of this, each time a new study is started, not only does a new group of animals need to be phenotyped, but all animals also need to be fully genotyped. The highly recombinant nature of these populations requires relatively large numbers of animals for sufficient statistical power [68–70] as well as high-density genotyping platforms [71].

2.2 Characterizing the Phenotypic Response of Commercial Inbred Mice and Collaborative Cross Mice to Experimental Periodontitis Using *Porphyromonas gingivalis* and *Fusobacterium nucleatum*

Periodontitis is the most common chronic inflammatory disease in humans, which results in destruction of tooth-supporting tissues and eventually leading to tooth loss. This process is characterized by destruction of the periodontal ligament, formation of periodontal pockets, and alveolar bone resorption [72]. The disease is initiated by periodontal pathogenic bacteria, which accumulate as sub gingival biofilm and stimulate an inflammatory response in the host gingiva [73]. An excessive or sustained response leads to chronic inflammation, which is a potent amplification system for recruiting humoral and cellular components of the immune system. Recently, several lines of evidence suggest that there is a significant genetic component associated with the susceptibility to chronic periodontitis [74, 75].

As a step towards identifying and subsequently cloning these genetic factors, our group has generated an (A/J × BALB/cJ) F2 mouse resource population, while A/J and BALB/cJ are the resistant and susceptible founders to the infection respectively. Oral mixed infection system (of the two anaerobic gram-negative bacteria *Porphyromonas gingivalis* and *Fusobacterium nucleatum*), as was previously described by Polak [76] was used to induce experimental periodontitis. The phenotype is measured as the residual alveolar bone volume, in mm^3, after infection and quantified by the microCT scan, which provides an accurate measurement of the bone volume around the animal teeth [77]. Initial phenotypic analyses showed a normal distribution among the 408 F2 mice populations, suggesting a polygenic trait (Fig. 4). Based on the phenotype of 408 mice, we performed a genome-wide search for QTL associated with the disease. To maximize our ability to detect QTL contributing to the variations in the responses to periodontitis, we genotyped only the phenotypic extremes of the F2 animals. Genome-wide association analysis detected two highly significant QTL on chromosomes 5 and 3, negative log 10 *P* value (log*P*) 4.79 and 3.93, respectively, with genome-wide significance (Fig. 5). Both peaks exceeded 95% genome-wide significance (log*P* = 3.5) based on log*P* per-mutation analysis. A third significant QTL was mapped on chromosome 1 (Fig. 2), with log*P* 2.47, at 50% genome-wide significance (log*P* = 2.3). The inheritance of QTL on chromosome 1 proved to be recessive, which means that the phenotype is seen only in the homozygote genotype, whereas the QTL on chromosomes 3 and 5 were additive, indicating that the phenotype strengthened from heterozygote genotype to homozygote genotype status. To our knowledge, this was the first report to map QTL associated with host susceptibility to periodontitis in mice. However, unfortunately, these QTL were mapped within large genomic interval of 30--50 MB (consists about 500 genes), as expected in F2 mapping [13].

In attempt to map a QTL with narrower intervals, we more recently assessed the phenotypic response of a total of 272 mice (103 females and 169 males) from 23 different CC lines (average:

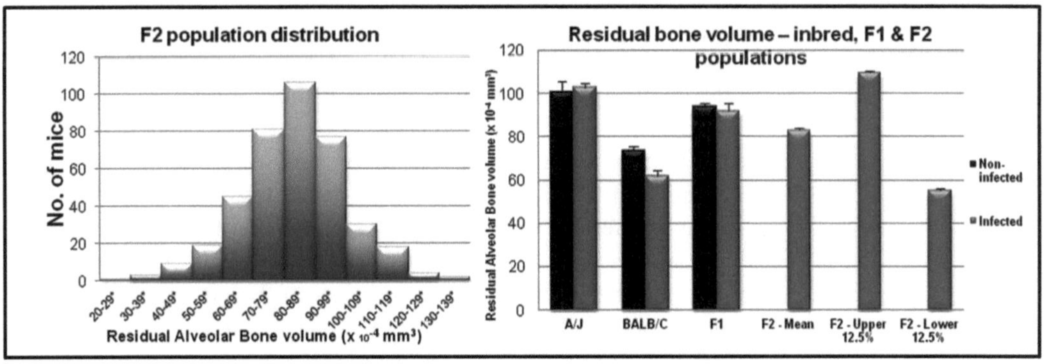

Fig. 4 Distribution of residual alveolar bone volume in 408 A/J × BALB/cJ F2 mice. (*Right*) Residual alveolar bone volume of inbred lines in F1 and F2 mouse populations. Each of the two inbred lines (A/J and BALB/cJ) and the F1 population (first generation) are represented by *two columns*. The *left* (*black*) *column* represents the control group (non-infected), and the *right* (*gray*) *column* represents the infected groups. The F2 infected mouse population is represented by *three single gray columns* [mean, high, and low (12.5% each)]. The F1 population shows a resistant phenotype. The F2 population displays a high diversity in the phenotypic response to the infection. Nos. of mice: *n* (A/J control) = 7; *n* (A/J infection) = 7; *n* (BALB/C control) = 6; *n* (F1 control) = 7; *n* (A/J control) = 7; *n* (F2 mean) = 395; and *n* (F2 12.5%) = 50 [13]

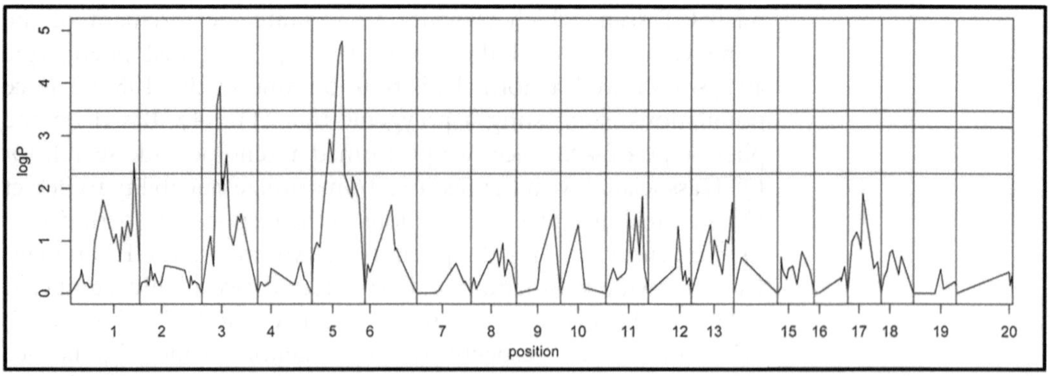

Fig. 5 Scan of periodontal infection in mice using F2 approach. The *x*-axis shows chromosome number, and the *y*-axis shows the statistical significance of the association, measured as –log (*p* value). The horizontal lines represent genome-wide significance thresholds of $E=0.50$ (*lower line*), 0.10 (*middle line*), and 0.05 (*upper line*), derived by permutation ($E=$percentage of permutations in which the genome-wide maximum did not exceed the threshold). Two highly significant QTL ($E<0.05$) were mapped on chromosomes 3 and 5 and one significant QTL ($E<0.5$) on chromosome 1 [13]

11.8 mice per line) to the *Porphyromonas gingivalis* and *Fusobacterium nucleatum* infection using the same methodology of disease induction and bone volume measurement (the data was already described by our group) [32]. The study design was to divide mice from each CC line into two groups, control and infected, and the differences between the bone volumes among these two groups were calculated and considered to be the value of the bone loss phenotype. In addition, the significance of difference was estimated using one way ANOVA and used to determine the susceptibility level of each different CC line. Figure 6 shows the mean control bone volume (CBV) , residual bone volume (RBV) after infection for the 23 CC lines. Of these, six lines (Il26, IL57, IL72, IL182, IL196, and IL711) showed a significant bone loss ($p<0.05$) and were considered to be susceptible to the infection, meaning that 25 % of the CC lines were highly susceptible. The remaining lines were probably a mixture of resistant lines and moderately susceptible lines that did not pass the significance threshold. Broad sense heritability was estimated for CBV, RBV, and LBV and found to be 0.4, 0.4, and 0.2 respectively. Based on these promising results, the number of the tested CC lines was extended to one hundred lines. Subsequently, the data of bone loss phenotype and genotype data of each line will be analyzed by HAPPY software and QTL mapping will be performed. Genomic interval of each QTL will be determined and candidate genes underlying these QTL will be identified (study in progress).

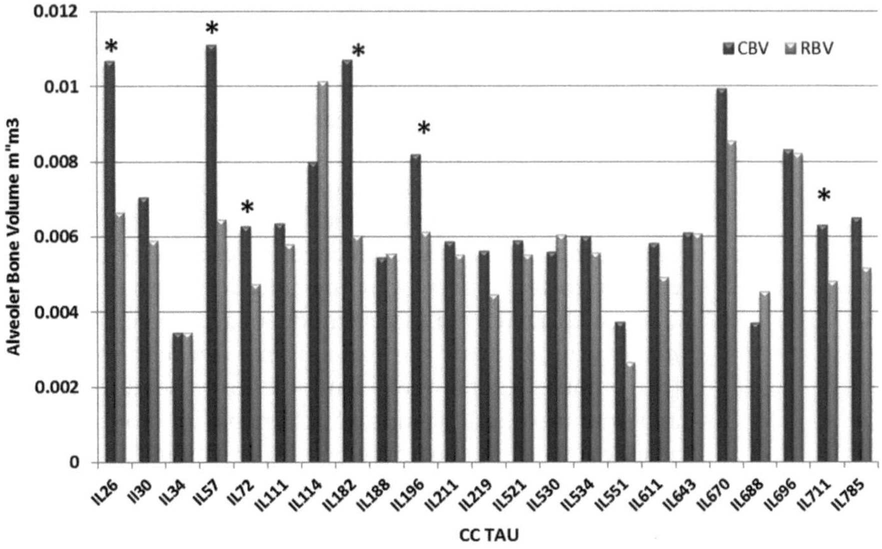

Fig. 6 The mean control bone volume (CBV) and residual bone volume (RBV) after infection for the 23 CC lines. Of these, six lines (Il26, IL57, IL72, IL182, IL196, and IL711) showed a significant bone loss ($p < 0.05$) and considered to be susceptible to the infection, i.e., 25 % of the CC lines were highly susceptible. The remaining *lines* were probably a mixture of resistant lines and moderately susceptible lines that did not pass the significance threshold. Axis *X* represents the CC Tau lines, whereas the *Y* axis represents the alveolar bone volume m″m3 [32]

2.3 Host Genetic Diversity Influences the Severity of Pseudomonas aeruginosa Pneumonia in the Inbred Mice and Collaborative Cross Mice

Pseudomonas aeruginosa is one of the major and dreaded sources of infections responsible for millions of cases each year and 10–15 % of all healthcare associated infections, with more than 300,000 cases annually in the EU, North US, and Japan. Patients at risk of acquiring *P. aeruginosa* are particularly the ones who are hospitalized in intensive care units (ICU) and who may develop ventilator-associated pneumonia (VAP) and sepsis [78]. The clinical outcome of *P. aeruginosa* infections may be extremely variable among individuals at risk and CF patients. In particular, heterogeneity in the severity of chronic bronchopulmonary *P. aeruginosa* infection is well documented in cystic fibrosis (CF), while it remains to be established in other patients [79]. According to clinical studies, the progression and severity of pulmonary disease in CF do not appear to correlate with the type of cystic fibrosis transmembrane regulator (CFTR) variant and rather seem to be largely dependent on secondary factors [80]. Much influence on disease outcome has been attributed more to different *P. aeruginosa* phenotypes rather than to host genetic background. Consistent with its larger genome size and environmental adaptability, *P. aeruginosa* contains the highest proportion of regulatory genes observed for a bacterial genome, which lead to large and complex phenotypic versatility. Thus, early studies from different groups [81–84] highlighted the responsibility of particular *P. aeruginosa* phenotypes for differential disease manifestations and pathogenesis. However, more recently, special interest has shifted toward understanding host genetic variation that alters the outcome of *P. aeruginosa* infection [85]. Identifying and tracking risk factors for *P. aeruginosa* infection remains one of the major research challenges.

As a first step toward the analysis of genetic traits influencing resistance and susceptibility to *P. aeruginosa* infection and the characterization of pathogenetic mechanisms, nine inbred mouse strains of differing ancestry were screened and chosen for the known differences in their ability to overcome infections with various pathogens. Using a characterized mouse model of acute infection with *P. aeruginosa* clinical strains and previous experience in this model system [86], mouse strains were identified for presenting deviant clinical and immunological phenotypes amenable for biological and genetic analyses. Nine different inbred mouse strains, including A/J, BALB/cJ, BALB/cAnNCrl, BALB/cByJ, C3H/HeOuJ, C57BL/6J, C57BL/6NCrl, DBA/2J, and 129S2/SvPasCrl were infected with 5×10^6 CFU of planktonic *P. aeruginosa* clinical isolate AA2 via intratracheal injection, and monitored for change in body weight and mortality over a period of 7 days. As shown in Fig. 7 a wide range of survival and weight loss was observed among different inbred mice, and different inbred murine strains were highly variable in their response to acute airway infection. During a time course analysis, a wide range response to *P. aeruginosa* infection has been observed both in the survival rate and body weight change of different inbred murine strains. Most notably, deviant clinical phenotypes were observed being the A/J, 129S2/SvPasCRL, and DBA/2J as the most susceptible while BALB/cAnNCrl and C3H/HeOuJ the most resistant murine strains. Other murine strains (BALB/cJ, BALB/cByJ, C57BL/6J, and C57BL/6NCrl) showed intermediate phenotype. These results confirmed that severity to *P. aeruginosa* infection is clearly affected by host genetic background, and also opened the question whether other mouse genetic background may better recapitulate the pulmonary abnormalities of CF patients during *P. aeruginosa* infection [87].

Recently, a total of 92 (50 males, 42 females) mice from 17 different CC mouse lines were assessed for their host response to *P. aeruginosa* infection. Briefly, mice were anesthetized and infected by intratracheal injection with a 10^6 colony forming unit (cfu) implanted into the lung via the cannula, with all lobes inoculated as described [88]. The CC lines showed a wide range of Survival time (ST) ranging from complete resistance (100% survival after seven days post-infection) to lethal disease (100% death after 1.5 days), while A/J mice showed an intermediate phenotype (30% of mortality rate after 7 days post-infection) (Fig. 8). Similarly, CC lines had a wide variation in body weight (BW) response to *P. aeruginosa* infection, ranging from a 23% decrease in BW after 3 days to those showing an almost total recovery of change in BW after five days (Fig. 8). A/J mice lost 16% of their change in BW after 3 days but

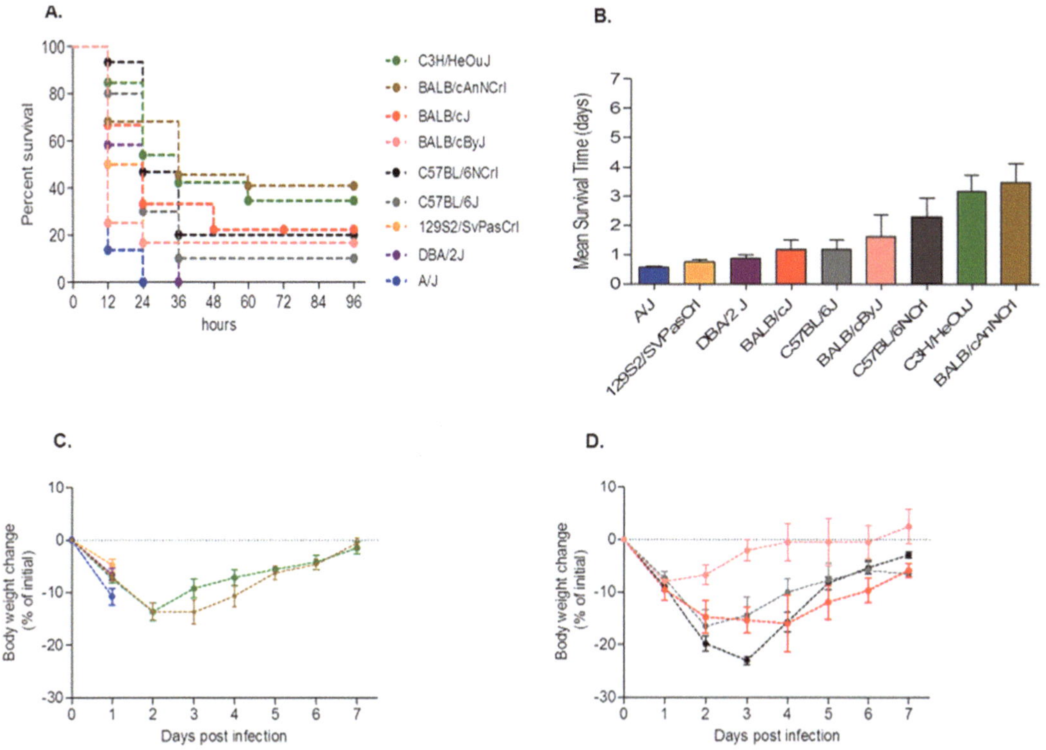

Fig. 7 Survival, body weight and mean survival time after *P. aeruginosa* infection in inbred mouse strains. A/J (*n*=22), BALB/cJ (*n*=9), BALB/cAnNCrl (*n*=8), BALB/cByJ (*n*=12), C3H/HeOuJ (*n*=26), C57BL/6J (*n*=10), C57BL/6NCrl (*n*=15), DBA/2J (*n*=12), and 129S2/SvPasCRL (*n*=12) mice were inoculated with 5×10^6 CFU of *P. aeruginosa* clinical isolate AA2, and monitored for survival (**a**) and weight change for a period of 7 days after infection (**c**, **d**). In addition, mean survival time was calculated based on the survival curve (**b**). *Bars* represent mean values and the error bars the standard error of the mean (SEM). The data are pooled from two to four independent experiments [87]

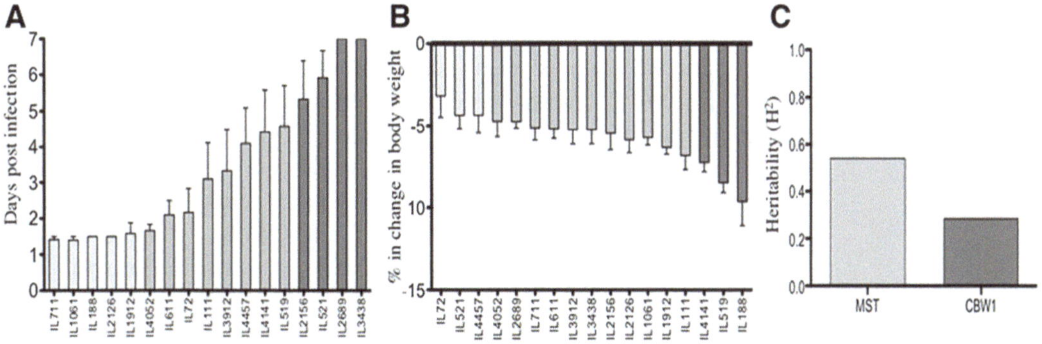

Fig. 8 Evaluation of mean survival times (MST), change in body weight (CBW1) and H2 of CC lines after P. aeruginosa airway infection. The CC mice resource population had a strong wide-response to *P. aeruginosa* airway infection in the MST (**a**) and CBW1 traits (**b**). MST and CBW1 of CC lines are arranged in increasing order of mean magnitude. Based on Bonferroni's Multiple Comparison Tests (BMCT) three distinct groups have been identified among 17 CC strains infected with *P. aeruginosa* and are indicated as scales of *gray*. Estimates of broad sense H2 (**c**) have been evaluated for MST and CBW1 [36]

did not recover completely after 7 days. At day 7 post-infection, bacterial cells were not recovered in the organs (blood, liver, and lung) of surviving mice (data not shown), indicating that differences in morbidity as assessed by recovery of body weight may be independent from bacterial clearance is independent from differences in morbidity as assessed by recovery of body weight. These results demonstrate that *P. aeruginosa* opportunistic infection has a wide range of disease phenotypes affected by multiple host genetic factors, such as multiple genetic loci or polymorphic variations. Broad sense heritability of both survival time after infection and body weight loss was estimated and found to be 0.54 and 0.28, respectively. These data strongly proved the influence of genetic profile rather than environmental factors among the CC lines during *P. aeruginosa* infection. Future mapping of key genetic loci/genes involved in the severity of *P. aeruginosa* infection will be carried out with the use of additional CC lines and further traits of disease phenotypes will be assessed [36].

2.4 Identify Genetic Factors (QTL) Associated With Klebsiella pneumoniae Infections Using the Collaborative Cross Mouse Population

Almost 90% of patients in intensive care units under ventilation become colonized in the upper respiratory tract with gram-negative bacteria. Gram-negative bacteria are causative agents for up to 70% of nosocomial pneumonia, a major cause of mortality in immunocompromised patients. *Klebsiella pneumoniae* (Kp) is a common nosocomial pathogen causing severe infections that can be considered a paradigm for gram-negative nosocomial pulmonary infections. Kp produces two major surface glycoconjugates, lipopolysaccharides (LPS, O-antigen) and capsular polysaccharides (CPS, K-antigen). Based on the molecular variability of the K and O-antigens, *Klebsiella* is serotyped into 77 capsular (K) and 9 (O) serotypes that differ significantly in pathogenicity and epidemiology. Because K2O1-like serotypes that are not recognized by C-type lectins cause infections exclusively in immunocompromised patients, it is conceivable that in immunocompetent hosts such serotypes are recognized by an as yet undefined innate immune mechanism.

Recently, a study was established in our labs and described by [29], which examined the response of four immune competent inbred strains and 73 Collaborative Cross lines to infection by virulent Kp K2O1-like serotypes. Indeed, ten females from each of BALB/CJ, DBA/2J, C3H/HeJ, and C57BL/6J) and 328 CC mice (184 females and 144 males) were included in this study and were determined for their phenotypic response to *Klebsiella* infection. Briefly, mice were challenged by intraperitoneally (IP) with 10^4 CFU of *Klebsiella pneumoniae* strain K2 (KP-2), and clinical assessment of susceptibility to infection during 15 days post-challenge was based primarily on survival time. All mice died during the infection, but with heritable variation in survival time. BALB/CJ mice were highly susceptible, DBA/2J and C3H/HeJ were highly resistant, and C57BL/6J was intermediate. Mean survival time of BALB/CJ, C57BL/6J, C3H/J, and DBA/2J was 2 days (s.d = 0.6), 2.8 days (s.d = 0.56), 3.8 days (s.d = 1.16), and 4 days (s.d = 1.88) respectively (Fig. 9). Based on statistical analysis, BALB/cJ was significantly different from the three other strains. The 328 CC mice also responded variably with mean survival time between 1 and 12 days. Differences in survival between the 73 CC lines were highly significant ($p < 0.0001$) and are shown in Fig. 9. Broad sense heritability was also estimated and found to be 0.17 (Table 2). Although the variation between lines was highly significant, indicating that the response to infection was heritable, there was also considerable variability within some lines. Mice that survived past the seventh day of infection tended to survive to the end of the experiment, suggesting this was a critical point in the disease progression.

Subsequently, alive/dead survival status for QTLs at different time points was tested to identify early- and late-acting QTLs. Figure 10 shows the three QTLs we found associated with host susceptibility to Kp infection at genome-wide $E < 0.5$ (FDR = 8.3 %). These QTL, named Kprl1–Kprl2 and Kprl3, (Klebsiella pneumonia-resistant locus) were located on chromosomes 4, 8, and 18, respectively, Kprl1 and 2 were mapped with 50 % confidence intervals (50 % CIs) of 0.48 and 0.51 Mb and 95 % CIs of 7.03 and 5.44 Mb.

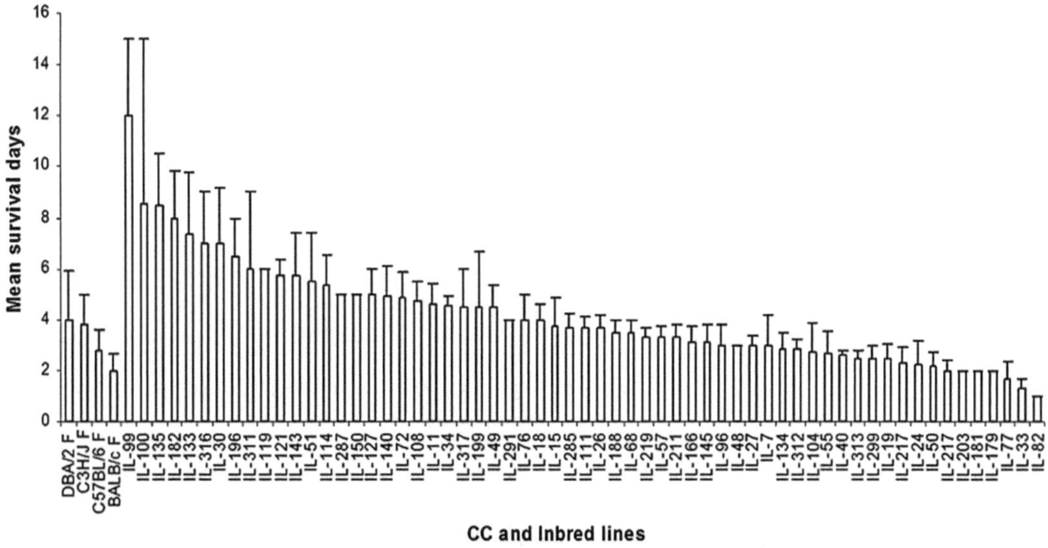

Fig. 9 Mean survival times after infection with *Klebsiella pneumoniae*. Mice from four classical inbred and representative Collaborative Cross (IL) lines were challenged and average survival times in days were calculated for each line, with standard errors indicated [29]

Table 2
Collaborative Cross, genetic variation among lines

Experiment	Trait	H^2	CV_g
PER	CBV $_{(23,139)}$	0.4	0.26
	RBV $_{(23,133)}$	0.4	0.24
	LBV $_{(23,117)}$	0.2	NA
KLEB$_{(60,345)}$	Mean survival (d)	0.17	0.33
LPS$_{(16,296)}$	Mean survival (Hr.)	0.19	0.158
P.A $_{(17,92)}$	Mean survival (d)	0.54	
	Body weight (%)	0.28	

Heritability and genetic coefficient of variation (CV_g) according to trait and experiment. In parentheses under column "Experiment," number of lines, number of mice. *PER* periodontal infection, *KLEB Klebsiella pneumoniae* infection, *LPS* lipopolysaccharide, *PA Pseudomonas aeruginosa*, *(d)* day, *H2* broad-sense heritability, CV_g genetic coefficient of variation

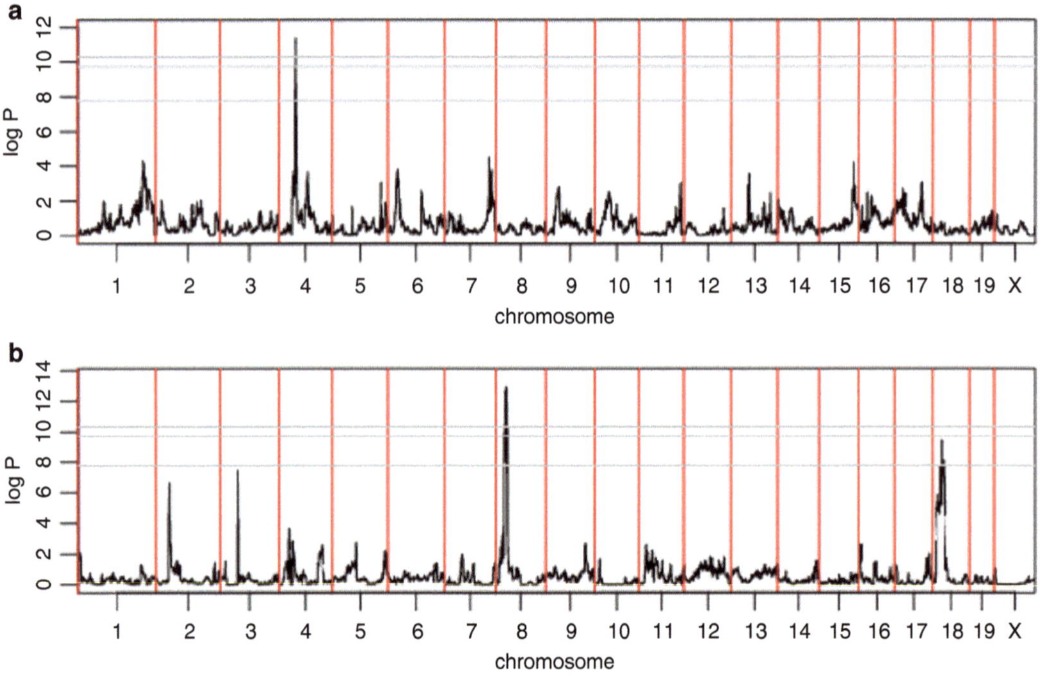

Fig. 10 Genome scans for days with significant QTLs. Three QTL associated with survival time were detected, on days 2 and 8 after infection with Kp. Experiment-wide thresholds of significance at 50, 90, and 95 % levels are log$P$=7.8, 9.7, and 10.3, respectively (e.g., the p% threshold means that in p% of permutations the genome wide maximum log$P$ across all analyses at different time points did not exceed the threshold). A. Scan at day 2 post infection showing QTL Kprl1 on chromosome 4 at day 2 of infection. B. Scan at day 8 post infection showing QTLs, Kprl2 and Kprl3 on chromosomes 8 and 18 [29]

2.5 Towards Identifying Genes Underlying Host Susceptibility to Sepsis Using Collaborative Cross Mice

Sepsis is a systemic inflammatory response syndrome arising from infection [89]. Despite the advances in antibiotic therapy and aggressive operative intervention, sepsis, severe sepsis, and its outcome are still reported to contribute to significantly high morbidity and mortality [90]. Much of the damage inflicted on the septic host is attributable to the host response to microbial toxins. One of the most important microbial toxins in the pathogenesis of sepsis is lipopolysaccharide (LPS) [91]. LPS is the major structural component of the outer membrane of gram-negative bacteria and accounts for approximately 70 % of the outer leaf. When bacteria multiply or die, LPS is released from their surface. The extracellular recognition of LPS by innate immune system triggers intracellular signal transduction that leads to activation of the transcription factor nuclear factor-kB (NF-kB) and induction of the acute phase response (APR) [92]. Similar pathogen-associated molecular pattern mediators exist in gram-positive bacteria and fungi that induce a potentially harmful host response during severe sepsis. The cell wall of gram-positive bacteria contains lipoteichoic acid (LTA) and peptidoglycan (PepG), which can activate leukocytes, stimulate the generation of proinflammatory cytokines, and hence, cause a moderate systemic inflammatory response syndrome [93]. Genetic epidemiologic studies suggest a strong genetic influence on the outcome of sepsis.

In a previous study, Iris Pinheiro et al. [94] used a backcross population of 140 mice by crossing (BxS) F1 hybrid females with C57BL/6 males, in order to identify loci responsible for LPS resistance of SPRET/Ei mice. All 140 offspring, of which 90 (64 %) survived the LPS challenge, were genotyped using 87 microsatellite markers evenly distributed over the genome, followed by a genome-wide linkage analysis. Two genome-wide significant QTL were identified, one on chromosome 2 (marker D2Mit510 at 65 cM; $\log 10(P) = 3.49$, Wald test) and one on the X chromosome (marker DXMit135 at 69 cM; $\log 10(P) = 3.4$, Wald test) spanning a 10 cM interval [94].

Recently, we initiated a study to characterize the genes underlying the host susceptibility to sepsis using the CC mouse model. At the time of writing this chapter, 16 lines of the Collaborative Cross mouse population were examined for their survival time following the challenge with LPS. Briefly, 296 mice from 16 different lines (age 11–12 weeks) were injected Intraperitoneally (IP) with LPS (15 mg/1 kg per mouse) dissolved in PBS in total volume of 100 μl. Based on our preliminary results (Fig. 11), the CC mice showed a high variation in their response to the LPS challenge. While some of the lines died after 30 h (on average), other lines survived the challenge. According to one way ANOVA analysis, there was a significant ($p < 0.05$) gender effect in survival after LPS injection. The mean survival of females and males separately is shown in Fig. 12. Moreover, our results show that all assessed CC lines developed hypothermia rather than hyperthermia (Fig. 13), confirming that the mice had acute inflammatory phase and that the challenge achieved its goal.

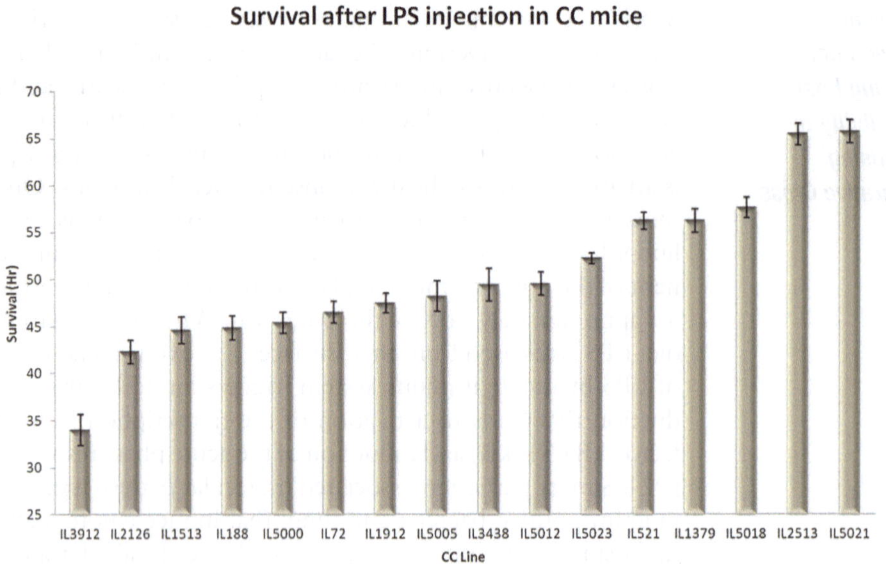

Fig. 11 Mean survival time (hours) in the different 16 CC lines (±SEM). Mice were injected intraperitoneally with LPS (15 mg/kg mouse). Survival data collected at different time points. *X*-axis—IL and *Numbers* represent the different CC lines, and CC population mean, while *Y* axis represents mean survival time with standard errors included. Significant variation was found between the different CC lines at $p<0.05$ (Unpublished)

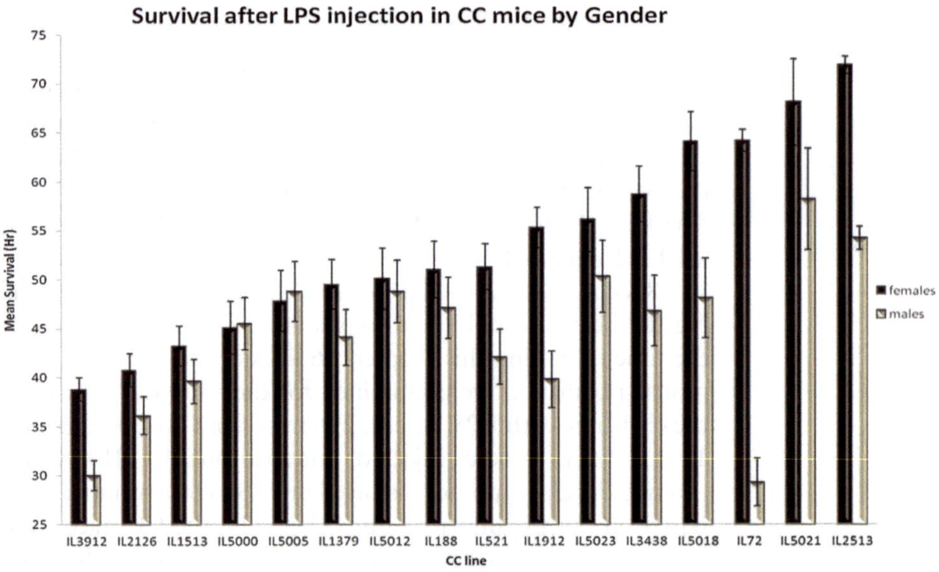

Fig. 12 Mean survival (hours) in the different CC lines separated by gender. Mice were injected intraperitoneally with LPS (15 mg/kg mouse). Survival data collected at the different time points. *X*-axis—IL and *Numbers* represent the different CC lines, and CC population mean, while *Y* axis represents mean survival time of male and female of each line with standard errors included. Significant variation was found between the different CC lines at $p<0.05$ (Unpublished)

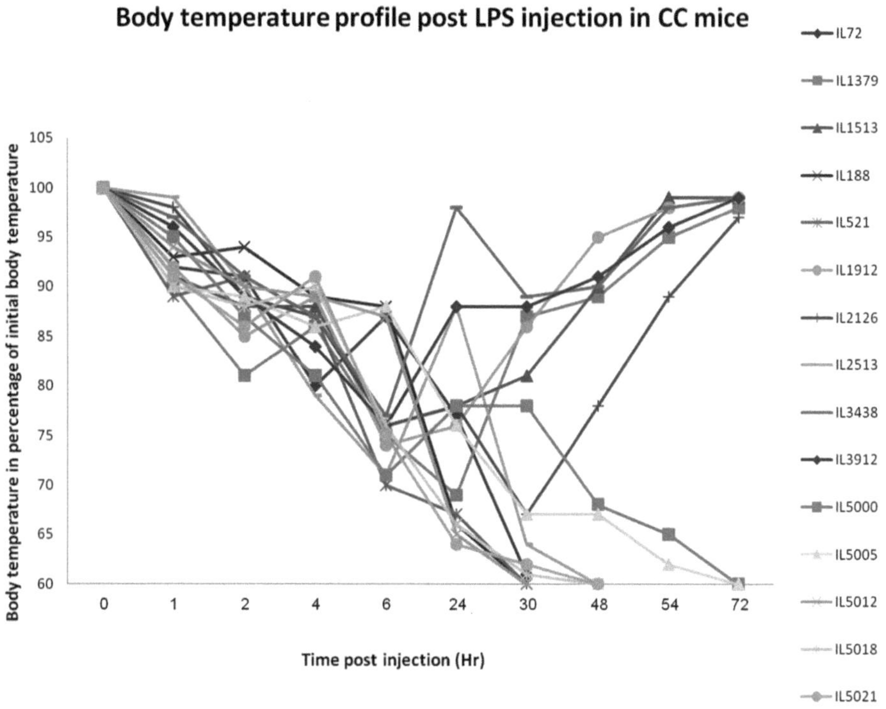

Fig. 13 Body temperature profile after LPS injection: mice were injected with LPS (15 mg/kg). Measurements of body temperature were performed at ten different time points (0, 1, 2, 4, 6, 24, 30, 48, 54, and 72 h) after LPS injection, using thermometer designed for small animals. Results presented as percentage of the initial body temperature (at time 0). *X*-axis represents the different time points where body temperature was measures of the tested CC lines, while *Y* axis represents the body weight of the tested CC lines at different time points, as percentage of the initial body (Unpublished)

The significant variance in the different CC lines was observed despite the fact that experiments were carried under the same environmental conditions, suggesting that there is a significant genetic effect on the response to LPS challenge. These results confirm the power of the CC as a genetic resource for dissecting complex trait. Assessment of more CC lines is ongoing and LTA challenge will also be performed. Once the phenotype data is available, genetic analysis and cloning of the orthologue genes can then be extended successfully to humans.

3 Conclusions

Infectious diseases are major constraints to human and livestock health worldwide. Studying the genetic factors of susceptibility to infectious diseases has become as a fundamental issue for our understanding of the pathogenesis of bacterial infections and its clinical variability between individuals infected with the same microbe. However, due to the complexity and limitations of carrying such investigations in humans, increasing efforts have been directed towards defining chromosomal regions responsible for the genetic variance of such complex traits as mapped quantitative trait loci (QTL) in experimental populations, especially the mouse resource. In the past two decades, a variety of approaches for mapping genes underlying QTL, based on crosses between genetically defined strains of mice, have been suggested by the scientific community. Yet for the most part, albeit with some important exceptions, the genes underlying these mapped loci remain unknown. However, with recent advances in the mouse genome project, high-throughput genotyping techniques, and new suggested models for mapping (QTL), such studies are becoming more feasible, although still daunting in terms of cost and scale.

Recombinant inbred (RI) strains, where the strains are generated by long-term inbreeding of the progeny of F2 crosses, became popular for the study of complex traits and biological systems in both medical and life sciences applications because genotyping is only required once (what has been described as the "genotype once, phenotype many times" paradigm), and replicate individuals can be produced with the same genotype at will, thereby allowing for optimal case–control and gene-by-environment designs [95]. The most advanced recombinant inbred lines (RIL) are the BXD recombinant inbred strains [45–47, 96, 97], and the newly generated Collaborative Cross population, which covers a genetic diversity twice as large as that of the human population and which enables high resolution mapping [30, 38, 40, 98, 99] (Table 1). The highly genetically diverse mouse resource population, Collaborative Cross (CC) mice, was designed to provide a new mouse resource for high resolution analysis of complex traits, with particular emphasis on traits relevant to human health in its broadest aspects [24, 100]. This unique reference genetic resource will eventually comprise a set of approximately 300 recombinant inbred lines (RIL) created from full reciprocal mating of eight divergent strains of mice: A/J, C57BL/6J, 129S1/SvImJ, NOD/LtJ, NZO/HiLtJ, CAST/Ei, PWK/PhJ, and WSB/EiJ . Controlled randomization was performed during the breeding process to break up large linkage disequilibrium blocks and to recombine the natural genetic variation present in these inbred strains with the aim to create a unique and inexhaustible resource of Recombinant inbred (RI) strains exhibiting a large phenotypic and genetic diversity [38]. Three founders of the CC (CAST/EiJ, PWK/PhJ, and WSB/EiJ) are wild-derived, representing the subspecies *M.m castaneus, M.m musculus*, and *M.m. domesticus*, respectively, and which contribute a large number of sequence variants not segregating among classical strains descended from *M.m. domesticus* (most classical strains differ from the reference C57BL/6J at about four million SNPs, whereas PWK and CAST each differ at about 17 million SNPs, and WSB at six million [37]. Consequently, quantitative trait locus (QTL) mapping using the CC is likely to uncover novel QTLs involving contrasts between the wild-derived strains. This was shown in a pilot experiment in which we fine-mapped QTLs associated with survival after infection by *Aspergillus fumigatus*; we mapped eight QTLs, five of which involved contrasts with wild-derived strains and which would not have been present in a cross between classical strains [30]. That study, and another by our US collaborators [40], further showed that by incorporating variation data from the genome sequences of the CC founders—available from the Sanger Mouse Genomes Project [37] and restricting attention to variants whose differences across the founders are consistent with the pattern of action of the QTL [101], the list of candidate genes under QTLs can be significantly refined. These studies confirm that by phenotyping a relatively modest number of CC lines (around 100 lines), with sufficient replication, it is possible to map QTLs to a resolution of about 1 Mb [41].

References

1. Dubos RJ (1950) Louis Pasteur, free lance of science. Little Brown, Boston, MA
2. Nicolle C (1937) Destin des maladies infectieuses. Alcan, Paris, p 301
3. Abel L, Dessein AJ (1998) Genetic epidemiology of infectious diseases in humans: design of population-based studies. Emerg Infect Dis 4:593–603. doi:10.3201/eid0404.980409
4. Burgner D, Levin M (2003) Genetic susceptibility to infectious diseases. Pediatr Infect Dis J 22:1–6. doi:10.1097/01.inf.0000043008.07700.6a
5. Kwiatkowski D (2000) Science, medicine, and the future: susceptibility to infection. BMJ 321:1061–1065
6. Hitzig WH (2003) The discovery of agammaglobulinaemia in 1952. Eur J Pediatr 162:289–304. doi:10.1007/s00431-003-1153-7
7. Bruton OC (1952) Agammaglobulinemia. Pediatrics 9:722–728
8. Allison AC (1954) Protection afforded by sickle-cell trait against subtertian malarial infection. Br Med J 1:290–294
9. Somech R, Amariglio N, Spirer Z, Rechavi G (2003) Genetic predisposition to infectious pathogens: a review of less familiar variants. Pediatr Infect Dis J 22:457–461. doi:10.1097/01.inf.0000068205.82627.55
10. Pharoah PDP, Antoniou A, Bobrow M, Zimmern RL, Easton DF, Ponder BAJ (2002) Polygenic susceptibility to breast cancer and implications for prevention. Nat Genet 31:33–36. doi:10.1038/ng853
11. Kemp SJ, Iraqi F, Darvasi A, Soller M, Teale AJ (1997) Localization of genes controlling resistance to trypanosomiasis in mice. Nat Genet 16:194–196. doi:10.1038/ng0697-194
12. Menge DM, Behnke JM, Lowe A, Gibson JP, Iraqi FA, Baker RL, Wakelin D (2003) Mapping of chromosomal regions influencing immunological responses to gastrointestinal nematode infections in mice. Parasite Immunol 25:341–349
13. Shusterman A, Durrant C, Mott R, Polak D, Schaefer A, Weiss EI, Iraqi FA, Houri-Haddad Y (2013) Host susceptibility to periodontitis: mapping murine genomic regions. J Dent Res 92:438–443. doi:10.1177/0022034513484039
14. Iraqi F (2000) Fine mapping of quantitative trait loci using advanced intercross lines of mice and positional cloning of the corresponding genes. Exp Lung Res 26:641–649. doi:10.1080/01902140150216729
15. Iraqi F, Clapcott SJ, Kumari P, Haley CS, Kemp SJ, Teale AJ (2000) Fine mapping of trypanosomiasis resistance loci in murine advanced intercross lines. Mamm Genome 11:645–648
16. Behnke JM, Iraqi FA, Mugambi JM, Clifford S, Nagda S, Wakelin D, Kemp SJ, Baker RL, Gibson JP (2006) High resolution mapping of chromosomal regions controlling resistance to gastrointestinal nematode infections in an advanced intercross line of mice. Mamm Genome 17:584–597. doi:10.1007/s00335-005-0174-0
17. Behnke JM, Menge DM, Nagda S, Noyes H, Iraqi FA, Kemp SJ, Mugambi RJM, Baker RL, Wakelin D, Gibson JP (2010) Quantitative

trait loci for resistance to Heligmosomoides bakeri and associated immunological and pathological traits in mice: comparison of loci on chromosomes 5, 8 and 11 in F2 and F6/7 inter-cross lines of mice. Parasitology 137:311. doi:10.1017/S0031182009991028

18. Hernandez-Valladares M, Naessens J, Gibson JP, Musoke AJ, Nagda S, Rihet P, Ole-MoiYoi OK, Iraqi FA (2004) Confirmation and dissection of QTL controlling resistance to malaria in mice. Mamm Genome 15:390–398. doi:10.1007/s00335-004-3042-4

19. Hernandez-Valladares M, Rihet P, ole-MoiYoi OK, Iraqi FA (2004) Mapping of a new quantitative trait locus for resistance to malaria in mice by a comparative mapping approach with human Chromosome 5q31-q33. Immunogenetics 56:115–117. doi:10.1007/s00251-004-0667-0

20. Wang M, Lemon WJ, Liu G, Wang Y, Iraqi FA, Malkinson AM, You M (2003) Fine mapping and identification of candidate pulmonary adenoma susceptibility 1 genes using advanced intercross lines. Cancer Res 63:3317–3324

21. Iraqi FA, Behnke JM, Menge DM, Lowe AM, Teale AJ, Gibson JP, Baker LR, Wakelin DR (2003) Chromosomal regions controlling resistance to gastro-intestinal nematode infections in mice. Mamm Genome 14:184–191. doi:10.1007/s00335-002-3049-7

22. Hanotte O, Ronin Y, Agaba M, Nilsson P, Gelhaus A, Horstmann R, Sugimoto Y, Kemp S, Gibson J, Korol A, Soller M, Teale A (2003) Mapping of quantitative trait loci controlling trypanotolerance in a cross of tolerant West African N'Dama and susceptible East African Boran cattle. Proc Natl Acad Sci U S A 100:7443–7448. doi:10.1073/pnas.1232392100

23. Silva MVB, Sonstegard TS, Hanotte O, Mugambi JM, Garcia JF, Nagda S, Gibson JP, Iraqi FA, McClintock AE, Kemp SJ, Boettcher PJ, Malek M, Van Tassell CP, Baker RL (2012) Identification of quantitative trait loci affecting resistance to gastrointestinal parasites in a double backcross population of Red Maasai and Dorper sheep. Anim Genet 43:63–71. doi:10.1111/j.1365-2052.2011.02202.x

24. Churchill GA, Airey DC, Allayee H, Angel JM, Attie AD, Beatty J, Beavis WD, Belknap JK, Bennett B, Berrettini W, Bleich A, Bogue M, Broman KW, Buck KJ, Buckler E, Burmeister M, Chesler EJ, Cheverud JM, Clapcote S, Cook MN, Cox RD, Crabbe JC, Crusio WE, Darvasi A, Deschepper CF, Doerge RW, Farber CR, Forejt J, Gaile D, Garlow SJ, Geiger H, Gershenfeld H, Gordon T, Gu J, Gu W, de Haan G, Hayes NL, Heller C, Himmelbauer H, Hitzemann R, Hunter K, Hsu H-C, Iraqi FA, Ivandic B, Jacob HJ, Jansen RC, Jepsen KJ, Johnson DK, Johnson TE, Kempermann G, Kendziorski C, Kotb M, Kooy RF, Llamas B, Lammert F, Lassalle J-M, Lowenstein PR, Lu L, Lusis A, Manly KF, Marcucio R, Matthews D, Medrano JF, Miller DR, Mittleman G, Mock BA, Mogil JS, Montagutelli X, Morahan G, Morris DG, Mott R, Nadeau JH, Nagase H, Nowakowski RS, O'Hara BF, Osadchuk A V, Page GP, Paigen B, Paigen K, Palmer AA, Pan H-J, Peltonen-Palotie L, Peirce J, Pomp D, Pravenec M, Prows DR, Qi Z, Reeves RH, Roder J, Rosen GD, Schadt EE, Schalkwyk LC, Seltzer Z, Shimomura K, Shou S, Sillanpää MJ, Siracusa LD, Snoeck H-W, Spearow JL, Svenson K, Tarantino LM, Threadgill D, Toth LA, Valdar W, de Villena FP-M, Warden C, Whatley S, Williams RW, Wiltshire T, Yi N, Zhang D, Zhang M, Zou F (2004) The Collaborative Cross, a community resource for the genetic analysis of complex traits. Nat Genet 36:1133–1137. doi:10.1038/ng1104-1133

25. Iraqi FA, Churchill G, Mott R (2008) The Collaborative Cross, developing a resource for mammalian systems genetics: a status report of the Wellcome Trust cohort. Mamm Genome 19:379–381. doi:10.1007/s00335-008-9113-1

26. Morahan G, Balmer L, Monley D (2008) Establishment of "The Gene Mine": a resource for rapid identification of complex trait genes. Mamm Genome 19:390–393. doi:10.1007/s00335-008-9134-9

27. Chesler EJ, Miller DR, Branstetter LR, Galloway LD, Jackson BL, Philip VM, Voy BH, Culiat CT, Threadgill DW, Williams RW, Churchill GA, Johnson DK, Manly KF (2008) The Collaborative Cross at Oak Ridge National Laboratory: developing a powerful resource for systems genetics. Mamm Genome 19:382–389. doi:10.1007/s00335-008-9135-8

28. Rogala AR, Morgan AP, Christensen AM, Gooch TJ, Bell TA, Miller DR, Godfrey VL, de Villena FP-M (2014) The Collaborative Cross as a resource for modeling human disease: CC011/Unc, a new mouse model for spontaneous colitis. Mamm Genome 25:95–108. doi:10.1007/s00335-013-9499-2

29. Vered K, Durrant C, Mott R, Iraqi FA (2014) Susceptibility to Klebsiella pneumoniae infection in Collaborative Cross mice is a complex trait controlled by at least three loci acting at different time points. BMC Genomics 15:865. doi:10.1186/1471-2164-15-865

30. Durrant C, Tayem H, Yalcin B, Cleak J, Goodstadt L, de Villena FP-M, Mott R, Iraqi FA (2011) Collaborative Cross mice and their power to map host susceptibility to Aspergillus fumigatus infection. Genome Res 21:1239–1248. doi:10.1101/gr.118786.110
31. Iraqi FA, Athamni H, Dorman A, Salymah Y, Tomlinson I, Nashif A, Shusterman A, Weiss E, Houri-Haddad Y, Mott R, Soller M (2014) Heritability and coefficient of genetic variation analyses of phenotypic traits provide strong basis for high-resolution QTL mapping in the Collaborative Cross mouse genetic reference population. Mamm Genome 25:109–119. doi:10.1007/s00335-014-9503-5
32. Shusterman A, Salyma Y, Nashef A, Soller M, Wilensky A, Mott R, Weiss EI, Houri-Haddad Y, Iraqi FA (2013) Genotype is an important determinant factor of host susceptibility to periodontitis in the Collaborative Cross and inbred mouse populations. BMC Genet 14:68
33. Iraqi FA et al (2012) The genome architecture of the Collaborative Cross mouse genetic reference population. Genetics 190:389–401. doi:10.1534/genetics.111.132639
34. Atamni HJA-T, Mott R, Soller M, Iraqi FA (2016) High-fat-diet induced development of increased fasting glucose levels and impaired response to intraperitoneal glucose challenge in the Collaborative Cross mouse genetic reference population. BMC Genet 17:10. doi:10.1186/s12863-015-0321-x
35. Kovacs A, Ben-Jacob N, Tayem H, Halperin E, Iraqi FA, Gophna U (2011) Genotype is a stronger determinant than sex of the mouse gut microbiota. Microb Ecol 61:423–428. doi:10.1007/s00248-010-9787-2
36. Lorè NI, Iraqi FA, Bragonzi A (2015) Host genetic diversity influences the severity of Pseudomonas aeruginosa pneumonia in the Collaborative Cross mice. BMC Genet 16:106. doi:10.1186/s12863-015-0260-6
37. Keane TM, Goodstadt L, Danecek P, White MA, Wong K, Yalcin B, Heger A, Agam A, Slater G, Goodson M, Furlotte NA, Eskin E, Nellåker C, Whitley H, Cleak J, Janowitz D, Hernandez-Pliego P, Edwards A, Belgard TG, Oliver PL, McIntyre RE, Bhomra A, Nicod J, Gan X, Yuan W, van der Weyden L, Steward CA, Bala S, Stalker J, Mott R, Durbin R, Jackson IJ, Czechanski A, Guerra-Assunção JA, Donahue LR, Reinholdt LG, Payseur BA, Ponting CP, Birney E, Flint J, Adams DJ (2011) Mouse genomic variation and its effect on phenotypes and gene regulation. Nature 477:289–294. doi:10.1038/nature10413
38. Roberts A, Pardo-Manuel De Villena F, Wang W, McMillan L, Threadgill DW (2007) The polymorphism architecture of mouse genetic resources elucidated using genome-wide resequencing data: implications for QTL discovery and systems genetics. Mamm Genome 18:473–481. doi:10.1007/s00335-007-9045-1
39. Beck JA, Lloyd S, Hafezparast M, Lennon-Pierce M, Eppig JT, Festing MFW, Fisher EMC (2000) Genealogies of mouse inbred strains. Nat Genet 24:23–25. doi:10.1038/71641
40. Aylor DL, Valdar W, Foulds-Mathes W, Buus RJ, Verdugo RA, Baric RS, Ferris MT, Frelinger JA, Heise M, Frieman MB, Gralinski LE, Bell TA, Didion JD, Hua K, Nehrenberg DL, Powell CL, Steigerwalt J, Xie Y, Kelada SNP, Collins FS, Yang IV, Schwartz DA, Branstetter LA, Chesler EJ, Miller DR, Spence J, Liu EY, Mcmillan L, Sarkar A, Wang J, Wang W, Zhang Q, Broman KW, Korstanje R, Durrant C, Mott R, Iraqi FA (2011) Genetic analysis of complex traits in the emerging collaborative genetic analysis of complex traits in the emerging Collaborative Cross. Genome Res 21:1213–1222. doi:10.1101/gr.111310.110
41. Valdar W, Flint J, Mott R (2006) Simulating the Collaborative Cross: power of quantitative trait loci detection and mapping resolution in large sets of recombinant inbred strains of mice. Genetics 172:1783–1797. doi:10.1534/genetics.104.039313
42. Kemp SJ, Darvasi A, Soller M, Teale AJ (1996) Genetic control of resistance to trypanosomiasis. Vet Immunol Immunopathol 54:239–243
43. Mashimo T, Lucas M, Simon-Chazottes D, Frenkiel M-P, Montagutelli X, Ceccaldi P-E, Deubel V, Guenet J-L, Despres P (2002) A nonsense mutation in the gene encoding 2′-5′-oligoadenylate synthetase/L1 isoform is associated with West Nile virus susceptibility in laboratory mice. Proc Natl Acad Sci U S A 99:11311–11316. doi:10.1073/pnas.172195399
44. Nganga JK, Soller M, Iraqi FA (2010) High resolution mapping of trypanosomosis resistance loci Tir2 and Tir3 using F12 advanced intercross lines with major locus Tir1 fixed for the susceptible allele. BMC Genomics 11:394. doi:10.1186/1471-2164-11-394
45. Boon ACM, deBeauchamp J, Hollmann A, Luke J, Kotb M, Rowe S, Finkelstein D, Neale G, Lu L, Williams RW, Webby RJ (2009) Host genetic variation affects resistance to infection with a highly pathogenic H5N1 influenza A virus in mice. J Virol 83:10417–10426. doi:10.1128/JVI.00514-09

46. Miyairi I, Tatireddigari VRRA, Mahdi OS, Rose LA, Belland RJ, Lu L, Williams RW, Byrne GI (2007) The p47 GTPases Iigp2 and Irgb10 regulate innate immunity and inflammation to murine Chlamydia psittaci infection. J Immunol 179:1814–1824. doi:10.4049/jimmunol.179.3.1814
47. Abdeltawab NF, Aziz RK, Kansal R, Rowe SL, Su Y, Gardner L, Brannen C, Nooh MM, Attia RR, Abdelsamed HA, Taylor WL, Lu L, Williams RW, Kotb M (2008) An unbiased systems genetics approach to mapping genetic loci modulating susceptibility to severe streptococcal sepsis. PLoS Pathog 4:e1000042. doi:10.1371/journal.ppat.1000042
48. Flint J, Mackay TFC (2009) Genetic architecture of quantitative traits in mice, flies, and humans. Genome Res 19:723–733. doi:10.1101/gr.086660.108
49. Parker CC, Palmer AA (2011) Dark matter: are mice the solution to missing heritability? Front Genet 2:32. doi:10.3389/fgene.2011.00032
50. Poltorak A, Smirnova I, He X, Liu M-Y, Van Huffel C, Birdwell D, Alejos E, Silva M, Du X, Thompson P, Chan EKL, Ledesma J, Roe B, Clifton S, Vogel SN, Beutler B (1998) Genetic and physical mapping of the Lps locus: identification of the Toll-4 receptor as a candidate gene in the critical region. Blood Cells Mol Dis 24:340–355. doi:10.1006/bcmd.1998.0201
51. Legare ME, Bartlett FS, Frankel WN (2000) A major effect QTL determined by multiple genes in epileptic EL mice. Genome Res 10:42–48
52. Yazbek SN, Buchner DA, Geisinger JM, Burrage LC, Spiezio SH, Zentner GE, Hsieh C-W, Scacheri PC, Croniger CM, Nadeau JH (2011) Deep congenic analysis identifies many strong, context-dependent QTLs, one of which, Slc35b4, regulates obesity and glucose homeostasis. Genome Res 21:1065–1073. doi:10.1101/gr.120741.111
53. Bryant CD, Kole LA, Guido MA, Sokoloff G, Palmer AA (2012) Congenic dissection of a major QTL for methamphetamine sensitivity implicates epistasis. Genes Brain Behav 11:623–632. doi:10.1111/j.1601-183X.2012.00795.x
54. Buchner DA, Geisinger JM, Glazebrook PA, Morgan MG, Spiezio SH, Kaiyala KJ, Schwartz MW, Sakurai T, Furley AJ, Kunze DL, Croniger CM, Nadeau JH (2012) The juxtaparanodal proteins CNTNAP2 and TAG1 regulate diet-induced obesity. Mamm Genome 23:431–442. doi:10.1007/s00335-012-9400-8
55. Shirley RL, Walter NAR, Reilly MT, Fehr C, Buck KJ (2004) Mpdz is a quantitative trait gene for drug withdrawal seizures. Nat Neurosci 7:699–700. doi:10.1038/nn1271
56. Stylianou IM, Christians JK, Keightley PD, Bünger L, Clinton M, Bulfield G, Horvat S (2004) Genetic complexity of an obesity QTL (Fob3) revealed by detailed genetic mapping. Mamm Genome 15:472–481. doi:10.1007/s00335-004-3039-z
57. Koudandé O, van Arendonk J, Iraqi F (2005) Marker-assisted introgression of trypanotolerance QTL in mice. Mamm Genome 16:112–119. doi:10.1007/s00335-004-2314-3
58. Parker CC, Sokoloff G, Leung E, Kirkpatrick SL, Palmer AA (2013) A large QTL for fear and anxiety mapped using an F2 cross can be dissected into multiple smaller QTLs. Genes Brain Behav 12:714–722. doi:10.1111/gbb.12064
59. Cheng R, Lim JE, Samocha KE, Sokoloff G, Abney M, Skol AD, Palmer AA (2010) Genome-wide association studies and the problem of relatedness among advanced intercross lines and other highly recombinant populations. Genetics 185:1033–1044. doi:10.1534/genetics.110.116863
60. Darvasi A, Soller M (1995) Advanced intercross lines, an experimental population for fine genetic mapping. Genetics 141:1199–1207
61. Parker CC, Cheng R, Sokoloff G, Lim JE, Skol AD, Abney M, Palmer AA (2011) Fine-mapping alleles for body weight in LG/J×SM/J F_2 and F(34) advanced intercross lines. Mamm Genome 22:563–571. doi:10.1007/s00335-011-9349-z
62. Lawson HA, Cady JE, Partridge C, Wolf JB, Semenkovich CF, Cheverud JM (2011) Genetic effects at pleiotropic loci are context-dependent with consequences for the maintenance of genetic variation in populations. PLoS Genet 7:e1002256. doi:10.1371/journal.pgen.1002256
63. Parker CC, Cheng R, Sokoloff G, Palmer AA (2012) Genome-wide association for methamphetamine sensitivity in an advanced intercross mouse line. Genes Brain Behav 11:52–61. doi:10.1111/j.1601-183X.2011.00747.x
64. Farber CR, Kelly SA, Baruch E, Yu D, Hua K, Nehrenberg DL, de Villena FP-M, Buus RJ, Garland T, Pomp D (2011) Identification of quantitative trait loci influencing skeletal architecture in mice: emergence of Cdh11 as a primary candidate gene regulating femoral morphology. J Bone Miner Res 26:2174–2183. doi:10.1002/jbmr.436
65. Jarvis JP, Cheverud JM (2011) Mapping the epistatic network underlying murine reproductive fatpad variation. Genetics 187:597–610. doi:10.1534/genetics.110.123505
66. Lawson HA, Lee A, Fawcett GL, Wang B, Pletscher LS, Maxwell TJ, Ehrich TH, Kenney-Hunt JP, Wolf JB, Semenkovich CF, Cheverud JM (2011) The importance of context to the genetic architecture of

diabetes-related traits is revealed in a genome-wide scan of a LG/J×SM/J murine model. Mamm Genome 22:197–208. doi:10.1007/s00335-010-9313-3

67. Lionikas A, Cheng R, Lim JE, Palmer AA, Blizard DA (2010) Fine-mapping of muscle weight QTL in LG/J and SM/J intercrosses. Physiol Genomics 42A:33–38. doi:10.1152/physiolgenomics.00100.2010

68. Mott R, Talbot CJ, Turri MG, Collins AC, Flint J (2000) A method for fine mapping quantitative trait loci in outbred animal stocks. Proc Natl Acad Sci U S A 97:12649–12654. doi:10.1073/pnas.230304397

69. Valdar WSJ, Flint J, Mott R (2003) QTL fine-mapping with recombinant-inbred heterogeneous stocks and in vitro heterogeneous stocks. Mamm Genome 14:830–838. doi:10.1007/s00335-003-3021-1

70. Mott R, Flint J (2002) Simultaneous detection and fine mapping of quantitative trait loci in mice using heterogeneous stocks. Genetics 160:1609–1618

71. Yalcin B, Flint J (2012) Association studies in outbred mice in a new era of full-genome sequencing. Mamm Genome 23:719–726. doi:10.1007/s00335-012-9409-z

72. Williams RC (1990) Periodontal disease. N Engl J Med 322:373–382. doi:10.1056/NEJM199002083220606

73. Wilson M (1995) Biological activities of lipopolysaccharides from oral bacteria and their relevance to the pathogenesis of chronic periodontitis. Sci Prog 78(Pt 1):19–34

74. Baker PJ, Dixon M, Roopenian DC (2000) Genetic control of susceptibility to Porphyromonas gingivalis-induced alveolar bone loss in mice. Infect Immun 68:5864–5868

75. Baker PJ, Roopenian DC (2002) Genetic susceptibility to chronic periodontal disease. Microbes Infect 4:1157–1167. doi:10.1016/S1286-4579(02)01642-8

76. Polak D, Wilensky A, Shapira L, Halabi A, Goldstein D, Weiss EI, Houri-Haddad Y (2009) Mouse model of experimental periodontitis induced by Porphyromonas gingivalis/Fusobacterium nucleatum infection: bone loss and host response. J Clin Periodontol 36:406–410. doi:10.1111/j.1600-051X.2009.01393.x

77. Wilensky A, Gabet Y, Yumoto H, Houri-Haddad Y, Shapira L (2005) Three-dimensional quantification of alveolar bone loss in Porphyromonas gingivalis-infected mice using micro-computed tomography. J Periodontol 76:1282–1286. doi:10.1902/jop.2005.76.8.1282

78. Gellatly SL, Hancock REW (2013) Pseudomonas aeruginosa: new insights into pathogenesis and host defenses. Pathog Dis 67:159–173. doi:10.1111/2049-632X.12033

79. Gibson RL, Burns JL, Ramsey BW (2003) Pathophysiology and management of pulmonary infections in cystic fibrosis. Am J Respir Crit Care Med 168:918–951. doi:10.1164/rccm.200304-505SO

80. Guillot L, Beucher J, Tabary O, Le Rouzic P, Clement A, Corvol H (2014) Lung disease modifier genes in cystic fibrosis. Int J Biochem Cell Biol 52:83–93. doi:10.1016/j.biocel.2014.02.011

81. Bianconi I, Milani A, Cigana C, Paroni M, Levesque RC, Bertoni G, Bragonzi A (2011) Positive signature-tagged mutagenesis in Pseudomonas aeruginosa: tracking patho-adaptive mutations promoting airways chronic infection. PLoS Pathog 7:e1001270. doi:10.1371/journal.ppat.1001270

82. Cigana C, Curcurù L, Leone MR, Ieranò T, Lorè NI, Bianconi I, Silipo A, Cozzolino F, Lanzetta R, Molinaro A, Bernardini ML, Bragonzi A (2009) Pseudomonas aeruginosa exploits lipid A and muropeptides modification as a strategy to lower innate immunity during cystic fibrosis lung infection. PLoS One 4:e8439. doi:10.1371/journal.pone.0008439

83. Bragonzi A, Paroni M, Nonis A, Cramer N, Montanari S, Rejman J, Di Serio C, Döring G, Tümmler B (2009) Pseudomonas aeruginosa microevolution during cystic fibrosis lung infection establishes clones with adapted virulence. Am J Respir Crit Care Med 180:138–145. doi:10.1164/rccm.200812-1943OC

84. Nguyen D, Singh PK (2006) Evolving stealth: genetic adaptation of Pseudomonas aeruginosa during cystic fibrosis infections. Proc Natl Acad Sci U S A 103:8305–8306. doi:10.1073/pnas.0602526103

85. Weiler CA, Drumm ML (2013) Genetic influences on cystic fibrosis lung disease severity. Front Pharmacol 4:40. doi:10.3389/fphar.2013.00040

86. Bragonzi A (2010) Murine models of acute and chronic lung infection with cystic fibrosis pathogens. Int J Med Microbiol 300:584–593. doi:10.1016/j.ijmm.2010.08.012

87. De Simone M, Spagnuolo L, Lorè NI, Rossi G, Cigana C, De Fino I, Iraqi FA, Bragonzi A (2014) Host genetic background influences the response to the opportunistic Pseudomonas aeruginosa infection altering cell-mediated immunity and bacterial replication. PLoS One 9:e106873. doi:10.1371/journal.pone.0106873

88. Lorè NI, Cigana C, De Fino I, Riva C, Juhas M, Schwager S, Eberl L, Bragonzi A

(2012) Cystic fibrosis-niche adaptation of Pseudomonas aeruginosa reduces virulence in multiple infection hosts. PLoS One 7:e35648. doi:10.1371/journal.pone.0035648

89. Levy MM, Fink MP, Marshall JC, Abraham E, Angus D, Cook D, Cohen J, Opal SM, Vincent J-L, Ramsay G (2003) 2001 SCCM/ESICM/ACCP/ATS/SIS international sepsis definitions conference. Crit Care Med 31:1250–1256. doi:10.1097/01.CCM.0000050454.01978.3B

90. Friedman G, Silva E, Vincent JL (1998) Has the mortality of septic shock changed with time. Crit Care Med 26:2078–2086

91. Van der Poll T, Opal SM (2008) Host-pathogen interactions in sepsis. Lancet Infect Dis 8:32–43. doi:10.1016/S1473-3099(07)70265-7

92. Tobias PS, Tapping RI, Gegner JA (1999) Endotoxin interactions with lipopolysaccharide-responsive cells. Clin Infect Dis 28:476–481. doi:10.1086/515163

93. Mattsson E, Verhage L, Rollof J, Fleer A, Verhoef J, van Dijk H (1993) Peptidoglycan and teichoic acid from Staphylococcus epidermidis stimulate human monocytes to release tumour necrosis factor-alpha, interleukin-1 beta and interleukin-6. FEMS Immunol Med Microbiol 7:281–287

94. Pinheiro I, Dejager L, Petta I, Vandevyver S, Puimège L, Mahieu T, Ballegeer M, Van Hauwermeiren F, Riccardi C, Vuylsteke M, Libert C (2013) LPS resistance of SPRET/Ei mice is mediated by Gilz, encoded by the Tsc22d3 gene on the X chromosome. EMBO Mol Med 5:456–470. doi:10.1002/emmm.201201683

95. Broman KW (2005) Mapping expression in randomized rodent genomes. Nat Genet 37:209–210. doi:10.1038/ng0305-209

96. Peirce JL, Lu L, Gu J, Silver LM, Williams RW (2004) A new set of BXD recombinant inbred lines from advanced intercross populations in mice. BMC Genet 5:7. doi:10.1186/1471-2156-5-7

97. Taylor BA, Wnek C, Kotlus BS, Roemer N, MacTaggart T, Phillips SJ (1999) Genotyping new BXD recombinant inbred mouse strains and comparison of BXD and consensus maps. Mamm Genome 10:335–348

98. Bottomly D, Ferris MT, Aicher LD, Rosenzweig E, Whitmore A, Aylor DL, Haagmans BL, Gralinski LE, Bradel-Tretheway BG, Bryan JT, Threadgill DW, de Villena FP-M, Baric RS, Katze MG, Heise M, McWeeney SK (2012) Expression quantitative trait Loci for extreme host response to influenza a in pre-Collaborative Cross mice. G3 (Bethesda) 2:213–221. doi:10.1534/g3.111.001800

99. Kelada SNP, Aylor DL, Peck BCE, Ryan JF, Tavarez U, Buus RJ, Miller DR, Chesler EJ, Threadgill DW, Churchill GA, Pardo-Manuel de Villena F, Collins FS (2012) Genetic analysis of hematological parameters in incipient lines of the Collaborative Cross. G3 (Bethesda) 2:157–165. doi:10.1534/g3.111.001776

100. Threadgill DW, Hunter KW, Williams RW (2002) Genetic dissection of complex and quantitative traits: from fantasy to reality via a community effort. Mamm Genome 13:175–178. doi:10.1007/s00335-001-4001-y

101. Yalcin B, Flint J, Mott R (2005) Using progenitor strain information to identify quantitative trait nucleotides in outbred mice. Genetics 171:673–681. doi:10.1534/genetics.104.028902

Chapter 28

The Collaborative Cross Resource for Systems Genetics Research of Infectious Diseases

Paul L. Maurizio and Martin T. Ferris

Abstract

An increasing body of evidence highlights the role of host genetic variation in driving susceptibility to severe disease following pathogen infection. In order to fully appreciate the importance of host genetics on infection susceptibility and resulting disease, genetically variable experimental model systems should be employed. These systems allow for the identification, characterization, and mechanistic dissection of genetic variants that cause differential disease responses. Herein we discuss application of the Collaborative Cross (CC) panel of recombinant inbred strains to study viral pathogenesis, focusing on practical considerations for experimental design, assessment and analysis of disease responses within the CC, as well as some of the resources developed for the CC. Although the focus of this chapter is on viral pathogenesis, many of the methods presented within are applicable to studies of other pathogens, as well as to case–control designs in genetically diverse populations.

Key words Use case, Infectious diseases, Collaborative Cross, Influenza

1 Introduction

A confluence of factors interacts to result in adverse infectious disease outcomes, including demographic, environmental, and genetic contributions from the host and pathogen. Given the many challenges of studying viral infections during primary human outbreaks, small animal model systems have been and continue to be essential for the assessment of host genes that drive differences in infection susceptibility and outcomes [1–5]. Differential immune regulation before and after infection is often modulated by complex genetic effects, such as gene-by-gene/gene-by-environment interactions, and allelic variation at individual genes (e.g., hypomorphs, deletions), as an increasing number of studies have begun to illustrate [6–9]. These complex effects are best uncovered and studied in the context of genetically diverse and multi-allelic systems. Therefore, in order to dissect the role of genetic variation on

Klaus Schughart and Robert W. Williams (eds.), *Systems Genetics: Methods and Protocols*, Methods in Molecular Biology, vol. 1488, DOI 10.1007/978-1-4939-6427-7_28, © Springer Science+Business Media New York 2017

host interactions with viruses and other pathogens, it is critical that novel frameworks are developed for the analysis of these complex traits within these genetically diverse systems.

Genetic reference populations (GRPs) have long proven to be powerful for studying complex traits and their underlying causal genetic variants. Many of the classical GRPs (e.g., the BxH [10–12], AxB [13] and BxD panels [14]), as well as classical backcrosses and intercrosses, have been critical in identifying polymorphic host genome regions that influence disease susceptibility and pathology. Building upon the utility of the classical systems, the Collaborative Cross (CC) GRP and the Diversity Outbred (DO) heterogeneous stock were created. These populations advanced the progress of complex trait studies in mice, while also modeling the genetic and allelic complexity present in naturally occurring populations [15–17]. Briefly, both the CC and the DO were derived from a common set of eight founder strains, which are comprised of the three major *Mus musculus* subspecies: *musculus*, *domesticus*, and *castaneus*. As a result of their breeding designs, both populations have high levels of genetic diversity (~45 million SNPs, and ~4 million indels) spread roughly evenly across the genome. Furthermore, in these GRPs, up to eight unique alleles may exist at any gene/locus, and novel epistatic (gene-by-gene) interactions have been introduced that are not present in any of the classical inbred laboratory mouse strains.

Concurrently with advances in the development and genetic characterization of GRPs, a variety of statistical and computational advances have been made. These have improved our ability to identify and characterize unique genetic variants driving differential traits. Improved power and precision for detecting QTL and causative underlying haplotypes, as evinced in refs. [18], have resulted specifically from: our enhanced ability to identify founder strain haplotypes [19]; the publication of annotated whole-genome sequences for the eight founder strains [20, 21]; and the development of powerful software packages for genetic mapping [22–24]. These advances have also enabled the narrowing of QTL regions down directly to candidate causative polymorphisms. Concurrently, RNA-seq and a variety of computational pipelines [25–27] allow for precise and accurate quantification of transcripts, allele-specific expression, and isoform expression within genetically heterogeneous populations. Together, these methods provide powerful new tools in the systems genetics arsenal.

To understand the contribution of host genetic effects on differential infectious disease responses, the CC recombinant inbred (CC-RI) lines and a variety of related populations, including the eight CC founder strains, the partially inbred incipient CC (preCC), the DO, and CC-F1 (recombinant inbred intercrosses, or CC-RIX), have been used in a number of recent studies. In the following sections, we summarize results from across these studies, and use them to provide a framework for researchers interested in using multiparent populations (MPPs) to study host responses to infection. Although we largely focus on viral pathogens, this guidance is equally useful for other pathogen systems, as well as for systems genetics studies using a case–control design.

Resources describing other uses of the CC and related populations are available in the accompanying chapters of this book, and are also referenced in the following reviews: ref. [28], which covers informatics resources for the Collaborative Cross; ref. [29], which discusses behavioral studies in complex genetic populations; ref. [30], which specifically deals with systems genetics of coronaviruses; and ref. [31], which reviews systems genetics and the utility of network modeling for inference. In addition, there have been several studies examining baseline immune status, autoimmunity, allergy, and inflammation in the CC [32–35]. While highlighting and expanding on the approaches described below, expansion to include specific autoimmune and allergic responses are beyond the scope of this chapter.

2 Methods

In this section, we provide some suggestions for experimental design of infectious disease studies in the CC. A general and useful basic guideline, as adapted from a chapter by [36] is as follows: (1) formulate statistical and biological hypotheses; (2) determine treatment variables, phenotypes of interest, and nuisance variables; (3) determine the population, selecting and/or excluding mouse lines, and simulate and estimate the number of mice that will be required; (4) decide on a randomization protocol; and (5) decide on tools for computational and statistical analysis; and we expand on these points below. We note that there are a large number of different approaches and goals for studying pathogens within the CC. These include, but are not limited to: identifying novel models of pathogenesis [37]; determining the effects, across genetic backgrounds, of variants at previously characterized genes of major effect (e.g., *Mx1* [38] or *Oas1b* [39]); and mapping genetic variants driving differential disease responses [38, 40]. We focus the methodology within in this section as if a researcher were interested in genetic mapping. The general principles and basic protocol are enumerated below:

1. Determine the range of phenotypes to be collected within the study. While many phenotypes are classically considered to be linked during viral infection in traditional inbred lines, it is likely that: (a) these phenotypes will become unlinked due to segregating variants within the CC; and (b) phenotypes causing severe pathology may be differentiated from those simply correlated with disease. Thus, collecting a variety of related phenotypes will

allow for better inferences about the pathways involved in disease pathogenesis. Additionally, in order to avoid confounding effects, potentially important baseline measures, prior to infection, should be considered and recorded. Genetically diverse mice also have phenotypically diverse baseline measures, such as body mass, coat color, litter size, susceptibility to spontaneous disease during aging, etc. Some of these measures may be important for causal inference of the effect of infection, or for clarifying misallocated sample identities, when the data is analyzed.

2. Consider the impact of genes of major effect in order to determine experimental design and/or select a subset of lines. For many pathogens, host genes or loci, e.g., MHC [41, 42], that exert major effects on control of viral disease have already been identified, e.g., *Cmv1* for cytomegalovirus, *Oas1b* for flaviviruses, *Mx1* for influenza, and *CCR5* for HIV. Furthermore, for *Oas1b* [39] and *Mx1* [38], there are both functional and nonfunctional variants segregating within the CC. Using the genetic sequence information available for CC-RIs, experimenters may wish to exclude specific lines from their experimental population. For example, a researcher interested in identifying genetic variants that enhance lung damage during influenza A virus infection might wish to exclude lines with a functional *Mx1* from their study.

   An analogous approach deals with those cases where reagents required to properly assess disease responses are genotype-sensitive or genotype-specific. One example includes a specific viral peptide or tetramer with a major histocompatibility complex (MHC) haplotype restriction. In this case, although CC lines with (e.g.) a C57BL/6J MHC haplotype should generate robust disease response data, CC lines with other founder haplotypes at the MHC locus might not be compatible with the reagent, and therefore accurate assessments of the antiviral states of these lines will not be possible. Thus, exclusion of specific lines, stratified analysis of all lines, or alternative experimental designs may be needed to address these issues. In both of these cases, the best approach to identify specific lines is to examine the founder-strain haplotypes at the genes/loci of interest. The CC status website (http://csbio.unc.edu/CCstatus/index.py) contains a variety of tools, reviewed in ref. [28], with which researchers can identify and visualize the haplotypes present in all available CC lines at given loci. In this way, specific lines can be identified, and subsequently included or excluded, based on the investigator's desired and required haplotypes. We note that while the DO might provide a greater number of genetically unique individuals for a study, the outbred nature of the DO and not being able to preselect animals with given haplotypes from the DO before purchase might strongly affect the ability to assess phenotypic variation if these haplotype-specific reagents and/or genes of major effect are present.

3. Assess a range of phenotypes in a preliminary subset of lines. Host responses to viral infections can differ in a variety of ways, including disease magnitude, kinetics, duration and infection dose responses. Depending on the question of interest, any number of study designs may be optimal for analysis. However, in all cases it is useful to understand the potential range of phenotypic variation being driven by host genetic variants in the CC. This can be achieved by assessing a preliminary subset of mouse lines. A common and useful approach is to screen the eight founder strains of the CC and DO, using a standard, well-characterized dose of virus and relatively long experimental timecourse. In this way, estimates of the range of variation in kinetics, magnitude, onset, and duration of disease can be obtained. Importantly, it is likely, due to transgressive segregation and allele shuffling, that some CC lines will express more extreme viral resistance or susceptibility phenotypes than the eight founder strains. Within these eight strains, one can collect data on the full range of viral pathogenesis phenotypes of interest (e.g., clinical disease, viral replication/dissemination, and tissue damage), following step (1), and determine the phenotypes which vary the most due to host genetic differences.

   In some cases, assessment of the founder strains will be insufficient for estimating phenotypic ranges within the CC. As mentioned above, prior knowledge may dictate that a specific founder strain haplotype should be included or excluded to accommodate experimental needs. In these cases, assuming that the founder haplotype distributions within the CC allow it, an initial screen may be performed using a subset of CC lines rather than the eight founder lines. To illustrate, if only one of the eight founder haplotypes is informative (e.g., seven of the founders are highly resistant to infection due to their allele at a major effect locus), screening several CC strains that contain the one haplotype may be preferable. In contrast, if seven or eight founder haplotypes are informative, screening the seven or eight founders of interest may be preferable to using a subset of CC lines.

4. Determine the batching/blocking and covariates to be used in the study. After an initial screen using the subset of lines in **step 3**, it will be useful to revisit and modify, as necessary, the experimental design for the larger CC study, including experimental block designs and specific covariate data collection, and design of the linear model to be used in the analysis. Again, such decisions are likely to be driven by the investigator's questions of interest, the infectious disease system, and experimental approaches that will be used. However, a few general rules may be helpful.

   One type of idealized experimental design might include an assessment of every treatment group, timepoint, and sex across multiple replicate animals in a single infection batch, with several full batches studied to confirm and generalize these results.

However, we note that even for those examining a single timepoint post-infection, a screen of replicate animals from the entire library of available CC lines might be logistically difficult. In such cases, some form of well-reasoned batching (or "blocking") is required to improve experimental feasibly, while still maintaining an ability to assess statistical significance. The investigator may also want to ensure that the characteristics of the various blocks are well-balanced with respect to the sample size and factors of interest.

There are a large number of ways to design blocking, and we suggest a few simple guidelines. First of all, attempt to randomize, where possible, such that if there is a choice to be made, mice of a given line and sex should be randomly selected from among those available. In order to simplify the screen, it may be preferable to assess and perform QTL mapping in a single sex, with follow-up studies of single lines or timepoints expanded into both sexes to broaden conclusions and to examine sex-specific differences. The inclusion of specific timepoints or subsets of lines will likely depend on the resources available and the phenotypes of interest, such as discovery of new models of previously restricted pathogens, genetic mapping of host variants affecting specific pathologic outcomes, or analysis of differentially expressed transcriptional pathways. For example, if genetic mapping at a single specific timepoint is critical, then ensuring that some lines are repeated across multiple batches, and that each batch contains lines that are repeated in other batches, can be useful for normalizing data across batches. In contrast, if examining the kinetics of differential transcriptional networks is the goal, batches should include all animals of each line in the experiment, with a subset of the total lines to be used. Most importantly, when mock samples are to be paired with samples from a specific timepoint post-infection (DPI, e.g., to study transcriptional differences at 2 DPI or to contrast immune cell infiltration into specific tissues), the mock animals and infected animals from each line should be assayed on the same day to explicitly control for any batch effects. To generalize, for a given contrast or factor of interest (sex, treatment/condition, dose, etc.), including all the levels of interest within each given batch (or even each cage), is preferable when feasible, so that the effect of confounding variables is reduced.

5. Collect phenotype data. Once an appropriate experimental blocking strategy is determined, the study should proceed following the investigator's appropriate infection protocols and design. We note that it is critical to carefully observe and record potentially important, yet previously undescribed disease responses. Such phenotypes might be useful for characterizing novel disease phenotypes in follow-up studies and/or for

improving disease classifications for transcriptional analysis. Be aware of and carefully annotate aberrant or unexpected phenotypes that might be useful as covariates in further analyses (e.g., tumors within tissues of importance that could impact immune phenotypes in those tissues).

6a. Examine the distribution of and correlation between phenotypes. Following data collection, quantify and visualize the within-strain means and variances, as well as the aggregate mean and variance for each phenotype. The use of a Box–Cox transformation on the raw pathogenesis phenotype data will ensure that the residuals follow a more normal phenotypic distribution, enabling a more robust array of statistical analyses. Once data are appropriately transformed, one may determine the genetic contribution to the variance in the data, otherwise known as the broad-sense heritability (H^2), and related measures.

$$\sigma_P^2 = \sigma_G^2 + \sigma_E^2$$

$$\sigma_G^2 = \sigma_A^2 + \sigma_D^2 + \sigma_I^2$$

$$H^2 = \frac{\sigma_G^2}{\sigma_G^2 + \sigma_E^2}$$

$$h^2 = \frac{\sigma_A^2}{\sigma_G^2 + \sigma_E^2}$$

$$g^2 = \frac{MS_B - MS_W}{MS_B + (2n-1)MS_W}$$

where σ_P^2 is the total phenotypic variance, σ_G^2 is the total variance attributable to genetics, and σ_E^2 is the remaining variance, attributable to environment and residual noise. σ_G^2 can be partitioned into additive (σ_A^2), dominance (σ_D^2), and epistatic (σ_I^2) components. Broad-sense heritability (H^2) is calculated as the ratio of the genetic variance (σ_G^2), to the total phenotypic variance. Narrow-sense heritability (h^2), which is a subset of H^2, is calculated as the ratio of the additive (σ_A^2) to the total phenotypic variance. The "coefficient of genetic determination" (g^2), which is used for estimating broad-sense heritability in inbred lines [35, 43], is a function of the number of animals tested per strain (n), and the between- and within-strain mean-squared errors (MS_B, MS_W).

Furthermore, a reexamination of the correlation structure of the disease phenotypes (both stratified by strain, and in aggregate) can help to clarify relationships between different aspects of viral pathogenesis, and can strengthen decision making regarding the phenotypes to be used for mapping causal loci. Simple packages such as corrplot (https://cran.r-project.org/web/packages/corrplot/index.html) and corrgram (https://cran.r-project.org/web/packages/corrgram/index.html) in R can be used to visualize the correlation and covariance structure of a phenotype matrix.

6b. Identify/select samples for transcriptional analysis. In some cases, researchers may wish to add whole-genome transcriptional analysis to further clarify the genes and pathways that are differentially expressed in concordance with specific phenotypes. In many cases, it will be cost-prohibitive to run transcriptional analyses on all samples. Transcriptional analyses that are focused on extreme phenotypic outcomes (e.g., contrast individuals with high vs. undetectable titers), such as is used in bulk segregant analysis, may provide increased power to identify transcripts associated with differential disease. This approach has been illustrated in ref. [44], where a combination of titer and weight loss extremes was used to identify reactive transcriptional networks differentiating the extreme phenotypic groups. Consider, additionally, whether banking a variety of specific immune-related tissues (bone marrow, lymph nodes, CNS, spleen), as well as "unrelated" control tissues may be helpful in follow-up studies, following transcriptional analysis or mapping. Also consider exploring other CC-related in vitro resources (cell-culture, such as mouse embryonic fibroblasts from CC-related mice) that would be useful for your study, especially in the follow-up stages.

7. Conduct genetic mapping. Once phenotypes with high heritability and sufficiently large variation have been identified, genetic mapping can be carried out. A number of software packages exist for multiparent mapping, including Bagpipe (http://valdar-lab.unc.edu/software.html), HAPPY (http://www.well.ox.ac.uk/happy/) [19], and DOQTL, a package for the R statistical computing environment [24], which also works for mapping in the CC. We currently recommend using the DOQTL package, as it is stably supported on Bioconductor (https://www.bioconductor.org/packages/release/bioc/html/DOQTL.html), and also has features to integrate SNP and gene variant features based on the Sanger Institute's resequencing of the eight founder strains of the CC, as described in refs. [17, 24]. For mapping in the CC or the DO, one uses a file(s) to describe the founder haplotype probabilities in the mapping population. This is used both for fully inbred lines, where probabilities should theoretically be 1.0 or 0.0 for any given founder haplotype at each locus, as well as for heterozygous populations, with fractional probabilities and haplotype uncertainties due to

recombination. A separate file is used, containing the transformed phenotypes and any covariates one might consider important (e.g., batch, sex, starting weight). The software uses a linear regression approach, asking whether there is a significant association between phenotypes and haplotype probabilities at each haplotype block along the genome. Significance thresholds are determined using permutation or false discovery rate approaches, both of which take into account the distribution of genotypes and phenotypes within the test population.

Covariates that can be included in the model may take a variety of forms, including demographic (age, sex), experimental (batches), and genetic (e.g., genes of major effect, previously discovered QTL). Proper accounting for these covariates often increases one's power to detect causative genetic polymorphisms underlying virus-induced disease. However, care must be taken in the analysis procedure to ensure that inclusion of covariates does not mask true genetic effects. For example, consider the scenario where two batches of CC lines are tested. In batch 1, all of the lines with a polymorphism enhancing viral titer are tested. In batch 2, all of the lines with a polymorphism repressing viral titer are tested. In this case, including a batch covariate will cause much of the polymorphism's effects to be attributed, incorrectly, to differences between experimental batches. We suggest that QTL scans are run both with and without inclusion of such covariates, with the investigator carefully examining the mapping results (maximum significance scores, significance thresholds, etc.) from all situations. Furthermore, if there are QTL that appear when scans are run without a covariate, but are absent when a covariate is included, we suggest follow-up studies that replicate some of the lines and timepoints that were previously split by covariates. Where possible, such studies will confirm and validate such QTL without the confounding of covariates. Additional, more rigorous model selection protocols for determining the inclusion/exclusion of covariates may be considered, but they are beyond the scope of this chapter.

8. Identify causative polymorphisms. Once QTL are identified, a number of approaches may be utilized to determine specific causative genetic variants. Toward this end, most multiparent mapping tools will generate allele effect plots. These plots display the estimated scaled effects of each founder allele on a trait of interest within a given genomic locus. In this way, one can distinguish groups of haplotypes that enhance, suppress, or have no effect on disease. By identifying SNPs and other genetic variants within the QTL that follow the allele effects patterns, one can narrowly focus on subsets of candidate genes or features that are likely to be causative for the phenotype of interest. For example,

at a given SNP, if both "high" and "low" phenotype groups share a founder allele, it is unlikely to be causative, whereas SNPs that contain alleles that segregate between the high- and low-responder strains are much more likely to be causative. Further integration of already available whole-genome expression data, or post-hoc gene expression analysis (e.g., qPCR) can help to narrow and refine candidates. For example, genes underneath a QTL locus which are differentially expressed between high and low groups, in a relevant tissue or compartment and at a relevant timepoint, will help lead to potential candidate genes and/or pathways impacting disease outcomes.

9. Consider alternate studies and experimental approaches. In the preceding text, we highlighted a case where follow-up studies might be useful in identifying genetic variants, i.e., where initial QTL scans suggested a locus, but the effect of that locus was confounded by a covariate such as experimental batch. Following the collection of initial data, there are a variety of other experiments which can help clarify and enhance the initial studies. One possibility is that only one or two CC lines show a desired or extreme disease response [37, 45]. Such outcomes may indicate either complex gene–gene interactions (epistasis) or de novo mutations arising in lines. In both cases, one should consider either a tailored follow-up genetic cross, such as an F2, as in refs. [45], or follow-up intensive expression analysis, as in refs. [37], to focus on likely causative loci or networks driving these unique disease responses. Another possible outcome is that a gene of major effect has been discovered. This would be a case where a QTL explains a large fraction (e.g., 50%) of trait variation for one or more of the pathology traits of interest. In these cases, it may be useful to either (as recommended above) redo analysis with the QTL of major effect as a covariate in the main QTL scans OR to subset your set of lines into those with the high versus low haplotypes at the QTL. These subpopulation style analyses can help in identifying further genetic variants that affect disease only in the context of a gene of major effect. For example, if a variant impacts viral dissemination from a primary tissue, its effect can be masked if there is an additional polymorphism that abrogates the viral receptor within the CC. Only by mapping with the receptor positive population will it be possible to identify the dissemination variant.

3 Expected Results

Given that there are a variety of possible genetic architectures underlying host responses to multiple aspects of viral infection, it is difficult to precisely predict outcomes for any given study type. However, based on the breadth of work conducted so far within the CC, the DO, and related populations, one can expect that there are genetic variants segregating within the CC system that will have impact on pathogenesis for any given virus. Indeed, for at least four viral pathogens, as well as for a variety of other bacterial and fungal pathogens, QTL have been identified that contribute to differential disease responses and pathogenesis. Taken as a whole, these QTL typically have shown modest effects on pathogenic traits (e.g., a summary of the results of [38, 40] show most QTL explaining 25 % of phenotypic variance for each trait). Furthermore, it is likely that transgressive segregation operates within the recombined genomes of the CC. That is, alleles driving extreme responses may come from a founder strain(s) that exhibits a mild or suppressed phenotype. Thus, only when the genetic structure of the founder strains has been rearranged will the true effects of alleles be identifiable. Lastly, it is likely that once QTL are identified, it will be possible to identify a set of high priority SNPs, based on the founder strain sequences, which act as the causative variants. Additional pathological and molecular phenotyping will be required for validation, but the integration of multiple allele effects, as well as sequence data, is a substantial improvement over classical positional cloning for identifying causal variants.

4 Lessons Learned

The use of the Collaborative Cross and related populations in studying infectious diseases is still in its nascent stages. Nevertheless, there are several important considerations, gleaned from the studies to date, that can specifically inform future studies and analysis of determinants of infectious disease susceptibility.

One clear lesson learned from virus infection studies in the CC and preCC so far is that phenotypic correlations present within any set of characterized founder strains or knockouts are likely to be broken apart within the CC, unless there are strong causal relationships between the correlated phenotypes. For example, a complete disassociation was seen between different aspects of SARS-coronavirus (SARS-CoV) induced pathology and disease within the preCC population [40]. Furthermore, QTL mapping will often show that distinct loci affect individual, distinct pathologic traits, as seen in both the influenza preCC and SARS-CoV preCC studies [38, 40]. These results highlight one main impetus for utilizing GRPs such as the CC (i.e., the discovery of novel phenotypic relationships and distinct genetic markers), but they also point to a critical consideration in the design and analysis of studies in these systems. Namely, the assessment of a wide variety of phenotypes, even those classically thought to be redundant, will be highly useful and enable a better understanding of disease pathogenesis.

It is well known that susceptibility and resistance genes of major effect are predominant within host–pathogen systems. These genes of major effect include, for example *Cmv1* for murine cytomegalovirus [46, 47]; *Oas1b* for flaviviruses [48]; and *Mx1* for influenza [49, 50]. Indeed, both functional and defective *Oas1b* and *Mx1* alleles circulate within the CC/DO [38, 39]. Given the genetic diversity present within the CC and DO, it is likely that other genes of major effect for specific pathogens will be found segregating within these populations. Although the presence of genes and alleles of major effect may appear to be an obstacle for discovery of novel regulators of disease, obscuring the contribution of genes or alleles of lesser effect size, new biological insight can still be obtained in the presence of these large effect alleles. For example, given the potential for up to eight alleles segregating within the CC at any locus, there may be several alleles, isoforms and/or transcriptional variants at a given causal locus. This was observed clearly in the influenza challenge of the preCC, where the antiviral and clinically protective effects of *Mx1* were disassociated via the presence of three unique alleles at the *Mx1* locus [38].

Furthermore, epistasis and transgressive segregation are at work within the CC population. Such segregation can most commonly reveal previously "hidden" genetic variation. For example, in the preCC study of SARS, the wild-derived founder strains (CAST, PWK, and WSB) all die of SARS-CoV infection at low doses [40]. In contrast, for the preCC lines that survived SARS-CoV infection, causative alleles at a variety of QTL are driven by wild derived parental alleles, and were therefore hidden in the context of super-susceptible parent founders. Thus, the allelic variants that affect immunopathology, viral replication, and immune infiltration were identified only in the context of disruption of founder haplotypes through recombination. Alternately, recombination driving reassortment of alleles may cause emergent phenotypes by introducing evolutionarily distinct allelic combinations. For example, it is only via this genetic reassortment across the CC that a severe Ebola virus (EBOV)-induced hemorrhagic fever was identified in mice, as this phenotype was not present in any of the CC founder strains [37].

There are several reasons for emphasizing the thoughtful use of mock controls in infectious disease studies in the CC. Firstly, given the novelty of the genetic backgrounds generated in the CC, the response to mock treatment in some lines may differ substantially from that of common inbred lines. Additionally, genetic loci that regulate baseline immune phenotypes may be quite distinct from those that regulate immune phenotypes after infection-induced pathways are upregulated or downregulated, hence QTL may be mapped for untreated or mock-treated animals as a complement to QTL mapped for infection response.

Finally, it should be noted that characterizing the variability or variance in a disease phenotype, both within-strain and between-strain, is worthwhile and may be critical for identifying genetic causes of differential disease. Identifying a strain or set of strains with increased variance may lead you to identifying a novel genetic factor or latent environmental variable that causes a substantial change in the phenotype of interest [51]. During the characterization of within-strain variation, you may be able to identify experimental issues that ought to be modeled or corrected (e.g., batch effects), or rare de novo genetic variants that substantially modulate your phenotype, which can be identified with additional genotyping or sequencing. In one recent study, using a diallel of the wild-derived CC founder mice and their F1 reciprocal crosses, gene expression was substantially altered in two mice, including one which was found to have a de novo duplication [25]. Thus, having well-characterized within- and between-strain variance estimates are critical for identifying novel genetic variants, for estimating statistical power, and for successful experimental design and analysis in the CC.

5 Further Considerations and Limitations

Although systems genetics approaches and genetically diverse study populations provide a powerful combination of tools to identify host genetic variants driving infectious disease, there are several caveats that ought to be considered in optimizing study design and analysis approaches: namely appropriate molecular phenotyping, disentangling complex phenotypic networks, and mechanistic insight into variant loci.

Omics analysis (transcriptomics, proteomics, metabolomics, etc.) is a cornerstone of the systems biology approach to research. One strong caveat for omics analysis is the dependence of these approaches on accurate assessment of genome sequences for utilized strains. The C57BL/6J genome has formed the backbone of mouse sequence analysis and annotation, however we know that the other CC founder strains, and therefore the CC themselves contain large numbers of polymorphisms, and more importantly structural variants and large insertion/deletions [52] (http://www.sanger.ac.uk/science/data/mouse-genomes-project). As described in more detail in this volume in a chapter by Green et al., integration of imputed genome sequences (pseudogenomes) for CC lines or DO animals [27] will substantially improve integration of these omics data within these GRPs.

A variety of factors, such as prior immune history, opportunistic coinfection, and microbiome influences on the immune system [53, 54] can influence host responses to viral infections. Furthermore, there is evidence for genetic variants within the CC affecting basal variation in immune populations [34]. Given the potential for host genetic variation to impact a variety of immune phenotype and the microbiome, it is likely that there will be complex causal networks underlying variation in the direct viral pathogen traits of interest of a researcher. While dissection of these networks may provide many years of fruitful study, they may present daunting obstacles to study within the CC. Careful design of experiments (e.g., cohousing animals from different CC lines; antibiotic pretreatment to limit bacterial coinfection) can help to ameliorate and control some of these effects and improve the ability to identify genetic variants directly affecting host responses to viral pathogens of interest.

Finally, we note that identification of genetic variants with a GRP affecting host responses to viral infection do little to identify the mechanisms and processes through which these variants act. While integration of a variety of phenotypes (e.g., pathological, immunological, and molecular responses) can help to highlight mechanisms and pathways of activity, it is only through classical (and phenotype-specific) manipulation and experimentation that a true understanding of these variants can be elucidated. Such approaches, often deemed "reductionist," are critically useful in transitioning broad systems-based responses with clear and actionable mechanistic processes.

6 Outlook

Small animal models for the host response to infectious disease pathogens are critical tools for the study of human susceptibility to disease, as well as for the development of novel prophylactics and therapeutics. Indeed, the utility of these systems for studying host–pathogen interactions appears to be persistent and critical. Notably, by varying the host genetic background in the study of infectious disease, we enable the detection of genetic variants that are important for disease across a population of genetically diverse individuals, improving our chances that variants are reproducible across experiments and, it is hoped, across species. Importantly, not only can these systems be used for identifying genetic susceptibility loci, but they can also be used to identify and develop of novel infectious disease models, using specific strains of CC mice as new resources for understanding severe disease, such as has been done in the recent development of CC mouse models of Ebola virus pathogenesis [37].

The CC is also useful for better understanding the genetic architecture of the host response to infection. It has been recognized that nonadditive genetic effects, such as dominance, epistasis, and parent-of-origin effects, may contribute substantially to quantitative traits, including the host immune system and infectious disease responses. In order to estimate, quantify, and explore such complex genetic interactions, and to quantify broad and narrow-sense heritability, future directions include characterizing infection phenotypes in F1 reciprocal crosses of the eight founder lines and of the CC lines (using CC-F1s). Such experiments will add to our knowledge about how disease susceptibility and resistance may be expressed and transmitted from parents to offspring, and this work may reveal important genetic complexities, hard to uncover in human studies. These complex genetic effects may be responsible for inhibiting our ability, at present, to identify candidate genes through GWAS and linkage mapping studies, which less often include rigorous screens for nonadditive effects. Finally, the experimental designs and phenotypic data sets that are being generated for systems genetics in the Collaborative Cross lend themselves to innovative statistical and quantitative genetics models. These new models and quantitative tools advance our understanding of human disease, and complement the variety of experimental tools being developed for the CC. Thus, infectious disease research in CC promises to advance our knowledge about complex host–pathogen interactions, and to enhance our ability to unravel and interpret increasingly complex biological networks in order to improve human health.

Acknowledgments

We acknowledge U19AI100625 to M.T.F. and 5T32AI007419-23 to P.L.M. for support.

References

1. Srivastava B, Błazejewska P, Hessmann M, Bruder D, Geffers R, Mauel S, Gruber AD, Schughart K (2009) Host genetic background strongly influences the response to influenza A virus infections. PLoS One 4(3):4857
2. Boon AC, deBeauchamp J, Hollmann A, Luke J, Kotb M, Rowe S, Finkelstein D, Neale G, Lu L, Williams RW, Webby RJ (2009) Host genetic variation affects resistance to infection with a highly pathogenic H5N1 influenza A virus in mice. J Virol 83(20):10417–10426
3. Bouvier NM, Lowen AC (2010) Animal models for influenza virus pathogenesis and transmission. Viruses 2:1530–1563
4. Boivin GA, Pothlichet J, Skamene E, Brown EG, Loredo-Osti JC, Sladek R, Vidal SM (2012) Mapping of clinical and expression quantitative trait loci in a sex-dependent effect of host susceptibility to mouse-adapted influenza H3N2/HK/1/68. J Immunol 188(8):3949–3960
5. Boon AC, Finkelstein D, Zheng M, Liao G, Allard J, Klumpp K, Webster R, Peltz G, Webby RJ (2011) H5N1 influenza virus pathogenesis in genetically diverse mice is mediated at the level of viral load. mBio 2(5):pii:e00171-11
6. Wei W-H, Hemani G, Haley CS (2014) Detecting epistasis in human complex traits. Nat Rev Genet 15(11):722–733
7. Lenz TL, Deutsch AJ, Han B, Hu X, Okada Y, Eyre S, Knapp M, Zhernakova A, Huizinga TW, Abecasis G, Becker J, Boeckxstaens GE, Chen WM, Franke A, Gladman DD, Gockel I, Gutierrez-Achury J, Martin J, Nair RP, Nöthen MM, Onengut-Gumuscu S, Rahman P, Rantapää-Dahlqvist S, Stuart PE, Tsoi LC, van Heel DA, Worthington J, Wouters MM, Klareskog L, Elder JT, Gregersen PK, Schumacher J, Rich SS, Wijmenga C, Sunyaev SR, de Bakker PI, Raychaudhuri S (2015) Widespread non-additive and interaction effects within HLA loci modulate the risk of autoimmune diseases. Nat Genet 47(9):4–7
8. Galson JD, Trück J, Fowler A, Clutterbuck EA, Münz M, Cerundolo V, Reinhard C, van

der Most R, Pollard AJ, Lunter G, Kelly DF (2015) B-cell repertoire responses to varicella-zoster vaccination in human identical twins. Proc Natl Acad Sci 112(2):500–505
9. Shin D-L, Hatesuer B, Bergmann S, Nedelko T, Schughart K (2015) Protection from severe influenza infections in mice carrying the *Mx1* influenza resistance gene strongly depends on genetic background. J Virol 89(19):01305–01315
10. Turcotte K, Gauthier S, Mitsos L-M, Shustik C, Copeland NG, Jenkins NA, Fournet J-C, Jolicoeur P, Gros P (2004) Genetic control of myeloproliferation in BXH-2 mice. Blood 103(6):2343–2350
11. Marquis J-F, LaCourse R, Ryan L, North RJ, Gros P (2009) Disseminated and rapidly fatal tuberculosis in mice bearing a defective allele at IFN regulatory factor 8. J Immunol 182(5):3008–3015
12. Berghout J, Langlais D, Radovanovic I, Tam M, MacMicking JD, Stevenson MM, Gros P (2013) Irf8-regulated genomic responses drive pathological inflammation during cerebral malaria. PLoS Pathog 9(7):e1003491
13. Hassan MA, Jensen KD, Butty V, Hu K, Boedec E, Prins P, Saeij JP (2015) Transcriptional and linkage analyses identify loci that mediate the differential macrophage response to inflammatory stimuli and infection. PLoS Genet 11(10):1005619
14. Nedelko T, Kollmus H, Klawonn F, Spijker S, Lu L, Heßman M, Alberts R, Williams RW, Schughart K (2012) Distinct gene loci control the host response to influenza H1N1 virus infection in a time-dependent manner. BMC Genomics 13(1):411
15. Churchill GA, Airey DC, Allayee H, Angel JM, Attie AD, Beatty J (2004) The collaborative cross, a community resource for the genetic analysis of complex traits. Nat Genet 36(11):1133–1137
16. Churchill GA, Gatti DM, Munger SC, Svenson KL (2012) The diversity outbred mouse population. Mamm Genome 23(9-10):713–718
17. Svenson KL, Gatti DM, Valdar W, Welsh CE, Cheng R, Chesler EJ, Palmer AA, McMillan L, Churchill GA (2012) High-resolution genetic mapping using the mouse diversity outbred population. Genetics 190(2):437–447
18. Aylor DL, Valdar W, Foulds-Mathes W, Buus RJ, Verdugo RA et al (2011) Genetic analysis of complex traits in the emerging collaborative cross. Genome Res 21(8):1213–1222
19. Mott R, Talbot CJ, Turri MG, Collins AC, Flint J (2000) A method for fine mapping quantitative trait loci in outbred animal stocks. Proc Natl Acad Sci 97(23):12649–12654
20. Keane TM, Goodstadt L, Danecek P, White MA, Wong K, Yalcin B, Heger A, Agam A, Slater G, Goodson M, Furlotte NA, Eskin E, Nellåker C, Whitley H, Cleak J, Janowitz D, Hernandez-Pliego P, Edwards A, Belgard TG, Oliver PL, McIntyre RE, Bhomra A, Nicod J, Gan X, Yuan W, van der Weyden L, Steward CA, Bala S, Stalker J, Mott R, Durbin R, Jackson IJ, Czechanski A, Guerra-Assunção AJ, Donahue LR, Reinholdt LG, Payseur BA, Ponting CP, Birney E, Flint J, Adams DJ (2011) Mouse genomic variation and its effect on phenotypes and gene regulation. Nature 477(7364):289–294
21. Adams DJ, Doran AG, Lilue J, Keane TM (2015) The mouse genomes project: a repository of inbred laboratory mouse strain genomes. Mamm Genome 26(9-10):403–412
22. Arends D, Prins P, Jansen RC, Broman KW (2010) R/qtl: high-throughput multiple QTL mapping. Bioinformatics 26(23):2990–2992
23. Zhang Z, Wang W, Valdar W (2014) Bayesian modeling of haplotype effects in multiparent populations. Genetics 198(1):139–156
24. Gatti SKL, Shabalin A, Wu LY, Valdar W, Simecek P, Goodwin N, Cheng R, Pomp D, Palmer A, Chesler EJ, Broman KW, Churchill GA (2014) Quantitative trait locus mapping methods for diversity outbred mice. G3 (Bethesda, MD) 4(9):1623–1633
25. Crowley JJ, Zhabotynsky V, Sun W, Huang S, Pakatci IK, Kim Y, Wang JR, MorganAP CJD, Aylor DL, Yun Z, Bell TA, Buus RJ, Calaway ME, Didion JP, Gooch TJ, Hansen SD, Robinson NN, Shaw GD, Spence JS, Quackenbush CR, Barrick CJ, Nonneman RJ, Kim K, Xenakis J, Xie Y, Valdar W, Lenarcic AB, Wang W, Welsh CE, Fu CP, Zhang Z, Holt J, Guo Z, Threadgill DW, Tarantino LM, Miller DR, Zou F, McMillan L, Sullivan PF, de Villena FP-M (2015) Analyses of allele-specific gene expression in highly divergent mouse crosses identifies pervasive allelic imbalance. Nat Genet 47(4):353–360
26. Zou F, Sun W, Crowley JJ, Zhabotynsky V, Sullivan PF, de Villena FP-M (2014) A novel statistical approach for jointly analyzing RNA-seq data from F1 reciprocal crosses and inbred lines. Genetics 197(1):389–399
27. Munger SC, Raghupathy N, Choi K, Simons AK, Gatti DM, Hinerfeld DA, Svenson KL, Keller MP, Attie AD, Hibbs MA, Graber JH, Chesler EJ, Churchill GA (2014) RNA-seq alignment to individualized genomes improves transcript abundance estimates in multiparent populations. Genetics 198(1):59–73
28. Morgan AP, Welsh CE (2015) Informatics resources for the collaborative cross and

related mouse populations. Mamm Genome 26(9):521–539
29. Mulligan MK, Williams RW (2015) Systems genetics of behavior: a prelude. Curr Opin Behav Sci 2:108–115
30. Schäfer A, Baric RS, Ferris MT (2014) Systems approaches to coronavirus pathogenesis. Curr Opin Virol 6(1):61–69
31. Civelek M, Lusis AJ (2014) Systems genetics approaches to understand complex traits. Nat Rev Genet 15(1):34–48
32. Kelada SNP, Aylor DL, Peck BCE, Ryan JF, Tavarez U, Buus RJ, Miller DR, Chesler EJ, Threadgill DW, Churchill GA, de Villena FP-M, Collins FS (2012) Genetic analysis of hematological parameters in incipient lines of the collaborative cross. G3 (Bethesda, MD) 2(2):157–165
33. Kelada SNP, Carpenter DE, Aylor DL, Chines P, Rutledge H, Chesler EJ, Churchill GA, de Villena FP-M, Schwartz DA, Collins FS (2014) Integrative genetic analysis of allergic inflammation in the murine lung. Am J Respir Cell Mol Biol 51(3):436–445
34. Phillippi J, Xie Y, Miller DR, Bell TA, Zhang Z, Lenarcic AB, Aylor DL, Krovi SH, Threadgill DW, de Villena FP-M, Wang W, Valdar W, Frelinger JA (2014) Using the emerging collaborative cross to probe the immune system. Genes Immun 15(1):38–46
35. Rutledge H, Aylor DL, Carpenter DE, Peck BC, Chines P, Ostrowski LE, Chesler EJ, Churchill GA, de Villena FP-M, Kelada SNP (2014) Genetic regulation of zfp30, cxcl1, and neutrophilic inflammation in murine lung. Genetics 198(2):735–745
36. Kirk RE (2009) The SAGE handbook of quantitative methods in psychology. Sage, Thousand Oaks, CA
37. Rasmussen AL, Okumura A, Ferris MT, Green R, Feldmann F, Kelly SM, Scott DP, Safronetz D, Haddock E, LaCasse R, Thomas MJ, Sova P, Weiss JM, Carter VS, Miller DR, Shaw GD, Korth MJ, Heise MT, Baric RS, de Villena FP-M, Feldmann H, Katze MG (2014) Host genetic diversity enables Ebola hemorrhagic fever pathogenesis and resistance. Science 346(6212):987–991
38. Ferris MT, Aylor DL, Bottomly D, Whitmore AC, Aicher LD, Bell TA, Bradel-Tretheway B, Bryan JT, Buus RJ, Gralinski E, Haagmans BL, McMillan L, Miller DR, Rosenzweig E, Valdar W, Wang J, Churchill GA, Threadgill DL, McWeeney SK, Katze MG, de Villena FP-M, Baric RS, Heise MT (2013) Modeling host genetic regulation of influenza pathogenesis in the Collaborative Cross. PLoS Pathog 9(2):e1003196
39. Graham JB, Thomas S, Swarts J, McMillan AA, Ferris MT, Suthar MS, Treuting PM, Ireton R, Gale M, Lund M (2015) Genetic diversity in the collaborative cross model recapitulates human West Nile virus disease outcomes. mBio 6(3):1–11
40. Gralinski LE, Ferris MT, Aylor DL, Whitmore AC, Green R, Frieman MB, Deming D, Menachery VD, Miller DR, Buus RJ, Bell TA, Churchill GA, Threadgill DW, Katze MG, McMillan L, Valdar W, Heise MT, de Villena FP-M, Baric RS (2015) Genome-wide identification of SARS-CoV susceptibility loci using the Collaborative Cross. PLoS Genet 11(10):e1005504
41. Blackwell JM, Jamieson SE, Burgner D (2009) HLA and infectious diseases. Clin Microbiol Rev 22(2):370–385
42. Sellers RS, Clifford CB, Treuting PM, Brayton C (2012) Immunological variation between inbred laboratory mouse strains: points to consider in phenotyping genetically immunomodified mice. Vet Pathol 49(1):32–43
43. Francis M, Festing W (1979) Notes on genetic analysis (Chapter 7). In: Inbred strains in biomedical research. Macmillan, New York, NY, pp 80–98
44. Bottomly D, Ferris MT, Aicher LD, Rosenzweig E, Whitmore A, Aylor DL, Haagmans BL, Gralinski LE, Bradel-Tretheway BG, Bryan JT, Threadgill DV, de Villena FP-M, Baric RS, Katze MG, Heise M, McWeeney SK (2012) Expression quantitative trait loci for extreme host response to influenza A in pre-Collaborative Cross mice. G3 (Bethesda, MD 2(2):213–221
45. Rogala AR, Morgan AP, Christensen AM, Gooch TJ, Bell TA, Miller DR, Godfrey VL, Villena FP-M (2014) The Collaborative Cross as a resource for modeling human disease: CC011/Unc, a new mouse model for spontaneous colitis. Mamm Genome 25(3-4):95–108
46. Scalzo AA, Fitzgerald NA, Simmons A, La Vista AL, Shellam GR (1990) Cmv-1, a genetic locus that controls murine cytomegalovirus replication in the spleen. J Exp Med 171(5):1469–1483
47. Scalzo AA, Yokoyama M (2008) Cmv1 and natural killer cell responses to murine cytomegalovirus infection. Curr Top Microbiol Immunol 321:101–122
48. Scherbik SV, Kluetzman K, Perelygin AA, Brinton MA (2007) Knock-in of the Oas1b(r) allele into a flavivirus-induced disease

susceptible mouse generates the resistant phenotype. Virology 368(2):232–237
49. Arnheiter H, Skuntz S, Noteborn M, Chang S, Meier E (1990) Transgenic mice with intracellular resistance to influenza. Cell 62:51–61
50. Staeheli P, Grob R, Meier E, Sutcliffe JG, Haller O (1988) Influenza virus-susceptible mice carry Mx genes with a large deletion or a nonsense mutation. Mol Cell Biol 8(10):4518–4523
51. Rönnegård L, Valdar W (2011) Detecting major genetic loci controlling phenotypic variability in experimental crosses. Genetics 188(2):435–447
52. Yalcin B, Wong K, Agam A, Goodson M, Keane TM, Gan X, Nellåker C, Goodstadt L, Nicod J, Bhomra A, Hernandez-Pliego P, Whitley H, Cleak J, Dutton R, Janowitz D, Mott R, Adams DJ, Flint J (2011) Sequence-based characterization of structural variation in the mouse genome. Nature 477(7364):326–329
53. Ichinohe T, Pang IK, Kumamoto Y, Peaper DR, Ho JH, Murray TS, Iwasaki A (2011) Microbiota regulates immune defense against respiratory tract influenza A virus infection. Proc Natl Acad Sci 108(13):5354–5359
54. Zhang D, Chen G, Manwani D, Mortha A, Xu C, Faith JJ, Burk RD, Kunisaki Y, Jang JE, Scheiermann C, Merad M, Frenette PS (2015) Neutrophil ageing is regulated by the microbiome. Nature 525(7570):528–532

Chapter 29

Using Systems Genetics to Understanding the Etiology of Complex Disease

Ramesh Ram and Grant Morahan

Abstract

Here, we discuss Systems Genetics applications for systematic evaluation of candidate causal genes together with follow-up bioinformatics pathway analysis. The aim of this chapter is to illustrate analytic procedures and we provide examples in the context of Type 1 diabetes (T1day), the risk of which is conferred by over 60 loci. We also describe the Type 1 Diabetes Systems Genetics website and provide a guide for its use and application to other diseases.

Key words Use case Type 1 diabetes, Nonsynonymous genetic modifiers, Variants that affect gene expression, *Cis*-regulatory genes, *Trans*-regulatory genes

1 Introduction

Genome-wide association studies have identified thousands of genetic variants that mediate dozens of complex traits and diseases [1]. However, many if not most of these variants are anonymous single nucleotide polymorphisms (SNPs) that reside in noncoding regions of the genome; their functional significance is unknown as well as the mechanisms by which they mediate disease risk. Systems genetics [2] can be employed to study complex human diseases. In principle, SNPs may mediate disease risk by encoding an amino acid change in a protein, or by affecting gene expression. It may be assumed the causal variants that lie outside protein-coding regions exert their effects on transcript levels, either directly by affecting the expression of genes in the region, or indirectly by influencing long-range chromatin conformations or affecting genetic networks and/or mediators such as micro-RNAs. Investigation of trait-associated loci with transcript-level measurement (from microarrays or RNA-Seq assays) would allow identification of candidate genes affected by causal variants.

In such expression quantitative trait locus (eQTL) analyses, genes can be identified whose expression is regulated by specific

Klaus Schughart and Robert W. Williams (eds.), *Systems Genetics: Methods and Protocols*, Methods in Molecular Biology, vol. 1488, DOI 10.1007/978-1-4939-6427-7_29, © Springer Science+Business Media New York 2017

SNP genotypes; these may be either in close proximity to (*cis*-acting SNPs) or at greater distances from, or on different chromosomes than, the regulated gene (*trans*-acting SNPs) [3]. Variation in promoters or enhancer elements could affect differential *cis*-regulation. The mechanisms involved in *trans*-regulation could include gene-mediated (e.g. transcription factors) or sterical effects such as "chromosome crosstalk"[4]. Some loci could have missense SNPs regulating either *cis*- and *trans*-genes. There have been two main reasons why it is difficult to identify potential gene regulatory effects. First, substantial correction for multiple testing is required if the interaction is analyzed in a genome-wide fashion. In a genome-wide 100K SNP set, for example, the *P*-value of an observed interaction would have to be in the range of $P=5\times10^{-7}$ per transcript before being considered significant. Second, substantial computation resources are needed to perform tests requiring billions of calculations. This problem has been resolved to some extent by the introduction of GPU-based computing resources.

Trait-associated SNPs are identified by "top-down" association analyses. In contrast, a "bottoms-up" systems genetic analysis of risk SNPs is achieved by interfacing the affected genotypes against transcriptomic datasets of cell line samples from genotyped subjects. Briefly, a systems genetics analysis of a complex genetic disease/trait may be accomplished in the following steps.

(a) All SNPs significantly associated with the disease or trait are compiled from various association studies [1].

(b) SNPs that are in strong linkage disequilibrium (LD) ($r^2>0.8$) with these SNPs are extracted from public domain databases such as Hapmap [5] and 1000 Genomes [6].

(c) The missense variants among these LD SNPs are isolated and analysis is performed using Polyphen-2 [7] to assess the impact of the protein-coding variation.

(d) Missense variants (if not genotyped already) are imputed to compare association *P*-values with reported peak SNPs.

(e) To test the impact of risk variants on gene expression, relevant cell lines are derived from samples of suitably genotyped affected and unaffected subjects and transcriptomic (microarray/RNA-Seq) scans are performed on these samples.

(f) The exported array data is normalized, corrected for batch effects and population structure effects.

(g) The corrected data is examined for differences between affected and unaffected subjects. If there is no significant difference, which is expected in the case of cell lines, the samples may be pooled for greater power.

(h) The data are then integrated against the reported peak (or best) SNPs and pairwise SNP-gene association tests are performed.

(i) Results are separated into *cis* and *trans* regulatory interactions based on the physical distances between the SNP and the gene whose expression it is associated with, and separate false discovery rate (FDR) corrections are applied.

(j) The lists of significant missense, cis- and trans-regulated genes are subject to bioinformatic network, pathway and enrichment analysis on a per locus basis as well as on the whole considering all candidate genes together.

In this chapter, we will focus on the data analysis procedures; the reader is referred elsewhere for sample preparation and laboratory procedures. We will go through detailed step-by-step methods for performing systems genetics analyses of complex genetic traits using Type 1 diabetes (T1D) as an example. To implement these methods, the reader should be familiar with the use of the R statistical package. For those who prefer to perform online analyses of curated samples, we will also provide instructions as to how to make use of our resource, the T1D Systems Genetics browser (accessible via this URL: http://www.sysgen.org/T1DGCSysGen/). This online tool permits visualization of SNP–gene interaction effects pertaining to T1D SNPs in four different cell types.

2 Materials

2.1 Datasets

1. Prepare a list of all T1D SNPs from various association studies [8–11] along with information such as chromosome band (e.g. 1p13.1), chromosome, base position, reported alleles, odds ratio, *P*-values, and genes of interest. If multiple best SNPs are reported at any one locus, retain the SNP that has the lowest reported *P*-value and eliminate the SNPs whose pairwise linkage disequilibrium (LD) with this SNP is $r^2 > 0.8$.
2. Download the latest pre-calculated LD data from Hapmap [5] and 1000 Genomes [6] for a relevant population type, e.g. the CEPH Utah CEU cohort samples. These data will be comprised of pair-wise LD calculations for exhaustive pairs of SNPs.
3. The next instructions pertain to Illumina HT-12 v4 expression beadchip arrays. Export microarray data using the Illumina Beadstudio software suite. Also export control probe profile data from Beadstudio. The probeset annotations file (HumanHT-12_V4_0_R2_15002873_B.txt) can be downloaded from the Illumina website. Prepare a sample information table with the following columns: part number, well position, analytical sample id, batch id (if known), date of scan, gender, and cell type.

3 Tools and Resources

1. Internet access.
2. Install Basic R package (cran.r-project.org). Install the following packages inside R: lumi, limma, MatrixEQTL, cluster.
3. PLINK (cog-genomics.org/plink2).
4. List of URLs.
   (a) T1D SNPs (www.sysgen.org/T1DGCSysGen/T1DSNPs.txt).
   (b) VAI (genome.ucsc.edu/cgi-bin/hgVai).
   (c) DAVID (david.ncifcrf.gov).
   (d) GATHER (gather.genome.duke.edu)
   (e) GENEMANIA (genemania.org)
   (f) DAPPLE (broadinstitute.org/mpg/dapple)

4 Methods

4.1 Analysis of Missense Variants in LD with T1D SNPs

1. Search the T1D SNP rsid's in the LD data files that were downloaded using *zgrep* (see **Note 1**). Place all the searched LD entries in one single file and apply a threshold of >0.80 to the R2 column. The LD entries remain are of those SNPs that are in strong LD with T1D SNPs. Obtain a unique set of these SNP ids.
2. In order to determine which of these LD SNPs are known missense variants, use the Variant Annotation Integrator (VAI) tool. Select the appropriate human genome assembly (as hg19) and paste the list of rs SNP ids in the box provided (*see* **Note 2**). Next, under options of "Database of Non-synonymous Functional Predictions", select PolyPhen-2. Under filters, only select CDS nonsynonymous option. Leave the other options as is and click on "get results".
3. The results screen will provide the list of missense variants that are in tight LD with T1D reported SNPs, gene name, the amino acid change with position in the protein sequence, and the corresponding Polyphen-2 prediction. Export results in a spreadsheet and for each missense SNP identify the T1D SNP to which there is LD. Also work out how many T1D loci harbor missense SNPs and how many of these are deleterious (*see* **Note 3**).
4. A comparison of association *P*-values between the T1D SNP and missense SNP is made next. For this, if the missense SNP was already genotyped in the GWAS study, then the *P*-value is noted. Otherwise, the SNP is imputed to determine the *P*-value association with the disease. Instances where the *P*-value of the missense SNP is better that the T1D SNP are noted. These would suggest the identification of candidate causal genes (*see* **Note 4**).

5. Additionally a conditional association (logistic regression) analysis of the missense SNPs is performed in PLINK [12] with reported best SNPs as covariates. A $P < 0.001$ indicates an independent secondary association signal at the missense SNP.
6. Compile the list of T1D SNPs, their missense variants, candidate gene names, Polyphen-2 predictions, and association *P*-values in the format of a table.

Additionally LD SNPs that are frameshift, stop gain/loss or in the splice-site/slice-region can be additionally identified using the VAI tool and reported separately.

4.2 Analysis of Effect of T1D SNPs on Gene Expression

1. Import the raw microarray data in R using the *lumi* [13] package. Append the control probe profile data to the array data object and perform background correction using *lumiB* with method as "bgAdjust." Follow this with variance stabilizing transformation (VST) and quantile normalization using functions *lumiT* and *lumiN*. Export the normalized expression data using function *exprs*. Replace the column headers with analytical ids of the samples. Repeat this process for all of the array data files exported using Beadstudio (*see* **Note 5**).
2. Merge the normalized expression data into a single file. In the case where more than one cell type/stimulation is investigated, create separate files per cell type/stimulation (*see* **Note 6**).
3. Next, batch effects are investigated. Perform principle components analysis (PCA) of the expression data. For this, compute the correlation matrix of the samples in R using the function *cor* and apply this matrix to function *eigen*. This should return the principal components (PCs). Plot PC1 against PC2 and color code the samples according to their batches (if known). If the samples cluster according to the batch, then batch effects are present and need to be removed. Also investigate PC1 against PCs 3–10 to check for sub-batch effects. Batched ids can be inferred by performing clustering analysis over PCs 1 and 2 using the *pam* (partition of medoids) function in the *cluter* R package.
4. To perform batch effect removal, first separate the samples into batches. Then, take each batch and center each row (gene) within it using the R function *scale* (*see* **Note 7**). Then merge all the batches together into one matrix. This matrix is now batch corrected and the PCA analysis should validate the correction (*see* **Note 8**). To calculate amplitude of each gene, calculate the mean expression of the gene in different batches and take a mean of the means. The amplitude value can be added row-wise to the batch corrected expression matrix (*see* **Note 9**).
5. Perform quality control next. Identify which genes have detection *P*-value <0.01 (in the raw Beadstudio exported files) in less than 10% of the samples and eliminate those. This will

eliminate genes whose expression is least detected. Also eliminate probesets with potential cross-hybridization problems. Do a *blastn* search of each probe sequence on a custom database on all 47,323 Illumina probeset sequences and only retain those with a single match.

6. Perform the exhaustive tests between gene expression matrix and T1D genotypes using the MatrixEQTL R package [14]. Format the T1D genotypes in an additive recode (0,1,2) using PLINK (*see* **Note 10**). Follow instructions in the MatrixEQTL manual to prepare the files. Prepare the files containing the SNP's chromosome and base positions and the probeset's chromosome and coordinates. Use ±1 Mbp as a conservative distance threshold for *cis* gene interactions. After running the association analysis, the results will be split into two files: *cis* and *trans* each with FDR corrected *P*-values. Eliminate the results where the SNP's position falls inside the probes start and end position. Calculate FDR *P*-values again after eliminating these transcripts whose expression may be incorrect due to differential hybridization to the SNP-containing probe.
7. To improve the detection of SNP-gene associations that may be confounded by population structure, expression derived Principal components can be used as covariates while performing the testing. For this, the first 20 PCs are derived from the batch corrected expression matrix. Then, one by one the PCs are sequentially added as covariates (eg. PC1, PCs1-2, PCs1-3, …) and MatrixEQTL analysis is repeated. At each of the 20 iterations, record the number of *cis* SNP–gene interactions detected at FDR $P<0.001$. The optimal number of PC's to correct is determined where maximum *cis* effects are recovered in these iterations (*see* **Note 11**).
8. Compile a table with the list of *cis* and *trans* genes associated with each of the T1D SNPs at different FDR P-value thresholds of 0.001, 0.01, and 0.05. The *cis* and *trans* genes detected below 0.001 are strong candidates for mediating T1D susceptibility.

4.3 Bioinformatic Follow-Up Analysis of the List of Candidate Genes Identified

1. Follow-up bioinformatic network, pathway, and gene ontology (GO) term enrichment analysis can be performed for the list of candidate genes identified in the above steps. This is done separately for each locus as well as for the combined set of genes taken from all T1D loci. For each analysis, compile a list of genes to be tested. Convert the gene ids to Entrez ids using the DAVID (david.ncifcrf.gov) gene id converter.
2. Perform analyses of the gene lists with the following online tools and compile the significant results reported: (*see* **Note 12**).
   (a) DAVID: enrichment of GO terms, Kegg Pathway.
   (b) GATHER: confirmatory enrichment of GO terms, Pathway.
   (c) GENEMANIA: Perform network analysis of all genes.

(d) DAPPLE: Perform deeper network analysis.

These results will provide functional insights into the possible mechanisms by which the SNPs mediate disease risk.

4.4 Use of Type 1 Diabetes Systems Genetic Browser

The T1D systems genetics browser is accessible via this URL: (www.sysgen.org/T1DGCSysGen). Here we present the effects of T1D SNPs on expression of genes in four cell types. An example of using the tools is illustrated in Fig. 1.

1. The browser presents various drop-down menus for choice of options. First, the user may select from four cell types available. Then depending on the choice of type of gene interaction as either *cis* or *trans*, the list of genes shown in the drop-down for the selected T1D SNP will vary. The results can be viewed as either box plots or gene network diagrams by clicking on the appropriate buttons.
2. In a box-plot view, the x-axis corresponds to the three genotypes of the T1D SNP and the y-axis corresponds to the normalized expression value. Based on the appearance of the boxes in the box plots, one may be able to tell if the SNP-gene effect is additive or dominant. It is also possible to mouse over the box plots to obtain the median and other information.
3. Additional boxplots display the comparison of SNP genotypes against normalized PC corrected gene expression sets that adjust for population structure (explained in Subheading 3.2) (*see* **Note 13**).
4. The *trans* regulatory gene interactions can be plotted in the form of a circular network diagram [15] as shown in Fig.1. Using these figures, it is possible to identify which SNPs are in genomic regions that mediate *trans* regulatory effects.
5. The same SNP–gene interaction pairs can be viewed across different cell types by changing the cell-type option. In doing so, it is possible to note the effect of a T1D SNP on a particular gene across different cell types. Where a SNP is associated with expression in multiple cell types, the effect directions should usually be consistent, i.e. the same allele is associated with increased expression in multiple cell types.

5 Notes

1. LD data downloaded will be in compressed format with separate .gz files for each of 24 human chromosomes. A fast method to search and extract LD SNPs in these files is to perform the Unix *zgrep* command. The usage of this command is:

   zgrep “search text” filename.gz

   where “search text” is replaced by the text the user wishes to find.

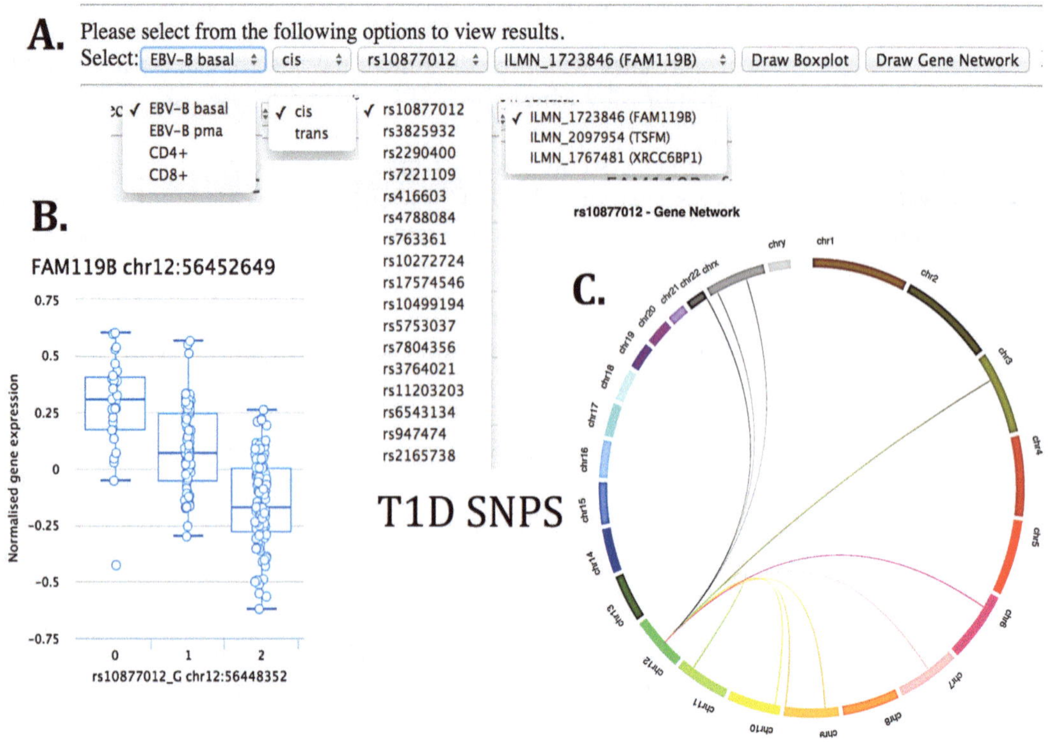

Fig. 1 Steps involved in making use of the Type 1 Diabetes Systems Genetics online browser. (**a**) Step 1. Select an input cell type. Step 2. Select the type of SNP–gene interaction as: *cis* or *trans*. Step 3. From the drop-down list of T1D SNPs, select a T1D SNP of interest. Step 4. From a drop-down list of candidate genes associated with the T1D SNP, select a gene of interest. Genes are sorted by association P-value. Step 5. To view, the SNP-gene association as a box plots, click "Draw Boxplot". (**b**) A figure of the box plot as shown will be displayed. Step 6. To view *trans* gene interactions associated with a T1D SNP, click "Draw gene network" button. (**c**) The result is displayed as a circos style circular network plot which is also referred to as a "genomagon" plot

2. The VAI tools are not capable of handling more than 100 SNPs in any one query. So it is recommended to split larger analyses into sets of 100 SNPs.
3. Assign each T1D locus an alias id such as (L1, L2, ...) for easier reference when summarizing the results.
4. Imputation can be performed with IMPUTE2 [16]. *P*-values are computed with the PLINK "–assoc" option.
5. The array data is imported in R using function *lumiR.batch* and control probe profile file is added to the imported data using *addControlData2lumi* function.
6. Repeat all the analysis procedures separately for each cell types.
7. Use the following option while calling the scale function: center = T and scale = F.
8. More advanced batch effect correction can be performed using COMBAT [17], however the methods presented here will suffice without over complicating the data.

9. Amplitude of a gene is simply the average level of expression at which a gene is usually detected in a particular cell type.
10. Using PLINK, the "–recodeA" option will return the additive recode of the SNP genotypes.
11. The number of PC's used for correction can be plotted against the number of *cis*-genes detected at FDR $P < 0.001$ to determine the optimal number of PCs.
12. The network and pathway analysis tools mentioned here provide a list of enriched annotations along with FDR adjusted *P*-value significances. Annotations with $P < 0.05$ in these can be treated as significant.
13. The boxplots of PC corrected expression sets are likely to show stronger association and tighter cluttering of the samples with the SNP genotypes compared to uncorrected data.

Acknowledgement

This work was supported by a grant from Diabetes Research Foundation of Western Australia, and Program Grant 1037321 and Project Grant 1069173 from the National Health and Medical Research Council of Australia.

References

1. Welter D, MacArthur J, Morales J, Burdett T, Hall P, Junkins H, Klemm A, Flicek P, Manolio T, Hindorff L, Parkinson H (2014) The NHGRI GWAS Catalog, a curated resource of SNP-trait associations. Nucleic Acids Res 42:D1001–D1006
2. Morahan G, Williams RW (2007) Systems genetics: the next generation in genetics research? Novartis Found Symp 281:181–188, discussion 188–191, 208–209
3. Veyrieras JB, Kudaravalli S, Kim SY, Dermitzakis ET, Gilad Y, Stephens M, Pritchard JK (2008) High-resolution mapping of expression-QTLs yields insight into human gene regulation. PLoS Genet 4, e1000214
4. Freedman ML, Monteiro AN, Gayther SA, Coetzee GA, Risch A, Plass C, Casey G, De Biasi M, Carlson C, Duggan D, James M, Liu P, Tichelaar JW, Vikis HG, You M, Mills IG (2011) Principles for the post-GWAS functional characterization of cancer risk loci. Nat Genet 43:513–518
5. The International HapMap Consortium (2003) The International HapMap Project. Nature 426:789–796
6. 1000 Genomes Project Consortium (2012) An integrated map of genetic variation from 1,092 human genomes. Nature 491:56–65
7. Adzhubei IA, Schmidt S, Peshkin L, Ramensky VE, Gerasimova A, Bork P, Kondrashov AS, Sunyaev SR (2010) A method and server for predicting damaging missense mutations. Nat Methods 7:248–249
8. Barrett JC, Clayton DG, Concannon P, Akolkar B, Cooper JD, Erlich HA, Julier C, Morahan G, Nerup J, Nierras C, Plagnol V, Pociot F, Schuilenburg H, Smyth DJ, Stevens H, Todd JA, Walker NM, Rich SS (2009) Genome-wide association study and meta-analysis find that over 40 loci affect risk of type 1 diabetes. Nat Genet 41:703–707
9. Rich SS, Akolkar B, Concannon P, Erlich H, Hilner JE, Julier C, Morahan G, Nerup J, Nierras C, Pociot F, Todd JA (2009) Overview of the type I diabetes genetics consortium. Genes Immun 10(Suppl 1):S1–S4
10. Morahan G, Mehta M, James I, Chen WM, Akolkar B, Erlich HA, Hilner JE, Julier C, Nerup J, Nierras C, Pociot F, Todd JA, Rich SS (2011) Tests for genetic interactions in type 1 diabetes: linkage and stratification analyses of 4,422 affected sib-pairs. Diabetes 60:1030–1040
11. Morahan G (2012) Insights into type 1 diabetes provided by genetic analyses. Curr Opin Endocrinol Diabetes Obes 19:263–270

12. Chang CC, Chow CC, Tellier LCAM, Vattikuti S, Purcell SM, Lee JJ (2015) Second-generation PLINK: rising to the challenge of larger and richer datasets. GigaScience 4:7
13. Du P, Kibbe WA, Lin SM (2008) lumi: a pipeline for processing Illumina microarray. Bioinformatics 24:1547–1548
14. Shabalin AA (2012) Matrix eQTL: ultra fast eQTL analysis via large matrix operations. Bioinformatics 28:1353–1358
15. Krzywinski M, Schein J, Birol I, Connors J, Gascoyne R, Horsman D, Jones SJ, Marra MA (2009) Circos: an information aesthetic for comparative genomics. Genome Res 19:1639–1645
16. Howie BN, Donnelly P, Marchini J (2009) A flexible and accurate genotype imputation method for the next generation of genome-wide association studies. PLoS Genet 5(6), e1000529
17. Johnson WE, Rabinovic A, Li C (2007) Adjusting batch effects in microarray expression data using Empirical Bayes methods. Biostatistics 8:118–127

INDEX

Klaus Schughart and Robert W. Williams (eds.), *Systems Genetics: Methods and Protocols*, Methods in Molecular Biology, vol. 1488, DOI 10.1007/978-1-4939-6427-7, © Springer Science+Business Media New York 2017

MIX
Papier aus verantwortungsvollen Quellen
Paper from responsible sources
FSC® C105338

If you have any concerns about our products,
you can contact us on
ProductSafety@springernature.com

In case Publisher is established outside the EU,
the EU authorized representative is:
Springer Nature Customer Service Center GmbH
Europaplatz 3, 69115 Heidelberg, Germany

Printed by Libri Plureos GmbH
in Hamburg, Germany